Selected Methods of Clinical Chemistry
Volume 9

**Selected Methods for the
Small Clinical Chemistry Laboratory**

Contributors to This Volume

Submitters

K. Owen Ash • Eugene S. Baginski • Mary Louise Baron • D. D. Bayse • Paige K. Besch • Gerald T. Biamonte • Epperson E. Bond • George N. Bowers, Jr. • Andy Burton • Thorne J. Butler • Roger R. Calam • Wendell T. Caraway • R. Thomas Chamberlain • Gerald R. Cooper • Genevieve C. Covolo • Collene J. Delaney • Murali Dharan • Albert A. Dietz • Basil T. Doumas • Patricia H. Duncan • Leonard K. Dunikoski, Jr. • Leila V. Edwards • Lilian M. Ewen • Frank W. Fales • Willard R. Faulkner • Craig C. Foreback • Arden W. Forrey • Christopher S. Frings • Richard H. Gadsden, Sr. • Patricia E. Garrett • Philip J. Garry • Samuel J. Gentile • Ronald L. Gillum • Thomas J. Giovanniello • Stanley E. Gordesky • Adrian Hainline, Jr. • Frank F. Hall • Keith B. Hammond • J. S. Hazlehurst • John W. Heinz • Thomas S. Herman • Richard L. Imblum • Mariet Iosefsohn • Gordon P. James • Brian J. Johnston • Alex Kaplan • Raymond E. Karcher • Richard A. Kaufman • Harvey L. Kincaid • Robert C. King • Bernard Klein • Thomas R. Koch • H. Peter Lehmann • Howard C. Leifheit • Stanley S. Levinson • John A. Lott • John N. Lukens • Slawa S. Marie • Alan Mather • Robert B. McComb • Samuel Meites • Dayton T. Miller • Samuel Natelson • Jane W. Neese • N. Vasudeva Paniker • Richard B. Passey • Joseph Pecci • Michael A. Pesce • Theodore Peters, Jr. • C. Ray Ratliff • Eugene W. Rice • James S. Roloff • Edward J. Rosenthal • Guilford G. Rudolph • Willie L. Ruff • John E. Sherwin • W. William Spencer • Herbert E. Spiegel • Thomas C. Stewart • Mildred Suszka • Lin-Nar L. Teng • Paul M. Tocci • Charles H. Tripp, Jr. • Kathie Turner • G. Russell Warnick • Seymour Winsten • Chien Yu • Bennie Zak

Evaluators

George J. Abraham • Ramesh C. Airan • John P. Aitchison • A. Akiyama • Nancy W. Alcock • A. Ernest Alexander • V. Thomas Alexander • Leroy H. Arnold • O. W. van Assendelft • Catherine Baker • Seymour Bakerman • Quentin C. Belles • Edward W. Bermes, Jr. • Lainie Blum • Thomas A. Blumenfeld • Geert J. M. Boerma • Epperson E. Bond • R. Bondar • June D. Boyett • Charles A. Bradley • Jean-Pierre Bretaudiere • Richard Brown • Thomas E. Burgess • Lester I. Burke • Angelo Burlina • Thorne J. Butler • Ronald J. Byrnes • R. Thomas Chamberlain • Daniel K. Y. Chan • Rui-Guan Chen • Mary H. Cheng • Samuel Y. Chu • Dafne Cimino • Francisco Civantos • Gerald E. Clement • Richard M. Corcoran • Shareen L. Cox • Steven P. Crause • Joseph S. Curcio • Fram R. Dalal • Manik L. Das • Irene Daza • A. De Leenheer • Albert A. Dietz • Patricia Dobbs • Richard F. Dods • James L. Driscoll • Kurt M. Dubowski • Sandra M. Durnan • Jack H. Edwards, Jr. • Leila V. Edwards • Robert C. Elser • E. Christis Farrell, Jr. • Ted W. Fendley • J. Ben Flora • Orlando R. Flores • Roger L. Forrester • Francis T. Fox • Robert B. Foy • Callum G. Fraser • Paul C. Fu • Suzanne M. Garszczynski • Melvin R. Glick • F. Goodland • Stephen H. Grossman • Michael Guarnieri • William L. Gyure • Montgomery C. Hart • Craig R. Hearne • A. R. Henderson • Thomas E. Hewitt • C. van der Heyden • Lindsay F. Hofman • Kap-Yong Hong • Walter E. Hordynsky • Norman Huang • Richard A. Jamison • Robert M. Johnson • Susan Johnson • Carl R. Jolliff • Raymond E. Karcher • John H. Kennedy • John W. Kennedy • Gerald Kessler • Arthur Kessner • Prabhaker G. Khazanie • Harry S. Kim • Fred A. Kincl • Sigrid G. Klotzsch • Man M. Kochhar • John A. Koepke • Richard S. Kowalczyk • M. L. Kuehneman • Ingeborg R. Kupke • Robert F. Labbé • Kwok-Wai Lam • Wayne Lambert • Betty J. Lash • Herbert K. Y. Lau • John R. Leach • F. Y. Leung • Sherwood C. Lewis • Sharon Shu-Hui Lin • Joseph W. Litten • Marion Littleman • Michael M. Lubran • Robert L. Lynch • A. H. J. Maas • P. A. Govinda Malya • Horace F. Martin • George M. Maruyama • Michael McAneny • Max E. McIntosh • Joyce A. McIntyre • Elizabeth A. McLister • Ruth D. McNair • Samuel Meites • Don S. Miyada • D. J. Moffa • Isaac D. Montoya • R. F. Moran • Philip G. Mullarky • H. J. Mulvihill • Shekhar Munavalli • Susan Myerow • Sheshadri Narayanan • Edwin W. Naylor • Gilbert H. Nelson • George Nix • Pieter J. Noordeloss • James W. North • Janet Y. Nutter • Karen Oates • T. Oeser • Adam P. Orfanos • Iris M. Osberg • Pennell C. Painter • David Palmer • John Pappas • Harry L. Pardue • Leslie Parker • Sharad Patel • B. Patterson • Richard Pavlovec • Charles A. Pennock • Mary Phillips • Russell A. Picard, Jr. • Donald R. Pollard • George T. Poole • Paul Pottgen • Rita Powers • C. P. Price • Nathan Radin • Ralston W. Reid • Robert Rej • Gustavo Reynoso • Paul F. A. Richter • Stephen E. Ritzmann • William R. Robertson • Karen Saniel-Banrey • Aaron Sarafinas • Irvin Schoen • Henry G. Schriever • Frank A. Sedor • Catherine Sherry • Leonard Sideman • Darshan Singh • William D. Slaunwhite • Jean M. Slockbower • John Southgate • Peter D. Spare • Willard A. Stacer • Bernard E. Statland • Bernard W. Steele • Lilla Sun • George H. Thomas • Marjorie Uhl • James L. Verdi • Kera Voigtlander • Nabil W. Wakid • Mary Kay Walker • G. Russell Warnick • Laken G. Warnock • Lester J. Watayi • John M. Waud • Diane Wawrzyniak • Nancy I. Weinberg • Graham M. Widdowson • Jannie Woo • C. Ying • S. Zeugner

Reviewers

O. W. van Assendelft • Stanley Bauer • D. Joe Boone • William D. Bostick • George N. Bowers, Jr. • Charles A. Bradley • Alan Broughton • Steven N. Buhl • Robert W. Burnett • Carl A. Burtis • Richard J. Carter • Gerald R. Cooper • Basil T. Doumas • James L. Driscoll • Patricia H. Duncan • Leila V. Edwards • Robert C. Elser • Bruce Lee Evatt • G. V. Freile-Gagliardo • Christopher S. Frings • Charles S. Furfine • Nathan Gochman • Frank A. Ibbott • Peter I. Jatlow • Roy B. Johnson, Jr. • Thomas R. Koch • A. H. J. Maas • Horace F. Martin • Alan Mather • Robert B. McComb • Henry C. Nipper • Billy W. Perry • Charles E. Pippenger • Edward A. Sasse • Rosanne M. Savol • F. W. Spierto • R. E. Vanderlinde • John Vasiliades • G. Russell Warnick • Harry F. Weisberg • James O. Westgard • David L. Witte

Selected Methods of Clinical Chemistry
Volume 9

Selected Methods for the Small Clinical Chemistry Laboratory

Editor-in-Chief
Willard R. Faulkner
Professor of Biochemistry, Emeritus
Vanderbilt University Medical Center
Nashville, TN 37212

Co-Editor
Samuel Meites
Director, Clinical Chemistry Laboratory
Children's Hospital
Columbus, OH 43205

Series Editor
Virginia S. Marcum

American Association for Clinical Chemistry
1725 K Street, N.W.
Washington, DC 20006

1982

BOARD OF EDITORS
Selected Methods of Clinical Chemistry

Gerald R. Cooper, Editor-in-Chief, Clinical Chemistry Division, Centers for Disease Control, Atlanta, GA 30333

Robert W. Burnett, Clinical Chemistry Laboratory, Hartford Hospital, Hartford, CT 06115

Willard R. Faulkner, Department of Biochemistry, Vanderbilt University Medical Center, Nashville, TN 37212

Martin Fleisher, Biochemistry Department, Memorial Sloan-Kettering Cancer Center, New York, NY 10021

Donald T. Forman, Clinical Chemistry, Department of Hospital Laboratories, University of North Carolina, Chapel Hill, NC 27514

Christopher S. Frings, Medical Laboratory Association, Birmingham, AL 35205

Richard H. Gadsden, Sr., Clinical Laboratories, Medical University of South Carolina, Charleston, SC 29416

Judith A. Hopkins, Bureau of Laboratories, Centers for Disease Control, Atlanta, GA 30333

Lawrence M. Killingsworth, Chemistry Laboratories, Sacred Heart Hospital, Spokane, WA 99204

Thomas R. Koch, Clinical Chemistry, University of Maryland Hospital, Baltimore, MD 21201

Samuel Meites, Clinical Chemistry Laboratory, Children's Hospital, Columbus, OH 43205

George N. Nichoalds, Department of Obstetrics and Gynecology, University of Tennessee, Memphis, TN 38163

Henry C. Nipper, Department of Pathology, University of Maryland Medical School, Baltimore, MD 21204

Tom C. Stewart, Medical Laboratories of Baton Rouge, Baton Rouge, LA 70806

Raymond E. Vanderlinde, Hahnemann Medical College and Hospital, Philadelphia, PA 19102

Library of Congress Catalog Card No. 80–66258
ISBN 0–915274–13–2

© 1982, by the American Association for Clinical Chemistry, Inc. All rights reserved. No part of this book may be reproduced nor transmitted, nor translated into a machine language without written permission from the publishers.

Printed in the United States of America.

This volume is dedicated to Gerald R. Cooper.

Gerald R. Cooper

Not many years ago, uncertainty prevailed in measuring physiologically important lipids. A multitude of non-validated methods existed, amid an absence of reliable standards and controls. Furthermore, there was no agency concerned with improving laboratory performance of lipid analysis. All that has now changed, due primarily to the dedicated, often herculean efforts of one man, Gerald R. Cooper. Working through the United States Public Health Service at the Center for Disease Control (CDC), Dr. Cooper has validated methods and standard reference materials for the measurement of cholesterol, triglycerides, lipoproteins, free fatty acids, and phospholipids. For this and other notable accomplishments he has received widespread acclaim from those who recognize the far-reaching impact of his work.

Though born in Scranton, South Carolina, young Cooper grew up and was educated in Durham, North Carolina. He received his A.B., M.A., and Ph.D. (1939) in chemistry, as well as his medical degree (1950), from Duke University. He then was engaged in research and teaching for two years in the Departments of Biochemistry and of Experimental Surgery and Medicine, at Duke School of Medicine.

In 1952 Dr. Cooper joined the staff of the Center for Disease Control, where for 28 years he has served in a number of different roles. In 1953, he was appointed Chief of the Clinical Chemistry, Hematology and Pathology Branch, Laboratory Division, and directed these complex and diverse laboratories for the next 20 years. In 1974 he became Chief of the Metabolic Biochemistry Branch, encompassing the sections of Lipid Reference, Lipid Metabolism Research, Clinical Trials Central Laboratory, and Clinical Chemistry Standardization. These sections, working collaboratively, administer programs to assist in calibration and standardization of lipid research laboratories; clinical trial laboratories of the National Heart, Lung, and Blood Institutes; and reference laboratories for manufacturers of in vitro clinical chemistry diagnostic products.

Dr. Cooper is currently the Medical Director of the World Health Organization Reference and Research Center for Blood Lipids, located at the CDC.

In 1978, Dr. Cooper retired from his commissioned status with the U.S. Public Health Service, but has continued to serve at the CDC in three capacities:

1. Research Medical Officer of the Clinical Chemistry Division.
2. Liaison Officer with the National Heart, Lung, and Blood Institute for laboratory standardization of clinical trials.
3. Research Consultant on lipids and lipoprotein reference materials and methods.

He served on the Council of Clinical Chemistry of the American Society of Clinical Pathologists as a member (1975–77), and as its chairman (1977).

Dr. Cooper has for 17 years participated in the affairs of the American Association for Clinical Chemistry. He has served on the Board of Editors, Selected (Standard) Methods of Clinical Chemistry, a member since 1967, and Chairman since 1970; the Publications Board and Board of Editors of *Clinical Chemistry;* the Board of Directors (1974–1976); and the Cholesterol and Triglyceride Reference Method Study Groups (Chairman since 1975).

A prolific worker in clinical chemistry, Dr. Cooper has published 108 scientific papers, written two chapters in books, written six manuals, and edited two volumes of *Selected Methods,* with another nearing completion. During the first three years of his membership on the Board of Editors of the *Selected Methods* series, he helped complete Volume 6. From 1970 to the present, as Editor-in-Chief, he supervised the production and publication of Volumes 7 and 8. These three volumes collectively comprise 70 chapters (totaling more than 800 pages) and involve the concerted efforts of 250 chemists. It was largely through his decisive influence that this volume, devoted to the small clinical chemistry laboratory, got off to an auspicious start in 1978, and was completed within three years. Volume 10 of the regular series is now on its way to completion.

Dr. Cooper's numerous significant contributions to clinical chemistry are reflected in the prestigious recognition he has received:

1954: Hektoen award for original work in electrophoresis.
1956: Billings award of the American Medical Association for an exhibit on diagnostic laboratory techniques for communicable diseases.
1964: Commendation medal of the U.S. Public Health Service, presented by the Surgeon General in recognition of his dedication to Public Health and in particular for his outstanding work on the Heart Disease Control program, which received worldwide acclaim.
1965: Superior Service Group Award of the U.S. Public Health Service for his outstandingly successful efforts in standardizing lipid methods.
1975: American Association for Clinical Chemistry award for outstanding contribution to clinical chemistry as a profession (Fisher Award).
1975: American Association for Clinical Chemistry award for innovations in quality-control techniques (Lecturer at the VIth International Symposium on Quality Control, Geneva, Switzerland).
1978: Distinguished Service Medal of the U.S. Public Health Services for sustained, dedicated service and outstanding contributions to the advancement of laboratory science in research on cardiovascular diagnosis.
1979: Gerulat Award of the New Jersey Section, AACC, for noteworthy activities in the field of clinical chemistry.

After all the factual information of Dr. Cooper's professional activities and attainments are stated, a question arises: what salient personality traits make him the leader he is? We note his boundless energy, his keen intellect and penetrating insight into the problems peculiar to clinical chemistry, and his ability to devise solutions for them. As we see it, however, there are also other, more personal, characteristics, which have summoned forth the willing support of many people, inducing them to work together so smoothly and effectively that they always reach their goals. Among these are enthusiasm, optimism, and goodwill to all.

We proudly dedicate this book to Dr. Gerald R. Cooper, our friend and mentor.

The Editors

Preface

This volume began in 1974 with an idea that Dr. J. Stanton King, Executive Editor for the American Association for Clinical Chemistry, expressed in a letter to Dr. Gerald R. Cooper, Editor, Selected Methods of Clinical Chemistry. Dr. King had found that the published volumes 1–7 of the *Standard* (now *Selected*) *Methods* series were being purchased in surprisingly large numbers by "generalists" in small laboratories (those with a workload equivalent to that in hospitals of 100 beds or less), many in developing foreign nations. This observation led him to conclude that many of these people might be better served by a volume written specifically to meet their needs. These needs, as we see them, are a set of sound methods of clinical chemistry, requiring little sophisticated equipment and no automation, written by authorities, and then evaluated thoroughly and independently by other chemists to ensure their reliability.

Although production of this special volume was discussed from time to time at meetings of the Selected Methods Editorial Board, with varying degrees of enthusiasm, it was not until July 1977 that the AACC Publications Board approved Dr. Cooper's proposal. Subsequently, we were invited to undertake the Editorship. Having accepted this task, we both set to work in the spring of 1978.

At the outset, we envisioned a text having these special features:

- It would contain basic clinical chemical determinations necessary for good patient care in any medical institution, regardless of size. This volume does, in fact, contain the methods that make up the heart of clinical chemistry, those most frequently used.
- It would avoid "automated" methods.
- It would minimize requirements for instruments and equipment. Only the following 12 basic instruments are needed:
    spectrophotometer (ultraviolet and visible ranges)
    pH/blood gas meter
    centrifuges, micro and regular size
    flame photometer
    microgasometer
    electrophoresis apparatus
    fluorometer, filter type
    osmometer
    bath, constant temperature, circulating
    chloride titrator
    refractometer
    analytical balance
- It would provide, whenever possible, alternative micromodifications of tests applicable to pediatric patients.

What is presented here is near to the state of the art, and the reader should keep in mind always that "working" or "field" methods—not "reference" methods—are described. No matter how perfect many reference procedures may appear, there is yet no approved system for validating field methods in an unequivocal manner. Fortunately, however, this important problem is receiving close attention by several concerned organizations.

For adequate operation, all clinical chemistry laboratories must do more than just perform analyses with approved methods. They also must be vitally concerned with such matters as quality control, safety, preventive maintenance of instruments, blood collection techniques, and proper transport of specimens. Accordingly, we have added chapters on these topics with the hope that they will prove helpful.

It was intended that each method be written by a chemist thoroughly familiar with it through practical experience, and also familiar with alternative procedures involving different principles. In each instance, we attempted to select an optimum technique, characterized by reliability, accuracy, precision, and ease of performance under practical conditions for the small clinical chemistry laboratory.

Each chapter was to be evaluated by at least one and preferably by three chemists working independently of the submitter. Chapters were then to be reviewed by a member of the AACC Committee on Standards or by a chemist working in its behalf. For the smooth, efficient accomplishment of this task, special thanks are due to Basil T. Doumas, then chairman of that group. In a few instances, we were unable to obtain the number of evaluators we wanted, or a reviewer: such chapters are labeled "Provisional," which is a reflection of our review process only, not of the reliability and usefulness of the method.

We acknowledge our indebtedness to the Publications Board for giving us the "go ahead" on this volume; to the Board of Editors, Selected Methods Series, for creating a climate of cooperation favorable to pursuing this endeavor; and to Scott Hunt for bringing to bear the formidable capabilities of the National Office in publicizing the plan to launch this book and for his never-failing enthusiasm and belief in the potential value of the *"Small Volume."* Finally, we are grateful to Virginia S. Marcum for her editorial refinement of each "completed" chapter.

This book is the result of Dr. King's concept, the response to Dr. Cooper's stimulus, and our synthesis of the concerted efforts of more than 300 clinical chemists working over a three-year period. How well these objectives have been attained will be judged by you, the reader. We welcome your comments, whether favorable or constructively critical, and especially your suggestions for future volumes.

Willard R. Faulkner
Editor-in-Chief

Samuel Meites
Co-Editor

February 25, 1981

Note: The naming of specific instruments, reagent brands, and manufacturers is for example or information only, and does not constitute approval or recommendation by the Editors, the Submitters, or the American Association for Clinical Chemistry.

Foreword

As the title indicates, this book is intended for the small clinical chemical laboratory, wherever it may be. By "small," we mean those laboratories that cannot afford the elaborate and elegant equipment, the staff of specialists, and the imposing array of tests offered in hospital, medical center, and commercial laboratories in metropolitan areas. Indeed, in the "underdeveloped" or "Third World" countries, as they are termed by the *Realpolitikers,* and even in some large cities there may be only clinical laboratories that are small by this definition. The reader is unlikely to have trouble in deciding for himself whether or not his laboratory is "small."

In any case, the objective of this book is to focus the expertise of the AACC on the problem of what methods provide the optimum trade-off between low cost of implementation, simplicity of operation, and validity of results obtained. The compilers gave special consideration to manually performed or inexpensively mechanized procedures, and to those analyses most commonly requested by physicians. For some analytes, alternative procedures are offered. Great pains have been taken to describe the procedures clearly, in plain English, so that the book will be self-sufficient; we assume that most small clinical laboratories will not have easy access to large medical libraries. Each procedure has actually been run in the author's laboratory, to guarantee that it works as described, and then has been evaluated independently in others.

This book, then, is rather different from preceding volumes in the *Selected Methods* series. It is aimed at a specific audience: those of you whose responsibility it is to run a clinical chemistry laboratory on a perilously small budget, who lack a highly expert staff, and who have no time to evaluate and compare methods, but who are expected to generate reliable and timely data.

There may be more clinical laboratories that fit this description than one might imagine from reading *Clinical Chemistry,* in which much that appears is aimed at the high-volume, highly automated, computerized laboratory.

We believe that workers in small laboratories around the world will find this book—which is, in effect, a manual expressly prepared for them by experts—indispensable and reassuring in their efforts to fulfill their obligations. It is a small contribution to the betterment of the human race, in a world that increasingly seems more interested in its own destruction.

J. Stanton King
AACC Executive Editor

Contents

Dedication: Gerald R. Cooper v
Preface vii
Foreword ix

GENERAL CONSIDERATIONS FOR LABORATORY OPERATION

BLOOD COLLECTION 3
Submitted by: Roger R. Calam
Evaluated by: Thomas A. Blumenfeld, Samuel Meites, Jean M. Slockbower, and Bernard E. Statland
Reviewed by: O. W. van Assendelft

TRANSPORTATION OF SPECIMENS . . . 11
Submitted by: Seymour Winsten and Stanley E. Gordesky
Evaluated by: Dafne Cimino, Robert M. Johnson, Fred A. Kincl, George T. Poole, Rita Powers, and Nancy I. Weinberg
Reviewed by: Frank A. Ibbott

QUALITY ASSURANCE: THEORETICAL AND PRACTICAL ASPECTS 17
Submitted by: Adrian Hainline, Jr.
Evaluated by: George J. Abraham, Charles A. Bradley, David Palmer, Joseph S. Curcio, A. De Leenheer, Richard A. Jamison, John W. Kennedy, Harry L. Pardue, Richard Pavlovec, Nathan Radin, and Irvin Schoen
Reviewed by: James O. Westgard

QUALITY ASSURANCE: PROFESSIONAL ORGANIZATIONS AND COMMERCIAL SYSTEMS 33
Submitted by: Thomas C. Stewart, Richard H. Gadsden, Sr., and Thomas S. Herman

PREVENTIVE MAINTENANCE OF INSTRUMENTS AND EQUIPMENT . . 37
Submitted by: Murali Dharan
Evaluated by: James L. Driscoll, Horace F. Martin, Leslie Parker, Russell A. Picard, Jr., Mary Kay Walker, and Jannie Woo
Reviewed by: Stanley Bauer

SAFETY 43
Submitted by: A. W. Forrey, Collene J. Delaney, and Willie L. Ruff
Evaluated by: Laken G. Warnock, Isaac D. Montoya, and John R. Leach
Reviewed by: R. E. Vanderlinde

SPECIFIC ANALYTES

ACID PHOSPHATASE ACTIVITY (THYMOLPHTHALEIN MONOPHOSPHATE SUBSTRATE) 59
Submitted by: Lilian M. Ewen
Evaluated by: A. Ernest Alexander, R. Bondar, and Kera Voigtlander
Reviewed by: George N. Bowers, Jr.

ACID PHOSPHATASE (PROVISIONAL) . . 65
Submitted by: Guilford G. Rudolph
Evaluated by: Herbert K. Y. Lau, James L. Verdi, and Kwok-Wai Lam

ALANINE AMINOTRANSFERASE, ALT (PROVISIONAL) 69
Submitted by: Thorne J. Butler
Evaluated by: Sigrid G. Klotzsch and Iris M. Osberg

ALCOHOL (ETHANOL) IN SERUM 75
Submitted by: Charles H. Tripp, Jr.
Evaluated by: R. Thomas Chamberlain, Susan Johnson, and James L. Verdi
Reviewed by: Charles A. Bradley and Christopher S. Frings

ALKALINE PHOSPHATASE, TOTAL ACTIVITY IN HUMAN SERUM 79
Submitted by: George N. Bowers, Jr., and Robert B. McComb
Evaluated by: Arthur Kessner, Robert Rej, and Jean-Pierre Bretaudiere

AMMONIA IN PLASMA, ENZYMIC PROCEDURE 85
Submitted by: C. Ray Ratliff and Frank F. Hall
Evaluated by: Sharon Shu-Hui Lin, Donald R. Pollard, and Darshan Singh
Reviewed by: D. Joe Boone

INTRODUCTION TO AMYLASE METHODS 91
Wendell T. Caraway

AMYLASE SCREENING TEST (STARCH–IODINE METHOD) 91
Submitted by: Wendell T. Caraway
Evaluated by: Ruth D. McNair
Reviewed by: Patricia H. Duncan

AMYLASE ACTIVITY IN SERUM AND URINE WITH A STARCH CHROMOGEN . . 95
Submitted by: Bernard Klein
Evaluated by: Steven P. Crause, Diane

Wawrzyniak, Roger L. Forrester, Lester J. Watayi, William R. Robertson, Thomas E. Hewitt, Gerald Kessler, Michael M. Lubran, Paul C. Fu, Richard Pavlovec, H. S. Kim, and Sheshadri Narayanan
Reviewed by: Patricia H. Duncan

ASPARTATE AMINOTRANSFERASE, AST (PROVISIONAL) 101
Submitted by: Thorne J. Butler
Evaluated by: Francisco Civantos, Bernard W. Steele, Ramesh C. Airan, and Sigrid G. Klotzsch

BARBITURATES, ULTRAVIOLET SPECTROPHOTOMETRIC METHOD . 109
Submitted by: Herbert E. Spiegel
Evaluated by: R. Thomas Chamberlain, Kurt M. Dubowski, and Ted W. Fendley
Reviewed by: Charles E. Pippenger

BILIRUBIN (TOTAL AND CONJUGATED) MODIFIED JENDRASSIK–GROF METHOD. 113
Submitted by: Thomas R. Koch and Basil T. Doumas
Evaluated by: Robert C. Elser, William L. Gyure, and Richard S. Kowalczyk
Reviewed by: Robert B. McComb

BILIRUBIN (DIRECT-REACTING AND TOTAL) MODIFIED MALLOY–EVELYN METHOD. 119
Submitted by: Samuel Meites
Evaluated by: Mary H. Cheng, Leroy H. Arnold, Shareen L. Cox, Gilbert H. Nelson, Willard A. Stacer, and Nabil W. Wakid
Reviewed by: Robert B. McComb

CALCIUM IN BIOLOGICAL FLUIDS . . . 125
Submitted by: Eugene S. Baginski, Slawa S. Marie, and Bennie Zak
Evaluated by: Nancy W. Alcock, Karen Saniel-Banrey, Richard F. Dods, Raymond E. Karcher, John Pappas, Paul Pottgen, and Frank A. Sedor
Reviewed by: George N. Bowers, Jr.

CARBON DIOXIDE CONTENT BY MICROGASOMETER 131
Submitted by: Samuel Natelson
Evaluated by: James W. North

CARBON MONOXIDE 139
Submitted by: Willard R. Faulkner
Evaluated by: Lester I. Burke, Patricia Dobbs, and Charles A. Pennock
Reviewed by: O. W. van Assendelft

CHLORIDES: INTRODUCTION TO THREE METHODS 143
Alan Mather

CHLORIDE, COLORIMETRIC METHOD . 143
Submitted by: Stanley S. Levinson
Evaluated by: George Nix, Sherwood C. Lewis, and Richard F. Dods
Reviewed by: Alan Mather

CHLORIDE, COULOMETRIC-AMPEROMETRIC METHOD. 149
Submitted by: Albert A. Dietz and Epperson E. Bond
Evaluated by: Catherine Baker, Edward W. Bermes, Jr., T. Oeser, and Daniel K. Y. Chan
Reviewed by: Alan Mather

CHLORIDE, DIRECT MERCURIMETRIC TITRATION (PROVISIONAL) 153
Submitted by: Alan Mather

INTRODUCTION TO ANALYSIS OF CHOLESTEROL, TRIGLYCERIDE, AND LIPOPROTEIN CHOLESTEROL: REFERENCE VALUES AND CONDITIONS FOR ANALYSES 157
Submitted by: Gerald R. Cooper
Reviewed by: Leila V. Edwards and G. Russell Warnick

CHOLESTEROL, ENZYMIC METHOD . . 165
Submitted by: Gerald R. Cooper, Patricia H. Duncan, J. S. Hazlehurst, Dayton T. Miller, and D. D. Bayse
Evaluated by: Graham M. Widdowson, H. J. Mulvihill, M. L. Kuehneman, Ingeborg Kupke, and S. Zeugner
Reviewed by: James L. Driscoll

CHOLESTEROL, LIEBERMANN–BURCHARD METHOD WITH 2-PROPANOL EXTRACTION 175
Submitted by: G. Russell Warnick and Chien Yu
Evaluated by: Samuel Y. Chu, Paul C. Fu, Michael M. Lubran, and Betty J. Lash
Reviewed by: James L. Driscoll

CHOLESTEROL, PAREKH–JUNG METHOD 179
Submitted by: Leila V. Edwards and Mildred Suszka
Evaluated by: Geert J. M. Boerma and Pieter J. Noordeloos
Reviewed by: Charles S. Furfine

CREATINE KINASE IN SERUM 185
Submitted by: John A. Lott and John W. Heinz
Evaluated by: A. H. J. Maas, C. van der Heyden, Thomas E. Burgess, and Lainie Blum
Reviewed by: Robert C. Elser

CREATINE KINASE ISOENZYMES IN SERUM. 191
Submitted by: John A. Lott
Evaluated by: E. Christis Farrell, Jr., D. J. Moffa, and Jannie Woo
Reviewed by: Robert C. Elser

CREATININE IN SERUM AND URINE WITH FULLER'S EARTH 201
Submitted by: Mariet Iosefsohn
Evaluated by: Manik L. Das, Iris M. Osberg,

Richard Pavlovec, Harry S. Kim, Sheshadri Narayanan, and Peter D. Spare
Reviewed by: Thomas R. Koch

CREATININE IN SERUM AND URINE, COLUMN-CHROMATOGRAPHIC METHOD. . . 207
Submitted by: John E. Sherwin
Evaluated by: Willard A. Stacer, Don S. Miyada, Orlando R. Flores, Lilla Sun, and Mary Phillips
Reviewed by: Thomas R. Koch

DRUG SCREENING. 213
Submitted by: R. Thomas Chamberlain
Evaluated by: John P. Aitchison, Francis T. Fox, Max E. McIntosh, and Joyce A. McIntyre
Reviewed by: Peter I. Jatlow

ELECTROPHORESIS OF SERUM PROTEINS ON CELLULOSE ACETATE 223
Submitted by: Michael A. Pesce and Genevieve C. Covolo
Evaluated by: Fram R. Dalal, Robert B. Foy, Carl R. Jolliff, and F. Y. Leung
Reviewed by: Basil T. Doumas and G. V. Freile-Gagliardo

ADDENDUM: ELECTROPHORESIS OF SERUM PROTEINS (PROVISIONAL) 232
Submitted by: Richard L. Imblum

ESTROGENS (TOTAL PREGNANCY) IN URINE. 235
Submitted by: Edward J. Rosenthal, Howard C. Leifheit, and Paige K. Besch
Evaluated by: Pennell C. Painter, and Norman Huang
Reviewed by: Alan Broughton

GLUCOSE, DIRECT HEXOKINASE METHOD. 241
Submitted by: Jane W. Neese
Evaluated by: Seymour Bakerman, Prabhaker G. Khazanie, Joseph W. Litten, and Thorne J. Butler
Reviewed by: Basil T. Doumas

GLUCOSE, o-TOLUIDINE METHOD . . . 249
Submitted by: Richard B. Passey, Ronald L. Gillum, and Mary Louise Baron
Evaluated by: Don S. Miyada, A. Akiyama, C. Ying, Orlando R. Flores, Paul F. A. Richter, Angelo Burlina, Callum G. Fraser, and Craig R. Hearne
Reviewed by: Edward A. Sasse

GLUCOSE-6-PHOSPHATE DEHYDROGENASE ACTIVITY IN ERYTHROCYTES . . . 255
Submitted by: N. Vasudeva Paniker, James S. Roloff, Andy Burton, and John N. Lukens
Evaluated by: V. Thomas Alexander and Angelo Burlina
Reviewed by: Carl A. Burtis

HEMATOCRIT (PACKED CELL VOLUME) 259
Submitted by: Eugene W. Rice
Evaluated by: John A. Koepke and O. W. van Assendelft
Reviewed by: Bruce Lee Evatt

HEMOGLOBIN 263
Submitted by: Eugene W. Rice
Evaluated by: John A. Koepke and O. W. van Assendelft
Reviewed by: Bruce Lee Evatt

IRON AND TOTAL IRON-BINDING CAPACITY 267
Submitted by: Thomas J. Giovanniello, Joseph Pecci, and Philip J. Garry
Evaluated by: Manik L. Das
Reviewed by: Richard J. Carter

LACTATE DEHYDROGENASE IN SERUM . 271
Submitted by: John A. Lott and Kathie Turner
Evaluated by: Quentin C. Belles, June D. Boyett, A. R. Henderson, Robert L. Lynch, and P. A. Govinda Malya
Reviewed by: Steven N. Buhl

MAGNESIUM IN BIOLOGICAL FLUIDS (PROVISIONAL) 277
Submitted by: Eugene S. Baginski, Slawa S. Marie, Raymond E. Karcher, and Bennie Zak
Evaluated by: Karen Saniel-Banrey

MAGNESIUM, TITAN YELLOW METHOD . 283
Submitted by: W. William Spencer and Samuel J. Gentile
Evaluated by: Michael McAneny, Marjorie Uhl, and Robert M. Johnson
Reviewed by: George N. Bowers, Jr.

OSMOLALITY. 287
Submitted by: Craig C. Foreback and Robert C. King
Evaluated by: Gerald E. Clement, John M. Waud, Rita Powers, Sharad Patel, Karen Oates, and Montgomery C. Hart
Reviewed by: Harry F. Weisberg, Roy B. Johnson, Jr., and John Vasiliades

pH AND BLOOD GAS ANALYSIS 293
Submitted by: Harvey L. Kincaid and Richard A. Kaufman
Evaluated by: R. F. Moran, Ronald J. Byrnes, Susan Myerow, Wayne Lambert, and Richard Brown
Reviewed by: A. H. J. Maas and Robert W. Burnett

PHENYLALANINE, FLUOROMETRIC METHOD. 305
Submitted by: Paul M. Tocci
Evaluated by: Mary H. Cheng, Rui-Guan Chen, Edwin W. Naylor, Adam P. Orfanos, and George H. Thomas
Reviewed by: F. W. Spierto

PHOSPHATE, INORGANIC. 313
    Submitted by: Eugene S. Baginski, Slawa S. Marie, and Bennie Zak
    Evaluated by: Elizabeth A. McLister, Philip G. Mullarky, Leonard Sideman, and Irene Daza
    Reviewed by: Basil T. Doumas

PROTEIN (TOTAL PROTEIN) IN SERUM, URINE, AND CEREBROSPINAL FLUID; ALBUMIN IN SERUM. 317
    Submitted by: Theodore Peters, Jr., Gerald T. Biamonte, and Basil T. Doumas
    Evaluated by: Sandra M. Durnan, Richard M. Corcoran, Lindsay F. Hofman, Man M. Kochhar, George T. Poole, Kap-Yong Hong, and Stephen E. Ritzmann
    Reviewed by: Richard J. Carter

PROTHROMBIN TIME AND ACTIVATED PARTIAL THROMBOPLASTIN TIME . 327
    Submitted by: Leonard K. Dunikoski, Jr.
    Evaluated by: Walter E. Hordynsky, Paul F. A. Richter, Henry G. Schriever, Catherine Sherry, Leonard Sideman, and Irene Daza
    Reviewed by: William D. Bostick

SALICYLATE 337
    Submitted by: Patricia E. Garrett
    Evaluated by: John H. Kennedy, Marion Littleman, John Southgate, and Charles A. Pennock
    Reviewed by: Henry C. Nipper

SODIUM AND POTASSIUM. 341
    Submitted by: Willie L. Ruff
    Evaluated by: Ronald J. Byrnes, Aaron Sarafinas, Wayne Lambert, Richard Brown, Jack H. Edwards, Jr., Suzanne M. Garszczynski, and George Nix
    Reviewed by: Billy W. Perry

SWEAT TEST FOR CYSTIC FIBROSIS. . . 347
    Submitted by: Keith B. Hammond and Brian J. Johnston
    Evaluated by: Melvin R. Glick, Gustavo Reynoso, and William D. Slaunwhite
    Reviewed by: Horace F. Martin

TRIGLYCERIDES IN SERUM, COLORIMETRIC METHOD. 353
    Submitted by: Christopher S. Frings
    Evaluated by: Stephen H. Grossman, B. Patterson, Janet Y. Nutter, Robert F. Labbé, Shekhar Munavalli, J. Ben Flora, F. Goodland, and C. P. Price
    Reviewed by: Gerald R. Cooper

UREA IN SERUM, UREASE–BETHELOT METHOD. 357
    Submitted by: Alex Kaplan and Lin-Nar L. Teng
    Evaluated by: Jack H. Edwards, Jr., and Betty J. Lash
    Reviewed by: David L. Witte

UREA IN SERUM, DIRECT DIACETYL MONOXIME METHOD. 365
    Submitted by: Frank W. Fales
    Evaluated by: June D. Boyett, Epperson E. Bond, Ralston W. Reid, Albert A. Dietz, Robert C. Elser, and Michael McAneny
    Reviewed by: David L. Witte

URIC ACID. 375
    Submitted by: Wendell T. Caraway
    Evaluated by: Lester I. Burke, Samuel Y. Chu, and George M. Maruyama
    Reviewed by: Nathan Gochman

URINALYSIS 379
    Submitted by: K. Owen Ash and Gordon P. James
    Evaluated by: William L. Gyure, Michael Guarnieri, and Charles A. Pennock
    Reviewed by: Rosanne M. Savol

APPENDIX I: CONVERSION FACTORS . . 391
    Submitted by: H. Peter Lehmann

APPENDIX II: ANALYTE REFERENCE INTERVALS 393

INDEX 395

GENERAL CONSIDERATIONS FOR LABORATORY OPERATION

Blood Collection

Submitter: Roger R. Calam, *Department of Pathology, St. John Hospital, Saint Clair Clinical Pathology Laboratories, Detroit, MI 48236*

Evaluators: Thomas A. Blumenfeld, *Babies Hospital, Columbia Presbyterian Medical Center, New York, NY 10032*

Samuel Meites, *Clinical Chemistry Laboratory, Children's Hospital, Columbus, OH 43205*

Jean M. Slockbower, *Department of Laboratory Medicine, Mayo Clinic, Rochester, MN 55901*

Bernard E. Statland, *Division of Pathology, University of California-Davis Medical Center, Sacramento, CA 95817*

Reviewer: O. W. van Assendelft, *Hematology Division, Center for Infectious Diseases, Centers for Disease Control, Atlanta, GA 30333*

Introduction

Long ago, the Greeks established the humoral theory in which black and green bile, phlegm, and blood were considered to be the four "humors" of the human body. This theory was responsible for the practice of bloodletting (phlebotomy). The ancients believed that blood removed from the veins would release the patient from the evil causes of disease. Today, phlebotomy, or venipuncture, has a much different meaning. The object of invading a vein with a needle attached to a syringe or evacuated collection tube is to obtain a blood specimen for the multiple analysis of its constituents in the clinical laboratory. The clinician will use the results to diagnose disease and to monitor the course of treatment for the patient. Blood collection also includes arterial and skin-puncture collection, the latter term being preferred to the less accurate one, "capillary" collection (1, 2).

The importance of specimen collection has recently been reviewed (3). There is a tendency to underestimate its importance, and therefore many laboratories assign blood collecting to inadequately trained personnel. Laboratorians must realize that, regardless of the merit of the technologist and the method used, the accuracy and precision of laboratory data depend upon the quality of the specimen collected.

Recent formation of the National Phlebotomy Association (4–6) should help promote the increasingly acknowledged importance of specimen collection. In addition, the National Committee for Clinical Laboratory Standards (NCCLS) has written or is in the process of writing several standards directed at the correct methods for collecting blood specimens (7–9). NCCLS has also published standards for evacuated blood-collection tubes (10) and for the handling and transfer of specimens (11).

The Patient

Ideally, for a test result to be meaningful, a blood specimen should reflect the patient's physiology at the time of collection. Some tests involve a challenge to the patient's metabolic response. A common example is the glucose-tolerance test: the patient should be on a carbohydrate diet, 1.75 g/kg of body weight, for three days preceding the test (12), then fast for 12–14 h before (and during) the test. Although the glucose-tolerance test and other tolerance tests (e.g., for lactose) require several fasting specimens, some tests require only one fasting specimen (single glucose, cholesterol, triglyceride).

Many tests do not require the fasting state, but it still is preferable that patients not have food for at least 4–6 h before venipuncture (13). This will reduce the possibility of chylomicron lipemia, which can interfere in many methods that are not appropriately blanked.

A greater problem than dietary interference is interference from medications (14). A comprehensive cross-referenced publication by Young et al. shows the current magnitude of this problem (15); this reference should be accessible to every clinical laboratory. When possible, known or potentially interfering medications should be discontinued for at least 48 h before blood collection.

Consumption of caffeine and alcohol and the use of tobacco are very common. Statland and Winkel have thoroughly reviewed physiological variation attributable to these materials (16). Caffeine increases the concentrations of free fatty acids in plasma, probably through hormonal changes. Nicotine can induce erythrocytosis and leukocytosis, changes mediated by an increase in plasma cortisol. Smokers also have higher carboxyhemoglobin content in their blood than do nonsmokers. Alcohol causes a pronounced increase in γ-glutamyltransferase activity, the increase being observable approximately three days after ingestion. The amount of this enzyme activity also depends upon the amount and the type of alcoholic drink.

Laboratorians should recognize that an interference can be methodological, physiological, or both. Only development of tests with greater specificity can overcome the methodological problem. Eliminating the undesired stimulus is the surest way to prevent physiological interference. The time may come when we will define a period of abstinence before the collection of blood for particular tests, and will be able to qualify results with the information reported on a patient's questionnaires. For example, noting a patient's recent strenuous exercise would be appropriate if apparently abnormal values for enzymes, particularly creatine kinase, were found; such values may reflect a normal physiological response to muscular stress (17). Fist-squeezing, or "hand pumping," during the collection process is now well recognized as causing abnormal potassium results (18).

Patients are categorized as outpatients or inpatients; these categories suggest individuals who are, respectively, ambulatory or for the most part recumbent. This may be an oversimplification, but it is known that variations in a patient's posture can affect test results (19, 20). An individual can have different results for certain tests, depending upon whether the individual is lying down, sitting, or standing at the time of venipuncture (21). The influence of posture on laboratory tests, in terms of inpatient/outpatient, is not totally new. As early as 1946, the average inpatient (recumbent) albumin result was approximately 3.7 g/dL, whereas the average outpatient (ambulatory) result was 4.2 g/dL. This difference is attributable to a normal physiological response. During a change from a supine to an upright position, there is a hemoconcentration effect, as body water shifts from the vascular compartment to the interstitium; hemodilution occurs in the recumbent position. Therefore, any substance that is transported in the blood by protein (e.g., drugs, hormones) will also be affected by this concentration/dilution effect. The physiological change is rapid, taking about 15 min. A notation on patient posture could be an important qualifier to a patient's results.

Tourniquet stasis causes a hemoconcentration effect similar to that caused by posture. Tourniquets used for venous collection should not be applied so tightly that the pulse of the radial artery cannot be felt. A tourniquet is meant to increase venous filling, causing the veins to become more prominent. Excessive application of the tourniquet (longer than 1 min) on the limb causes hemoconcentration. The effect is more evident with some analytes than others. If a tourniquet is removed and repositioned without allowing 2–3 min of blood flow before repositioning, results may be unreliable. Specimens for some analytes (e.g., lactic acid, pH, arterial gases) should not be drawn with use of a tourniquet. Bokelund et al. have documented the tourniquet as a contributor to variable results (22).

Collection Sites

Venous Specimens

Venipuncture is the most frequent mode of collection, and veins of the anticubital fossa (median cubital, cephalic) are most commonly used. These veins are usually large and easily palpated for selecting the puncture site. Inspection of both arms may help reduce the frequency of unsuccessful puncture. Veins of the wrist, ankle, and back of the hand are also sites for venipuncture.

Arterial Specimens

The radial artery at the wrist and the brachial artery of the arm are the sites most commonly chosen for arterial puncture. Tourniquets are not necessary with arterial collection because the arteries have sufficient hemostatic pressure to be distended; moreover, the pulse helps locate the position of an artery and differentiates it from a vein. Local anesthetic at the puncture site, procaine or xylocaine, is sometimes advocated, but may cause additional problems and discomfort; abandoning this practice should be seriously considered. It is mandatory that strong positive pressure be applied directly to the puncture site for at least 5 min after the collection, and perhaps longer if a patient's hemostasis is to any extent compromised. The pressure should be applied by the person collecting the specimen, not by the patient. If bleeding persists, a physician must be informed.

Skin-Puncture Specimens

There are several skin-puncture sites: earlobe, palmar surface of the fingertip, and the plantar surface of the big toe or the heel. The most common site for the newborn and young infants is the heel. Recent investigation, however, has demarcated a heel zone that should *never* be used for skin-puncture collection because of the potential hazards of penetrating the calcaneous and causing necrotizing chondritis (23, 24). There are only two areas where heel puncture should be permitted: *(a)* medial to a line drawn posteriorly from the middle of the great toe to the heel, and *(b)* lateral to a line drawn posteriorly from between toes 4 and 5 to the heel (2, 8, 23). Penetration of the calcaneous is a very real possibility when the puncture is deeper than 2.4 mm. The Bard-Parker scalpel blade should not be used as a puncturing device. Phlebotomists should use a lancet that cannot puncture deeper than 2.4 mm. Several commercial lancets and their tip dimensions have been listed (2). A site previously used, or one that is unhealed, infected, or edematous, should never be reused.

After the skin puncture, it is important that the blood flow freely. Discard the first drop of blood after a puncture because of contamination from the damage caused by the puncture itself. Squeezing or "milking" the area will presumably contaminate the specimen with tissue juices. Perhaps adequately warming the collection site to 40–42 °C for approximately 3 min with a warm, moist towel, like that used for pH/blood gas collections, should be used for all heel punctures; this procedure can increase the blood flow as much as sevenfold (1). Skin-puncture collection is also useful with obese patients and others whose veins may be difficult to locate. Skin puncture is often the only recourse for burn patients and for those on chemotherapy.

Catheters

Catheters can be inserted into the arteries and veins of the arm or leg. They are also inserted into the umbilical artery of the newborn. When blood is obtained through a catheter, the first few milliliters of blood are disposed of because the blood is diluted by the anticoagulant used to flush the catheter lock. Results from this aliquot would be unreliable. In the case of newborns, some neonatologists draw off 1–2 mL of anticoagulated blood, which is then put back into the patient after an additional aliquot (for the laboratory analyses) is taken with a second syringe. Umbilical collection is obtained by a physician.

Patients Receiving Intravenous Treatment

For patients receiving medication or intravenous nutrition, avoid drawing a specimen from the limb with the intravenous insertion. Drawing distal to the insertion does not guarantee a representative specimen unless a tourniquet is placed between the intravenous needle and the puncture site, and unless the test requested is not glucose when glucose is being administered (25). If there is no other choice (consider the other arm), it is sometimes possible to have the attending physician shut off the intravenous flow for approximately 3 min, and then draw the specimen. Regardless, all such test results should be qualified on the laboratory report.

Irrespective of the mode of collection—venipuncture, arterial, or skin puncture—edematous sites or hematomas should be avoided, because results from these sites are invalid.

Collection Apparatus

Blood-collecting apparatus consists of the following: tourniquets, cleansing swabs, gauze, needles, adapters, evacuated tubes, syringes, and micro collecting devices (e.g., glass capillaries, lancets).

Tourniquets. Tourniquets should be kept clean, because they may carry infectious agents. Irreversibly stained tourniquets should be replaced, as should those used on patients with infectious disease.

Cleansing swabs. The low cost of disposable, individually wrapped, 70% (700 mL/L) isopropanol swabs should replace all standing containers of sponges or cotton balls soaking in 70% alcohol. Use of these individual swabs gives greater assurance that the puncture site is aseptic. When collecting a specimen for an alcohol determination, use Zephiran chloride (benzalkonium chloride), a non-alcoholic preparation, for cleansing.

Povidone-iodine is commonly used to cleanse the puncture site when collection is for microbiology specimens. Avoid collecting any specimen for the chemistry laboratory after the use of this agent, because it may lead to falsely high results for certain tests (potassium, phosphorus, uric acid) (26, 27).

Allow the intended puncture site to dry after cleansing. This prevents hemolysis by the cleansing agent. A dry site can also reduce the pain of venipuncture (28).

Needles. The multiple-draw needle, designed to eliminate dripping blood during the collection of multiple tubes from a single venipuncture site, has evolved as the commonest needle in current use. All needles are available in several gauges (gauge number indicates needle size; the larger the gauge, the smaller the needle). Gauges 18 through 22 are most commonly used. Look at the tip of the needle to determine that it is open and free of barbs or hooks that could be painful to the patient. Sterile, disposable needles are marketed. A needle cutter should be used before disposing of a needle.

Evacuated tubes. The Vacutainer Tube[1] evacuated collection tube, in use now for several years, was the first of its kind to be commercially available (29). More recently, other evacuated collection tubes have been introduced, the more advanced containing a gel material and a clotting activator that facilitates specimen processing (SST[1], Corvac[2]). Collection tubes come in several different sizes with or without anticoagulants and preservatives. The smaller tubes are for pediatric collection. Virtually all are siliconized, including the stopper. The silicone coating enhances wiping characteristics of the inner surface, thus reducing clot-cell hangup.

Stoppered evacuated collection tubes should not be used to collect specimens for zinc or copper determinations, because these trace elements are often present in the rubber stopper (30–32). A plasticizer,

---

[1] Vacutainer™, SST™, Microtainer™: Becton-Dickinson, Division of Becton, Dickinson & Co., Rutherford, NJ 07070.
[2] CORVAC™: Monojet, Division of Sherwood Medical, A Brunswick Co., St. Louis, MO 63103.

tributoxyethyl phosphate (TBEP), an interferent in gas-chromatographic procedures for toxicology, is also leached from rubber stoppers as well as from plastic syringes *(33, 34)*. Until recently, acid-washed glass syringes had to be used for the collection of blood for trace metals or toxicology determinations. A Vacutainer Tube (royal blue stopper, with or without heparin) is now available that contains minimal trace-element contamination (e.g., zinc, copper, iron); it is also claimed that TBEP interference has been eliminated from this tube *(35)*. Another new tube (brown stopper) has minimal lead content (less than 0.1 µg/tube).

Currently, sterile and nonsterile evacuated collection tubes are available; increasingly, sterile collection tubes are being used. In 1975 Katz et al. *(36)* published a paper documenting a case of septicemia attributable to "backflow hazard" from a nonsterile evacuated blood-collecting tube; additional publications have also dealt with this problem *(37, 38)*. Since this alert, all companies manufacturing collection tubes have issued specific directions (package inserts) for blood collection. All laboratorians responsible for and involved with specimen collection should consult these instructions to make sure that the correct procedure is carried out. Eventually, *all* evacuated collection tubes will be sterile, which will obviate many of the difficulties one encounters when attempting to use exacting collection techniques.

Syringes. Although the idea of the hypodermic syringe was conceived in the 1800s, many decades passed before this concept was applied to collecting blood specimens. Until early 1900, blood was collected by performing a surgical "cut-down" to expose the vein, followed by venesection. Large amounts of blood were required because micromethods were virtually nonexistent. After its introduction, the syringe was the principal device for blood collection until the advent of Vacutainer Tubes (late 1940s). In Europe, syringe collection remains the main technique.

Syringes are still commonly used to collect specimens for pH/blood gases determinations and for toxicology. For arterial collection, glass syringes are favored because the plunger responds to arterial pressure, enabling the phlebotomist to know when the vessel has been entered. Newer plastic syringes, designed specifically for collection of specimens for pH/blood gas determinations, also have plungers that respond to arterial pressure. Petty et al. have reported on an innovative method for blood gas collection involving a sampling device without a plunger *(39)*. Plastic should not be used for the collection of toxicology specimens *(33, 34)*. A detailed standard for arterial puncture is currently being drafted by NCCLS *(9)*.

Microcollection devices. Microcollection devices are used after skin puncture. The blood may be collected dropwise into a small tube or by capillary action into a capillary tube. In the case of Microtainer Tubes[1] blood passes through a capillary into a small plastic tube *(40)*. The Microtainer Tube containing EDTA is designated for hematology blood collections. Several glass capillaries are commercially available with or without anticoagulant. The anticoagulant exclusively used is ammonium heparin. Capillary devices are sealed with either a cap or a claylike sealant. A NCCLS standard (PSH-14) is currently being drafted for microcollection devices *(41)*.

Miscellaneous

1. Liquid crystals: Davison et al. have written an excellent publication dealing with venous physiology and the visualization of arm veins by use of liquid crystal technology *(42)*.

2. Arteriolization: The process of warming the skin-puncture site to 40–42 °C for approximately 3 min is commonly used and necessary for the collection of newborn or pediatric blood gas/pH specimens *(43)*. It should also be considered when the desired volume of blood to be collected could be a problem. An inexpensive and very effective method is to use a towel that has been warmed and moistened with hot tap water. If the temperature of the towel is not too hot to be held in the collector's hand, it will not burn the patient and will adequately warm the puncture site. The younger the patient, the more careful the collector must be with the towel temperature—burn damage to the site is possible. Recently, a new disposable product, T-Pak[3], has been introduced *(2)* that provides controlled heat at the puncture site. The T-Pak chemical reaction, when triggered, is instantly and constantly at 40 °C for 30 min. Use of T-Pak allows better control of the arteriolization process, but at added expense.

Anticoagulants and Preservatives

To obtain plasma or whole blood rather than serum, an anticoagulant is used. Except for heparin, which inhibits prothrombin, thrombin, and platelet aggregation, anticoagulants chelate or bind calcium. Obviously, nonheparinized plasma cannot be used for most calcium determinations.

There are several collection tube systems. Many phlebotomists select their tubes by the color of the stopper. However, there are now several different additive concentrations and combinations in tubes with the same colored stoppers. It is important that the person collecting the blood specimen know what concentration of anticoagulant or preservative should be used. It is also important to recognize that

---

[3] T-Pak™: Kay Laboratories, Inc., Medical Division, San Diego, CA 92138.

the concentration of an additive in a tube must be appropriate for a particular amount of blood. Too often, incomplete collections are made because of collecting difficulties, attempts to conserve blood, or to save time. The laboratorian should consider whether there will be any influence on the analyte to be measured or on the method to be used by having the additive present in a concentration greater than intended.

Heparin. Heparin is a sulfated mucopolysaccharide available as the sodium, lithium, or ammonium salt. A salt form that will interfere with the analyte to be measured should not be used. Heparin is very effective at 0.2 mg/mL of blood. Unfortunately, the anti-clot activity does not persist and the plasma develops fibrin when stored longer than 24 h. Heparinized plasma is recommended for potassium determinations (44). Potassium concentrations in serum will be increased by approximately 0.2–0.4 mmol/L because of the clotting process. Other advantages and disadvantages of serum vs plasma have been explored by several investigators (44–47). One must be especially careful in syringe collection of pH/blood gas specimens; too much heparin will dilute the blood and falsify the results. Heparin is expensive; otherwise, it would probably be used exclusively in place of other anticoagulants.

Oxalate/fluoride. Potassium oxalate is frequently used in combination with sodium fluoride. The oxalate acts as the anticoagulant, the fluoride (a weak anticoagulant) as a preservative for glucose, inhibiting glycolysis. Sodium fluoride is also a urease inhibitor and cannot be used with urease methods for determining urea nitrogen. Potassium oxalate is more often used than the sodium salt because of its greater solubility. The usual concentration for each, oxalate and fluoride, is 2.0 mg/mL of blood. Unfortunately, both agents alter erythrocyte membrane permeability, resulting in a slightly hemolyzed plasma.

Iodoacetate. Sodium iodoacetate, which is not an anticoagulant, has been reported to be effective as an antiglycolytic agent (48). The hemolysis observed with fluoride is eliminated when the proper iodoacetate concentration is used. It does not have anti-urease activity (48), but does inhibit creatine kinase (49). Instability of iodoacetate limits its reliability.

EDTA. Ethylenediaminetetraacetic acid (EDTA) is a calcium chelating agent, effective at a concentration of 1 to 2 mg/mL of blood. The dipotassium salt is commonly used, being more soluble than the disodium salt; tripotassium EDTA (liquid) is also in use. EDTA is widely used as an anticoagulant in hematology because it does not produce changes in erythrocyte volume.

Citrate. Sodium citrate (3.8% w/v; 38 g/L) is effective as a liquid anticoagulant when used in a ratio of one part of citrate to nine parts of blood. It is the preferred anticoagulant for coagulation studies because it appears to best preserve labile procoagulants. Citrate chelates calcium.

The Collection

Primum non nocere—"The first thing is to not inflict damage." Use of the correct procedure for venipuncture, skin-puncture, and arterial blood collection is critical. Detailed procedures have recently been published (7–9). The possibilities for serious problems at a site of multiple punctures, the infant's heel, have already been discussed. The availability of the veins commonly used for venipuncture must be preserved. Veins for the administration of therapeutic agents must be kept patent.

Phlebotomists should remember that they are representatives of the laboratory. Personal neatness and cleanliness are very important. The patient should perceive in the phlebotomist an image of friendliness and self-confidence. For child or adult, inpatient or outpatient, anxiety should not be a part of the collection process. Be truthful and inform the patient that there will be discomfort, but only momentarily (28). Keep in mind that the patient does have the right to refuse blood collection. If the patient cannot be persuaded to cooperate, inform his or her physician.

Identification. Correct identification of the patient is crucial. Matching the name on the patient's wrist band with the name on the request slip is the surest way for positive identification. Excellent data from the laboratory are of no value if they are reported on the wrong patient, and could result in improper care. Correct labeling of the specimen collected is mandatory.

Isolation. Collecting blood from a patient in isolation is potentially hazardous not only to the phlebotomist but also to all laboratory personnel who will subsequently handle the specimen. Appropriate alert precautions (labeling) should be systematized within the institution. Any other laboratory to which the specimen may be sent for additional analyses must, of course, be informed of possible hazards.

Timed tests. Some specimens must be obtained at timed intervals because metabolic challenge (e.g., glucose tolerance test), therapeutic drug monitoring (e.g., digoxin), or biological rhythm (e.g., iron, cortisol) may be involved. It is important to collect these specimens on time and to label them correctly. Intra-individual variation (within-day and day-to-day) has been documented (50).

Order of collection. When several specimens are needed for different determinations, the order of drawing different tubes can be important. The usual recommended order is as follows: sterile tubes for blood culture, plain tubes, coagulation tubes, and tubes with additives. The order should remain the same even when all tubes are sterile.

Transport. Transport specimens to the laboratory without delay after collection. When necessary (as for determinations of pH/gases, ammonia, etc.), inhibit cellular metabolism within the collected specimen by using ice chips *in water*. If access to ice is not possible, commercial cooling aids are available (T-Pak).[3] The concept of prechilling the collection device may become common in the future as new analytical methods for more labile substances become available. Although specimens for the most part are carried to the laboratory by hand, pneumatic-tube and gravity transport systems have been utilized *(51–52)*.

Specimen Processing

Process specimens rapidly and remove the sample from the cells within 2 h after the specimen is collected *(59)*. Allow a period of 20 to 30 min before processing blood for serum. The clotting process, if not complete, will reduce expected yield and possibly delay the availability of the sample for analysis. One of the major advantages of the new integrated collection/processing tubes (Corvac, SST) is the rapid clotting that occurs (10–15 min), which is induced by including an activator (glass or silica particles) in the tube itself.

Glucose, the most frequently requested clinical chemistry determination, in adults remains stable when in contact with the cells for approximately 24 h at 25 °C and up to 48 h at 4 °C, if fluoride is a part of the collected specimen. In the absence of a glycolytic inhibitor, glucose concentration is significantly reduced within 2 h. The rate of decrease is even faster in newborn blood because of the greater metabolic activity of the newborn's erythrocytes. Even with an inhibitor, there is no ideal way to completely inhibit glycolysis in the blood from newborns or in blood from some adults *(60)*. Without an inhibitor it is mandatory that serum or plasma be separated from the cells as rapidly as possible. The newer processing devices involving gels are quite useful, but the formed barrier must completely separate the cells from the serum.

Certain blood collections are cooled in a mixture of ice and water because of analyte instability. These specimens require rapid processing, and refrigerated centrifugation is appropriate. Laboratorians should recognize a prevailing contradiction: often, a specimen that has been transported in ice is put into a warm centrifuge. All laboratories should have at least one refrigerated centrifuge for labile specimens. Bench-top refrigerated centrifuges are now available. The additional expense is justified. Generally, refrigerated centrifuges are set to maintain 20–25 °C. Cooler temperatures can be used when indicated. Although refrigeration enhances analyte stability, potassium specimens should not be centrifuged at 4 °C; potassium moves out of the cell faster at this temperature than at 25 °C *(61)*.

The centrifuge is critical for proper specimen processing. It is very important that the centrifuge operate according to the manufacturer's specifications. Proper operation can be critical for the gel-separating devices. If the *g*-force is not adequate, proper gel separation does not take place. Many laboratories still confuse rotational speed (rpm) with relative centrifugal force (RCF), i.e., *g*-force. Given the radius between the axis of rotation and the center of the tube being centrifuged, the RCF can be calculated as follows:

$$RCF = 1.118 \times 10^{-5} \times r \times (rpm)^2$$

RCF = relative centrifugal force, in *g* units
1.118×10^{-5} = a constant
r = the radius, in centimeters, between the axis of rotation and the center of the tube being centrifuged
rpm = revolutions per minute

Today's clinical chemistry laboratory, except for pH/blood gas analyses, infrequently makes determinations on whole-blood specimens. Whole-blood sodium and potassium determinations may increase in the future, however, particularly for pediatric patients, in view of the ion-selective electrode technology that has recently become available *(57)*. Regardless, most blood specimens for clinical chemistry will continue to be processed, with serum or plasma harvested and used as the sample in the analytical system. A detailed and structured approach that addresses pre-analytical factors and their influence on variation in results has been published by Statland and Winkel *(16)*, and the NCCLS is currently preparing a standard procedure for processing blood specimens *(58)*.

In 1977 Calam comprehensively reviewed processing devices *(3)*; Corvac and SST integrated collection/separation tubes were then considered to be the most technically advanced of the processing devices. Except for slight material modifications, no significant changes in processing technology have occurred since then. The Corvac and SST tubes now contain revised gel formulations made necessary because of tube sterilization procedures. Clinical trials conducted in our laboratory, comparing the new Corvac with the old, indicate no statistical or clinically significant differences in results when using one tube rather than the other. The Sure-Sep[4] device also has a revised silicone gel to facilitate the centrifugal delivery of the gel.

In the selection of a processing device, several questions should be considered:

---

[4] Sure-Sep™: General Diagnostics, Division of Warner-Lambert Co., Morris Plains, NJ 07950.

1. Can hemolysis caused by excessive handling be eliminated through the use of less serum centrifugation and decanting, or less pipetting? Less handling also reduces identification errors, and the time saved means that the test can be done sooner.

2. Can the stopper remain in place, avoiding aerosol formation? Possible evaporation effects would also be eliminated.

3. Is the yield of serum greater than that obtained by conventional processing? Perhaps fewer tubes of blood need to be drawn, thus conserving the patient's blood.

4. Does the device interfere with any of the laboratory's methods? Beware of interference from plastic *(33, 34, 56)*.

Specimen collection involves more than physically procuring a tube of blood. Laboratorians must not forget that the patient is the "alpha and the omega" of all analysis: the alpha, because the patient is where everything begins—without the patient there would be no specimen, no sample for analysis; the omega, because test results are not meant to embellish laboratory statistics, but rather to contribute to the care and health of the patient. The laboratory must continue to meet the challenge as advances in blood-collection technology occur. Excellence in laboratory medicine is dependent upon the quality of the specimen collected.

References

1. Blumenfeld, T. A., Hertelendy, W. G., and Ford, S. H., Simultaneously obtained skin-puncture serum, skin-puncture plasma, and venous serum compared, and effects of warming the skin before puncture. *Clin. Chem.* **23**, 1705–1710 (1978).
2. Meites, S., and Levitt, M. J., Skin-puncture and blood-collection techniques for infants. *Clin. Chem.* **25**, 183–189 (1979).
3. Calam, R. R., Reviewing the importance of specimen collection. *J. Am. Med. Technol.* **39**, 297–302 (1977).
4. Crawford, D. C., Forming the National Phlebotomy Association. *Lab World* **29**, 102 (1978).
5. Moore, D., Improving the quality of performance. *Lab World* **29**, 66 (1978).
6. Tyslan, D. R., Phlebotomist draws blood in battle for recognition and certification. *Lab World* **29**, 82 (1978).
7. NCCLS, ASH-3 (approved standard). Standard procedures for the collection of diagnostic blood specimens by venipuncture. NCCLS, 711 E. Lancaster Ave., Villanova, PA, 1978.
8. NCCLS, TSH-4 (tentative standard). Standard procedures for the collection of diagnostic blood specimens by skin-puncture. NCCLS, Villanova, PA, 1978.
9. NCCLS, PSH-11 (proposed standard). Standard for percutaneous collection of arterial blood for laboratory analysis. NCCLS, Villanova, PA, 1978.
10. NCCLS, ASH-1 (accepted standard). Standard for evacuated tubes for blood specimen collection. NCCLS, Villanova, PA, 1978.
11. NCCLS, TSH-5 (tentative standard). Standard procedure for the handling and transport of diagnostic medical specimens and etiologic agents. NCCLS, Villanova, PA, 1978.
12. Tietz, N. W., Ed., *Fundamentals of Clinical Chemistry*, 2nd ed. W. B. Saunders Co., Philadelphia, PA, 1976, p 254.
13. Annino, J. S., and Relman, A. S., The effect of eating on some of the clinically important chemical constituents of the blood. *Am. J. Clin. Pathol.* **31**, 155–159 (1959).
14. Caraway, W. T., and Kammeyer, C. W., Chemical interferences by drugs and other substances with clinical laboratory test procedures. *Clin. Chim. Acta* **41**, 395–434 (1972).
15. Young, D. S., Pestaner, L. C., and Gibberman, V., Effects of drugs on clinical laboratory tests. *Clin. Chem.* **21**, ID–432D (1975).
16. Statland, B. E., and Winkel, P., Effects of preanalytical factors on the intra-individual variation of analytes in the blood of healthy subjects: Consideration of preparation of the subject and time of venipuncture. *CRC Crit. Rev. Clin. Lab. Sci.* **8**, 105–144 (1977).
17. King, S., Statland, B. E., and Savory, J., The effects of a short burst of enzyme activity values in sera of healthy subjects. *Clin. Chim. Acta* **72**, 211–218 (1976).
18. Romano, A. T., and Young, G. W., Mild forearm exercise during venipuncture, and its effects on potassium determinations. *Clin. Chem.* **23**, 303–304 (1977).
19. Dixon, M., and Paterson, C. R., Posture and the composition of plasma. *Clin. Chem.* **24**, 824–826 (1978).
20. Kelly, A., Munan, L., PetitClerc, C., and Billon, B., Impact of posture on reference ranges for serum calcium and protein. *Clin. Chem.* **23**, 1938 (1977). Letter.
21. Statland, B. E., Bokelund, H., and Winkel, P., Factors contributing to intra-individual variation of serum constituents: 4. Effect of posture and tourniquet application on variation of serum constituents in healthy subjects. *Clin. Chem.* **20**, 1513–1519 (1974).
22. Bokelund, H., Winkel, P., and Statland, B. E., Factors contributing to intra-individual variation of serum constituents: 3. Use of randomized duplicate serum specimens to evaluate sources of analytical error. *Clin. Chem.* **20**, 1507–1512 (1974).
23. Blumenfeld, T. A., Turi, G. K., and Blanc, W. A., Recommended site and depth of heel puncture in newborns based on anatomical measurements and histopathology. *Lancet* **i**, 230–233 (1979).
24. Lilien, L. D., Harris, V. J., Ramamurthy, R. S., and Pildes, R. S., Neonatal osteomyelitis of the calcaneous: Complication of heel puncture. *J. Pediatr.* **88**, 478–480 (1976).
25. Ong, Y. Y., Boykin, S. F., and Barnett, R. N., You can draw blood from the "IV arm" below the intravenous needle if you put a tourniquet in between. *Am. J. Clin. Pathol.* **72**, 101–102 (1979).
26. Hait, W. N., Snepar, R., and Rothmen, C., False-positive hematest due to povidone-iodine. *N. Engl. J. Med.* **297**, 1350–1351 (1977).
27. Van Steirteghem, A. C., and Young, D. S., Povidone-iodine (Betadine) disinfectant as a source of error. *Clin. Chem.* **23**, 1512 (1977).
28. Philips, P. J., Pain, R. W., and Brook, G. E., The pain of venipuncture. *N. Engl. J. Med.* **294**, 116 (1976).
29. Barnard, R. D., Gordon, G. B., and Weitzner, H. A., The preservation of thromboplastin suspensions in the frozen state. *NY State J. Med.* **49**, 1057–1058 (1949).
30. Healy, P. J., Turvey, W. S., and Willats, H. G., Interference in estimation of serum copper concentrations resulting from use of silicone-coated tubes for collection of blood. *Clin. Chim. Acta* **88**, 573–576 (1978).
31. Helman, E. Z., Wallick, D. K., and Reingold, I. M.,

Vacutainer contamination in trace element studies. *Clin. Chem.* **17**, 61–62 (1971).
32. Hughes, R. O., Wease, D. F., and Troxler, R. G., Collection of blood uncontaminated with Ca, Cu, Mg, or Zn, for trace-metal analysis. *Clin. Chem.* **22**, 691–692 (1976).
33. Dusci, L. J., and Hackett, L. P., Interference in Dilantin assays. *Clin. Chem.* **22**, 1236 (1976).
34. Missen, A. W., and Dickson, S. J., Contamination of blood samples by plasticizer in evacuated tubes. *Clin. Chem.* **20**, 1247 (1974).
35. Lin, F. C., Fay, A., Iannuzzelli, V. F., and Narayanan, S., Evaluation of an improved Vacutainer Tube for trace metals and routine clinical chemistry determinations. *Clin. Chem.* **24**, 1019 (1978).
36. Katz, L., Johnson, D. L., Neufeld, P. D., and Gupta, K. G., Evacuated blood collection tubes—the backflow hazard. *Can. Med. Assoc. J.* **113**, 208–213 (1975).
37. Baer, K., Ed., Contaminated vacuum tubes pose hazards. *Hosp. Infect. Control* **2**, 141–145 (1975).
38. Baer, K., Ed., Attention focused on vacuum tube sterility. *Hosp. Infect. Control* **3**, 129–131 (1976).
39. Petty, T. L., Bailey, D., and Best, C., A new device for arterial blood gas sampling. *J. Am. Med. Assoc.* **239**, 2016–2017 (1978).
40. Hicks, J. M., Rowland, G. L., and Buffone, G. J., Evaluation of a new blood-collecting device (Microtainer) that is suited for pediatric use. *Clin. Chem.* **22**, 2034–2036 (1976).
41. NCCLS, PSH-14 (proposed standard). Devices for the collection of skin puncture blood specimens. NCCLS, Villanova, PA, 1979.
42. Davison, T. W., Studebaker, E., and Blessington, J., Minimizing difficult venipuncture situations using venous physiology and liquid crystals. *Clin. Lab. Prod.* **7**, 20–36 (1978).
43. Gambino, S. R., Collection of capillary blood for simultaneous determinations of arterial pH, CO_2 content, p_{CO_2} and oxygen saturation. *Tech. Bull. Regist. Med. Technol.* **31**, 1–9 (1961).
44. Lum, G., and Gambino, S. R., A comparison of serum vs heparinized plasma for routine chemistry tests. *Am. J. Clin. Pathol.* **61**, 108–113 (1974).
45. Kubasik, N. P., and Sine, H. E., Results for serum and plasma compared in 15 selected radioassays. *Clin. Chem.* **24**, 137–139 (1978).
46. Ladenson, J. H., Tsai, L. M. B., Michael, J. M., Kessler, G., and Joist, J. H., Serum versus heparinized plasma for eighteen common chemistry tests. *Am. J. Clin. Pathol.* **62**, 545–552 (1974).
47. Hoke, S. R., and Trent, J. L., Non-effect of heparin on serum electrolyte values. *Clin. Chem.* **22**, 1540 (1976).
48. Marbach, E. P., McLean, M., Scharn, M., and Jones, T., Sodium iodoacetate as an antiglycolytic agent in blood samples. *Clin. Chem.* **21**, 1810–1812 (1975).
49. Moore, J. J., Sax, S. M., and Narayanan, S., Efficacy of iodoacetate as a glycolytic inhibitor and its utility in clinical chemistry procedures. *Clin. Chem.* **24**, 998 (1978).
50. Winkel, P., Statland, B. E., and Bokelund, H., Factors contributing to intra-individual variation of serum constituents: 5. Short-term day-to-day and within-hour variation of serum constituents in healthy subjects. *Clin. Chem.* **20**, 1520–1527 (1974).
51. Lapidus, B. M., and Dehner, G. F., Gravity delivery of laboratory specimens. *Am. J. Clin. Pathol.* **64**, 127–135 (1975).
52. Lapidus, B. M., and Mac Indoe, R. C., A slow-speed gravity delivery system for laboratory specimens. *Am. J. Clin. Pathol.* **68**, 243–249 (1977).
53. Lapidus, B. M., and Mac Indoe, R. C., A multi-station gravity delivery system. *Am. J. Clin. Pathol.* **69**, 73–76 (1978).
54. Pragay, D. A., Edwards, L., Toppin, M., Palmer, R. R., and Chilcote, M. E., Evaluation of an improved pneumatic-tube system suitable for transportation of blood specimens. *Clin. Chem.* **20**, 57–60 (1974).
55. Weaver, D. K., Miller, D., Leventhal, E. A., and Tropeano, V., Evaluation of a computer-directed pneumatic-tube system for pneumatic transport of blood specimens. *Am. J. Clin. Pathol.* **70**, 400–405 (1978).
56. Brown, H. H., Vanko, M., and Meola, J. M., Interference from serum separators in drug screening by gas chromatography. *Clin. Chem.* **20**, 919 (1974).
57. Ladenson, J. H., Direct potentiometric measurement of sodium and potassium in whole blood. *Clin. Chem.* **23**, 1912–1916 (1977).
58. NCCLS, PSH-18 (proposed standard). Standard for the processing of blood specimens. NCCLS, Villanova, PA, 1980.
59. Laessig, R. H., Indriksons, A. A., Hassener, D. J., Paskey, T. A., and Schwartz, T. H., Changes in serum chemical values as a result of prolonged contact with the clot. *Am. J. Clin. Pathol.* **66**, 598–604 (1976).
60. Meites, S., and Saniel-Banrey, K., Preservation, distribution, and assay of glucose in blood, with special reference to the newborn. *Clin. Chem.* **25**, 531–534 (1979).
61. Oliver, T. K., Young, G. A., Bates, G. D., and Adamo, J. S., Factitial hyperkalemia due to icing before analysis. *Pediatrics* **38**, 900–902 (1966).

Transportation of Specimens

Submitters: Seymour Winsten and Stanley E. Gordesky, *Chemistry Section, Division of Laboratories, Albert Einstein Medical Center, Philadelphia, PA 19141*

Evaluators: Dafne Cimino, *Boston Medical Laboratory, Inc., Waltham, MA 02215*

Robert M. Johnson, *Department of Pathology, St. Joseph Mercy Hospital, Ann Arbor, MI 48106*

Fred A. Kincl, *The College of Staten Island, St. George Campus, Staten Island, NY 10301*

George T. Poole, *The Hospital of St. Raphael, New Haven, CT 06511*

Rita Powers, *Special Chemistry, St. John's Hospital, Springfield, IL 62701*

Nancy I. Weinberg, *Bartholomew County Hospital, Columbus, IN 47201*

Reviewer: Frank A. Ibbott, *Department of Research, Bio-Science Laboratories, Van Nuys, CA 91405*

Introduction

In the middle of this century, when clinical chemistry was still in its infancy, most clinical laboratories performed all requested tests "in-house." However, with the development of sophisticated analytical techniques for measuring certain compounds, it became common even for large medical and referral center laboratories to send their samples to facilities specially dedicated to these analyses. We have all had the experience of requesting a determination from a large central laboratory, only to discover that the sample has been referred to yet another facility. Because many of these tests are performed infrequently, it is quite difficult to obtain reliable information about the stability of the chemical constituents. Information has usually come from the research laboratory that developed the technique, where the problem of stability of samples related only to in-house storage and not to the difficulties that might be created by transporting the samples to other laboratories perhaps as much as a continent away.

The purpose of this chapter is to present methods for handling specimens that are to be sent to referral laboratories for analysis of constituents not normally assayed in the laboratory of a hospital of 50 to 150 beds. To avoid a cumbersome presentation, we have made some assumptions about what the small laboratory can do. For example, a modern small laboratory should be capable of measuring electrolytes and performing the analyses included in a commercial 12- or 18-test profile (Table 1). Such a laboratory should have an accurate ultraviolet spectrophotometer and therefore should be able to perform a number of tests, available in kit form, in which the reagents can be stored for some period of time at a minimal cost. Assays such as amylase and lipase, for example, must be performed "in-house" in the modern clinical laboratory because of the urgency with which the result is needed: a serum amylase value needed today is of no clinical relevance if one gets the result tomorrow. Other clinically relevant information must be gained by sending a test to a referral laboratory.

The storage data presented in our tables are based on the assumption that the tests to be sent out are dispatched by mail or some other common carrier, and not by a direct courier to a local laboratory. In the event that samples are sent to a laboratory by courier, more latitude is available to both the sending and receiving laboratories in the manner of specimen handling. The sender should communicate with the reference laboratory about necessary guidelines for transporting samples.

The information presented herein was obtained from critical reviews of sample-handling techniques provided by several referral laboratories, from an analysis of some of the pertinent literature on this subject, and from our own experience.

Table 1. On-Site Tests [a]

Alanine aminotransferase	Creatinine
Albumin	γ-Glutamyltransferase
Alkaline phosphatase	Glucose
Amylase	Lactate dehydrogenase
Aspartate aminotransferase	Lipase
Bilirubin	Potassium
Calcium	Sodium
CO_2	Total protein
Chloride	Urea nitrogen
Creatine kinase	

[a] Given the technology currently available, these are chemistry tests that every hospital laboratory should be capable of performing.

General Rules for Handling Specimens

1. Instruct laboratory personnel in the safe and proper techniques for each of the samples to be collected. It is advisable to prepare a manual of written procedures containing instructions for proper collection of samples and general rules for handling these samples.

2. Handle samples as promptly as possible. When cells are to be separated from serum or plasma, the separation should generally take place within 2 h of the time the specimen was drawn. Except as noted in the tables, store the serum or plasma at 4 °C. In cases where the specimen must be frozen, use a freezer capable of maintaining a temperature of −20 °C. Frost-free freezers are not suitable for storing specimens because they have a freeze/thaw cycle that allows the temperature of the specimen to increase and then refreeze. This may cause some constituents, particularly enzymes and other proteins, to deteriorate. In unusual cases, such as storage of estrogen-receptor protein, it is advisable to use a freezer capable of maintaining −70 °C; these types of specimens should ultimately be shipped in solid CO_2.

3. Containers adequate for the shipping these specimens must be available in the laboratory. It is imperative to use proper containers that adequately protect the sample being shipped. For example, a cardboard carton is not adequate to ship a sample that must be kept deep-frozen for a long time; in this case, a styrofoam shipping package is preferable. All shipping containers must provide space for proper identification of the sample contained. If neither the sending nor the receiving laboratory is directly involved in the transmission of the sample, then personnel of the sending laboratory must determine the exact conditions under which the specimen will be transported: *this cannot be overemphasized.* Specimens carried by the postal service over long distances are usually shipped by air. The temperature in the cargo hold of modern aircraft is approximately −40 °C, which is excessively cold for specimens that should not be frozen. One may avoid this problem by using other types of carriers. When time is not pressing, common carriers such as United Parcel Service, Air Express, or other courier services may be employed to carry specimens. Further, almost all airlines have a small-package service in which they will hand-carry packages for a fee.

4. When a rarely measured compound is to be assayed, it is advisable to obtain an adequate sample so that repeat studies may be performed. One should split the sample, store a portion under optimum conditions "in-house," and send the other fraction of the sample to the referral laboratory. The stored sample should be retained until the referral laboratory has completed its report. This precaution is especially important in metabolic studies on patients who have had several timed specimens drawn, because interpretation of the data depends on changes occurring in these samples.

5. In cases where the referral laboratory has advocated transport procedures for a specimen, the sending laboratory should attempt to verify whether these procedures do, in fact, preserve the sample. This may be done by comparing stability data supplied by the referral laboratory with similar data obtained from independent sources, or may be determined experimentally in one's own laboratory.

6. Do not collect a specimen and send it to a referral laboratory if it is impossible to obtain a clinically valid result, owing to circumstances beyond the control of either the referral laboratory or the sender. A case in point is the rapid increase with time of ammonia in serum. Remember, an incorrect result may be worse than no result.

7. A recent publication by the National Committee of Clinical Laboratory Standards (NCCLS, Tentative Standard No. 5) generally summarizes currently acceptable procedures for handling samples, and should be used as a primary source for answers to this problem *(1)*.

Identification of Transported Samples

The laboratory should establish clerical procedures for ensuring proper handling of specimens to be transported. These procedures will be influenced by the size, staffing, and degree of automation of the laboratory, but certain basic rules must be followed:

1. Include with every sample the patient's name, an appropriate reference number, the test(s) to be performed, and the name and address of the sending laboratory.

2. Also keep on hand sufficient information about referred tests. At a minimum, these records should include patient identification, the date and time of acquisition of the sample, the date the sample was sent, the laboratory to which the sample was sent, the method of shipping the sample if different from laboratory courier, the name of the referring physician, the condition of the sample (lipemic, hemolyzed, etc.), and a record of the return of the data. A very good mechanism for keeping this information readily accessible is to develop a log book. Not only is this an easy way to collate the above information, but also periodic review of the log will allow one to determine when results are overdue and thus help trace lost specimens.

3. Many specimens sent to referral laboratories are not medically urgent and may involve time-consuming procedures. It is advantageous to inform hospital personnel and the referring physician that the sample has been transmitted to an outside facility and that the data will be returned to the laboratory

as soon as possible. A simple "sticker" attached to the patient's chart can indicate this information.

Most laboratories and physicians appear to obtain satisfactory results for analysis of routine chemical constituents of serum when they use the shipping technique currently recommended by each referral laboratory. However, one must keep in mind any additional conditions that may influence shipment. For example, samples being transported by truck or car service in summer should be placed in refrigerated containers that can maintain a temperature of approximately 15 °C during transit. In winter, one must be careful not to transport the specimens too close to the heater in the vehicle used.

Although Laessig et al. (2) indicate that many chemical substances are not affected when serum remains on a clot, it is still a useful general rule to separate sera or plasma before transit. Whole blood or clotted blood should be transmitted only when cells are required for analysis. Tables 2 and 3 summarize currently accepted requirements for the preservation, during shipment, of some of the less-frequently analyzed constituents of blood and urine.

Guidelines on the stability of the more commonly measured biological constituents have been reported (3–8). Much of the information presented in the tables in this chapter was obtained from manuals prepared by Bio-Science Laboratories and Nichols Institute, although manuals from other referral laboratories have also been reviewed.

The data from Bio-Science Laboratories, Van Nuys, CA, and from Nichols Institute, San Pedro, CA, were obtained through the cooperation of Carl Alper and John Langan, respectively.

Table 2. Blood Constituents Requiring Special Conditions[a]

Constituent	Sample	Condition
Acetone, acetoacetate [b,c]	S	Freeze
Acid phosphatase [d]	S	Add citrate (10 mg/mL); freeze
ACTH (corticotropin)	P	Add heparin (0.2 mg/mL); freeze within 15 min of collection
Alcohol	S	Add NaF (10 mg/mL) to serum
Aldolase	S	Freeze
Aldosterone	P,S	Add boric acid (25 mg/mL) or freeze
Amino acids	P	Add heparin (0.2 mg/mL); freeze
Androstenedione	S	Draw sample in a.m. (0000 to 1200 hours)
Ascorbic acid	S	Freeze
Barbiturates	S	Do not use heparin
Bile acids	S	Freeze
C peptide	S	Draw from fasting patient; freeze
Calcitonin	S	Freeze
Carcinoembryonic antigen (CEA)	P	Add EDTA (2 mg/mL)
Cholinesterase	B	Add heparin (0.2 mg/mL)
Citric acid	S	Freeze
Complement	S	Freeze
Compound S	P	Add heparin (0.2 mg/mL), separate immediately
Cortisol	P	Add heparin (0.2 mg/mL), separate immediately
Creatine	S	Freeze
Creatine kinase isoenzymes	S	Freeze
Creatinine	S	Freeze
Cryoglobulins	S	Keep above 20 °C
Digitoxin	P,S	Draw 6–12 h post-administration
Digoxin	S	Draw 8 h post-administration
Estradiol	P	Add heparin (0.2 mg/mL); freeze
Estrogen receptors	Tumor tissue	Freeze at −70 °C
Fatty acids (free and esterified)	P	Add heparin (0.2 mg/mL); freeze
Fibrinogen	P	Do not use heparin [blue-top tube (citrate) is commonly used]
Fluoride	S,B	Do not collect in glass container
Folate (tetrahydrofolate)	P	Freeze
Galactose	B	Add heparin (0.2 mg/mL) and NaF (10 mg/mL)
Gastrin	S	Collect from fasting patient; freeze
Glucose-6-phosphate dehydrogenase [b]	B	Add EDTA (2 mg/mL)
HDL cholesterol	P	Collect after 12–14 h fast, add EDTA (2 mg/mL), and freeze
Hemoglobins (quantitative and electrophoresis samples)	B	Add EDTA (2 mg/mL)
Histidine	S	Freeze
17-Hydroxyprogesterone	S	Draw between 0900 and 1100 hours
Insulin	S	Collect from fasting patient; freeze
Isocitric dehydrogenase	S	Freeze
Lactic acid	B	Immediately dilute with an equal volume of 5% (50 g/L) perchloric acid; shake and mix
Lipoprotein phenotyping	S	Freeze
Lysozyme	S	Freeze
Lead	B	Collect in lead-free tube containing heparin

(continued on next page)

Table 2. (cont.)

Constituent	Sample	Condition
Magnesium	S	Separate immediately
Parathyroid hormone (PTH)	S	Freeze
Pepsinogen	S	Freeze
Placental lactogen	S	Freeze
Prolactin	S	Freeze
Prostaglandin $F_2\alpha$	S	Freeze
Pyridoxal phosphate (vitamin B_6)	P	Add EDTA (2 mg/mL) and protect from light
Pyruvic acid	B	Immediately dilute with an equal volume of 5% (50 g/L) perchloric acid; shake and mix
Renin	P	Chill during collection, centrifuge, and add EDTA (2 mg/mL)
Vitamin A	S	Protect from light
Zinc	S	Use only acid-washed glass; avoid hemolysis

P = plasma, S = serum, B = whole blood, EDTA = ethylenediaminetetracetate.

[a] All specimens listed, except those to be assayed for cryoglobulins, should be transported at temperatures below 15 °C and preferably at 4 °C, except as noted. It is further advisable to transport specimens in plastic instead of glass, to minimize the chances of breakage.
[b] There is some question whether this constituent remains stable even with this procedure.
[c] A sealable styrofoam container with "freezer packs" is usually sufficient to keep samples frozen for 12 h. Solid CO_2 is necessary for longer periods.
[d] It is advisable, when no harm can come to the patient, to add preservative to the tube before the addition of blood. Blood must be gently but thoroughly mixed with preservative (2 min of gentle inversion).

Table 3. Urine Constituents Requiring Special Conditions[a]

Constituent	Conditions
Aldosterone	Bring pH to 3–5 with boric acid or glacial acetic acid
Amino acids	Bring pH to 3–5 with 6 mol/L HCl
δ-Aminolaevulinic acid (ALA)	Bring pH to 1–2 with 6 mol/L HCl; protect from light
Bilirubin	Protect from light; freeze
Calcium	Bring pH to < 3.0 with 6 mol/L HCl
Catecholamines	Bring pH to 1–2 with 6 mol/L HCl
Choriogonadotropin (hCG)	Add boric acid[b]
Compound S	Add boric acid[b]
Cortisol	Freeze
Creatine	Freeze
Creatinine	Freeze
Cystine	Bring pH to 2–3; freeze
Estriol	Add boric acid[b]
Estrogens	Add boric acid[b]
N-Formiminoglutamic acid (FIGLU)	Freeze (Use 1- to 8-h timed specimen)
Heavy metals	Bring pH to 2 with 6 mol/L HCl or 6 mol/L HNO_3
5-Hydroxyindoleacetic acid (5-HIAA)	Bring pH to 1–2 with 6 mol/L HCl
Hydroxycorticoids	Add boric acid[b]
Hydroxyproline	Bring pH to 2 with concd HCl
Homogentisic acid	Freeze
Ketogenic steroids	Add boric acid[b]
Ketosteroids	Add boric acid[b]
Lysozyme	Freeze
Mercury	Bring pH to 2 with concd HNO_3
Metanephrines	Bring pH to 2 with 6 mol/L HCl
Osmolality	Freeze
Phenylpyruvic acid	Freeze
Porphobilinogen	Freeze; protect from light
Porphyrins	Protect from light; add EDTA (6 mg/10 mL of urine)
Pregnanediol	Add boric acid[b]
Pregnanetriol	Add boric acid[b]
Tetrahydro-compound S	Add boric acid[b]
Vanillylmandelic acid (VMA)	Bring pH to 1–2 with 6 mol/L HCl
Xylose	Add NaF (10 mg/mL)

[a] Tests requiring a 24-h or other timed collection can usually be aliquoted to be sent out, as long as the original volume is noted, e.g., 100 mL of a 1600-mL collection.
[b] 10 g/L of specimen.

References

1. NCCLS Tentative Standards: TSH-5. Standard procedures for the handling and transport of diagnostic medical specimens and etiologic agents. National Committee for Clinical Laboratory Standards, Villanova, PA; revised July, 1978, 15 pp.
2. Laessig, R. H., Indricksons, A. A., Hassemer, D. J., Paseley, T. A., and Schwartz, T. H., Changes in serum chemical values as a result of prolonged contact with clot. *Am. J. Clin. Pathol.* **66,** 598–604 (1976).
3. Wilding, P., Zilva, J. F., and Wilde, C. E., Transport of speciments for clinical chemistry analyses. *Ann. Clin. Biochem.* **14,** 301–306 (1977).
4. Winsten, S., Collection and preservation of specimens. *Stand. Methods Clin. Chem.* **5,** 1–15 (1965).
5. Young, D. S., Pestaner, L. C., and Gibberman, V., Effects of drugs on clinical laboratory tests. *Clin. Chem.* **21,** 1D-432D (1975).
6. Meites, S., Ed., *Pediatric Clinical Chemistry,* 1st ed., American Association for Clinical Chemistry, Washington, DC, 1977, Appendix 3, pp 236–245.
7. Alper, C., Specimen collection and preservation. In *Clinical Chemistry, Principles and Technics,* 2nd ed., R. J. Henry, D. C. Cannon, and J. W. Winkelman, Eds. Harper and Row, Hagerstown, MD, 1974, pp 374–388.
8. Bermes, E. W., Jr., and Forman, D. J., Collection and handling of specimens. In *Fundamentals of Clinical Chemistry,* 2nd ed., N. W. Tietz, Ed. W. B. Saunders, Philadelphia, PA, 1976, pp 47–53.

Quality Assurance: Theoretical and Practical Aspects

Submitter: Adrian Hainline, Jr., *Clinical Chemistry Division, Center for Environmental Health, Centers for Disease Control, Atlanta, GA 30333*

Evaluators: George J. Abraham, *Coney Island Hospital, Brooklyn, NY 11235*
Charles A. Bradley and David Palmer, *Clinical Chemistry Laboratories, Vanderbilt University Medical Center, Nashville, TN 37232*
Joseph S. Curcio, *Warner Lambert Co., Morris Plains, NJ 07950*
A. De Leenheer, *Laboratorium voor Medische Biochemie en Klinische Analyse, Rijks-Universiteit, Akademisch Ziekenhuis, 9000 Gent, Belgium*
Richard A. Jamison, *Educational Services, Diagnostics Division, Fisher Scientific Co., Orangeburg, NY 10962*
John W. Kennedy, *Medistat Consultants, Plainsboro, NJ 08536*
Harry L. Pardue, *Department of Chemistry, Purdue University, West Lafayette, IN 47907*
Richard Pavlovec, *Laboratory Metropolitan Hospital, New York, NY 10029*
Nathan Radin, *Clinical Chemistry Division, Center for Environmental Health, Centers for Disease Control, Atlanta, GA 30333*
Irvin Schoen, *Automated Laboratory Services, Van Nuys, CA 91406*

Reviewer: James O. Westgard, *University of Wisconsin, Madison, WI 53792* (on behalf of the AACC Committee on Standards)

This chapter presents a summary of some of the things that must be done to assure quality performance in the clinical chemistry laboratory. Concepts of error and error measurement are briefly described. The system of quality control recommended is based upon duplicate analyses of one or more control pools. This system involves use of statistical control charts to evaluate precision and accuracy. The principles involved are not affected by the size or output of the laboratory.

Quality assurance is a broad term encompassing the ideas and goals of quality control. For the purpose of this text, quality assurance includes all things that are done to make analytical measurements reliable. *Quality control* is the system by which acceptable performance is maintained. In the analytical laboratory, acceptable performance is maintained by measuring the variability of the analytical process, by evaluating the results of the measurement against certain criteria or specifications, by correcting excess variability when necessary, and by documenting these actions. Quality control should be considered to be an important servant rather than the master of the laboratory.

The concept of quality control has come into increasing use in the twentieth century. After statistical systems of quality control for manufacturing were developed by such pioneers as W. A. Shewhart, analytical chemists recognized the merit of the statistical approach for measuring and monitoring error in the analytical laboratory. The application of Shewhart's statistical quality-control chart for the analytical chemistry laboratory was first described by Wernimont (1). Ten years later, Levey and Jennings (2) proposed the control chart as a useful tool for the clinical chemist. Since then, much progress and interest in clinical chemistry laboratory quality control have been demonstrated and several techniques are now in use.

Assurance of Quality

Before one can control and maintain quality, it must first be achieved. Quality performance is assured by the effective use of instrumentation, management, methods, personnel, reagents, standards, samples, and a quality-control system. These considerations are fundamental to achieving accurate and reliable results and are not related to the size or output of the laboratory.

The quality of *instrumentation* and *equipment* is an essential element in producing quality work. All equipment must be in proper operating order, chemically clean, dry, and safe for use. Daily checks of temperatures of refrigerators, baths, and incubators must be made and recorded. Ease of repair and maintenance of equipment and availability of service must be considered at the time of purchase. Construction and design of some instruments may prevent or make difficult certain adjustments that may be necessary for optimizing performance. Continued use of an improperly functioning instrument perpetuates unsatisfactory performance. Unscratched, matched cuvets are essential. Flowthrough cuvets must function properly: test their reproducibility and photometer response by replicate readings of the same colored solution, such as a solution made from pure dyes or colored salts. Wavelength calibrations and colorimeter filters need frequent, scheduled inspections. Laboratories must have adequate floor space, lighting, sound control, and ventilation to encourage optimum performance by personnel.

Management is a crucial component of quality assurance. Effective quality assurance requires as much management skill as technical talent. Management attitudes that relegate quality to a lower priority than quantity are a major obstacle to quality performance. Good management provides the laboratory staff with well-written guidelines, procedures, and instructions that describe each operational detail. The effective laboratory director must maintain familiarity with these details through regular periodic review. The management system should provide a feedback mechanism through which analysts and director regularly communicate about performance. Any laboratory that employs two or more persons needs written job descriptions for each position to describe who does what and how all workers are kept informed. Schedules must be available and in force for performing and documenting regular preventive maintenance and equipment evaluation. Quality performance does not just happen—it evolves from planning, preparation, training, and careful execution. A good manager will assure that these things are done.

Methods (including procedures available in kits) must be evaluated before being used to analyze patients' specimens. Information must be available as to how the method will perform within the specific laboratory where it will be used. Any modification of a method essentially produces a new method, which also requires evaluation. Documented measurements of precision, bias, linearity, and specificity are required. Analysts must develop familiarity with and skills for use with each method, and maintain an ongoing evaluation of its performance.

Well-trained, skilled *personnel* are essential for the performance of quality analyses. Training is necessary for all new personnel in the day-to-day operations of a laboratory, regardless of their previous experience. The loss of one analyst from the staff of a small laboratory produces a void that may be proportionately larger than the same loss in a large laboratory. This situation does not, however, diminish the need to train replacement workers properly before allowing them to analyze patients' specimens. In-service training should be concluded only after a worker has been evaluated and shown to be competent. Continuing education is vital for each member of the staff of every laboratory, to maintain familiarity with advances in instrumentation and methods, and each technical worker should be provided with an opportunity to participate in professional meetings and workshops at least once yearly. Attendance at seminars and discussion groups should be encouraged. There should be ready access to scientific literature in the laboratory or in a nearby library.

Reagents. Quality cannot be assured unless reagents are prepared properly and remain stable throughout the period of analysis. Volumetric glassware of high quality must be available with which to prepare quantitative reagents. The necessary quality reagents and kits must be purchased; grade or quality must not be compromised, because the user must rely upon the manufacturer's integrity and skills. When possible, assume continuity of performance level by overlapping analyses to check new lots of reagents against old reagents. Reagents should be dated when opened and stored acccording to specifications. Mixing must be complete whenever water or solvents are added to dissolve dry components. Appropriate quality of water is essential for accurate analyses.

Standards serve as the basis for accurate analysis. The user of purchased standard solutions or calibrators inherits any errors introduced by the manufacturer. If biased procedures are used to assign values to calibrators, the biases will be included in the calibrator values. Whenever possible, calibrators should have values assigned through use of reference methods or methods generally recognized as specific. Single-point calibration is often uncertain and inadequate; if possible, additional calibrators and control materials of different concentrations should be analyzed to verify the calibration at additional points.

The laboratory that cannot prepare its own standard solutions is at a disadvantage. It must rely upon kit manufacturers and other sources to provide and assure the validity of these bases of accuracy. Overlap studies are essential when replacing standards. The accuracy of a standard material must not be taken for granted.

The *sample* must accurately represent the system from which it is collected. Proper collection of a specimen begins with the proper preparation of the patient. Selecting and using the appropriate type and amount of anticoagulant and preservatives, and protecting the specimen from exposure to abuse and adverse environmental conditions during collection, transport, and analyses are the direct responsibilities of the laboratory. Therefore, written directions and requirements that instruct anyone involved with specimen collection should be prepared.

The laboratory must assure by means of a defined identification and accession system that specimens

Table 1. Potential Sources of Analytical Error

Control sample: bacteria, contamination, evaporation, inhomogeneity, instability, matrix, reconstitution, turbidity.
Equipment: calibration, contamination, maintenance, maladjustment, pipette, repair, speed, temperature, timing, volume.
Data processing: confusion in specimen accession, clerical error, miscalculation, misreading, digit preference, incorrect units.
Instrumentation: calibration, cuvets, electrical noise, optical filters, linearity, maintenance, power supply, stray light.
Laboratory environment: air supply, design, fumes, housekeeping (cleanliness, orderliness), humidity, lighting, power supply, stability, temperature, utilities, ventilation.
Method: accuracy, bias, efficiency, interference, instructions, nonlinearity, precision, sensitivity, specificity, stability, timing.
Personnel: attitude, conduct, skill, training, workload.
Reagents: composition, contamination, container, deterioration, evaporation, inhomogeneity, potency, preparation, purity, stability, storage, turbidity.
Sample or specimen: anticoagulant, blood volume, circadian rhythm, collection, container, contamination, denaturation, diet, dilution, drugs, evaporation, fibrin, glycolysis, hemolysis, homogeneity, identity, interferants, interaction, lipemia, matrix, measurement, medication, mixing, patient preparation, reconstitution, sample preparation, preservatives, sequencing, storage, timing, tourniquet application, turbidity.
Standards: calibration factor, contamination, cross reaction, incorrect concentration, evaporation, dilution, linearity, matrix, measurement, purity, single-point calibration, stability, storage.

or analytical results from different patients are not confused. Results of analyses must be correctly aligned with specimens. The appropriate blood fraction must be harvested and maintained under correct storage conditions until analysis can begin. After refrigerated storage or prolonged standing, specimens must be well mixed before sampling.

A *quality-control system* is a necessary part of the quality-assurance program. Instructions for making computations, reporting, and filing results should be specific. It is essential that analytical error be measured, evaluated, and documented according to written guidelines and specifications. Causes of analytical error that exceed predescribed action limits must be corrected promptly, and the corrective actions documented.

Some analysts attempt to assure quality performance without using a defined quality-control system. Such people may take great pains to perform properly and reproducibly. Unfortunately, they may have no tangible measures of quality performance by which to evaluate their efforts. Visibility is the most important contribution made by a quality-control system and, at its best, provides anyone concerned—whether analyst, director, or laboratory inspector—with information by which to evaluate performance.

There is no appropriate distinction between the quality-control system suitable for the small laboratory and that which is suitable for the large laboratory. Analytical results need to be equally accurate whether from a small or large laboratory. Procedures for attaining accuracy are not related to the size or the number of analyses performed. The small laboratory may actually bear a proportionally heavier burden than the large laboratory in maintaining an appropriate quality-control system.

Variability in Measurement

Analytical error. Each quantitative laboratory measurement is performed for the purpose of estimating the concentration or activity of some constituent in the specimen. The perfect measurement would not deviate from the truth and would have no error. In practice, a number of factors, some inherent in the measurement process and some related to the specimen, produce deviation from the actual or true value.

Table 1 lists a number of items, each of which may be a problem or the potential source of analytical error if not properly used. Considerations must be given to assuring that proper conditions exist and to eliminating adverse conditions.

Measurement of variability. In the typical quality-control procedure, precision is measured by analyzing a homogeneous, stable solution (e.g., a control serum) a number of times. Figure 1 is a histogram or bar graph that portrays the distribution of 102

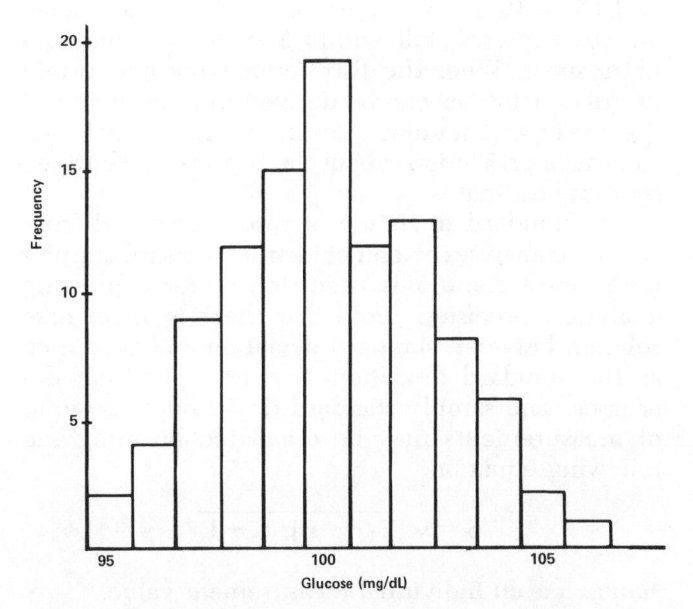

Fig. 1. Histogram illustrating the distribution of 102 glucose control results

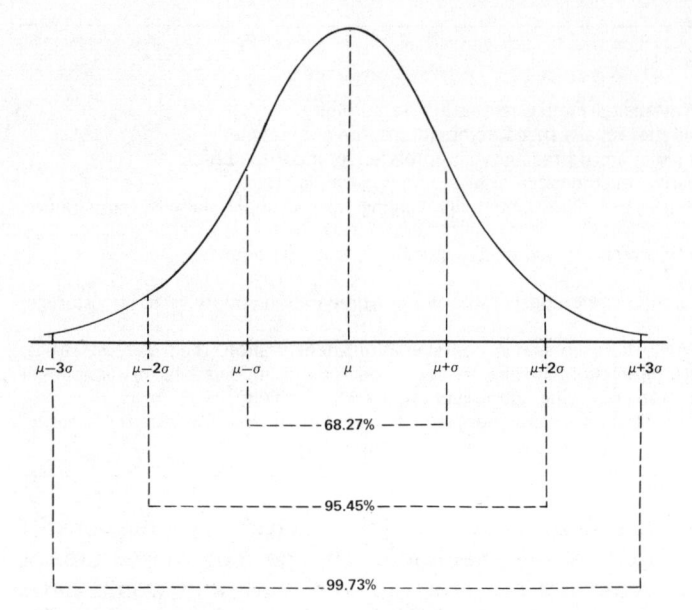

Fig. 2. The normal frequency distribution and proportionate areas included within various multiples of σ

values obtained by analyzing a control serum for glucose. More values cluster near and about the middle than in the extremes on either side.

Symmetry is a feature of the normal frequency distribution (Figure 2), which represents the distribution of an infinite number of measurement values. An infinite number of normal curves are possible, each characterized by a mean, μ, which locates the central value, and a standard deviation, σ, which indicates the spread of values about μ. When the distribution is normal (gaussian), 68.34% of its values fall within an area bounded by 1 standard deviation on either side of the mean; 95.45% of the values will fall within 2 standard deviations of the mean, and 99.73% will fall within 3 standard deviations of the mean. When the distribution is normal, certain useful conclusions can be derived from estimates of the mean and standard deviation, from which one can make predictions about the process represented by those estimates.

The standard deviation of results obtained from repeated analyses of a set of identical control samples is the most commonly used statistic for expressing analytical precision. Note that there is an inverse relation between standard deviation and precision: as the standard deviation increases, precision decreases. The sample standard deviation of a group of measurements may be calculated by using the following equation:

$$S = \sqrt{[\Sigma(x - \bar{x})]/(n - 1)}$$

where: x is an individual measurement value.
Σ is a symbol of summation.
$\bar{x}$ (pronounced "x-bar") is the mean of n measurement values and is an estimate of μ, the true population mean.[1]
n is the number of measurement values, and n − 1 equals degrees of freedom.
S is the sample standard deviation, which is an estimate of σ, the population (or true) standard deviation.[1]

Variability may also be expressed as the coefficient of variation (CV), or relative standard deviation. CV is the standard deviation expressed as a percentage of the mean concentration or activity and is calculated from this formula: $CV = 100S/\bar{x}$. CV is useful for expressing the relative precision over a wide range of values.

The average range of replicate determinations is another useful expression of precision. The range is the difference between the largest and the smallest values in a group of measurements. The average range, $\bar{R}$, of analyses performed in duplicate within runs on a pooled serum, is calculated by summing each difference, R, between pairs of results and dividing by the number of pairs, N.

$$\bar{R} = (\Sigma R)/N$$

$\bar{R}$ is an index of within-run precision, being based upon differences between paired results within each run. The average successive mean range, $\bar{R}_s$, is determined from the difference between the means of pairs in successive runs and includes both within-run and between-run error components. An example of the use of $\bar{R}$ and $\bar{R}_s$ to evaluate a precision problem is given on page 29.

Random error. It is important to consider the components of analytical error. Random error affects precision and is the basis for disagreement between repeated measurements; it results from variation of the measurement process. Analytical random error will have within-run and between-run components.[2] With care to reproduce exactly each analytical step every time, random error can be reduced. There is usually a point, however, beyond which it may not be economically feasible to reduce random error.

Systematic error, or bias, arises from factors that contribute a constant influence, either positive or negative, and directly affect the estimate of the mean. In some analyses, contributions to bias can be corrected by measuring a blank. Systematic error may be defined as the average deviation from a true value. The true value is not usually known in clinical chemistry. Instead, one must rely upon a reference value, which is the average value obtained with a reference method of established reliability. For more information, see references *3* and *4*.

---

[1] The population mean and standard deviation are hypothetical values of an infinite-sized population. Samples selected from that population form the basis for calculating $\bar{x}$ and S. The reader may wish to consult a statistics text for more information.
[2] Note the example given in Table 4.

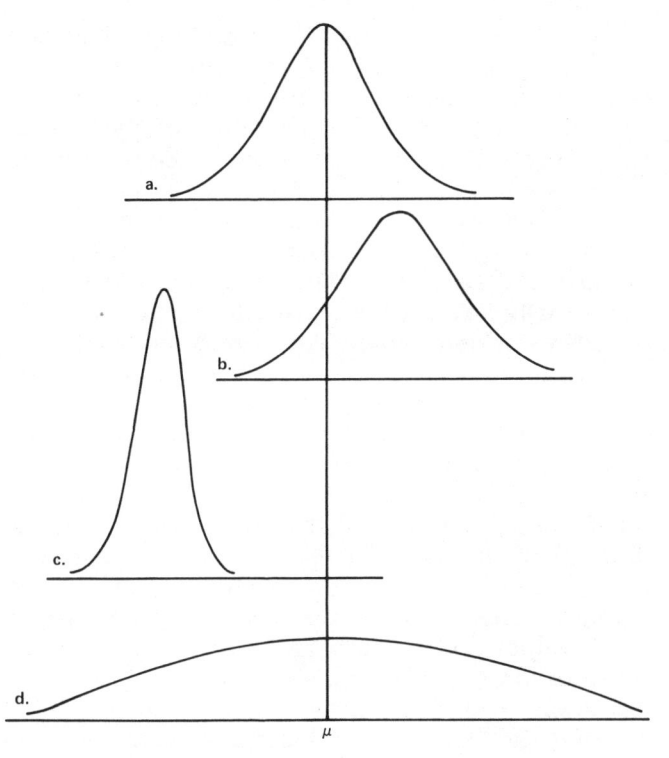

Fig. 3. Accuracy—a matter of bias and precision: *a,* accurate; *b,* biased; *c,* biased, high precision; *d,* inaccurate, imprecise

Personal error. In addition to controlling analytical error, the clinical laboratory must be vigilant to avoid the contribution to error produced by personal negligence. This type of error is not statistically predictable, neither in magnitude nor in rate of occurrence. *Clerical error* is a common cause of widely aberrant results and arises from mistakes performed in writing down and processing results. It may also arise because of confusion of samples, miscalculations, or the use of incorrect calculation formulas, factors, or concentration units. The *analytical mistake* may also contribute to the number and magnitude of discrepant values known as "outliers" or "wild values." Outliers may arise as a result of such mistakes as incorrect volumes of reagents, incorrect timing, incorrect incubation temperatures, improper sequence of adding reagents, inadequate mixing, faulty analytical techniques, omission of steps, and unauthorized alterations of procedure. An occasional outlier may occur in any set of results and does not necessarily arise because of an easily identifiable cause. Any increase in the number of outliers should be a cause for concern and must be eliminated. Personal errors may be detected and the number reduced by repeating all analyses of specimens with abnormal results. Clerical error and analytical mistakes do not frequently occur in a well-managed laboratory staffed with experienced, trained technologists.

Accuracy is the degree of agreement between a measurement value and the actual or true value. Precise methods that have low bias are required to obtain accurate measurements. Figure 3 illustrates this principle with four different combinations of precision and bias. The first method, *a,* has no bias and low variation. The second method, *b,* has the same standard deviation as method *a,* but its bias indicates that most of the values differ from the true value, μ. Much emphasis has been placed upon the need for repeatability of analyses performed in the clinical laboratory. Precise measurements, however, do not assure accurate measurements if the analytical process is excessively biased. The third method, *c,* is very precise but is so biased that none of its results agree with the true value. The fourth method, *d,* has no bias on the average, but its wide standard deviation indicates that many individual results differ from the true value by a large amount. Although method *d* might be judged accurate if only its mean were considered, it is not accurate because of its imprecision.

The Statistical Control Chart

General considerations. The statistical control chart is a graphic device used to evaluate and moni-

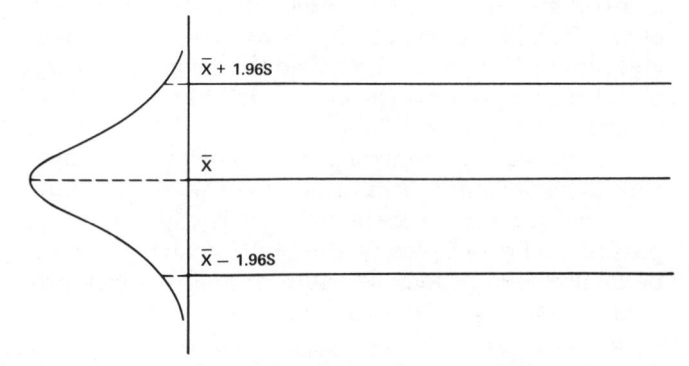

Fig. 4. Conceptual relationship between limit lines on the quality-control chart and the normal curve

tor analytical variability. It provides a pictorial record to facilitate the comparison of each result obtained from the analysis of a reference sample (control sample) with results that precede or follow it. The chart is considered by many to be superior to recording control results in a log because it enables the reviewer to see patterns of performance as they develop. The chart helps the analyst decide whether performance is satisfactory and whether the analysis process should continue or be stopped.

Statistical control charts are constructed with limits based upon probability theory. A relationship between a normal distribution curve and a control chart can be demonstrated by turning the normal curve 90° counterclockwise and extending lines to the right (Figure 4). The center line represents the mean and the lines located 1.96 standard deviations from the mean are the 95% control limits.[3] Statistical control charts are usually constructed with mean and limit lines and with graphic coordinates to represent concentration on the y-axis and time or run numbers along the x-axis.

Such charts are called x or $\bar{x}$ charts, depending upon whether x or $\bar{x}$ is the variable, and are used for controlling analytical variability arising both from changes in systematic error and from random error. A range chart is used to monitor random error and is not influenced by systematic error. The range chart is thus a more sensitive indicator of changes in the overall magnitude of random error than the $\bar{x}$ chart. Two types of range charts complement the $\bar{x}$ chart, one based upon the average range, $\bar{R}$, and the other based upon the average successive mean, $\bar{R}_s$.

Before preparing a set of control charts, One must obtain the necessary data by analyzing samples of control material. One or more pools (5) of biological materials, such as serum, are obtained in sufficient quantity to last about a year. The laboratory director may choose to control at more than one concentration. Many laboratories use a normal pool and an high-concentration pool for quality-control analyses. Control materials are selected to represent effects of the sample matrix as closely as possible.[4] Bowers and others (5, 6) have described the preparation and use of serum pools by the clinical laboratory for quality-control purposes.

Samples of each control pool are analyzed in duplicate in each routine run under conditions that simulate the analysis of patients' samples as closely as possible. The samples of the control material must be analyzed in at least 20 runs[5] to give a sufficiently valid estimate of the variability.

When sufficient data have been obtained, calculate the overall sample mean, $\bar{\bar{x}}$, the standard deviation of individual analyses, S; the means of individual pairs, $\bar{x}$; and the standard deviation of the means, $S_{\bar{x}}$. Average range, $\bar{R}$, and average successive range, $\bar{R}_s$, are also calculated.

Should any evidences of loss of control appear during this period, as described under *Interpretation of Statistical Control Charts*, corrective action must be taken before analyses are allowed to continue. Results from an out-of-control period and from outliers should not be included in calculating the mean, range, or standard deviation. Identify and eliminate outliers from the statistical calculations. Including outliers distorts the statistical values and may widen the control limits excessively. On the other hand, injudicious discarding of data may lead to narrow limits that may be difficult to meet, and may lead to an excessive number of rejected runs. Care and common sense must be used when precision is very good; values that are otherwise acceptable may be considered for rejection because of a small standard deviation.

Limits for the control charts are calculated with the appropriate factors given in Table 2. Probability values for control limits are given in the left column. To calculate the 99% probability or control limits for $\bar{x}$, multiply the standard deviation, $S_{\bar{x}}$, by 2.58. This value is then added to the overall mean, $\bar{\bar{x}}$, to obtain the upper control limit (UCL), or subtracted from the mean to obtain the lower control limit (LCL). The upper and lower warning limits are similarly obtained from the product of $S_{\bar{x}}$ multiplied by 1.96. (For ease in calculation one may round 1.96 to 2; however, do *not* round 2.58 to 3.) There are no lower limits on range because perfect range is zero, which is the lowest point on the range chart. There can be no negative range.

---

[3] So named because the probability is such that 95% of the values in a normal distribution will fall in an area between bounds of 1.96 standard deviations on either side of the mean. Likewise, the probability is such that 99% of the values will fall between bounds of 2.58 standard deviations from the mean. In practice, 1.96 is rounded to 2 standard deviations and 2.58 is rounded by some laboratories to 3. However, the probability value of 3 standard deviations is actually 99.7%, not 99.0%.

[4] Several manufacturers of clinical laboratory control sera offer quality-control services in which statistical calculations and interpretations of results are supplied for a wide variety of analyses. In addition to saving the labor of performing many calculations and preparing reports, these programs allow laboratories to compare themselves with their peers. In general, however, the lag times between submitting results and receiving a report tend to diminish some of the effectiveness of these programs. The laboratory staff must therefore be prepared to do some of the evaluations themselves as the occasion arises.

[5] The small-volume laboratory may have problems in accumulating sufficient control analysis data if specimens are not analyzed each day. To obtain the data, it may be necessary to analyze runs composed of only control samples if patients' samples are not available. Leftover portions of patients' samples may be reanalyzed to simulate typical runs, if desired. Control samples should be analyzed as if they were patients' samples.

Table 2. Calculation of Control Limits

p	Limit	Mean	Range	Successive range
95%	Upper warning (UWL)	$\bar{\bar{x}} + 1.96 S_{\bar{x}}$	$2.51\bar{R}$	$2.51\bar{R}_s$
99%	Upper control (UCL)	$\bar{\bar{x}} + 2.58 S_{\bar{x}}$	$3.27\bar{R}$	$3.27\bar{R}_s$
95%	Lower warning (LWL)	$\bar{\bar{x}} - 1.96 S_{\bar{x}}$	—	—
99%	Lower control (LCL)	$\bar{\bar{x}} - 2.58 S_{\bar{x}}$	—	—

Limits provide certainty to the evaluation process and enable laboratory directors to specify performance. Temporary limit lines can be approximated with which to evaluate performance and exert some control during the preliminary measurements. For example, one can estimate by eye a line for the mean that can be used until sufficient data are accumulated to provide a reliable value. Drifts, shifts, and precision changes can be detected by visual estimation before control limits are added.

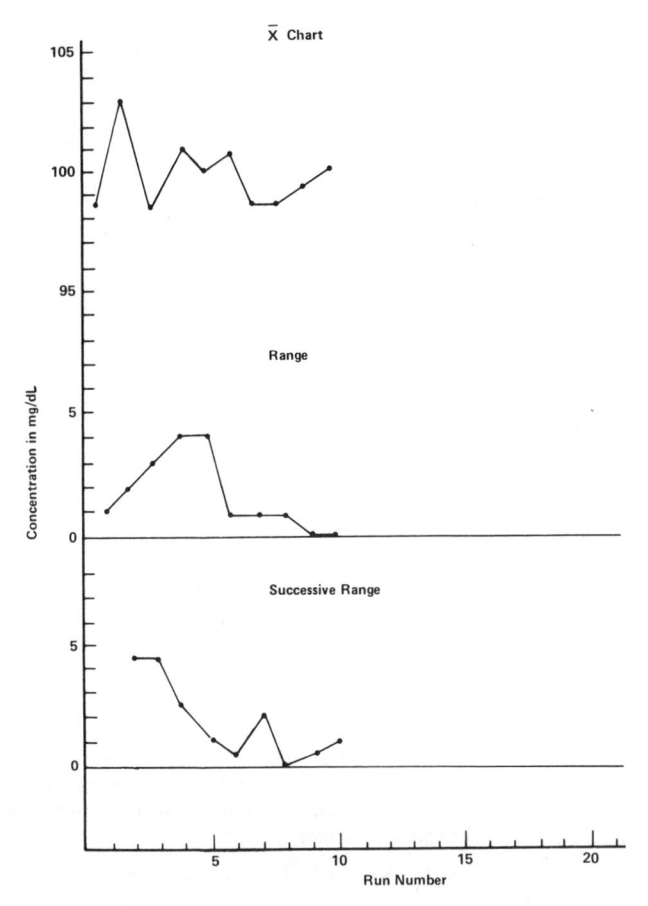

Fig. 5. Example of a statistical control chart under preparation

Results are plotted on the chart as they are accumulated, which provides opportunities to make evaluations until limit lines can be added

Temporary limits on precision can be based upon prior performance of the same analyst. Control lines can also be arbitrarily assigned by the laboratory director. For example, a certain fraction of the normal-range interval (e.g., ¼), or a percentage (e.g., 5% of the mean) might be specified. Limits with a statistical basis, however, are the most useful because they are based on actual performance and allow prediction for the future.

Construction. Control charts are usually constructed on commercially available graph paper. The size of the grid is optional. Ordinarily, one seeks to display results so that they can be easily read and clear notations can be made. It is usually advantageous to prepare charts so that data from at least two or more months can be entered and displayed simultaneously. Charts can be kept in a loose-leaf notebook but should always be readily accessible to the analyst at the workbench as well as to the laboratory supervisor.

The data from the analyses should be promptly entered on the control chart as they are obtained (Figure 5). All results should be recorded on the chart except for those obtained as a consequence of a recognized mistake in procedure (e.g., use of a wrong reagent, incorrect volume delivered, or improper timing). Comments are entered as appropriate to indicate changes in reagents, standards, or personnel and corrective actions. Because of space restrictions on the chart, more detailed notes may be kept in a log.

Figure 6 illustrates a portion of a completed control chart. The means of 30 duplicate analyses are plotted on the $\bar{x}$ chart. The differences between the duplicates, R, are plotted on the range chart. The differences between the successive means of duplicates are plotted on the successive range chart. The median of each range is drawn on its range chart. Calculations are summarized in Table 3.

Table 3. Condensed Instructions for Processing Quality-Control Data Based on Analyses of Duplicate Aliquots of a Control Material[a]

Instructions:

1. Record control analyses x_1 and x_2 by day or by run.
2. Calculate the average, $\bar{x}$, for each pair; record.
3. Calculate the difference, R, between x_1 and x_2 of each pair; record.
4. Calculate the difference, R_s, between successive $\bar{x}$'s, beginning after the second run; record.
5. Plot $\bar{x}$, R, and R_s on the control chart immediately; compare with previous values.
6. When at least 25 runs have been made, sum $\bar{x}$, R, and R_s. Divide each sum by the number of pairs, N, to obtain $\bar{\bar{x}}$, $\bar{R}$, and $\bar{R}_s$. (For the first set, divide R_s by $N-1$, the actual number of differences.) Determine the median of each range by selecting the middle value.
7. Calculate the overall standard deviation, S, of all single values.
8. Calculate the standard deviation, $S_{\bar{x}}$, of the means of the paired values. If one value is missing, drop that pair from the calculation.

Calculations:

$n = 60$ $N = 30$ $\bar{\bar{x}} = (\Sigma \bar{x})/N = 2999.5/30 = 99.98$
$\bar{R} = (\Sigma R)/N = 63/30 = 2.1$ $\bar{R}_s = (\Sigma R_s)/(N-1) = 48.5/29 = 1.67$
$S = \sqrt{[\Sigma(x-\bar{x})^2]/(n-1)}$ $S_{\bar{x}} = \sqrt{[\Sigma(\bar{x}-\bar{\bar{x}})^2]/(N-1)} = 1.689$

9. Calculate control limits on $\bar{x}$ chart
   $1.96 S_{\bar{x}} = 3.3$ $2.58 S_{\bar{x}} = 4.4$
   95% UWL: $\bar{\bar{x}} + 1.96 S_{\bar{x}} = 100 + 3.3 = 103.3$
   95% LWL: $\bar{\bar{x}} - 1.96 S_{\bar{x}} = 100 - 3.3 = 96.7$
   99% UCL: $\bar{\bar{x}} + 2.58 S_{\bar{x}} = 100 + 4.4 = 104.4$
   99% LCL: $\bar{\bar{x}} - 2.58 S_{\bar{x}} = 100 - 4.4 = 95.6$
10. Calculate control limits on range chart (within-run variability)
    95% WL: $2.51 \bar{R} = 2.51 \times 2.1 = 5.3$
    99% CL: $3.27 \bar{R} = 3.27 \times 2.1 = 6.9$
11. Calculate control limits on successive range chart (between-run variability)
    95% WL: $2.51 \bar{R}_s = 2.51 \times 16.7 = 4.2$
    99% CL: $3.27 \bar{R}_s = 3.27 \times 1.67 = 5.5$

[a] Example given is for 30 runs of glucose analyses. See Figure 6.

A Recommended System of Quality Control

The system of quality control recommended in this chapter is suitable for any clinical chemistry laboratory. It requires a sizeable input of time, money, and effort, and the proportionate burden is large for the small laboratory, but this input is necessary if quality is to be assured. The predominant system of statistical quality control currently in use is based

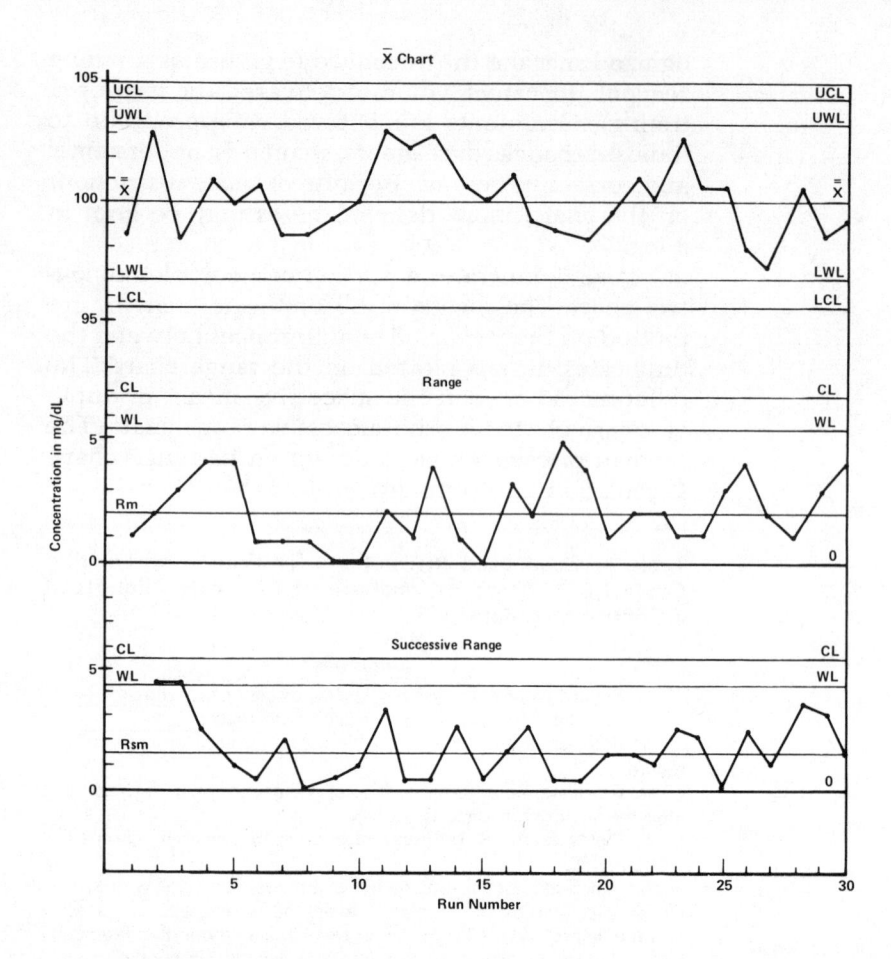

Fig. 6. A completed statistical control chart showing results from 30 runs of glucose analyses; means of duplicates and ranges are plotted and limits drawn

$\bar{\bar{x}} = 100$, $S = 1.69$, $\bar{R} = 2.1$, $\bar{R}_s = 1.67$

on single analyses. Single-analysis quality control, however, lacks the sensitivity and certainty afforded by the duplicate-analysis system described here.

The recommended system is based upon information obtained from analysis of a duplicate pair of control samples in each analytical run. This system thereby allows the use of two range charts as well as the $\bar{x}$ chart and permits rapid evaluation of both within-run and between-run precision. The instructions that follow describe the use of one pool. If a second concentration of control material is analyzed, a second set of charts is prepared. For additional information about multiple controls, see Westgard et al. (7). The exact design of data forms, control charts, and reports is left to each laboratory. It is assumed that an electronic calculator is available to ease the load in calculating the statistics required for quality control.

The system of quality control discussed in this chapter is based upon the assumption that a hypothetical population of an infinite number of measurement values performed on a control material will have a normal distribution. It is also assumed that the magnitude of the error estimated from the analyses of control samples is representative of the error in the analyses of patients' specimens.

1. Analyze each control pool in duplicate with each run.[6] Place one sample of the control at or near the beginning of the run. Place its duplicate at a variable location in each run among the samples. A suggested arrangement is to locate the second control sample at a run position number corresponding either to the day of the week or month or to the run number, so that each day the control will be in a different location. For example, if there is an average of 10 samples in a run, place the controls at positions 1 and 3 on Monday, 1 and 4 on Tuesdays, 1 and 5 on Wednesdays, and so on.

2. Accumulate results from at least 20 runs. Record results for the controls in a log kept especially for quality control. Immediately when obtained, plot the mean of each pair of results on a control chart prepared on graph paper (for example, 10 squares per inch or per 25 mm). Plot the difference between each pair, R, on the portion reserved for range. Calculate and plot the absolute differences between the means, R_s, for each successive run on the successive range chart. For example, the first successive range

---

[6] For the purpose of this discussion, a run is a set of analyses usually grouped and analyzed together without stopping and traceable to a standard or group of standards analyzed in the same set. The size limits for a run are defined by the individual laboratory and the conditions under which it operates.

is the difference between the first and second means: $R_{s1} = \bar{x}_1 - \bar{x}_2$. The second is the difference between the second and third means: $R_{s2} = \bar{x}_2 - \bar{x}_3$, and so on. Note that there is no value for R_s for run 1 because there was no prior run (see Table 4). Connect the points for run means, and connect range points and successive range points on the range charts for each run. There is no negative range.

Table 4. Example of Change in Precision for Cholesterol Analyses (mg/dL)

Run	x_1	x_2	R	$\bar{x}$	R_s
1	295	296	1	295.5	—
2	294	288	6	291	4.5
3	290	292	2	291	0
4	290	296	6	293	2
5	285	286	1	285.5	7.5
6	300	305	5	302.5	17
7	280	284	4	282	20.5
8	303	298	5	300.5	18.5

3. Inspect the data in the chart for evidence of drift, shift, or poor precision as new points are plotted. An approximate and temporary midline can be fitted visually after 10 or 12 sets of data have been entered.[7] If performance does not meet acceptable criteria or is not satisfactory at any time, institute appropriate correction procedures. Data that form the basis for control chart calculations must come from runs that are under control and free from excess error. If variability of performance is considered to be marginal by the laboratory director, additional data should be collected by extending the collection period.

4. When data from at least 20 runs have been accumulated, compute the overall mean, $\bar{\bar{x}}$, by summing the means of the pairs, $\bar{x}$'s, and dividing the total by the number of pairs (Table 3 shows a completed data summary).

5. Calculate the standard deviation of the means of duplicates, $S_{\bar{x}}$.

6. Calculate the mean range, $\bar{R}$, by summing the R's and dividing by the numbers of pairs. In a similar manner, calculate $\bar{R}_s$ by dividing the summation of R_s by the number of differences.

7. Examine the individual values for outliers. This can be best done by preparing a histogram of the control values such as the one shown in Figure 1. If any result is obviously much different from the others, calculate the mean and overall standard deviation, S, for single values both with and without that result. If the questionable value exceeds 3.3S from the mean, discard it. Do not include unpaired results in the final calculation of $\bar{\bar{x}}$ and $S_{\bar{x}}$: discard the other mate to the pair if one of the values is missing or dropped. If one or more outliers must be dropped, accumulate additional data until there are at least 20 pairs. Recalculate $\bar{\bar{x}}$, $S_{\bar{x}}$, $\bar{R}$, and $\bar{R}_s$.

8. Calculate limits for the mean and for the range, using the factors listed in Table 2. Determine the median range.[8]

9. Draw lines for limits, means, and range medians on control charts and reinspect for trends, shifts, or poor precision. Suggestion: for easier chart inspection, use colored lines such as blue for 95% limits and red for 99% limits.

10. Continue recording and plotting control values as they accumulate, evaluating the results on the basis of how the mean of each pair of values relates to previous values and to the overall mean. It is important to record control values as soon as possible, to make interpretations at or near the time of analysis while memories are fresh. Institute appropriate corrective actions when performance is interpreted as unacceptable. Plot all values, including any that are "out of control."[9] Identify and label out-of-control values and note corrective actions on the chart.

11. Label the chart with any appropriate information for control. For example, the analyst's initials should appear on the first value of a new contol plotted by that analyst. A replacement analyst should also initial the chart. Indicate any actions taken in response to out-of-control results. Changes in reagent lots, new calibrator solutions, instrument adjustments, etc., should be briefly indicated on the control chart as well as detailed in their respective logs. The system of record keeping should be designed to inform the analyst and supervisor and thereby facilitate evaluation and decision-making.

12. During the initial period when analyses are being performed to set values on the new pool, analyze any accuracy-control samples available to the laboratory, such as previous control pools, special materials with assayed values, or samples from another laboratory. Use every opportunity during the characterization of control pools to analyze and compare values with any other reference material that may be available.

---

[7] Analysts are cautioned not to make readings that are prejudiced by this estimate of the mean. Prejudiced readings, based on a desire to "agree" with the mean, will distort the calculation of random error and falsely narrow the precision limits. Any bias of readings introduced at this point will persist when the final limits are calculated and will diminish the effectiveness of the chart. Such limits will affect future operations and result in excessive rejection of runs.

[8] The median range, R_m, is preferable to the mean range, $\bar{R}$, for establishing the center of the range chart. To determine the median range, R_m, arrange each set of range values in an array from the lowest to the highest, then choose the middle value. If the total number of values is an even number, select the two middle values, add them together, and divide by two to obtain the median.

[9] It is appropriate to record "out-of-control" values except those values obtained as a consequence of a recognized mistake.

13. If possible, avoid changes in personnel, standards, reagents, and procedures during the period when a new pool is being characterized. It is wise to avoid any change on the first day that a new control is put into service, especially if there has been no opportunity to overlap properly with the old control. If at all possible, a new supply of control material should be overlapped for at least a month with the old control before being completely replaced. This is necessary to establish a reliable mean and standard deviation and a relation between the new and the old values. If possible, never allow the old control supply to be completely used up until the new control is established.

14. When an additional 20 to 30 runs have been completed, combine the results with those of the first 20 runs and calculate the final $\bar{\bar{x}}$, $S_{\bar{x}}$, $\bar{R}$, and $\bar{R}_s$. Calculate the final limits and prepare the final control charts. Calculate the overall standard deviation, S, and the CV, which may be used to state the variability of single values when patients' analyses are performed as singles.

15. At the end of each month, pool all new and old results of each control that correspond to accepted runs and for which results for patients were reported, and calculate a new overall mean and standard deviation for the past month. Plot these summary statistics on a separate monthly statistical control chart. Such calculations should provide statistics which permit comparison of (a) the previous month, (b) the past month, and (c) the entire period that the control has been in use.[10]

16. Laboratory directors should schedule to meet at least monthly with the analysts to review control charts and other quality-control data. More frequent meetings should occur when indicated by performance difficulties.

Corrective Action Rules

Written instructions should be provided in the laboratory operation manual detailing specific actions to be taken, including notification of the supervisor and necessary steps for documentation. The laboratory director may wish to specify additional limits and actions. As each control value is obtained and plotted on the control chart, the results on the control chart are interpreted as satisfactory; possibly indicative of a problem (warning); or out of control. The analytical procedure is considered to be out of control when the following events occur, and the analysis should stop:

[10] Laboratories electing to subscribe to one of the commercial quality-control services will have this laborious task performed for them. These statistics are regularly available to participants in reports of certain commercial quality-control programs. Nevertheless, if results appear to be questionable for an analyte, the analyst should be prepared to calculate the appropriate statistics to help perform necessary evaluations without waiting.

1. A single daily $\bar{x}$ value falls above the upper 99% limit or below the lower 99% limit.
2. Two successive daily $\bar{x}$ values exceed the 95% limits. When one value exceeds the 95% limit, a warning condition exists, and the analyst should be alert to the possibility of loss of control.

Corrective actions are indicated by the $\bar{x}$ chart when:

1. Seven consecutive $\bar{x}$ values fall so that all are either above or below the mean line.
2. Seven or more successive $\bar{x}$ values progressively increase or decrease from previous values (drift).
3. The difference between $\bar{\bar{x}}$ and the mean of eight consecutive values is equal to or greater than 1 standard deviation (shift).
4. A pattern develops that is indicative of shift, drift, or imprecision, as illustrated in Figure 7.

Corrective actions are indicated by the range chart when:

1. A single R or R_s value exceeds the 99% limit.
2. Two successive R or R_s values exceed the 95% limit.
3. Seven R or R_s values in succession exceed the median. Whenever more values exceed the median than fall below it, random error has increased.

Actions. When a *warning condition* exists, the analyst should:

1. Check readings and calculations for errors.
2. Review control charts and records, including those regarding maintaining the instruments and preparing the reagents and standard solutions used in the analysis in question.
3. Stop analysis if an obvious cause is found, and apply the appropriate remedial action.
4. Report to the supervisor the value(s), the results of the review, and any actions taken. (The order of steps may vary as specified by the supervisor or director.)

If an *out-of-control condition* exists, the analyst should:

1. Stop analysis of specimens. Report the situation to supervisor.
2. Follow the steps 1–3 listed for a warning condition.
3. Repeat the analysis of the control (if only a single value is in question) in duplicate from a newly opened vial.
4. If the mean of the repeated analysis is still out of control, institute corrective procedures stepwise as appropriate to the problem revealed on the control chart. Avoid applying simultaneous multiple corrections unless necessary.
5. Recalibrate. Analyze precision-control and accuracy-control samples as required, to determine whether performance is now acceptable.
6. When control has returned, as indicated by acceptable results for controls, reanalyze the patient specimens analyzed during out-of-control run.

7. Report actions and results of actions to the supervisor.

8. Record all actions on control charts and in logs.

Interpretation of Statistical Control Charts

The control chart is not intended to be used in isolation from the other sources of information that should be available to the laboratory. Logs of reagent and standard preparation and instrument maintenance and personnel records are valuable adjuncts to the control chart. Liberal notes should be made of items and events that may affect or be related to performance. The chart should be indexed with times and dates to permit ready referral to other records.

Limit lines should not be considered as rigid bounds that can never be crossed. On the average,

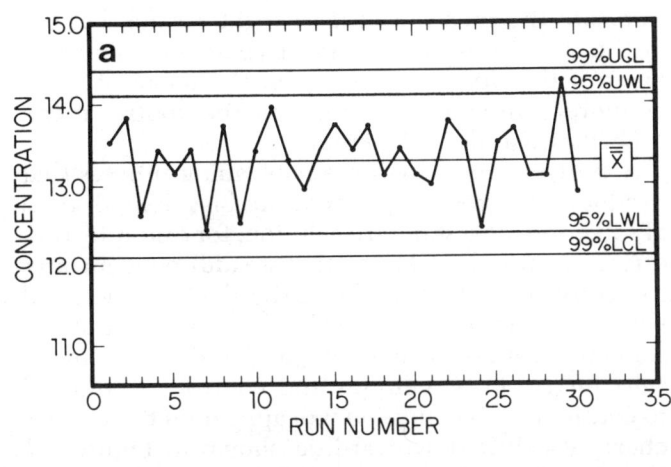

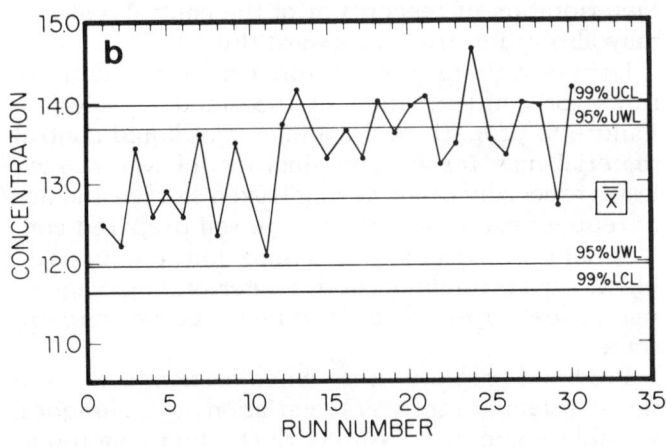

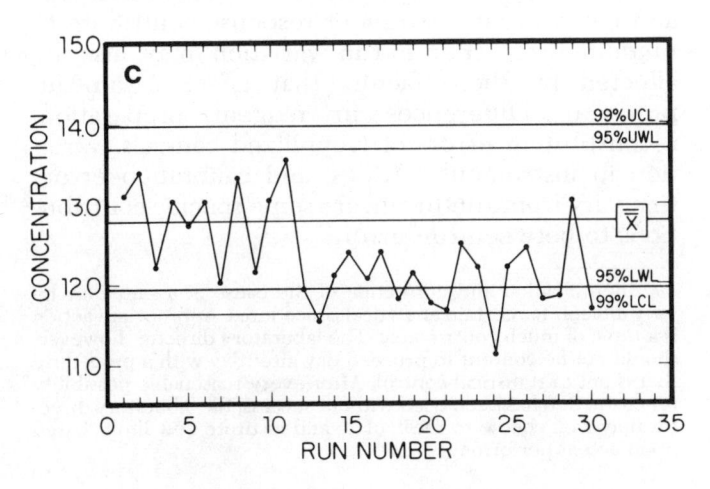

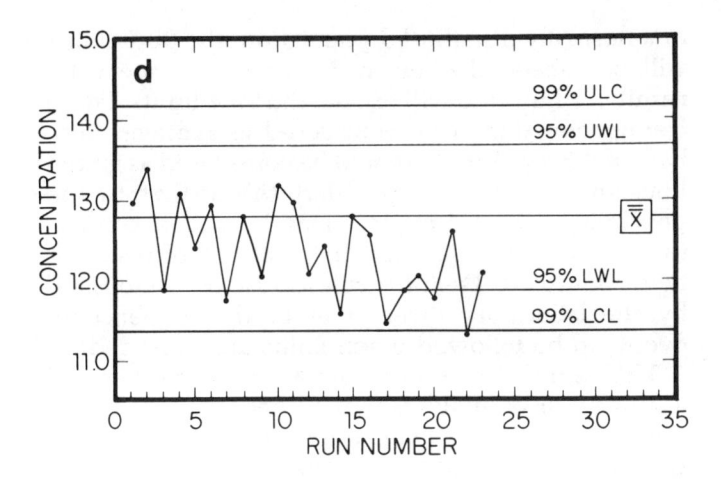

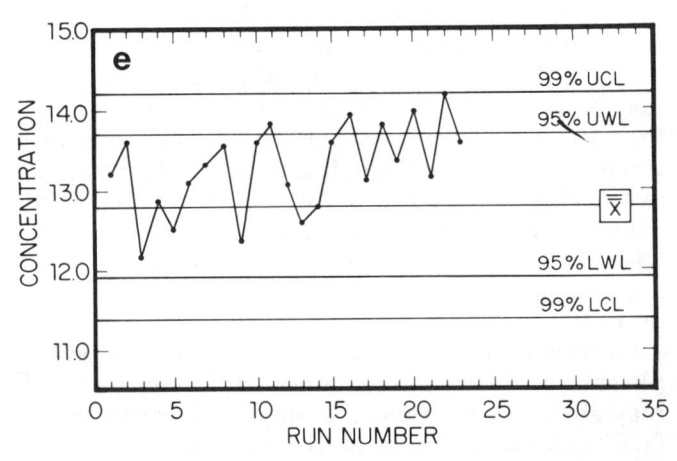

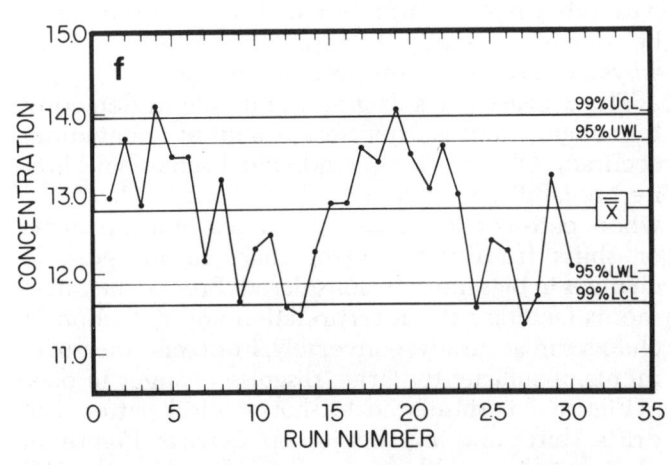

Fig. 7. Quality-control chart: *a*, acceptable performance; *b*, shift upward of the mean; *c*, shift downward of the mean; *d*, drift downward of the mean; *e*, drift upward of the mean; *f*, loss of precision

Source: Radin, N., Quality assurance in clinical chemistry. I. Workbook. Centers for Disease Control, Atlanta, GA, 1971.

in a normally distributed population, the 95% limits will be exceeded once in 20 times and one valid result in a hundred will exceed the 99% limits. These frequencies should be considered as average numbers, not fixed. Limits should be considered as guidelines that, when crossed, alert the analyst to the possibility that the process may be out of control, in the sense that another "population" of results is being sampled. Definite rules must be established by the laboratory director as to the sequence of events to be followed when limits are exceeded.

The result of a single analysis or of the mean of a single pair of analyses that exceeds a control limit may be difficult or even impossible to evaluate in the absence of additional data. Some laboratories may choose to reanalyze a control sample, to obtain additional data, if the control result exceeds the 99% limits. If the repeated analysis is acceptable, the run may be declared acceptable. The reader should note that, although commonly performed, this procedure is not necessarily reliable; it is theoretically possible, even when a shift has occurred, for the repeated value to fall "in control" on a random or chance basis. To be on the safe side, an additional repeated analysis of the control sample should be made later on in the run.

Observing the patterns formed by results of multiple analyses plotted in sequence on the control chart may aid interpretation. The statistical control chart facilitates evaluation of individual points in context with other points. Imprecision, drifts, and shifts may be apparent visually. Interpretation of charts is always less certain when precision is poor.

The $\bar{x}$ chart plays an important role in detecting bias change and is, therefore, useful in maintaining accuracy. Observations of individual values have limited reliability for detecting bias changes. However, when patterns in successive values indicate drifts or shifts from the control chart mean, possible changes in bias must be considered. Precise measurements facilitate the interpretation and detection of changes in accuracy; conversely, imprecise measurements, if sufficiently large, disguise changes in bias.

Figure 7 is simulated to show typical patterns of drifts, shifts, and imprecision on $\bar{x}$ charts. Figure 7a illustrates acceptable performance for a hypothetical analyte. In practice, corrective actions would be taken as early as possible and the period of unacceptable performance would not be allowed to continue as shown on the other charts.

Ordinarily, the visual pattern of the points on the chart should help distinguish a drift from a shift. Shifts occur abruptly, a result of some specific cause that continues to produce a constant effect until corrected. It is appropriate to consult records to determine whether some change was made at the time or date of the shift: a variation in procedure or the introduction of a new standard, reagent, or item of equipment. Such conditions might be avoided by performing overlapping studies of the new with the old. Instrumentation problems such as a defective thermostat, a change in volume delivered by an automatic pipette or a pump, and incorrect temperature settings may also cause a shift.

Sometimes a shift[11] will occur in such a way that all values collect on one side of the mean. Such a change in bias is indicated when seven values in a series fall above or below the mean. When many, but not all, points, are on one side of the mean, it may not be clear whether a change in bias has occurred. To determine whether a shift has occurred, calculate the mean of at least eight results that fall in the suspicious area. Corrective action should be taken if the difference between the mean of eight or more consecutive results and the control mean exceeds 1 standard deviation.

A drift is caused by some continuing process that produces progressively increasing bias. Evaporation of solvent from a standard solution, for example, produces a continuing change that steadily concentrates the constituent to more than the declared nominal value. Such an increase in the concentration of the standard will result in a comparable decrease in results for patients' samples and controls. If allowed to continue, this condition will appear on the control chart as a drift downward, as shown in Figure 7d. Deterioration of reagents or of the control material may also cause drift downward (Figure 7e).

Drift upward may result from progressive deterioration or contamination of the standard solution. Failure to properly seal containers of liquid control material may result in gradual loss of water. Such losses especially occur during frozen storage and are a frequent cause of variability in self-prepared control sera. A steadily increasing but unmeasured blank, such as might occur in a deteriorating reagent, may cause upward drift under certain circumstances.

Precision problems arise when certain factors are not constant (Figure 7f). Variation in techniques, variable volumes measured, fluctuating conditions, and inconsistent instrument response contribute to random error. Run-to-run variation may also be affected by those factors that affect within-run precision. Differences in reagent preparation, reconstitution errors of lyophilized controls, variation in instrument settings, and calibration errors that vary from run-to-run are some specific contributions to between-run error.

---

[11] Identification and correction of the cause of a shift may be very difficult for certain analytical procedures. Appropriate action is a topic of much controversy. The laboratory director, however, should not be content to proceed day after day with a procedure that is out of statistical control. After every reasonable possibility for correction has been tried without success, the laboratory director may well choose to recalculate and institute new limits based upon actual performance.

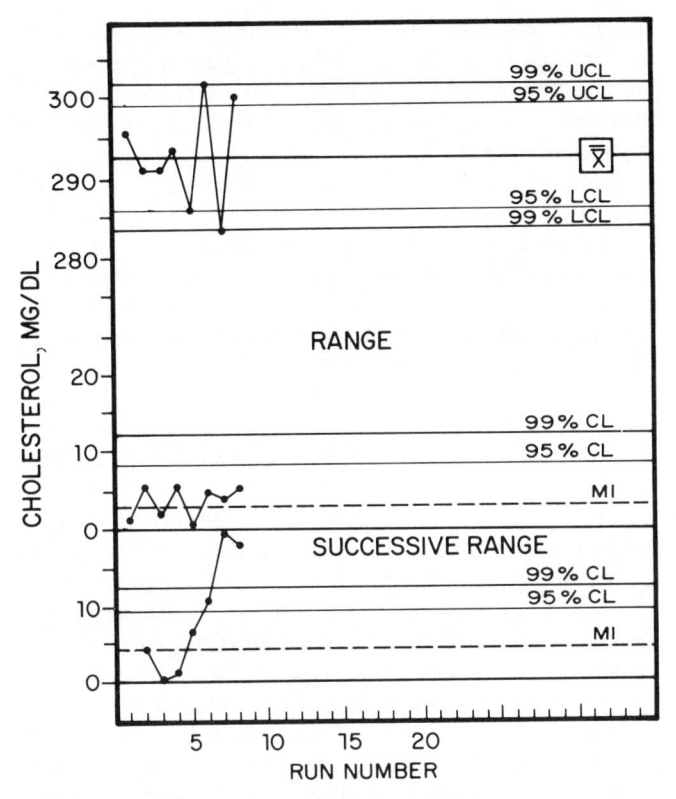

Fig. 8. Control chart for cholesterol, illustrating a precision change

The successive range chart is especially helpful for monitoring between-run error if it is used with the within-run range chart. Figure 8 illustrates the use of quality-control charts to diagnose the cause of increased variability in the cholesterol analysis. Data for this period are assembled in Table 4. The $\bar{x}$ chart shows an increase in variability, starting with the fifth run. The nature of the imprecision is clarified by comparing the range chart with the successive range chart. An examination of records showed that at Run 5 a different technologist took over the daily reconstitution of the lyophilized control serum for cholesterol. According to procedure, aliquots for each daily duplicate were taken from the same vial. The run-to-run variation indicates that each vial was reconstituted to a different volume. It is evident that the second technologist needs training in reconstitution.

The average range, $\bar{R}$, from the first four runs is $15/4 = 3.75$. The average successive range, $\bar{R}_s$, is $6.5/3 = 2.2$. For the second four runs, $\bar{R} = 15/4 = 3.75$. However, $\bar{R}_s$ is $58/4 = 15.875$, indicating a marked increase in the run-to-run variability, although the within-run error did not change. The mean of the first four runs is 292.625, and so is the mean of the second four runs.

Achievement of Accuracy

Analyzing an unassayed control material and using a control chart do not assure accuracy. Accuracy is assured by analyzing standards and reference materials with a reliably assigned value. Most clinical laboratories must rely upon assayed commercial reference materials.

Severe limitations are placed upon the laboratory by certain methods such as those involving enzymic reagents and requiring serum calibrators. Reliability of a particular manufacturer may be the only guide for choice of a calibrator.

Assayed reference materials must be selected with caution. Checking with materials obtained from more than one supplier is usually recommended as a help in obtaining reliability. If there is a choice, the analyst should select materials that have been analyzed by reference procedures.

Participation in quality-control programs and in proficiency testing programs offers opportunities to compare results with other laboratories. Unfortunately, the summarized results for some programs may be limited and may represent a combination of the results of either a variety of methods or variations of methodologies, some with positive biases and others with negative biases. Moreover, the mean of values of all the participants' values who use a common approach is not always as reliable as may be desired. If there is no better point of reference, however, participants may derive some deserved comfort when their results approach the mean values obtained by the others. Certainly, the opposite situation—when results do not agree with those of others—may be appropriate cause for questioning one's method.

Analyzing different concentrations of standard solutions and controls helps to monitor analytical linearity. Excess variation may occur at either extreme of the analytical range. Adverse signal-to-noise ratios may increase the standard deviation at low concentrations. Both accuracy and precision are influenced at the high-concentration range by adequacy of reagent composition and potency. Methods that produce S-shaped response curves should be corrected or avoided. Perhaps the system that relies upon a single-point calibration without adequate control is the greatest potential contributor to excess error at either the low or the high range. Inaccuracy may arise from the analyst's ignorance of the actual nature of the curve. Systems set with single-point calibrators must be carefully controlled with matrix-based controls at concentrations other than the set-point value, to assure linearity.

An Alternative Method of Quality Control

Another system of quality control is especially useful when no satisfactory control materials are available; this system is based upon the repeated (paired) analyses of patients' specimens. Specimens are held and reanalyzed from the same day or from previous days' analyses if stability permits.

By appropriate placement of samples, within-run or day-to-day precision may be estimated from the results of these analyses when control pools are not available. When the second sample is analyzed in the same run as the first, information is provided about within-run precision. Analyzing the second sample in a second run or on a second day provides information about between-run or between-day precision.

The difference, d, is obtained by subtracting the lesser of two values in a pair from the greater. The standard deviation of the differences may be calculated by using this relationship:

$$S = \sqrt{(\Sigma d^2)/2N}$$

A control chart similar to a range chart can be used for monitoring the difference. Limits for the chart can be set upon the average difference, $\bar{d}$, just as limits are set upon average range. However, this procedure lacks the statistical reliability that can be attained when one material is analyzed repeatedly. Because it does not provide a control upon the mean, the procedure does not detect drifts and shifts.

The value of the information may be increased if the identity of the repeat specimen is unknown to the analyst. However, if specimens are submitted "blind" with fictitious names, the laboratory must be careful not to deliver the reports to patients' record files.

Kit and Method Selection and Evaluation

It is necessary to select good methods or kits. The small-volume laboratory, like many other laboratories, may choose a kit largely because of convenience. A kit is a packaged method, however, and must be given just as much consideration in selection and during use as the evaluation of any method. Several of the references provide very good descriptions of kit evaluation procedures (8–12).

In selecting a method, much thought must be given to the principles used, the propriety of any modifications, and what level of accuracy and precision the user might expect. Such information comes from carefully reading the literature, either in published papers or in the package inserts for the kits. Kit manufacturers should provide very concise descriptions of the results of evaluation studies as well as information about their use.

The manufacturer of an acceptable kit is expected to make appropriate evaluations about accuracy. The literature accompanying the kit should summarize results from a number of experiments designed to evaluate both bias and precision of the kit. These experiments should be identical in nature to those any developer of a method should perform before using or reporting a new, prospective method for analyses of patients' samples. For example, the manufacturer should be expected to perform recovery studies, which test linearity as well as accuracy at high concentrations.

Adequately described evaluation studies will contain results of the measurements of variability. Means and standard deviations should be stated in the kit literature as measured on a number of samples covering a range of concentrations. Under no circumstances should the laboratory use the standard deviations listed by the kit manufacturers to assign control limits. In some instances, these values may have been pooled from data from several laboratories and may be too wide for acceptable performance. The user must determine how the method or kit performs under the conditions of the laboratory in which the kit is to be used. Information should be presented in the method or kit literature about the usable analytical range for which linearity can be expected. The type of expected response should be stated, if other than linear (as may occur in certain enzymic and immunochemical procedures).

When evaluating a new method or kit, one should make some preliminary analyses of standards and controls, to become familiar with the new method. Sometimes a method will be rejected because of obvious shortcomings that are apparent early. If the kit survives early testing or if its deficiencies can be corrected, precision should be measured by analyzing reference pools of at least three different concentrations spanning the intended analytical range (low, medium, and high). Control samples should be analyzed in duplicate in at least 20 runs, and control charts should be constructed on which to plot the values and make an evaluation. Means, ranges, and standard deviations should be calculated for each pool and evaluated before making the final decision regarding the suitability of the kit.

Special Concerns of the Small Clinical Chemistry Laboratory

Quality control may represent a greater relative expenditure of committed resources for the small laboratory than for the large laboratory. The ratio of the number of controls and standards to patients' samples is high and may exceed 1:1 in the small-volume laboratory. Thus, the relative cost per specimen of small-volume laboratory analyses may be expected to be higher than that for larger laboratories. Nevertheless, quality analyses require that standards and controls be placed in every run, regardless of length—without the information they supply, the laboratory is operating blindly.

Pressures may encourage management to overextend the laboratory's technical capability or available time. Backup instrumentation and backup methods are likely to be missing from the small laboratory, or if available, may not always be capable of producing acceptable results. It is not necessary for each laboratory to offer every procedure available in the field to provide good patient care; certain tests may be equally well performed by another laboratory. Emphasis on high-quality performance in whatever is done should be the foremost consideration.

There are other handicaps of being small, although these problems may be experienced by any laboratory. In a small staff, absences of individuals are felt acutely whenever they occur. As a result, directors of small laboratories may be reluctant to provide and schedule continuing-education opportunities for their personnel.

To summarize, a small size or a small-volume workload does not exempt a laboratory from the need to exert intensive efforts to assure quality performance. High-quality analyses result from consistent application of care, skill, and effort. Quality-control procedures, aided by quality-control charts, provide the necessary visibility to the analytical process by revealing whether performance is being maintained at acceptable levels. The information obtained from quality-control procedures assures the physician as well as the laboratory of the reliability of the measurements, and is of ultimate benefit to the patient.

References

1. Wernimont, G., Use of control charts in the analytical laboratory. *Anal. Ind. Eng. Chem.* **18**, 587–592 (1946).
2. Levey, S., and Jennings, E. R., The use of control charts in the clinical laboratory. *Am. J. Clin. Pathol.* **20**, 1059–1066 (1950).
3. Buttner, J., Borth, R., Boutwell, J. H., and Broughton, P. M. G., IFCC Committee on Standards, Provisional recommendations on quality control in clinical chemistry. Part 1. General principles and terminology. *Clin. Chem.* **22**, 532–540 (1976).
4. Buttner, J., Borth, R., Boutwell, J. H., and Broughton, P. M. G., IFCC Committee on Standards, Provisional recommendations on quality control in clinical chemistry. Part 2. Assessment of analytical methods for routine use. *Clin. Chem.* **22**, 1922–1932 (1976).
5. Bowers, G. N., Jr., Burnett, R. W., and McComb, R. B., Preparation and use of human serum control materials for monitoring precision in clinical chemistry. *Clin. Chem.* **21**, 1830–1836 (1975).
6. Burnett, R. W., Accurate estimation of standard deviations for quantitative methods used in clinical chemistry. *Clin. Chem.* **21**, 1935–1938 (1975).
7. Westgard, J. O., Groth, T., Aronsson, T., Falk, H., and deVerdier, C. H., Performance characteristics of rules for internal quality control: Probabilities for false rejection and error detection. *Clin. Chem.* **23**, 1857–1867 (1977).
8. Barnett, R. N., and Youden, W. J., A revised scheme for the comparison of quantitative methods. *Am. J. Clin. Pathol.* **54**, 454–462 (1970).
9. Broughton, P. M. G., Evaluation of analytical methods in clinical chemistry. In *Progress in Clinical Pathology*, 8, M. Stefanini, A. A. Hossaini, Sr., and H. D. Isenberg, Eds., Grune and Stratton, New York, NY, 1978.
10. Lloyd, P. H., A scheme for the evaluation of diagnostic kits. *Ann. Clin. Biochem.* **15**, 136–145 (1978).
11. Nielsen, L. G., and Ash, K. O., A protocol for the adoption of analytical methods in the clinical chemistry laboratory. *Am. J. Med. Technol.* **44**, 30–37 (1978).
12. Westgard, J. O., deVol, D. J., Hunt, M. R., Quam, E. F., et al., Concepts and practices in the evaluation of clinical chemistry methods. I. Background and approach. *Am. J. Med. Technol.* **44**, 290–300 (1978). II. Experimental procedures. *Ibid.*, pp 420–429. III. Statistics. *Ibid.*, pp 552–571. IV. Decisions of acceptability. *Ibid.*, pp 727–741.

Quality Assurance: Professional Organizations and Commercial Systems

Submitters: Thomas C. Stewart, *Medical Laboratory of Baton Rouge, Baton Rouge, LA 70806*
Richard H. Gadsden, Sr., *Division of Clinical Chemistry, Medical University of South Carolina, Charleston, SC 29425*
Thomas S. Herman, *Memorial Medical Center, Savannah, GA 31403*

Introduction

It is impossible to reproduce exactly any human act, including the performance of laboratory analysis. Quality control provides the laboratorian with data regarding the reproducibility of his performance of laboratory analysis. We have to live with some variation from test to test, day to day, but any variation greater than a certain minimum may reach the point of unacceptability, and cause an erroneous report. When the value of a component in a quality-control pooled specimen lies outside of acceptable limits, there is a good chance that a *systemic* error, rather than a *random* error, is the cause. Areas to suspect for the cause of the error include specimen collection and handling, instrument reliability, reagent reliability, personnel integrity, and laboratory management.

Presented here is a procedure by which individuals in a small clinical laboratory can efficiently and economically initiate and maintain a program for quality assurance. We do not present a discussion on the theories of quality control or quality assurance. A detailed presentation on these topics is found in the preceding chapter and throughout this book.

Personnel

A laboratorian must be qualified, experienced, and highly motivated. It is not enough that analysts know how to perform a procedure; they must also recognize whether the procedure is working properly, and must be able to troubleshoot and correct problems when they occur. Match the employee to the job on the basis of the individual's capabilities. Failure to do so will create problems in providing useful data, as well as problems in morale and personnel management.

Encourage continuing education. With the rapid changes taking place in clinical chemistry, employees should keep current by attending meetings and seminars and by reading current literature in their particular fields. Up-to-date reference books and journals should be available in the laboratory.

Documentation

A detailed procedure manual and records on instruments are essential in developing a good quality-control program. The organization of the laboratory should be outlined, indicating the responsibility and authority of each employee. Each procedure must be so described that a qualified "outsider" could perform it reliably. The package inserts of many commercial products now fulfill this need and could be incorporated into the procedure manual. Any changes in procedure in the laboratory must be reviewed and initialed by a responsible supervisor.

Keeping detailed performance and maintenance records is necessary to recognize instrument problems. It is often difficult to trace an instrumental problem (e.g., loss in speed of a centrifuge, or wavelength drift of a spectrophotometer) when accumulated data are not available. Repairing an instrument may be virtually impossible if a previous modification or change has not been recorded. Because a sudden shift in values can often be attributed to new reagents, record the performance of each new "lot number" of reagents, using controls run with both the old and the new lot numbers. If a standard curve is not calculated each time the procedure is run, check the linearity of the procedure frequently, using at least three values (low, normal, and high).

Check critical temperatures daily. These include the storage temperatures of specimens, controls, and reagents. The stated outdating of controls and reagents is correct only if correct storage temperatures are used. Heating block and waterbath temperatures, and the reaction temperature built into automated equipment, control the rate of reaction and final results of many procedures. Check and record these temperatures, making sure that they are within acceptable limits.

Intralaboratory Checks

One of the more valuable checks in the small laboratory is *repeat* analysis. A specimen that has been previously analyzed for a component is properly stored and re-analyzed along with the next run. The value given by the second analysis should be within the coefficient of variation of that particular procedure [CV = (1 standard deviation/mean) × 100, expressed as percent]. If not, a careful study should be undertaken to determine why not, and corrective action taken.

A *delta check* is sometimes useful when a patient is having repeat analysis done daily or otherwise for a particular component. It is doubtful, for example, that any patient would have a 100-fold change (delta) of any component in a 24-h period. The limits of change vary from component to component, and must be derived from a cooperative effort of the laboratory and the physician.

Be aware of *absurd values*. It is unlikely that a patient who walked into the laboratory to have blood drawn would have a blood urea nitrogen value of 150 mmol/L or a glucose of 420 mg/L. Many laboratories find it useful to prepare an "absurd-value" chart and post it where the analyst can see it.

Pattern recognition is also a valuable check. Gross abnormality in one electrolyte without corresponding abnormality in another electrolyte is suspicious. Specific-organ enzymes usually are increased to above-normal values as a group. A marked increase in uric acid, blood urea nitrogen, or creatinine usually results in an abnormality in one of the other analytes. Hemoglobin concentration has a numerical value of about one-third the hematocrit value. Many other patterns can be found.

Interlaboratory Quality Control with Commercial Sources

Although reference (calibrator) and control materials are most often obtained commercially, the individual laboratory bears the responsibility for determining the validity of using these materials. Historically, tests of laboratory performance throughout the world have shown that variations exist between laboratories even when the methods, conditions for reaction, and instrumentation used are identical. These efforts have highlighted a general lack of accuracy and precision in quality control on an interlaboratory basis. Once this problem had been identified, professional agencies (e.g., College of American Pathologists, Institute of Clinical Scientists, American Association of Bioanalysts) and governmental agencies (federal and state) have intensified efforts to improve laboratory performance. They have made materials available for assay purposes and have compared the results on a regional to national basis. This has been a major contribution toward assessment of analytical improvement in the clinical laboratory.

Recognizing a deficiency in performance and taking proper corrective action must be timely. Often, in the small clinical laboratory this recognition must be obtained through external assessment; therefore, the more frequent the assessment, the sooner the recognition and corrective action. External assessment may assure that analytical accuracy and, often, precision are being maintained within acceptable limits.

Several quality-assurance programs are offered by industries and through professional organizations related to the clinical laboratory. The common factor among all these programs is the availability of a material (primarily lyophilized serum or urine) that is analyzed repeatedly in the laboratory over a specific period. The results obtained are regularly (e.g., monthly) returned to the providing agency. The provider collates results from the participating laboratories according to the analytical interlaboratory variations, i.e., methods (and conditions) and instrumentation, and this collated information is forwarded to the participating laboratories for their use in assessing their performance. The individual laboratories involved are not identified.

Many of the programs are flexible to suit the needs of the individual laboratory. Assessment of accuracy may be the most desirable information sought. Programs are available for "run-to-run" and "day-to-day" precision evaluation. Most users desire participation of multiple laboratories analyzing the same "pool" because the data obtained are useful for both intra- and interlaboratory assessment. The cost to the laboratory varies with the amount of material used. The usual basis is cost per milliliter of specimen, number of analytes chosen, and the frequency of data points per analyte, with computer cost included. In choosing a program, the user must understand clearly whether precision or both accuracy and precision are being evaluated by the individual program offered.

Listed below are U.S.-based agencies that have quality-control programs available for laboratory use. We recognize that the list is not exhaustive, nor do we endorse any in particular, but they are all reputable. For a full description of the programs, contact each organization.

The following companies or organizations use computer-based programs to provide long-term intra- and interlaboratory precision evaluation of serum constituents. Accuracy is generally evaluated on a consensus basis. Data are listed and collated by method, reagent source, and instrumentation. Some programs include the source of controls used by the reporting laboratories. Lyophilized sera are provided for analysis.

Bio-Rad
Environmental Chemicals Specialty Division
3700 East Miraloma Ave.
Anaheim, CA 02807
Tel: (714) 630–6400

College of American Pathologists
P.O. Box 1234
Traverse City, MI 49684
Tel: (800) 253–1790

American Dade
P.O. Box 520672
Miami, FL 33152
Tel: (800) 372–1572

Fisher Scientific
Diagnostics Division
526 Route 303
Orangeburg, NY 10962
Tel: (800) 241–8912

General Diagnostics
Division of Warner/Lambert Co.
201 Taber Rd.
Morris Plains, NJ 07950
Tel: (800) 241–9616

Hyland Diagnostics
One Baxter Parkway
Deerfield, IL 60015
Tel: (800) 323–1472

Ortho Diagnostics Systems, Inc.
Raritan, NJ 08869
Tel: (800) 241–2549

The following companies or organization provide materials and computer-based programs for assessing long-term precision analysis of more specialized constituents, e.g., radio-ligands, therapeutic drug monitoring, pH and blood gases, lipids and lipid fractions, heavy metals, and enzymes.

American Dade
Fisher Scientific
General Diagnostics

Hyland Diagnostics
Ortho Diagnostics

American Association for Clinical Chemistry
1725 K. St., N.W.
Washington, DC 20006
Tel: (202) 833–3590
Note: TDM program only

Instrumentation Laboratories
Biomedical Division
113 Hartwell Ave.
Lexington, MA 02173
Tel: (617) 861–0710
Note: pH–blood gas program only

Summary

The whole point of a quality-assurance program is to attempt to maintain or to improve the quality of the laboratory results reported each day. Adequate quality control does not consist of merely assaying these materials daily and charting the results; or even reporting the results to the computer center, and then glancing at the results when they are returned. It primarily consists of taking appropriate action when results fall outside of the established limits. This means reviewing results daily, just as soon as they are available.

What do you do when results are outside of the limits? No single answer can be given. First, look for an obvious source of error. Did the technologist use the wrong filter or the wrong wavelength? Did the analyst, or a clerk, or a nurse mislabel the specimen or the standards? Is there any other obvious reason for the error? Were control results for other tests off in the same direction on the same day, suggesting that something is wrong with that particular vial of control material? In any case, you must have a logical and defensible plan of action, and it must be routinely instituted in your laboratory when the results for quality-assurance procedures exceed the accepted tolerance limits.

Preventive Maintenance of Instruments and Equipment

Submitter: Murali Dharan, *Mallinckrodt, Inc., Bohemia, NY 11716*

Evaluators: James L. Driscoll and Horace F. Martin, *Rhode Island Hospital, Providence, RI 02902*
Leslie Parker, *The Washington Hospital, Washington, PA 15301*
Russell A. Picard, Jr., *St. Agnes Hospital, Fond du Lac, WI 54935*
Mary Kay Walker, *New England Deaconess Hospital, Boston, MA 02215*
Jannie Woo, *Department of Pathology and Laboratory Medicine, The University of Texas, Houston, TX 77025*

Reviewer: Stanley Bauer, *Technicon Instruments Corp., Tarrytown, NY 10591*

Introduction

All quantitative chemical analyses performed in the clinical laboratory involve the use of one or more instruments. They may be mechanical, optical, electronic, or a combination of these. All of them need maintenance for a smooth and error-free operation. Mechanical equipment may need oiling or lubrication and some general cleaning. Delicate instruments need periodic adjustments of various parts and calibration besides general cleaning.

All these preventive maintenance procedures take time. But the time spent is worthwhile in terms of long life of instruments and equipment and increased accuracy and precision in laboratory test results (1).

In this chapter, the Submitter has reviewed the various preventive maintenance measures required for instruments and equipment. The very general discussion will be by class and not by brand name. Therefore, readers must consult the operators' manuals of their instruments for more specific details of preventive maintenance procedures.

Spectrophotometers

Spectrophotometers are perhaps the most commonly used instruments in the clinical chemistry laboratory that can influence the accuracy and precision of analytical determinations. It is therefore important that spectrophotometers receive regular and careful preventive maintenance (2).

All spectrophotometric measurements are based on Beer's law, which states that the concentration of a substance being analyzed is proportional to the absorbance of the final "colored" product. Mathematically, Beer's law is expressed as $c = A/ab$, where c is the concentration of the analyte, A is absorbance, a is absorptivity, and b is length of the light path, in centimeters. According to this equation, the concentration of the substance being measured depends on the absorbance, the absorptivity (a function of wavelength), and the length of the light path (the internal diameter of the cuvette). In other words, accurate analytical determination with a spectrophotometer depends on the accuracy in measuring these three components (3).

Wavelength Calibration

Wavelength calibration should be checked each week and recalibrated when needed. For spectrophotometers having a spectral band width of 20 nm or more, a didymium filter may be used for calibration. A didymium filter has its highest absorbance at 585 nm. If the wavelength calibration knob is slowly turned from 580 nm to 590 nm with this filter in place, the absorbance should slowly increase, giving the highest reading at 585 nm, then decrease steadily from 585 nm through 590 nm. If the absorbance does not peak at 585 nm, perform a wavelength calibration. For further details of calibration procedures of your particular instrument, consult the operator's manual.

For spectrophotometers with a spectral band width of 2 to 8 nm, use a holmium oxide filter (peak absorbance at 360 nm) to determine the need for wavelength calibration.

Calibration of Absorbance

Absorbance calibration should be carried out every six months and checked at least monthly. Some spectrophotometer manufacturers supply one or more absorbance calibrators. Otherwise, some calibrating solutions may be prepared as follows (1, 4):

1. Weigh out accurately 50 mg of potassium dichromate ($K_2Cr_2O_7$) and transfer all of it into a clean 1-L volumetric flask. Dissolve the dichromate crystals in about 300 mL of 10 mmol/L sulfuric acid, then dilute to 1 L with the acid. Mix well and store in a clean, tightly stoppered brown bottle. Stable for six months, this solution should give an absorbance of 0.535 ± 0.002 at 340 nm, read against 10 mmol/L sulfuric acid.

2. Weigh out 20.0 g of copper sulfate, and transfer this into a clean 1-L volumetric flask. Dissolve the crystals in about 300 mL of distilled water. Add 10 mL of concd. sulfuric acid and mix well. Then dilute to 1 L with distilled water. Mix well once again and store in a clean, dry bottle. Mark an expiration date of six months. This solution should give an absorbance of 0.527 ± 0.002 at 650 nm.

3. Weigh 40.0 mg of potassium chromate and transfer it carefully to a 1-L volumetric flask. Dissolve the crystals in about 800 mL of distilled water. Add 3.30 g of potassium hydroxide and mix well until totally dissolved. Dilute to 1 L with distilled water. Mix well and store in a clean and dry polypropylene bottle. Stable for three months, this solution should give an absorbance of 0.396 ± 0.002 at 400 nm.

Sealed vials of absorbance standards are available from the National Bureau of Standards in Washington, DC, and also from some commercial sources. Calibration standard solutions made in the laboratory should be double-checked before any major adjustments are made on the spectrophotometer. Always calibrate wavelength before attempting to calibrate absorbance.

Spectrophotometers should also be cleaned periodically. The outside should be wiped clean of spills and dust with a partly damp cloth at least once a week. The interior should be cleaned free of dust with an air gun every three months. Also, use lens-cleaning paper to clean the light source every three months.

Summary

1. Wipe the outside with partly damp cloth once a week. Clean up major spills immediately.
2. Clean the interior of the instrument with an air gun or vacuum at least once every three months to eliminate dust.
3. Clean the light source once every three months, using lens-cleaning paper.
4. Check wavelength calibration weekly and recalibrate as needed.
5. Check absorbance calibration weekly and recalibrate as needed. If absorbance calibration is not possible, perform the manufacturer's recommended procedure to correct the situation or contact the manufacturer's service personnel.

Thermometers

Mercury-in-glass thermometers are used in the laboratory to measure temperatures of water baths, incubators, refrigerators, heating blocks, and freezers. A mercury thermometer consists of a graduated capillary tube and a bulb containing mercury. Accuracy of a thermometer depends of the integrity of the mercury column, the thermal quality of the glass used in its construction, and the accuracy of standardization.

Check the mercury column monthly. A broken mercury column—one with gaps between sections containing mercury—can show completely erroneous temperatures. Rapid heating and cooling of the thermometer bulb, especially when the capillary is very narrow, can cause jerky rather than smooth movement of the mercury through the column and sometimes produces broken mercury columns.

Do not use thermometers with broken mercury columns. Regenerate the continuity of the column by cooling the bulb to very low temperatures so as to withdraw all the mercury back into the bulb. When the bulb warms up, mercury will slowly rise into the capillary in one continuous column. This technique may not work if the column is dirty or moist; in such cases, discard the thermometer. Examine the mercury columns in thermometers for continuity at least once a month.

Thermal quality of the thermometer glass is important in maintaining year-to-year accuracy in measurement. Good thermometer bulbs are made of a special thermal-quality glass so that they regain their original volume rapidly after exposure to high and low temperatures. The thermometer bulb may sometimes expand or contract into a distorted shape, causing an undesirable change in volume that can lead to erroneous readings. If the temperature shown by the thermometer deviates by more than five divisions from the true temperature, suspect a problem with the bulb and discard the thermometer.

Standardize thermometers at least once every three months with a National Bureau of Standards certified thermometer. After standardization, tag or mark the tested thermometer. A thermometer that is used in the 37 °C water bath, for example, should be calibrated at that temperature: if a true 37 °C water bath shows 37.2 °C on the thermometer being calibrated, write 37 °C = 37.2 on the thermometer tag. If the thermometer shows a temperature that deviates by more than five divisions from the true temperature, consider it defective and discard.

During calibration, as well as during a regular measurement of temperature, use an ascending mercury column, not a descending one. For example, if calibrating for 60 °C, the thermometer should be reading below 60 °C before it goes up to 60 °C.

Summary

1. Check the thermometer liquid column for continuity at least once a month. If the column is broken, try to regenerate a continuous column. If the column remains broken, discard the thermometer.

2. Standardize the thermometer at least once every three months. If the temperature shown by the thermometer is more or less than the true temperature by more than five divisions, discard the thermometer.

Refrigerators and Freezers

The refrigerator is an essential piece of equipment in any laboratory. It stores perishable reagents and patients' samples to prevent spoilage. Refrigeration prevents deterioration of reagents and samples in three ways:

1. It prevents or retards microbial growth. Microorganisms may grow very rapidly in reagents and samples at room temperature, especially when nutrients are present. Proteins in patients' serum samples are fairly good nutrients for microorganisms. Growth of microorganisms may be visible to the unaided eye in sera left at room temperature in the laboratory for one to four days; however, the same samples may stay without significant microbial growth in the refrigerator at 2 to 8 °C for two to four weeks or in the freezer at −20 °C for one to two years. Many reagents and chemicals can themselves function as nutrients, making room-temperature storage undesirable. Addition of antimicrobial agents to serum and reagents without proper testing for interference is not usually a recommended procedure for preventing microbial growth. Refrigeration of samples and reagents is both easy and convenient.

2. It retards the decomposition of reagents. Many clinical laboratory chemicals and reagents, especially enzymes and coenzymes, decompose at all temperatures. When the temperature is lowered, the rate of decomposition is also lowered. At the refrigerated temperature such decomposition is considerably decreased. On the other hand, malfunction of a refrigerator can often lead to the loss of expensive reagents or erroneous test results from use of spoiled reagents.

3. It retards the reaction between the various ingredients within a single reagent. Many reagents undergo a certain degree of reaction between the various constituents of the reagent. This reaction can be slowed down considerably by refrigeration or by lowering the temperature of the reagent. When such reagents are not properly refrigerated, they may become unusable or cause erroneous test results.

Thus unexpected refrigeration failure or improper functioning may result in expensive losses as well as inconveniences in the smooth operation of the laboratory; sometimes reagent loss may even affect proper patient care in the hospital. For good and uninterrupted service from a refrigerator, carry out the following maintenance procedures at the given intervals:

1. Check and record the temperatures of refrigerators and freezers daily. Place a thermometer in a suitable location inside each refrigerator and freezer. Hang the thermometer vertically. Use a mercury-in-glass thermometer for refrigerators and an alcohol-in-glass thermometer for freezers.

2. Adjust the temperature setting once a week if necessary. The temperature of refrigerators should be between 4 and 8 °C. If the average for the week was not 6 °C, adjust the setting. Freezer temperature should be between −30 and −20 °C.

3. Defrost refrigerators at least once every three months, and freezers once every six months.

4. Vacuum or remove any dust from the condensing coil (condenser) at the back of the refrigerator at least once a year. Some units have a condenser fan; make sure the fan is moving properly. When a refrigerator fails to cool properly, 95% of the time a dusty condenser or a poor fan ventilator is at fault.

5. Clean up any spillage in the refrigerator immediately.

Note: Always set the back of the refrigerator at least 5 in. (13 cm) away from the wall to provide adequate air circulation to cool the compressor unit; this will minimize the chance of compressor "burn out" or loss of cooling effect. Never use an extension cable on a freezer or refrigerator because improper cable size will immediately cause compressor damage or fire.

Analytical Balances

Analytical balances are used in the clinical laboratory to determine accurately the weights of chemicals or other materials. Many types of analytical balances are available, but all of them need some occasional care for proper functioning. Because all respond to force acting on the weighing pans, the analyst should keep all extraneous forces such as air currents, heating effects, changes in relative humidity, magnetic influences, vibrations, fingerprints, etc. away from the weighing pans and other parts of the balance.

Error-free operation of an analytical balance depends also on the proper placement of the balance. It should be away from any draft, air vents, and direct sunlight, and should be placed very level on a vibration-free stand or table. The balance should be kept dustfree; whenever it is not in use, keep it covered with a dust cover. Regularly follow these maintenance procedures for analytical balances:

1. Return the scale of the balance to zero after each use.
2. Clean up spillage in the balance after each use.
3. Keep the area near the balance dry, clean, and dust-free.
4. Check the accuracy of the balance every three months with calibrated weights from the National Bureau of Standards.

Flame Photometers

Flame photometers are used to measure the concentration of alkali metals (sodium, potassium, and lithium) in biological fluids. The intensity of the color imparted by the metals to the flame is proportional to the concentration and is measured photometrically. To perform a good analysis, one should have a clean, uniform, and steady flame with no distortion or discontinuity. The filters used in the photometer section should be clean, and the amount of sample aspirated into the flame per unit of time should always be the same. To maintain accuracy in the analysis, follow these preventive maintenance procedures for the flame photometer:

1. Check the aspiration rate and adjust it weekly if necessary.
2. Check flame conditions weekly and adjust the gas and oxygen supply if required.
3. Clean encrusted salts from the burner head and ignition block, when necessary. Check weekly.
4. Check the air/oxygen and gas pressure daily.
5. Remove and clean the chimney once every month.
6. If they are removable, clean the optical filters with an air gun or lens-cleaning paper monthly.

Centrifuges

Centrifuges are used extensively in clinical laboratories to separate cells from blood in preparing serum or plasma, to clarify fluids, to separate suspended solid particles from solution, to concentrate and purify various biological and chemical agents, and to perform certain analyses involving the quantitative separation of solids from liquids.

An object revolving at a high speed in a circle having a certain radius generates a centrifugal force that is proportional to the speed as well as the radius. Centrifugal force is, therefore, expressed both in terms of revolutions per minute (rpm) or gravity (*g*-value). Many analytical procedures involve centrifugation at a certain speed. For the centrifuge to function and reproduce speed or *g*-force accurately, the following preventive maintenance procedures should be carried out at regular intervals (5):

1. Clean up major spills and broken tubes immediately.
2. Once a month wipe the interior and exterior with a damp cloth.
3. Lubricate the bearings every three months, if the unit is not permanently lubricated and sealed; follow the manufacturer's recommendation.
4. Every three months check that the unit is cushioned well on the floor or table, that it is balanced and free of vibration, and that the accessories are well balanced.
5. Check the brushes and replace if they are worn to $5/16$-in. (8 mm). Do this every three months. Follow the manufacturer's recommendation.
6. Check the oil level in the reservoir every six months.
7. Match centrifuge tube carriers of the same "stamped" weight in the opposite position.
8. Check the rpm with a tachometer once every three months.

Water Baths

Water baths are used in the clinical laboratory to carry out certain analytical reactions. A reaction required in the analysis might be too slow at room temperature, but will be faster at a higher temperature. The most commonly used temperature for a water bath is 37 °C, but other temperatures are also used. In all cases, the specified temperature is important because, for a particular analysis, too much or too little reaction may occur if the temperature is not maintained within the tolerance limits. Inconsistent reactions cause falsely increased or decreased test results. For proper functioning of the unit and to assure the accuracy of the temperature used, use the following preventive maintenance procedures at regular intervals:

1. Check and record the temperature of water baths daily. Adjust the setting if the temperature is outside the specified limits. Leave a properly tagged thermometer in the water bath.
2. Check the water level daily. If it is too low, fill the bath up to the appropriate mark with distilled water.
3. Clean the bath thoroughly and fill it with distilled water once a month and whenever there is a spillage.
4. Thymol (1 g/L) can be added to the water to inhibit bacterial growth.

Conclusion

In general, oiling the moving parts, changing bulbs, and maintaining clean surfaces need no elaboration except to emphasize their importance. Instruments and equipment manufactured by different companies may vary slightly in their maintenance schedule; therefore, follow the instructions given in the operator's manual before planning a preventive maintenance program.

INSTRUMENT:		MODEL NO:		
MANUFACTURER:		SERIAL NO:		

SERVICEMAN:

PREVENTIVE MAINTENANCE:

1.

2.

3.

4.

DATE	MAINTENANCE	PERFORMED BY	REMARKS	DATE OF NEXT MAINTENANCE

Fig. 1. A form for documentation of a preventive maintenance program

It is a good practice to *record everything done* in preventive maintenance on forms such as the one shown in Figure 1. A preventive maintenance file should contain one such form for each instrument. Such forms also help in management problems by having the instrument model number, serial number, address and telephone number of the manufacturer and the serviceman, etc. in one place. It should be useful to the inspectors from government and professional accrediting groups. Any problems or troubles involving instrument malfunction should be recorded in a separate book for future reference if trouble-shooting is necessary.

References

1. Dharan, M., *Total Quality Control in the Clinical Laboratory.* C. V. Mosby Co., St. Louis, MO, 1977.
2. Winstead, M., *Instrument Check System.* Lea & Febiger, Philadelphia, PA, 1971.
3. Keitges, P. W., Standardization and calibration of spectrophotometers. American Society of Clinical Pathology Technical Improvement Service, Chicago, IL, 1971.
4. Rand, R. N., Practical spectrophotometric standards. *Clin. Chem.* 15, 839–863 (1969).
5. Tenczar, F. J., and Kowlski, T. A., The centrifuge: Maintenance and calibration. American Society of Clinical Pathologists Technical Improvement Service, Chicago, IL, 1973.

Safety

Submitters: A. W. Forrey, *Veterans Administration Medical Center, Seattle, WA 98108*
Collene J. Delaney, *Veterans Administration Medical Center, Seattle, WA 98108, and University of Washington, Department of Laboratory Medicine, Seattle, WA 98102*
Willie L. Ruff, *Clinical Laboratories, Howard University Hospital, Washington, DC 20060*

Evaluators: Laken G. Warnock, *Department of Biochemistry, Vanderbilt University, and Research Laboratories, Veterans Administration Hospital, Nashville, TN 37203*
Isaac D. Montoya, *Medicus Systems, Houston, TX 77006*
John R. Leach, *Safety Operations Section, Occupational Safety and Health Branch, National Institute of Health, Bethesda, MD 20205*

Reviewer: R. E. Vanderlinde, *Clinical Laboratories, The Hahnemann Medical College and Hospital of Philadelphia, Philadelphia, PA 19102*

Chapter Outline

Safety Goals and Perspectives
  A. Agencies concerned with laboratory safety
  B. Practical objectives
  C. Understanding agency scope and authority
Hazards and Their Control in General Laboratory Operations
  A. Pipetting
  B. Eye protection
  C. First aid
  D. Working alone
  E. Eating, smoking, applying cosmetics
  F. Safety equipment
  G. Emergency situations and plans
  H. Safe work practices
  I. Waste disposal
  J. Centrifuges
  K. Lifting
Use of Toxic and Hazardous Substances
  A. Understanding the properties of chemicals
  B. Planning the use of chemicals
  C. Hazardous vapors
  D. Chemical and solvent waste storage and disposal
    1. Insoluble waste
    2. Organic solvents
    3. Sodium azide
    4. Nontoxic waste chemicals
    5. Toxic and heavy metal wastes
  E. Specific hazards
    1. Mercury
    2. Phenol
    3. Other vapor hazards
Solvents, Fire Hazards, and Explosion
  A. Solvents and solvent storage
  B. Fire and explosion hazards
  C. Fire and explosion protection systems
Biohazards
  A. Handling infectious agents
  B. Specimen processing
Radioactivity
  A. Requirements and regulations
  B. Information required by the Nuclear Regulatory Commission
Compressed and Cryogenic Gases
Electrical Equipment, and General Safety Systems and Services
  A. Instrument usage
  B. Electrical circuitry
  C. Gas
  D. Vacuum
  E. Drains
  F. Communications
Ventilation, Air Conditioning, and Environmental Control
  A. Special local ventilation
  B. Fume hoods
  C. Biological safety cabinets and laminar-flow hoods
Laboratory Design, Maintenance, and Housekeeping
  A. Access, security, and layout
  B. Furniture
  C. Storage
  D. Repair of instrumentation, equipment, and apparatus
  E. Maintenance of the physical plant
  F. Laboratory staff responsibilities for housekeeping
  G. Noise, lighting, and work and walking surfaces
Designing a Laboratory for Safety

In the Occupational Safety and Health Act (OSHA) of 1970, the U.S. Government recognized the responsibility of employers to provide a safe working environment for their employees. The Toxic Substances Control Act of 1976 (TOSCA) was also a step in recognizing the need to control human exposure to hazardous chemicals. Although several professional societies have promoted safety in laboratory design and practice, only in the last few years have individual clinical and industrial chemists become safety-conscious. We now recognize that in terms of safety all members of laboratory professions are prone to oversight, forgetfulness, and the pressures of expediency, leading to personal risk-taking that cannot be condoned. This situation is as prevalent in the small clinical laboratory as in the most advanced research facility.

Certain potential hazards are more prevalent in the small laboratory than in the large one. However, because each facility differs, no single hazard can be ruled out. Because the topics selected cannot be reviewed completely, we have provided a compact bibliography for the reader's convenience. References 1–10 are good general references on laboratory safety.

Safety Goals and Perspectives

Table 1 displays the mosaic of federal, state, local, and private organizations concerned with safety in clinical laboratories in the United States. OSHA is the primary enforcer and may delegate its authority to approved state programs. Nevertheless, there is an interlocking network of related agencies concerned with different and specific aspects of safety.

A major difficulty in promoting safety in the laboratory is that many groups have an interest in pieces of the program, but none is responsible for a completely integrated program, or even a substantial part of the program. For this reason local, state, and federal agencies may compete. Such competition sometimes hinders rational solutions of safety problems in an institution.

The purpose of this chapter is to assist laboratorians in identifying concepts that may be used in obtaining effective and economic solutions to specific safety problems.

Each of the groups listed in Table 1 affect, in some way, the three major facets of any laboratory safety program: (a) an operational document of safety policy for the facility, (b) an inspection program, and (c) an educational program. Each of these activities depends on an effective, dedicated safety group

Table 1. Agencies Concerned with Laboratory Safety

Govt. Regulatory/Research	Standards/Criteria	Professional Societies/ Affected User Groups	Informational
Federal			
U.S. Dept. of Labor	Natl. Council Radiation Protection & Meas.	Am. Chem. Soc.	Natl. Safety Council
OSHA		Am. Assoc. Clin. Chem.	
Fed. Radiation Council	Internatl. Council of Radiation	Am. Soc. Med. Tech.	
U.S. Dept. of Energy	ANSI	Am. Soc. Microbiol.	
Nucl. Reg. Comm.	NFPA	College Am. Pathol.	
U.S. Dept. of HHS	Western Fire Chiefs Assoc.	FASEB	
Center for Dis. Control/NIOSH	Int. Assoc. Pl. & Mech. Off.	Am. Soc. Biol. Chem.	
Bur. Radiol. Health	Int. Assoc. Bldg. Off.	Am. Inst. Nutrition	
Soc. Sec. Admin.	OSHPAC	Am. Physiol. Soc.	
U.S. Off. Ed.	Am. Pub. Health Assoc.	Am. Assoc. Immunol.	
Envir. Protection Agency	Am. Conf. Gov. Indus. Hyg.	Am. Soc. Exp. Pathol.	
U.S. Dept. of Transportation	Am. Acad. Indus. Hyg.	Am. Soc. Pharmacol. & Exp. Ther.	
U.S. Dept. of Commerce	Am. Acad. Occup. Med.	Campus Safety Assoc.	
Cons. Prod. Safety Comm.	Am. Indus. Hyg. Assoc.	Compressed Gas Assoc.	
Natl. Bur. of Standards	Am. Occup. Med. Assoc.	Manuf. Chem. Assoc.	
Fire Res. Center	Am. Soc. Safety Engineers		
U.S. Fire Admin.	Health Physics Soc.		
	Joint Comm. Accred. Hospitals		
	ASHRAE		
State (e.g., Wash. St.)			
Wash. St. Dept. of Labor & Indus.		ACS, Puget Sound Sec.	Evergreen Safety Council
WISHA		AACC, Pacific N.W. Sec.	Wash. St. Hosp. Assoc.
Wash. St. Dept. Soc. & Health Serv.			
Dept. of Laboratories			
State Board of Health			
Radiation Safety			
Wash. St. Hosp. Plan. Comm.			
Insurance Commissioner, St. of Washington			
State Fire Marshall			
County/City (e.g., King/Seattle)			
Fire Dept.			
Building Dept.			
Metro (sewer)			

ANSI, American National Standards Institute
ASHRAE, American Society of Heating, Refrigeration and Air Conditioning Engineers
FASEB, Federation of American Societies for Experimental Biology
OSHA, Occupational Safety and Health Administration

OSHPAC, Occupational Safety and Health Accrediting Committee
NIEHS, National Institute of Environmental Health Sciences
NFPA, National Fire Protection Association
WISHA, Washington Industrial Safety and Health Administration

within the laboratory. In the larger laboratory, full-time safety positions, as well as committees, may be created, but in the small laboratory it is more appropriate to assign this role as a side duty. Nevertheless, the steps to be taken are identical, and the responsibilities are undiminished.

Practical Objectives

The major goal of any laboratory safety program is the *elimination of accidents or illness* resulting from unsafe working conditions. Any reasonable step directed towards this end, regardless of whether it diverges from conventional practice, is valid. Such steps provide an incentive for innovation in safety and should be communicated to established professional groups as part of the continuing process of compiling valid safety measures in the laboratory. The practical objectives of any laboratory program are:

1. To establish and review periodically a document of policy containing the steps taken in each area of hazard.

2. By using this document, to establish for the facility a practical, systematic, and comprehensive inspection plan, including check-off lists, that will reveal unsafe conditions within the laboratory, and to make concerned individuals aware of the basis for the hazard.

3. To hold regular instruction on safety pertinent to the facility. The instructor should know which agencies and regulations affect the laboratory. This provides an insight into the issues to be addressed during the instruction of both new and long-term employees. Currently, there are no comprehensive statistics on accidents or illnesses prevalent within laboratories in general. Therefore, each clinical laboratory should recognize its own problem areas.

4. To facilitate investigation by the supervisor of accidents or incidents, to determine their cause and prevent recurrence.

To guide the small clinical chemistry laboratory in establishing an effective and economic safety program, we emphasize the need for developing an understanding, based on Table 1, of the scope and authority of each of the groups involved. It is important to recognize where and how each group develops the emphasis that it brings to the small laboratory in the course of executing its particular duties. Each laboratory, however, has unique problems despite a similar spectrum of hazards. Also, if the emphasis on how to assure safe conditions is incorrectly placed, one must learn how to resolve conflicts and to clarify communications.

Note: Usually, state and local statutes are based on established standards and codes such as the NFPA or Uniform Building and Fire Codes. Many of these documents are not always current with respect to modern clinical laboratories, even though updated frequently. Accepted approaches and supporting data may change markedly and negate a provision or an entire statute. This is especially true in management of exposure to potentially toxic chemicals. Currently, the only response to an apparently capricious ruling is via persuasion based on established professional consensus that, in turn, is backed by broad experience or experimental evidence. It is therefore important to note the code or selected sources used as a basis for local codes, and to note to which professional society additional data on the issue in question should be presented. The arguments rebutting a specific policy should be soundly *based on fact* and should state the problem addressed, the statements of various standards, and (or) professional consensus. The solutions proposed should be accompanied by the supporting data for each alternative.

It is always to the advantage of the laboratorian to challenge apparently improper applications of codes. It must also be realized that in a specialized field such as the clinical chemistry laboratory, documentation of a balanced approach to the problem may not have been developed or, more likely, may never have reached the personnel who inspect and accredit laboratories. An informed challenge will show this lack of information. The dialogue can then be beneficial for all, but only when it is based on discussion of the *facts*. In this dialogue it is well to remember that, as in science, no set of assembled facts is immutable and, in the specific area of safety, several alternative approaches are usually possible, once the problem is clearly stated. Even if a given appeal should fail after the subject is communicated to a professional society (e.g., AACC, ASCP, ASMT, ACS, ASM), representatives on policy-setting bodies can debate the issues and articulate new approaches. These positions can then be brought to bear on local codes. In building a safety program, the small clinical laboratory that adopts and pursues the above philosophy will be able to deal more effectively with the problems that are certain to occur.

Hazards and Their Control in General Laboratory Operations

Certain hazards are common to all laboratories *(11–14)*. Specific safety products are commercially available to deal with each hazard. A list of vendors is given in the references *(15–18)* and should be consulted.

Pipetting

Pipetting by mouth is dangerous because hazardous materials or their vapors (e.g., radioactive, infectious, or toxic) may be drawn into the mouth, or because the mouth portion of the pipette may be contaminated from dirty fingers or bench tops. Therefore, use one of the many mechanical aids now available: speed is not sacrificed, and accuracy is improved, yet the devices are economical.

The greatest hurdle is attitude. Mechanical pipettors may not always be necessary; however, in most clinical laboratories hazardous materials must at some time be pipetted. Manual pipetting habits are so ingrained that, in switching from innocuous to hazardous substances, manipulators revert to their habitual procedures, and mistakes occur. Learn to use a good mechanical pipettor (e.g., Pipetteman, Eppendorf, etc.), and strongly discourage all oral pipetting. Resist arguments for exceptions to the rule; they can invariably be shown to increase risk, whether or not a serious accident is the immediate result.

Case Report

A technologist in a hospital laboratory was mouth-pipetting a serum sample for routine serological analysis obtained from a patient with active syphilis. The sample had been properly identified with warning labels, but the technologist was momentarily interrupted and drew the sample contents into her mouth. This incident occurred in spite of procedures that forbade oral pipetting of infectious samples. Fortunately, the employee did not become infected, although she became quite anxious, and time and expense for precautionary follow-up were required.

Eye Protection

Inexpensive, convenient, commercially available devices can prevent eye damage from splashed corrosive liquids, flying debris, dangerous radiation, or combinations of these hazards (see list of vendors, *15–18*). There is no valid reason for laboratory workers to suffer eye damage while mixing reagents or while viewing analytical results with ultraviolet or other hazardous radiation; nor should workers be the victim of flying debris. Use a face shield when splash hazards exist. Be sure that eyewash fountains are located conveniently in the working area. Do not wear contact lenses in the laboratory if you work with hazardous materials.

Case Report

This incident occurred in a small community hospital during handling of x-ray film developing fluid, a situation not usually regarded as hazardous. Alkaline developing fluids, however, are hazardous if splashed. During transfer of a large batch of this fluid, some was accidentally splashed onto the worker's face. Despite immediate application of copious quantities of water, much time elapsed before the worker's vision returned to normal. The lesson to be learned: label conspicuously *all* hazardous solutions, including glassware washing solutions, and ensure that eye and face protection is available for use.

First Aid

Maintain sufficient supplies to care for an injury until it can receive proper medical attention. Because small injuries such as needle sticks, cuts, or scratches by potentially infective objects can seem minor, they may be casually treated and not reported until infection appears. Require treatment and reporting of *every* injury. A list of supplies might include:

triangular bandages
gauze pads
bandage compresses
wound wipes
splinter forceps
thimerosal (Merthiolate) swabs
adhesive tape

Case Report

This incident occurred in a toxicology laboratory where contaminated glassware was washed. The laboratory aide broke a glass item, causing a severe gash and immediate copious bleeding. The possible serious consequences of this accident were exacerbated by the lack of immediately available first aid supplies for treating the injury before transporting the victim to the emergency room. Fortunately, the judicious, immediate use of hand pressure on the severed blood vessels during transport prevented more serious consequences. This example illustrates the need to identify potential accidents and to provide appropriate first aid supplies that will be required before more help can be obtained, or until the injured worker can be transported. *The mere fact of being in a hospital does not remove the need for maintaining appropriate first aid supplies.*

Working Alone

In a small clinical laboratory there is the strong probability of working alone on late shifts, weekends, or holidays. From the safety viewpoint, this may be quite hazardous. Arrangements should be made to have such lone workers checked on periodically, at the least.

Case Report

A worker was alone on a weekend in a laboratory located in an isolated corridor. During the time that the lightly clad worker was inside a cold storage room, its door jammed shut. Fortunately, a temperature-alarm system had been installed that the worker was able to activate from the inside. However, there was some delay as security personnel tried to locate the source of the alarm. The worker experienced mild hypothermia, which could have had led to grave consequences if assistance had been further delayed. The lesson learned: have well-defined procedures for personnel working alone, communicate alarm sources to security staff, and keep instances of working alone to a minimum.

Eating, Smoking, Applying Cosmetics

The major hazard from eating, drinking, smoking, storing food, and applying cosmetics in laboratories is that toxic or infectious material may be ingested. The recommended policy is to forbid such activities, and to provide separate rest facilities for laboratory personnel. Recently proposed regulations may make these steps mandatory. Many laboratories have desk and writing areas integrated within working spaces, and many older facilities, attempting to make economic use of areas designed for nonlaboratory uses, do not provide adequate rest facilities. Hence, this becomes a difficult policy to enforce. This problem is a significant justification for redesigning or remodelling work areas and nonwork areas. Provide handwashing facilities within the laboratory and encourage their frequent use. Personal items should be stored outside the laboratory. Forbid the introduction of domestic conveniences or decorations because these may create additional hazards.

Safety Equipment

One frequently noted problem, aside from general laboratory design, layout, and equipment (see later section), is the lack of an adequate amount of safety equipment. This includes strategically placed safety showers and eyewash fountains with adequate drains. Plan and provide for periodic testing of these devices. Locate showers in the vicinity of the expected hazard and away from electrical outlets. Eyewash/hand showers can be easily located at major sinks where corrosive liquids may be handled. Locate safety shields, devices for eye and face protection, fire blankets, and extinguishers near the expected hazard.

Emergency Situations and Plans

A close scrutiny of laboratory activities during periodic safety inspections reveals those areas that have the potential for creating major emergencies: electrocution hazards, eye damage, fire, etc. Plans for coping with emergencies should include contingencies for fire, flood, storms, and power outages. The staff should discuss a written plan involving immediate and follow-up action by designated personnel (with provisions for alternates). Some have insisted upon a complete emergency procedure book, but this is undesirable because true emergencies are probably rare. The instructions should be simple, straightforward, and widely known. Comprehensive, complicated instructions are difficult to follow and should be avoided. Conduct first-aid training and qualify laboratory personnel in cardiopulmonary resuscitation.

Safe Work Practices

A number of tasks in the laboratory require using equipment that is potentially dangerous. For this type of equipment it is important to establish safe practices and teach them to each user. Vacuum pumps as sometimes used with desiccators are one such hazard, owing to the mechanical hazard of using unguarded pulleys and the implosion hazard of unprotected desiccators, filter flasks, and other vessels under reduced pressure. Use goggles and face shields when working with desiccators or with corrosive liquids. Provide pulley and belt guards on pumps and mechanical equipment.

Use protective shields and goggles when working with vessels and systems that operate under relatively high pressure or that might quickly generate large volumes of vapor or gas, such as distillation apparatus, boiling liquids, or rapid chemical reactions (e.g., H_2SO_4 plus small amounts of water). Post signs regarding potential hazards, and establish procedures regarding which personnel are permitted to use the laboratory in the absence of the regular staff. Prepare and post a policy regarding visitors in the laboratory.

Case Report

A worker was using a vacuum desiccator for drying samples. The desiccator was untested, and a small flaw caused it to implode during a drying operation. Two technicians nearby were lacerated, although no eye damage occurred. Pretesting of desiccators and high-pressure/reduced-pressure vessels would largely prevent such accidents. Locating such vessels in a protected area with at least minimal barriers would eliminate the hazard.

Waste Disposal

Exposure of laboratory workers to waste produced in the laboratory is a hazard to health. Release of this waste, however small, into the general environment can endanger public health. It seems inevitable that at some time the controlled disposal of all laboratory wastes will be required. Potentially infectious biowastes, including sample containers, needles, syringes, sample cups, and processing materials, will require isolation and sterilization. These materials will require identifying labels and containers as well as special handling instructions. Radioactive waste already requires separate, labeled waste containers, disposal procedures, and monitoring. Chemical wastes will now require the same procedures. The public sewer systems no longer will be available for dumping toxic substances in an activated state. Discharging reagents containing heavy metal, for example, even if only in low concentrations, will most likely be prohibited where collection at the source is possible. These contaminants can be easily precipitated and concentrated for delivery to appropriate receivers *(19–25)*. In many cases, chemical compounds can be reacted in such a way as to convert them into nontoxic biodegradable forms. The small laboratory should be prepared to deal with this disposal problem. Environmental monitoring methods and inspection and accreditation procedures will make it impossible to dump chemical waste down the sink.

Broken-glass waste, syringe needles, and other discarded objects also must be placed into special puncture-proof, labeled containers.

Centrifuges

Centrifuges at almost any speed can be the source of aerosol and missile hazards. Laboratories using high-speed centrifuges or ultracentrifuges must be especially careful to maintain these instruments regularly, and to avoid using unbalanced rotors. Ultracentrifuge rotors, if used, should be downrated for maximum speed at regular intervals.

Lifting

Identify the locations of objects that may have to be lifted by laboratory personnel and, if possible, either design the laboratory to provide mechanical lifts or ensure that the location of heavy objects is such that a proper lifting procedure can be used easily. Back injuries caused by improper lifting of such objects as distilled-water bottles, solvent cans, and medical equipment are avoidable hazards.

Use of Toxic and Hazardous Substances

Understanding the Properties of Chemicals

Clinical chemistry laboratories increasingly use chemical substances that are potentially hazardous because of flammability, violent reactivity with water, explosiveness, sensitivity to temperature, chemical toxicity, carcinogenicity, or teratogenicity. It is important to understand the chemical properties of every chemical in use and to label each container with these hazards. Labeling regulations for laboratory chemicals are under development insofar as they relate to manufacturers. However, each user of chemicals must also appropriately label all reagents prepared in the laboratory. At the present this means hand-labeling (see the labels available now from laboratory safety supply houses, which use the NFPA 704M system) *(15–18)*. New warning codes and standards may be forthcoming to help simplify and unify the procedures for this task and ensure that there are no "unknown" containers in the laboratory.

Case Report

This incident took place in a large hospital clinical laboratory where reagents were being prepared. A container holding a toxic chemical cracked and the technologist handling it received a deep cut, which allowed the chemical to enter the wound. Though the dose was not lethal, the technologist suffered symptoms of toxicity for several days after the accident. This situation illustrates the need not only to discard defective glassware but also to recognize the hazards of the chemical being used. In this case neither adequate warning nor hand protection was provided. Other precautions for dealing with the chemical—such as preventing splashing, skin absorption, or inhalation of vapors—should also be taken.

Planning the Use of Chemicals

Nothing will substitute for knowing what you are working with: Reference *26* provides a guide to potential hazards when new uses are contemplated. Based on the intrinsic hazardous properties and the potential reactivity with other compounds, the use and storage of chemicals can be determined individually so that reactive materials are stored well separated. Avoid storage of chemicals solely by alphabetical order. Store oxidizing substances and oxidizable substances apart from each other. Store perchloric acid bottles in nonorganic drip containers, and do not accumulate excessive amounts. Use polyvinylchloride-coated containers for storing acids, bases, or other hazardous substances.

Store explosive substances in appropriately protected spaces. For example, peroxide forms in ether upon storage; hence, plan short turnover of such solvents and perform appropriate tests for the presence of peroxides. Date peroxide formers when received, and again when opened; discard the container six months after opening.

Store volatile substances in well-ventilated spaces. If vapor from one substance will react with another, separate these substances.

Store corrosive substances, particularly mineral acids (HCl and HNO_3), organic acids (acetic or formic acid), and any of the volatile bases (NH_4OH or volatile amines), in special ventilated, corrosion-resistant spaces. Keep such substances separated to prevent reaction. Always transport corrosive acids in special carriers (available from laboratory safety supply vendors) to prevent breakage and spillage.

Hazardous Vapors

When working with solvents and other chemicals emitting hazardous fumes or vapors, carry out the work under adequate ventilation. This is particularly important for reagent preparation and acid-washing glassware. Consult the section on ventilation to plan the design of a laboratory so that adequate general ventilation is provided as well as the required special ventilation for the expected hazards.

Chemical and Solvent Waste Storage and Disposal

Insoluble wastes. Although the volume of solvent waste in the small laboratory is not great, insoluble waste cannot be discharged into the sewer system.

Organic solvents. Store nonhalogenated solvent waste (e.g., acetone) in one container and halogenated (e.g., $HCCl_3$) solvent waste in another. Use ventilated, approved waste-disposal cans for this purpose. Solvent waste is frequently incinerated, but halogenated compounds must be handled by special procedures. Store the separate containers in a solvent cabinet (see below) and, when they are full, move the containers to solvent bulk storage, if available, until the waste can be removed for disposal. Proper labeling, naming the types of solvents pooled, will aid in their proper disposal.

Sodium azide can form explosive compounds with lead or copper piping. Though its use is now being avoided when possible, it has been added as a preservative in diluent solutions. Use copious amounts of water when disposing of such fluids in the sanitary drains.

Nontoxic waste chemicals. Be careful in disposing

of nontoxic waste chemicals via the public sewer system. Collect toxic chemical waste from analytical instruments and deactivate them before disposal. Several references *(19–25)* exist to guide this process.

Toxic and heavy metal wastes. Never dispose of heavy-metal-containing reagents and solutions into the sewer system, but precipitate them and turn over the treated reagent to resource-recovery firms. Carefully plan these procedures, when using them for large volumes, to avoid any chemical incompatibilities. Planning will allow regular pooling of defined waste streams. Small volumes of waste, infrequent test waste, and other low-volume, toxic waste chemicals may be disposed of safely by dilution. However, it is advisable to plan the collection of all toxic waste and its deactivation before disposal: the monitoring of chemicals in sewage will undoubtedly increase, particularly in those areas with limited water supplies, and capture of all toxicants at the source will become more widely required.

Specific Hazards

Mercury. This metal is a component of pressure manometers, thermometers, and certain kinds of electrical switches found in the laboratory. Free mercury establishes a vapor phase in equilibrium with the liquid. At 25 °C this is 0.00184 mm (245 mPa), which is 150- to 200-fold the threshold limit value. A considerable amount of mercury can be vaporized in a large space. Mercury under water establishes an equilibrium with mercury in solution, which is also in equilibrium with the air. Consequently, mercury in sink traps and floor drains can eventually saturate the air in a closed room. In conjunction with other in-house clinical facilities that also use mercury-forming devices, obtaining a mercury "vacuum" for cleaning up all mercury spills may be justified. Mercury-spill kits are commercially available. Alternatively, contracting with a consultant to conduct surveys and to clean up any spilled mercury at appropriate intervals may be more justifiable for the small laboratory.

Case Report

In a small hospital, broken thermometers and manometers were stored in an open cabinet in a closed and poorly ventilated room. Employees from time to time spent extended periods repairing equipment and taking inventory of various items stored there. When the room was monitored for mercury, the air was found to be nearly saturated. The employees who had used it had received, over the course of several hours, about 100-fold the threshold limit value of mercury vapor. Mercury in the traps of the sink and floor drains as well as free mercury in the cabinet contributed to the saturation. The lesson: enclose all free mercury and monitor frequently all areas where mercury is used, particularly those not well ventilated.

In a second case, a technician in a small laboratory was using a drying oven containing a mercury thermometer. The thermometer was heated beyond its upper limit and broke. When the technician opened the oven to retrieve his samples, the escaping mercury vapor subjected him to a severe toxic excess and saturated the room for several hours. The solution to this problem: use *alcohol* rather than mercury thermometers and bimetalic regulators rather than mercury switches.

Phenol. This chemical is used in several reagents in the clinical laboratory. Its threshold limit value is 5 ppm in air (5 µL/L), but its major danger is through the splashing of phenol-containing solutions onto the skin or oral ingestion during pipetting. As with other corrosive or caustic substances, store phenolic solutions no higher than eye level to prevent falling and splashing, and use safety goggles and protective clothing when working with phenolic solutions. Restrict phenolic vapors, and clean up spills quickly with a disposable absorbant. Store pure phenol in ventilated, well-marked areas (either with solvents or organic acids) where they will be handled with care.

Other vapor hazards. Common volatile chemicals such as those listed below should also be stored and used with adequate ventilation.

Compounds	Formula	Threshold limit value, ppm (µL/L)
Nitric acid	HNO_3	2
Hydrochloric acid	HCl	10
Carbon tetrachloride	CCl_4	10[a]
Chloroform	$CHCl_3$	10[a]
Benzene	C_6H_6	10[a,b]
Toluene	$C_6H_5CH_3$	100
Xylene	$C_6H_4(CH_3)_2$	100
Acetone	CH_3COCH_3	1000
Diethyl ether	$(C_2H_5)_2O$	400
Glutaraldehyde	$CHO(CH_2)_3CHO$	0.2
Formaldehyde	HCHO	2

[a] Carcinogenic hazard.
[b] 1 ppm (µL/L) proposed threshold limit value.

This list is not exhaustive and refers primarily to acute toxic hazards and only secondarily to the chronic toxic hazards. The threshold limit values are currently being revised significantly downward for a number of substances *(27)*.

Solvents, Fire Hazards, and Explosion

Solvents and Solvent Storage

The small clinical laboratory does not handle large volumes of organic solvents. If the laboratory is in a building housing patients, the requirements for solvent storage are contained in NFPA 56C *(10)*, whereas if it is in a detached building, the requirements are controlled by the less stringent NFPA 45 *(9)*. Maximum container size for class 1A, 1B, and 1C solvents (e.g., ethers, ethanol, and heavy solvents) is 1 pt., 1 qt., and 1 gal. (473 mL, 946 mL, and 3.78

L), respectively, under NFPA 56C. If "working" storage of solvents on bench tops is minimized, and all bulk storage is kept in a storage cabinet meeting the NFPA specifications, the requirements are easily met. Keep larger volumes of acetone and alcohols in safety cans (stainless steel, when purity is essential). A maximum size of 5 gal. (18.9 L) is permitted in laboratories complying with NFPA 45, whereas 2 gal. (7.57 L) is the maximum in laboratories that must comply with NFPA 56C.

Remember the constraints on ventilation when using solvents in work areas. Use a hood, if available, though it is not essential if the areas used are well ventilated and ignition sources are absent. Ether used for sample extractions is a particular hazard.

Do not store solvents in refrigerators (modified or unmodified commercial models), but only in the specifically designed products now available (15–18). Keep foods away from solvent storage areas and refrigerators; toxic fumes may be taken up by food solids and lipids.

Fire and Explosion Hazards

Most fires in laboratories arise from ignition of solvent vapors. Prohibit smoking in all laboratory working areas, and also rigidly control smoking in nonlaboratory areas adjacent to areas where solvents are used. Give due attention to heaters, open flames, or sparking electrical apparatus; to the placement of combustibles; to the use of azide in copper, lead, or zinc pipes or vessels (such as waterbaths); and to the drainage into the sewer of such solutions in large volumes over a prolonged time. Be careful in the use of oxidizing, temperature-sensitive, or water-reactive substances. A general rule is: know the properties of the substances with which you are working! Sources of information can be found in NFPA 49, Properties of hazardous substances (28); NFPA 491M, Hazardous chemical reactions (26); the Merck Index (23); the Manufacturing Chemists' safety manual (7); and the large chemical-supply companies' catalogs. See also references 29–31.

Case Report

Paper chromatograms were being dried in an oven in a hospital laboratory. The solvent system was volatile and flammable. When the dried chromatograms were to be removed and the oven door (not a chromatographic oven) was opened, a spark caused a fire and explosion.

In a second incident a research laboratory separated from the hospital was being remodeled. Ether bottles, one of which had a defective cap, were left unnoticed in a disconnected hood. Welders, brazing pipe, entered and began their operations. Because the hood was shut down, vapors reached a nearby welder's torch, causing a tremendous explosion and fire. Fortunately, no injuries accompanied the extensive damage.

Obviously, all volatile and flammable solvents must be handled with extreme caution. Informed personnel who know the properties of the solvents must be present when repair or housecleaning is done by uninformed personnel.

Fire and Explosion Protection Systems

For most laboratories, the portable fire extinguisher is the major fire-fighting device, though sprinkling systems might be installed for some purposes. The laboratory fume hood is usually the most appropriate housing for potentially violent and fume-producing chemical reactions. Use bench and face shields as well as goggles during these operations. It is unlikely that a clinical laboratory will have to contend with blast and explosion protection but, should this be a consideration, NFPA 45 (9) contains appropriate guidelines and references to other publications. Drench showers are probably more effective than fire blankets and can be used for chemical splash emergencies.

Biohazards

Handling Infectious Agents

Infectious agents are a major hazard in any clinical laboratory. Such agents enter the body through cuts or pricks in the skin and often originate from specimen handling. Inadvertent entry by inhalation or by mouth is also a major hazard, hence the prohibition of eating, drinking, smoking and application of cosmetics in the laboratory. Because of this major hazard, prohibit storage of foods in the labs and permit them only in specially designed and furnished employee lounges. (Please note the design implications of this requirement.) Use self-adhesive labels and automatic pipetting devices for processing all specimens. Provide convenient hand-washing facilities and establish procedures to encourage the practice of using them. Provide germicidal hand-scrubs and knee- or pedal-operated devices on sinks.

Wear protective clothing while handling specimens and provide for safe storage of soiled linen. Contaminated clothing should not leave the laboratory, except for sterilization and proper laundering, and special racks must be provided for hanging garments within the laboratory. Provide eye protection if splash hazards exist. Store contaminated waste in specially labeled containers. Wear disposable gloves at all times when handling specimens.

Decontaminate work areas daily after thorough cleaning to remove all residue from specimens or reagents. Chlorine bleach (20-fold diluted) is recommended for all except surfaces sensitive to corrosive substances; these require chemically toxic 2% (20 mL/L) glutaraldehyde. Prepare new batches of disinfectant on a weekly basis. Post biohazard warning signs in all specimen-handling areas. Autoclave all reusable glassware and centrifuge carriers after soaking in a 20-fold diluted chlorine bleach solution once daily. Also wipe centrifuge bowls with glutaraldehyde disinfectant. Collect biowaste in autoclavable bags and then autoclave them before disposal.

Report all accidents whether they are as seemingly insignificant as needlesticks or are the more involved cuts or spills. Infection-control personnel must monitor the sequence of all such accidents after appropriate medical treatment. If a large area is contaminated by such accidents as broken sample tubes containing infected specimens, evacuate the area until aerosols are dispersed and then institute thorough decontamination steps. All personnel involved in the accident should be immediately reported and checked. A similar procedure must apply to drawing samples from patients in isolation. Establish a rigid procedure for labeling specimens from patients known to be infected. Train all personnel in biologic decontaminization procedures. Store all needles and glassware in receptacles that cannot be broken or punctured by the waste. Needle cutters or benders should not be used because of the potential aerosol hazard.

Case Report

In a hospital laboratory, a culture of highly pathogenic organisms (*Coccidioides immitis*) was dropped during handling, thereby creating an aerosol—an extreme hazard. The personnel involved immediately evacuated the space, which was sealed and decontaminated over several days. The technologists were immediately given skin tests and kept under surveillance for several weeks. Fortunately, positive cases were not reported.

Specimen Processing

During specimen processing, monitoring is required to prevent aerosol production while opening tubes (for example, be sure to twist the caps rather than popping them open). Before centrifuging, inspect tubes for cracks to prevent breakage. Use germicidal solutions to wipe down the centrifuge and buffer the tubes during spinning. To remove supernates, use transfer pipettes rather than decantation. Autoclave specimens before their final disposal. Dispose of effluent from automatic instruments only in authorized drains that are thoroughly washed daily with germicidal solution. See also references *32–34* for specific suggestions.

Radioactivity

Requirements and Regulations

Only approved investigators or facilities under the direction of such persons are authorized to handle radioactive isotopes (Federal Regulations: 10 CFR 31.35). Carefully check each area that has been approved by such authorized persons to ensure that procedures for storage, work space, monitoring programs, waste disposal, record keeping, and laboratory operations are adequate for the work proposed. Highly hazardous areas require special consideration *(34–38)*. An adequate sink, monitoring instruments, and specially designated and protected work areas are required. A flat, stainless-steel pan with a ½-in. (approx. 1 cm) lip and a disposable, absorbant, impervious-backed pad is adequate. Specially designated refrigerators or freezers with appropriate lead-shielding for storing isotopic substances are also required. Use appropriate shielding when handling or transporting containers of radioisotopic solutions. Post all areas with proper warning signs. The small clinical laboratory will undoubtedly use only limited amounts of radioactive substances for radioimmune, radiometric, or radioenzymic assays. For these functions safety precautions are easily met.

Information Required by the Nuclear Regulatory Commission

Record the receiving and disposal of all isotopes, including estimates of radioactive waste. The ordering procedures will require checking, usually by the suppliers before shipment, of the appropriate license that permits use of the substances ordered. Nevertheless, attention to this bookkeeping detail will ensure that requirements of this aspect of radiologic safety are met. An outline of the required information is shown in Table 2.

Empty radioactive aqueous waste into a sealed container filled with vermiculite or similar absorbant. Radioactive solvent waste should be treated with the same caution as general solvent wastes except that liquid-scintillation vials may be stored in the "egg crate" cartons in which they are supplied. These crates may be stored with other solvent waste while awaiting disposal of the radioactivity, but keep them in specially segregated areas that are always properly labeled. Store solid radioactive waste materials in labeled plastic-lined cardboard cartons, seal tightly when full, label as to isotopic content, and store in a segregated, labeled solid-waste storage area until they can be picked up.

Monitor waste areas regularly and keep a log of the readings. If "hotspots" occur, quickly decontaminate and discard all "hot" waste as above. Highly hazardous areas, such as for radioiodinations, are not likely in the small laboratory, but should your facility have such areas, consult reference *38*.

Compressed and Cryogenic Gases

In the small laboratory the major compressed gases used are likely to be fuel (acetylene, propane) and oxidant (air, oxygen, nitrous oxide) gases for flame spectrometers (atomic emission, atomic absorption, and flame-ionization detectors for gas–liquid chromatography), carrier gases (nitrogen, helium) for gas-chromatographic units, calibration gas (oxygen, car-

Table 2. Information Required (by the Nuclear Regulatory Commission) To Be Kept on File

General (updated yearly)
1. Facility name
2. Authorized (licensed) user
3. Department/position
4. Phones in radioisotopic laboratories
5. Mailing address of the laboratory
6. List of laboratory rooms (see laboratory room records)
7. Names of laboratory personnel using radioisotopes (see individual personnel records)
8. Instrumentation
9. Monitoring pattern
10. Waste-disposal procedures
11. Isotope inventory by isotope (updated quarterly)
    A. Isotope
    B. Authorized maximum inventory level
    C. Current usage
    D. Projected usage

Laboratories (a record for each laboratory room) (updated yearly)
1. Room identification
2. Name of responsible person
3. Facilities: number of sinks, room security, monitoring equipment
4. Regular use, type
5. Isotope inventory (updated quarterly)
6. Special uses

Laboratory personnel (for each laboratory worker) (updated yearly or upon job change)
1. Name
2. Birthdate
3. Date record was last verified
4. Employment history (for each department worked in by employee: date employed, date left, isotopes exposed to or worked with in this dept.)
5. Lab rooms used in present position (all facilities and spaces regularly worked in and the dates)
6. Current work, job description, supervisor
7. Exposed record (film badge readings, dates) (updated at appropriate intervals)

---

bon dioxide, nitrogen) for blood-gas machines, and anerobic gases (nitrogen, carbon dioxide) for microbiological incubators.

Secure all cylinders in use or in storage firmly to a solid structure to prevent tipping and avoid damage to the valve. Escaping high-pressure gas converts any cylinder into a deadly missile. Take care to use proper fittings and regulators with each cylinder, and prohibit the use of adapter fittings because they promote dangerous mismatches of the fittings to the gas used. Use only oxygen-type regulators with oxygen cylinders; they are designed to remain nonflammable in oxygen-enriched atmospheres. Such atmospheres create hazards of extreme flammability in the presence of any potentially combustible material, which can then flash at low ignition temperatures. Note that cryogenic gases such as liquid N_2 can condense oxygen on cold surfaces, creating the same hazard.

Transport the gas cylinders only when properly secured on a dolly. When they are being used, open valves carefully and double-check the regulators. A regulator should always be "backed off" after use, before closing the main valve. Transfilling operations are particularly hazardous and should be prohibited in the small facility. Leaks or accidents at high pressure, as noted above, create missile hazards. If only small amounts are needed, obtain small cylinders from suppliers, to diminish storage problems.

Segregate reserve storage tanks in the laboratory into fuel, oxidizing, and inert gas categories, maintaining a 20-ft. distance between fuels and oxidants. Store acetylene and hydrogen reserve tanks outside the building with the main reserve gas storage. Keep full and empty tanks separated, labeled as to status, and firmly secured. Store cryogenic gases in a well-ventilated area to prevent the possibility of asphyxiation from accumulations of nonoxygen gases.

Return the used large tanks to the supplier and dispose of the small disposable tanks in clearly marked receptacles (for subsequent disposal in accordance with the manufacturer's instructions), *not* in the regular trash. See also references *39–41*.

Electrical Equipment and General Safety Systems and Services

Instrument Usage

The design of a laboratory should provide adequate areas for instruments, ensuring that they are properly grounded and not exposed to corrosive vapors or to water, which may cause the insulation to deteriorate and lead to the potential danger of electrocution. Establish clear procedures to make sure that each instrument is used as intended (see refs. *42–44*). Placing instruments in special settings such as cold rooms, refrigerated boxes, or incubators requires continual surveillance to enforce this policy and to prevent the unexpected occurrence of electrical hazards. Check each newly received instrument for safety features such as proper grounding, location of switches and plugs, and a reasonable design that segregates the electrical circuits from the pathways for liquids or physical access. If not already in place, install a proper safety plug when the instrument is received and checked by the local electrician. Apply the same criteria to "loaner" instruments and noninstrument appliances.

Case Report

An employee in a small hospital laboratory brought a radio into the laboratory to use during work hours. The cord was frayed and the plug defective. Water was spilled on a desk top and engulfed the radio. As the employee reached to shut it off, he suffered a severe electric shock and needed emergency treatment from other staff members. The lesson to be learned: ground *all* instruments properly, using three-wire plugs. Grounded equipment is much less susceptible to shock hazard.

Circuitry

Circuits in the laboratory should be designed so that a cord no longer that 2 m (about 6 ft.) is required for any equipment. Longer cords cause tripping hazards. Insist that all receptacles are located on work surfaces such that water or other liquids cannot contact them in a way that creates a conductivity path. Scrutinize the proximity of receptacles to sinks, safety showers, and eyewash fountains.

Make certain that frayed cords or defective devices are quickly repaired and that electrical fixtures are properly maintained. Check electrical panels and switches to see that they are properly labeled and are not blocked by laboratory equipment. Inform the staff of the locations of these panels for use in emergencies but discourage access to them under routine conditions.

Gas

Install regulators for flammable gases in conformity with the NFPA 50A, 54A, 58, 59A (see ref. 9) for piped fuel gases, and maintain as described. Regularly check switches and outlets and use them only as intended. Compressed air systems should conform with NFPA-50. Nonflammable medical gases are regulated by NFPA-56F.

Vacuum

Vacuum systems are described in the Uniform Plumbing Code. Built-in vacuum systems must be properly trapped to prevent the introduction of water, solvents, or corrosive liquids into the system. For emergency purposes, clearly label each system, including the backup valves to the laboratory. Piping systems should conform to NFPA and ANSI standards.

Drains

All drainage systems should be readily apparent in the laboratory, and installed according to conventional building codes and standards (i.e., Uniform Plumbing and Mechanical Code). Clearly label all cleanouts, and install systems that allow easy recovery of items inadvertently carried into traps. Drains serving instrumentation that produces corrosive effluent must be installed with corrosion-resistant piping. Run water into each drain weekly to prevent the trap from drying up, which would allow the accompanying escape of toxic or flammable gases into the workspace.

Communications

Consider your capacity to communicate quickly in an emergency, and to activate local alarms, in addition to the convenience and efficiency of such systems in the usual work routine. Suitable systems are described in NFPA No. 45, Section A 11–6.11 (9).

Ventilation, Air Conditioning, Environmental Control

The ventilation system should be given major attention in the design of a laboratory. The small laboratory deals with a large enough variety of substances that ventilation should be carefully considered at renovation or new construction. NFPA No. 45 (9) and 56C (10) provide general principles of guidance, whereas the technical details can be found in their cited references. The first requirement is that laboratory ventilation must be separate from other institutional ventilation systems. A separate exhaust and supply system must be installed, balanced so that airflow is directed into the laboratory from supply and corridors. Do not rely on fumehoods to carry the exhaust load, but locate the hoods so that good general airflow takes place within the laboratory space, with no "dead spots." There should be six to seven room-volume changes of air per hour.

Locate the hoods away from doors and traffic areas. Place storage and other fixtures so that exhaust grills are not blocked. Inspect to ensure that this condition exists.

Special Local Ventilation

For special procedures in which large volumes of fumes are generated, design spot ventilation to draw these fumes off before they reach the general room circulation. Special slots, canopies, and other techniques are generally used (39).

Fume Hoods

Fume hoods are one type of special ventilation. A properly designed hood confines all fumes inside the cabinet and exhausts them from the laboratory regardless of the work activities at the face of the hood (9, 10). Do not use hoods for storage. Provide special ventilated storage areas for corrosive or odorous liquids, for radioactive substances or solvents. Stored materials invariably disrupt the hood airflow, allowing the substances that should be contained to escape into the room. Most hoods are designed to contain both fire and low-level explosive hazards. Perchloric acid hoods are specially designed, are corrosion-resistant, and have wash-down systems. Annually inspect and label each hood with the height at which the sash should be kept to ensure adequate airflow.

Biological Safety Cabinets and Laminar-Flow Hoods

Biological safety cabinets have an *inward* flow of air to contain hazardous agents and are designed to sweep hazardous aerosols into a filtration or incineration system. Laminar-flow hoods—"clean cabinets"—are designed to provide an *outward* flow of clean air onto the work surface to prevent contamination of delicate or sensitive objects or procedures (such as cell cultures). In special cases the two might be combined to supply clean air, free from contaminating organisms, and also to prevent the hazardous organisms from escaping into the environment. Do not substitute these devices for fume hoods used in containing fire and explosion in the chemical laboratory, and do not use flammable solvents in recirculating devices of this type.

Regularly check the safety features of these different types of hood.

Laboratory Design, Maintenance, and Housekeeping

Access, Security, and Layout

Some general principles relating to safe design will be outlined here; more comprehensive treatments have been published elsewhere *(7, 45)*. Surround each laboratory unit with 1-h rated firewalls and doors *(9, 10)*. Provide additional exits based upon the floor area. Plan sensibly for exits in case of emergency, regardless of the minimum requirements (see NFPA No. 101, "Life Safety Code") in references *9* and *10*. Locate hoods where they will not block exits. Locate storage, utility services, light, ventilation, and furniture with respect to the expected hazards and the location of and access to safety equipment. Providing for security should be consistent with the requirements for access, including windows, locking devices, and door swings (they should swing outward from laboratory exits). Laboratories should be easily lockable when responsible personnel are not present.

Furniture

Select laboratory fixtures not only for functional convenience but also for safety in providing "working" and "backup" storage; in particular, try to impede the accidental knocking off of bottles. For storage located above eye level, make sure there are physical restraints to keep stored items from falling. Arrange services in modules, so as to avoid chemical incompatibilities and to provide unobstructed work surfaces. Fixtures must be easily cleaned and maintained. Locate sitting and desk areas away from workbench areas to discourage eating and smoking, but retain convenience for data-recording.

Storage

A major hazard *created* in laboratories is clutter, caused by nonexistent or badly planned storage of supplies and equipment. This leads to stacking materials high overhead without proper restraints or cramming them in underbench knee holes, causing spilling or breakage of containers of chemicals and reagents. Provide adequate space for supplies and unused instruments and apparatus. Make appropriate use of this space a matter of policy and sustain the policy at every inspection. If the total amount of space is inadequate, increase it. If the distribution is poor, reorganize it to provide flexible, modular, convenient storage, in which hazardous and incompatible substances can be kept separate. Ensure that access to emergency equipment or services is never blocked by the storage of any item. Justify each increase of space on the basis of *safety* rather than convenience. In an uncluttered laboratory, enforced housekeeping procedures and maintenance lead to greater productivity as well as safety. The importance of this is not made sufficiently clear to nonlaboratory maintenance and housekeeping personnel, or to administration.

Repair of Instrumentation, Equipment, and Apparatus

Attention by the laboratory staff to this task will ensure fewer accidents from explosions, electrocutions, and falls caused by malfunctioning equipment. Institute a regular plan of preventive maintenance (see preceding chapter) and of assessing the adequacy of equipment, and follow through with appropriate correction of deficiencies.

Maintenance of Physical Plant

This activity includes not only repair of architecturally fixed structures and services but also the day-to-day housekeeping. To interact properly with nonlaboratory staff, laboratory staff must correctly label all storage space and waste receptacles and develop *explicit* instructions. If the maintenance and housekeeping staff is to carry out its function, the laboratory staff must clearly describe what is to be done and whom to contact if there are questions. Moreover, laboratory staff should be present when repairs are being made. Make sure that equipment is decontaminated before maintenance is performed. Failure to attend to these details will invariably expose some people to hazards.

Laboratory Staff Responsibilities for Housekeeping

Because laboratory staff members know the principles of the procedures they perform and are trained in the use of equipment, they are responsible for its cleanliness and basic maintenance, *and that of the surrounding work areas.* This responsibility should be explained to every laboratory worker as a condition for continued employment. Negotiate and enforce a housekeeping procedure whereby each individual is assigned a segment of the required duties. This act alone will greatly reduce the opportunity for the creation of hazardous situations. In an uncluttered laboratory environment the chances of an accident's occurring should be significantly reduced, although never completely eradicated. "Eternal vigilance is the price of safety!"

Noise, Lighting, and Work and Walking Surfaces

Distractions from noise and poor lighting create an environment for accidents. Insist on adequate attention to these details in the design and maintenance of a laboratory. Likewise, provide nonskid walking surfaces; injuries can be caused by water on freshly waxed tile. Work surfaces selected on the basis of appearance alone rather than safety may lead to slipping and falling accidents.

Designing a Laboratory for Safety

In designing or renovating a laboratory where functions are well defined, a primary problem is to find the appropriate documents (regulations, standards, codes) that define approaches to reducing perceived hazards. To be truly effective, those individuals who form the safety organization for the laboratory must approach the building committee or similar group within the institution when the plans for a laboratory are in their infancy. Once the functional plan for the laboratory is sketched out, the safety group must consult several sources to understand the current technology available via the existing standards and codes. Usually architects, building engineers, safety engineers, or consultants are sought out, in addition to a series of documents. We enclose a bibliography of useful sources. It is important also to consult chemists, microbiologists, etc. who have had experience with the potential hazards noted for the proposed facility. As is well known, solutions proposed by lay personnel frequently turn out to be nonfunctional. Quite often, however, information to this effect is never fed back to the designers, architects, and engineers, unless laboratorians turn to their colleagues with experience. Laboratorians must also communicate design errors and solutions by publications in the literature on safety, and by communications to their professional societies.

Summary

We have attempted to outline both the procedural and design issues regarding hazardous areas in an integrated fashion. Thus the approach to the handling of a particular hazard is consistent with the approach to another hazard when a common laboratory facility is in use. For example, ventilation must be considered along with procedural aspects of operations, solvents, explosive operations, and biohazards; storage and access; and hazardous radioactive chemical and biological waste storage and disposal. Numerous alternatives exist in dealing with hazards either through design or by appropriate policies and procedures. It is desirable, therefore, when designing a facility, to detail decisions and how the result will be manifested in the design. The ramifications of these decisions will be evident in the laboratory safety procedural document on which the inspection and education program will be built. Involvement of the laboratorians in this decision process, especially in the small facility, is essential, particularly if the institution has already been inspecting the laboratory for some time and conducting periodic education of the staff. An involved staff can easily spot alternative approaches and point out the difficulties with proposed solutions in that facility. Rejection of inadequate alternatives can, in the long run, save the institution time, money, and human resources. Thus, it is important at the earliest stages to get this concept accepted by the administrative or managerial authorities involved in planning new facilities.

References

General References

1. Muir, G. D., *Hazards in the Chemical Laboratory*, 2nd ed., Chemical Society, Letchworth, Herts., U.K., 1977.
2. Steere, N. V., *Handbook of Lab Safety*, 2nd ed., Chemical Rubber Co. Press, Cleveland, OH, 1971.
3. Flury, P. A., and Deluca, K., *Environmental Health and Safety in the Hospital Laboratory*, Charles C Thomas, Springfield, IL, 1978.
4. Hartree, E., and Booth, V., *Safety in Biological Laboratories*, Spec. Publ. 5, Biochemistry Society, London, 1977.
5. Pecsok, R. L., Chapman, K., and Ponder, W. H., *Chemical Technology Handbook*, American Chemical Society, Washington, DC, 1975.
6. Henry, R. J., Olitsky, I., Lee, N. D., Walker, B., and Beattie, J., *Safety in the Clinical Laboratory*, Bio-Science Enterprises, Van Nuys, CA, 1976.
7. Manufacturing Chemists Association, *A Guide for Safety in Chemical Laboratories*, 2nd ed., Van Nostrand, Reinhold Press, New York, NY, 1972.
8. Green, M. E., and Turk, A., *Safety in Working with Chemicals*, MacMillan, New York, NY, 1978.
9. Safety in laboratories using chemicals. National Fire Protection Association publ. 45, Boston, MA, 1981.
10. Safety in health-related institutions. National Fire Protection Association publ. 56C, Boston, MA, 1980.

Laboratory Practices

11. Steere, N. V., *Safety in the Chemical Laboratory*, I–III, J. Chem. Ed., Easton, PA, 1968, 1971, 1974.
12. Chapter 3 in ref. *5*.
13. Chapters 2 and 3 in ref. *3*.
14. Chapters 2, 3, 4, and 6 in ref. *8*.

Safety Supplies Catalogs

15. Lab Safety Supply Co., P.O. Box 1368, Janesville, WI 53545.
16. Nalge Co., Nalgene Labware Dept., P.O. Box 365, Rochester, NY 14602.
17. Fisher Scientific Co., 711 Forbes Ave., Pittsburgh, PA 15219.

18. Instruments for Research and Industry, 108 Franklin Ave., Cheltenham, PA 19012.

Toxic Substances

19. Gaston, P. J., *The Care, Handling, and Disposal of Dangerous Chemicals*, Northern Publishers, Aberdeen, U.K., 1970.
20. *The Design of Laboratories for Radioactive and Other Toxic Substances*, Koch-Light Laboratories, Colnbrook, U.K., 1978.
21. Aldrich Chemical Catalog, Aldrich Chemical Co., Inc., Milwaukee, WI, 1979.
22. How to deal with hazardous chemicals. British Drug Houses, Poole, U.K., 1976. Wall chart.
23. *Merck Index*, 9th ed., Merck & Co., Rahway, NJ, 1976.
24. *Occupational Diseases, A Guide to Their Recognition*, NIOSH publ. 77–81, US Govt. Printing Office, Washington, DC, 1977.
25. Chapter 5 in ref. *8*.
26. Hazardous chemical reactions. National Fire Protection Association publ. 491M, Boston, MA, 1975.
27. Threshold limit values for chemical substances in the workplace. *National Safety News* **119**, 65 (1979).
28. Properties of hazardous substances. National Fire Protection Association publ. 49, Boston, MA, 1975.

Solvents, Fire, and Explosion

29. Chapter 4 in ref. *5*.
30. Chapter II–3, 4 in ref. *6*.
31. Chapters 4 and 5 in ref. *3*.

Biohazards

32. Chapter 9 in ref. *3*.

33. Chapter II–8 in ref. *6*.
34. Chapter 5 in ref. *4*.

Radioactivity

35. Chapter 7 in ref. *4*.
36. Chapter II–9 in ref *6*.
37. Chapter 6 in ref. *8*.
38. A guide for preparation of applications for medical programs. N. REG 338, Rev. 1–77, Nuclear Regulatory Commission, Bethesda, MD, 1977.

Compressed Gases

39. Pamphlets P-1, P-2, Compressed Gas Association, New York, NY, 1974.
40. The handling of compressed gas cylinders. British Oxygen, London, U.K., 1974.
41. Chapter 9 in ref. *5*.

Electrical and Laboratory Services

42. Chapter 2 in ref. *4*.
43. *Industrial Ventilation: A Manual of Recommended Practice*, 13th ed., American Conference Governmental Industrial Hygienists, P.O. Box 453, Lansing, MI, 1973.
44. Standard for laboratory fumehoods. Scientific Apparatus Makers Association, Washington, DC, 1975.

Laboratory Design

45. Ferguson, W. R., *Practical Laboratory Planning*, Halstead Press, Div. of Wiley and Sons, New York, NY, 1973.

SPECIFIC ANALYTES

Acid Phosphatase Activity (Thymolphthalein Monophosphate Substrate)

Submitter: Lilian M. Ewen, *Department of Clinical Biochemistry, Royal Columbian Hospital, New Westminster, B.C., V3L 3W7, Canada*

Evaluators: A. Ernest Alexander, *Department of Laboratories, St. Paul's Hospital, Vancouver, B.C., V6Z, 1Y6, Canada*
R. Bondar and Kera Voigtlander, *Worthington Diagnostics, Freehold, NJ 07728*

Reviewer: George N. Bowers, Jr., *Clinical Chemistry Laboratory, Department of Pathology, Hartford Hospital, Hartford, CT 06115*

Introduction

Measurement of the activity of acid phosphatase [orthophosphoric monoester phosphohydrolase (acid optimum); EC 3.1.3.1] in serum has been used as a confirmatory test for prostatic cancer for more than 40 years since Gutman et al. described increased activity in the serum of patients with metastatic carcinoma of the prostate gland *(1, 2)*. But acid phosphatases are found in a wide variety of tissues and cells, including liver, kidney, spleen, platelets, erythrocytes, and bone marrow, as well as prostate, and acid phosphatase in normal serum is derived from several of these sources. Measurement of total acid phosphatase activity in serum is not, therefore, a measure of an enzyme derived from the prostate exclusively, and unfortunately prostatic carcinoma has frequently metastasized before the fraction derived from this gland is great enough to be distinguished as an increase beyond the range of acid phosphatase activities usually found in the serum of either sex.

Efforts over the years have been directed, therefore, towards improving the sensitivity and specificity of tests, in an attempt to detect prostatic cancer earlier, particularly to detect it while the carcinoma is still localized in the pelvic area and potentially curable by surgery. Recent efforts to increase the specificity of the assay have involved immunological principles, using antibody produced against purified prostatic enzyme for this purpose. Current nonisotopic assays and radioimmunoassays *(3, 4)* hold promise as diagnostic aids for earlier detection of prostatic carcinoma, but the reagents and equipment required for these methods are not universally available, and a widely applicable method is still required both for routine use and as a reference against which new advances can be measured.

Acid phosphatases hydrolyze many orthophosphoric monoesters, and several different substrates have been used for measuring acid phosphatase activity in serum. Two approaches have been taken to distinguish acid phosphatase of prostatic origin from that derived from other cells and tissues. These are *(a)* use of methods that include inhibitors with some specificity for various forms of the enzyme (e.g., formaldehyde, tartrate) and *(b)* use of substrates with greater specificity for acid phosphatase of prostatic origin. The method described below, in which thymolphthalein monophosphate is the substrate, falls into the latter category.

This substrate was first used for acid phosphatase in 1971 *(5)* and subsequently developed further *(6)*. Although thymolphthalein monophosphate is not completely specific for prostatic acid phosphatase, it is the most specific substrate known at this time. The method is simple and can be performed in any laboratory.

Principle

Acid phosphatase hydrolysis of the substrate thymolphthalein monophosphate produces thymolphthalein and phosphate. These products are colorless at the acid pH used during hydrolysis, but thymolphthalein becomes a self-indicating chromogen when the reaction is terminated by addition of alkali. Measurement of the absorbance increase at 595 nm over a measured time interval is proportional to liberated thymolphthalein, thus giving a direct measure of enzyme activity. The equation of the reaction catalyzed by acid phosphatase is:

Final reaction conditions are shown in Table 1.

Table 1. Final Reaction Conditions

Temperature	37 °C
pH	5.4
Acetate buffer	0.15 mol/L
Thymolphthalein monophosphate	1.0 mmol/L
Brij-35	1.5 g/L
Volume fraction of sample	1:12

Materials and Methods

Reagents

Thymolphthalein monophosphate, disodium salt. This substrate is sold by a number of manufacturers but is not, however, of uniform quality. It is vital that the material used be of consistent and appropriate quality, meeting the specifications detailed in the Appendix to this chapter. Of the products evaluated, those supplied by Regis Chemical Co., Morton Grove, IL 60053, and J. B. L. Chemical Co., San Luis Obispo, CA 93401, met these specifications and gave similar and consistent activities.[1] Brij-35, available as a 300 g/L solution from B. D. H. Chemicals, Montreal, Canada, or through Gallard Schlesinger Chemical Manufacturing Corp., Carle Place, New York, NY 11514, is suitable. All other chemicals and water should be of the highest purity available, and should meet ACS reagent grade specifications.

1. *Acetate buffer, 5 mol/L, pH 5.4 (25 °C).* Dissolve 68.0 g of sodium acetate trihydrate in water in a 100-mL volumetric flask, dilute to volume, and mix well. Measure 28.87 mL of glacial acetic acid into a 100-mL volumetric flask, dilute to volume, and mix well. Add the diluted acetic acid to the sodium acetate until the pH of the mixture is 5.4 at 25 °C. Store well-stoppered at room temperature. Stable at least three months.

2. *Acetate buffer, 0.25 mol/L, pH 5.4 (25 °C).* Pipet 5 mL of the acetate buffer (Reagent 1) into a 100-mL volumetric flask, dilute to volume with reagent grade water, and mix well. Store refrigerated. Stable at least one month.

3. *Buffered substrate reagent.* Dilute Brij-35 solution to 3.24 g/L with reagent grade water; i.e., dilute 10.8 mL of Brij-35 solution (300 g/L) to 1 L. For 100 mL of final buffered substrate solution measure 50 mL of the diluted Brij-35 solution and dissolve in it the appropriate weight of disodium thymolphthalein monophosphate to give a concentration of 1.1 mmol/L when the reagent is brought to volume. For example, for thymolphthalein monophosphate containing 18% (by weight) water of crystallization, which is equivalent to 6.75 mol of H_2O, 74.4 mg of disodium thymolphthalein monophosphate should be used.[2] Add 1.92 g of sodium acetate trihydrate to the solution, mix until dissolved, and adjust the pH of the solution to 5.4 (25 °C) with 0.1 mol/L HCl. Bring the total volume to 100 mL with reagent grade water. Store refrigerated and discard after two months or when the blank values become unacceptable, i.e., exceed the specification given in the Appendix.

4. *Alkaline solution for color development.* Dissolve 10.6 g of anhydrous Na_2CO_3 and 4.0 g of NaOH in reagent grade water and dilute to 1 L (final concentration, per liter: 0.1 mol each of Na_2CO_3 and NaOH). Mix well and store in a polyethylene container at room temperature. Stable at least three months.

5. *Thymolphthalein standard solution (3 mmol/L).* Dissolve 129.2 mg of thymolphthalein in *n*-propanol/water (70/30 by vol) to make a total volume of 100 mL. Well-stoppered, the solution is stable at room temperature for at least three months.

Note: High-quality thymolphthalein, recrystallized to constant absorptivity, should be used. Material available from J. B. L. Chemical Co. was used by the Submitter to determine the molar absorptivity.

Apparatus

Volumetric glassware should meet the National Bureau of Standards (NBS) Class A specifications, or be individually calibrated before use. Tubes used should be borosilicate glass and chemically clean.

For measuring reagents, repetitive dispensors (e.g., Oxford type) may be used, and for serum specimens mechanical pipettes with disposable tips are suitable, but check the accuracy and precision of each type before use.

Calibrate the pH-measuring device at 25 °C against reference buffers from an authoritative source (NBS or IUPAC).

A water bath with continuous agitation capability is recommended to maintain uniform conditions throughout the bath. Temperature should be controlled at the set point ± 0.05 °C.

Note: Developmental work for this assay was performed at 37 °C. A temperature of 29.77 °C (with a 1-h incubation period) is being used in the study of the NBS Standard Reference Material 909 by the Joint NBS and Cooperating Laboratories Enzyme Study Group (7).

Collection and Handling of Specimens

Preparation of tubes for sample preservation. Pipet 25 μL of acetate buffer (5 mol/L, pH 5.4, Reagent

---

[1] Bowers, G. N., Jr., Onoroski, M., Schifreen, R. S., Brown, L. R., Klein, R. E., and Ewen, L. M., ms. in preparation.

[2] Formula weight of anhydrous thymolphthalein monophosphate = 554.5; F.W. of 6.75 mol of H_2O = 18 × 6.75 = 121.5; F.W. of thymolphthalein monophosphate containing 6.75 mol of H_2O per mole = 676. Therefore, thymolphthalein monophosphate (1.1 mmol/L) requires 676 × 1.1 = 743.6 mg/L (74.4 mg/dL).

1) into tubes. Evaporate in an oven at 45 °C for 7–8 h, or overnight if necessary. Stopper and store at room temperature until required.

Note: Check the tubes on initial preparation and periodically during storage to ascertain that patients' specimens will be properly acidified. Add 0.5 mL of serum to several tubes selected randomly, agitate gently to dissolve the acetate salts, and measure the pH, which should be approximately 5.6–5.7.

Preservation of patients' specimens. Collect blood with minimal venous stasis. Serum, or plasma obtained by collection into sodium heparin (0.2 g/L), may be used. Centrifuge whole blood 10 min at 3000 × g (25 °C); immediately after centrifugation add 0.5 mL of the serum or plasma to a tube containing dried acetate buffer salts. Agitate gently to dissolve the acetate salts, and store the specimens frozen (−20 °C) if they are not to be assayed on the same day.

Reconstitution of lyophilized control samples for assay. Use acetate buffer (0.25 mol/L, pH 5.4, Reagent 2) to reconstitute lyophilized control specimens. Aliquot and freeze (−20 °C) for use on subsequent dates. Samples thus prepared and stored retain activity for at least three months.

Procedure

1. For each specimen to be assayed label two tubes, "test" and "blank."
2. Pipet into each 550 µL of buffered substrate (Reagent 3).
3. Bring the "test" solutions to the temperature of assay in a water bath; at convenient intervals (e.g., 30 s) add 50 µL of each specimen to the appropriate "test" tube, mix gently, and allow to incubate at reaction temperature for exactly 30 min.
4. Add 1.0 mL of alkaline solution for color development (Reagent 4) to each tube and mix.
5. While the "test" samples are incubating, add 1.0 mL of the alkaline solution to each of the "blank" tubes, then 50 µL of the appropriate patient's specimen, and mix.
6. Read the absorbances of "test" and "blank" solutions at 595 nm. Subtract "blank" from "test" absorbances and determine acid phosphatase activity by one of the methods described under *Standardization*.

Note: For specimens with an acid phosphatase activity yielding a color exceeding the absorbance limits of the spectrophotometer (up to an absorbance of 3.5) either dilute the specimen and re-assay, or: repeat the assay with a 5-min incubation period, read the absorbance at 595 nm, subtract the absorbance of the "blank," determine the activity as usual, and multiply by 6.

Standardization

Activity of acid phosphatase can be determined by reference to a standard curve, by calculation with the molar absorptivity of thymolphthalein, or by preparing a standard with each batch of assays.

1. *Preparation of a standard curve (to 50 U/L).* Pipet into six test tubes 0.0, 0.1, 0.2, 0.3, 0.4, or 0.5 mL of thymolphthalein standard solution (Reagent 5). Add 1.0, 0.9, 0.8, 0.7, 0.6, or 0.5 mL of *n*-propanol/water mixture (70/30 by vol) to make a total volume of 1.0 mL and mix well.

Into another set of six test tubes labeled 0, 10, 20, 30, 40, and 50 dispense 550 µL of buffered substrate (Reagent 3) and 1.0 mL of alkaline solution for color development (Reagent 4). Pipet 50 µL of each diluted standard into these tubes. Mix well and read the absorbances of these solutions with a spectrophotometer at 595 nm; use the tube labeled 0 as blank. Under the conditions of the method the absorbances obtained will be equivalent to those obtained by the action of 10, 20, 30, 40, and 50 U of acid phosphatase per liter of serum. Plot these absorbances against U/L on linear graph paper.

2. *Calculation of acid phosphatase activity, using the molar absorptivity of thymolphthalein under the conditions of the assay.*

$$\text{Activity (U/L)} = (\Delta A/t) \cdot (V/\epsilon b v) \cdot 10^6$$

where ΔA is the change in absorbance at 595 nm during the incubation period of the assay (A "test" − A "blank"), t is the reaction time in minutes, V is the reaction volume in liters, ϵ is the molar absorptivity of thymolphthalein under the final reaction conditions (L · mol^{-1} · cm^{-1}), b is the pathlength of the cuvette in centimeters, and v is the sample volume in liters; that is:

$$\text{Activity (U/L)} = \frac{\Delta A}{30} \times \left(\frac{1.6 \times 10^{-3} \times 10^6}{40.15 \times 10^3 \times 1 \times 5 \times 10^{-5}} \right)$$

$$= \Delta A_{595 \text{ nm}} \times 26.57$$

Note: The number 40.15×10^3 was obtained in the Submitter's laboratory. It should be checked by each laboratory under the conditions there, for thymolphthalein solutions of known concentration prepared as under step 1 above, *Preparation of a standard curve*; e.g., the concentration of thymolphthalein in the 20 U/L standard is 1.875×10^{-5} mol/L. The molar absorptivity = $[A_{595 \text{ nm}} (20 \text{ U/L standard}) - A_{595 \text{ nm}} (0 \text{ U/L standard})] \times (1.875 \times 10^{-5})^{-1}$

$$= \Delta A_{595 \text{ nm}} \times 5.33 \times 10^4$$

3. *Standardization with each batch of specimens.* Prepare two standards, e.g., 20 U/L and 0 U/L, as described in step 1 of standardization. Determine their absorbances at 595 nm and calculate the difference in absorbance between them. Calculate specimen activities directly from these standards, using the formula

$$\frac{\Delta A_{595 \text{ nm}} \text{ "test"}}{\Delta A \text{ between standards}} \times \text{"activity difference of standards"} = \text{U/L}$$

i.e., $\dfrac{\Delta A_{595\text{ nm}} \text{"test"} \times 20}{\Delta A \text{ between standards}} = \text{U/L}$

Reference Intervals and Precision of Assay

Acid phosphatase activities in sera of individuals without prostatic disease (measured at 37 °C) ranged from 0.5 to 1.9 U/L ($\bar{x} \pm 2$ SD, n = 48) in the Submitter's laboratory. Activities were similar in both sexes.

The within-run CV for this method was 2.1% at an activity of 1.7 U/L and 0.8% at 5.0 U/L. Between-day assays showed CVs of 5.0% and 2.1%, respectively, for these two activities.

Discussion

Concentrations of reagents in the reaction mixture are as previously described (6): acetate buffer, 0.15 mol/L; thymolphthalein monophosphate, 1.0 mmol/L; and Brij-35, 1.5 g/L. The serum/reaction volume ratio is 1/12. The reaction conditions were chosen after evaluating the effects of buffer type and concentration, surfactant concentration, and serum/reaction volume ratio on the activity of prostatic acid phosphatase. Because citrate concentrations adequate to buffer the reaction were shown to decrease prostatic acid phosphatase activity, acetate buffer is preferred. The concentration of surfactant required to give maximum activity is influenced by the serum/reaction volume ratio, as is also the range of linearity. The importance of high-quality substrate, meeting all of the specified criteria of purity, cannot be overemphasized. Because of the different storage conditions prevailing in different laboratories, water content is particularly important. Substrates with greater hydration have been observed to be more stable and are recommended. Under the conditions chosen, the range of linearity of the method extends to an absorbance of about 3.5, in excess of the capabilities of most spectrophotometers.

The fact that the range of activities of acid phosphatase found in sera of normal females is similar to that of males is evidence that the substrate thymolphthalein monophosphate is not absolutely specific for prostatic acid phosphatase. In some patients platelet acid phosphatase could contribute to an increase in total activity, e.g., in individuals with thrombocytopenia; use of heparinized platelet-free plasma will eliminate this problem.

Thymolphthalein monophosphate is more specific than other substrates used for colorimetric assay of acid phosphatase (5). With less specific substrates such as p-nitrophenyl phosphate and α-naphthyl phosphate, prostatic acid phosphatase is differentiated from nonprostatic by addition of inhibitors such as tartrate to inhibit preferentially prostatic enzyme activity, and some "approximation" results. Inhibition of prostatic forms is not complete—stated to be about 95% (8)—and not exclusive to the prostatic form of the enzyme (9). The proportion of acid phosphatase of nonprostatic origin is usually much greater than that of prostatic origin and a small difference between two larger numbers is used to calculate activity. In addition, techniques involving inhibitors have the disadvantage of requiring an extra tube for each assay. The molar absorptivity of thymolphthalein is more than twice that of p-nitrophenol and one-third greater than α-naphthol, the chromogens measured in assays where inhibitors are used to promote specificity. The method presented here therefore gives a more sensitive endpoint per micromole of substrate transformed. This method is being used in modified form (29.77 °C, 1-h incubation) by the Joint NBS and Cooperating Laboratories Enzyme Study Group (7) to provide a standard material with an assayed activity.

In summary, measurement of acid phosphatase with thymolphthalein monophosphate offers the advantages of simplicity with a minimum of sample tubes, greater sensitivity than other substrates, and ready applicability to any laboratory. Immunochemical methods show promise, but additional work in development and standardization of these techniques is still needed (10). Comparison of the diagnostic value of methods involving determination of endpoints based on measurement of *quantity* of enzyme, as in counterimmunoelectrophoresis and radioimmunoassay, and on *activity* of enzyme, as in immunofluorescence, must still be assessed. Until one of these techniques shows greater efficacy for earlier detection of prostatic carcinoma, the method described here provides a useful interim procedure and a reference against which the sensitivity and specificity of newer immunochemical techniques can be measured.

References

1. Gutman, E. B., Sproul, E. B., and Gutman, A. B., Increased phosphatase activity of bone at site of osteoblastic metastases secondary to carcinoma of prostate gland. *J. Urol.* **28**, 485–495 (1936).
2. Gutman, A. B., and Gutman, E. B., An "acid" phosphatase occurring in the serum of patients with metastasing carcinoma of the prostate gland. *J. Clin. Invest.* **17**, 473–478 (1938).
3. Lee, C., Wang, M. C., Murphy, G. P., and Chu, T. M., A solid-phase fluorescent immunoassay for human prostatic acid phosphatase. *Cancer Res.* **38**, 2871–2878 (1978).
4. Foti, A. G., Herschman, H., and Cooper, J. F., A solid-phase radioimmunoassay for human prostatic acid phosphatase. *Cancer Res.* **35**, 2446–2452 (1975).
5. Roy, A. V., Brower, M. E., and Hayden, J. E., Sodium thymolphthalein monophosphate: A new acid phosphatase substrate with greater specificity for the prostatic enzyme in serum. *Clin. Chem.* **17**, 1093–1102 (1971).
6. Ewen, L. M., and Spitzer, R. W., Improved determination of prostatic acid phosphatase (sodium thymol-

phthalein monophosphate substrate). *Clin. Chem.* **22**, 627–632 (1976).
7. Bowers, G. N., Jr., Cali, J. P., Elser, R., Ewen, L. M., McComb, R. B., Rej, R., and Shaw, L. M., Activity measurements for seven enzymes in lyophilized human serum SRM 909. *Clin. Chem.* **26**, 969 (1980). Abstract.
8. Nigam, V. N., Davidson, H. M., and Fishman, W. H., Kinetics of hydrolysis of the orthophosphate monoesters of phenol, *p*-nitrophenol, and glycerol by human prostatic acid phosphatase. *J. Biol. Chem.* **234**, 1550–1554 (1959).
9. Hennebemy, M. O., Engel, G., and Grayhack, J. T., Acid phosphatase. *Urol. Clin. North Am.* **6**, 629–641 (1979).
10. Brice, A. W., Mahan, D. E., Morales, A., Clark, A. F., and Belville, W. D., An objective look at acid phosphatase determinations. A comparison of biochemical and immunological methods. *Br. J. Urol.* **51**, 213–217 (1979).

Appendix

The quality of thymolphthalein monophosphate varies among manufacturers. Substrates used should therefore meet the reagent specifications detailed below. Variation in the number of molecules of water of crystallization also occurs, and materials with greater hydration, being less hygroscopic, are less subject to change on storage or error in weighing.

A certificate of analysis should accompany each lot.

Formula: $C_{28}H_{29}O_7PNa_2 \cdot xH_2O$
Anhydrous formula weight: 554.5
Appearance: pale yellow powder
H_2O content (by Karl Fischer titration): 160–260 mg/g, or 5.6–9.1 molecules of H_2O per mole.
Solubility: an aqueous solution (25 g/L) should be a clear pale yellow, with a pH between 8.2 and 8.4
Molar absorptivity: 2150–2500 at 445 nm and 25 °C (see no. 1 below)
Free thymolphthalein: less than 5 mmol/mol of thymolphthalein monophosphate (see no. 2 below)
Storage: at 4 °C or −20 °C, as recommended by the manufacturer
Analysis by "high-performance" liquid chromatography: > 97.5% of total peak area at 254 nm from thymolphthalein monophosphate

1. To check molar absorptivity, dissolve an accurately weighed amount (20–25 mg) of thymolphthalein monophosphate, disodium salt, in 0.1 mol/L Na_2CO_3 (10.6 g/L) in a 100-mL volumetric flask and dilute to volume with 0.1 mol/L Na_2CO_3. Measure the absorbance at 445 nm in a temperature-controlled cuvet at 25 °C with Na_2CO_3 as reference. Calculate the molar absorptivity (ϵ) from the following equation:

$$\epsilon = \left(\frac{A_{445\ nm}}{mg}\right) \times \left(\frac{554.5}{10}\right) \times \left(\frac{100}{100 - \%\ moisture}\right)$$

Note: Thymolphthalein monophosphate exhibits definite thermochromic properties. A solution of the above concentration will show an increase in absorbance of about 0.004 per °C increase in temperature between 25 °C and 37 °C. Temperature must therefore be controlled carefully. In addition, at a constant temperature of 25 °C, there is a very slow decrease in absorbance over time (about 0.010 per hour). The absorbance should therefore be measured as soon as the solution is made.

2. The mole fraction of free thymolphthalein in thymolphthalein monophosphate should be less than 0.005. To test this, add 1.0 mL of alkaline reagent (Reagent 4) to 0.5 mL of buffered substrate. The $A_{595\ nm}^{25\ °C,\ 1\ cm}$ of this solution vs alkaline reagent as blank should be less than 0.075 (calculated from the molar absorptivity of thymolphthalein at 595 nm as being 40.15×10^3).

Acid Phosphatase (Provisional)[1]

Submitter: Guilford G. Rudolph, *Department of Biochemistry and Molecular Biology, Louisiana State University Medical Center, School of Medicine in Shreveport, Shreveport, LA 71130*

Evaluators: Herbert K. Y. Lau, *Sisters of Charity Hospital, Department of Laboratory, Buffalo, NY 14214*
James L. Verdi, *Sparks, NV 89431*
Kwok-Wai Lam, *Department of Biochemistry, Albany Medical College Union University, Albany, NY 12208*

Introduction

Acid phosphatases [orthophosphoric-monoester phosphohydrolases (acid optimum), EC 3.1.3.2] are enzymes that catalyze the hydrolysis of an orthophosphoric monoester to give an alcohol and phosphate; their optimum reaction is at pH values below 7.0. The acid phosphatase of greatest clinical importance is that in the prostate gland, where it is found in high concentration. These enzymes are found also in the stomach, liver, kidney, muscle, spleen, platelets, erythrocytes, and bone marrow.

Acid phosphatase can hydrolyze a variety of phosphate esters. Some of the substrates used over the years include β-glycerol phosphate *(1–3)*, phenyl phosphate *(4, 5)*, p-nitrophenyl phosphate *(6, 7)*, α-naphthyl phosphate *(8)*, and thymolphthalein monophosphate *(9,10)*.

Because the prostatic enzyme is the acid phosphatase of greatest clinical interest in serum, the method used should be capable of distinguishing acid phosphatase of prostatic origin from those of erythrocyte, platelet, and other tissue origins. In the method described here, inhibition by tartrate is used to allow quantitation of prostatic acid phosphatase activity in the presence of other acid phosphatases in the reaction mixture. In certain other colorimetric methods, certain substrates (e.g., α-naphthyl phosphate and thymolphthalein monophosphate), reported to be more specific than β-glycerol phosphate, are used to distinguish prostatic from other phosphatases. Unfortunately none of these techniques is completely specific for the acid phosphatase of prostatic origin.

Principle

Enzymic hydrolysis of p-nitrophenyl phosphate, the substrate in the assay described here, results in the production of p-nitrophenol and phosphate.

$$O_2N-C_6H_4-O-PO_3^{2-} + H_2O \xrightarrow[\text{(pH 4.85)}]{\text{acid phosphatase}} O_2N-C_6H_4-O^- + HPO_4^{2-} + H^+$$

L-Tartrate strongly inhibits prostatic acid phosphatase, whereas acid phosphatases from other sources are largely unaffected *(11–14)*. Serum is incubated with substrate in the presence and in the absence of L-tartrate. The liberated reaction product, p-nitrophenol, is stabilized with sodium hydroxide and then measured spectrophotometrically *(7, 11, 13–15)*. The difference in absorbances, compared with the absorbances of p-nitrophenol standards, gives a measure of the enzymic activity, which is expressed in U/L.

Materials and Methods

Reagents

1. *Citrate buffer, 90 mmol/L, pH 4.85 (37 °C).* Solution *a*: 90 mmol/L citric acid. Dissolve 18.91 g of citric acid ($C_6H_8O_7 \cdot H_2O$) in reagent-grade water and dilute to 1 L. Solution *b*: 90 mmol/L sodium citrate. Dissolve 26.46 g of sodium citrate ($Na_4C_6H_5O_7 \cdot 2H_2O$) in reagent-grade water and dilute to 1 L. Mix 180 mL of solution *a* with 320 mL of solution *b*. The pH should be 4.85 at 37 °C. If necessary, adjust by adding solution *a* to lower or solution *b* to raise the pH. Add a few drops of chloroform as a preservative.

2. *Citrate buffer, 81 mmol/L, pH 4.85 (37 °C), containing tartrate, 20 mmol/L.* Dissolve 30.0 g of L-tartaric acid ($C_4H_6O_6$) in about 300 mL of reagent-grade water in a 500-mL volumetric flask. Adjust the pH to 4.85 with sodium hydroxide (about 350 mmol, or 35 mL of 10 mol/L NaOH solution). Dilute to 500 mL. Transfer exactly 50.0 mL of the Na tar-

[1] *Editors' Note:* This chapter is marked "Provisional," indicating that it may not have been completely evaluated, and has not met the arbitrary criteria of our reviewing process. On the other hand, the method is used daily in the Submitter's laboratory, and is not only clinically useful, but has met current standards of quality for many years.

trate solution into a 500-mL volumetric flask and dilute to 500 mL with citrate buffer (reagent 1). Add a few drops of chloroform as a preservative.

3. *Substrate, 36 mmol/L.* Dissolve 134 mg of disodium *p*-nitrophenyl phosphate hexahydrate ($C_6H_4O_6NPNa_2 \cdot 6H_2O$) in 10 mL of reagent-grade water.

Note: p-Nitrophenyl phosphate hexahydrate obtained either from Sigma Chemical Co., St. Louis, MO 63178, or from Calbiochem-Behring Corp., San Diego, CA 92112, has been satisfactory.

Note: The substrate solution when stored in a refrigerator at 2–4 °C shows a slow hydrolysis but is useful as a substrate for about a week. At room temperature, hydrolysis is more rapid, making the reagent useful for only about 8 h.

4. *Sodium hydroxide, 100 mmol/L.* Prepare a carbonate-free solution by dissolving about 70 g of NaOH in water, diluting to 100 mL, and allowing to stand for a week. This solution will be about 17 mol/L. Dilute 5.9 mL of the saturated solution to 100 mL with water. Titrate with standard acid and adjust to 100 mmol/L by adding water.

5. *Stock p-nitrophenol standard, 1 mmol/L.* Dissolve 139.1 mg of high-purity *p*-nitrophenol ($C_6H_5NO_3$) in reagent-grade water. Dilute to 1 L in a volumetric flask at 25 °C and mix thoroughly. When stored in the dark, this reagent is useful for several months. The absorptivity was found in the Submitter's laboratory to be 18 600 L · mol^{-1} · cm^{-1} under these conditions.

Using a volumetric pipette, transfer 4.00 mL of 1 mmol/L *p*-nitrophenol stock standard at 25 °C into a 100-mL volumetric flask. Dilute to the mark with 10 mmol/L NaOH and mix well. The absorbance of this 40 µmol/L *p*-nitrophenol solution at 402 nm in a 10-mm cuvet should be 0.736 after subtraction of the blank.

Note: Sources of the p-nitrophenol are Sigma Chemical Co.; Calbiochem-Behring Corp.; and J. T. Baker Chemical Co., Phillipsburg, NJ 08865.

6. *Acetate buffer, 5 mol/L, pH 5.4 (25 °C).* Dissolve 68 g of sodium acetate trihydrate ($C_2H_3O_2Na \cdot 3H_2O$) in reagent-grade water and dilute to 100 mL with reagent-grade water. Measure 28.9 mL of glacial acetic acid ($C_2H_4O_2$) into a 100-mL volumetric flask and dilute to the mark with reagent-grade water. To this solution add the sodium acetate solution until the pH is 5.4 at 25 °C *(10)*.

Collection and Handling of Specimens

Collect blood so as to minimize hemolysis and separate serum from cells soon after clot formation occurs. Use of hemolyzed serum is not desirable, because the specimen will be contaminated with acid phosphatases from the erythrocytes. However, tartrate inhibition, as used in this procedure, minimizes the effect. Because acid phosphatase is unstable at neutral and alkaline conditions, the pH must be kept below 6.5, by adding to 1 mL of serum either 1 mg of disodium citrate *(17)* or 50 µL of acetate buffer (reagent 6) *(10)*. Either method keeps the acid phosphatase stable for two days at room temperature, or two to three months frozen. If the assay is not to be performed within one day, the specimen should be frozen.

Do not use fluoride or oxalate as anticoagulants; they inhibit acid phosphatase activity.

Procedure

1. Add 0.50 mL of substrate solution (reagent 3) to each of three tubes labeled 1, 2, and 3.
2. Add as follows:
   Tube 1, 0.50 mL of citrate-tartrate buffer (reagent 2)
   Tube 2, 0.50 mL of citrate buffer (reagent 1)
   Tube 3, 0.50 mL of citrate buffer (reagent 1)
3. Place in 37 °C water bath for 5 min to equilibrate.
4. Add 0.20 mL of serum to tubes 1 and 2. Mix and return to water bath. Note the time the serum was added.
5. Incubate for exactly 30 min.
6. At exactly 30 min after adding the serum sample, add 5.0 mL of 100 mmol/L NaOH (reagent 4) to all three tubes.
7. Add 0.20 mL of serum to tube 3.
8. Transfer the solutions to cuvets; read and record the absorbances at 402 nm with water as the reference.

Calculation

Subtract the absorbance of the solution in tube 1 from that of tube 2 to calculate prostatic acid phosphatase activity. Subtract the absorbance of the solution in tube 3 from that of tube 2 to calculate the total acid phosphatase activity.

Calculate enzymic activity as units per liter where a unit (U) is µmol/min.

$$U/L = \left(\frac{\Delta A_{402}}{t}\right) \cdot \left(\frac{V}{\epsilon l v}\right)$$

ΔA_{402} is change in absorbance at 402 nm
t is the reaction time in minutes
V is the final volume (L)
v is the sample volume (L)
ϵ is the micromolar absorptivity of *p*-nitrophenol (0.0184 L · µmol^{-1} · cm^{-1}) in 10 mmol/L NaOH *(16)*
l is the pathlength of the cuvet in centimeters

$$U/L = \frac{\Delta A_{402} \cdot 0.0062}{30 \cdot 0.0184 \cdot 0.0002}$$
$$= \Delta A_{402} \cdot 56.1$$

Conversion factors:
  1 U/L = 16.67 nkat/L [kat (katal) is mol/s]
  1000 nkat/L = 60 U/L

Sample calculation:
  Absorbance set at 0 with water
  Sample readings at 402 nm
  Tube 1, A = 0.15
  Tube 2, A = 0.22
  Tube 3, A = 0.02

Total acid phosphatase, U/L = (0.22 − 0.02) × 56.1 = 11.2

Prostatic acid phosphatase, U/L = (0.22 − 0.15) × 56.1 = 3.9

Discussion

Methods for determination of serum acid phosphatase activities have been published in previous volumes of this series (3, 5, 7). With the substrate p-nitrophenyl phosphate, the method is standardized with p-nitrophenol, and determinations require a minimum of manipulations.

Because of the clinical importance of increased serum activities of acid phosphatase in the diagnosis of prostatic cancer, it is desirable to differentiate between the activities of the prostatic phosphatase and of the non-specific forms. The enzyme from the prostate is inhibited by L-tartrate ions (12, 18).

Immunochemical methods have been recommended as the most specific for prostatic acid phosphatase (19, 20). A recent report describes the identical antigenicity between some of the isoenzymes of leukocytes and that of the prostate (21). Yam has reviewed the significance of human acid phosphatases (22).

Reference Intervals

The total and tartrate-sensitive enzyme activities are not significantly different in men and women (23, 24). Total serum acid phosphatase activity in normal adults is 2–11 U/L (30–180 nkat/L).

Prostatic (tartrate-sensitive) acid phosphatase in normal adults is 2–3 U/L (30–50 nkat/L). Higher activities of prostatic (tartrate-sensitive) acid phosphatase are seen in patients with metastases. In one report, however, 25% of the patients with progressive metastatic disease from prostatic cancer had normal serum acid phosphatase activities (25).

References

1. Bodansky, A., Phosphatase studies. II. Determination of serum phosphatase. Factors influencing the accuracy of the determination. J. Biol. Chem. 101, 93–104 (1933).
2. Shinowara, G. Y., Jones, L. M., and Reinhart, H. L., The estimation of serum inorganic phosphate and "acid" and "alkaline" phosphatase activity. J. Biol. Chem. 142, 921–933 (1942).
3. Kaser, M. M., and Baker, J., Alkaline and acid phosphatase. Stand. Methods Clin. Chem. 2, 122–131 (1958).
4. King, E. J., and Armstrong, A. R., A convenient method for determining serum and bile phosphatase activity. Can. Med. Assoc. J. 31, 376–381 (1934).
5. Carr, J. J., Alkaline and acid phosphatase. Stand. Methods Clin. Chem. 1, 75–83 (1953).
6. Bessey, O. A., Lowry, O. H., and Brock, M. J., A method for the rapid determination of alkaline phosphatase with five cubic millimeters of serum. J. Biol. Chem. 164, 321–329 (1946).
7. Berger, L., and Rudolph, G. G., Alkaline and acid phosphatase. Stand. Methods Clin. Chem. 5, 211–221 (1965).
8. Babson, A. L., and Read, P. A., A new assay for prostatic acid phosphatase in serum. Am. J. Clin. Pathol. 32, 88–91 (1959).
9. Roy, A. V., Brower, M. E., and Hayden, J. E., Sodium thymolphthalein monophosphate: A new acid phosphatase substrate with greater specificity for the prostatic enzyme in serum. Clin. Chem. 17, 1093–1102 (1971).
10. Ewen, L. M., and Spitzer, R. W., Improved determination of prostatic acid phosphatase (sodium thymolphthalein monophosphate substrate). Clin. Chem. 22, 627–632 (1976).
11. Jacobsson, K., The determination of tartrate-inhibited phosphatase in serum. Scand. J. Clin. Lab. Invest. 12, 367–380 (1960).
12. Abul-Fadl, M. A. M., and King, E. J., Properties of the acid phosphatases of erythrocytes and of the human prostate gland. Biochem. J. 45, 51–60 (1949).
13. Fishman, W. H., and Lerner, F., A method for estimating serum acid phosphatase of prostatic origin. J. Biol. Chem. 200, 89–97 (1953).
14. Fishman, W. H., Bonner, C. D., and Homburger, F., Serum "prostatic" acid phosphatase and cancer of the prostate. N. Engl. J. Med. 255, 925–933 (1956).
15. Andersch, M. A., and Szczypinski, A. J., Use of p-nitrophenyl phosphate as the substrate in determination of serum acid phosphatase. Am. J. Clin. Pathol. 17, 571–574 (1947).
16. Bowers, G. N., Jr., and McComb, R. B., A continuous spectrophotometric method for measuring the activity of serum alkaline phosphatase. Clin. Chem. 12, 70–89 (1966).
17. Doe, R. P., Mellinger, G. T., and Seal, U. S., Stabilization and preservation of serum prostatic acid phosphatase activity. Clin. Chem. 11, 943–950 (1965).
18. Nigam, V. N., Davidson, H. M., and Fishman, W. H., Kinetics of hydrolysis of the orthophosphate monoesters of phenol, p-nitrophenol, and glycerol by human prostatic acid phosphatase. J. Biol. Chem. 234, 1550–1554 (1959).
19. Cooper, J. F., and Foti, A., A radioimmunoassay for prostatic acid phosphatase. I. Methodology and range of normal male serum values. Invest. Urol. 12, 98–102 (1974).
20. Cooper, J. F., Foti, A., Herschman, H. H., and Finkle, W., A solid phase radioimmunoassay for prostatic acid phosphatase. J. Urol. 119, 388–391 (1978).
21. Lam, W. K. W., Yam, L. T., Wilbur, H. J., Taft, E., and Li, C. Y., Comparison of acid phosphatase isoenzymes of human seminal fluid, prostate, and leukocytes. Clin. Chem. 25, 1285–1289 (1979).
22. Yam, L. T., Clinical significance of the human acid phosphatases. Am. J. Med. 56, 604–616 (1974).
23. Dow, D., and Whitaker, R. H., Prostatic contribution to normal serum acid phosphatase. Br. Med. J. iv, 470–472 (1970).

24. Whitmore, W. F., Jr., Bodansky, O., Schwartz, M. K., Ying, S. M., and Day, E., Serum prostatic acid phosphatase levels in proved cases of carcinoma or benign hypertrophy of the prostate. *Cancer* 9, 228–233 (1956).

25. Johnson, D. E., Scott, W. W., Gibbons, R. P., Prout, G. R., Schmidt, J. D., and Murphy, G. P., Clinical significance of serum acid phosphatase levels in advanced prostatic carcinoma. *Urology* 8, 123–126 (1976).

Alanine Aminotransferase, ALT (Provisional)[1]

Submitter: Thorne J. Butler, *Southern Nevada Memorial Hospital, Las Vegas, NV 89102*

Evaluators: Sigrid G. Klotzsch, *Union Carbide Corporation, Wallingsford, CT 06492*
Iris M. Osberg, *Pediatric Microchemistry, University of Colorado Medical Center, Denver, CO 80262*

Introduction

Alanine aminotransferase (ALT), L-alanine:2-oxoglutarate aminotransferase, EC 2.6.1.2, is an enzyme that catalyzes the transfer of an amine group from alanine to oxoglutarate, forming glutamate and pyruvate.[2] As with the clinically useful asparate aminotransferase (AST), the enzyme is widely distributed in all tissues but has a particularly high concentration in the liver. Consequently, ALT assays are useful in evaluating hepatic disease. Many schemes involving ratios of ALT to AST have been used as an approach to staging and classifying hepatic disorders (1).

In spite of the long clinical experience with serum ALT activity as an indicator of acute hepatic necrosis, there has been relatively little effort to investigate optimum conditions for ALT assays, compared with numerous investigations of AST assays.

Principle

The principles of kinetic assays of enzyme activity are applicable to ALT determination. The amount of pyruvate formed is quantitated indirectly; pyruvate is reduced to lactate and NADH is oxidized to NAD$^+$ in the presence of excess lactate dehydrogenase (L-lactate:NAD$^+$ oxidoreductase, EC 1.1.1.27). The assay system has two reactions: the primary ALT reaction and the indicator lactate dehydrogenase reaction.

---

[1] *Editors' Note:* This chapter is marked "Provisional," indicating that it may not have been completely evaluated, and has not met the arbitrary criteria of our reviewing process. On the other hand, the method is used daily in the Submitter's laboratory, and is not only clinically useful, but has met current standards of quality for many years.

[2] Nonstandard abbreviations used: ALT, alanine aminotransferase; AST, aspartate aminotransferase; Tris, tris(hydroxymethyl)methylamine.

Primary alanine aminotransferase reaction:

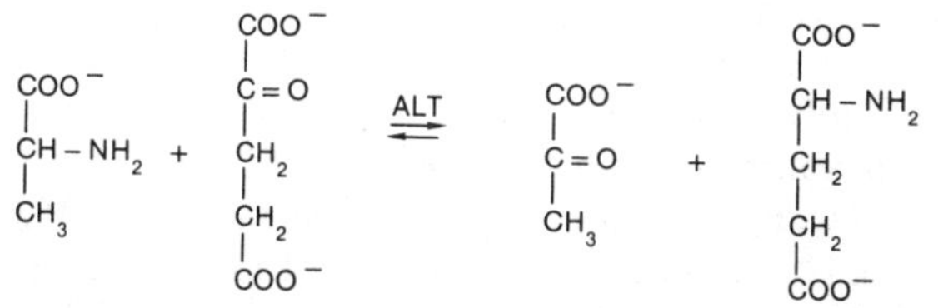

$$\text{Alanine} + \alpha\text{-Ketoglutarate} \underset{}{\overset{\text{ALT}}{\rightleftharpoons}} \text{Pyruvate} + \text{Glutamate}$$

Indicator reaction:

$$\text{Pyruvate} + \text{NADH} + \text{H}^+ \underset{}{\overset{\text{lactate dehydrogenase}}{\rightleftharpoons}} \text{Lactate} + \text{NAD}^+$$

Because the indicator reaction equilibrium lies far to the right, the equilibrium of the primary reaction is not relevant: pyruvate reacts immediately in the indicator reaction.

One observes the decrease in absorbance of the NADH at 340 nm until a rectilinear phase (zero-order kinetics) occurs. The change in absorbance per unit time ($\Delta A/\Delta t$) represents the activity of the ALT enzyme.

The presence of other keto acids such as pyruvate within the sample requires an appropriate lag phase so that these contaminants are consumed before the steady state, or zero-order kinetic condition, occurs. Glutamic dehydrogenase, which can appear in serum samples, catalyzes the amination of oxoglutarate to form glutamate, with the concomitant oxidization of NADH to NAD$^+$. To avoid this possible interference, ammonium sulfate suspensions of lactate dehydrogenase should not be used in the reagent.

Materials and Methods

Because the equipment and procedure used for the determination of ALT are identical with those used for AST, the reader should refer to the chapter on AST for a description of the apparatus and procedure used.

Reagents

1. *Stock ALT buffer:* 100 mmol/L Tris, pH 7.5, and DL-alanine, 465 mmol/L. In a plastic beaker

with a 1-L calibration mark, add 42 g of DL-alanine (no. A-7502; Sigma Chemical Co., St. Louis, MO 63178), 12. 1 g of tris(hydroxymethyl)methylamine (Tris) base (Sigma no. T-1378), and 800 mL of deionized or distilled water.

Note: Reagents prepared in-house are convenient to use and relatively inexpensive, compared with lyophilized reagents, and are stable for long periods. Moreover, only the volume required for the workload need be prepared.

Note: Evaluator I. M. O. states that when using a Tris buffer, the pH of the reagent should be measured at the reaction temperature.

Mix until dissolved. With a pH electrode in the stirring solution, add 2.5 mol/L HCl until the pH is 7.5. Transfer to a 1-L volumetric flask, and with washings of the beaker, bring to volume with water. Add 1.0 g of sodium azide, NaN_3 (no. 1953; Mallinckrodt Chemical Co., Hazelwood, MO 63042). Aliquot into capped 100-mL glass or plastic bottles. Autoclave in the bypass mode for 4 min. Allow to cool and store at 2 to 6 °C. Stable for six months; if stored frozen, the shelf life is increased to 18 months.

Note: If an autoclave is not available, then bring the mixture to a light boil for 1 min and aliquot as stated above. Allow to cool, cap, and store at 2 to 6 °C.

2. *Stock nicotinamide adenine dinucleotide (NADH): 11.3 mmol in aqueous ethylene glycol.*[3] Add 187 mg of NADH (Sigma no. N-8129) into a 25-mL flask. Make to volume with 30% ethylene glycol AR (Mallinckrodt no. 5001) in de-ionized water (300 mL of ethylene glycol per liter of solution). Store in a freezer at −10 to −15 °C.

3. *Stock lactate dehydrogenase (about 1.2 MU/L), in phosphate-buffered 50% glycerol, pH 7.4.* Weigh a 5-mg aliquot of lyophilized salt-free lactate dehydrogenase (Sigma no. L-1254) into a capped 5.0-mL glass bottle. Add 5.0 mL of buffered glycerol diluent (see below). Mix. Store at −10 to −15 °C. This reagent will not freeze and has a shelf life of one month.

Note: Lactate dehydrogenase solutions are relatively unstable. Loss of activity prolongs the lag phase in the ALT assay.

Into a 1-L flask add 1.179 g of KH_2PO_4 (Mallinckrodt no. 7100), 4.303 g of Na_2HPO_4 (Mallinckrodt no. 7917), and 0.05 g of NaN_3. Make up to volume with de-ionized water and mix to dissolve. Store in a refrigerator. Prepare 250 mL of working phosphate-buffered glycerol diluent solution by mixing 125 mL of glycerol AR (Mallinckrodt no. 5092) and 125 mL of phosphate buffer. Store in a refrigerator.

[3] The use of ethylene glycol as a stabilizing agent of biological materials at low temperatures is covered by U.S. patents no. 3,876,375 and 4,121,905 held by Beckman Instruments, Inc., Fullerton, CA 92364.

Note: A convenient source of phosphate buffer is the pH 7.384 buffer used as a calibrator in blood-gas instruments. The buffered 50% glycerol is an excellent vehicle for stabilizing enzyme solutions in the freezer.

4. *Stock 2-oxoglutarate, 1.4 mol, and 82 mmol of Tris per liter of 30% (300 mL/L) aqueous ethylene glycol.* To 2.0 g of 2-oxoglutaric acid (Sigma no. K-1750) and 1.0 g of Tris base, add 10 mL of the aqueous ethylene glycol solution. Mix; store in freezer at −10 to −15 °C.

5. *Working ALT assay reagent: pH 7.5, 100 mmol of Tris, 400 mmol of L-alanine, 14 mmol of oxoglutarate, 0.1 mmol of NADH, and 2400 U of lactate dehydrogenase per liter.* To 25 mL of stock ALT buffer add 0.25 mL of stock oxoglutarate, 0.3–0.5 mL of NADH stock, and 0.05 mL of lactate dehydrogenase stock. This is stable for 24 h at room temperature and 48 h at 2 to 6 °C.

Note: The volume of added NADH will vary with each batch of stock NADH. The actual volume required should give the final working solution an absorbance, read against water at 340 nm in 1-cm light path, of between 1.100 and 1.300.

Collection and Handling of Specimens

Freshly drawn and separated serum from a 12-h fasted individual is the preferred sample, collected in either a plain glass syringe or a Vacutainer Tube® (Becton Dickinson, Rutherford, NJ 07070). Plasma samples collected with ethylenediaminetetraacetate, fluoride, or heparin do not affect ALT activity *(2)*. However, plasma samples and nonfasted serum samples tend to be turbid, causing erratic absorbance changes. Oxalate inhibits enzyme activity and should therefore not be used as an anticoagulant.

ALT activity in separated serum is stable for 48 h at room temperature, seven days at 2 to 6 °C, and at least eight months at −20 °C. ALT activity preserved in 30% aqueous ethylene glycol at −15 to −20 °C is stable up to 18 months.[3]

Procedure

The procedure for the determination of ALT activity is identical with that described in the chapter for AST. Refer to the AST chapter for details.

Standardization, Controls, and Calculations

Refer to the chapter on AST.

Discussion

Liver cells have a high concentration of cytoplasmic ALT activity. Two isoenzymes of approximately equal activity are reported in serum, a cationic-migrating form (isoenzyme A) and an anionic-migrating form (isoenzyme B) *(3)*. In hepatic cells, isoenzyme B has 16-fold more activity than isoenzyme A. With liver cell necrosis, the increased activity of serum ALT is primarily due to the increased presence of isoenzyme B. There have been suggestions that the change in ratio of the isoenzymes in the serum may be a better indicator of hepatic cellular necrosis *(4)*. This approach may prove useful in defining active necrosis in the presence of underlying chronic hepatic disease. From a practical point of view, ALT increases have a high degree of specificity as an indicator of acute hepatic cellular necrosis.

Note: Evaluator I. M. O. states that very high ALT activities are seen in patients with muscular dystrophy who show no evidence of hepatic or cellular damage, i.e., normal values for gamma-glutamyltransferase.

The assay system for determining ALT by an indirect measurement of the pyruvate formed can involve a variety of techniques such as colorimetry, ultraviolet spectrophometry, fluorometry, and chromatography. The colorimetric method of Reitman and Frankel *(5)* remains popular, but is limited by the reaction of diphenylhydrazine with other keto acids (causing high blanks), the nonoptimal concentration of substrates, and the nonlinearity of the reaction. The scheme elucidated by Karmen *(6)* for the AST assay was adapted by Henly and Pollard *(7)* to ALT measurement. Modifications by Henry et al. *(2)* have been widely accepted for several years. The conditions of the assay are designed to permit zero-order kinetics. Procedures that are practical and applicable to the clinical laboratory are "one step": serum added to a premixed reagent initiates the reaction.

There are few recommendations on the composition of ALT assay reagents. In 1974, the Scandinavian Society for Clinical Chemistry and Clinical Physiology published a recommended procedure *(8)*. Wilkinson et al. recommended a reference method for ALT involving use of a phosphate buffer *(9)*. Because phosphate inhibits transaminase reactions, phosphate buffer systems are to be avoided; the one included here is used only in the indicator reaction. Bergmeyer et al. have recommended an optimized ALT assay *(10)*, and state that 500 mmol/L DL-alanine is necessary.

Note: The Submitter has found that 400 mmol/L DL-alanine gives an excellent dynamic range and linearity to 550 U/L. The reagent composition presented in this chapter allows sufficient flexibility in the volume fraction of sample to permit the use of a wide variety of manual and semi-automated pipettor-dilutor devices.

The importance of pyridoxal phosphates in transaminase reactions is undisputed. The suggestion that pyridoxal phosphate plays a role in the reactivation of ALT is similar to the recommendation for AST assays *(11)*.

Note: The submitted method does not include the addition of pyridoxal phosphate. Serum specimens assayed without the addition of that coenzyme show good reproducibility, have adequate clinical correlations, and show excellent accuracy when examined in proficiency surveys.

Note: The Submitter has compared the activities of samples with and without the addition of pyridoxal phosphate. Although the addition of pyridoxal phosphate will increase the observed activity of samples with high ALT activity, the effect upon normal-range samples is minimal, and not significant. Furthermore, pyridoxal phosphate must be added to the serum at least 10 min before determination of activity to obtain the maximum activating effect. This requirement causes operational difficulties, one of which is a loss of the advantage of having a one-step assay (adding a sample to a reagent containing all ingredients). Pyridoxal phosphate also imparts a yellow color to the reagent and thereby significantly increases the blank reagent absorbance. The initial absorbance is a valuable indicator in detecting substrate depletion, an effect that could be masked by adding pyridoxal phosphate.

The method described here has the following advantages: *(a)* linear reaction rate, *(b)* primary spectrophotometric standardization (based on the molar absorptivity of NADH), *(c)* constant removal of pyruvate from the reaction mixture (pyruvate is an inhibitor of ALT activity), and *(d)* adequate sensitivity.

The assay conditions as presented are selected with the following considerations:

1. Obtaining maximum ALT activity in the sample with an adequate range practical for the clinical laboratory.
2. Achieving clinically acceptable accuracy and precision.
3. Providing conditions such that minor variations in methodological approach, or in composition of reagents, will cause only minimal differences in assayed activities.

The final reaction mixture has the following composition:

pH (at measurement temperature), 7.5
Volume fraction of sample, 0.052 (1/19)
Tris, 86 mmol/L
DL-Alanine, 400 mmol/L
NADH, 18–20 mmol/L
Lactate dehydrogenase, 2400 U/L
Temperature, 30 °C

Several commercial sources supply reagents, lyophilized or liquid, that give results comparable with those obtained by using an in-laboratory reagent made up to the above specifications. However, many commercial sources continue to use a phosphate rather than a Tris buffer and therefore can be criticized for using a known inhibitor to transaminase assays.

Note: The Submitter prepares a nonlyophilized (liquified) reagent having excellent long-term stability and, once reconstituted as a working reagent, at least 24-h stability at room temperature. The composition of the final reaction mixtures presented by the Submitter has sufficient flexibility to permit additional dilution with water when using semi-automated pipettor-dilutors without jeopardizing analytical requirements.

Calibration of the Spectrophotometer

In the Submitter's experience, the major contributor to interlaboratory variability is failure to calibrate the spectrophotometer. Both wavelength peak and absorbance must be checked against appropriate calibrators. Most instrument manufacturers supply filters for this purpose. See the chapter on AST for further details.

Analytical Variables

See the chapter on AST for a discussion of the importance of the initial absorbance value, the $\Delta A/\Delta t$ intervals, and the linearity of the assay.

Note: For many reasons, variation in sample-to-reagent volume ratio will occur. Differences in that ratio cause changes in assayed activity, but the difference is clinically insignificant.

Reagent and Serum Blanks

There are two types of blank rates. The reagent blank rate represents a linear change in absorbance in the reaction mixture with a water sample replacing serum. The sample blank rate represents a linear change in absorbance in a reaction containing the serum sample but with one of the substrates, DL-alanine or 2-oxoglutarate, removed.

Note: The Submitter's experience indicates that the blank rate for serum is almost always less than 2 U/L and for the reagent less than 1 U/L.

The reagent blank is routinely assayed to assure than no exogenous contamination has occurred during reagent preparation.

Precision Studies and Reference Intervals

A review of the College of American Pathologists (CAP) Special Enzyme Survey *(12)* indicates a broad consensus for the upper limits of normal for ALT of 24 U/L at 30 °C. In the Submitter's laboratory an analysis of 60 sera selected without conscious bias from adults in apparent good health gave a range of 5–19 U/L.

Precision data are based upon assaying two serum specimens, one with normal activity and the other highly active, for 20 consecutive times on each of three different days within one week (Table 1). A lyophilized plasma pool reconstituted daily with

Table 1. Precision Data

	ALT, U/L at 30 °C, mean ± 1 SD
Within-run precision (20 replicates, ×3)	"Normal" serum, 14 ± 3.0
	"High" serum, 255 ± 8.0
Day-to-day precision, (N = 300)	"Normal" pool, 14 ± 4.5
	"High" pool, 60 ± 5.0

sterile, de-ionized water produced greater variability.

Note: Minor variants of the submitted method were used by about 250 participants in the CAP survey. Despite many different instruments and variations in methods used, the reported activities for ALT were in very close agreement.

The sensitivity of this manual enzymic method is limited by the change of absorbance detected during a specific measuring period. Because the spectrophotometer can detect an absorbance change of 0.001/min, sensitivity will be between 3 and 5 U/L. The greater the volume of serum used in the assay, the better the sensitivity.

The Submitter thanks Dr. Bonita Cornett, Mrs. Martha Kerby, and Miss Rosa Tang for their skilled technical assistance, and Mrs. Jane E. Russell for preparation of the manuscript.

References

1. DeRitis, E., Coltorti, M., and Guisti, G., Serum transaminase activity in liver disease. *Lancet* i, 685–687 (1972).
2. Henry, R. J., Chiamori, M., Golub, O. J., and Berkman, S., Revised spectrophotometric methods for the determination of glutamic oxalacetic transaminase, glutamic pyruvic transaminase, and lactic acid dehydrogenase. *Am. J. Clin. Pathol.* 34, 381–398 (1960).
3. Chen, S. H., Giblett, E. R., Anderson, J. E., and Fossum, B. L. G., Genetics of glutamic–pyruvic transaminase; its inheritance, common and rare variants, population distribution, and differences in catalytic activity. *Ann. Hum. Genet.* 35, 401–409 (1972).
4. Ortanos, A. P., Gabriel, E. R., and Pragay, D. A., Separation of glutamic–pyruvic transaminase isoenzymes in human serum. *Res. Commun. Chem. Pathol. Pharmacol.* 1, 266–376 (1970).
5. Reitman, S., and Frankel, J. A., A colorimetric method for the determination of serum glutamic oxalacetic and glutamic pyruvic transaminase. *Am. J. Clin. Pathol.* 28, 56–63 (1957).
6. Karmen, A., A note on the spectrophotometric assay of glutamic oxalacetic transaminase in human blood serum. *J. Clin. Invest.* 34, 131–133 (1955).
7. Henly, K. S., and Pollard, H. M., A new method for the determination of glutamic oxalacetic and glutamic pyruvic transaminase in plasma. *J. Lab. Clin. Med.* 46, 785–789 (1955).
8. The Committee on Enzymes of the Scandinavian Society for Clinical Chemistry and Clinical Physiology,

Recommended methods for determination of four enzymes in blood. *Scand. J. Clin. Lab. Invest.* **33**, 291 (1974).
9. Wilkinson, J. H., Baron, D. N., Moss, D. W., and Walker, P. G., Standardization of clinical enzyme assays: A reference method for aspartate and alanine transaminases. *J. Clin. Pathol.* **25**, 940 (1972).
10. Bergmeyer, H. U., Scheibe, P., and Wahlefeld, A. W., Optimization of methods for aspartate aminotransferase and alanine aminotransferase. *Clin. Chem.* **24**, 58–73 (1978).
11. Lustig, V., Activation of alanine aminotransferase in serum by pyridoxal phosphate. *Clin. Chem.* **23**, 175–177 (1977).
12. College of American Pathologists, Special Enzyme Survey, VA through VD. CAP, Skokie, IL, 1979.

Alcohol (Ethanol) in Serum

Submitter: Charles H. Tripp, Jr., *Toxicology Laboratory, Office of the Medical Examiner, Travis County, Austin, TX 78767*

Evaluators: R. Thomas Chamberlain and Susan Johnson, *Clinical Chemistry and Toxicology Sections, Veterans Administration Medical Center, Memphis, TN 38104*

James L. Verdi, *Clinical Laboratory Section, Veterans Administration Hospital, Reno, NV 89520*

Reviewers: Charles A. Bradley, *Clinical Chemistry Laboratory, Vanderbilt University, Nashville, TN 37232*

Christopher S. Frings, *Cunningham Pathology Associates, Birmingham, AL 35256*

Introduction

The most frequently requested toxicological test received by the clinical laboratory has traditionally been ethyl alcohol (ethanol). Either alone or in combination with other drugs, it may produce coma or death and is a contributory factor to many accidents. In 56 accidental deaths occurring in Travis Co., TX, during the first half of 1980, ethyl alcohol was detected in 30 of them. The values obtained ranged from 0.2 to 3.9 g/L, with a mean value of 1.8 g/L. The analyst must realize that the results of an ethyl alcohol analysis may be used for medical as well as legal purposes. Therefore, good laboratory technique, strict adherence to established quality-control measures, and the maintenance of adequate records are mandatory.

Several methods have been used to determine ethyl alcohol in biological specimens. One of the more popular approaches involves the oxidation of alcohol and subsequent determination of the amount of oxidizing agent consumed. In these redox procedures the alcohol is often separated by diffusion or distillation [1]. Breath analysis for ethyl alcohol is popular among law enforcement agencies because no venipuncture is required and the derived values for concentrations in blood correlate well with more direct quantitation [2].

Gas chromatography, although considered by many toxicologists as the "method of choice," is not available to many clinical laboratories because of the expensive and sophisticated instrumentation required.

Several methods involving the enzyme alcohol dehydrogenase (ADH; EC 1.1.1.1; alcohol:NAD$^+$ oxidoreductase) have already gained popularity in Europe and are becoming more widely used in the United States [1]. The advantages of these enzymic methods for the small clinical laboratory are *(a)* specificity, *(b)* speed of analysis, and *(c)* use of simple equipment.

The procedure selected for the determination of ethyl alcohol is based on the work of Bonnichsen and Theorell [3], in which absorbance of the NADH formed is measured at 340 nm. The methodology and calculations follow the procedure outlined in Sigma Technical Bulletin No. 331-UV [4].

Principle

$$\text{Ethyl alcohol} + \text{NAD}^+ \xrightarrow{\text{ADH}} \text{Acetaldehyde} + \text{NADH}$$

NAD$^+$ = nicotinamide adenine dinucleotide
NADH = nicotinamide adenine dinucleotide, reduced

In the above reaction, alcohol dehydrogenase catalyzes the oxidation of ethyl alcohol to acetaldehyde, which, in turn, forms a complex with the "trapping agent," semicarbazide. If semicarbazide is not added, the reaction proceeds at a slower rate.

The formation of acetaldehyde is favored when the pH of the reaction is approximately 9. An increase in absorbance at 340 nm occurs when NAD$^+$ is converted to NADH, and is directly proportional to the concentration of ethyl alcohol present [4]. In contrast to NADH, NAD$^+$ has maximum absorption at 260 nm.

Material and Methods

Reagents

Reagents may be purchased commercially or prepared in the laboratory. Purchased reagents should not be used after the stated expiration dates. Reagents prepared in the laboratory are stable for 12 months, properly stored.

Note: Manufacturers of reagents for ADH analysis of ethyl alcohol include Sigma Chemical Co., Boehringer-Mannheim Corp., Calbiochem-Behring, and Worthington Diagnostics Corp.

1. *NAD⁺–ADH single assay vial*. This contains 1.8 µmol of NAD⁺ and 150 Sigma units of alcohol dehydrogenase (yeast) in 3.0 mL. Store desiccated vials of NAD⁺ and ADH below 0 °C. Use the NAD⁺–ADH within 1 h after reconstitution with pyrophosphate buffer solution (reagent 3).

Note: 1 Sigma unit converts 1.0 µL of ethyl alcohol to acetaldehyde per minute at pH 8.8 at 25 °C.

2. *Ethyl alcohol standard solution*, 0.8 g/L. Store capped at refrigerator temperature in a glass bottle. For use, bring to room temperature and cap immediately after use to prevent loss of ethyl alcohol.

3. *Pyrophosphate buffer solution*. This contains, per liter, 75 mmol of tetrasodium pyrophosphate ($Na_4P_2O_7 \cdot 10H_2O$), 75 mmol of semicarbazide hydrochloride, and 22 mmol of glycine. Store refrigerated. Close immediately after use, to prevent contamination from ethyl alcohol or other volatiles.

4. *Trichloroacetic acid (TCA) solution*, 62.5 g/L. Store refrigerated in a brown glass bottle.

Collection and Handling of Specimens

Clean the site of venipuncture with a suitable aqueous disinfectant such as benzalkonium chloride (Zephiran), thimerosal (Merthiolate), or mercuric chloride *(5)*. *Never use alcohol or other volatile disinfectants.*

Because well-stoppered and refrigerated specimens may show changes in ethyl alcohol values with time, a preservative should be used. Several kinds of preservatives are adequate, but a mixture of sodium fluoride and potassium oxalate is most preferable *(5)*.

Procedure

1. Pipet into a 15-mL screw-cap culture tube 2.0 mL of TCA solution (reagent 4).
2. Swirl the tube while slowly adding 0.5 mL of test sample or standard. Screw on the top immediately and shake vigorously to disperse any clumps.

Note: Because of the legal implications of alcohol analyses, the Submitter runs all analyses in duplicate.

3. Centrifuge blood samples for 5 min at $1100 \times g$ or until a clear supernate is obtained.
4. While samples are being centrifuged, label the appropriate number of NAD⁺–ADH single assay vials: blank, sample₁, sample₂, standard, etc.
5. Pipet 3.0 mL of pyrophosphate buffer solution (reagent 3) into each vial. Cap and gently invert to dissolve contents.
6. To the vial labeled blank, add 50 µL of distilled or de-ionized water and 50 µL of TCA reagent (reagent 4).
7. To each vial labeled test or standard, add 0.10 mL of clear supernate prepared in steps 1–3.
8. Cap the tubes and invert gently.

9. Allow to stand 45 min at room temperature or place in a 37 °C water bath for 20 min.

Note: The incubation period is not critical. The times indicated are sufficient to permit the reaction to go to completion. As long as the vials remain capped, the reaction mixture may be left as much as 30 min longer than the time indicated.

10. Measure and record the absorbance at 340 nm for the blank and each sample or standard vs distilled or de-ionized water in a 1.0-cm cuvet.

Note: The absorbance of the blank vs distilled water in a 1.0-cm cuvet should be less than 0.10. A higher reading may indicate contamination of the blank or one of the reagents.

Calculations

1. By using a narrow-bandwidth spectrophotometer (<20 nm)—such as the Beckman DU, DK, 25, 34, or 35 series; or the Cary, Zeiss, or Gilford spectrophotometers—and an absorption cell with a 1-cm lightpath, one may calculate the ethyl alcohol concentration as follows:

Ethyl alcohol, g/L = $A_{340nm} \times 1.15$, which is derived from

$$\text{Ethyl alcohol, g/L} = \frac{A_{340nm} \times 3.1 \times 4.6 \times 10^{-5} \times 1000}{6.22 \times 0.02}$$

Where A = absorbance
 3.1 = liquid in absorption cell, mL
 4.6×10^{-5} = 1 µmol of ethyl alcohol, g
 1000 = converts g/mL to g/L
 6.22 = absorptivity of a 1 µmol/mL solution of NADH
 0.02 = mL of specimen in test

Example: In an assay the blank solution read $A = 0.080$. The sample, 1.0 mL of supernate, read $A = 1.020$. Therefore, the absorbance of the sample = 1.020 minus 0.080 = 0.940. Ethyl alcohol (g/L) = $0.940 \times 1.15 = 1.1$ g/L.

2. If a wide-bandwidth spectrophotometer such as a Bausch and Lomb Spectronic 20 or the Coleman Jr. II is used, a calibration curve must be prepared, as follows:

a. Mix together 1.0 mL of water, 1.0 mL of the TCA solution, and 0.5 mL of ethyl alcohol standard solution. Cap this until used.

b. Prepare and number five NAD⁺–ADH vials as indicated:

Vial	Pyrophosphate buffer	Standard from Step *a*	Ethyl alcohol content, g/L
	mL	mL	
1 (blank)	3.10	0.00	0.0
2	3.05	0.05	0.4
3	3.00	0.10	0.8
4	2.95	0.15	1.2
5	2.90	0.20	1.6

c. Allow these tubes to stand for 45 min at room temperature or place in a 37 °C water bath for 20 min.

d. Read and record the absorbance at 340 nm for each standard, using vial 1 as the blank.

e. Using linear graph paper, plot the absorbance values obtained vs the corresponding ethyl alcohol concentration. The plot obtained may not be a straight line but should pass through the origin.

Note: The calibration curve should be prepared any time an analytical parameter is changed, such as use of a new reagent or replacement of a spectrophotometer source bulb.

Note: The reaction mixture is designed to assay ethyl alcohol values of 1.6 g/L or less. If higher values are observed, repeat the assay, using 50, 25, or 10 µL of the supernatant fluid, as appropriate, and multiply the corresponding results by 2, 4, or 10.

Quality Control

1. Run all analyses in duplicate. Values obtained on duplicates should not vary by more than 0.1 g/L. If greater variability is obtained, reanalyze the sample.

2. Using at least 20 assays, establish an acceptable range (mean ± 2SD) for the 0.8 g/L ethyl alcohol standard solution. If the values of the standard for a particular group of analyses falls outside the acceptable range, the values of the analyses should not be reported. Once the source of error for the out-of-range results is discovered and corrected, repeat the analyses of the samples and standard. If the value of the standard is now within the acceptable range, the values of the samples may be reported.

3. If the absorbance of the blank vs distilled or de-ionized water (in a 1.0-cm cuvet) exceeds 0.10, the blank or one of the reagents may be contaminated. The values of the analyses should not be reported. Find and remove the source of the contamination and repeat the analyses; if the blank vs water absorbance value is now less than 0.10, the results of the analyses may be reported.

Interpretation

The phrase "under the influence" of alcohol is often used to describe the effects produced by ethyl alcohol, and may be defined as a state in which an individual has been deprived of the clearness of intellect and self-control that he or she might normally possess *(6)*. It is conceivable, however, that an individual may be "under the influence" of alcohol and not display the behavior one traditionally associates with drunkenness, i.e., slurred speech, stumbling, etc. *(2)*.

The Uniform Vehicle Code, a model law developed jointly by the National Safety Council and the American Medical Association, states the following concerning persons under the influence of intoxicating liquor: At an alcohol concentration of 0.5 g/L or less, an individual is not considered to be affected by alcohol. If the alcohol concentration is 0.5–1 g/L, no presumption is made either way; i.e., other information such as absence or presence of slurred speech, staggering gait, etc. is needed before a decision can be made. At an alcohol concentration of 1 g/L or higher, all individuals are considered to be under the influence of alcohol *(7)*.

Note: Evaluator R. T. C. states that these "presumptions" are "rebuttable presumptions" that are listed in most states' driving statutes; therefore, they really deal with driving under the influence of alcohol. Moreover, these presumptions may vary from jurisdiction to jurisdiction.

At 2.5 g/L or higher, unconsciousness and a marked decrease of response to stimuli are observed. At 4.0 g/L or more, the individual is comatose, body temperature may be subnormal, and death may ensue *(4)*.

The method described is not entirely specific for ethyl alcohol. The alcohols listed in Table 1 are also oxidized by ADH to the degree of reactivity indicated (ethyl alcohol has a reactivity of 100). The other alcohols, however, are usually not present in blood at appreciable concentrations *(4)*.

Table 1. Comparative Reactivity of Alcohols in ADH Assay for Ethanol

Ethyl alcohol	100.0
n-Propyl alcohol	36.0
n-Butyl alcohol	17.5
n-Amyl alcohol	12.5
Isopropyl alcohol	6.6
Methyl alcohol	3.0
Ethylene glycol	3.9
Octyl alcohol	1.3
Glycerol	0.4

Table 2 shows the analytical recovery of alcohol from reference solutions. Values were obtained by the Submitter and by Evaluator J. L. V. Within-day and day-to-day precision are shown.

Table 2. Precision of ADH Assay for Ethyl Alcohol

	n	Range	Mean	SD	CV, %
Within-day					
Evaluator J.L.V.	20	2.43–2.55	2.50	0.03	1.2
Between-day					
Evaluator J.L.V.	20	2.40–2.66	2.52	0.08	3.17
Submitter	118	0.70–0.79	0.75	0.02	2.67
Submitter	36	0.95–1.05	1.00	0.02	3.00

Evaluator J. L. V. analyzed each of 20 samples once by gas chromatography (standard reference method) and three times each over a three-day period by the ADH method. A comparison of the average of the three ADH values and the gas-chromatographic values produced a correlation coefficient (r^2) of 0.994.

References

1. Lundquist, F., The determination of ethyl alcohol in blood and tissues. *Methods Biochem. Anal.* **3**, 218–248 (1957).
2. Harger, R. N., Ethyl alcohol. In *Toxicology, Mechanisms and Analytical Methods,* **2,** C. R. Stewart and A. Stolman, Eds. Academic Press, New York, NY, 1961, pp 137–147.
3. Bonnichsen, R. K., and Theorell, H., An enzymatic method for the microdetermination of ethanol. *Scand. J. Clin. Lab. Invest.* **3,** 58 (1951).
4. Technical Bulletin No. 331-UV, Ethyl alcohol (ethanol), Sigma Chemical Co., St. Louis, MO, revised 1978.
5. Jones, D., Gether, L. P., and Drell, W., A rapid enzymatic method for estimating ethanol in body fluid. *Clin. Chem.* **16,** 402–406 (1970).
6. Steffani vs. State (1935). 45 Ariz. 210, 42 P.2d 615, as quoted on p 266, Arizona Revised Statutes annotated Vol 9A, Titles 28 & 29.
7. *Alcohol and the Impaired Driver.* American Medical Association, Chicago, IL, 1968.
8. Kaye, S., *Handbook of Emergency Toxicology,* Charles C Thomas, Springfield, IL, 1961, pp 172–181.

Addendum

Sigma Chemical Co. has developed a new enzymic assay kit for ethanol (no. 332-UV) that differs from the method reported here, in that glycine replaces pyrophosphate in the buffer, and hydrazine replaces semicarbazide as the trapping agent (*Clin. Chem.* **27**: 1106, 1981, abstract 423). Results reportedly correlate well with those by this method, with either a single-beam ($r = 0.996$) or a double-beam ($r = 0.998$) spectrophotometer (n = 21).

Alkaline Phosphatase, Total Activity in Human Serum

Submitters: George N. Bowers, Jr., and Robert B. McComb, *Clinical Chemistry Laboratory, Hartford Hospital, Hartford, CT 06115*

Evaluators: Arthur Kessner, *Clinical Chemistry Laboratory, Hartford Hospital, Hartford, CT 06115*
Robert Rej and Jean-Pierre Bretaudiere, *New York State Department of Health Laboratories, Albany, NY 12201*

Introduction

Measurements of alkaline phosphatase [international classification: orthophosphoric monoester phosphohydrolase (alkaline optimum), EC 3.1.3.1] *(1)* activity have been used in medicine for more than 70 years and are the subject of well over 10 000 publications *(2)*. Currently, this enzyme ranks just behind the aminotransferases in number of tests ordered annually in clinical laboratories in the United States *(2)* and in Europe (M. Hørder, personal communication). Alkaline phosphatase activity measurements are used either to help in the diagnosis of bone, hepatobiliary, and other diseases or to monitor the treatment of these diseases *(2)*.

Reference values are extremely important for interpretation of alkaline phosphatase results for two reasons. First, because of normal bone growth, the activity values are routinely higher in infants, children, and adolescents than in adults. This age-dependent contribution of skeletal isoenzyme is added to a liver isoenzyme to give the total activity (see Figure 1). Second, the magnitude of the numerical results can vary as much as 100-fold when different methods and scale units are used. For example, various studies (Figure 2, A–L) suggest that healthy newborns one to six days old frequently have alkaline phosphatase activity as much as 2.5-fold that of adults; note, however, that the numerical values range from 1 to 8 Bessey–Lowry–Brock units at 37 °C to as much as 100 to 1000 U/L by the Scandinavian method, also at 37 °C. Therefore, alkaline phosphatase methods for clinical use must be accompanied by *reference values* that clearly take into account the physiological changes observed in the young vs adults.

The procedure given below is an adaptation of the Manual Reference Method used by the joint National Bureau of Standards (NBS) and cooperating laboratories Enzyme Study Group formed with the purpose of assigning a value for alkaline phosphatase activity to NBS/SRM 909 human serum *(3)*. This method differs from that of the study group only in that a two-point absorbance measurement option may be used if equipment for continuous recording of absorbance is not available.

Principle

The formation of the yellow product 4-nitrophenol is measured with a spectrophotometer at 402 nm as the alkaline phosphatase(s) in serum catalyzes the hydrolysis of the colorless substrate 4-nitrophenyl phosphate. This carefully controlled enzymic reaction is maintained in a transphosphorylating buffer, 2-amino-2-methyl-1-propanol (2A2M1P), the final reaction conditions being:

4-Nitrophenyl phosphate	16 mmol/L
2A2M1P	1.0 mol/L
Magnesium	1.0 mmol/L
Reaction pH$_{29.77°C}$	10.5
Reaction temperature	29.77 °C
Volume fraction (sample/total)	0.0164 (1/61)

The reaction rate is monitored only within the first 2.5 min to avoid the effects of subsequent slower rates, thought to be due to the removal of zinc from the enzyme by chelation to 2A2M1P and(or) inhibitory compounds in the buffer.

Materials and Methods

Reagents

Published specifications for 4-nitrophenol *(4)* and 4-nitrophenyl phosphate *(5)* should be met. All other chemical reagents should be American Chemical Society (ACS) reagent grade, and water should meet National Committee for Clinical Laboratory Standards Specifications C-3 *(6)*.

1. *Magnesium solution (3 mmol/L).* Dissolve 650 mg of magnesium acetate · 4H$_2$O in water in a 1-L volumetric flask. Dilute to volume and mix well. This solution is stable indefinitely at 4 °C.

2. *2A2M1P (1.5 mol/L, pH 10.50 at*

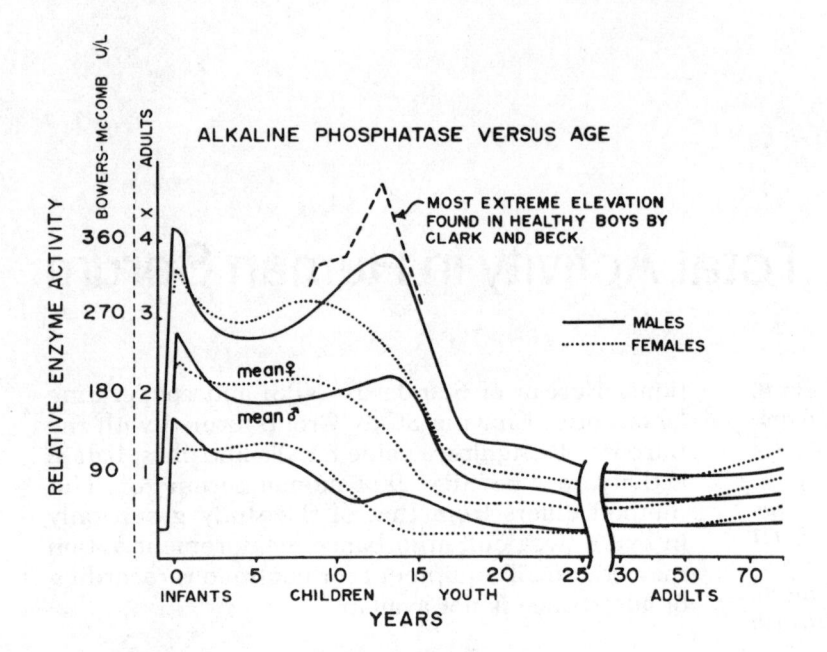

Fig. 1. Age- and sex-dependent reference limits for serum alkaline phosphatase by the method described

The extreme increase for healthy boys (dashed line) was from interpolation of data from L. C. Clark, Jr., and E. Beck, J. Pediatr. 36, 335 (1950)

$29.77\,°C$). Warm 2A2M1P at 30–35 °C until it is completely liquefied. Weigh 135 g of this liquid directly into a 1-L volumetric flask. Add about 500 mL of reagent-grade water, and mix. Transfer 190 mL of 1.000 mol/L HCl to the flask and allow the solution to cool at 25 ± 1 °C. Adjust to volume with reagent-grade water and mix well. At 29.77 °C, this buffer has a pH of 10.50 ± 0.05 when measured after calibrating the pH meter with special "Tris-type" electrodes and NBS buffers. When protected from atmospheric CO_2 and stored at room temperature, this buffer may be used for one month.

Note: 2A2M1P has increasingly been reported to contain alkaline phosphatase "inhibitors"; therefore, for best results each new batch should be tested as described by various authors (7–9, and Evaluator R. R.).

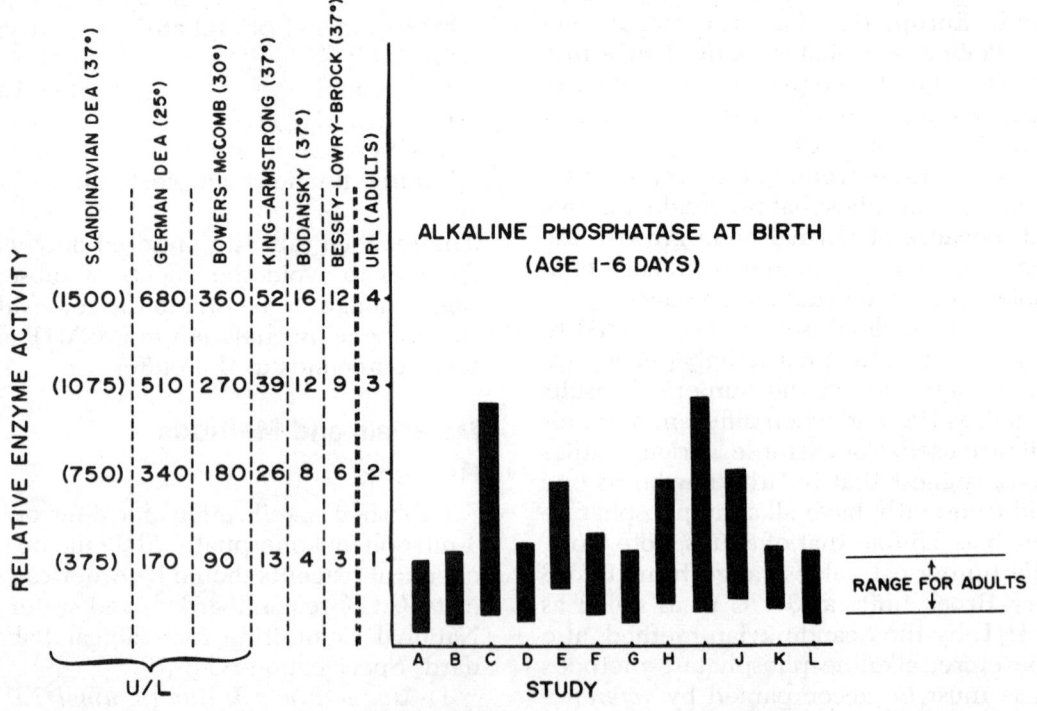

Fig. 2. Ranges of alkaline phosphatase activities determined in neonates one to six days old

The studies A–L are from the literature (see page 533 of reference 2 for details). The six methods listed on the ordinate demonstrate how numerical values can differ by one or two orders of magnitude yet have the same pathophysiological meaning when normalized (see page 532 of reference 2 for more details of methods). URL, upper range limit

Furthermore, the pH measurements at 10 and above in aminopropanol solutions require special "Tris-type" electrodes. These can be obtained from either Sigma Chemical Co., St. Louis, MO 63178, as "Trisma" electrode no. E4878, or from the Beckman Instrument Co., Fullerton, CA 92634, as described in their Applications Research Technical Report No. 542.

3. *4-Nitrophenyl phosphate (24.5 mmol/L in 2A2M1P buffer at pH 10.50).* For each 1 mL of buffered substrate desired, dissolve 9.1 mg of disodium 4-nitrophenyl phosphate hexahydrate in buffer (Reagent 2). Make fresh solutions daily.

Note: The quality of the substrate, disodium 4-nitrophenyl phosphate hexahydrate salt, can vary from lot to lot and from manufacturer to manufacturer. We use the following reagent specifications (see reference 5 for greater details):

Molar absorptivity: At 311 nm and 25 °C in 10 mmol/L NaOH the molar absorptivity, $\epsilon_{311\,nm}^{4NPP}$, should be 9867 L · mol⁻¹ · cm⁻¹. Experience has shown that differences in water content of this hydrated salt can vary by 5%, leading to similar variations in the measured molar absorptivity. Therefore, the actual concentration of 4-nitrophenyl phosphate should be determined by the absorbance of this substrate at 311 nm in NaOH, not by the formula weight of 371.1, because the actual relative molecular mass can be as low as 335 in excellent materials *(5)*.

Test for 4-nitrophenol: The mole fraction of 4-nitrophenol in 4-nitrophenyl phosphate

(mol 4-nitrophenol/mol 4-nitrophenyl phosphate) × 100

should be less than 0.3%. The $A_{432\,nm}$ at 25 °C of a 100 mmol/L solution of 4-nitrophenyl phosphate should be less than 0.300 *(5)*.

Test for inorganic phosphate (Pi): The mole fraction of Pi in 4-nitrophenyl phosphate

(mol Pi/mol 4-nitrophenyl phosphate) × 100

should be less than 1%. This test is easily made by measuring the inorganic phosphate in a 100 mmol/L substrate solution by most methods used daily in a clinical chemistry laboratory to determine Pi in human serum.

4. *Stock standard of 4-nitrophenol solution (1 mmol/L).* Dissolve 139.1 mg of high-purity 4-nitrophenol in water. Dilute to volume in a 1-L volumetric flask with water at 25 ± 1 °C and mix thoroughly. When protected from light, this solution is stable for many months.

Note: High-purity 4-nitrophenol that has been recrystallized from water, dichloromethane, and water *(4)* has been found to have a molar absorptivity, at 401 nm and 25 °C in 10 mmol/L NaOH, 10.00 mm pathlength, $\epsilon_{401\,nm,\,25\,°C}^{4NP}$, equal to 18 380 ± 90 L · mol⁻¹ · cm⁻¹. This can be tested by transferring 25.00 mL of 1 mmol/L 4-nitrophenol stock solution (Reagent 4) to a 1-L flask and diluting to volume at 25 ± 1 °C with 10 mmol/L NaOH. The $A_{401\,nm}^{1\,cm,\,25\,°C}$ of this 25 μmol/L solution after blank correction is 0.460.

5. *Working standard solution of 4-nitrophenol (25 μmol/L in 2A2M1P, pH 10.50 at 29.77 °C).* Place 900 mL of buffer (Reagent 2) into a 1-L volumetric flask and transfer 25.00 mL of stock standard solution (Reagent 4) into this flask. With contents at 25 ± 1 °C, dilute to volume with water and mix. The absorbance of this buffered solution at 402 nm (29.77 °C, 10.00-mm cuvet) is 0.476, corrected for absorbance of the blank (Reagent 2/H₂O, 9/1 by vol). This is equivalent to a molar absorptivity of 19 050 L · mol⁻¹ · cm⁻¹. If the value obtained varies by more than 1% from this value, use the observed value to calculate the activity, if the difference can be rationally explained. The standard in 2A2M1P has been shown to be stable at least for eight weeks.

Apparatus

Any of a number of moderately priced spectrophotometers are suitable for this assay. Minimum requirements are: a spectral bandwidth of less than 8 nm and temperature control within the cell compartment of ± 0.1 °C. A pH meter is needed to establish the buffer pH to ± 0.05 pH units (see Note to Reagent 2, above). Glassware should meet the NBS Class A specifications or be individually calibrated before use. Absorbance changes can be monitored with either a recorder or an accurate stop watch.

Note: The gallium melting point at 29.772 °C has been recommended as the reference temperature for reference methods in clinical enzymology *(10, 11)*. In the United States both 30 and 37 °C have been endorsed as the reaction temperature for assaying the catalytic activity concentration (U/L) *(12)*. Temperature factors for alkaline phosphatase in pooled human sera by this reference method at 25, 30, and 37 °C, relative to the activity measured at the gallium melting point at 29.77 °C, are 0.79, 1.01, and 1.33, respectively, when no correction is made for thermal expansion or temperature effects on absorbance. Similar temperature conversion factors have been found with a crude preparation of alkaline phosphatase from human liver and added to a human serum pool.

Collection and Handling of Specimens

A single venipuncture should be made after a fast of at least 8 h; the subject should be in a sitting position for at least 15 min before venipuncture. Tourniquets may be used if the period of venous occlusion does not exceed 30 s. Heparinized plasma and serum give similar results, but all other anticoagulants or metabolic inhibitors should be avoided *(2)*.

Cells should be separated from serum within 1–2 h after collection. The presence of hemoglobin or bilirubin does not normally influence alkaline phosphatase activity, but the high absorbance blanks associated with these two materials may cause problems in the spectrophotometric measurements *(2)*.

Because alkaline phosphatase activities in human serum increase slowly with time *(13)*, the analysis should be performed as quickly as possible after blood collection. This change is temperature dependent, and if the sample must be stored for periods longer than 2–3 h, storage at low temperature is preferred. Refrigeration of samples at 4 °C is satisfactory for an overnight storage, but storage at −20 °C or lower is preferred for longer periods. Thawing may cause a marked increase in activity with time, an effect called "reactivation." Lyophilized reference materials are even more susceptible to this "reactivation" phenomenon and should be assayed immediately after reconstitution or deliberately held until "reactivation" is complete.

Procedure

1. Into a test tube containing 1.0 mL of Reagent 1, add 50 µL of sample. A calibrated mechanical diluter/pipette or a "to contain" pipette may be used to deliver the sample, followed by two washes of the pipette. After thorough mixing, place the diluted sample in a water bath at 30 °C for exactly 5 min so that magnesium activation and temperature equilibrium can occur before the rate measurement is made.

Note: Volumes of sample and reagents can be reduced by a factor of 2.5 (20-µL sample) without impairing the precision if spectrophotometers with micro-sipper cells are used.

2. Immediately add 2.0 mL of buffered substrate (Reagent 3) previously warmed to 30 °C to the sample/magnesium solution and mix thoroughly.

3. Immediately transfer the reaction mixture to a cuvet (10.00-mm light path). This cell should be capable of bringing the reaction mixture to 29.77 ± 0.05 °C within 30 s after filling and should hold this temperature for the remaining 120 s of the reaction.

4. Read the absorbance change vs time at 402 ± 1 nm against a blank of 2 mL of buffered substrate plus 1 mL of water. Either one of two time-measuring options is available:

(a) Record the absorbance vs time at 402 nm continuously with a recorder or at 10-s intervals for a total of 150 s.

(b) Record the absorbance at 30 s and again at 150 s.

Note: The method is relatively insensitive to variation in wavelength setting or spectral bandwidth when measurements are made with spectrophotometers in which stray light is low. A wavelength-setting error of ± 3 nm with a spectral bandwidth of 5 nm results in an absorbance error of less than 1% *(14)*. Over the past 20 years we have used numerous types of spectrophotometers to follow the time course of the alkaline phosphatase catalyzed reaction: Beckman Instrument Co. Models DU, B, DK-1, and 24; Cary Instruments (Varian Associates) Models 16, 210, and 219 (the last equipped with temperature-controlled sipper cuvet); Gilford Instrument Co. Models 240, 2000, 300N, System 5, and 3500 (the latter three with temperature-controlled sipper cuvet); LKB Instrument Co. Model 8600; and Perkin-Elmer Co. Models 202, 1000, KA150, and Coleman 571. The major difficulty with different instruments has been to set and control temperature within 0.1 °C of the reaction temperature set point.

5. Determine the rate of absorbance change per minute ($\Delta A_{402\ nm}$/min) between 30 and 150 s from the spectrophotometric data by graphic, mathematical, or electronic means.

Calculations

The catalytic activity concentration *(15)* of alkaline phosphatase in human serum measured at 29.77 °C by the above procedure is calculated *(16)* in terms of the International Union of Biochemistry unit (U) per unit volume as follows:

$$U/L = (\Delta A/\Delta t) \cdot (V/\epsilon l v)$$

where $\Delta A/\Delta t$ is the change in absorbance at 402 nm per minute, V is final reaction volume in liters, v is sample volume in liters, ϵ is the molar absorptivity of 4-nitrophenol under the final reaction conditions (19 050 L · mol⁻¹ · cm⁻¹), and l is path length of the cuvet in centimeters.

Therefore,

$$U/L = \frac{\Delta A_{402nm}}{min} \cdot \frac{3.05 \cdot 10^{-3}\ L}{19.05 \cdot 10^3\ [L \cdot mol^{-1} \cdot cm^{-1}]} \times \frac{1}{1[cm] \cdot 5 \cdot 10^{-5}[L]}$$

$$= \frac{\Delta A_{402nm}}{min} \cdot 3202$$

Note: At present in the United States the IUB unit (micromole per minute, U) is the dominant way of reporting catalytic activity, and the catalytic activity concentration in serum is calculated in terms of units per liter (U/L). The IFCC recommends use of the katal (moles per second, kat) to report catalytic activity, and the catalytic activity concentration in serum is calculated in terms of katal per liter (kat/L). The relationship between these two calculations of catalytic activity concentration is U/L = 16.67 nkat/L.

Reference Intervals

The reference intervals for healthy individuals, determined with this method at 29.77 °C, are as follows:

Age, sex	Catalytic activity, U/L
Adults	25–100
Newborns < 1 week	Up to 250
Girls	
1–12 yr	Up to 350
>15 yr	Same as adults
Boys	
1–12 yr	Up to 350
12–15 yr	Up to 500
>20 yr	Same as adults

Discussion

This method is a direct adaptation of the Manual Reference Method for Alkaline Phosphatase, which has been used to assign the enzymic activity value of NBS SRM 909 (human serum) *(3)*. The assay differs slightly from an earlier Selected Method *(17)* in the final reagent conditions. The buffer pH has been raised to 10.5, the substrate concentration has been increased to 16 mmol/L, and the magnesium concentration to 1 mmol/L. The procedural steps, however, have been *markedly altered* so that the serum or other sample is preincubated in 3 mmol/L magnesium solution for at least 5 min and then the activity is measured promptly to avoid prolonged exposure of the enzyme to 2A2M1P *(18)*. "Round-robin" testing in six laboratories established the precision of this method to be 3% (CV) under ideal conditions at 75 U/L (three-fourths the upper reference limit for adults). The precision of daily service analysis with this adaptation will probably be double this figure. Using the two-point measurement of ΔA (option *4b* above), one chemist in our laboratory achieved a within-day precision of 6% (CV) on a serum control pool ($\bar{x} = 44.9$ U/L, n = 10).

The apparent simplicity of this reaction is misleading because of numerous subtle activation and inactivation phenomena *(2)*. Incorrect standardization and failure to maintain proper temperature control also contribute to inaccuracy and lack of precision in this as well as any other enzyme assay.

Many commercial kits are available for alkaline phosphatase assay, but the use of these kits does not circumvent the problems cited above. Indeed, with careful attention to experimental detail, we feel that the procedure described above is a satisfactory method for determining alkaline phosphatase in a small clinical laboratory at minimum cost.

Perhaps the most troublesome aspect of this procedure is the need to stabilize the reaction temperature quickly during the time the absorbance readings are made. A model yet pragmatic system for rapid and accurate temperature control at 29.77 °C ($\pm$ 0.05 °C) that can easily be achieved in any size laboratory has been described *(11)*, and we recommend its wider use in clinical enzymology.

Much clinical confusion in the interpretation of alkaline phosphatase results has been caused when the appropriate reference limits were either not available or ignored. Interpretations of results can be made more meaningful by using the proper reference limits for the patient's age and sex and for the specific method used.

References

1. *Enzyme Nomenclature.* Recommendations (1978) of the Nomenclature Committee of the International Union of Biochemistry, Academic Press, New York, NY, 1979.
2. McComb, R. B., Bowers, G. N., Jr., and Posen, S., *Alkaline Phosphatase*, Plenum Press, New York, NY, 1979.
3. Bowers, G. N., Jr., Cali, J. P., Elser, R., Ewen, L. M., McComb, R. B., Rej, R., and Shaw, L. M., Activity measurements for seven enzymes in lyophilized human serum SRM No. 909. *Clin. Chem.* **26**, 969 (1980). Abstract 048.
4. Bowers, G. N., Jr., McComb, R. B., Christensen, R., and Schaffer, R., High-purity 4-nitrophenol: Purification, characterization, and specifications for use as a spectrophotometric reference material. *Clin. Chem.* **26**, 724–729 (1980).
5. Bowers, G. N., Jr., McComb, R. B., and Upretti, A., 4-Nitrophenyl phosphate—characterization of high-purity materials for measuring alkaline phosphatase activity in human serum. *Clin. Chem.* **27**, 135–143 (1981).
6. Specifications for reagent water used in the clinical laboratory. Publication C3-A, National Committee for Clinical Laboratory Standards, Villanova, PA 19085, 1980.
7. Derks, H. J. G. M., Borrias-van Tongeran, V., Terlingen, J. B. A., and Koedam, J. C., Elimination of inhibitors of alkaline phosphatase from 2-amino-2-methyl-1-propanol. *Clin. Chem.* **27**, 318–321 (1981).
8. Williamson, J. A., and Thompson, J. C., An impurity in the buffer 2-amino-2-methyl-1-propanol, which correlates with depression of measured alkaline phosphatase activity. *Clin. Chem.* **24**, 1611–1613 (1978).
9. Pekelharing, J. M., Noordeloos, P. J., and Leijnse, B., Influence of zinc ions addition to different lots of 2-amino-2-methyl-1-propanol (AMP) buffer on the alkaline phosphatase activities. *Clin. Chim. Acta* **98**, 61 (1979).
10. Mangum, B. W., The gallium melting-point standard: Its role in our temperature measurement system. *Clin. Chem.* **23**, 711–718 (1977).
11. Bowers, G. N., Jr., and Inman, S. R., The gallium melting-point standard: Its application and evaluation for temperature measurements in the clinical laboratory. *Clin. Chem.* **23**, 733–737 (1977).
12. Enzyme assay conditions: Choice of standard reaction temperature for assaying activity values of enzymes in human sera. Publication C10-CR, National Committee for Clinical Laboratory Standards, Villanova, PA 19085, 1980.
13. Massion, C. G., and Frankenfeld, J. K., Alkaline phosphatase: Lability in fresh and frozen human serum and lyophilized control material. *Clin. Chem.* **18**, 366–373 (1972).
14. Schifreen, R. S., and Burnett, R. W., Effects of wavelength error and spectral bandwidth on the measurement of alkaline phosphatase activity. *Clin. Chem.* **25**, 429–431 (1979).
15. Bowers, G. N., Bergmeyer, H. U., Hørder, M., and Moss, D. W., Approved recommendation (1978) on general considerations concerning the determination of the catalytic concentration of an enzyme in the blood serum or plasma of man. *Clin. Chim. Acta* (IFCC Section) **98**, 163F–174F (1979).
16. Provisional Guidelines (1980) for the Calculation and Evaluation of Reaction Rates in Clinical Chemistry. IFCC draft document of August 2, 1980, from Expert Panels on Instrumentation and Expert Panel on Enzymes.
17. Bowers, G. N., Jr., and McComb, R. B., Measurement of total alkaline phosphatase activity in human serum. *Clin. Chem.* **21**, 1988–1995 (1975).
18. Bowers, G. N., Jr., McComb, R. B., and Kelley, M. L., Effect of magnesium on alkaline phosphatase activity in human serum. *Clin. Chem.* **23**, 1149–1150 (1977). Abstract.

Ammonia in Plasma, Enzymic Procedure

Submitters: C. Ray Ratliff, *Department of Pathology, Quillen-Dishner College of Medicine, East Tennessee State University, Johnson City, TN 37601*
Frank F. Hall, *International Clinical Laboratories, Dallas, TX 75247*

Evaluators: Sharon Shu-Hui Lin, *Clinical Chemistry Laboratory, The Childrens' Hospital, Columbus, OH 43205*
Donald R. Pollard and Darshan Singh, *Doctors Hospital, Columbus, GA 31901*

Reviewer: D. Joe Boone, *Clinical Chemistry and Toxicology Section, Centers for Disease Control, Atlanta, GA 30333*

Introduction

High concentrations of ammonia in blood have been associated with severe liver dysfunction for many years; hyperammonemia has been reported in hepatic encephalopathy and in coma resulting from cirrhosis, severe hepatitis, Eck's fistula, and drug hepatotoxicity. Increased blood ammonia has also been reported in cardiac failure, azotemia, and pulmonary emphysema (1, 2). More recently, blood ammonia measurements have gained attention in evaluating abnormalities of urea-cycle enzymes, ketotic hyperglycinemia, ornithinemia, and effects of parenteral nutrition (3, 4). Also, blood ammonia concentration is used increasingly in the diagnosis (and prognosis) of Reye's syndrome.

The full value of blood ammonia measurements has not been realized in the past, perhaps because of equivocal results obtained in some instances, variations in individual tolerances to ammonia intoxication, or poor methodology. With the improved method described in this chapter, we anticipate that increased plasma ammonia values can be better correlated with the pathological condition of the patient, which should enhance the status of this test as a creditable laboratory procedure.

Although determination of plasma ammonia may be a "low-volume" test in many laboratories, it should be made available in all acute-care, diagnostic, newborn, and pediatric facilities because of specimen lability and other problems in transporting specimens to reference laboratories.

Principle

The enzymic determination of ammonia is based on the following reaction, catalyzed by glutamate dehydrogenase [GLDH; L-glutamate:NAD(P)$^+$ oxidoreductase (deaminating), EC 1.4.1.3]:

$$NH_4^+ + \alpha\text{-ketoglutarate} + NADPH \underset{}{\overset{GLDH}{\rightleftharpoons}} \text{L-glutamate} + NADP^+ + H_2O$$

The equilibrium of the reaction lies far to the right, so that when all reactants are in excess except NH_4^+, the amount of NADPH oxidized to $NADP^+$ is stoichiometrically related to the amount of ammonium ion (ammonia) present. The "endpoint" quantity of $NADP^+$ produced is determined by monitoring the decrease in absorbance at 340 nm and by applying the well-characterized absorptivity of NADPH at 340 nm (6.22 L · μmol^{-1} · cm^{-1}). The inclusion of ADP (adenosine 5'-diphosphate) accelerates the reaction toward completion by stabilizing the GLDH at the alkaline pH of the triethanolamine buffer.

This procedure for plasma ammonia is based on a method described by Da Fonseca-Wollheim (5) and is commercially available in kit form from Bio-Dynamics/*bmc*, Indianapolis, IN 46250 (Reagent Set™, cat. no. 125857).

Materials and Methods

Blood Collection

Collect blood from a stasis-free vein into an evacuated tube containing EDTA (Becton Dickinson No. 4204Q Vacutainer Tubes, or a comparable brand), mix gently, place on ice, and deliver immediately to the laboratory. Release the residual vacuum in the collection tube after the blood is drawn, because decreased CO_2 tension in the sample may increase measurable ammonia (6). Avoid hemolysis, and separate plasma from the cells without delay. Plasma may be stored (tightly sealed) up to 2 h in a refrigerator, although the analysis should be done within 30 min. If the procedure cannot be performed within 2 h, freeze the plasma and store in the coldest available freezer; determine the ammonia content as soon as possible. However, under these conditions, an abnormally high value for ammonia is suspect, and analysis should be repeated with a fresh sample at the first opportunity.

Reagents

Note: Although few laboratories will prepare "in-house" reagents for this enzymic procedure, the reagents are outlined here. General biochemical suppliers can provide the reactive ingredients. However, the Submitters' sole experience with this enzymic ammonia procedure has been with reagents supplied in the kit (7).

1. *NADPH (bottle 1 in BMC kit)*. Each bottle contains 0.24 μmol of NADPH. Store refrigerated.

2. *Buffer/substrate (bottle 2 in BMC kit)*. Per liter: 152 mmol of triethanolanine buffer adjusted to pH 8.6 with HCl, 15.2 mmol of α-ketoglutarate, and 1.52 mmol of ADP. Store refrigerated.

3. *Glutamate dehydrogenase (bottle 3 in BMC kit)*. Greater than 750 kU (25 °C) of bovine liver GLDH per liter of glycerol/water (50/50 by vol). Store refrigerated. Use undiluted.

4. *Working reagent*. Reconstitute the entire contents of reagent 1 with 2.5 mL of reagent 2. This working solution should be near room temperature (20–25 °C) before use; it is stable for 24 h if stored refrigerated in a tightly closed container.

5. *Ammonia control solutions*. Frozen plasma pools are not satisfactory for control samples because their ammonia content continuously increases, even if the samples are maintained at super-cold temperatures. However, ammonia in aqueous solutions (5 g/dL) of bovine or human serum albumin (Fraction V), which have been supplemented with ammonium sulfate to make the desired ammonia concentrations, are stable indefinitely at −20 °C or lower (8). Aqueous standard solutions of ammonium sulfate may be used as quality-control specimens and are stable if stored in the freezer.

Note: The Submitters prefer standard solutions of ammonium sulfate for use in quality control of ammonia determinations, at concentrations within and up to threefold the upper limit of the expected range of plasma ammonia. These solutions may be used to check (a) the accuracy of the method, (b) the precision of the method, by means of standard quality-control statistical treatment, (c) the background contamination of reagents with environmental ammonia, and (d) the absorptivity of NADPH at 340 nm.

Note: Ammonium sulfate standard solutions (20 mg of nitrogen per 100 mL) may be obtained from the College of American Pathologists, 7400 N. Skokie Blvd., Skokie, IL 60076, or can be prepared in the laboratory from desiccated, analytical grade ammonium sulfate. Use ammonia-free water and volumetric techniques for all dilutions.

6. *Ammonia-free water*. A supply of amonia-free water must be available for making up standard solutions, diluting specimens, and rinsing any glassware used in the procedure. The current laboratory supply of distilled or de-ionized water may be satisfactory if from a large ion-bed demineralizer or doubly distilled. This enzymic ammonia procedure can be used to check for excessive ammonia contamination of the water source by substituting water for the specimen. Ammonia > 2.5 μmol/L indicates excessive ammonia contamination, and the water should be treated to remove ammonia, as follows: Shake 5–10 g of Permutit (Fisher Scientific) or Dowex-50W (Dow Chemical Co.) with each liter of the best source of distilled or de-ionized water in the laboratory. Allow the resin to settle in the container and use the water from above the settled resin. Use tightly capped, glass containers to store supplies of ammonia-free water. Plastic bottles are unsatisfactory.

Equipment

1. Pipettes or pipetting devices capable of accurately delivering 2.5, 0.5, and 0.02 mL.
2. Cuvets, 1-cm, square, suitable for making spectrophotometric readings at 340 nm.
3. Timing device.
4. A spectrophotometer capable of measuring absorbance at 340 nm.

Procedure

1. Appropriately label three matched cuvets and dispense 2.5 mL of working reagent into each. Pipet 0.5 mL of plasma, and normal and high-concentration ammonia control solutions, into the respective cuvets. Mix by inversion (do not use finger to cap cuvet), and equilibrate at room temperature for at least 10 min but not longer than 15 min. If a sufficient number of matched 1-cm square cuvets are unavailable, one may alternatively add specimens and controls directly to the bottles containing working reagent (reconstituted "bottle 1"). At appropriate times, the contents can be transferred to a single cuvet and absorbance read in the usual manner. Between absorbance readings, pour the cuvet contents back into each respective "bottle 1."

2. After step 1, measure the absorbance (A_1) of each sample at 340 nm against an air blank.

3. Pipet 20 μL of GLDH from reagent 3 into each cuvet, cap, and rapidly mix by inversion.

Note: It is important to measure absorbance after at least 10 min but no longer than 15 min after the GLDH is added to the cuvet. The interval used must be exactly the same for all cuvets within a run.

4. After 10–15 min, read the absorbance (A_2) of each sample again.

5. Make a second addition of 20 μL of GLDH from reagent 3 to all cuvets and measure the absorbance (A_3) at 340 nm again after an additional 10–15 min.

Note: This absorbance value corrects for any changes caused by clearing of lipids by glycerol in lipemic specimens after mixing with the GLDH solutions. The Submitters have found that this step is unnecessary unless the plasma is noticeably turbid or hazy in appearance.

6. The ΔA_S (change of absorbance) of each specimen or ammonia control solution is calculated as follows:

$\Delta A_S = (A_1 - A_2)$ (if no turbidity correction is made)
$\Delta A_S = (A_1 - A_2) - (A_2 - A_3)$
     (if turbidity correction is necessary)

Note: A change in absorbance of a reagent blank (ΔA_{RB}) must be calculated for each different lot no. of (kit) ammonia reagents, to correct for any endogenous (background) ammonia in the reagents. To do this, perform a determination exactly as described in steps 1 through 4 and 6 above, but without adding a specimen or an ammonia control solution to the cuvet. Calculate ΔA for specimen blanks as follows: $\Delta A_{BLK} = (A_{BLK_1} - A_{BLK_2})$. Calculate reagent-blank corrected absorbance as follows: $\Delta A_{RB} = \Delta A_{BLK} \times 0.83$. The factor 0.83 corrects for the change in dilution because no specimen is added to the blank. The Submitters have found this blank value to be significant in all BMC ammonia kits used so far. However, for a given lot of ammonia reagents, a single determination of ΔA_{RB} will usually suffice until a new lot of reagents is obtained. Also, we use two primary ammonia standards for quality control; if the calculated values for ammonia (after reagent blank correction) in these standards are *not* significantly higher than expected values, we assume that the determined ΔA_{RB} is satisfactorily correcting for any background ammonia in the reagents.

7. Calculate the concentration of ammonia in each specimen and ammonia control solutions as follows:

$$\frac{(\Delta A_S - \Delta A_{RB}) \times V_T \times 1000}{\epsilon \times D \times V_S} = NH_3, \mu mol/L$$

where ΔA_S and ΔA_{RB} are obtained as described above, and

V_T = total assay volume (3.02 mL)
1000 = conversion to liter volume
ϵ = molar absorptivity of NADPH at 340 nm
    (6.220 L · μmol^{-1} · cm^{-1})
D = length of lightpath (1 cm)
V_S = specimen volume (0.5 mL)

After substitution, the simplified calculation is:
$(\Delta A_S - \Delta A_{RB}) \times 971 = NH_3, \mu mol/L$

Note: Evaluator S. S.-H. L. used the following calculation: $(\Delta A_S - \Delta A_{RB}) \times 835 = NH_3, \mu mol/L$.
The factor 835 was obtained empirically, from measurements of a series of ammonia standards. The factor 971, above, is based on the theoretical characteristics of NADPH at 340 nm.

Precautions on procedure

1. Round cuvets are unsatisfactory.
2. Keep all tubes, cuvets, and vials covered with Parafilm and store all reagents refrigerated in tightly capped bottles to prevent absorption of ammonia from the environment. Do not perform the procedure in an atmosphere likely to be high in ammonia—from tobacco smoke, concentrated ammonium hydroxide, or ammonia-containing reagents or specimens, or where 24-h urines are measured and poured.

3. Rinse all glassware thoroughly with ammonia-free water.
4. The spectrophotometer used to measure the change in absorbance of the coenzyme NADPH at 340 nm must be in close calibration (within 2%) of the theoretical response to a known coenzyme concentration if accurate ammonia determinations are to be obtained (7). In the Submitters' laboratories, this means that (a) the spectrophotometer is giving satisfactory quality-control values for other kinetic enzyme procedures that depend upon changes in concentration of NADH or NADPH measured at 340 nm, and (b) the values for the primary ammonia control solutions are within 5% of theoretical values.
5. If plasma specimens have ammonia values exceeding 400 μmol/L, dilute them fivefold (1+4) with ammonia-free water and reassay. Multiply results by 5.

Discussion

Method Considerations

Blood and plasma ammonia tests based on Conway's microdiffusion technique date back to 1935 (9), but have suffered from lack of precision and from other technical difficulties. Volatile amines diffusing at the same rate as ammonia are included in the results, and alkaline hydrolysis of protein produces additional ammonia.

Ion-exchange methods, arising from the work of Dienst (10) in 1961, are more accurate than microdiffusion and are relatively easy to perform. Indeed, the development of accurate ion-exchange procedures has contributed significantly to the study of hyperammonemia in defects of the urea cycle.

Enzymic procedures, with their better specificity for circulating ammonia, reportedly eliminate some problems of the microdiffusion and ion-exchange methods (3, 6, 11, 12). The BMC reagent kit reported here is based on NADPH instead of NADH as cofactor for GLDH, thereby reducing preincubation time and the risk of spurious ammonia from conversion of oxidizable substrates by plasma enzymes that use the more general NADH as cofactor (12). Furthermore, the inclusion of ADP in the test system accelerates the reaction toward completion by stabilizing GLDH at the alkaline pH of the triethanolamine buffer (5). Several investigators have reported satisfactory agreement between enzymic and ion-exchange methods (3, 8).

Evaluator S. S.-H. L. evaluated the procedure with a Stasar II spectrophotometer (Gilford Instrument Labs., Oberlin, OH 44074), using a 1-cm lightpath quartz nonthermostated cuvet. Day-to-day and within-day precision are shown in Table 1. Recovery of added ammonia in human serum albumin and fresh plasma was slightly higher than expected [104%: ammonia added to Permutit-treated 5 g/dL

Table 1. Precision of the Method

Material	No.	Mean (SD), µg/dL	CV, %
Day-to-day			
Std., 50 µg/dL	20	47 (3.5)	7.5
Std., 300 µg/dL	20	296 (4.0)	1.4
HSA, 5 g/dL	20	209 (6.4)	3.1
HSA, 5 g/dL + 100 µg/dL std.	20	307 (7.2)	2.3
HSA, 5 g/dL + 400 µg/dL std.	20	596 (6.5)	1.1
Within-day			
Std., 100 µg/dL	20	95 (4.0)	4.0
HSA, 5 g/dL	20	206 (4.8)	2.3
HSA, 5 g/dL + 400 µg/dL std.	20	594 (4.7)	0.8
Std., 200 µg/dL	10	206 (3.4)	1.7
Std., 200 µg/dL	10	205 (2.5)	1.2
HSA, 5 g/dL + 100 µg/dL std.	9	307 (3.8)	1.2

Std., standard; HSA, human serum albumin.

human serum albumin (Fraction V), 99 µg/dL; measured value (n = 6), 103 µg/dL; ammonia added to fresh plasma from donor, 100 µg/dL; measured value (n = 5), 105 µg/dL].

Evaluators D. R. P. and D. S. adapted the procedure to the ABA-100 (Abbott Diagnostics, S. Pasadena, CA 91030). With this method, described in Table 2, 30 tests can be determined in about 20 min. This adaptation is especially advantageous because it reduces specimen (and reagents) consumption to micro-quantities, a most important consideration in laboratories serving pediatric patients.

Table 2. ABA-100 Plasma Ammonia Procedure

ABA-100 instrument settings:

Power	On
Incubator	37 °C
Mode Selector	Rate
Reaction Direction	Down
Analysis Time	5 min
Carousel Revolution	4
Filters	340/380
Zero	Air (0000)
Decimal Setting	0000
Calibration factor	1097

Operation:

Add 250 µL of distilled water into cuvet position 01 and 250 µL of reagent into subsequent cuvets.

Use cuvets 02 and 03 as reagent blanks and add 50 µL of standards or plasma into subsequent cuvets.

Press Run button. In the second revolution, as soon as each cuvet crosses the light path, add 2.0 µL of GLDH, using the SMI pipettor. Mix well and add an additional 2.0 µL of GLDH to the cuvets during the next revolution. The instrument will print on revolutions 3 and 4.

Calculation:

Subtract reading no. 2 from reading no. 1 for each cuvet. Average duplicates.

Multiply Blank reading by 0.83 to obtain adjusted Blank value.

Subtract the adjusted Blank value from all cuvets to yield ammonia concentration in µg/dL.

Notes on Methodology

Although improved accuracy and specificity have been obtained with newer methods, problems associated with specimen collection, stability, physiological conditions, and ammonia contamination remain common to all methods. Increased circulating ammonia may result from exercise and other muscular contractions, either voluntary or involuntary (convulsions). Venous ammonia will not always correlate with arterial values because of ammonia uptake by tissues *(6)*. After collection of the specimen, ammonia in fresh blood rapidly increases on standing; therefore, immediate separation of plasma is necessary. Rapid analysis reduces this source of error and the potential for absorption of environmental ammonia.

The in vivo action of the substances listed in Table 3 may increase ammonia by contributing to the NH_4^+ pool, by interfering with ammonia metabolism, or by reacting with circulating ammonia. Ammonia production (and absorption) by bacteria in the gut are decreased by antibiotics.

Table 3. Drugs and Metabolites Affecting Plasma Ammonia In Vivo (13)

Increased results	Decreased results
Ammonium salts	Arginine
Asparaginase	Glutamic acid
Barbiturates	Isocarboxazid
Blood transfusions	Kanamycin (gut)
Chlorothiazide	*Lactobacillus acidophilus* (gut)
Ethanol	Lactulose
Glucose	Neomycin (gut)
Isoniazid	Potassium
Morphine	Sodium
Oral resins	Tetracycline (gut)
Tetracycline (intravenous)	
Thiazides	

Ammonia concentration of the laboratory environment, glassware, and reagents must be avoided. Traces of formaldehyde remaining after disinfecting de-ionizer components will affect the Berthelot reaction for determining ammonia but not enzymic reactions *(8)*. Plastic bottles should not be used for reagents because ammonia may be leached from or absorbed through the walls of these containers.

Reference Values

There is some variation in reference values reported for plasma ammonia determined by enzymic procedures (see Table 4).

Clinical Considerations

In cirrhotics with gastrointestinal bleeding, increased blood ammonia may be seen, whereas in noncirrhotics with gastrointestinal bleeding the am-

Table 4. Reference Values for Plasma Ammonia (Enzymic Method)

μmol/L (Reference)		μg/dL[a]
	1–35 (8)	17–60
	5–35 (14)	8.5–60
	18–46 (12)	1–78
	10–47 (15)	17–80
	12–47[b]	20–80
Males	16.6–47.3 (7)	28.2–80.4
Females	11.5–38.0 (7)	19.5–64.6
Children (16)		
Newborn[c]	to 171	to 290
"Post-natal"	to 43	to 73
Infant and child	to 91	to 155

[a] Calculated: μmol/L × 1.7 = μg/dL.
[b] Submitters' laboratories.
[c] Values are higher in premature and jaundiced newborns than in full-term and nonjaundiced newborns.

monia concentration is usually normal (2). Psychotic patients when euphoric may have increased blood ammonia and decreased concentrations when depressed. High ammonia values in peritoneal fluid suggest a strangulated, perforated, or lacerated bowel (2).

Hepatic encephalopathy. Hepatic encephalopathy is a disturbance in cerebral function resulting from liver disease and defects in nitrogen metabolism. Increased concentrations of ammonia in blood and cerebrospinal fluid are common to encephalopathy associated with liver disease. Studies of patients with hepatic encephalopathy have correlated hyperammonemia with the degree of neuropsychologic status.

Circulating ammonia in normal individuals is relatively low, even though ammonia is continuously being produced from dietary and tissue amino acid metabolism. Ammonia is liberated from several tissues, including the kidneys and muscle, but most of it arises in and is absorbed from the gastrointestinal tract. It is estimated that more than 4 g of ammonia per day is produced in the jejunum, the colon, and their fecal content (17). In normal individuals, ammonia absorbed from the intestinal tract is transported to the liver for urea synthesis. However, in patients with hepatic disease, gut ammonia reaches the systemic circulation through portal-systemic collateral vessels and increases to toxic values because of impaired liver function (17, 18).

Experimental rats with acute ammonia intoxication have shown a marked decrease in ATP and phosphocreatine in the brainstem, along with development of coma. These studies indicate that brain energy stores are depleted during ammonia intoxication, and the relationship of the mechanism of high-energy phosphate depletion to glutamine synthesis remains under investigation (18). Ammonia uptake by the brain, proportional to arterial concentration, commences when the ammonia in blood exceeds approximately 60 μg/dL, thereby initiating increased synthesis of glutamine. Concentrations of spinal fluid glutamine have been shown to correlate with ammonia concentrations in patients with hepatic encephalopathy, but spinal fluid glutamine remains normal in patients with nonhepatic coma (19). Therefore, measurement of spinal fluid glutamine may assist in the laboratory differentiation of ammonia intoxication in liver disease from other disorders. A small amount of metabolic ammonia is used in glutamine synthesis, but most is removed by conversion to urea in the Krebs–Henseleit cycle (Figure 1).

Other substances implicated in hepatic encephalopathy include methionine, tryptophan metabolites, and short-chain fatty acids (18).

Inborn errors of metabolism—diseases of the urea cycle. The development of the more accurate ion-exchange procedure for plasma ammonia greatly assisted in the discovery and elaboration of inborn errors of metabolism that affect the urea cycle. These studies have played an important role in reinstating the importance of ammonia determinations.

Figure 1 outlines the reaction sequences in the biosynthesis of urea (20). The rate-limiting steps in the synthesis are reactions 1, 3, and 4. Diseases of the urea cycle are associated with almost every step in this pathway. Syndromes in this group that may lead to severe mental retardation are associated with the following steps in the urea cycle:

Reaction 1: Carbamoyl-phosphate synthetase (EC 6.3.4.16) deficiency

(1) $\text{Ammonia} + CO_2 + 2\text{ATP} \xrightarrow[\text{carbamoyl-phosphate synthetase}]{\text{acetylglutamate; Mg}^{++}} \text{carbamyl phosphate} + 2\text{ADP} + P_i$

(2) $\text{Carbamyl phosphate} + \text{ornithine} \xrightarrow[\text{transcarbamylase}]{\text{ornithine}} \text{citrulline} + P_i$

(3) $\text{Citrulline} + \text{aspartate} \xrightarrow[\text{argininosuccinate synthetase (condensing enzyme)}]{\text{ATP} \quad \text{AMP} + PP} \text{argininosuccinic acid}$

(4) $\text{Argininosuccinic acid} \xrightleftharpoons[\text{(cleavage enzyme)}]{\text{argininosuccinate lyase}} \text{arginine} + \text{fumaric acid}$

(5) $\text{Arginine} + H_2O \xrightarrow[\text{Mn}^{++}]{\text{arginase}} \text{urea} + \text{ornithine}$

Fig. 1. Reaction sequences in the urea cycle, showing sites of enzyme action where metabolic blocks induce hyperammonemia
Arginase, EC 3.5.3.1; see text for EC identification of other enzymes

Reaction 2: Ornithine transcarbamylase (EC 2.1.3.3) deficiency

Reaction 3: Argininosuccinate synthetase (EC 6.3.4.5) deficiency

Reaction 4: Argininosuccinate lyase (EC 4.3.2.1) deficiency

In all of these syndromes except argininosuccinic aciduria, greatly increased concentrations of blood ammonia have been found, and the measurement of plasma ammonia is now considered an important diagnostic test in these diseases. Another condition frequently associated with hyperammonemia in the urea cycle is citrullinemia (a defect in reaction 3). In patients with chronic ammonia intoxication, blood urea concentration usually remains normal, even in the presence of markedly increased ammonia (4, 20).

Plasma ammonia has become an established procedure in the laboratory diagnosis of urea-cycle enzymopathies, monitoring the patient during intoxication episodes, and assessing therapeutic results.

Reye's syndrome. We are seeing an increasing number of requests for plasma ammonia determinations in the laboratory diagnosis of Reye's syndrome. This condition was reported in 1963 by Reye et al. (21) in a group of children with acute and usually fatal encephalopathy; then Huttenlocher et al. in 1969 described ammonia intoxication in children with this syndrome (22). A prodromal illness, considered to be of viral etiology with severe liver involvement, Reye's syndrome has been epidemiologically linked to influenza in recent reports from the Centers for Disease Control, and is now recognized as an important cause of morbidity and mortality in children under 18 years of age (23). In cases with markedly increased ammonia in blood, therapy is monitored by serial determinations of plasma ammonia. In this disease, results of aspartate aminotransferase and other liver tests become abnormal, except for bilirubin, which consistently remains normal or only slightly increased. This factor, along with ammonia determinations, helps the laboratory differentiate Reye's syndrome from other liver problems.

References

1. Bessman, S. P., and Bessman, A. N., The cerebral and peripheral uptake of ammonia in liver disease with an hypothesis for the mechanism of hepatic coma. *J. Clin. Invest.* 34, 622–628 (1955).
2. Zimmerman, H. J., Tests of hepatic function. In *Clinical Diagnosis by Laboratory Methods,* I. Davidsohn and J. B. Henry, Eds., 15th ed., W. B. Saunders, Philadelphia, PA, 1974, pp 817–818.
3. Wu, J., Ash, K. O., and Mao, E., Modified micro-scale enzymatic method for plasma ammonia in newborn and pediatric patients; comparison with a modified cation-exchange procedure. *Clin. Chem.* 24, 2172–2175 (1978).
4. O'Brien, D., Ibbott, F. A., and Rodgerson, D., Determination of blood ammonia nitrogen. In *Laboratory Manual of Pediatric Micro-Biochemical Techniques,* Harper & Row, Hagerstown, MD, 1968, pp 39–41.
5. Da Fonseca-Wollheim, F., The significance of the hydrogen ion concentration and the addition of ADP in the determination of ammonia with glutamate dehydrogenase: An improved enzyme determination of ammonia. *J. Clin. Chem. Clin. Biochem.* 11, 421–425 (1973).
6. Kingsley, G. R., and Tager, H. S., Ion-exchange method for the determination of plasma ammonia nitrogen with the Berthelot reaction. *Stand. Methods Clin. Chem.* 6, 115–126 (1970).
7. Package Insert, Reagent Set™ Ammonia, Biodynamics/bmc, Indianapolis, IN, 1977.
8. Doumas, B. T., Hause, L. L., Sciacca, R. D., Jendrzejczak, B., Foreback, C. C., Hoover, J. D., Spencer, W. W., and Smock, P. L., Performance of the DuPont *aca* ammonia method. *Clin. Chem.* 25, 175–178 (1979).
9. Conway, E. J., Apparatus for the microdetermination of certain volatile substances; the blood ammonia, with observations on normal human blood. *Biochem. J.* 29, 2755–2772 (1935).
10. Dienst, S. G., An ion-exchange method for plasma ammonia concentration. *J. Lab. Clin. Med.* 55, 149–155 (1961).
11. Van Anken, H. C., and Schiphorst, M. E., A kinetic determination of ammonia in plasma. *Clin. Chim. Acta* 56, 151–157 (1974).
12. Munir, P. I., Kumar, S., and Willis, C. E., Enzymatic determination of plasma ammonia: Evaluation of Sigma and BMC kits. *Clin. Chem.* 24, 2044–2046 (1978).
13. Young, D. S., Pestaner, L. C., and Gibberman, V., Effects of drugs on clinical laboratory tests. *Clin. Chem.* 21, 252D–253D (1975).
14. Morzdac, A., Ehrlich, G. E., and Seegmiller, J. E., An enzymatic determination of ammonia in biological fluids. *J. Lab. Clin. Med.* 66, 526–533 (1965).
15. Technical Bulletin No. 170-UV, Sigma Chemical Co., St. Louis, MO, 1977.
16. Meites, S., Ed., *Pediatric Clinical Chemistry,* 2nd ed., Am. Assoc. Clin. Chem., Washington, DC, 1981, p 95.
17. Summerskill, W. H. J., and Wolpert, E., Ammonia metabolism in the gut. *Am. J. Clin. Nutr.* 23, 633–639 (1970).
18. Walker, C.O., and Schenker, S., Pathogenesis of hepatic encephalopathy with special reference to the role of ammonia. *Am. J. Clin. Nutr.* 23, 619–632 (1970).
19. Steigmann, F., Kazimi, F., Dubin, A., and Kessane, J., Cerebrospinal fluid glutamine in the diagnosis of hepatic coma. *Am. J. Gastroenterol.* 40, 378–386 (1963).
20. Efron, M. L., Diseases of the urea cycle. In *The Metabolic Basis of Inherited Diseases,* J. B. Stanbury, J. B. Wyngaarden, and D. S. Fredrickson, Eds., 2nd ed., McGraw-Hill, New York, NY, 1960, pp 393–408.
21. Reye, R. D. C., Morgan, G., and Baral, J., Encephalopathy and fatty degeneration of the liver. *Lancet* ii, 749 (1963).
22. Huttenlocher, P. R., Schwartz, A. D., and Klatskin, G., Reye's syndrome. Ammonia intoxication as a possible factor in the encephalopathy. *Pediatrics* 43, 443–454 (1969).
23. Centers for Disease Control, Morbidity and Mortality Weekly Report, 1979, Vol. 28 (4), Atlanta, GA.

Introduction to Amylase Methods

Two methods are presented for the determination of amylase activity. In the starch–iodine procedure soluble starch is the substrate. The reagents are reasonably stable and can be prepared easily in the laboratory. The method is rapid, easy to perform, requires no centrifugation, and is readily adapted to microliter volumes of sample. A major disadvantage is the lack of a suitable reference standard. Commercially available starches may vary from lot to lot and from one manufacturer to the next.

The Cibachron Blue amylase procedure (Amylochrome®) is commercially available as a ready-to-use reagent set. Substrate is provided in the form of stable tablets, along with a buffered diluent and stock standard. This method is also rapid and easy to perform, requires one centrifugation, and is designed for use with small sample volumes.

Because both procedures are technically easy to perform, the choice of method depends largely on other factors. With the starch–iodine method, cost of reagents is negligible and the analyst has control over their preparation. The Amylochrome method has the convenience of prepackaged reagents and includes a standard to ensure consistent results from lot to lot. Either method provides reliable values that the physician may use with confidence as aids in diagnosis and management.

—*Wendell T. Caraway*

Amylase Screening Test (Starch–Iodine Method)

Submitter: Wendell T. Caraway, *Laboratories of McLaren General Hospital and St. Joseph Hospital, Flint, MI 48502*

Evaluator: Ruth D. McNair, *Department of Pathology, Providence Hospital, Southfield, MI 48075*

Reviewer: Patricia H. Duncan, *Clinical Chemistry Division, Center for Environmental Health, Centers for Disease Control, Atlanta, GA 30333*

Introduction

The determination of amylase activity in serum and urine is useful as an aid in the diagnosis of acute pancreatitis. α-Amylase (1,4-α-D-glucan glucanohydrolase; EC 3.2.1.1) catalyzes the hydrolysis of starch to yield primarily intermediate dextrins and maltose. The rate of the reaction may be monitored by measuring either the disappearance of starch by use of the starch–iodine reaction (amyloclastic) or the amount of reducing sugars formed (saccharogenic).

Several dye-labeled substrates have been prepared and introduced commercially. Most of these are synthesized by linking amylose or amylopectin to various dyes. The substrates are attacked by amylase to produce small, water-soluble, dye-containing fragments that can be measured colorimetrically after being separated by centrifugation from the insoluble unreacted substrate.

Continuous-monitoring-type assays (kinetic) have also been developed. Maltotetraose, for example, is split by amylase into two molecules of maltose, which can be coupled to three successive enzymic reactions to generate NADH. The rate of appearance of the latter, after a suitable lag period, is a function of amylase activity.

Results obtained by the amyloclastic and saccharogenic methods usually show fairly good agreement over a wide range of amylase activity (1). Correlation with the dye-labeled or kinetic methods is less satisfactory. Because results by various methods may not be interchangeable, each laboratory should select one procedure and use that method exclusively. The following procedure involves starch as the substrate and is designed for laboratories that wish to prepare their own reagents. For those that do not, a reagent set is available from Harleco, Division of EM Industries, Inc., Gibbstown, NJ 08027, or from other manufacturers that also supply kits based on the amyloclastic procedure. These reagents may be substituted in this method only if the concentrations of reagents are identical to those presented here.

Principle

Starch is hydrolyzed in the presence of amylase to intermediate dextrins and maltose. When serum or urine is incubated with starch substrate under controlled conditions, subsequent addition of iodine produces a blue color with the remaining starch. The decrease in color, compared with that obtained in the absence of the enzyme, provides a measure of amylase activity (1, 2). The basic procedure has been evaluated and published as a Standard Method (3).

Materials and Methods
Reagents

1. *Buffered starch substrate, pH 7.0.* Dissolve 26.6 g of anhydrous disodium phosphate (Na_2HPO_4), 1.75 g of sodium chloride (NaCl), and 8.6 g of benzoic acid (C_6H_5COOH) in 500 mL of water in a large beaker. Bring to boiling. Mix separately 400 mg of soluble starch in 10 mL of cold water and add it, with stirring, to the boiling mixture. Rinse the residual starch into the beaker with added cold water. Continue boiling for 1 min. Allow the solution to cool to room temperature, then dilute to 1000 mL with water and mix well.

This solution, if not contaminated, is stable at room temperature and should remain water-clear. Stability is monitored by noting the absorbance of the standard with each set of determinations. Use care to avoid contamination with saliva or other body fluids. Because starches vary in their molecular mass and the degree of branching within the molecule, results may vary for different starches. The Submitter recommends using soluble starch that meets American Chemical Society (ACS) specifications.

2. *Stock iodine solution, 50 mmol/L.* Dissolve 3.567 g of potassium iodate (KIO_3) and 45 g of potassium iodide (KI) in 800 mL of water. Add slowly and with mixing 9.0 mL of concentrated hydrochloric acid (12 mol/L), and dilute to 1000 mL with water. Store in a brown glass bottle at room temperature. The solution is stable for one year.

3. *Working iodine solution, 5 mmol/L.* Dilute 100 mL of stock iodine solution to 1000 mL with water. Store in a brown glass bottle at 4–8 °C. The solution is stable for two months.

Collection and Handling of Specimens

Serum or heparinized plasma may be used in the assay. Citrate, oxalate, and ethylenediaminetetraacetate reportedly inhibit amylase activity, presumably by complexing with free calcium ion, which is needed for optimum amylase activity. Separate the serum or plasma and refrigerate (4–8 °C) if analysis cannot be completed within 6 h. With the following method, amylase values appear to increase as much as 100 to 200 units when serum is stored at room temperature for three days (see *Calibration* for discussion of amylase units used). This is not a true increase, but is related to an interference with the starch–iodine reaction that occurs with aged serums. Amylase values are stable for at least two weeks when serum is stored at 4 °C.

Amylase in urine is reported to be stable for two weeks at room temperature. Urine may be collected at timed intervals ranging from 1 to 24 h; however, a timed 2-h collection is recommended (4). Measure the volume because the interpretation is based on the total units of amylase excreted per hour.

Procedure

1. Use two tubes graduated at 50 mL (these are available as Folin–Wu N. P. N. tubes from Fisher Scientific Co., Pittsburgh, PA 15238); mark "Test" and "Standard."

Note: The "Standard" in this assay is actually a reagent blank equivalent to 800 units of amylase activity per 100 mL.

2. Pipet 5.0 mL of buffered starch substrate to each tube.

3. Place the tubes in a water bath at 37 °C for 5 min to warm contents.

4. Pipet exactly 0.10 mL of serum or urine into the tube marked "Test," mix, and incubate exactly 7.5 min in the water bath.

Note: An automatic micropipette delivering 100 µL (0.1 mL) is recommended. A glass 0.1-mL Mohr pipette, which delivers between two marks, may also be used. Avoid contaminating the reaction mixture with saliva.

5. After 7.5 min remove both tubes from the bath and immediately add, with mixing, 5.0 mL of working iodine solution. Dilute both tubes to the 50-mL mark with water without delay, cap, and mix well by inversion.

6. Measure the absorbance *(A)* of each solution without delay. Set the spectrophotometer to zero absorbance with water at 660 nm and measure the absorbance of the test and standard with minimal time variation between the two readings.

7. Calculations:

$$[(A_{standard} - A_{test})/A_{standard}] \times 800$$
$$= \text{amylase activity in units/dL}.$$

If the amylase activity exceeds 400 units/dL, repeat the analysis with a fivefold (1 + 4) dilution of serum or urine in water. Multiply the final value by 5. For timed urine collections:

$$[(\text{units/dL}) \times \text{volume, mL}]/(100 \times \text{hours collected}) = \text{units/h}.$$

Calibration

The ratio $(A_{standard} - A_{test})/A_{standard}$ gives the fraction of starch digested by the enzyme during incubation. The amylase unit used here is defined as the digestion of 10 mg of starch in 30 min at 37 °C. In the above method, 5 mL of reagent containing 2 mg of starch is incubated for 7.5 min with 0.1 mL of serum; this is equivalent to incubating 8000 mg of starch with 100 mL of serum for 30 min, so digestion of all the starch would correspond to 800 units/dL. Because the relation between absorbance and activity in the reaction departs from linearity when more than one-half of the starch is digested, the test must be repeated on diluted samples when the absorbance of the test is less than one-half that of the standard, i.e., when the activity exceeds 400 units/dL.

Discussion

All volumes may be reduced proportionately without affecting results. Thus, 1.0 mL of buffered starch substrate, 20 µL of serum, 1.0 mL of working iodine solution, and 8.0 mL of water may be used in the procedure.

As noted above, commercially available lyophilized serum and patient's serum that has been stored at room temperature up to three days may show results 100 to 200 units/dL higher than the true amylase value. This effect can be corrected by including a "serum blank" with zero incubation time as follows: Set up the test as described above, add serum, then *immediately* add working iodine solution and dilute to the mark. The calculation then becomes:

$$[(A_{serum\ blank} - A_{test})/A_{standard}] \times 800 = \text{units/dL}$$

Serum must be added to the starch before the iodine solution.

With fresh or refrigerated sera, the serum blank correction is usually negligible and can be ignored. Values given for commercially assayed control sera are highly method-dependent and may also be related to the source of enzyme and the variability of starch substrate; hence, it is often impossible to relate values between two methods. Each laboratory should establish and use normal reference values for its own procedure.

Maximum absorbance for the starch–iodine complex occurs near 590 nm. At 660 nm the absorbance from iodine is negligible, and readings on the standard are brought closer to center scale. Good agreement with Beer's law is obtained at 590, 660, or 700 nm. Choice of wavelength can be varied, depending on cuvet diameters. There is a slight increase with time in absorbance of the final solution; hence, the test and standard should be set up and read in sequence, with minimum variation in the interval between the two readings.

Amylase activity is proportional to the difference in absorbance between the standard and test. In the low-normal range, this difference between two high absorbance values is subject to more uncertainty, and results in lower precision. Precision improves at higher values and is best in the above-normal range.

Note: Evaluator R. D. M. used 20 µL of serum in the micro-modification with two lyophilized commercial controls. Over a seven-month period, the day-to-day precision studies with one control showed a mean value of 250 (SD 14.7) units/dL (CV 5.9%). For a control with a higher activity, the mean value was 350 (SD 16.5) units/dL (CV 4.7%). Within-day values on three specimens were as follows:

N	Mean (SD), units/dL	CV, %
20	80.2 (5.8)	7.3
24	155.6 (6.7)	4.3
20	218 (9.1)	4.2

Results by the starch–iodine method were found to agree closely with those obtained by the Somogyi saccharogenic method *(1)*; however, comparisons with methods based on dye-labeled or synthetic substrates are generally less satisfactory.

Reference Intervals

"Normal" values. A high percentage of amylase determinations performed in the average clinical laboratory fall within the range of "normal" results. Tabulation of all results for about 200 determinations provides a frequency distribution from which a "normal range" can be estimated. On the basis of four reported studies, the normal range for serum amylase by the above procedure is 40 to 180 units/dL *(1–3, 5)*. Values are not related to age or sex.

> *Note:* Evaluator R. D. M. generally accepts an upper limit of normal of 160–180 units/dL, although values exceeding 200 units/dL have occasionally been observed in patients with no signs or symptoms of pancreatic disease.

Based on a timed 2-h urine collection, normal excretion of amylase is less than 300 units/h *(6)*.

Abnormal values. Salt and Schenker have reviewed the clinical significance of amylase *(7)*. The highest values for serum amylase, up to 5000 units/dL, are encountered in patients with acute pancreatitis. Values increase within 24 to 48 h of the acute onset and generally return to normal within three to five days. Serum amylase activity is usually normal during the quiescent phase of chronic pancreatitis.

Serum amylase may be above normal in renal insufficiency but seldom more than twice the upper limit of "normal." Salivary gland lesions, including mumps or obstruction of the gland ducts, may result in increased serum amylase values averaging around 500 units/dL. Other diseases associated with serum amylase values of up to 1000 units/dL include cholecystitis, cholelithiasis, perforated peptic ulcer, intestinal obstruction, and diabetic ketoacidosis. Increased serum amylase has been found in normal pregnant women during the second and third trimesters, with values as great as twice the upper limit of normal.

Urine amylase excretion normally parallels serum amylase activities but may remain increased for several days after serum amylase activity has returned to normal. Several studies suggest that amylase in a 2-h specimen, expressed as units per hour, is a more sensitive index of the presence of pancreatitis than is serum amylase *(6)*.

In macroamylasemia an amylase–globulin complex is formed that is too large to be excreted readily by the kidneys. This results in an increased serum amylase, but normal or low urinary amylase excretion. The prevalence of this condition in the population is probably between 1 and 2%. This disorder, which is not associated with any clinical symptoms, should be considered in any patient with above-normal serum amylase whose urine amylase excretion (with unimpaired renal function) is normal.

References

1. Van Loon, E. J., Likins, M. R., and Seger, A. J., Photometric method for blood amylase by use of starch–iodine color. *Am. J. Clin. Pathol.* **22**, 1134–1136 (1952).
2. Caraway, W. T., A stable starch substrate for the determination of amylase in serum and other body fluids. *Am. J. Clin. Pathol.* **32**, 97–99 (1959).
3. McNair, R. D., Iodometric measurement of amylase (Caraway). *Stand. Methods Clin. Chem.* **6**, 183–188 (1970).
4. Saxon, E. I., Hinkley, W. C., Vogel, W. C., and Zieve, L., Comparative value of serum and urinary amylase in the diagnosis of acute pancreatitis. *Arch. Intern. Med.* **99**, 607–621 (1957).
5. Flick, A. L., Bark, C. J., and Harrell, J. H., Serum amylase values by the Caraway method in hospitalized patients and normal controls. *Am. J. Clin. Pathol.* **53**, 458–461 (1970).
6. Gambill, E. E., and Mason, H. L., Urinary amylase excretion per hour in 100 individuals without gastrointestinal disease or renal insufficiency. *J. Lab. Clin. Med.* **63**, 173–176 (1964).
7. Salt, W. B., II, and Schenker, S., Amylase—its clinical significance: A review of the literature. *Medicine* **55**, 269–289 (1976).

Amylase Activity in Serum and Urine with a Starch Chromogen

Submitter: Bernard Klein, *Department of Diagnostic Research and Product Development, Hoffmann-La Roche Inc., Nutley, NJ 07110* [Present address: *Department of Laboratory Medicine, Albert Einstein College of Medicine, Bronx, NY 10461*]

Evaluators: Roger L. Forrester, Lester J. Watayi, and William R. Robertson, *Clinical Chemistry Laboratory, Mercy San Juan Hospital, Carmichael, CA 95608*

Thomas E. Hewitt and Gerald Kessler, *Department of Biochemistry, The Jewish Hospital of St. Louis, St. Louis, MO 63110*

Michael M. Lubran and Paul C. Fu, *Harbor-UCLA Medical Center, Torrance, CA 90509*

Richard Pavlovec, H. S. Kim, and Sheshadri Narayanan, *Department of Pathology, Metropolitan Hospital Center, New York, NY 10029*

Diane Wawrzyniak and Steven P. Crause, *J. Spencer Love Clinic, Chapel Hill, NC 27514*

Reviewer: Patricia H. Duncan, *Clinical Chemistry Division, Center for Environmental Health, Centers for Disease Control, Atlanta GA 30333*

Introduction

It has long been known that α-amylase (EC 3.2.1.1, 1,4-α-D-glucan glucanohydrolase) activity in serum is usually increased in acute pancreatitis and parotitis, and the determination of serum α-amylase activity has been extensively studied and accepted as a reliable procedure for the diagnosis of pancreatic disease (1, 2). Most methods available to the clinical chemist for this assay depend on various physical and chemical properties of starch, the natural substrate; similar straight- or branch-chain carbohydrate polymers; or the oligomeric products of the amylolytic process. These methods have been evaluated (3, 4) and chapters on the assay of serum amylase have been published in previous volumes of this series (5, 6).

About 10 years ago, the synthesis of several amylase substrates, by covalently linking "reactive" dyes to starch or its components, amylose and amylopectin, created a new approach to the assay of serum and urinary amylase. These new substrates include Remazol brilliant blue R–starch (7), Cibachron Blue F3GA–cross-linked-starch (8, 9), Reactone Red 2B–amylopectin (10, 11), Cibachron Blue F3GA–amylose (12, 13), and Procion Brilliant Red M2BS–amylopectin (14). The assay reaction involves the hydrolytic scission by α-amylase of the water-soluble or insoluble dye–carbohydrate polymer to form soluble, colored oligosaccharide fragments of as yet unknown structure. Under the controlled conditions of the assay, the absorbances of these soluble colored substances are proportional to the α-amylase activity.

Principle

This procedure is based on the use of the commercially available substrate, Cibachron Blue F3GA–amylose (13). The blue tablet, containing the buffered substrate and sodium chloride, is suspended in water, treated with an aliquot of serum or urine, and incubated at 37 °C for 15 min. The reaction is stopped with an acidic diluent, the products are centrifuged, and the absorbance of the clear-blue supernatant liquid is measured at 625 nm. The amylase activity, expressed as dye units per deciliter, is determined by reference to a calibration curve prepared by diluting the stock dye standard solution. Tablet variation between lots is adjusted by a correlation factor.

Materials and Methods

Reagents (see note)

1. *Synthesis of the substrate (12).* Make a slurry of 500 g of amylose (Superlose; Avebe America Inc., Union, NJ 07083) in 2 L of water, then gradually dilute with an additional 3 L of water with good mechanical stirring. Warm the suspension to 50 °C and treat with a solution of 50 g of Cibachron Blue F3GA (Ciba-Geigy Chemical and Dye Co., Fairlawn, NJ 07410) in 5 L of water. Add 1 kg of anhydrous sodium sulfate gradually over a 15-min period, then

add a warm solution (50 °C) of 50 g of trisodium phosphate in 759 mL of water. Continue heating at 50 °C for an additional 75 min, allow the dyed-amylose to settle, and remove the supernatant fluid. Resuspend the residue in about 5 L of water with vigorous mechanical stirring; allow it to settle and remove the supernate again. Continue washing until the washings are colorless. Then wash the residue similarly with methanol until the washings are colorless. Air-dry. When the product is thoroughly dried, mill it to a 60–200 mesh (yield, 400–450 g of dyed-amylose).

Two hundred milligrams of the substrate are used in each assay.

2. *Phosphate buffer (40 mmol of phosphate and 20 mmol of NaCl per liter), pH 7.0.* Dissolve 5.52 g of $NaH_2PO_4 \cdot H_2O$ and 1.17 g of NaCl in 500 mL of de-ionized or distilled water, adjust the pH at 25 °C with dilute NaOH (0.1 mol/L) and dilute to 1 L at room temperature. The solution is stable for at least two months at 4–8 °C.

3. *Buffered diluent, 0.1 mol/L, pH 4.2–4.3.* Dissolve 13.8 g of $NaH_2PO_4 \cdot H_2O$ in de-ionized or distilled water and dilute to 1 L at 25 °C. The solution is stable for at least two months at 4–8 °C.

4. *Stock standard.* The solution contains 0.2 mg of Cibachron Blue F3GA per liter. The solution is stable for at least four years.

Note: All the reagents described above are available as Amylochrome®, Roche Diagnostics, Div. of Hoffmann-La Roche Inc., Nutley, NJ 07110. Each tablet contains 200 mg of Cibachron Blue–amylose, 40 mmol of phosphate buffer, pH 7.0, and 20 mmol of NaCl. The tablets are stable at room temperature for at least four years.

Apparatus

No special apparatus or equipment other than a reliable incubating bath and spectrophotometer is required.

Note: Evaluators D. W. and S. P. C. encountered problems with a water bath incubator at a temperature 2–4 °C less than the required 37 °C; this, of course, affected the precision of their results. Evaluators M. M. L. and P. C. F. claim a heating block is unsatisfactory because of wide hole-to-hole temperature variation; they recommend a water bath. The Submitter has had considerable experience with both types of incubators and has found them equally satisfactory. Obviously, the temperature variation of any incubating device must be monitored carefully and continuously to ensure accurate and reproducible results.

The Submitter and Evaluators M. M. L. and P. C. F. agree that the use of automatic pipetters for repetitive delivery of fixed volumes of liquids improves the precision of the assay. Delivery of the 100-μL test samples with a piston-displacement-type micropipette is recommended. The Submitter has used Eppendorf (Brinkman Instruments, Inc., Westbury, NY 11590) and MLA (Medical Laboratory Automation, Inc., Mt. Vernon, NY 10550) pipettes; Evaluators M. M. L. and P. C. F. prefer the SMI (SMI, Berkeley, CA 94710), and claim that the Oxford pipette, with disposable plastic tips, is unsatisfactory for the accurate delivery of serum. Evaluators D. W. and S. P. C., however, found the latter acceptable.

All micropipettes should be recalibrated periodically, the frequency depending on the amount of usage.

Collection and Handling of Specimens

Test specimens should be collected in the customary manner for enzyme analysis. Serum and urinary amylase are reportedly stable for one week at room temperature and for several weeks at 4 °C if kept free of bacterial contamination *(15)*.

To obtain a specimen for measurement of urinary amylase activity:
1. Have the patient void.
2. Give the patient 240 mL (8 oz.) of water.
3. Collect and pool all urine voided during the following precisely timed 2-h period. Measure the total volume voided. If urine is collected for less or more than the 2 h, note the exact duration of collection.
4. Perform the amylase assay with 0.1 mL of urine.

Procedure

1. Drop one blue substrate tablet into 1.9 mL of de-ionized or distilled water in a test tube. The tablet will disintegrate rapidly. Mix the resulting slurry thoroughly by vortex-mixing gently and warm to 37 °C.

Notes: If bulk dyed-amylose is used, substitute 200 (± 5) mg of accurately weighed substrate for the tablet and suspend it in 1.9 mL of pH 7.0 buffer (Reagent 2).

The time required to bring the slurry to incubation temperature will have to be determined by the analyst. In the Submitter's laboratory 5 min was usually sufficient.

A 15 × 125 mm tube is recommended. Any excess grainy substrate on the test tube wall is normal and will not interfere with the assay.

2. Add 0.1 mL of serum (or urine) and mix again by vortex-mixing gently.

Note: Take care when pipetting the sample and reagents not to contaminate the test with saliva. Use of mechanical pipetting devices is recommended. If more than one specimen is to be analyzed, allow 0.5 to 1 min between sample additions.

3. Incubate this mixture exactly 15 min at 37 °C.
4. Stop the reaction by adding 8.0 mL of buffered diluent to the tube, mix well, and centrifuge to obtain a clear supernatant fluid. Transfer this clear supernate into a glass cuvet (12 × 75 mm is recommended).

Notes: Evaluators T. E. H. and G. K. recommend that the supernate be transferred to a clean tube with a

Pasteur pipette before measuring the absorbance in a flow-through cuvet. This prevents carryover of interfering particulates.

The minimum centrifugation time required to obtain a clear solution should be determined by the analyst.

Evaluators M. M. L. and P. C. F. used square glass cuvets with a 1.0-cm light path. The Submitter and Evaluators M. M. L. and P. C. F. agree that the disposable square plastic cuvets should not be used. Other Evaluators did not specify the size or type of cuvet they used.

The supernate (separated from precipitate) is stable indefinitely at room temperature.

5. Read the absorbance of the test fluid at 625 nm (red filter) vs the reagent blank. Determine the amylase activity from the calibration curve, including the appropriate correlation factor (see below).

Preparation of a calibration curve. Pipet 1, 2, and 3 mL of the stock standard solution into 10-mL volumetric flasks, dilute to the 10-mL mark with diluent, and mix well. To prepare standards at concentrations of less than 200 dye units/dL, dilute the stock standard appropriately; e.g., for 100 dye units/dL, pipet only 0.5 mL of stock standard into the 10-mL volumetric flask before diluting.

Read absorbance of the three standards against the diluent at 625 nm.

Plot absorbance of the standards against their equivalent concentrations of 200, 400, and 600 dye units/dL on linear graph paper. Typically, with a 1.0-cm cuvet, the standards containing 200, 400, and 600 dye units/dL give absorbance readings of 0.220 ± 0.020, 0.440 ± 0.020, and 0.660 ± 0.020 (range), respectively.

Note: Evaluators M. M. L. and P. C. F. stored diluted standards in the dark at room temperature. They found that absorbance increased by 0.7% for every °C as the refrigerated standard solutions warmed from 10 to 23 °C.

Preparation of reagent blank. Incubate one substrate tablet with 2.0 mL of distilled water for 15 min at 37 °C, add 8.0 mL of diluent solution, and centrifuge. The resulting supernatant fluid is used to adjust the photometer to zero. This solution is stable for two weeks at room temperature.

Note: Refrigeration is not recommended because the solution will slowly turn turbid.

Preparation of a serum (or urine) blank. The absorbance of a serum or urine specimen blank should be measured, especially in the case of a turbid, hemolyzed, or discolored specimen.

Note: Evaluators T. E. H. and G. K. measured the serum blank with each assay and found a mean absorbance of 0.023 (SD 0.0055). This confirms the Submitter's experience; most patients' sera tested did not exceed 0.030, equivalent to about 30 dye units/dL.

To prepare a serum or urine blank, drop one substrate tablet (or the preweighed bulk substrate) into 1.9 mL of distilled water. Mix the resultant slurry by vortex-mixing. Incubate this mixture for exactly 15 min at 37 °C. Add 8.0 mL of diluent to the tube, mix, then add 0.1 mL of serum or urine. Mix well and centrifuge. The resulting supernate is the serum or urine blank.

Calculate results as follows: test absorbance − serum or urine blank absorbance = corrected absorbance. Use the corrected absorbance to obtain amylase activity from the calibration curve.

Use of the Correlation Factor (see note)

The following example and sample calculation demonstrate the application of the reference substrate correlation factor. Assume the lot of tablets in use is found to have an activity of 107% of the reference substrate. Multiplying the results of the assay by the correlation factor of 0.93 (i.e., 100/107) correlates the test assay with the reference substrate assay. Sample calculation:

Test result × lot correlation factor = corrected value

200 dye units/dL × 0.93 = 186 dye units/dL

Note: Each lot of tablets will have its own reference substrate correlation factor. Check each kit before use to make certain that the lot number of the tablets and the lot number of the accompanying correlation factor agree. When a new lot of tablets is received, attach the slip with the new correlation factor for that lot to the calibration reference curve before any assays are attempted. There is no change in the preparation of the calibration curve because it is not influenced by the substrate assay value. Follow the instructions for preparation of the calibration curve as given above.

Calculation of Results

Results are obtained either from the calibration curve or from Beer's law calculation. Thus,

(Concn of standard)/(absorbance of standard) = (concn of unknown)/(absorbance of unknown)

For example, if the absorbance of the 400 dye units/dL standard is 0.440 and the absorbance of the unknown test assay is 0.220, then by the formula: $400/0.440 = x/0.220$; or $x = 0.220 \times 400/0.440$. The concentration of the test specimen would therefore be 200 dye units/dL. These calculations are performed automatically by direct readings from the calibration curve.

Note: If the amylase activity exceeds 600 dye units/dL, repeat the test with 0.1 mL of a saline-diluted specimen. Multiply the activity of the diluted sample (dye units/dL) by the dilution factor to obtain the correct amylase activity.

Urinary amylase results. Use the calibration curve

to determine the dye units liberated by the urine specimen. Results are calculated as follows:

(dye units/mL) × vol (mL) of 2-h urine pool/time in min = dye units/min

Sample calculation: Assume 240 mL of urine pooled from patient has 336 dye units/dL determined from curve (see note), corresponding to 3.36 dye units/mL of urine:

(3.36 × 240)/120 = 6.72 dye units/min

Note: This represents the final corrected dye units/dL. It incorporates any calculations necessary if the specimen required dilution and takes into account the current reference substrate correlation factor.

Quality Control

Either locally prepared human serum controls (16) or commercial enzyme control sera at clinically normal and clinically abnormal concentrations should be used to monitor test performance. The analyst should determine, in replicate, the dye equivalents of each lot of such control sera by the above procedure, including a determination of the serum blank. All subsequent amylase control activity values should be compared with the values so obtained. Quality control records may be expressed in terms of dye equivalents. The serum pool properly subdivided and labeled should be stable for several months under refrigeration.

Note: Evaluators T. E. H. and G. K. found a loss of 12 to 20% of amylase activity of a patients' serum pool stored 30 days at 4 °C. It was not clear whether the serum pool had been subdivided into smaller aliquots as described in reference 16.

Discussion

Quantitative determination of serum α-amylase is an important diagnostic aid not only in acute pancreatitis but also in other disorders involving abdominal areas adjacent to the pancreas, such as perforated peptic and gastric ulcers and obstruction of the intestine, or pancreatic and bile ducts.

Use of traditional serum α-amylase substrates has several deficiencies, notably, long reaction times, interference by glucose, unstable endpoint colors, and poor reproducibility. Developers of the dyed-carbohydrate polymer substrates wanted a more uniform synthetic substrate with characteristics and products that could be more precisely controlled.

Structure of Cibachron Blue F3GA–Amylose

Until recently (17), the structure of the reactive dye, Cibachron Blue F3GA, was proprietary information. The reactive halogen of the substituted monochlorotriazine probably reacts with at least one primary alcohol in the repeating 1 → 4 anhydroglucose unit of amylose (Figure 1). The degree of substitution is probably temperature related (18). Studies repeatedly demonstrate zero-order kinetics in this heterogeneous reaction, as well as a linear relation between the amount of soluble blue products formed and the amount of enzyme used in the assay. The structures of the colored products are unknown, but hydrolytic scission has been shown to occur at the 1,4-anhydro linkage in true amylolytic fashion (12).

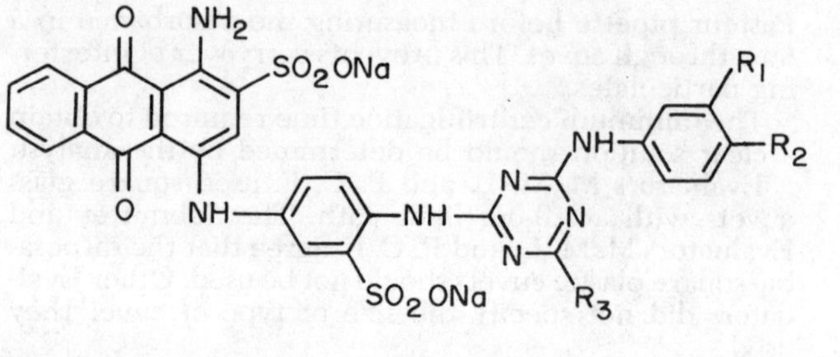

The reaction is stopped with an acidic buffer rather than an alkaline solution to produce lower reagent blank values. Sodium hydroxide solutions, used in similar assays (8, 18) to quench the reaction, convert the sulfonic acid groups to sulfonate anions; this renders the molecule more soluble and hence produces a higher blank.

R_1 = H or SO_2ONa R_3 = Cl = Cibachron Blue F3GA®
R_2 = SO_2ONa or H R_3 = O-Amylose = Amylochrome®

Fig. 1. Dye structures

Units

Uniform expression of serum α-amylase activity is still unsettled. Each technique has its particular unit, and there are no direct relationships among them. The popular Somogyi unit, based on a saccharogenic assay, is the amount of enzyme capable of generating reducing substances, expressed as milligrams of glucose per 100 mL of specimen in 30 min at 40 °C. However, as has often been demonstrated, the short-term reducing product of α-amylase on starch is maltose, which has about half the reducing power of glucose.

The Cibachron Blue F3GA–amylose (Amylochrome) amylase assay is reported in dye units/dL. Early efforts to standardize the assay in terms of the Somogyi unit disclosed that the dye unit was statistically equivalent to approximately 0.63 Somogyi units (13) and correlated well with the Somogyi assay. When attempts were made to calibrate and control the quality of the assay with commercial enzyme control sera containing hog pancreatic amylase, the hog pancreatic amylase yielded only half the soluble colored product given by human pancreatic amylase, leading to gross errors. For this reason, calibration with a dye solution of exactly known composition was instituted and has served well.

Reference Correlation Factor

Synthesis of substrate and tablet manufacture are rigidly controlled, but small lot-to-lot variations in substrate response do occur. To ensure uniform substrate response, especially in quality-control serum assays, the activity of each lot of tablets is compared with a reference substrate, and an activity ratio—the reference correlation factor—is determined. A specific correction factor accompanies each lot of tablets to ensure the reproducibility of results and allow the same dye calibration curve to be used for all lots.

Specimens

Serum and heparinized plasma yield identical results. Plasma containing Ca^{2+}-chelating or precipitating agents such as citrate, oxalate, or ethylenediaminetetraacetate may give lower results because α-amylase is activated by small amounts of Ca^{2+}. Icteric serum may impart a greenish color to the test solution but will not interfere with the validity of the results.

Characteristics of the Method

Precision and reproducibility. In the Submitter's laboratory, recent studies on the precision and reproducibility of the assay gave the following results: At clinically normal concentrations of amylase, within-day precision (mean, SD, and CV) was 91 ± 3 dye units/dL, 3.3%; day-to-day reproducibility, 82 ± 9 dye units/dL, 11.0%. At clinically abnormal concentrations, within-day precision was 269 ± 7 dye units/dL, 2.6%; day-to-day reproducibility, 243 ± 25 dye units/dL, 10.3%.

Note: Results of precision and reproducibility studies by the Evaluators are given in Tables 1 and 2.

Recovery studies. Three evaluating laboratories determined the analytical recovery of amylase added to serum; their results are given in Table 3. One laboratory reported a mean recovery of 97.6% with specimens ranging in activity from 50 to 1024 units/dL. The second laboratory, in a more limited study, reported recovery of 99 and 100% of the amylase activity added. The third laboratory added various amounts of diluted purified hog pancreatic amylase to preassayed control serum and recovered 102, 99.4, and 103.4% of the enzyme activity added.

Correlation studies. Two evaluating laboratories correlated the results of serum assays by the dyed-amylose procedure with results obtained by a coupled-enzymic assay involving a maltotetraose substrate *(19)*. They obtained correlation coefficients *(r)* of 0.984 and 0.934, respectively. Another laboratory correlated the dyed-amylose assay results with those by a nephelometric procedure *(20)* and obtained $r = 0.961$. A fourth laboratory, correlating the results of dyed-amylose assays with those by a modified saccharogenic procedure *(21)* obtained $r = 0.996$.

Table 1. Within-Day Precision Studies

	Mean (SD) dye units/dL	CV, %
Evaluators R. L. F., L. J. W., W. R. R.[a]	126.9 (4.6)	3.6
	212.5 (7.1)	3.4
	325.5 (11.3)	3.5
Evaluators T. E. H., G. K.[b]	132–136 (2.3–10.7)	1.6–7.0
	284–354 (4.3–16.8)	1.3–5.5
Evaluators M. M. L., P. C. F.[c]	240	3.5
	412	3.5
Evaluators D. W., S. P. C.[d]	98.5 (5.9)	6.0
	253.8 (12.7)	5.0
	199.7 (6.8)	3.4

[a] n = 20 assays at each concentration.
[b] Range of values found when eight to 10 specimens were assayed daily on 11 successive days; lower concentration, n = 107; higher concentrations, n = 104.
[c] Precision calculated from measured absorbances; n = 20 each.
[d] Results obtained at each concentration on each of three test days varied with the water-bath temperature; n = 15 each.

Table 2. Day-to-Day Reproducibility

	Mean (SD) dye units/dL	CV, %
Evaluators R. L. F., L. J. W., W. R. R.	111[a] (8.9)	8.0
	203[b] (14.3)	7.0
	291[b] (26.7)	9.2
Evaluators T. E. H., G. K.	144[c] (10.1)	7.0
	315[d] (21.1)	6.7
Evaluators M. M. L., P. C. F.	240	4.8
	412	5.2
Evaluators D. W., S. P. C.	85.6[e] (5.4)	6.3
	231[e] (11.2)	4.9
	191[e] (10.7)	5.6

[a] n = 28. [b] n = 30. [c] n = 107. [d] n = 104. [e] n = 45.

Table 3. Results of Recovery Studies

	Dye units/dL		
Evaluator	Expected	Found	Recovery, %
R. L. F.	51.2	48.7	95.1
	102.4	98.5	96.2
	204.8	207.7	101.4
	409.5	433.5	105.9
	819.1	806.0	98.4
	1023.9	903.7	88.3
			Mean 97.6
T. E. H.	119	121	102
	119	118	99.4
	119	123	Mean 101.6
R. P.	0.275[a]	0.273[a]	99
	0.303[a]	0.303[a]	100
			Mean 99.5

[a] Absorbances measured at 625 nm.

Two laboratories correlated the results of urine assays by the dyed-amylose procedure with the results obtained by the coupled-enzymic assay and calculated $r = 0.990$ and 0.976, respectively.

Linearity. Evaluators D. W. and S. P. C. found good assay linearity to 900 dye units/dL, well beyond the 600 units/dL claimed for the assay. In another laboratory Evaluators R. P., H. S. K., and S. N. found the absorbances produced by assays of diluted sera to be 99.7% of the theoretical absorbances calculated by least-squares regression. They recommended careful dilution of specimens having amylase activity greater than 600 units/dL for accurate results.

Evaluators T. E. H. and G. K. investigated the linearity of assays when a high-activity serum pool was diluted with a low-activity pool or with saline. The experimentally measured activities compared well with the expected or calculated results, indicating good linearity with either method of dilution.

Reference Intervals

Among adults in good health and without a history of pancreatic disease, the range of serum amylase activity was 45 to 200 dye units/dL *(13)*. Normal excretion of urinary amylase has not been established, and random sampling may show wide disparities in values recovered. In a study in which urine was collected in the manner of a renal clearance, the excretion rate ranged from 0.66 to 5.43 dye units/min *(22)*.

Results of this method of collection and calculation can be used directly to calculate amylase clearance/creatinine clearance ratios. The ratios, which are significantly increased in acute pancreatitis, are considered a most sensitive test, though not clearly diagnostic in the presence of renal impairment *(23)*.

References

1. Janowitz, H. D., and Dreiling, D. A., The plasma amylase. Source, regulation and diagnostic significance. *Am. J. Med.* **27**, 924–935 (1959).
2. Schwartz, M. K., and Fleisher, M., Diagnostic biochemical methods in pancreatic disease. *Adv. Clin. Chem.* **13**, 113–159 (1970).
3. Wilding, P., and Dawson, H. F., Human serum amylase: A brief biochemical evaluation. *Clin. Biochem.* **1**, 101–109 (1967).
4. Searcy, R. L., Wilding, P., and Berk, J. E., An appraisal of methods for serum amylase determination. *Clin. Chim. Acta* **15**, 189–197 (1967).
5. Young, N. F., and Kaser, M. M., Amylase. *Stand. Methods Clin. Chem.* **1**, 8–10 (1953).
6. McNair, R. D., Iodometric measurement of amylase (Caraway). *Stand. Methods Clin. Chem.* **6**, 183–186 (1970).
7. Rinderknecht, H., Wilding, P., and Haverback, B. J., A new method for the determination of α-amylase. *Experientia* **23**, 805 (1967).
8. Ceska, M., Hultman, E., and Ingelman, B. C. A., A new method for determination of α-amylase. *Experientia* **25**, 555–556 (1969).
9. Ceska, M., Birath, K., and Brown, B., A new and rapid method for the clinical determination of α-amylase activities in human serum and urine. Optimal conditions. *Clin. Chim. Acta* **26**, 437–444 (1969).
10. Babson, A. L., Kleinman, N. M., and Megraw, R. E., A new substrate for serum amylase determination. *Clin. Chem.* **14**, 802–803 (1968). Abstract.
11. Babson, A. L., Tenney, S. A., and Megraw, R. E., New amylase substrate and assay procedure. *Clin. Chem.* **16**, 39–43 (1970).
12. Klein, B., Foreman, J. A., and Searcy, R. L., The synthesis and utilization of Cibachron Blue–amylose. A new chromogenic substrate for the determination of amylase activity. *Anal. Biochem.* **31**, 412–425 (1969).
13. Klein, B., Foreman, J. A., and Searcy, R. L., New chromogenic substrate for determination of serum amylase activity. *Clin. Chem.* **16**, 32–38 (1970).
14. Sax, S. M., Bridgwater, A. B., and Moore, J. J., Determination of serum and urine amylase with use of Procion Brilliant Red M-2BS–amylopectin. *Clin. Chem.* **17**, 311–315 (1971).
15. Caraway, W. T., Clinical and diagnostic specificity of laboratory tests; effect of hemolysis, lipemia, anticoagulants, medications, contaminants, and other variables. *Am. J. Clin. Pathol.* **37**, 445–464 (1962).
16. Bowers, G. N., Jr., Burnett, R. W., and McComb, R. B., Preparation and use of human serum control materials for monitoring precision in clinical chemistry. *Clin. Chem.* **21**, 1830–1836 (1975).
17. Böhme, H.-J., Kopperschläger, G., Schulz, J., and Hofmann, E., Affinity chromatography of phosphofructokinase using Cibachron Blue F3GA. *J. Chromatogr.* **69**, 209–214 (1972).
18. Ewen, L. M., Synthesis of Cibachron Blue F3GA–amylose with increased sensitivity for determination of amylase activity. *Clin. Chim. Acta* **47**, 233–248 (1973).
19. Pierre, K. J., Tung, K. K., and Nadj, H., A new enzymatic kinetic method for determination of α-amylase. *Clin. Chem.* **22**, 1219 (1976). Abstract.
20. Zinterhofer, L., Wardlow, S., Jatlow, P., and Seligson, D., Nephelometric determination of pancreatic enzymes. I. Amylase. *Clin. Chim. Acta* **43**, 5–12 (1973).
21. Henry, R. J., and Chiamori, N., Study of the saccharogenic method for the determination of serum and urine amylase. *Clin. Chem.* **6**, 434–452 (1960).
22. Klein, B., and Foreman, J. A., Urinary amylase excretion as measured by Amylochrome assay. *Clin. Chem.* **19**, 1226–1227 (1973).
23. Levitt, M. D., Rapoport, M., and Cooperband, S. R., The renal clearance of amylase in renal insufficiency, acute pancreatitis, and macroamylasemia. *Ann. Intern. Med.* **71**, 919–925 (1969).

Aspartate Aminotransferase, AST (Provisional)[1]

Submitter: Thorne J. Butler, *Southern Nevada Memorial Hospital, Las Vegas, NV 89102*

Evaluators: Francisco Civantos, Bernard W. Steele, and Ramesh C. Airan, *Department of Pathology, Jackson Memorial Hospital, Miami, FL 33101*

Sigrid G. Klotzsch, *Medical Products Division, Union Carbide Corporation, Wallingsford, CT 06492*

Introduction

Transaminase reactions, of major interest in cellular metabolism for many years, burst into clinical importance when LaDue, Wroblewski, and Karmen reported in 1954 that serum glutamic oxalacetic transaminase (SGOT) activity is increased after myocardial infarction *(1)*. This enzyme, since renamed aspartate aminotransferase (AST; L-aspartate:2-oxoglutarate aminotransferase, EC 2.6.1.1) catalyzes the transfer of an amine group from L-aspartate to oxoglutarate, forming oxalacetate and L-glutamate.[2] The enzyme is widely distributed in tissues, and has particularly high concentrations in heart muscle, liver cells, and skeletal muscle. The enzyme occurs in two isoenzyme forms that have been purified from human liver tissue and located as cytoplasmic and mitochondrial forms *(2)*. The isoenzymes have different affinities for the assay substrates *(3)*. Human serum contains predominantly cytoplasmic isoenzyme. However, in pathological states, particularly with necrosis of cells, there can be a marked increase of the mitochondrial form *(4)*. An increase of AST activity in the serum generally suggests damage predominantly to heart muscle or liver. However, cellular damage to the kidney and pancreas has also been reported to markedly increase AST activity.

Previous editions of *Selected Methods* have offered both the ultraviolet *(5)* and colorimetric techniques *(6)* for AST analysis. Although several adaptations of the colorimetric method originally described by Reitman and Frankel *(7)* and later modified by others *(8)* have been popular for many years in clinical laboratories, there are so many substantive criticisms of these assay methods that we recommend their abandonment. For the determination of AST, the ultraviolet techniques originally described by Karmen *(9)* have withstood the test of time. Since the original publications by Karmen, a number of other methods have been recommended. In the course of developing these methods, almost every parameter has been varied to obtain optimum assays. Several international groups, including the American Association for Clinical Chemistry (AACC), the International Federation of Clinical Chemistry (IFCC), and the Scandinavian Society of Clinical Chemistry and Clinical Pathology, have extensively studied this assay and have made various recommendations on procedure and reference materials. Apparently, the formulations recommended by the IFCC and the AACC are capable of producing clinically acceptable and reproducible assay data.

Principle

The method presented here is based on recently published information thoroughly summarized in the Second International Symposium on Clinical Enzymology, 1976 *(10)*. The reagents proposed for the method have been used daily in the Submitter's laboratory for the past decade. They are adapted to both an automated (centrifugal analyzer) and a manual system.

As previously stated, several international organizations have chosen the principles of the Karmen method in which the reaction is monitored spectrophotometrically at 340 mm as NADH is converted to NAD^+. The assay system depends on two reactions: the primary AST reaction and the indicator malate dehydrogenase (MD; L-malate:NAD^+ oxidoreductase, EC 1.1.1.37) reaction.

---

[1] *Editors' Note:* This chapter is marked "Provisional," indicating that it may not have been completely evaluated, and has not met the arbitrary criteria of our reviewing process. On the other hand, the method is used daily in the Submitter's laboratory, and is not only clinically useful, but has met current standards of quality for many years.

[2] Nonstandard abbreviations used: AST, aspartate aminotransferase; MD, malate dehydrogenase; LD, lactate dehydrogenase; Tris, tris(hydroxymethyl)methylamine.

Primary aspartate aminotransferase reaction:

$$\underset{\text{Aspartate}}{\begin{array}{c}COO^-\\|\\CH-NH_2\\|\\CH_2\\|\\COO^-\end{array}} + \underset{\substack{\alpha\text{-Keto-}\\\text{glutarate}}}{\begin{array}{c}COO^-\\|\\C=O\\|\\CH_2\\|\\CH_2\\|\\COO^-\end{array}} \underset{\longleftrightarrow}{\overset{AST}{}} \underset{\substack{\text{Oxalo-}\\\text{acetate}}}{\begin{array}{c}COO^-\\|\\C=O\\|\\CH_2\\|\\COO^-\end{array}} + \underset{\text{Glutamate}}{\begin{array}{c}COO^-\\|\\CH-NH_2\\|\\CH_2\\|\\CH_2\\|\\COO^-\end{array}}$$

Indicator reaction:

$$\underset{\substack{\text{Oxalo-}\\\text{acetate}}}{\begin{array}{c}COO^-\\|\\C=O\\|\\CH_2\\|\\COO^-\end{array}} + NADH + H^+ \underset{\longleftrightarrow}{\overset{MD}{}} \underset{\text{Malate}}{\begin{array}{c}COO^-\\|\\CHOH\\|\\CH_2\\|\\COO^-\end{array}} + NAD^+$$

Because the equilibrium of the indicator reaction lies far to the right, the equilibrium of the primary reaction is not relevant: oxalacetate reacts immediately in the indicator reaction.

Materials and Methods

Apparatus

Use a spectrophotometer with the following characteristics:
Parallel-faced cuvet, 10-mm internal light path.
Flow-through feature requiring 0.5–1.0 mL of sample.
Thermoregulated (within 0.1 °C) cuvet.
8-nm bandpass from 320 to 700 nm.

Notes:
1. Evaluator S. G. K. comments that an 8-nm bandpass is at the upper limit of tolerance. Thermoregulated cuvets are absolutely necessary in this type of assay.
2. There is some argument over the selection of temperatures even though the enzyme is stable between 30 and 37 °C. Whether or not to add lactate dehydrogenase (LD) is still disputed. For this method, the Submitter and Evaluators take a neutral position; the potential of adding other enzyme contaminants (such as AST) in the LD reagent is a distinct possibility, and therefore may reduce precision.
3. An electronic monitor, an attractive additional feature for a spectrophotometer, is programmable and can follow the absorbance change per unit time, multiply that change by an appropriate factor, and print the activity in desired units. The improvement in precision and the reduction in analyst fatigue may merit the expense. Attachments manufactured by Syva Co., Palo Alto, CA 94304, and Gilford Instrument Laboratories, Inc., Oberlin, OH 44074, are suitable for these purposes.

Reagents

1. *Stock AST buffer: 90 mmol/L Tris, pH 7.8, and L-aspartic acid, 225 mmol/L.* In a plastic beaker add: 11 g of tris(hydroxymethyl)methylamine (Tris) base (no. T-1378; Sigma Chemical Co., St. Louis, MO 63178), 30 g of L-aspartic acid (Sigma no. A-9250), with approximately 800 mL of de-ionized (distilled) water. Mix until dissolved. With a pH electrode in the solution, add with stirring 2.5 mol/L NaOH until the pH is 7.8. Continue stirring for 30 min. Transfer to a 1-L volumetric flask and, with washings of the beaker, bring to volume with water. Add 1 g of sodium azide, NaN_3 (no. 1953; Mallinckrodt Chemical Co., Hazelwood, MO 63042). Aliquot into capped 100-mL glass or plastic bottles. Autoclave in the bypass (100 °C) mode for 4 min. Allow to cool, and store at 2 to 6 °C. Stable for six months; if stored frozen, shelf life is increased to 18 months.

Notes:
1. If an autoclave is not available, then bring the mixture to a light boil for 1 min, and aliquot as stated above.
2. Evaluator S. G. K. advises discarding any Tris base that is yellowish.
3. Reagents prepared in-house are convenient to use and relatively inexpensive, compared with lyophilized reagents, and are stable for long periods. Additionally, only the volume required for the workload need be prepared. Several commercial sources supply reagents, lyophilized or liquid, that provide results comparable with those obtained with reagents prepared in-house.

2. *Stock nicotinamide adenine dinucleotide (NADH): 11.3 mmol/L in aqueous ethylene glycol.*[3] Add 187 mg of NADH (Sigma no. N-8129) into a 25-mL flask. Make to volume with 30% ethylene glycol AR (Mallinckrodt no. 5001) in de-ionized water (300 mL of ethylene glycol per liter of solution). Store in a freezer at −10 to −15 °C.

3. *Stock malate dehydrogenase (about 500 kU/L) in phosphate-buffered 50% glycerol, pH 7.4.* Into a 1-L flask add 1.179 g of KH_2PO_4 (Mallinckrodt no. 7100), 4.303 g of Na_2HPO_4 (Mallinckrodt no. 7917), and 0.05 g of NaN_3. Make up to volume with de-ionized water and mix to dissolve. Store in refrigerator. Prepare 250 mL of working phosphate-buffered 50% glycerol diluent solution by mixing 125 mL of glycerol AR (Mallinckrodt no. 5092) and 125 mL of the phosphate buffer. Store in refrigerator.

Note: A convenient source of phosphate buffer is the pH 7.384 buffer used as a calibrator of pH in blood-gas instruments. The buffered 50% glycerol is an excellent vehicle for stabilizing enzyme solutions to be stored in the freezer.

To a 20-mL flask or cylinder add 10 kU of MD

---

[3] The use of ethylene glycol as a stabilizing agent of biological materials at low temperatures is covered by U.S. patents no. 3,876,375 and 4,121,905 held by Beckman Instruments, Inc., Fullerton, CA 92364.

in phosphate-buffered 50% glycerol (Sigma no. M-2634). Make to volume with diluent solution. Mix and store at −10 to −15 °C in a freezer. This reagent will not freeze, and its shelf life is six months.

4. *Stock lactate dehydrogenase (about 1.2 MU/L) in phosphate-buffered 50% glycerol, pH 7.4.* Weigh a 5-mg aliquot of lyophilized salt-free LD (Sigma no. L-1254) into a capped 5.0-mL glass bottle. Add 5.0 mL of buffered glycerol diluent. Mix. Store at −10 to −15 °C. This reagent will not freeze, and its shelf life is one month.

5. *Stock 2-oxoglutarate, 1.4 mol, and 82 mmol of Tris per liter of aqueous ethylene glycol.* To 2.0 g of 2-oxoglutaric acid (Sigma no. K-1750) and 1.0 g of Tris base, add 10 mL of the aqueous ethylene glycol (300 mL/L) in de-ionized water. Mix; store in a freezer at −10 to −15 °C.

6. *Working AST assay reagent, pH 7.8: per liter, 87 mmol of Tris, 217 mmol of L-aspartate, 14 mmol of 2-oxoglutarate, 0.18 mmol of NADH, 2000 U of MD, and 2400 U of LD.* To 25 mL of stock buffer add: 0.25 mL of stock 2-oxoglutarate, 0.30–0.50 mL of NADH stock, 0.1 mL of MD stock, and 0.05 mL of LD stock. This reagent is stable for 18 h at room temperature and 36 h at 2 to 6 °C.

Notes:
1. Evaluator S. G. K. urges rechecking the pH of the working AST reagent each time it is prepared.
2. The volume of added NADH will vary with each batch of stock NADH. The actual volume required should give the final working solution an absorbance, when read against water at 340 nm with a 1-cm light path, of between 1.100 and 1.300.
3. Evaluators F. C. and R. C. A. found 0.17 mL of NADH stock to be sufficient.

Collection and Handling of Specimens

Freshly drawn and separated serum from a 12-h fasted individual is the preferred sample, collected in either a plain glass syringe or a Vacutainer Tube® (Beckton Dickinson, Rutherford, NJ 07070). Ethylenediaminetetraacetate, fluoride, and heparin *do not* affect AST activity *(11).* However, plasma samples tend to be more turbid than serum samples, and cause erratic absorbance changes. Oxalate inhibits enzyme activity and therefore should not be used as an anticoagulant.

AST activity in separated serum is stable to 48 h at room temperature, seven days at 2–6 °C, and up to eight months at −20 °C. AST activity preserved in the aqueous ethylene glycol at −15 to −20 °C is stable up to 18 months.[3]

Avoid hemolysis because intracellular AST can increase sample activity up to 10-fold. Serum allowed to stand in contact with the clot apparently elutes reactants from the cells, causing an increased delay or lag time to remove those endogenous reactants (see *Discussion*, linearity) and thereby increases the likelihood of error (Figure 1).

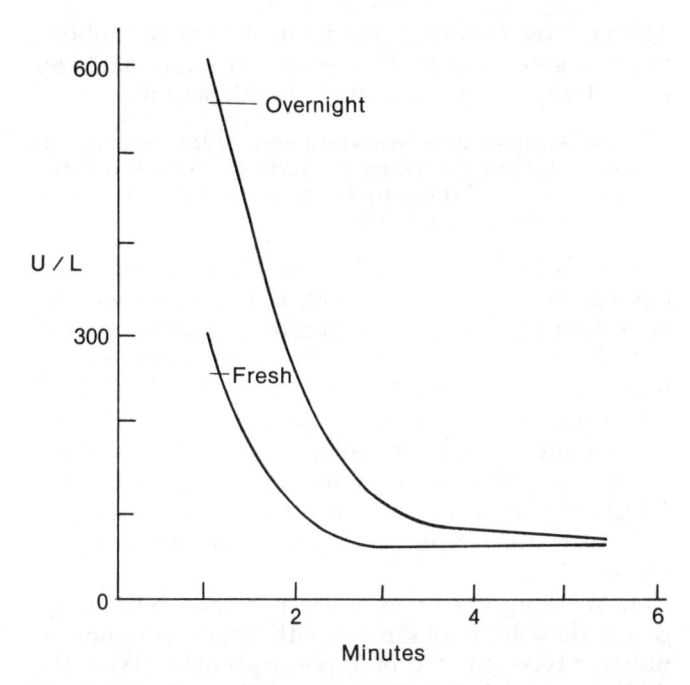

Fig. 1. Time for endogenous reactants to be consumed
Fresh, serum removed soon after clotting; *overnight*, serum in contact overnight with clot

Procedure

This procedure is recommended for a spectrophotometer with a flow-through thermoregulated cuvet.

1. Into three labeled (blank, control, unknown) 1.5- to 2.0-mL polystyrene AutoAnalyzer-type cups (Croan Engineering, Huntington Beach, CA 92649), add with a pipettor-dilutor 60 μL of sample with 100 μL of water rinse. Use water for the blank.

2. Set the spectrophotometer to read between 0 and 200 absorbance units (0.000–0.200) with a water blank at 340 nm. Set the cuvet thermoregulator to 30 °C. (See *Discussion* for calibration of the spectrophotometer.)

3. Working with one cup at a time, add with a pneumatic plunger-type pipettor 1000 μL of premixed reagent. Invert with Parafilm and aspirate the contents into the spectrophotometer cuvet.

4. After 150 s (2.5 min) record the absorbance (see *Discussion* for interpreting this absorbance), then record absorbance every 30 s for the next 180 s (3.0 min).

Note: Evaluators F. C. and R. C. A. found they could begin measuring after only 90 s when using a Gilford Stasar III. This shortening of the lag phase would potentially lengthen the linearity range.

5. Calculate the change in absorbance (ΔA) for each of the 30-s intervals. Select two adjacent 30-s intervals that differ by not more than two absorbance units (0.002) and add; the sum represents the change in absorbance per minute. Multiply the sum by the appropriate factor (see *Calculations*) to obtain total

AST activity. Subtract activity in the blank to obtain the net activity (see *Discussion* on blank activity): net activity = total activity − blank activity.

Note: Samples demonstrating zero-order kinetics will have intervals for which the third to sixth absorbance increments will differ by less than ± 0.002 (see *Discussion* on absorbance intervals.)

The following procedure is recommended for laboratories having spectrophotometers without flow-through features or thermoregulated cuvets:

1. Set the spectrophotometer at zero absorbance with water at 340 nm. Set the heating block at 30 °C.
2. Into cuvets labeled blank, control, and unknown, add an adequate volume of premixed reagent to fill the cuvets while maintaining a final ratio of 50 µL of sample to 1000 µL of reagent.
3. Place the cuvets in the heating block for at least 5 min.
4. Working with one cuvet at a time, add an appropriate volume of sample with either a pneumatic plunger-type pipette or pipettor-dilutor. (With the dilutor, maintain the ratio of sample volume to water rinse at 1/1.) Mix by inversion.
5. Leave the cuvet in the heating block for 150 s (2.5 min).
6. Transfer the cuvet to the spectrophotometer, then read and record the absorbance.
7. Immediately return the cuvet to the heating block.
8. Repeat, determining absorbance every 30 s for a total of six measurements.
9. Calculate the change in absorbance (ΔA) for each of the 30-s intervals. Select two adjacent intervals that differ by not more than 0.002 and add; the sum represents the change in absorbance per minute. Multiply the sum by an appropriate factor (see *Calculations*) to obtain the total activity in the sample. Subtract the activity in the blank to obtain the net activity: net activity = total activity − blank activity.

Standardization

Because standard reference materials for enzyme assays are nonexistent, the molar absorptivity of NADH is the standard for transaminase assays. To assure a quantity of NADH adequate to maintain linearity for a practical range of activity, the absorbance of the blank reagent (reagent plus water blank) should read between 1.100 and 1.300 at 340 nm in a 10-mm cuvet. The concentration of NADH in the stock NADH reagent is not exactly known. Therefore, the analyst must determine experimentally the volume of NADH stock needed to add to the mixed reagent to assure proper absorbance.

Controls

Commercial control materials usually contain normal and two- to threefold the normal range of transaminase activity. The controls are moderately turbid and tend to have more endogenous substrates, thus requiring more preincubation time to reach "zero-order" kinetics than do fresh serum samples.

A control may be prepared to test the dynamic range of assay for values greater than 250 U/L. Obtain lyophilized AST (Sigma no. G-7005) and reconstitute in phosphate-buffered 50% glycerol (see reagent 3); this gives a stock control having an activity of approximately 5×10^4 U/L. Reconstitute a normal control material with phosphate-buffered 25% glycerol (phosphate buffer/glycerol, 3/1 by vol) and add an appropriate volume of stock control to give a working control within the range of 200–300 U/L. Store in a freezer.

Note: Reconstituting control materials in phosphate-buffered 25% glycerol and storing them in a freezer prolongs their shelf life for up to 10 weeks.

Calculations

The International Union of Pure and Applied Chemistry (IUPAC) and others recommend the reporting of enzyme activity in the SI unit termed katal (kat): 1 kat = 1 mol of substrate converted per second per liter of sample. The so-called International Union of Biochemistry unit (U) of enzyme activity per liter of sample is also SI-based and is widely accepted in clinical chemistry: 1 U/L = 1 µmol of substrate converted per minute per liter of sample. With the following formula, the activity of the sample is calculated.

$$U/L = (\Delta A/\min) \cdot (TV/SV) \cdot 10^{-6}/\epsilon$$

where TV = Total volume of reaction
SV = Sample volume
ϵ = Molar absorptivity of NADH
   = 6.21 L · cm^{-1} · mol^{-1}

Note: Evaluator S. G. K. considers a value of 6.3 L · cm^{-1} · mol^{-1} more accurate.

Alternatively, calculate activity as follows:

$$U/L = (\Delta A/\min) \cdot F, \text{ where } F = (TV/SV) \cdot 10^{-6}/\epsilon$$

If absorbance is reported in whole numbers (i.e., 0.010 is expressed as 10), then F = (TV/SV) · 6.21. Convert U/L into katals as follows:

1 U/L = 16.67 nkat
(1 U = katals · 10^{-6}/60 = 16.67 · 10^{-9} kat)

Discussion

The conditions of the assay are designed to cover a reasonably wide range of substrate concentration and enzyme activity while maintaining "zero-order" kinetics. The following conditions will assure zero-order kinetics for the determination of AST activity.

The original method of Karmen, as modified by

Henry *(11, 12)* and recommended by the IFCC *(13, 14)*, requires mixing the serum with reagents not containing oxoglutarate (α-ketoglutarate), and later adding the oxoglutarate to initiate the reaction. Most laboratories find this two-step addition of reagents inconvenient, and prefer a procedure involving a pre-mixed reagent to which the serum sample is added. In the original recommendation secondary reactions involving other keto acids depleted those acids before the oxoglutarate was added. The presence of glutamate dehydrogenase in serum specimens may lead to false increases of measured AST activity *(15)*. Glutamate dehydrogenase catalyzes amination between the ammonium ion and α-ketoglutarate to form glutamate. By incubating the sample with reagents lacking aspartate, it is possible to deplete the assay system of substrates sensitive to that enzyme. Because glutamate dehydrogenase requires ammonium ion as an essential substrate, assay systems should *not* include ammonium sulfate suspensions of enzymes. To assist in the rapid depletion of contaminating keto acids that occur, either in the reagent mixture or in the patient's specimen, most investigators now add an excess of LD.

Adding the controversial pyridoxal phosphate coenzyme has been shown to be important in transaminase reactions involving alpha-amino acids. Pyridoxal phosphate is reversibly transformed from its free aldehyde form to the aminated pyridoxamine phosphate, resulting in an increase in enzyme activity *(16, 17)*. However, the method presented here excludes pyridoxal phosphate. In the Submitter's experience, results with sera assayed without the coenzyme show good reproducibility, are accurate in terms of proficiency surveys, and have adequate clinical correlation.

Note. The Submitter has compared the activities of samples with and without the addition of pyridoxal phosphate. Although the addition of pyridoxal phosphate will increase the observed activity of samples with high AST activity, the effect upon normal-range samples is minimal, and not significant. Furthermore, pyridoxal phosphate must be added to the serum at least 10 min before determination of activity to obtain the maximum activating effect. This requirement causes operational difficulties, one of which is a loss of the advantage of having a one-step assay (adding a sample to a reagent containing all ingredients). Pyridoxal phosphate also imparts a yellow color to the reagent and thereby significantly increases the blank reagent absorbance. The initial absorbance is a valuable indicator in detecting substrate depletion, an effect that could be masked by adding pyridoxal phosphate.

In general, the concentration of the ingredients in the proposed method are only slightly different from those recommended by the AACC and IFCC. However, there is a significant difference in concentrations of LD and MD.

The selected method described here has the following advantages: *(a)* linear reaction rate, *(b)* primary spectrophotometric standardization (the molar absorptivity of NADH), *(c)* constant removal of oxalacetate from the reaction mixture (oxalacetate is an inhibitor of AST activity), and *(d)* adequate sensitivity. The assay conditions presented were selected with the following considerations:

1. To obtain maximum AST activity in the sample with an adequate range practical for the clinical laboratory.
2. To achieve clinically acceptable accuracy and precision.
3. To provide conditions such that minor variations in method or in composition of reagents will cause only slight difference in assayed activities.

Calibration of the Spectrophotometer

A major cause of interlaboratory variability is the failure to calibrate the spectrophotometer. Both wavelength peak and absorbance must be checked against appropriate calibrators. Most instrument manufacturers supply filters for this purpose.

Note: The Submitter prefers a holmium oxide filter (H-153; National Bureau of Standards, Washington, DC) for peak calibration, because it gives a sharp peak at 360 nm. Absorbance calibration is more difficult, even though manufacturers' neutral-density filters perform fairly well. The Submitter suggests using an accepted absorptivity of 10.7 ± 0.1 L · g^{-1} · cm^{-1} for acid dichromate solutions at 350 nm as a preferred method. Preparation and values for these solutions are well reviewed in papers by Rand *(18)* and Chamran and Keiser *(19)*.

Analytical Variables

Initial absorbance

The initial absorbance is an all-important warning of a potential problem in the assay. A low initial absorbance (less than 0.800) suggests a highly active sample and the possibility of substrate depletion. With substrate depletion the $\Delta A/1$-min intervals usually become smaller with each increment, demonstrating non-zero-order kinetics. If activity is extremely high, initial absorbance may be near total substrate exhaustion and the ΔA/min for each interval may appear factitiously constant. Therefore, whenever the initial absorbance is less than 0.800, the assay should be repeated with a fivefold (1 + 4) dilution of the sample.

A high initial absorbance (greater than 1.600) is generally caused by turbid or jaundiced samples. If the background absorbance is sufficiently great, the fact that a sample has high activity may be masked even after substrate depletion has occurred. The phenomenon is relatively uncommon but the analyst should be alert to the possibility. In this situation the Submitter recommends repetition of the assay with a fivefold (1 + 4) dilution of the sample.

Absorbance intervals

Ideally, all $\Delta A/\Delta t$ interval readings should remain within ± 0.001; however, this does not occur in routine practice. One waits for the reaction velocity to become constant. Unfortunately, with samples of low activity, ΔA of ± 0.002 occurs between intervals, thereby causing a calculated change of ± 3–8 U/L between intervals. To minimize this variation, select two adjacent intervals in which absorbances do not differ by more than 0.002/30 s, and sum these two values.

Wide variance between intervals suggests dirty cuvets, a bubble in the light path, turbid samples, or a highly active sample. Samples containing substantial endogenous reactants will take longer to reach equilibrium and in some cases may never show true zero-order kinetics.

Linearity

Linearity in the enzyme assays refers to the capacity of the assay system to give a response that is linear with activity over a wide range of sample activities. Check linearity by diluting a highly active sample (e.g., 200–400 U/L) with a normal serum. Plot observed activity against the percentage of dilution. Although the plot will have a negative x-intercept, the points should form a straight line. Reagent problems can be quickly detected by this procedure. Periodic linearity checks assure linear responses to enzyme activity with this assay system.

Note: For many reasons variations in the sample-to-reagent volume ratio will occur. Differences in that ratio cause significant changes in assayed activity; fortunately, however, the difference is often *clinically insignificant*. Table 1 summarizes the effect of changing the sample-to-reagent volume ratio for two sera—one with normal activity, and the other with high activity. These observations are supported by the report of Bergmeyer et al. (20).

Reagent and Serum Blanks

There are two types of blank rates. The reagent blank rate represents a linear change in absorbance in the reaction mixture, with a water sample replacing serum. The sample blank rate represents a linear change in absorbance in a reaction containing the serum sample but with one of the substrates, either L-aspartate or 2-oxoglutarate, removed.

Note: The blank rates for serum are almost always less than 2 U/L, and the rate for the reagent blank is less than 1 U/L.

The reagent blank is routinely assayed to assure that no exogenous contamination has occurred during reagent preparation.

Precision Results and Reference Intervals

A review of the College of American Pathologists (CAP) Special Enzyme Survey, 1979 (21), indicates a broad consensus for the upper limits of normal AST values in adults 27 U/L at 30 °C. In the Submitter's laboratory an analysis of 155 sera selected without conscious bias from adults in apparent good health gave a range of 10–29 U/L.

Precision studies were made on two sera, one having normal activity and the other having increased activity. Analyses were performed on these sera 20 times on three different days, within one week. A lyophilized plasma pool, reconstituted daily with sterile, de-ionized water, was also analyzed and showed greater variability (Table 2).

Table 1. Effect of Sample/Reagent Volume Ratio on AST Activity Detected

Sample/Reagent[a]	Ratio	AST, U/L Normal acty	High acty
20/100	0.0178	16	320
30/100	0.0265	15	318
50/100	0.0434	14	310
60/100	0.0517	14	308
80/200	0.0625	13	303
100/300	0.0769	13	303

[a] All volumes in microliters. Reagent volume = 1000 µL.

Table 2. Precision of the Method

	Within-run				Day-to-day			
	N	$\bar{x}$ (U/L at 30 °C)	SD	CV, %	N	$\bar{x}$ (U/L at 30 °C)	SD	CV, %
Submitter								
Normal serum	3	22.0	1.5	6.8	300	18.0	2.25	12.5
High serum	3	158.0	3.5	2.2	300	91.0	3.00	3.3
Evaluators F. C. and R. C. A.								
Normal serum	20	18.2	1.8	9.8	21	16.6	2.28	13.7
High serum	20	66.1	2.4	3.6	21	62.0	5.67	9.1

Minor variations of the submitted method were used by about 250 participants in the CAP survey. Despite many different instruments and probable variation in methods, the reported activities for AST agreed closely.

Note: Evaluators F. C. and R. C. A. report average analytical recovery of 96% for 0.1-mL aliquots of a 635 U/L AST solution added to 1 mL of a "normal" or an abnormal serum. When they compared results by this method (y) with results with an Abbott ABA 100 (x), the linear regression equation was $y = 0.87\ x + 4.01$. Evaluator S. G. K. compared this method with results with a Union Carbide CentrifiChem (x) and found that $y = 12.5\ x - 2.0\ (r = 0.9967)$.

The sensitivity of this manual enzymic method is limited by the change of absorbance detected during a specific measuring period. Because the capability of the spectrophotometer can detect an absorbance change of 0.001/min, this will result in a sensitivity of between 3 and 5 U/L. The greater the volume of serum used in the assay, the better the sensitivity.

The Submitter thanks Dr. Bonita Cornett, Mrs. Martha Kerby, and Miss Rosa Tang for their skilled technical assistance, and Mrs. Jane E. Russell for preparation of the manuscript.

References

1. LaDue, J. S., Wroblewski, F., and Karmen, A., Serum glutamic oxaloacetic transaminase in human acute transmural myocardial infarction. *Science* **120**, 497–499 (1954).
2. Wilkinson, J. H., *Isoenzymes*, J. B. Lippincott Co., Philadelphia, PA, 1966, pp 95–103.
3. Rej, R., and Vanderlinde, R. E., Evaluation of assays of aspartate aminotransferase activity with use of human enzyme reference materials. *Clin. Chem.* **21**, 1028 (1975). Abstract.
4. Schmidt, E., and Schmidt, F. W., In *Methods of Enzymatic Analysis*, H. U. Bergmeyer, Ed., 2nd English ed., Academic Press, New York, NY, 1974, pp 7–8.
5. Friedman, M. M., and Taylor, T. H., Transaminase. *Stand. Methods Clin. Chem.* **3**, 207–216 (1961).
6. Sax, S. M., and Moore, J. J., Glutamic oxaloacetic transaminase (colorimetric). *Stand. Methods Clin. Chem.* **6**, 149–158 (1970).
7. Reitman, S., and Frankel, J. A., A colorimetric method for the determination of serum glutamic oxaloacetic and glutamic pyruvic transaminase. *Am. J. Clin. Pathol.* **28**, 56–63 (1957).
8. Babson, A. L., Shapiro, P. O., Williams, R. A. R., and Phillips, G. E., The use of a diazonium salt for the determination of glutamic-oxaloacetic transaminase in serum. *Clin Chim. Acta* **7**, 199–205 (1962).
9. Karmen, A., A note on the spectrophotometric assay of glutamic oxaloacetic transaminase in blood serum. *J. Clin. Invest.* **34**, 131–133 (1955).
10. Tietz, N. M., Weinstock, A., and Rodgerson, D. O., *Proceedings of the Second International Symposium of Clinical Enzymology.* American Association for Clinical Chemistry, Washington, DC, 1976, pp 91–104.
11. Henry, R. J., Chiamori, M., Golub, O. J., and Berkman, S., Revised spectrophotometric methods for the determination of glutamic oxaloacetic transaminase, glutamic pyruvic transaminase, and lactic acid dehydrogenase. *Am. J. Clin. Pathol.* **34**, 381–398 (1960).
12. Demetriou, J. A., Drewes, P. A., and Gin, J. B., Glutamic-oxalacetic transaminase. In *Clinical Chemistry, Principles and Technics,* R. J. Henry, D. C. Cannon, and J. W. Winkelman, Eds., 2nd ed., Harper and Row, New York, NY, 1974, pp 873–893.
13. Bergmeyer, H. U., Bowers, G. N., Jr., Hørder, M., and Moss, D. W., Provisional recommendations on IFCC methods for the measurement of catalytic concentrations of enzymes. *Clin. Chem.* **23**, 887–899 (1977). Special Report.
14. Bergmeyer, H. U., Hørder, M., and Moss, D. W., Provisional recommendations on IFCC methods for the measurement of catalytic concentrations of enzymes. *Clin. Chem.* **24**, 720–721 (1978). Special Report.
15. Rodgerson, D. O., and Osberg, I. M., Sources of error in spectrophotometric measurement of aspartate aminotransferase and alanine aminotransferase activities in serum. *Clin. Chem.* **20**, 43–50 (1974).
16. Hørder, M., Moore, R. E., and Bowers, G. N., Jr., Aspartate aminotransferase activity in human serum. Factors to be considered in supplementation with pyridoxal-5'-phosphate in vitro. *Clin. Chem.* **22**, 1876–1883 (1976).
17. Hørder, M., and Bowers, G. N., Jr., Biological variability in aspartate aminotransferase activity in serum of healthy persons, and effect of in-vitro supplementation with pyridoxal-5'-phosphate. *Clin. Chem.* **23**, 551–554 (1977).
18. Rand, R. N., Practical spectrophotometric standards. *Clin. Chem.* **15**, 839–863 (1969).
19. Chamran, M. S., and Keiser, B. S., Maintaining optimum spectrophotometer performance. *Lab. Med.* **8**, 33–39, 9, 35–44 (1978).
20. Bergmeyer, H. U., Scheibe, P., and Wahlefeld, A. W., Optimization of methods for aspartate aminotransferase and alanine aminotransferase. *Clin. Chem.* **24**, 58–73 (1978).
21. College of American Pathologists, Special Enzyme Survey, VA thru VD. CAP, Skokie, IL, 1979.

Barbiturates, Ultraviolet Spectrophotometric Method

Submitter: Herbert E. Spiegel, *Department of Clinical Laboratory Research, Hoffmann-La Roche, Inc., Nutley, NJ 07110*

Evaluators: R. Thomas Chamberlain, *Clinical Chemistry and Toxicology Sections, Veterans Administration Medical Center, Memphis, TN 38104*
Kurt M. Dubowski, *University of Oklahoma, College of Medicine, Oklahoma City, OK 73190*
Ted W. Fendley, *Pathology and Cytology Laboratories, Inc., Medical Heights, Lexington, KY 40503*

Reviewer: Charles E. Pippenger, *Department of Neurology, Columbia University College of Physicians and Surgeons, New York, NY 10032*

Introduction

Hypnotics are drugs that by acting on the central nervous system can produce a depressed state resembling sleep. In smaller doses these drugs produce drowsiness, and when used for this purpose are called sedatives. Barbiturates represent one of the most important classes of hypnotics. They are used in combination with analgesics for pain, as antidotes for stimulant drugs and convulsive disorders, and as anesthetics. Overdoses of these drugs, which can be habit forming and addicting, can lead to death (1).

The clinical chemistry laboratory is frequently required to analyze concentrations of barbiturates in blood and urine to monitor therapy when these drugs are used as anticonvulsants, and in emergency toxicology cases. The ultraviolet spectrophotometric method described here is particularly useful for routine use in a community hospital laboratory (2–4).

Principle

Serum or blood containing barbiturates is extracted with chloroform. The organic layer is back-extracted with an alkaline solution, which is then separated and divided into two equal aliquots. To one aliquot is added more alkali; to the other is added ammonium chloride, which lowers the pH to 10.5. Lowering the pH causes a shift in the absorbance peak of the barbiturate to a shorter wavelength (hypsochromatic shift), the ionized form of the barbiturate at pH 10.5 having maximum absorbance at 240 nm. In addition, a hyperchromic effect (increase in absorbance) is produced at 240 nm at the lower pH. A spectrum of two aliquots of the same specimen on both sides of the light beam, at different pHs, is obtained with a double-beam spectrophotometer. This spectrum, called a difference spectrum, shows well-defined negative and positive peaks. If the less-alkaline solution is placed in the reference beam of the spectrophotometer, then the 260 nm peak of the less-alkaline sample is positive to the baseline and the 240 nm peak is negative. If the identity of the barbiturate is known, the peak at 260 nm is used for its quantitation.

Materials and Methods

Reagents

1. Chloroform, spectro-grade: Burdick and Jackson Laboratories, Inc., Muskegon, MI 49442.
2. Ammonium chloride.
3. Sodium hydroxide, 0.45 mol/L.
4. Whatman 41 H paper (Whatman, Inc., Clifton, NJ 07014).
5. Stock barbiturate standard: 1 g/L in 0.45 mol/L sodium hydroxide.
6. Working reagents: Add 0.1, 0.2, 0.3, 0.4, and 0.5 mL of stock solution to separate 10-mL volumetric flasks. Fill each flask to the mark with buffered 0.45 mol/L NaOH solution. The concentrations of these standards are respectively 10, 20, 30, 40, and 50 mg/L.

Instrumentation

The Beckman DBG (DB) (Beckman Instruments, Inc., Fullerton, CA 92634) or any automatic ratio-recording ultraviolet spectrophotometer with a suitable logarithmic recorder can be used.

Procedure

1. Pipet 3 to 5 mL of serum, plasma, or blood into a 60-mL glass-stoppered bottle or separating funnel and add 50 mL of chloroform (use a Propipette or some similar device).
2. Shake the mixture on a mechanical shaker for 5 min.
3. Aspirate the aqueous layer, using a pipette fitted with a bulb.
4. Filter the organic phase and place a 40-mL aliquot in another 60-mL bottle.
5. Add 8.0 mL of 0.45 mol/L NaOH.
6. Shake the mixture for 5 min, then remove the organic phase by aspiration (a Pasteur pipette attached by tube to a aspirator with reduced pressure is suitable).
7. Centrifuge the upper phase to separate any residual chloroform, which is then aspirated.
8. Place two 3-mL aliquots of the aqueous phase into two clean, stoppered silica cuvets.
9. To one aliquot, add 0.5 mL of NH_4Cl (160 g/L).
10. Cap both aliquots and mix by inversion.

Note: Treat urine or gastric washings by the same procedure as blood but use 0.1 mol/L HCl to acidify the sample before the initial extraction with chloroform. Wash the chloroform layer with 5 mL of 0.1 mol/L phosphate buffer, pH 7, and discard after washing. Process a 40-mL aliquot of the filtered chloroform as described above.

11. Adjust the spectrophotometer to read zero against air, which allows the entire spectrum to be on scale.
12. Place the more-alkaline solution in the sample compartment and the less-alkaline solution in the reference compartment.
13. Record the difference spectrum from 340 nm to 220 nm. For barbiturates, a two-peak curve results in a positive deflection from baseline in the 260 nm region and a negative deflection in the 240 nm region. A crossover point occurs near 250 nm. Use the reading above the baseline at 260 nm to quantitate the barbiturate if the barbiturate ingested is known. The presence of the barbiturate is indicated by the shape of the difference spectrum.

Calculation

Calculate barbiturate concentration as follows:

concn of barb. unknown (mg/L)

$$= \frac{\text{absorbance difference 260 nm of unknown}}{\text{absorbance difference 260 nm of standard}}$$

$$\times \text{ concn of barb. standard (mg/L)}$$

or read concentration from a calibration curve made by plotting the absorbance difference of each standard at 260 nm vs the concentration of each standard.

Discussion

The spectrophotometric method is a useful way to determine the presence or absence of a barbiturate. Identifying the barbiturate when the ingested drug is unknown depends on the analytical skill of the technician as well as the quality of the sample. Identification by more sophisticated means involves use of "high-pressure" liquid chromatography, thin-layer chromatography, and gas–liquid chromatography. Knowing the identity of the barbiturate is important because its half-life in blood and hence the duration of any toxic symptoms are structure-dependent.

An example of a long-acting barbiturate is phenobarbital; a medium-acting barbiturate is amobarbital, and a short-acting barbiturate is secobarbital. To further identify and quantify the barbiturate detected by the ultraviolet spectrophotometric analysis, perform a recovery experiment by adding a known barbiturate to an aliquot of the specimen, then analyzing this sample with another aliquot of the patient's specimen to which no additions have been made.

A correlation of clinical condition with blood concentrations of selected barbiturates is presented in Table 1.

Table 1. Relationship of Average Blood Concentrations of Barbiturate to Clinical Condition

Stage	Phenobarbital	Amobarbital	Secobarbital	Pentobarbital
1. Awake, mild sedation	35	15	8	
2. Sedated, reflexive	44	25	15	20
3. Comatose, reflexive	65	47	25	20
4. Comatose, reflexive	100	62		
5. Comatose, with circulatory and respiratory difficulties		86		

Concn, mg/L[a]

[a] Averages of four to six nontolerant patients entering each stage. A "nontolerant" patient is one who has not taken the drug long enough to become resistant to its pharmacological effects.

The reader is encouraged to consult textbooks of medical pharmacology for information concerning the relationship of structure to half-life of each barbiturate in blood *(1)*. Moreover, when interpreting results, one should be aware of tolerance phenomena (enzyme induction) in epilepsy patients and the potentiation effects of concomitant medication.

References

1. *The Pharmacological Basis of Therapeutics*, 5th ed., L. S. Goodman and A. Gilman, Eds., MacMillan, New York, NY, 1975, pp 102–121.
2. Sunshine, I., Barbiturate. *Stand. Methods Clin. Chem.* 3, 46–54 (1961).
3. Goldbaum, L. R., Analytical determination of barbiturates. *Anal. Chem.* 24, 1604–1607 (1952).
4. Broughton, P. M. G., Ultraviolet method for the determination and identification of barbiturates in biological material. *Biochem. J.* 63, 207 (1956).
5. Sunshine, I., *Methodology for Analytical Toxicology*, CRC Press, Inc., Boca Raton, Fl, 1975, pp 34–36.

Bilirubin, Total and Conjugated, Modified Jendrassik–Grof Method

Submitters: Thomas R. Koch, *Department of Pathology, University of Maryland School of Medicine, Baltimore, MD 21201*

Basil T. Doumas, *Department of Pathology, Medical College of Wisconsin, Milwaukee, WI 53226*

Evaluators: Robert C. Elser, *Department of Pathology, York Hospital, York, PA 17405*

William L. Gyure, *Chemistry Laboratory, North Central Bronx Hospital, Bronx, NY 10467*

Richard S. Kowalczyk, *Northern Michigan Hospitals, Petoskey, MI 49770*

Reviewer: Robert B. McComb, *Hartford Hospital, Hartford, CT 06115*

Introduction

Bilirubin was first measured in serum by van den Bergh and Snapper in 1913 (1), using the diazo reaction described by Ehrlich (2). The differentiation of bilirubin conjugated with glucuronic acid ("conjugated" or "direct" bilirubin) from unconjugated bilirubin was described three years later (3). Since then, many methods have been proposed for the quantitation and fractionation of this important analyte in human serum.

Conjugated bilirubin reacts promptly with the diazo reagent at an acid pH. Under the same conditions, unconjugated bilirubin reacts very slowly and the reaction goes to completion only in the presence of an accelerator. A number of methods in current use are modifications of that of Malloy and Evelyn (4), in which methanol is used to speed the reaction of unconjugated bilirubin, and the absorbance of the colored product (azobilirubin) is measured at pH 1.5. This method was greatly improved by Meites and Hogg (5, 6) and has been described in two chapters of *Standard Methods of Clinical Chemistry* (7, 8).

In 1938, Jendrassik and Grof (9) developed a method for total bilirubin in which caffeine and benzoate accelerate the reaction of unconjugated bilirubin at a pH near 6.5, and the absorbance of azobilirubin is measured after alkalinization to pH 13. This method has been used in many manual and automated versions, and has been published as a Standard Method (10). The main advantages of the Jendrassik–Grof method are: (a) the molar absorptivity of azobilirubin is highly reproducible and its value is not affected by the protein matrix (11); (b) interference by hemoglobin, at concentrations likely to be encountered in serum, is negligible, and can be eliminated by the use of ascorbic acid; (c) color development is very rapid; and (d) the color developed adheres to Beer's law, up to an absorbance of 1.7 and perhaps higher (11).

The quantity of conjugated bilirubin is estimated by allowing the coupling reaction to proceed in the absence of an accelerator for a fixed time. We now know that what is called conjugated bilirubin comprises several molecular species. Because pure preparations are unavailable for standardization, the accuracy of all methods for conjugated bilirubin is unknown.

Determination of total and conjugated bilirubin is a useful diagnostic tool in a variety of clinical conditions. Total bilirubin values are used in neonatal jaundice as a guide to intervention (phototherapy or exchange transfusion) for preventing kernicterus; in this case, accurate analysis is required in microsamples (frequently, hemolyzed samples) at bilirubin concentrations between 10 and 20 mg/dL (170–340 μmol/L). In adults and children, total and conjugated bilirubin determinations are used to diagnose liver disease, obstruction of the common bile duct, and hemolytic disorders. To obtain precise bilirubin measurements (desired for early detection of some diseases), sometimes a large sample volume is used. To meet all these needs, we have presented two versions of the method of Jendrassik and Grof: a micromethod, with a standard curve that is linear to at least 30 mg/dL (513 μmol/L), involving a small serum volume, and subject to minimum interference from hemolysis; and a macromethod, requiring a larger serum volume to provide good precision in the adult reference interval.

Principle

For determination of total bilirubin, serum is mixed with an acetate-buffered solution of caffeine and sodium benzoate, which accelerates the reaction of unconjugated bilirubin. Diazotized sulfanilic acid splits bilirubin at the central methylene carbon, forming two molecules of red azobilirubin. The mechanism of this reaction has been described by Overbeek et al. *(12)* and Landis and Pardue *(13)*. The reaction is completed rapidly; after 10 min, no further increase in absorbance is detected. Ethylenediaminetetraacetate (EDTA) is added to the caffeine reagent to prevent formation of heavy metal complexes with azobilirubin *(14)*.

Alkalinization to pH 13 with alkaline tartrate solution converts the azobilirubin to its blue form (it appears green because of the yellow background color of the reagent). The absorbance of the alkaline azobilirubin is measured at the absorption maximum of 600 nm.

Materials and Methods
Reagents

All chemicals are reagent-grade unless otherwise specified. De-ionized water with a specific conductivity of at least 1 $\mu\Omega^{-1}cm^{-1}$ (preferably meeting Type I specifications of the College of American Pathologists) is used throughout.

1. *Caffeine reagent.* Dissolve 56 g of anhydrous sodium acetate, $NaC_2H_3O_2$ (or 93 g of $NaC_2H_3O_2 \cdot 3H_2O$); 56 g of sodium benzoate, $NaC_7H_5O_2$; and 1 g of disodium EDTA in about 500 mL of water. Add 38 g of caffeine. Dissolve without heating and dilute to 1 L. Mix well and filter. Store at room temperature in a polyethylene bottle; the reagent is stable for at least six months.

2. *Sulfanilic acid.* Add 5.0 g of sulfanilic acid ($NH_2C_6H_4SO_3H \cdot H_2O$) and 15.0 mL of concd HCl to 500 mL of water in a 1-L volumetric flask. Dissolve and dilute to 1 L with water. Store at room temperature in a polyethylene bottle; this solution is stable for at least six months.

3. *Sodium nitrite.* Dissolve 0.50 g of sodium nitrite ($NaNO_2$) in water and dilute to 100 mL. Store in a Pyrex bottle at 4 °C. This reagent is stable for at least two weeks and may be stable for more than nine months *(11)*. A slight yellow color indicates reagent deterioration.

4. *Diazo reagent.* Combine and mix sodium nitrite and sulfanilic acid reagents in the proportions of 0.10 mL of $NaNO_2$ to 4.0 mL of sulfanilic acid. Although this reagent has been reported to have longer stability *(10, 11)*, we recommend use within 5 h.

5. *Alkaline tartrate.* Dissolve 75 g of sodium hydroxide and 263 g of sodium tartrate ($Na_2C_4H_4O_6 \cdot 2H_2O$) in water and dilute to 1 L. Store in a polyethylene bottle; this is stable for at least six months. Sodium potassium tartrate, $KNaC_4H_4O_6 \cdot 4H_2O$ (323 g/L), may be substituted for sodium tartrate.

Note: A single batch of reagents was prepared and used over a four-month period by Evaluator R. S. K.; no deterioration was detected. Evaluator W. L. G. found reagents 1, 2, and 5 to be stable for five months, reagent 3 stable for two months at room temperature, and reagent 4 stable for at least 5 h at 4 °C.

Apparatus

1. *Automatic pipettor* (Eppendorf type), 100 μL for the micromethod or 500 μL for the macromethod.

2. *Spectrophotometer.* Any spectrophotometer with a spectral bandpass of 10 nm or less and equipped with 1-cm cuvets or a flow cell is suitable. The use of round cuvets is not recommended. The Submitters use a Stasar II spectrophotometer (Gilford Instruments, Inc., Oberlin, OH 44074).

Specimen

Serum or heparinized plasma is suitable for analysis. Protect specimens from light (daylight or fluorescent lights) at all times to prevent photo-oxidation of bilirubin. Make every effort to avoid hemolysis during blood collection, especially from neonates (see the other chapter on bilirubin determination in this volume).

Bilirubin in serum or plasma is reportedly stable for four to seven days at 4 °C *(15)*, or for at least three months if kept frozen *(16)*. We recommend that samples be kept for no more than one day at 4 °C, to avoid hydrolysis of conjugated bilirubin.

Micromethod
Total Bilirubin Procedure

1. For each control or unknown, label tubes for Test and Blank. If a flow cell is used, set up a reagent blank tube (see *Note*). This blank solution will be used to condition the flow cell.

2. Add 100 μL of serum to each tube.

3. Add 1.0 mL of caffeine reagent to each tube.

4. Add 0.5 mL of freshly prepared diazo reagent to each Test tube. Mix well and allow to react for 10 min.

5. During this time, add 0.5 mL of sulfanilic acid reagent to each Blank tube. Mix well.

6. Add 1.0 mL of alkaline tartrate to each tube. Cover the tubes with Parafilm and mix thoroughly by inversion; inadequate mixing of the viscous final solution is a common source of error.

7. Set wavelength to 600 nm and set to zero absorbance for water.

8. If a flow cell is used, aspirate the reagent blank solution to pre-treat the flow cell.

9. Read in order all Blank solutions, then all Test solutions in a 1-cm cuvet. (This order minimizes effects of any carryover that may occur.) The alkaline azobilirubin color is stable for at least 30 min.

Note: A reagent blank is prepared exactly as the Test except that water is substituted for serum. Despite the yellow color, its absorbance against water at 600 nm is essentially zero.

Calculations. Subtract each Blank absorbance

from that of the respective Test to obtain the corrected absorbance of the Test. Using the corrected absorbance of the Test, calculate the concentration of total bilirubin from the calibration curve (see *Standardization*).

Conjugated Bilirubin Procedure

We cannot endorse a Selected Method for the measurement of conjugated bilirubin, for the reasons mentioned in the *Introduction*. For those who wish to perform conjugated bilirubin analyses, we suggest the following modification of the method described by Nosslin *(17)*:

1. Add 100 µL of serum to tubes labeled Test. Add 1.0 mL of 50 mmol/L HCl to each tube.
2. Add 0.5 mL of diazo reagent to the Test tubes and mix. After 5 min, add 50 µL of ascorbic acid solution (40 g/L; 0.2 g in 5 mL of water) and mix well (see *Notes*).
3. Add 0.5 mL of sulfanilic acid and 50 µL of ascorbic acid to the tubes labeled Blank.
4. Add 1.0 mL of alkaline tartrate to each tube. Mix well and read at 600 nm vs water as described above for total bilirubin.

Calculations. Subtract the absorbance of each Blank from that of the corresponding Test, then use the corrected Test absorbance to calculate the concentration of conjugated bilirubin from the calibration curve.

Notes:
1. The 5-min reaction time is not critical because the reaction is essentially completed even when the concentration of conjugated bilirubin is high ($\simeq$ 10 mg/dL).
2. With 50 mmol/L HCl as diluent, the interference by unconjugated bilirubin is negligible. When bilirubin standards with concentrations of 2, 5, 10, and 20 mg/dL were analyzed, values for direct-reacting bilirubin were 0.06, 0.09, 0.13, and 0.18 mg/dL, respectively.
3. The ascorbic acid solution should be prepared freshly daily.

Standardization

Acceptable bilirubin preparations *(18)* are available from several sources. A certified bilirubin is available from the National Bureau of Standards, Washington, DC 20234, as Standard Reference Material (SRM) No. 916, at a rather high cost. The Submitters examined bilirubins from Pfanstiehl Labs., Inc., Waukegan, IL 60085 (Reference Grade), and Harleco Division, American Hospital Supply Corp., Philadelphia, PA 19143, and found their purity acceptable. However, each lot of such materials must be shown to yield an acceptable molar absorptivity value for alkaline azobilirubin (see below).

Preparation of Stock Bilirubin Solution, 20 mg/dL (342 µmol/L)

The preparation of the standard is carried out in subdued light.

Weigh 20.0 ($\pm$ 0.1) mg of bilirubin in a small Weigh-Boat (Scientific Products, Inc., McGaw Park, IL 60085, cat. no. B2045-5) and transfer into the bottom of a 100-mL volumetric flask. Rinse the boat with 1.0 mL of dimethylsulfoxide directly into the flask. Swirl the flask until a homogeneous suspension is obtained, and then add 2.0 mL of 0.1 mol/L (10 g/L) Na_2CO_3. Mix by swirling until all bilirubin is dissolved; a *clear* red-orange solution should be obtained. Add about 80 mL of 40 g/L aqueous bovine serum albumin solution, previously adjusted to pH 7.4 ($\pm$ 0.1). Add 2.0 mL of 0.1 mol/L HCl and dilute to 100 mL with the bovine serum albumin solution. After mixing the solution by inversion, centrifuge an aliquot (6–10 mL) at 2000 rpm for 10 min. The presence of an orange sediment at the bottom of the tube indicates that the bilirubin had not been completely dissolved before the addition of the albumin. If this occurs, a new solution must be prepared.

Notes:
1. Use the stock bilirubin solution (see below) immediately after preparation, or within 2 to 4 h; store at 4 °C until used.
2. Because the bovine serum albumin solution is used to prepare working standards, make an adequate volume ($\simeq$ 200 mL).
3. Bovine serum albumin, Cohn Fraction V, can be obtained from Reheis Chemical Co., Division of Armour Pharmaceutical Co., Phoenix, AZ 85077, stock no. 2293-01; Sigma Chemical Co., St. Louis, MO 63178, cat. no. 4503; or Miles Labs., Inc., Research Products Division, Elkhart, IN 46515, code no. 81-066-1.

Working Standards and Calibration Curve for Total Bilirubin

Pipet the following volumes of stock bilirubin standard (20 mg/dL) and bovine albumin solution (40 g/L) into test tubes.

Tube	Bovine serum albumin, mL	Stock std., mL	Bilirubin concn, mg/dL
1	—	5.0	20.0
2	1.0	3.0	15.0
3	3.0	3.0	10.0
4	7.0	3.0	6.0
5	8.0	2.0	4.0
6	9.0	1.0	2.0
7	5	0	0 (Blank)

Analyze the six standards and the Blank for total bilirubin in duplicate. (Do not run Blanks with sulfanilic acid.) Measure the absorbance of all solutions at 600 nm against water. Subtract the absorbance of the Blank (albumin) from that of the working standards. Plot on linear graph paper the absorbance of the standards against concentration. A straight line should be obtained. Along with the standards, also analyze any commercial bilirubin controls or calibrators routinely used in your laboratory. Compare your values with those assigned by the manufacturer. If there is a disagreement, use your own values. To avoid frequent standardization, have on hand a year's supply of controls with one lot number. If you have used a spectrophotometer with good photometric accuracy *(19)*, cuvets with 1.0-cm pathlength, and volumetric pipettes, you should expect to obtain a molar absorptivity of between 74 000 and 76 000 L · mol^{-1} · cm^{-1}. Calculate the molar absorptivity (ϵ) as follows:

$$\epsilon = 584.7 \times 2.6 \times \text{absorbance of standard (20 mg/dL)}/0.02$$

where:

584.7 = relative molecular mass of bilirubin
2.6 = volume of the reaction mixture, in milliliters
0.02 = milligrams of bilirubin in 100 μL of stock standard

Notes:

1. For the National Bureau of Standards' bilirubin a mean ϵ value of 74 700 L · mol^{-1} · cm^{-1} has been obtained in several laboratories. In accord with the convention used in most publications, this constant is calculated by using the relative molecular mass of bilirubin instead of that of the absorbing species (azobilirubin).

2. The calibration curve may also be used for calculating conjugated bilirubin values. The 2% error, introduced by the slight increase in volume (2.60 vs 2.65 mL), is tolerable for conjugated bilirubin analyses. If greater accuracy is desired, the conjugated bilirubin values obtained from the curve should be multiplied by 1.02.

3. If you intend to use the macromethod, prepare a calibration curve by using the volumes of reagents and sample indicated for this method (see below). The concentration of the highest working standard should not exceed 5.0 mg/dL.

4. The stock and working bilirubin standards can be stored for up to six months at −70 °C without appreciable deterioration. When stored at −20 °C, the loss of bilirubin is about 1.5% per month *(11)*. The remainder of the bovine serum albumin solution can be stored indefinitely at −20 °C.

Macromethod for Total Bilirubin

Some laboratories may wish to use a method that is highly sensitive near the reference intervals for normal adults. The macromethod provides excellent sensitivity and precision, and yields the same molar absorptivity as the micromethod.

Procedure

1. Add 500 μL of serum to tubes labeled Test and Blank.
2. Add 1.0 mL of caffeine reagent to each tube.
3. Add 0.5 mL of freshly prepared diazo reagent to the Test tubes. Mix well and allow to react for 10 min.
4. Add 0.5 mL of sulfanilic acid to each Blank tube.
5. Add 1.0 mL of alkaline tartrate to each tube. Mix well by inversion and read at 600 nm, as described in the micromethod for total bilirubin.

Calculations. Calculate as described for the micromethod for total bilirubin, using the calibration curve for the macroprocedure.

Note: Evaluator W. L. G. finds the method unsatisfactory because it is too sensitive, requires a large volume of sample, and is subject to hemoglobin interference; he recommends using a 200-μL sample. According to Evaluator R. C. E. the method has excellent precision, better than that of the micromethod.

Discussion

Standardization

Inadequate attention to accurate standardization remains the predominant source of error in bilirubin determinations. Meites and Traubert *(20)* described the preparation of a stable bilirubin standard in chloroform that produced the same azobilirubin molar absorptivity as bilirubin in serum with the Meites–Hogg method *(5)*. However, such a standard is usable only in methods in which alcohol is an accelerator.

A sound approach for the preparation of bilirubin standards in serum was described by a Joint Committee of several professional organizations in 1962 *(18)*. The procedure we use is similar to that of the Joint Committee except that *(a)* a small amount of dimethylsulfoxide is used to speed dissolution, *(b)* the alkaline sodium carbonate is neutralized with HCl, and *(c)* a purified protein (bovine serum albumin), free of bilirubin, is used instead of the "acceptable" serum diluent.

Although daily use of freshly prepared standards in albumin is the preferred approach to standardization, it is not practical for most clinical laboratories.

The use of a calibration curve or a factor derived from it is an acceptable approach, provided its validity can be ascertained. To do this, include bilirubin controls in every run. For the micromethod the use of control sera at bilirubin concentrations near 1.5 and 20 mg/dL is recommended.

Many laboratories find the preparation of bilirubin standards difficult and prefer to use commercially available lyophilized preparations as standards. In the past, such products were unreliable *(5, 11)*, but there has been substantial improvement since then. The Submitters have recently examined bilirubin calibrators by the Jendrassik–Grof method and found some accurate. Such products are available from Hyland Diagnostics, Deerfield, IL 60015; Beckman Instruments, Inc., Fullerton, CA 92634; and perhaps other sources.

Evaluators R. C. E. and R. S. K. feel that the section on standardization is beyond the expertise of small laboratories and recommend the use of high-quality secondary standards. Because that argument is valid, the Submitters have indicated sources where such calibrators can be obtained.

Precision

All evaluators studied both the within-run and day-to-day precision of the micromethod at one or two concentrations of bilirubin, and reported the following data:

Within-run. R. C. E.: $\bar{x} = 0.38$ mg/dL, CV = 4.6%, n = 20; R. S. K.: $\bar{x} = 1.2$ mg/dL, CV = 3.7%, n = 24, and $\bar{x} = 4.7$ mg/dL, CV = 1.5%, n = 24; W. L. G.: $\bar{x} = 1.7$ mg/dL, CV = 4.1%, n = 20, and $\bar{x} = 4.7$ mg/dL, CV = 1.9%, n = 20.

Day-to-day. R. C. E.: $\bar{x} = 1.5$ mg/dL, CV = 5.5%, n = 20, and $\bar{x} = 4.9$ mg/dL, CV = 2.4%, n = 20; R. S. K.: $\bar{x} = 1.7$ mg/dL, CV = 5.1%, n = 20, and $\bar{x} = 9.2$ mg/dL, CV = 2.2%, n = 20; W. L. G.: $\bar{x} = 1.7$ mg/dL, CV = 5.9%, n = 30, and $\bar{x} = 4.7$ mg/dL, CV = 3.2%, n = 30.

The day-to-day precision of the macromethod was evaluated in the laboratory of Submitter B. T. D. over an eight-month period with use of a frozen serum pool. The mean bilirubin value was 0.5 mg/dL (CV 3%, n = 24).

Linearity

With the micromethod, Beer's law is reported to be followed to a bilirubin concentration of 30 mg/dL (R. C. E.), 19.7 mg/dL (R. S. K.; highest concentration tested), and 25.6 mg/dL (W. L. G.). The linearity of the macromethod extends to at least 5 mg of bilirubin per 100 mL.

Analytical Recovery

We tested the analytical recovery of bilirubin added to many preparations of human serum albumin (Cohn Fraction V), bovine serum albumin, and a large number of sera from hospitalized individuals. The recovery was between 99.5 and 100.5%. This quantitative recovery demonstrates and confirms a previous report (11) that the molar absorptivity of azobilirubin by the Jendrassik–Grof method is not dependent upon the protein matrix. That is why bovine serum albumin can be substituted for serum in the preparation of bilirubin standards. Evaluator W. L. G. reported a recovery of between 97 and 103%.

Interference

Diazo methods in an acid pH are subject to interference by hemoglobin. For reasons that are not well understood, the interference by hemoglobin in the Jendrassik–Grof method is minimal.

Evaluator W. L. G. found no interference with the micromethod with hemoglobin concentration as great as of 3320 mg/L. With the macromethod, the presence of 1660 mg of hemoglobin per liter of serum lowered the bilirubin concentration by 0.6 mg/dL (15%). Our findings agree with his. Note that the suppression of bilirubin values does not depend on the bilirubin/hemoglobin ratio (identical in both methods) but rather on the hemoglobin concentration in the reaction mixture.

Meites and Lin (21) measured hemoglobin in a large number of specimens from newborns (0 to 13 days old). They found that less than 3% of the values exceeded 100 mg/dL; the highest value was 147 mg/dL. In view of these data, hemoglobin interference will rarely be encountered with the micromethod. In a case of gross hemolysis the interference by hemoglobin can be eliminated or greatly decreased by adding 50 μL of ascorbic acid solution (40 g/L) before adding the alkaline tartrate (22).

Sodium azide blocks the coupling reaction and should not be added as a preservative to protein solutions that might be used for preparing or diluting standard bilirubin solutions.

Consult the publication by Young et al. (23) for more information on interfering substances; these are very few and unlikely to be found in blood at concentrations that could interfere.

Color Stability

In our experience, the color is stable for at least 1 h; however, we recommend that absorbance measurements be made within 10 to 15 min after adding the alkaline tartrate, to prevent the appearance of slight turbidity seen occasionally in some serum samples. The use of a serum blank eliminates this error because the degree of turbidity is the same in both Test and Blank.

Evaluator W. L. G. reports color fading when solutions are read 70 min after alkalinization; he recommends reading all solutions within 30 min. He also found that results are not affected when the coupling reaction is prolonged for 10 to 30 min.

We have decreased the concentration of the caffeine reagent to 75% of its original strength to eliminate the possibility of caffeine precipitation in the final solution (11).

Evaluator R. C. E. feels that serum blanks are not necessary with nonturbid specimens because the Blank absorbance is equivalent to < 0.2 mg of bilirubin per 100 mL. Although his observation is correct, we disagree because this approach is not a sound laboratory practice. Furthermore, turbidity is present in most lyophilized (freeze-dried) controls and, as mentioned above, may develop sometimes even when specimens are clear.

Sensitivity

Most methods for total bilirubin that involve the diazo reaction provide similar values for the molar absorptivity of azobilirubin. A typical value reported for the Meites–Hogg method, for example, is 70 300 L · mol^{-1} · cm^{-1} (5), compared with 74 700 for the method recommended here. Thus, the sensitivity of diazo methods can be increased substantially only by *(a)* increasing the serum volume relative to the total volume in the reaction mixture, or *(b)* increasing the pathlength.

Laboratories wishing to use smaller serum volumes than suggested here should reduce proportionately the reagent volumes and use microcuvets (note that the same cuvets must be used for standardization). To decrease the serum volume without decreasing the reagent volumes would diminish the test sensitivity and result in poor precision.

Reference Intervals

Evaluator W. L. G. has studied 150 apparently healthy subjects (ages 18–73 years) on four occasions over a period of four years. Based on his data, the reference interval for total bilirubin is 0.1 to 1.1 mg/dL. For Evaluator R. S. K. the upper limit of normal is 1.2 mg/dL. Both Submitters use this same range in their laboratories.

For a detailed discussion of reference intervals, see the following chapter on bilirubin by S. Meites.

References

1. van den Bergh, A. A. H., and Snapper, J., Die Farbstoffe des Blutserums. *Dtsch. Arch. Klin. Med.* **110**, 540–561 (1913).
2. Ehrlich, P., Sulfodiazobenzol, ein Reagens auf Bilirubin. *Centr. Klin. Med.* **4**, 721–723 (1883).
3. van den Bergh, A. A. H., and Muller, P., Uber eine direkte und eine indirekte Diazoreaction auf Bilirubin. *Biochem. Z.* **77**, 90–103 (1916).
4. Malloy, H. T., and Evelyn, K. A., The determination of bilirubin with the photoelectric colorimeter. *J. Biol. Chem.* **119**, 481–490 (1937).
5. Meites, S., and Hogg, C. K., Studies on the use of the van den Bergh reagent for determination of serum bilirubin. *Clin. Chem.* **5**, 470–478 (1959).
6. Hogg, C. K., and Meites, S., A modification of the Malloy and Evelyn procedure for the micro-determination of total serum bilirubin. *Am. J. Med. Technol.* **25**, 281–286 (1959).
7. Kingsley, G. R., Getchell, G., and Schaffert, R. R., Bilirubin. *Stand. Methods Clin. Chem.* **1**, 11–15 (1953).
8. MacDonald, R. P., Bilirubin (modified Malloy and Evelyn)—Provisional. *Stand. Methods Clin. Chem.* **5**, 65–74 (1965).
9. Jendrassik, L., and Grof, P., Vereinfachte photometrische Methoden zur Bestimmung des Blutbilirubins. *Biochem. Z.* **297**, 81–89 (1938).
10. Gambino, S. R., Bilirubin (modified Jendrassik and Grof)—Provisional. *Stand. Methods Clin. Chem.* **5**, 55–63 (1965).
11. Doumas, B. T., Perry, B. W., Sasse, E. A., and Straumfjord, J. V., Jr., Standardization in bilirubin assays: Evaluation of selected methods and stability of bilirubin solutions. *Clin. Chem.* **19**, 984–993 (1973).
12. Overbeek, J. T. G., Vink, C. L. J., and Deenstra, H., Kinetics of the formation of azobilirubin. *Recl. Trav. Chim. Pays-Bas.* **74**, 85–97 (1955).
13. Landis, J. B., and Pardue, H. L., Kinetics of the reactions of unconjugated and conjugated bilirubins with *p*-diazobenzenesulfonic acid. *Clin. Chem.* **24**, 1690–1699 (1978).
14. Holtz, A. H., and van Dreumel, H. J., Stabilization of azobilirubin solution by EDTA. *Clin. Chim. Acta* **20**, 355–357 (1968).
15. Henry, R. J., Golub, O. J., Berkman, S., and Segalove, M., Critique on the icterus index determination. *Am. J. Clin. Pathol.* **23**, 841–853 (1953).
16. Winkleman, J. W., Cannon, D. C., and Jacobs, S. L., Liver function tests, including bile pigments. In *Clinical Chemistry, Principles and Technics*, R. J. Henry et al., Eds., Harper and Row, Hagerstown, MD, 1974, p 1059.
17. Nosslin, B., The direct diazo reaction of bile pigments in serum, experimental and clinical studies. *Scand. J. Clin. Lab. Invest.* **12**, Suppl. 49, 1–176 (1960).
18. Joint Committee Report, Recommendation on a uniform bilirubin standard. *Clin. Chem.* **8**, 405–407 (1962).
19. Rand, R. N., Practical spectrophotometric standards. *Clin. Chem.* **15**, 839–863 (1969).
20. Meites, S., and Traubert, J. W., Use of bilirubin standards. *Clin. Chem.* **11**, 691–699 (1965).
21. Meites, S., and Lin, S. S.-H., Hemolysis in plasma samples from skin puncture of children. *Clin. Chem.* **26**, 987 (1980). Abstract.
22. Michaëlsson, M., Bilirubin determination in serum and urine; studies on diazo methods and a new copperazo pigment method. *Scand. J. Clin. Lab. Invest.* **13**, Suppl. 56, 1–79 (1961).
23. Young, D. S., Pestaner, L. C., and Gibberman, V., Effects of drugs on clinical laboratory tests. *Clin. Chem.* **21**, 1D–432D (1975). Special issue.

Addendum

The mechanism of hemoglobin interference with the methods of Malloy and Evelyn, and Jendrassik and Grof has been explained in two papers by B. Schull et al. since this chapter was written. See *Clin. Chem.* **26**, 22–25, 26–29 (1980).

Bilirubin, Direct-Reacting and Total, Modified Malloy–Evelyn Method

Submitter: Samuel Meites, *Clinical Chemistry Laboratory, Children's Hospital, Columbus, OH 43205*

Evaluators: Mary H. Cheng and Leroy H. Arnold, *Clinical Chemistry Laboratory, Children's Hospital of Los Angeles, Los Angeles, CA 90054*
Shareen L. Cox, *Clinical Chemistry Laboratory, Children's Hospital, Columbus, OH 43205*
Gilbert H. Nelson, *Miami Valley Hospital, Dayton, OH 45409*
Willard A. Stacer, *Harper-Grace Hospitals, Detroit, MI 48201*
Nabil W. Wakid, *Department of Laboratory Medicine, American University Hospital, Beirut, Lebanon*

Reviewer: Robert B. McComb, *Clinical Chemistry Laboratory, Hartford Hospital, Hartford, CT 06115*

Introduction

The Meites–Hogg modification (1) of the Malloy–Evelyn procedure (2) for measuring direct-reacting (conjugated) and total (direct-reacting and indirect-reacting) bilirubin in serum or plasma was published in 1965 as a provisional selected (standard) method (3). Now it is presented as a manual semi-micromethod that may be readily scaled upward to a macromethod or, less easily, downward to an ultramicromethod.

An important advantage of the acid–diazo analytical method described here is that the total bilirubin determination is relatively easy to standardize and control by using pure standards and reference sera. Moreover, the azobilirubin formed in the analysis of total bilirubin has been optimized to give a stable color that reaches its maximum development within 10 min. A principal disadvantage is that, for unexplained reasons, hemoglobin interferes with the determination. Knowledge of the apparent concentration of direct-reacting bilirubin is highly useful clinically; because there is no pure standard, however, its assay is far from acceptable.

Principle

Bilirubin in plasma or serum is coupled with diazotized sulfanilic acid in acid medium (pH 1.2) to produce a purplish product (azobilirubin, absorption maximum at 560 nm), the intensity of which is proportional to bilirubin concentration, and which is measured spectrophotometrically. For total bilirubin 50% aqueous methanol (water/methanol, 50/50 by vol) serves as the accelerator, without causing protein precipitation, whereas direct-reacting (conjugated) bilirubin is measured in aqueous solution alone. Methanolic dilutions of a stock standard of bilirubin in chloroform are used to standardize measurement of both forms of bilirubin.

Materials and Methods

Reagents

Note: Use ACS-grade chemicals, unless otherwise indicated. See the note with methanol, in particular.

1. *Diazo A reagent.* Dissolve 5.0 g of sulfanilic acid in 60 mL of concd. HCl (376 g/L, sp. gr. 1.19, about 12 mol/L), mix, and carefully dilute to 1 L with water. This reagent is stable indefinitely at room temperature.

2. *Sodium nitrite stock, 200 g/L.* Dissolve 20 g of sodium nitrite in water and dilute to 100 mL. Store at 2–10 °C in a brown, glass-stoppered bottle. Discard when the reagent becomes tinged with yellow color (nitrate; this may take many months).

3. *Diazo B reagent.* Dilute the stock nitrite 10-fold with water (1 + 9) to prepare a 20 g/L solution. Prepare daily.

4. *Diazo blank.* Carefully dilute 60 mL of concd. HCl to 1 L with water; this reagent is stable indefinitely at room temperature.

5. *Diazo reagent.* Add 0.3 mL of Diazo B reagent to 10 mL of Diazo A reagent, and mix. This reagent is unstable at room temperature.

Note: Prepare this reagent just before use. There is disagreement about the duration of its stability when stored at 4 °C. Evaluator N. W. W. found it to be stable at −17 to −20 °C for 72 h.

6. *Methanol, absolute.* This solvent must meet ACS specifications, with purity of at least 995 mL/L. It must be stored in a glass container.

Note: Use of methanol of only slightly lesser grade, e.g., methanol, anhydrous (980 mL/L) may cause pronounced turbidity. Storage in metal containers may also cause this problem.

7. *Reference sera, controls, and calibrators.*

Note: The Submitter uses Versatol Pediatric and Validate-A (both from General Diagnostics, Morris Plains, NJ 07950) with each run. When reconstituted, these sera are dispensed into 1-mL polyethylene tubes, and stored frozen at −10 to −20 °C. Contents of one or more tubes are used daily for five consecutive days, but no longer. Evaluators G. H. N. and N. W. W. have used other commercial sera and found some values by this method to be less than the label values. For an explanation of these discrepancies, see No. 3 in *Notes on Standardization and the Use of Reference Sera.* N. W. W. has found that aliquots of Ortho Abnormal Control Serum (Ortho Diagnostics Inc., Raritan, NJ 08869) can be stored for eight consecutive days at −17 to −20 °C, and Ortho Elevated Bilirubin Control Serum for at least 15 days.

Method

General notes

1. Add the reagent and diluted plasma (serum) *in the order stated*, to avoid turbidity. Add the diluted plasma *last*.

2. The final color in total bilirubin assay is stable for at least 2 h for concentrations of unconjugated bilirubin as great as 30 mg/dL. Dilute higher concentrations with water.

3. Because sunlight destroys bilirubin, protect samples and reaction vessels from it. Fluorescent light is acceptable, but not harmless. If the serum cannot be analyzed within 1 h of collection, store it in the dark.

4. Avoid using hemolyzed plasma; hemolysis decreases values.

5. The Submitter collects the patient's blood into blue-tinted, lithium-heparinized, 550-mL polyethylene microcentrifuge tubes with caps (KEW Scientific, Inc., Columbus, OH 43227) and determines bilirubin in plasma. The *opaque* tubes shield the blood from light, and the use of plasma avoids the clotting process, which may contribute some hemolysis to serum. Make every effort to avoid hemolysis during blood collection. For skin puncture of infants, this means be careful to avoid excess pressure at the puncture site (heel or finger), to dry the area of puncture with sterile gauze, and to use dry collecting tubes. For further details on blood collection and materials, see reference 4. Recent evidence indicates that with proper skin-puncture technique, hemolysis can be virtually eliminated as an interference in bilirubin determination (5).

6. If you prefer a macroprocedure, use multiples of each volume listed, and use them *in the order given.*

Procedure and calculations

1. Pipet 1.9 mL of de-ionized water into a 12 × 75 mm tube.

2. Add 100 μL of plasma or reference serum. Mix.

Note: The Submitter uses an automatic pipette (Micromedic Systems, Inc., Philadelphia, PA 19044) first picking up the sample, and then washing it into the tube with water.

3. Pipet in the exact order given the following volumes of reagents and diluted plasma or reference serum into appropriately labeled tubes. Mix thoroughly after each step.

Direct-Reacting

Plasma test	Plasma blank
0.50 mL of water (Submitter uses a Lancer pipette, Sherwood Medical Instruments, Inc., St. Louis, MO 63103)	same
0.10 mL of diazo reagent (Mohr pipette)	0.10 mL of diazo blank
0.40 mL of diluted plasma or reference serum [Submitter uses a Fixed Finn pipette (Labsystems Oy, Helsinki, Finland) to dispense the diluted plasma]	same

Allow to stand 5 min in capped tubes, at room temperature (23–27 °C)

Total

0.50 mL of methanol	same
0.10 mL of diazo reagent (Submitter uses a Micromedic pipette, first picking up the diazo reagent or blank, then washing out with methanol)	0.10 mL of diazo blank
0.40 mL of diluted plasma or reference serum	same

Allow to stand 10 min in capped tubes, at room temperature (23–27 °C)

Note: The Submitter uses 1.5-mL polyethylene centrifuge tubes with caps.

4. Read the absorbance at 560 nm with water as the blank. Subtract the plasma blank absorbance from the plasma test absorbance to obtain the net absorbance.

Note: Submitter uses a Stasar II spectrophotometer (Gilford Instruments Inc., Oberlin, OH 44074) with flow-through microcuvet, requiring 0.25 mL volume per reading. Multiple readings are made to ensure that bubbles are not trapped in the sampling system.

5. Convert net absorbance reading to concentration of total or direct-reacting bilirubin (mg/L) by use of a standard curve. To calculate free or "indirect" bilirubin, subtract the concentration of direct-reacting (5-min) bilirubin from the concentration of the total bilirubin.

Standardization

Stock standard, 10 mg/dL, in chloroform

1. Weigh exactly into a tared 20-mL beaker about 10 mg of bilirubin. Use this weighed value in calculating the concentration of the stock bilirubin in methanol, and of the working methanol standards. (See *Notes on Standardization,* No. 1.)

2. Using a glass funnel, transfer quantitatively the precisely weighed bilirubin, with several washings of chloroform, into a 100-mL volumetric flask, and dilute to the mark with chloroform. The chloroform may be warmed slightly to help dissolve the bilirubin, but this is generally unnecessary.

3. Store the stock standard in a brown glass bottle at -10 to -20 °C. Warm an aliquot of the solution to room temperature before use. The chloroform solution is stable for at least three months.

Note: Evaluator G. H. N. found no change in bilirubin concentration for this standard stored in 10-mL aliquots at -70 °C for at least nine months.

Stock standard, 1 mg/dL, in methanol

Dilute 10.0 mL of stock standard to 100 mL in a volumetric flask with methanol. *Use this methanolic standard immediately (within 20 min)* because bilirubin will slowly precipitate.

Note: The Submitter delivers the stock standard solution with a premarked volumetric pipette prepared as follows: deliver 10 mL of *water* from the pipette. Mark the position of the remaining drop of water with a glass-marking file or diamond pencil. Deliver the stock standard between the marks on the pipette. Calibrate the pipette this way because chloroform drains *completely* from the pipette, whereas the pipette is calibrated to deliver water.

Working standards

1. Pipet the following volumes of methanolic bilirubin standard (10 mg/L) and methanol into 1.5- to 2.0-mL tubes, in duplicate:

Tube	Methanol standard, mL	Methanol, mL	Bilirubin, mg/dL
1	0.60	0.30	30.0
2	0.50	0.40	25.0
3	0.40	0.50	20.0
4	0.30	0.60	15.0
5	0.20	0.70	10.0
6	0.10	0.80	5.0
7 (blank)	0.00	0.90	0

Note: Submitter uses 1.5-mL polyethylene disposable centrifuge tubes with caps.

2. *Immediately* add 0.10 mL of diazo reagent to each tube, mix, and allow to stand 10 min.

3. Read absorbances at 560 nm with water as the blank. To obtain net absorbances, subtract the absorbance of the reagent blank, tube 7, from all other readings, and average the duplicates. Construct a curve on rectangular coordinate paper, plotting absorbance vs milligrams of bilirubin per 100 mL.

Notes on standardization and the use of reference sera

1. Bilirubin obtained commercially may be impure. Unless packaged in an inert atmosphere, it may deteriorate upon aging. For best results use the National Bureau of Standards (Washington, DC 20234) Standard Reference Material no. 916, Bilirubin. (See also the previous chapter in this text on total bilirubin by Koch and Doumas.)

2. Standardization of total bilirubin in methanol, as described above, gives a slightly greater volume than that obtained after addition of diluted plasma to methanol–diazo reagent mixture, in the test procedure. The addition of 0.40 mL of diluted plasma to the 0.60 mL of mixture results in a final volume somewhat less (3%) than 1.00 mL because of contraction of the solution. This contraction is not ordinarily taken into consideration in standardization.

3. Discrepancies between labeled values of bilirubin concentration in reference sera and those obtained with this modification are based on differences in methods of standardization and preparation of the parent material. The Meites–Hogg modification allows the use of a definable standard bilirubin in chloroform. The values obtained from bilirubin added to serum are the same as those obtained from chloroform–methanol standardization *(6).* The standard preparation of bilirubin in albumin described by Koch and Doumas (previous chapter, this text) should help resolve differences between commercial reference sera, as well as between the method of Jendrassik–Grof and the Meites–Hogg modification presented here.

Studies of Precision, Recovery, and Method Comparison

Table 1 shows the precision of this modification. Previous experience indicates that the precision is lower at lower concentrations of bilirubin, and increases as the values increase above the reference (normal) range. This is shown in the day-to-day data for CV in this table.

Analytical recovery studies performed by Evaluators M. H. C. and L. H. A. for 12 different bilirubin concentrations added to serum, giving ranges of 14 to 231 mg/L, averaged 97% (range, 87–102%).

Method comparisons were made by Evaluators M. H. C. and L. H. A., using a modified Jendrassik–Grof method (American Monitor Corp., Indianapolis, IN 46268). The difference between means obtained by the two methods on 44 sera ranging in bilirubin values from 2 to 314 mg/L is statistically significant (95% confidence limit), the Jendrassik–Grof method giving higher results. On the other hand, linear re-

Table 1. Precision Studies, Total Bilirubin

Evaluator	Material analyzed	No.	Mean (SD), mg/L	CV,%
			Within-run	
M.H.C. and L.H.A.	serum	19	43.7 (0.86)	2.0
			Day-to-day	
	serum P$_1$[a]	18	4.7 (0.72)	15.3
	serum P$_2$[a]	20	42.9 (1.5)	3.5
	serum BC[a]	19	179 (2.6)	1.5
W.A.S	serum 1	26	21.1 (1.1)	5.2
	serum 2	26	59.4 (2.6)	4.4
S.M.	serum 1[b]	28	190 (9.3)	4.9
	serum 2[b]	28	40.7 (2.8)	6.9

[a] Data obtained by one analyst.
[b] Data obtained by seven analysts over a 30-day period.

gression shows a strong correlation between the two methods:

Student's t-test

Jendrassik–Grof bilirubin (x): mean, 8.47 mg/dL; SD, 0.75 mg/dL; standard error of estimate, 0.42 mg/dL; df, 43; t-value, 8.6937.

Modified Malloy–Evelyn bilirubin (y): mean, 7.66 mg/dL; SD, 0.71 mg/dL; standard error of estimate, 0.45 mg/dL.

Linear regression

	y on x	x on y
No. of pairs	44	44
Correlation coefficient	0.9982	0.9982
Slope of regression line	1.0596	0.9403
y-intercept	0.3574	−0.308

Reference (Normal) Values

Reference values of total bilirubin cited for adults generally place lower limits at 0.1 mg/dL (10 mg/L) and upper limits at 1.0 to 1.5 mg/dL (10–15 mg/L). The upper limits for direct-reacting (conjugated) bilirubin in the method described here are about 0.10 to 0.30 mg/dL (1.0–3.0 mg/L), with lower limits at 0. Bear in mind, however, that for the Meites–Hogg modification, 5 to 10% of the unconjugated (indirect, free) bilirubin reacts in aqueous medium. When the concentration of free bilirubin is markedly increased, as in pathological jaundice, the "conjugated" fraction is necessarily high. If free (indirect-reading) bilirubin is at 15 mg/dL, for example, the direct-reacting fraction could be as high (falsely) as 1.5 mg/dL in the absence of *any* conjugated bilirubin. Interpretation, therefore, must be based on the relative increase of direct-reacting bilirubin in relation to the indirect-reacting bilirubin.

Figure 1 shows the physiological change in bilirubin concentration with age in the newborn according to birth weight (7). In 10 infants the indirect-reacting bilirubin is 1–3 mg/dL in umbilical cord serum, and increases at a rate of less than 5 mg/

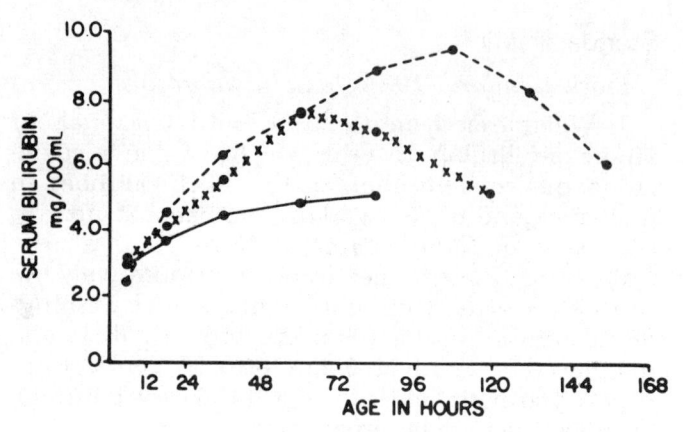

Fig. 1. Mean serum bilirubin concentration as related to age in three groups of infants
— — —, pre-term, birth wt. < 2500 g (appropriate for gestational age); xxx, full-term, birth wt. < 2500 g (small for gestational age); ———, full-term, birth wt. > 2500 g (appropriate for gestational age). Reprinted with permission from *Neonatology* (7)

dL per 24 h. The peak value (5 to 6 mg/dL) is usually reached between the second and fourth days, and then decreases to less than 2 mg/dL between the fifth and seventh day of life. In the premature newborn, however, the peak is not reached until the fifth to seventh day (8–12 mg/dL). According to a standard textbook (8), diagnosis of physiological jaundice in the *pre-term* infant can be established by excluding known causes of jaundice on the basis of history, and clinical and laboratory findings. In general, a search to determine the cause of jaundice should be made if (a) it appears in the first 24 h of life; (b) serum bilirubin increases faster than 5 mg/dL per 24 h; (c) serum bilirubin exceeds 12 mg/dL in full-term or 14 mg/dL in pre-term infants; (d) jaundice persists after the first week of life; or (e) direct-reacting bilirubin exceeds 1.0 mg/dL at any time.

Discussion

The assumption that conjugated bilirubin in aqueous medium reacts quantitatively as if it were unconjugated bilirubin in the diazo methods is clinically useful, and has been solidly established in the diagnosis of several cases of pathogenic conjugated hyperbilirubinemia, particularly in infants with infectious, metabolic, anatomic, and cholestatic disorders. However, in the absence of a pure bilirubin diglucuronide preparation with which to standardize the direct-reacting bilirubin, the use of any quantitative values obtained with this method is at least open to doubt. Because of the absence of the diglucuronide for study, its chemical and physical properties are only partly known. When the direct-reacting fraction is genuinely increased in plasma, its *exact* contribution to "total" bilirubin is unknown; hence, the total value is questionable (9).

Discrepancies in total bilirubin values between various commercial reference sera are well known. Resolution of these differences made significant progress with the appearance of the National Bureau of Standards preparation of high purity bilirubin for standardization. Differences between the Jendrassik–Grof and the modified Malloy–Evelyn methods have been significantly reduced with the appearance of an acceptable standard bilirubin in serum matrix, useful in both procedures (10).

When infrequent turbidity is noted in the final mixture, three causes may be suspected: reversal of steps in the procedure, the use of impure methanol, and lipemia. The first two causes are easily corrected; and use of the serum blank generally helps subtract the effects of lipemia.

A principal advantage of this method is its relative ease of standardization and control. The chloroform-methanol standards are highly "definable"; moreover, they are independent of a protein matrix, and of the ambiguities caused by variations in the composition of serum used as the medium for dissolving bilirubin. A principal disadvantage of the method is that dissolved hemoglobin, particularly when greater than 1000 mg/L, causes decreased values; hence, one must avoid the use of visibly hemolyzed specimens. Indeed, hemolyzed plasma specimens not only should occur rarely, but also should be avoided for most chemical analyses in the clinical laboratory.

References

1. Meites, S., and Hogg, C. K., Studies on the use of the van den Bergh reagent for the determination of serum bilirubin. *Clin. Chem.* **5**, 470–478 (1959).
2. Malloy, H. T., and Evelyn, K. A., The determination of bilirubin with the photoelectric colorimeter. *J. Biol. Chem.* **119**, 481–490 (1937).
3. MacDonald, R. P., Bilirubin (modified Malloy and Evelyn), Provisional. *Stand. Methods Clin. Chem.* **5**, 65–74 (1965).
4. Meites, S., and Levitt, M. J., Skin-puncture and blood collecting techniques for infants. *Clin. Chem.* **25**, 183–189 (1979).
5. Meites, S., and Lin, S. S.-H., Hemolysis in plasma samples from skin puncture of children. *Clin. Chem.* **26**, 987 (1980). Abstract.
6. Meites, S., and Traubert, J. W., Use of bilirubin standards. *Clin. Chem.* **11**, 691–699 (1965).
7. Behrman, R. E., Ed., *Neonatology*, C. V. Mosby Co., St. Louis, MO, 1973, p 221.
8. *Nelson Textbook of Pediatrics*, 11th ed., V. C. Vaughn, R. J. McKay, Jr., and R. E. Behrman, Eds., W. B. Saunders Co., Philadelphia, PA, 1979, pp 442–445, 1109–1119.
9. Meites, S., Quantitative total bilirubin determination: A part-time illusion. *Clin. Chem.* **25**, 1981 (1979). Letter.
10. Recommendation on a uniform bilirubin standard. *Clin. Chem.* **8**, 405–407 (1962).

Addendum

Evaluator S. L. C. has adapted the manual procedure to the ABA-100 and VP Bichromatic Analyzers (Abbott Laboratories, Dallas, TX 75247). The settings and procedure for each instrument are presented separately, then followed by data on precision and accuracy.

ABA-100 Settings

Power	On
Incubator	Heater Off (room temperature)
Mode Selector	Endpoint
Reaction Direction	Up
Analysis Time	5 Min
Carousel Rev	2
Filters	550/650
Syringe Plate	1:26
Sample Size	10 µL
Decimal Setting	000.0
Zero	0000
Calibration Factor	Conc. Stds.

Procedure

1. Fill the syringe plate with the total bilirubin reagent (below). Add 50 µL of water to cup 01 and 50 µL of STDS$_1$, STDS$_2$ (see *Note* below), controls, and patients' samples to subsequent cups, to no. 15. Prime the syringe plate four times by placing the sample probe in the reagent reservoir, using the manual Dispense switch. Position the sample probe on the boom arm, and depress the Run button.

2. As soon as 15 samples are pipetted, remove the syringe plate, and wash it 15 times.

3. Depress the Stop button at the end of the first revolution (5 min).

4. Fill the syringe plate with direct bilirubin reagent (below) and prime the plate as described above.

5. Remove the Multicuvette, and turn it clockwise so that positions 00 and 01 are empty.

6. Position sample probe on the boom arm, and depress the Run button.

7. As soon as the 15 samples are pipetted, depress the Stop button. When running batches of fewer than 10 samples, do not stop until this revolution is complete, to maintain a constant time interval.

8. Remove the Multicuvette and turn it counter-clockwise so that the Total Bilirubin water blank is in the 01 position.

9. Place the empty sample cups in positions 17 to 31, or as in the case of smaller batches, at the positions where the direct-reaching bilirubins were pipetted.

10. Adjust the instrument to zero with the water blank and calibrate it with the Total Bilirubin calibrator (STDS$_1$).

11. Turn the carousel forward to position 00 and depress the Run button. When the carousel has moved to the 01 position, turn it forward to the 00 position to activate the print out (Rev. 2).

Total bilirubin will be printed in positions 01 to 15. Direct-reacting bilirubin will be printed in positions 17 to 31.

Note: The calibrating serum (STDS$_1$) used for this procedure is Versatol Pediatric. To attain at least two reference points for calibrating this instrument, dilute the calibrating serum with an equal volume of diluent, to make STDS$_2$. For serum bilirubin values exceeding 24 mg/dL, the sample must be diluted and rerun. *However, the calibrators and samples cannot be diluted with water.* Dilute with a control serum of low bilirubin content, e.g., 0.4 to 0.6 mg/dL. The corresponding value of bilirubin in the diluting serum can be subtracted from the results, but this is usually unnecessary. *This type of dilution is mandatory* for both the ABA-100 and the VP analyzers.

ABA-100 Reagents

1. Total bilirubin: Mix 5.5 mL of methanol, 4.5 mL of water, and 1.0 mL of diazo reagent.
2. Direct-reacting bilirubin: Mix 10 mL of water and 1.0 mL of diazo reagent.

VP Settings

Test Name	Bili T
Temperature	25 °C
Filter Value	550/650
Units of Measurement	mg/dL
Dilution Ratio Setting	1:26
Rev Time	2
Aux Disp	NO
FRR	NO
Reaction Direction	Up
High Std	Conc of STD$_1$
Lo Std	Conc of STD$_2$
Test Type	EP
Bgn Prt Rev	2
No. Prt Rev	1
Max Absorbance	0.80

Note: Reagents and calibrators are same as described in the previous section.

Procedure

1. Prime the instrument with the total bilirubin reagent.
2. Enter the new test number.
3. Add 50 µL of water to cup 01 and 50 µL of STDS$_1$, STDS$_2$, controls, and patients' samples to subsequent cups, to no. 15. Depress E.
4. Depress the Stop button at the end of the first revolution.
5. Remove the Multicuvette and turn it clockwise so that no. 31 and 00 are empty.
6. Enter 001 to Wash, then 003 W/P, and prime with the direct-reacting bilirubin reagent.
7. Enter new test number, then E. As soon as the 15 samples are pipetted, depress the Stop button.
8. Remove the Multicuvette, and turn it counterclockwise so that the Total Bilirubin water blank is in position 01.

9. Enter test no. 03 Dispense Off; then depress E.
10. Enter new test number for Bili T; then depress E. Total bilirubin will be printed in positions 01 to 15. Direct-reacting bilirubin will be printed in positions 18 to 00.

Study of Precision, ABA-100

The following table shows within-run and day-to-day precision:

	Within-run for samples			
	1	2	3	4
Insert value, mg/L	5.0	27	47	190
Mean, mg/L	6.4	28.2	48.8	194
SD, mg/L	0.50	1.1	0.80	2.7
CV, %	7.8	4.0	1.6	1.4
No.	60	30	30	20

	Day-to-day	
	ABA-100	Manual
Insert value, mg/L	190	190
Mean, mg/L	191	188
SD, mg/L	2.4	4.3
CV, %	1.3	2.3
No.	10	10

Method Comparison

The following statistical studies (linear regression and *t*-tests) were made on the total bilirubin determination, comparing the manual method with each of the analyzers. Patients' sera used were selected on the basis of showing little to no direct-reacting component.

	ABA-100 (y); manual (x)		VP (y); manual (x)	
	x on y	y on x	x on y	y on x
y-intercept	0.4768	−0.3743	0.6208	−0.4590
Slope	0.9913	0.9983	0.9693	1.0092
r	0.9948	0.9948	0.9922	0.9928
Standard error of estimate, mg/L	5.767		1.874	
No.	31	31	27	27
Bilirubin, range, mg/L	4.0–196		34–192	
t-test results	2.433[a]		2.865[a]	

[a] Significant at $p = 0.05$.

Although the two instrumental methods correlate well with the manual method, a small but significant bias occurs, as indicated by the *t*-test. The cause for this has not yet been ascertained.

In the absence of a true standard for determining direct-reacting bilirubin, studies of correlation and *t*-tests show only that there is no strong basis for comparison of methods. Correlation is much weaker, and bias is far greater than that observed in the methods for total bilirubin.

Calcium in Biological Fluids

Submitters: Eugene S. Baginski and Slawa S. Marie, *Chemistry Laboratory, Doctors Hospital, Detroit, MI 48207*
Bennie Zak, *Wayne State University School of Medicine, Detroit, MI 48201*

Evaluators: Nancy W. Alcock, *Clinical Physiology–Renal Laboratory, Memorial Hospital, New York, NY 10021*
Karen Saniel-Banrey, *Clinical Chemistry Laboratory, The Children's Hospital, Columbus, OH 43205*
Richard F. Dods, *Pathology Department, Louis A. Weiss Memorial Hospital, Chicago, IL 60640*
Raymond E. Karcher, *Clinical Pathology Department, William Beaumont Hospital, Royal Oak, MI 48072*
John Pappas, *Advance Medical and Research Center, Inc., Pontiac, MI 48057*
Paul Pottgen, *Med-Check Laboratories, Inc., Pittsburgh, PA 15238*
Frank A. Sedor, *Clinical Chemistry Department, Duke University Medical Center, Durham, NC 27710*

Reviewer: George N. Bowers, Jr., *Clinical Chemistry Laboratory, Hartford Hospital, Hartford, CT 06115*

Introduction

Calcium plays an important role in human physiology; its determination in serum serves as a useful diagnostic indicator of disturbances in calcium metabolism. Both hypercalcemia and hypocalcemia are often life-threatening, and there is need for a rapid and reliable method for determining serum calcium. Inasmuch as hypocalcemia is often encountered in neonates, the method of choice is also the one that requires a small volume of specimen.

The method described here involves use of 20 µL of serum or plasma, requires no deproteinization, and is complete within 5 min. The technique can be applied also to other biological fluids, notably cerebrospinal fluid and urine.

Principle

In a strongly basic medium, calcium forms a complex with cresolphthalein complexone, with maximum absorbance at 575 nm. Interference by magnesium is circumvented by incorporating 8-hydroxyquinoline in the medium, and the presence of other metals is masked by cyanide.

Dimethyl sulfoxide serves as a solubilizing agent in the preparation of the cresolphthalein complexone reagent, and also helps keep the final solution clear when moderate lipemia is encountered. However, highly chylous or turbid sera make the final mixture opalescent, causing light-scattering. This is a minor problem that can be easily corrected as described later.

The presence of dimethyl sulfoxide also eliminates a "protein effect," a phenomenon responsible for a proportional change in slope (a matrix effect) when standards are prepared in the presence of protein, compared with the slope of a standard curve based on aqueous standards. Dimethyl sulfoxide also appears to repress the ionization of cresolphthalein complexone, lowering the absorbance of the blank.

Materials and Methods

Reagents

1. *Hydrochloric acid, concd.*
2. *Stock calcium standard, 100 mg/dL.* Transfer 250.0 mg of anhydrous calcium carbonate into a 100-mL volumetric flask. Dissolve in a small volume of concd. HCl, and dilute to the mark with metal-free water.

Note: Calcium standard solution can be prepared from calcium carbonate ($CaCO_3$), issued and certified by the National Bureau of Standards (SRM 915). Reviewer G. N. B. urges that this be used as one of the approaches to greater accuracy. Dry this reference material for 4 h at 200 °C and cool to room temperature in a desiccator before use.

3. *Working calcium standard, 10 mg/dL.* Dilute 10.0 mL of the calcium stock standard to 1 dL with water.
4. *Dimethyl sulfoxide, reagent grade.*
5. *8-Hydroxyquinoline.*
6. *Cresolphthalein complexone (CPC).* Dissolve 2.5 g of 8-hydroxyquinoline in 100 mL of dimethyl sulfoxide. Add 40.0 mg of cresolphthalein complexone and shake until all is dissolved. Add about 300 mL of water; mix. Add 1.0 mL of concd. HCl, mix

again, and finally dilute to 1 L with water. The acid must be present; otherwise, the chemical will reprecipitate upon dilution.

7. *Potassium cyanide, reagent grade.* (Danger! Extremely toxic!)

8. *Diethylamine buffer solution.* Dissolve 0.5 g of potassium cyanide in water, add 40.0 mL of diethylamine, and dilute to 1 L with water. Store in a polyethylene bottle.

9. *Ethylene glycol bis-(2-aminoethyl ether) tetraacetic acid (EGTA).* Prepare a 5 g/L aqueous solution to be used for blanking chylous specimens. If necessary, add a few drops of concd. KOH solution to facilitate solution.

Apparatus

Perform the test in 4.0-mL polystyrene cups (AutoAnalyzer™; Technicon Instrument Corp., Tarrytown, NY 10591) with polyethylene caps. Pipettes should have polypropylene tips.

Specimen Requirement

Serum, heparinized plasma, cerebrospinal fluid, or acidified 24-h urine may be used.

Procedure

1. Pipet a 20-µL sample of the specimen into one 4-mL polystyrene cup and 20 µL of the working calcium standard into another.

2. Add 1.0 mL of CPC reagent to each of the first two cups and to a third cup to be used as a reagent blank.

3. Add 1.0 mL of diethylamine buffer to each of the three cups, cap them, and mix by inversion.

4. After about 5 min determine absorbances of the sample and the standard against the reagent blank with a spectrophotometer set at 575 nm.

5. When turbid or chylous specimens are encountered, proceed as follows: Measure the absorbance of the tube containing the sample (Step 4) against the reagent blank. Add 50 µL of EGTA solution to each, mix, and determine the absorbance of the sample vs the blank again. Use the difference in absorbances for calculations.

Calculations

$$(A_S/A_{ST}) \times 10.0 = Ca, mg/dL$$

where A_S = absorbance of sample
A_{ST} = absorbance of standard

To convert mg/dL to meq/L, divide the above by 2; to convert to mmol/L, multiply meq/L by 0.5, or divide mg/dL by 4.

Reference Intervals

Serum: 8.5–10.5 mg/dL (4.3–5.3 meq/L, 2.1–2.6 mmol/L).

Cerebrospinal fluid: 4.2–5.4 mg/dL (2.1–2.7 meq/L).

Urine: 50–150 mg/24 h (may vary widely with diet).

Note: Evaluator P. P. submitted the following normal serum calcium range, established in his laboratory: 8.6–10.4 mg/dL.

Evaluator K. S.-B. provided the following list of normal serum calcium values from reference *10:*

Premature:	First wk:	6.0–10.0 mg/dL (1.5–2.5 mmol/L)
Full-term:	First wk:	7.0–12.0 (1.75–3.00)
Child:		8.0–11.0 (2.0–2.75)
Adult:		8.5–11.0 (2.13–2.75)

Evaluator F. A. S. reported these serum calcium values from a large number of patients: outpatients (1500), 8.8–10.9 mg/dL; inpatients (2500), 8.5–10.5 mg/dL.

Discussion

The ubiquitous presence of calcium, especially on laboratory glassware, and the high sensitivity of this method make it necessary to use plasticware for storage of reagents and for containers in which the test is performed. Both polyethylene and polystyrene appear satisfactory.

Note: According to Evaluator R. F. D., the amount of blank absorbance depends on the purity of cresolphthalein.

It may be convenient to prepare a "one-piece" reagent by premixing the CPC and diethylamine reagents shortly before use. However, such a mixture has limited stability, whereas individual reagents stored in polyethylene bottles are stable.

Hemolysis, jaundice, or moderate lipemia does not appear to interfere with the test *(1, 2)*. However, grossly lipemic or turbid specimens may pose a problem, which can be easily resolved as follows: After performing the procedure as described and determining the absorbance of the sample, add a small amount of EGTA solution to the tube containing the sample and to the reagent blank. Measure the absorbance again and subtract from the value obtained previously. Use this "corrected" sample absorbance, obtained after EGTA addition, in the calculations. When a spectrophotometer equipped with a flow-through cuvet is used, the solution cannot be retrieved; therefore, perform the test in duplicate, with and without EGTA. The chelating agent, having a higher affinity for calcium than does cresolphthalein complexone, binds the metal and disrupts the colored complex. The amount of scattered light is presumed to be the same before and after addition of EGTA.

Note: According to Evaluator K. S.-B., 10 µL of EGTA can be used when the method is scaled down to 10 µL of sample, 0.5 mL of CPC, and 0.5 mL of diethylamine, by using a Micromedic pipette for sample pickup and dilution of CPC.

The correction for the presence of turbidity as described here is possible only because of the large dilution used in the method. In most cases, however, the final solution is clear and the correction is unnecessary, even though the original specimen may appear moderately lipemic. When in doubt, perform the simple correction; otherwise, the turbidity may substantially affect the test, causing falsely high values.

Physiological and Clinical Considerations

Most body calcium is mineralized in the skeleton, and can be quickly mobilized to maintain a constant concentration in the blood. Approximately 41% of serum calcium is bound to albumin, a small amount is chelated, and the rest exists as ions. It is the latter form that plays a vital role in muscle contraction, in maintenance of cardiac rhythmicity, and in membrane stability. Calcium is also a vital co-factor in many enzymic activities. Consequently, maintenance of calcium homeostasis is of utmost importance in preserving normal physiology.

Calcium metabolism seems to be directly controlled by the parathyroid hormone (parathyrin; PTH) synthesized in the parathyroid gland. The hormone functions to mobilize calcium from the skeleton by affecting bone resorption, a process that is opposed by calcitonin, synthesized in the "C" cells of the thyroid gland. Parathyrin also affects calcium absorption from the intestine, if vitamin D is available, and increases reabsorption of calcium and decreases reabsorption of phosphate in the renal tubules. Hence, in a typical case of hyperparathyroidism, the calcium concentration in blood is increased, while the phosphate concentration decreases. The opposite situation exists in hypoparathyroidism. In patients with chronic uremia, hyperphosphatemia is often seen, but the serum calcium is usually normal.

Hypocalcemia in the neonatal period is common (3), especially in premature infants, infants born after complicated pregnancies, infants born to diabetic mothers, and others. These infants, as well as those born under normal conditions, are characterized by having a higher serum concentration of inorganic phosphate than do adults (4). Hypocalcemia in infants may be also secondary to hypomagnesemia (5, 6). Prolonged hypocalcemia may lead to disorders of bone mineralization, causing rickets in children and osteomalacia in adults.

Hypocalcemia is often associated with dietary deficiency or malabsorption, and with vitamin D deficiency or resistance to it. It is also found in renal tubular syndrome, chronic renal disease, and hypophosphatasia. A moderate form of hypocalcemia may be seen in patients with decreased serum protein; conversely, hyperproteinemia is often accompanied by hypercalcemia.

Hypercalcemia is also a well-recognized complication of malignant neoplastic disease, most often associated with osteolytic metastases but occasionally

Table 1. Evaluators' Studies of Precision

Evaluator	Material[a]	Within-day Mean (and SD), mg/dL	CV, %	Day-to-day Mean (and SD), mg/dL	CV, %
J. P. and R. E. K.	Serum	10.06 (0.143)	1.4	10.69 (0.407)	3.8
	Serum	–	–	7.45 (0.211)	2.8
N. W. A.	CSF	4.9 (0.084)	1.7	5.0 (0.12)	2.4
	Serum	9.7 (0.133)	1.4	9.7 (0.26)	2.7
	Urine	1.4 (0.058)	4.1	–	–
	Urine	5.2 (0.153)	3.0	5.2 (0.16)	3.1
P. P.	Serum	7.8 (0.17)	2.2	7.8 (0.28)	3.6
	Serum	10.2 (0.15)	1.5	10.2 (0.33)	3.2
	Serum	12.5 (0.18)	1.4	12.5 (0.31)	2.5
R. F. D.	Serum[b]	–	4.3–5.8	–	–
K. S.-B	Serum	6.62 (0.124)	1.9	6.73 (0.10)	1.5
	Serum	9.92 (0.206)	2.1	9.61 (0.18)	1.9
	Serum	14.45 (0.302)	2.1	14.55 (0.28)	1.9
	Serum	–	–	10.68 (0.21)	2.0
	Serum	–	–	9.51 (0.35)	3.7
	Serum	–	–	13.16 (0.38)	2.9
F. A. S.	Serum[c]	8.10 (0.21)	2.6	8.34 (0.31)	3.7
	Serum[c]	11.49 (0.14)	1.2	11.73 (0.34)	2.9

[a] n = 20 for each entry, except where indicated. CSF, cerebrospinal fluid.
[b] Five sera, 10 assays each.
[c] Within-day, n = 24; day-to-day, n = 25.

Table 2. Evaluators' Studies of Analytical Recoveries

Evaluator	Sample (no.)	Range of added Ca, mg/dL	Range of recovery, %
J.P. and R.E.K.	Serum (6)	6.0–10.0	99.4–101.4
N.W.A.	Urine (5)	0.9–9.4	89.9–98.3
	Serum (5)	9.4–10.7	100.1–104.1
	CSF (18)	3.9–7.5	97.6–100.6
P.P	Serum (4)	1.2–6.0	97–101
K.S.-B.	Serum (9)	2.5–12.5	102–104
F.A.S.	Serum (6)	1.9–5.1	91–108

with osteoblastic lesions *(7)*. An increased concentration of calcium in serum may also develop in patients who are immobilized for a long time, and is a common complication in multiple myeloma, sarcoidosis, hyperthyroidism, breast carcinoma, vitamin D toxicity, and other diseases.

Digitalis glycosides are known to increase calcium uptake by the myocardium, and hypercalcemia often induces digitalis toxicity. Hypocalcemia, on the other hand, may cause resistance to digitalis glycosides *(8)*. Anticonvulsant drugs, especially a combination of phenobarbital and phenytoin, may interfere with hepatic hydroxylation of calciferol to 25-hydroxycalciferol and thus lead to hypocalcemia *(9)*.

Precision, Recovery, and Method Comparison

The good precision of this method is presented in Table 1. Within-day precision studies of the Evaluators show that the CV varies between 1.4 and 5.8%, and that day-to-day precision is between 1.9 and 3.7%. Analytical recoveries (Table 2) range from 89.9 to 108%. Results by this method strongly correlate with those by atomic absorption *(11–14)* (Table 3).

References

1. Baginski, E. S., Marie, S. S., Clark, W. L., and Zak, B., Direct microdetermination of serum calcium. *Clin. Chim. Acta* **46**, 49–54 (1973).
2. Zak, B., Epstein, E., and Baginski, E. S., Review of calcium methodologies. *Ann. Clin. Lab. Sci.* **5**, 195–215 (1975).
3. Root, A. W., and Harrison, H. E., Recent advances in calcium metabolism. II. Disorders of calcium homeostasis. *J. Pediatr.* **88**, 177–199 (1976).
4. Root, A. W., and Harrison, H. E., Recent advances in calcium metabolism. I. Mechanisms of calcium homeostasis. *J. Pediatr.* **88**, 1–18 (1976).
5. Anast, C. S., Mohs, J. M., Kaplan, S. L., and Burns, T. W., Evidence for parathyroid failure in magnesium deficiency. *Science* **177**, 606–608 (1972).
6. Suh, S. M., Tashjian, A. H., Jr., Matsuo, N., Parkinson, D. K., and Fraser, D., Pathogenesis of hypocalcemia in primary hypomagnesemia: Normal end-organ responsiveness to parathyroid hormone, impaired parathyroid gland function. *J. Clin. Invest.* **52**, 153–160 (1973).
7. Lynwood, H. S., and Riggs, B. L., Clinical and laboratory considerations in metabolic bone disease. *Ann. Clin. Lab. Sci.* **5**, 252–256 (1975).
8. Chopra, D., Janson, P., and Sawin, C. T., Insensitivity to digoxin associated with hypocalcemia. *N. Engl. J. Med.* **296**, 917–918 (1977).
9. Anast, C. S., Anticonvulsant drugs and calcium metabolism. *N. Engl. J. Med.* **292**, 587–588 (1975).
10. Meites, S., Normal total plasma calcium in the newborn. *CRC Crit. Rev. Clin. Lab. Sci.* **6**, 1–18 (1975).
11. Cali, J. P., Bowers, G. N., Jr., and Young, D. S., A referee method for the determination of total calcium in serum. *Clin. Chem.* **19**, 1208–1213 (1973).
12. National Committee for Clinical Laboratory Standards. Tentative Standard T5C-14, NCCLS, 771 E. Lancaster Ave., Villanova, PA 19085.
13. Cali, J. P., Mandel, J., Moore, L., and Young, D. S., A referee method for the determination of calcium in serum. NBS Spec. Publ. 260–36, US Dept. of Commerce, National Bureau of Standards (Supt. of Docu-

Table 3. Evaluators' Comparison of Methods

Evaluator	Comparison method and material (no.)	Slope	y-intercept	r	Comments
J.P. and R.E.K.	AA, serum (60)	1.05	−0.33	0.963	$t = 3.55$ ($p = 0.05$)
	SMAC, serum (60)	0.91	0.97	0.972	$t = 4.43$ ($p = 0.05$)
	AA, serum (9–20)				F-test = 2.37, within-day ($p > 0.05$)
					F-test = 3.67, day-to-day, Level 1 ($p = 0.05$)
					F-test = 8.45, day-to-day, Level 2 ($p = 0.05$)
N.W.A.	AA, serum (25)	0.9856	0.2216	0.9692	
	AA, CSF (20)	0.9744	0.2508	0.9668	
	AA, urine (20)	1.0185	0.4066	0.9958	
P.P.	AA, serum (70)	1.04	0.29	0.98	Student's t-test, $p > 0.05$; F-test, $p > 0.05$
R.F.D.	AA, serum (19)	0.857	1.36	0.873	$t = 1.16$ ($p > 0.05$); F-test = 1.14 ($p > 0.05$)
F.A.S.	AA, serum (40)	0.986	0.098	0.998	$t = 1.184$ ($p > 0.05$)

AA, atomic absorption, SMAC, Technicon trade mark.

Table A1. Day-to-Day Precision of the Methods[a]

Insert value, mg/dL	Method 1 Mean (SD), mg/dL	CV, %	Method 2 Mean (SD), mg/dL	CV, %	Method 3 Mean (SD), mg/dL	CV, %
6.5	6.7 (0.10)	1.5	6.7 (0.07)	1.0	6.7 (0.18)	2.7
6.8	7.3 (0.18)	2.5	7.2 (0.16)	2.2	7.3 (0.19)	2.6
8.2	8.2 (0.21)	2.6	7.9 (0.15)	1.9	8.1 (0.15)	1.9
9.1	9.6 (0.16)	1.7	9.7 (0.11)	1.1	9.7 (0.22)	2.3
9.6	9.5 (0.38)	4.0	9.6 (0.23)	2.4	9.3 (0.30)	3.2
10.2	10.6 (0.45)	4.2	10.8 (0.18)	1.7	10.4 (0.23)	2.2
12.6	13.2 (0.35)	2.7	13.0 (0.21)	1.6	12.7 (0.41)	3.2
13.6	14.6 (0.30)	2.1	14.7 (0.20)	1.4	14.4 (0.56)	3.9

[a] Eight commercial reference sera were used, with n = 15 over a 42-day period.

ments, US Govt. Printing Office, Washington, DC 20802, cat. no. C13.10:260-36), 1972.
14. Tietz, N. W., A model for a comprehensive measurement system in clinical chemistry. *Clin. Chem.* **25**, 833–839 (1979).

Addendum: Modifications of the Method for Pediatrics

Evaluator K. S.-B. reports the following modifications:

Manual Procedure, Method 1.

1. Pick up 10 µL of sample with a Micromedic Automatic Pipette (Model 25000), and dilute with 0.5 mL of CPC reagent (stored in a brown glass bottle). Deliver into a 1.5 mL polyethylene centrifuge tube (with cap) (KEW Scientific, Inc., Columbus, OH 43227, or Beckman Instruments, Inc., Fullerton, CA 92634).
2. Add 0.5 mL of diethylamine reagent [Dispensette (Brinkmann Instruments, Inc., Westbury, NY 11590) pipetting device, on plastic bottle].
3. Cap the tubes and mix with a vortex-type mixer. Allow to stand 5 min. Read absorbance at 575 nm with a spectrophotometer.

Notes:
1. Wrap the plastic lines of the automatic pipette with aluminum foil to reduce exposure of the CPC to light.
2. For turbid specimens, use 10 µL of the EGTA, rather than 50 µL originally recommended.
3. Day-to-day precision (n = 30): Serum 1: mean 9.27 (SD 0.26) mg/dL; CV, 2.8%. Serum 2: mean 12.6 (SD 0.31) mg/dL; CV, 2.4%. See also Table A1.

Automated Procedures

Use the following settings for an ABA-100 Bichromatic Analyzer (Abbott Diagnostics, South Pasadena, CA 91303):

Power	On
Incubator	37 °C
Mode Selector	Endpoint
Reaction Direction	Up
Analysis Time	5 min
Carousel Rev	2
Filter	550/650
Syringe plate	1:101
Sample Size	5 µL
Zero	0000.
Decimal	000.0
Cal Factor	(Concn std/A_d) × 0.5

2. With the above settings, the procedure may be used in either of two ways (Methods 2 and 3). After premixing equal volumes of CPC and diethylamine reagents, proceed as follows:

Method 2:

1. Fill the syringe plate with the premixed reagent. Add 50 µL of water to cup 01, and 50 µL of standards, controls, and sera to subsequent cups.
2. Position the probe in the reservoir, recycle the reagent four times through the syringe and back into the reagent reservoir, using the Manual Dispense switch. Position the probe on the boom arm, and push the Run button. At the end of Revolution 1, press the Stop and then the Test buttons.
3. Place cup 01 and corresponding cuvet receptacle in the light path; using the Fine Scaling dial, adjust the 10 standard to 010.0. Push Stop and Run buttons, and when the carousel rotates to 01 position, advance the carousel one complete revolution to position 00.
4. The carousel will now begin the second revolution, in which the calcium values will be printed out in mg/dL.

Method 3:

1. Prime the premixed reagent into a Micromedic Automatic Pipette set to deliver the same sample and reagent volume as the ABA-100 syringe plate. Deliver the sample dilutions directly into the ABA-100 cuvet, and allow to stand about 5 min.
2. Proceed as in Step 3, Method 2.

Carbon Dioxide Content by Microgasometer

Submitter: Samuel Natelson, *Department of Environmental Practice, College of Veterinary Medicine, University of Tennessee, Knoxville, TN 37901*

Evaluator: James W. North, *Clinical Instruments Division, Beckman Instruments Inc., Brea, CA 92621*

Introduction

The significance of total carbon dioxide content (total CO_2) determination has been reviewed in detail *(1, 2)*. Recently, it has been emphasized again that estimation of total CO_2 affords information that cannot be substituted for by the measurement of other blood components *(3)*.

CO_2 content of blood was estimated, originally, gasometrically. This was done either volumetrically *(4)* or manometrically *(5)* with instruments developed by Van Slyke. Micro versions of these instruments were developed subsequently *(6, 7)*. Of these, the microgasometer of Natelson has been used most widely, and continues to be used as a reference method for CO_2 content of serum or plasma.

Other techniques used for this purpose include titrimetric procedures *(8)*, gas chromatography *(9)*, and infrared measurement of CO_2 absorbance *(10)*. In the continuous-flow method for total CO_2, the CO_2 is liberated by acidification and its concentration estimated colorimetrically by the change in color of a buffered solution of indicator *(11)*. Alternatively, the CO_2 liberated is estimated by potentiometry *(12)* or by infrared spectrometry *(10)*.

An indirect method has been used widely, in which total CO_2 is calculated from the Henderson–Hasselbalch equation, after the determination of the p_{CO_2} and pH of whole blood *(13)*. This procedure has come under criticism, because the pK' of the equation is not constant in disease *(14)* and varies with blood pH *(15)*.

The microgasometer has significant advantages for the small laboratory. Compared with other instruments, it is relatively inexpensive. There are no electronics associated with the instrument. As a result, there is less "down" time with the microgasometer, and when malfunctioning, it can be readily repaired by the technician.

A significant advantage is that the measurement is absolute, that is, independent of standards, and dependent only on the calibration of the glassware; the other methods cited require calibration with standards each time the test is run. The sample size is small (30 µL or less), the technique is simple, and the result is available within 3 min. The microgasometer has been used routinely in many laboratories for the past 25 years, and is still used in some pediatric laboratories, especially as a backup for an automated instrument. It also serves instrument manufacturers as a reference procedure for the calibration of other instruments.

Principle

The manometric measurement of CO_2 is based on the fact that the gram molecular volume of CO_2 at 0 °C and 760 mmHg pressure is 22.2 L. Therefore, 1 mmol of CO_2 occupies 22.2 mL at standard conditions, and the volume of CO_2 liberated by a measured volume of serum or plasma can by simple calculation be reported in mmol/L.

Note: 1 mmHg ≃ 133 Pa.

In the manometric procedure, the CO_2 is liberated from the sample by the addition of acid into an evacuated volume of reduced pressure created in a mercury manometer. The pressure developed, in mmHg, is read on the manometer. The CO_2 is reabsorbed in alkali and the pressure is read again. The difference is the pressure due to the CO_2. Normally, the volume sampled is 30 µL. The volume at which the gas is measured is 120 µL. By Boyle's Law *(PV = P'V')*, the volume at 760 mmHg, multiplied by 0.12 mL, is equal the measured pressure *(P')* multiplied by the unknown volume. This yields the volume of CO_2 at 760 mmHg and ambient temperature. A temperature correction is then applied, by Charles' Law *(TV = T' V')*, to adjust to 0 °C, where T is in kelvins. The temperature correction factors are given in the handbooks of chemistry and physics. The volume obtained is the volume of CO_2, at standard conditions, in 30 µL of sample. This is readily converted to mmol/L as described above.

In actual practice, the calculations merely require

the multiplication of the measured pressure by a factor, determined from a table listing the factor at different temperatures.

Note that the readings are independent of the ambient barometric pressure. This is not true for gasometers, which measure the gas volumetrically.

Materials and Methods

Reagents

1. CO_2 reference standards: Dry some sodium carbonate (Na_2CO_3, anhydrous, ACS certified; Fisher Scientific Co., Pittsburgh, PA 15219) at 100 °C overnight in a drying oven. Cool and keep in a desiccator over anhydrous $CaSO_4$. Dissolve 50 mmol (5.300 g) in water and dilute to 1 L. Dilute again to make 25 and 12.5 mmol/L solutions. Keep in well-stoppered polyethylene bottles in a refrigerator. Distribute these solutions in 10-mL polyethylene bottles for use with the gasometer. While these solutions are stable for many months, it is advisable that they be made fresh monthly, comparing the new solutions with the old as to their CO_2 content. Standardized stock Na_2CO_3 solutions are available from many suppliers, e.g., Fisher or Beckman.

2. Lactic acid (1.25 mol/L): Dilute 90 mL of lactic acid (850 g/kg, ACS certified, sp. gr. 1.246, 13.83 mol/L; Fisher Scientific) to 1 L. Stable for at least one month when refrigerated.

3. NaOH (3 mol/L): Dissolve 120 g of NaOH (Fisher Scientific, ACS certified) and dilute to 1 L. Stable indefinitely at room temperature in a well-stoppered polyethylene bottle.

4. Reagents prepared for assay: Into five 5-mL test tubes or vials fitted with screw tops place 2 mL of Hg. Add 2 mL of each of the following reagents to each of the labeled tubes: lactic acid (1.25 mol/L), NaOH (3 mol/L), and the 12.5, 25, or 50 mmol/L Na_2CO_3 standards. To a 10-mL tube or vial with screw top, add 5 mL of Hg and 4 mL of distilled water. Cover the lactic acid with 0.5 mL of caprylic alcohol. Keep these solutions in a rack near the microgasometer, and renew them from stock solutions as they are used up in the test. They are stable at room temperature for about a week if kept tightly stoppered when not in use.

Apparatus

Microgasometer: This instrument is sold as the Natelson microgasometer by several supply houses (e.g., Fisher Scientific). There is a manual model, in which shaking is performed manually, and a motorized model, in which a motor moves the piston back and forth and stirs with a magnetic stirrer. The procedure is the same for both models. The essential features are a pipette for sampling when the piston is moved back, a reaction chamber where the gas is released, and a manometer connected by a Y-tube

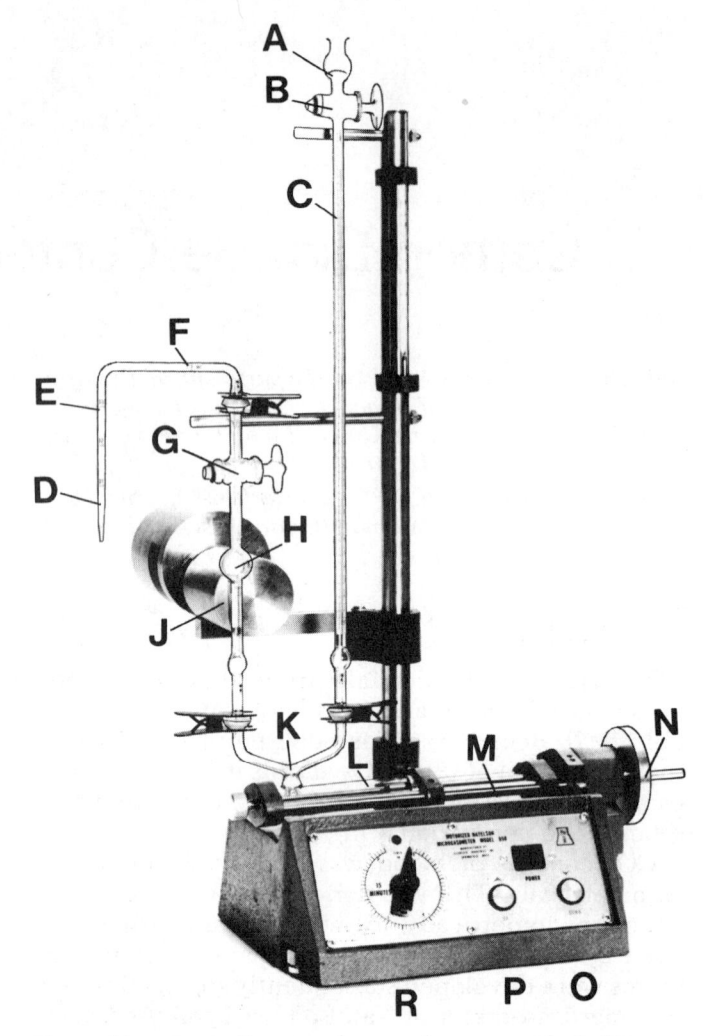

Fig. 1. Motorized microgasometer: *A*, mercury seal; *B*, stopcock; *C*, reaction bulb, 3 mL; *D*, pipette; *E*, mark at 30 µL; *F*, mark at 0.1 mL; *G*, stopcock; *H*, reaction bulb, 3 mL; *J*, magnetic stirrer; *K*, Y-tube; *L*, syringe barrel, *M*, piston; *N*, handle to adjust piston manually; *O*, control for piston down; *P*, control for piston up; *R*, timer

to the piston and manometer, so as to adjust and measure the pressure (Figure 1).

Assemble the microgasometer according to the instructions of the manufacturer, and fill it with clean Hg. As the instrument is used, Hg accumulates in the waste, but can be cleaned and re-used as follows: Transfer the Hg to a 500-mL beaker and run tap water over it in the sink, until the reagents are removed. Fill the beaker to the top with 10% HNO_3 (100 mL/L). Stir with a glass rod or pipette until the Hg looks clean. The water should be acid to litmus. Run tap water over the Hg, with stirring, until the HNO_3 is washed away. The pH should be about 5. Decant as much water as practicable. Rinse once with distilled water, and decant again. Dry the Hg with surgical gauze, to remove the last traces of water. Pour the Hg onto a filter paper held in a funnel. Puncture a small hole with the edge of a paper clip at the apex of the filter paper. The clean Hg will come through, and the last traces of water will be absorbed by the paper. Some oxidized Hg and dust will remain in the funnel; save it to be cleaned with the next batch of Hg.

Keep the microgasometer on a luncheon tray, or its equivalent, to facilitate collecting any spilled Hg. Keep the tray in a well-ventilated hood, to avoid the hazards of breathing the Hg vapor. The working reagents are kept on the tray with the microgasometer. It is also advisable that the operator work with disposable, thin plastic gloves.

Procedure

1. Blood is drawn into a heparinized syringe or evacuated blood-collection tube from the vein or artery. Blood from a fingertip or heel prick is collected into Natelson blood-collecting tubes or polyethylene microcentrifuge tubes, 250-µL capacity with caps (Kew Scientific, Columbus, OH 43227).

2. Transfer the blood from the syringe to Natelson blood-collecting tubes. Cap the bottom of the blood-collecting tube with a plastic cap (Criticaps; Scientific Products, Division of American Hospital Supply Co., Evanston, IL 60085). This seal must be tight to avoid leaking. Preferably, the blood is centrifuged in the refrigerated centrifuge, and the blood-collecting tubes can be placed in test tubes containing crushed ice. Alternatively, the blood can be centrifuged at room temperature, if there is no delay in assaying with the microgasometer. The blood from the syringe may also be transferred to test tubes. If they are not to be assayed promptly, seal the containers and store in the refrigerator at 4 °C. In any case, samples must be assayed within 1 h after they are drawn.

3. Draw 30 µL of serum or plasma into the pipette of the microgasometer by pulling back the piston; then draw in 30 µL of Hg *(see Note no. 1 for details)*.

4. Draw in 30 µL of lactic acid, followed by 30 µL of Hg.

5. Draw in 0.1 mL of water, followed by enough Hg to bring the Hg meniscus to the 0.12-mL mark in the reaction chamber, beyond the stopcock.

6. Close the stopcock, and pull back the piston until the aqueous meniscus is below the midpoint of the 3-mL bulb.

7. Shake for 2 min or let stand for 2 min with occasional shaking. Shaking should be gentle to avoid fragmenting the Hg. Pull back the piston so that the Hg meniscus is at the 3-mL mark.

8. Advance the piston, and bring the aqueous meniscus (caprylic alcohol) to the 0.12-mL mark. Read the pressure on the manometer promptly. This is P_1.

9. Advance the piston slowly until the aqueous layer approaches the stopcock and a slight resistance is felt. Open the stopcock. Advance the piston until a drop of Hg protrudes from the tip of the pipette.

10. Immerse the tip of the pipette in the NaOH solution and draw up 30 µL, followed by enough Hg to bring the NaOH solution into the reaction chamber and the Hg meniscus to the 0.12-mL mark.

11. Close the stopcock and pull back the piston to bring the solution below the middle of the reaction bulb. Shake for about 10 s and bring the Hg meniscus back to the 3-mL mark. Let stand for 1 min.

12. Advance the aqueous meniscus to the 0.12-mL mark, and read the pressure. This is P_2. $P_1 - P_2$ is the net pressure due to the CO_2.

13. To clean the instrument, release the pressure by bringing the aqueous meniscus to the stopcock and opening the stopcock. Advance the piston to eject the aqueous contents. Sample 0.1 mL of lactic acid followed by 30 µL of caprylic alcohol. Bring this to the 3-mL mark, without following with Hg. Eject the aqueous layer and take up 0.1 mL of water, as for the lactic acid wash. Bring to the 3-mL mark below the reaction chamber and then eject the water. Advance the Hg until the bubble protrudes. The instrument is ready for the next specimen.

14. Calculate the result with the following equation:

$$(P_1 - P_2) \times \text{Factor (see Table 1)} = CO_2 \text{ content, in mmol/L}$$

Notes:

1. Care must be taken not to introduce air into the instrument while sampling. This is achieved by advancing the Hg until a bubble protrudes from the pipette. Dip the pipette tip into the test tubes containing the reagents. The protruding Hg drop at the tip of the pipette is removed by touching it to the Hg at the bottom of the container. The reagent is then sampled, the tip is dipped again into the Hg, and 30 µL of Hg is sampled as a seal. To sample serum or plasma, take 30 µL of sample, and advance the piston until a convex meniscus forms at the tip of the pipette. Dip this into Hg contained in a separate test tube, and plug with 30 µL of Hg, as for the reagents.

2. To sample from the Natelson blood-collecting tube, place it into a test tube and centrifuge. Cut off and discard about 1 cm from the narrow tip. Now cut at the interface between the plasma and cells, to isolate the plasma. Tilt so that the plasma runs to the end of the tube. Now press your thumb at the opposite end of the cut tube. The tube can now be held in a vertical position, with the plasma at the top. Bring it to the tip of the sampling pipette of the microgasometer so that the tip dips into the plasma and is supported by the walls of the cut tube. Release the thumb from the bottom of the tube. The plasma remains suspended, by capillary pressure. Sample 30 µL of plasma and then 30 µL of Hg from a test tube, proceeding with the test as described. We recommend 30 µL of Hg as a plug between reagents; lesser amounts may be used, as long as the same amount is used for all specimens.

3. At the end of each run, introduce 0.5 mL of excess Hg into the instrument. It is advisable to mark the location of the piston with a marking pencil at the beginning of the run, and to bring the piston back to this position at the end of the test, by expelling the excess Hg, so

that the same area of the piston is used each time, and the instrument is not overloaded with Hg.

4. Caprylic alcohol is used to prevent foaming. It is not necessary to sample this material because the lactic acid solution should be saturated with caprylic alcohol. In addition, caprylic alcohol is introduced into the instrument during the washing procedure, and some adheres to the walls of the reaction chamber. *Commercial anti-foaming agents should be avoided, as many contain amino compounds, which react with CO_2.*

5. Hg adhering to the stopcock in the measuring chamber indicates that the stopcock has been greased inadequately. Rotate the stopcock a few millimeters to locate a new spot on the stopcock. If the Hg still remains suspended below the stopcock, warm the gas above the suspended Hg with the thumb and forefinger so as to expand the gas and force the Hg down.

6. The test is described for plasma. This is advantageous because the same sample can also be used for related measurements such as p_{O_2} and pH. Serum can also be used in a similar manner. Other biological fluids, such as cerebrospinal fluid and urine, can also be used.

7. If the blood is collected in evacuated tubes (e.g., Vacutainer Tubes, Becton Dickinson Co., Rutherford, NJ 07070), results for normal serum or plasma will be about 7% lower than those obtained if the blood is drawn with a syringe *(16)*. It is assumed that if the evacuated tube is filled to capacity, then any released CO_2 will redissolve. However, there is usually a residual volume, which contributes to the error. This error can be larger if the tube is not filled to capacity with blood. It is recommended, where practicable, that the blood be drawn into a syringe, the needle removed, and the syringe capped, placed in an ice bath, and then sent to the laboratory. Keep in mind when evaluating the results that fingertip blood is about 93% arterial blood *(17)*; it should be drawn into heparinized tubes, capped, and sent to the laboratory in ice, especially if the blood gases are also to be assayed.

8. When bringing the volume to the 0.12-mL mark, read the pressure promptly, because some of the CO_2 will tend to dissolve as the pressure is increased.

9. The factors of Table 1 are derived by applying pressure and temperature corrections to arrive at 0 °C and 760 mmHg pressure. In addition, a factor is applied for the solubility of CO_2 in water, and for the fact that some CO_2 redissolves as the meniscus is brought to mark to be read. These are derived from the equation developed by Van Slyke *(5, 18)*.

$$V_{0, 760} = a \times \left(\frac{P}{760}\right) \times \left(\frac{1}{1 + 0.00384t}\right) \times \left[1 + \left(\frac{S}{A-S}\right)\alpha'\right] i$$

Where a = the measured volume (0.12 mL), $P = P_1 - P_2$ (measured), t = temperature in °C. Note that the temperature factor, $273/(273 + t)$, simplifies to $1/(1 + 0.00363t)$. An additional adjustment is introduced to correct for the expansion of the mercury and glass with temperature. A = 3 mL, the volume of the chamber, and S is the volume of the aqueous phase (0.16 mL). α' is the solubility of CO_2, in milliliters, in 1 mL of H_2O, or the partition coefficient for CO_2 at 760 mmHg pressure of CO_2. This is also known as the distribution coefficient. A factor i, added to correct for the reabsorption of CO_2 as the volume is brought to 0.12 mL, has been calculated to be 1.017. To convert to millimoles per liter, multiply the volume *(V)* by 1000/0.03, to obtain the volume of gas in 1 L of plasma, and divide by 22.2.

10. Because the factory calibration of the glassware may be inaccurate, it is advisable to prepare one's own set of factors by repeatedly analyzing the 50, 25, and 12.5 mmol/L standards. If the $P_1 - P_2$ readings are reproducible, and if P_2 remains fairly constant, then the machine is functioning correctly, and a factor can be calculated for the temperature at which the assay is run. If the temperature were 25 °C, for example, and a factor of 0.240 yielded the correct CO_2 concentration in the standards, when multiplied by $P_1 - P_2$, then the other temperature factors can be corrected, for this particular set of glassware, by multiplying each factor by 0.240/0.232 (the 0.232 is the correction factor for 25 °C from Table 1). If new glassware is installed in the instrument, the new factors need to be redetermined.

Table 1. Temperature Correction Factors

°C	Factor	°C	Factor
20	0.237	27	0.230
21	0.236	28	0.229
22	0.235	29	0.228
23	0.234	30	0.227
24	0.233	31	0.225
25	0.232	32	0.224
26	0.231		

Discussion

The CO_2 is liberated in the microgasometer at a pH of about 2. Under these conditions, the CO_2 is from HCO_3^-, undissociated H_2CO_3, dissolved CO_2, and small amounts of CO_3^{2-}. Carbamino compounds will hydrolyze at acid pH to liberate CO_2, but there is not enough time for this to occur during this determination. Thus CO_2 content, as performed manometrically, does not measure all the CO_2 that could be derived from the plasma.

About 94% of the CO_2 in blood exists as HCO_3^-, at pH 7.4. At pH 7.2, only about 87% of the total CO_2 exists as HCO_3^-. Thus the measurement of CO_2 content is not a direct measure of bicarbonate.

Various techniques have been recommended for sampling. One method is to draw the sample, and allow the heparinized blood to stand in the sealed syringe in a 37 °C air bath, until the cells settle. The supernatant plasma is then transferred to a test tube and sampled promptly. The rationale behind this procedure is that CO_2 diffuses into the erythrocytes as the temperature is lowered. If the sample is cooled from 37 °C to 25 °C, the results will be lower by 0.5 mmol/L *(18)*.

Another technique is the one recommended above, where the blood is drawn and the syringe is sealed and placed in an ice bath to stop metabolism. When the sample reaches the laboratory, it is transferred to a test tube and centrifuged (sealed) in a refrigerated centrifuge. This is also readily done with blood-collecting tubes, centrifuging the sealed tubes in ice, and separating the plasma by cutting the tube. Under these conditions blood can be kept for at least 1 h without a change in CO_2 content. If incubated at 37 °C, glucose concentrations decrease and CO_2 content increases substantially. These problems are eliminated if the blood is analyzed promptly.

The most commonly used procedure is to collect the blood into evacuated tubes. As pointed out above, it is important to fill these tubes when the sample is drawn. Some analysts, however, then pour the serum into cups on a turntable; as a result, the last specimen stands exposed for as much as 40 min to 1 h. The results obtained in this manner have little significance, especially for patients with a respiratory acidosis. If samples are kept at 4 °C, in sealed tubes or in syringes, results will be valid for at least 2 h after the blood is drawn.

Historically, blood for this purpose was drawn under oil and allowed to clot at room temperature. After centrifugation, the serum and oily layer were poured into another test tube. This system was used widely before the advent of evacuated collection tubes. This procedure has been criticized, with the claim that mineral oil extracts the CO_2, causing low results (19). Table 2 illustrates results found with the microgasometer for blood samples from 10 healthy men, ages 23 to 28 years, comparing the different methods of sampling the blood.

When comparing the differences between the mean values by Student's *t*-test, no significance could be adduced at the 5% level. The *F*-test, comparing the coefficients of variation, also did not show a significant difference at the 5% level. However, in every case, the value obtained with the Vacutainer Tubes was lower, by at least 1 mmol/L, than the lowest value obtained with the other procedures. This was true even though only those samples were used where the Vacutainer Tube had been filled completely at the time of drawing. This difference decreased as the blood was allowed to stand. At the end of 4 h at room temperature, however, the CO_2 content of the blood in the evacuated tubes showed a mean value of 29.5 mmol/L. Thus there is a compensation of errors with the evacuated collection tubes, CO_2 content increasing with time until analysis and decreasing from extraction by the reduced pressure in the tubes.

Table 2 justifies the collection of blood under oil. However, the oil tends to cling to the glass, and presents some problems in sampling into the microgasometer, some oil getting into the reaction chamber at times.

When drawing blood from the vein or artery, the Submitter prefers drawing the blood into a syringe containing heparin. The needle is removed, after it has been ascertained that the blood extends to the tip, and the syringe tip is capped. The syringe is placed in an ice bath and delivered to the laboratory. There, blood from the syringe is transferred to the Natelson blood-collecting tubes, which are capped and centrifuged. The blood in the syringe is injected into the gas analyzer for pH, p_{CO_2}, and p_{O_2} determinations. The plasma in the blood-collecting tubes, along with that taken separately from the fingertip, is separated by cutting the tube as described in *Notes*, and sampled into the microgasometer for estimation of CO_2 content. Thus the same technique is used for venous, arterial, and fingertip blood with the microgasometer.

CO_2 content of venous blood, for infants and adults, is substantially higher than that observed with fingertip blood. In a study involving 52 healthy newborn infants, the Submitter found that fingertip plasma yielded a mean value of 24.8 mmol/L, compared with 26.1 mmol/L observed in plasma taken from the scalp vein (Table 3). Similar observations were made in adults.

The reproducibility attainable with the microgasometer depends on the skill of the operator in keeping air out of the instrument. It is also important that the piston be returned to its original position at the end of each determination, expelling the excess Hg. P_2, obtained after reabsorption of the CO_2, should be constant within 1 mmHg; this indicates

Table 2. CO_2 Content Obtained with Different Methods of Collecting Venous Blood

Technique	CO_2 content, mmol/L	CV, %
Plasma from blood maintained at 37 °C	27.3	5.5
Plasma from blood cooled in ice bath	26.5	4.2
Plasma from Vacutainer Tube at room temperature	24.9	7.1
Serum from blood under oil at room temperature	27.6	3.9

All specimens were analyzed 1 h from time of blood collection.

Table 3. Reference Intervals for CO_2 in Adults and Newborn Infants

	In adults, mmol/L		In newborns, mmol/L	
Sample	Mean	Range	Mean	Range
Venous plasma	27.1	25.1–28.3	25.8	24.3–26.9
Venous serum	27.9	26.1–29.1	26.1	24.9–27.3
Arterial plasma	24.1	22.5–26.7	23.5	21.9–25.5
Arterial serum	24.7	23.1–26.9	24.7	22.3–26.0
Fingertip plasma	25.2	24.3–27.4	24.8	23.1–26.2

Table 4. Reproducibility of the Microgasometer CO₂ Content Assay, as Measured with 50, 25, and 12.5 mmol/L Standards

	mmol/L	
Standard	Mean found	CV, %
12.5	12.41	1.6
25.0	25.0[a]	1.4
50.0	50.1	1.2

[a] Factors were adjusted, as described in text, to make this standard read 25.0 mmol/L as a correction for the error in glassware calibration for the particular microgasometer used.

that the serum and reagents have been sampled properly. With these precautions, Table 4 represents the results of a study of aqueous CO_2 standards designed to test the reproducibility of results from the microgasometer. The factors having been derived by using the 25 mmol/L standard, as described above, this is also a test of accuracy for the 12.5 and 50 mmol/L standards as well. From the data it is apparent that within the 95% confidence limits a reproducibility of ± 3.2% can be achieved readily.

The reference values listed in Table 3 have been compiled from data collected during the years 1955–1978, comprising four major studies conducted by the Submitter, each involving more than 50 individuals, and obtained with the experimental procedure recommended in this report.

Evaluator J. W. N. compared the microgasometer with the Beckman CO_2 Analyzer (12) for correlation and precision. CO_2 results obtained from 184 patients by both methods ranged from 8.6 to 40.8 mmol/L. Figure 2 shows the regression line comparing the two methods. No significant error in either proportional or constant statistical parameters was observed (slope = 1.005, y-intercept = 0.20). The standard error of estimate (uncertainty about the regression line) was 0.89 mmol/L. From the data, no significant difference was demonstrated between the two procedures as to accuracy or precision.

A second correlation plot (Figure 3) is the result of a study during the CAP Verification of Performance Claims on the Beckman Rate pH method. In this study, blood from 171 patients was analyzed. All analyses were carried out in groups of five samples to minimize the effect of CO_2 loss from the samples. The range of results for CO_2 was 7.1 to 40.2 mmol/L. Of the patients in the study, 44 had values < 20 mmol/L, 100 values were between 20 and 30 mmol/L, and 27 values were greater than 30 mmol/L. This correlation study showed a proportional difference between the microgasometer and the Beckman procedure of less than 1.6%. The y-intercept was −0.6 mmol/L. By the regression statistics, the computation for bias at 20 mmol/L was −0.3 mmol/L and at 30 mmol/L, −0.26 mmol/L. These data confirm the data in Figure 2, and Figures 2 and 3 are in general agreement with the data in Table 4.

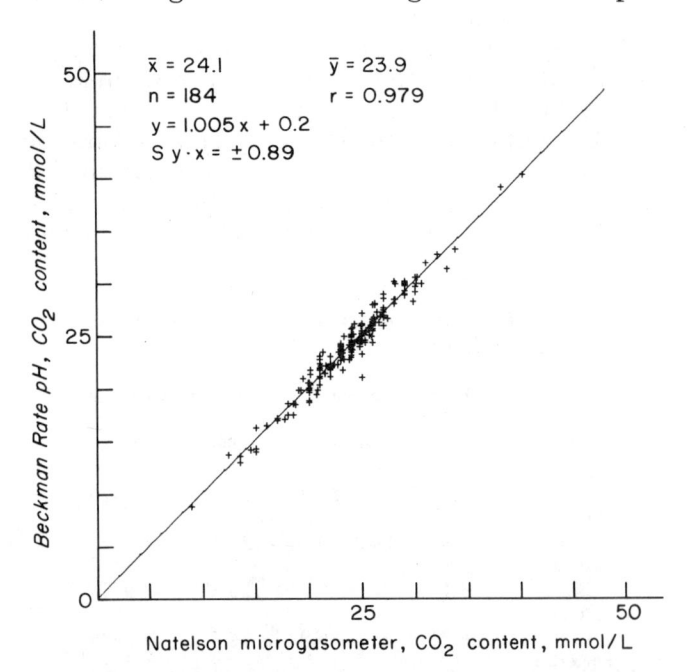

Fig. 2. Comparison of Beckman CO_2 Analyzer with the Natelson Microgasometer

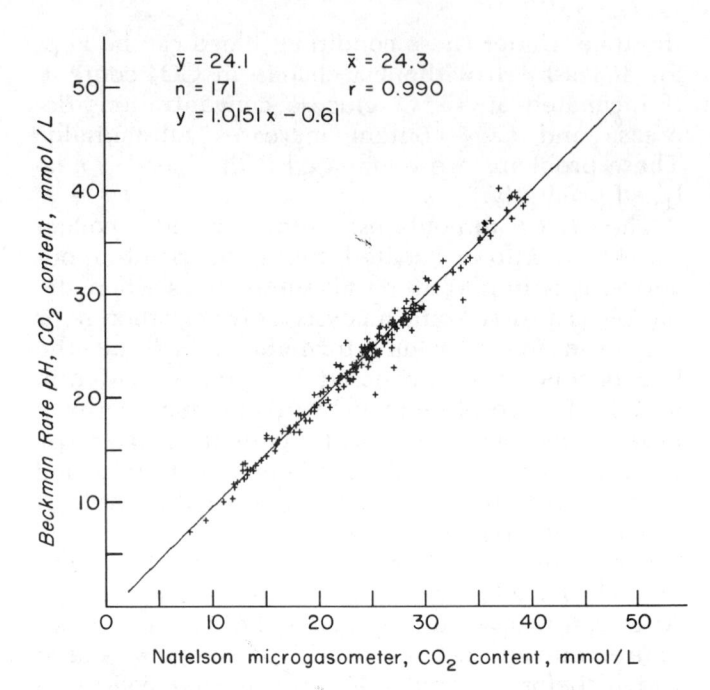

Fig. 3. Regression line obtained comparing Natelson Microgasometer with Beckman CO_2 Analyzer during performance study by the College of American Pathologists

References

1. Natelson, S., and Natelson, E., Abnormal blood pH (acidosis and alkalosis). In *Principles of Applied Clinical Chemistry*, **1**, *Maintenance of Fluid and Electrolyte Balance*. Plenum Press, New York, NY, 1975, pp 213–246.

2. Tietz, N. W., and Siggaard-Andersen, O., Acid–base and electrolyte balance. In *Fundamentals of Clinical Chemistry*, N. Tietz, Ed., W. B. Saunders, Philadelphia, PA, 1976, pp 945–974.
3. Natelson, S., Miletich, D. J., Seals, C. F., Visintine, D. J., and Albrecht, R. F., Clinical biochemistry of epilepsy. II. Observations on two types of epileptiform convulsions induced in rabbits with corticotropin. *Clin. Chem.* **25**, 898–913 (1979).
4. Van Slyke, D. D., II. A method for the determination of carbon dioxide and carbonates in solution. *J. Biol. Chem.* **30**, 347–368 (1917).
5. Van Slyke, D. D., and Neill, J., The determination of gases in blood and other solutions by vacuum extraction and manometric measurement. I. *J. Biol. Chem.* **61**, 523–573 (1924).
6. Scholander, P. F., and Irving, L., Micro blood gas analysis in fractions of cubic millimeters of blood. *J. Biol. Chem.* **169**, 561–569 (1947).
7. Natelson, S., Routine use of ultra-micro methods in the clinical laboratory. I. A scheme of analysis for estimation of Na, P, Cl, protein and hematocrit. II. Easily constructed micropipets. III. Estimation of CO_2 content of serum with a practical microgasometer. *Am. J. Clin. Pathol.* **21**, 1153–1172 (1951).
8. Segal, M. A., Rapid electrometric method for determining CO_2 combining power in plasma or serum. *Am. J. Clin. Pathol.* **25**, 1212–1216 (1955).
9. Natelson, S., and Stellate, R. L., Apparatus for extraction of gases for injection into the gas chromatograph. *Anal. Chem.* **35**, 847–851 (1963).
10. Rabinow, B. E., Geisel, A., Webb, L. E., and Natelson, S., Automated system for total CO_2 analysis of plasma contained in capillaries by infrared spectrometry. *Clin. Chem.* **23**, 180–185 (1977).
11. Kenny, M. A., and Cheng, M. H., Rapid automated simultaneous determination of serum CO_2 and chloride with the AutoAnalyzer I. *Clin. Chem.* **18**, 352–354 (1972).
12. Hicks, J. M., Aldrich, F. T., and Josefsohn, M., An evaluation of the Beckman chloride/carbon dioxide analyzer. *Clin. Chem.* **22**, 1868–1871 (1976).
13. Sinclair, M. J., Hart, A., Pope, H. M., and Campbell, E. J. M., The use of the Henderson–Hasselbalch equation in routine medical practice. *Clin. Chim. Acta* **19**, 63–69 (1968).
14. Natelson, S., and Nobel, D., Effect of variation of pK' of the Henderson–Hasselbalch equation on values obtained for total CO_2, calculated from p_{CO_2} and pH values. *Clin. Chem.* **23**, 767–769 (1977).
15. van Steklenburg, G. J., Valk, C., and Donckerwolcke, R. A. M. G., Variation of carbonic acid pK'_{1g} in blood and urine during $NaHCO_3$ infusion and NH_4Cl loading. A study of two renal acidotic patients. *Clin. Chem.* **26**, 60–65 (1980).
16. Gambino, S. R., Heparinized vacuum tubes for determination of plasma pH, plasma CO_2 content and blood oxygen saturation, with an extensive discussion of pH methodology. *Tech. Bull. Regist. Tech.* **29**, 123–131 (1959).
17. Natelson, S., and Menning, C. M., Improved methods of analysis for oxygen, carbon monoxide and iron on fingertip blood. *Clin. Chem.* **1**, 165–179 (1955).
18. Astrup, P., Simple electrometric technique for determination of carbon dioxide tension in blood and plasma. *Scand. J. Clin. Invest.* **8**, 33–43 (1956).
19. Paulsen, L., Comparison between total CO_2-content (total CO_2) in plasma/serum from clotted blood collected with or without paraffine oil. *Scand. J. Clin. Invest.* **9**, 402–405 (1957).

Carbon Monoxide

Submitter: Willard R. Faulkner, *Department of Biochemistry, Vanderbilt University Medical Center, Nashville, TN 37212*

Evaluators: Lester I. Burke, *Holy Cross Hospital, Chicago, IL 60629*
Patricia Dobbs and Charles A. Pennock, *Paediatric Chemical Pathology, Bristol Maternity Hospital, University of Bristol, Bristol BS2 8EG, U.K.*

Reviewer: O. W. van Assendelft, *Hematology Division, Center for Infectious Diseases, Centers for Disease Control, Atlanta, GA 30333*

Introduction

Carboxyhemoglobin (HbCO) can be measured in any of several different ways.[1] One major approach involves the liberation of carbon monoxide bound to hemoglobin, followed by quantitation of the gas. Methods such as these include microdiffusion *(1)*, volumetric and manometric procedures *(2)*, gas chromatography *(3)*, infrared spectroscopy *(4)*, and measurement of the thermal change after oxidation of carbon monoxide in the presence of Hopcalite *(5)*. With all of these procedures, hemoglobin must also be measured so that its degree of saturation with carbon monoxide can be calculated. Microdiffusion is time-consuming and is only semiquantitative. The other methods all require special equipment, and some require skills that are attainable only through long experience.

Alternatively, HbCO can be measured by spectrophotometric procedures without liberating the carbon monoxide. This type of procedure is based on the measurement of absorbances at two or more wavelengths chosen in such a manner that the concentrations of each of two or three components in the sample may be calculated through the solution of simultaneous linear equations in two or three unknowns. Procedures based on direct spectrophotometry are simple, rapid, and sufficiently accurate for clinical purposes. The method presented here is based on the method of Tietz and Fiereck *(6)*.

---

[1] Abbreviations: HbCO, carboxyhemoglobin; HbO$_2$, oxyhemoglobin; Hb, deoxygenated hemoglobin; SHbCO, sulfcarboxyhemoglobin.

Principle

In dilute alkaline solution (pH about 11), oxyhemoglobin (HbO$_2$) and HbCO have similar double absorption bands in the visible region (450–650 nm), as shown in Figure 1, whereas deoxygenated hemoglobin (Hb) has a single broad band around 555 nm (Figure 2), Addition of sodium hydrosulfite (Na$_2$S$_2$O$_4$), a reducing agent, readily deoxygenates HbO$_2$ to Hb but has relatively little, if any, effect on HbCO. After Na$_2$S$_2$O$_4$ treatment of an HbCO-containing dilute blood sample, the ratio of absorbances at 541 nm and 555 nm is a linear function of the percent saturation of hemoglobin with carbon monoxide *(1)*. Figure 3 shows a calibration graph developed from blood specimens with 0, 50, and 100% saturations. These observations form the basis for the method presented here.

Materials and Methods

Reagents

1. *Ammonium hydroxide reagent, 4 g/L.* Dilute 15.9 mL of concd. ammonium hydroxide (29% NH$_3$;

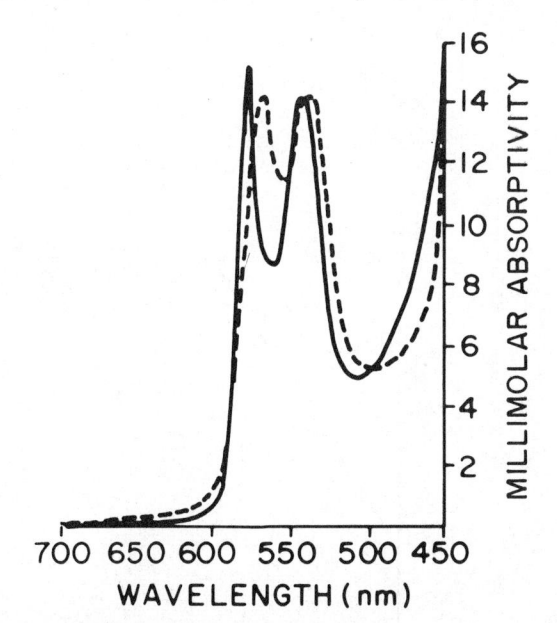

Fig. 1. Spectral curves of oxyhemoglobin (———) and carboxyhemoglobin (---)
Adapted from van Assendelft *(7)*, with permission. Absorptivity values are based on definition of hemoglobin molecule as one iron-containing heme group

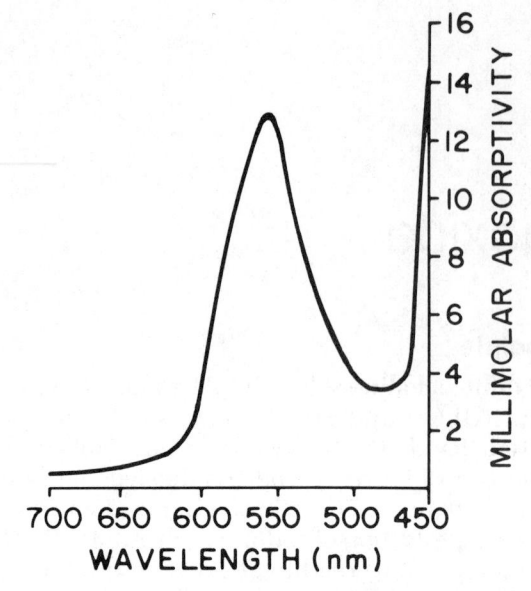

Fig. 2. Spectral curve of deoxygenated hemoglobin
Source and absorptivity definition in Fig. 1

290 mL/L) to 1 L with water. This reagent is stable indefinitely.

2. *Sodium hydrosulfite (sodium dithionite) powder ($Na_2S_2O_4$).* Preweigh 10-mg portions of this reagent into individual 10 × 75 mm tubes and cover with Parafilm™ (Fisher Scientific, Pittsburgh, PA 15219). When 10 mg is added to 3 mL of the ammonium hydroxide reagent, the solution should be clear. If turbidity develops, use a different lot of hydrosulfite.

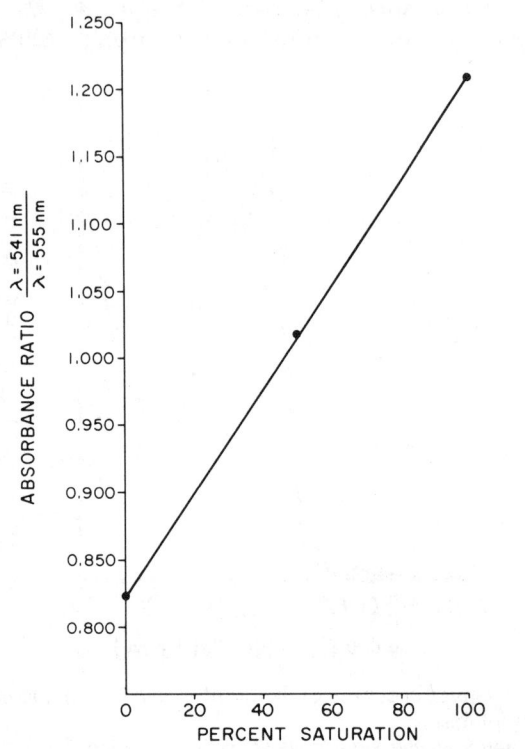

Fig. 3. Standard plot of ratios of absorbances (A_{541}/A_{555}) at hemoglobin saturations from 0 to 100%

3. *Carbon monoxide,* 99% pure.

Note: For greatest safety keep the cylinder secured to the wall in a fume hood.

4. *Oxygen,* 99.6% pure.
5. *Nitrogen,* 99.99% pure.

Note: The Submitter obtained all of the above gases from Price-Bass Co., Nashville, TN, a distributor of products of Linde Division, Union Carbide Corporation, Florence, SC 29501.

6. *Mercury, ACS grade* (Fisher Scientific).

Equipment

1. Two 12.5-mL separatory funnels or a tonometer.
2. Erlenmeyer flasks, 50 mL.
3. A spectrophotometer having a narrow bandpass (less than 4 nm in the range of 325—700 nm).
4. Volumetric pipettes, 0.5, 1.0, and 25-mL sizes.
5. Lang–Levy pipettes, 100 µL.
6. Two 3-mL syringes with metal caps (no Luer-Lok).
7. A properly functioning fume hood.

Procedure

1. Pipet 25.0 mL of ammonium hydroxide reagent into a 50-mL Erlenmeyer flask, and with a Lang–Levy pipette add 100 µL of a well-mixed blood specimen. Mix and allow to stand 2 min.

Note: Evaluators P. D. and C. A. P. prefer Sahli blood pipettes because of their ease of handling and excellent precision.

2. Transfer 3 mL of the diluted specimen into a suitable cuvet. Fill the reference cuvet with ammonium hydroxide reagent.

3. Add 10-mg portions of sodium hydrosulfite powder to both cuvets. Cover the cuvets with Parafilm and mix by inverting 10 times. Allow to stand exactly 5 min.

4. Read the absorbances at 541 and 555 nm.

5. Calculate the ratios of absorbances, A_{541}/A_{555}, and read the percent saturation of the hemoglobin in the blood specimen from a standard curve as shown in Figure 3.

Note: Evaluators P. D. and C. A. P. observed no difference in absorbance at either wavelength when the cuvets were allowed to stand from 30 s to 10 min after adding the sodium hydrosulfite. They did note a small decrease in absorbance after 10 min, which did not affect the ratio. After 30 min there was a marked decrease in the ratio, producing unreliable results.

These Evaluators recommend the inclusion of a 100% carboxyhemoglobin control with each run; its absorbance will be about 1.2. This can be prepared by bubbling carbon monoxide into a small portion of blood in a test tube.

Note: If a sample cannot be analyzed immediately, stopper the container tightly and place it in a refrigerator where it will be stable for several days. However, blood that has been stored for over a week or postmortem blood may contain interfering pigments, making it unsuitable for analysis by this method *(6)*.

Standardization

1. Draw into a heparinized tube 12 mL of blood from a healthy nonsmoker.

Note: Although most chemists use heparin, there is no evidence that other anticoagulants are not suitable.

2. Place 5-mL portions of the fresh blood in each of two 125-mL separatory funnels.

3. Treat one sample with pure oxygen and the other with pure carbon monoxide for 15 min, while rotating the funnels gently. *Do this in a fume hood.*

Note: The Submitter used an IL Model 237 tonometer (Instrumentation Laboratory, Inc., Lexington MA 02173) and equilibrated the blood for 15 min at 300 mL/min gas flow at 37 °C. Three milliliters was then drawn into a syringe, the gas expelled, the cap filled with mercury, and the tip pressed into the cap. The blood was thoroughly mixed within the syringe before sampling.

4. After treating the samples with the gases, close the separatory funnels and rotate gently for an additional 15 min.

5. Analyze the fully saturated samples immediately in triplicate by the procedure given above.

6. Calculate the ratios A_{541}/A_{555} at 0 and 100% saturation with carbon monoxide. Plot the points on linear coordinate paper and draw a line connecting them.

7. For intermediate points, fill the funnel containing 100% HbCO with nitrogen and rotate for 5 min. (This removes the physically dissolved CO and a small amount of CO bound to hemoglobin.) Continue with steps 1 through 4 under *Procedure*. Calculate the absorbance ratio and read the percentage saturation from the plot prepared in step 6 above. Prepare intermediate standards by mixing in small vials appropriate proportions of the nitrogen-treated sample with the oxygen-treated one. Calculate the percentage saturation by multiplying the percentage of CO saturation in the N_2-treated sample by the appropriate factor:

N_2-treated sample, mL	O_2-treated sample, mL	% CO-saturation factor
0.5	1.5	0.25
0.5	0.5	0.50
1.5	0.5	0.75

8. Analyze the above mixtures in triplicate by the *Procedure*, steps 1 through 4. Plot the calculated concentrations vs the corresponding absorbance ratios obtained. These points should fall on the line or close to it.

Reference Intervals

Blood was collected over a period of six months from 43 adult employees of the Vanderbilt University Medical Center and analyzed by the method presented here. This group of individuals included light, medium, and heavy smokers, as well as nonsmokers. The mean value for all the subjects was 3.15% CO saturation (range, 0–10%).

Blood of the nonsmoking employees had a carbon monoxide saturation range of 0–2%, whereas that of several heavy smokers was as much as 10% saturated. Heavy automobile traffic in urban areas may result in a HbCO saturation of 0–4%. Heavy tobacco smoking may cause values typically in the 5–7% range, with occasional concentrations as high as 14% *(8, 9)*.

Discussion

The method described in this chapter requires measurements at 541 and 555 nm in an alkaline solution. At 555 nm, Hb has an absorption maximum and HbCO has a minimum. At 541 nm, these compounds do not have an isobestic point. Of the commonly encountered hemoglobin derivatives (Hb, HbO_2, HbCO, and methemoglobin), only methemoglobin is pH-dependent over a wide range. This derivative is reduced by sodium hydrosulfite ($Na_2S_2O_4$) to Hb, and therefore does not interfere in the test.

A potential, though not likely, source of error is the production of sulfcarboxyhemoglobin (SHbCO) from treatment of the initial specimen with hydrosulfite. This derivative has a broad peak with a maximum at 614 nm *(10)*, which is considered to be a slight disadvantage in measuring HbCO in a Hb/HbCO mixture containing $Na_2S_2O_4$. However, van Kampen and Zijlstra *(10)* have compared a method of this type with one involving $Na_2S_2O_4$ and have concluded that only slight errors are introduced by formation of SHbCO.

Note: Evaluator L. I. B. found no difference in the ratio A_{541}/A_{555}, using from 10.5 to 30.4 mg of the hydrosulfite at an HbCO concentration of 1%.

Within-day precision for the method presented here was reported by Evaluators P. D. and C. A. P.: for 20 samples the mean was 46.3% saturation and the SD was 2.08% saturation, yielding a CV of 4.5%.

These Evaluators report that when a blood sample saturated with CO was kept in a half-filled 2-mL heparinized container overnight at 4 °C there was no loss of CO. When the container was one-fourth full, there was a loss of about 10%. Overnight storage of a full container kept at room temperature resulted in a loss of 2–3% of CO, which is within the error of the method.

Clinical Interpretation

The toxic effects of carbon monoxide are attributable primarily to tissue hypoxia, caused by inability of the blood to transport sufficient oxygen. The affinity of CO for Hb is between 200- and 250-fold that of O_2 for Hb, and for this reason even a low CO concentration in the air breathed (0.05%, 500 ppm, 0.5 mL/L) results in a significant concentration of HbCO in the blood (20–25%). In carbon monoxide poisoning, O_2 transport is impaired not only through blocking a portion of the Hb but also by causing a marked shift of the O_2 dissociation curve to the left, resulting in a further decrease in the availability of O_2 to the tissues.

The signs and symptoms of CO poisoning (Table 1) are related to the HbCO concentration in the blood and length of time of exposure to the CO. Toxicity is affected by such factors as the general state of health of the victim, the amount of physical activity engaged in at the time, and the Hb concentration. Signs and symptoms may be inconsistent or even absent at moderately low concentrations of CO in the atmosphere and may not follow a specific sequence. With respect to this possible inconsistency, it is important to realize that the percentage of HbCO can be a misleading guide, because it is the actual amount of HbO_2 that remains available to O_2 transport rather than the percentage. If this amount is low, as in anemia, clinical symptoms are likely to appear at a lower concentration of HbCO than in an individual with a normal Hb concentration.

Because of the low affinity of Hb for O_2, compared with that for CO, the latter is not readily replaced from Hb by O_2 except at high oxygen tensions. Under normal breathing conditions at one atmosphere of pressure (101 kPa), 4 h is required to lower a given HbCO concentration by one-half. At 100% O_2, the time required is reduced to about 40 min. With hyperbaric O_2 the time can be further decreased to 20–30 min.

In patients treated with O_2 before a blood specimen was collected, the measured HbCO may not reflect the degree of CO poisoning.

Table 1. Signs and Symptoms of Carbon Monoxide Poisoning *(11)*

	% HbCO saturation
No symptoms	0–10
Tightness across forehead, possibly slight headache, dilitation of cutaneous blood vessels, shortness of breath	10–20
Headache, throbbing in temples, irritation, disturbances of judgment	20–30
Severe headache, weakness, dizziness, dimness of vision, nausea and vomiting, collapse	30–40
Same as above with greater possibility of collapse or syncope, increased respiratory and pulse rates	40–50
Syncope, coma with intermittent convulsions, Cheyne–Stokes respiration	50–60
Coma with intermittent convulsions, depressed heart action and respiration, possible death	60–70

References

1. Conway, E. J., *Microdiffusion Analysis and Volumetric Error*, 5th ed., The MacMillan Co., New York, NY, 1958.
2. Peters, J. P., and Van Slyke, D. D., *Quantitative Clinical Chemistry, 2, Methods*, Baltimore, MD, 1932, pp 229–442.
3. Collison, H. A., Rodkey, F. L., and O'Neal, J. D., Determination of carbon monoxide in blood by gas chromotography. *Clin. Chem.* **14**, 162–171 (1968).
4. Gaensler, E. A., Cadigan, J. B., Jr., Ellicott, M. F., Jones R. H., and Marks, A., A new method for rapid precise determination of carbon monoxide in blood. *J. Lab. Clin. Med.* **49**, 945–957 (1957).
5. Linderholm, H., Sjödstrand, T., and Söderström, B., A method for determination of low carbon monoxide concentration in blood. *Acta Physiol. Scand.* **66**, 1–8 (1966).
6. Tietz, N. W., and Fiereck, E. A., The spectrophotometric measurement of carboxyhemoglobin. *Ann. Clin. Lab. Sci.* **3**, 36–42 (1973).
7. van Assendelft, O. W., *Spectrophotometry of Hemoglobin Derivatives*, Royal Vangorcum, Ltd., Assen, The Netherlands, 1970.
8. *Newsletter*, October-November 1973, Pittsburgh Poison Center, 125 DeSoto St., Pittsburgh, PA 15213.
9. Lawther, P. J., and Commins, B. T., Cigarette smoking and exposure to carbon monoxide. *Ann. N. Y. Acad. Sci.* **174**, 135–147 (1970).
10. van Kampen, E. J., and Zijlstra, W. G., Determination of hemoglobin and its derivatives. *Adv. Clin. Chem.* **8**, 141–187 (1965).
11. Goodman, L. S., and Gilman, A., Eds., *The Pharmacological Basis of Therapeutics*, 5th ed. MacMillan Publishing Co., Inc., New York, NY, 1975, p 902.

Chlorides: An Introduction to Three Methods

Three quite different analytical approaches to the determination of chloride in serum are presented to meet various laboratory situations with respect to workload and available instrumentation. All are based upon the very slight solubility (or low ion product) constants for the salts of Group I metallic ions, notably silver and mercury. Titrimetric analyses depend upon the insolubility or only slight dissociation of their chlorides; color reactions for chloride, and many of the indicators used in its titration, rely upon replacement reactions among various anions of silver and mercuric salts, a number of which have even lower ion product constants than do the chlorides. In common with most chloride methods, all three methods presented are subject to various interferences by other halides, but only bromide presents a problem with clinical samples; the coulometric titration, unlike the other methods, is not interfered with by iodide but can be affected by certain reducing constituents that are seldom seen clinically.

Coulometric determination of metallic ions is a classical technique, and still one of the few "definitive" measurements in analytical chemistry. Except for the initial investment in instrumentation, it is still one of the most economical, rapid, precise, and easily performed analyses in the clinical laboratory. However, the "cell constant" (consisting of the "background" titration to bring the concentration of excess titrant to a detectable endpoint and the lag-time to bring the mixture to the sensing electrodes) must be averaged and subtracted by some means; the more recent appearance of faster (higher current) instruments has increased the time and variability of this "zero correction," which becomes troublesome only when high precision and accuracy are required.

Colorimetry, based upon the replacement by free chloride ion of thiocyanate, from mercuric thiocyanate, has been widely used in automated systems. Although the manual version requires somewhat more knowledge and care in setting up and monitoring the reaction, colorimetry may be preferable for the final measurement for some available instrumentation, particularly those in which a part or all of the process may be carried out in sequentially automated or semiautomated systems. The concentration range over which the absorbance is linear depends upon reagent composition, and the calibration should be checked at several concentrations.

A simple "direct" mercurimetric titration has been included for use in laboratories where a low work volume does not justify special instrumentation or maintenance of a colorimetric method. Either argentometric or mercurimetric titrimetry of the chloride in a precisely performed protein-free filtration (20-fold or greater dilution in tungstic acid) eliminates problems with the determination of the visual endpoint and is a more widely accepted procedure for defining accuracy. However, carefully performed direct titration of serum with mercuric salts produces apparently unbiased results. Difficulties with the detection of the endpoint arise from competition of protein for the mercury cation and from highly jaundiced specimens; these effects can be minimized, and at least they do not include the fairly rapid reversal of endpoints common to direct titrations with silver salts.

In recent surveys of more than 700 laboratories engaged in interstate commerce, 52% of participants used mercuric thiocyanate (preponderantly automated) color measurements, 22% used coulometric titrations, 10% used mercurimetric titrations, and 13% used ion-selective electrode analysis.—*Alan Mather*

Chloride, Colorimetric Method

Submitter: Stanley S. Levinson, *Department of Laboratory Medicine, Sinai Hospital of Detroit, Detroit, MI 48235* (former address: *Clinical Laboratory, Brookline Hospital, Brookline, MA 02146*)

Evaluators: George Nix, *Blood Gas Section, Department*

of Pathology, Wishard Memorial Hospital, Indiana University Medical Center, Indianapolis, IN 46202

Sherwood C. Lewis, *Department of Pathology and Laboratory Medicine, St. Francis Hospital and Medical Center, Hartford, CT 06105*

Richard F. Dods, *Clinical Biochemistry Section, Pathology Department, Louis A. Weiss Memorial Hospital, Chicago, IL 60640*

Reviewer: Alan Mather, *Clinical Chemistry Division, Center for Environmental Health, Centers for Disease Control, Atlanta, GA 30333*

Introduction

Chemical–mercuric (1, 2) and coulometric–amperometric titrating (3) methods for assaying chloride in biological fluids have previously been described in the *Selected (Standard) Methods* series. Other techniques for assaying chloride include the AutoAnalyzer® (Technicon Instruments Corp., Tarrytown, NY 10591) and ion-selective electrodes (4). Disadvantages of these approaches include slowness, poor reproducibility of manual chemical titrations owing to subjective endpoint determinations, and the need for costly equipment to obtain rapid, reliable results for only a limited number of specimens in the small laboratory.

Colorimetric techniques, another way of assaying chloride, may be used with instrumentation already available in most small laboratories. Chloride in water has long been assayed colorimetrically by measuring the amount of red color produced when ferric ion reacts with thiocyanate (5). Two modifications of this method for the manual analysis of serum chloride have been published (6, 7). The method described here is similar to one of these modifications (6), except that the volumes of the reagents have been adjusted for use with microsamples and with modern spectrophotometers that can accurately measure up to 1.500 absorbance units in small volumes (8). These reagent adjustments enable rapid, reproducible assay of multiple samples of "stats" with minimum interference from bilirubin or lipids. The method involves inexpensive chemicals that can be purchased already prepared and that have a long shelf-life. Although high-quality spectrophotometers are required, to provide precision comparable with the coulometric–amperometric, AutoAnalyzer, and ion-electrode techniques, this type of instrumentation is found in most small laboratories.

Principle

The principle of the assay is given by the reactions:

$$2\text{ Cl}^- + \text{Hg(SCN)}_2 \rightleftharpoons \text{HgCl}_2 + 2\text{ SCN}^-$$
$$\text{SCN}^- + \text{Fe}^{3+} \rightleftharpoons \text{FeSCN}^{2+} \text{ (red)}$$

Chloride quantitatively displaces thiocyanate from mercuric thiocyanate. Liberated thiocyanate ion reacts with ferric ion from ferric nitrate to form the red color that is measured spectrophotometrically in the range of 460 to 480 nm.

The sensitivity and linearity of the reaction are adjusted by the amount of mercuric ion included in the solution, because mercuric nitrate reacts preferentially with chloride over thiocyanate.

$$2\text{Cl}^- + \text{Hg}^{2+} \rightleftharpoons \text{HgCl}_2$$

Depending on whether more or less mercuric nitrate ion is added, less or more chloride, respectively, will be available to react with thiocyanate, thereby regulating the intensity of the color.

Materials and Methods

Reagents

Stock solutions[1]

1. Mercuric thiocyanate solution, saturated. Suspend 2.0 g of mercuric thiocyanate in 1 L of distilled water. Allow the solution to stand at room temperature for 48 h with occasional stirring. Decant and filter the solution through Whatman no. 2 filter paper (this solution is similar to Technicon formula T01-0026).

2. Ferric nitrate solution, 202 g/L. Dissolve 202 g of ferric nitrate in a 1-L volumetric flask containing 500 mL of distilled water. Add 31.5 g of nitric acid to the flask and mix. Dilute with distilled water to a final volume of 1 L (similar to Technicon formula T01-0028).

3. Mercuric nitrate solution, 68.5 g/L. Dilute 20.8 g of nitric acid with 800 mL of distilled water in a 1-L volumetric flask. Add 68.5 g of mercuric nitrate and shake until dissolved. Dilute with distilled water to a final volume of 1 L (similar to Technicon formula T01-0027).

4. Working solution: Add 100 mL of ferric nitrate solution to 1 L of mercuric thiocyanate, satd. solution; then add 0.85 mL of mercuric nitrate solution.[2]

The working solution takes no more than 20 min to prepare and is stable after preparation for at least two months at room temperature.

Note: Evaluator R. F. D. found it difficult to reproduce the working solution to give the same absorbance from batch to batch; he suggests titrating each new batch with mercuric nitrate solution or mercuric thiocyanate

---

[1] The stock solutions are the same as those used for the Technicon AutoAnalyzer "chloride by thiocyanate" method and may be purchased already prepared from Fisher Scientific, Pittsburgh, PA 15219. The commercially prepared solutions currently cost less than 1¢/mL and are stable for at least one year when stored at room temperature in brown bottles.

[2] The mercuric nitrate solution must be added precisely, with a pipette that is at least as accurate as a 1 in 0.01-mL Mohr-type glass pipette (yellow coded, Class A with an accuracy range of ± 0.002 mL; Fisher Scientific Co.).

solution until a linear relationship between the standard and absorbance is obtained. If the absorbance of the lowest standard is too high, add mercuric nitrate solution; if too low, add thiocyanate solution.

The assay has been performed for more than three years in the laboratory of the Submitter; this problem has not been observed when commercially purchased reagents are used. It is possible that reagents prepared "in house" exhibit less batch-to-batch consistency than does the commercial counterpart. If so, titration, as suggested by R. F. D., is a way to secure a usable working solution without having to remake it.

5. Chloride standards. Dissolve 4.38, 5.84, and 7.3 g of dried sodium chloride (ACS grade) with 1 L of distilled water in separate 1-L volumetric flasks: this provides 75, 100, and 125 mmol/L standards. Commercially prepared 80 and 125 mmol/L standards are available from Hycel, Houston, TX 77306, and 100 mmol/L standards, traceable to the National Bureau of Standards, can be obtained from several biomedical distributors, including Scientific Products, McGaw Park, IL 60085 (Harleco prod. no. 23831–16).

Apparatus

It is desirable to use a spectrophotometer that can assay accurately up to 1.500 absorbance units (usually digital) at wavelengths between 460 and 480 nm. For results that are comparable in speed and accuracy with those obtained with automated analyzers, use direct concentration readout when available.

Note: Instruments with a 500/600 filter such as the ABA-50 (Abbott Diagnostics Division, S. Pasadena, CA 91030) can also be used.

An alternative back-up type procedure, described later, involves spectrophotometers that read accurately up to only 0.80 absorbance units, such as the Coleman Junior II (Perkin-Elmer Corp., Oak Brook, IL 60521).

Collection of Specimens

Collect blood in silicone-coated red-top glass tubes without additive. The size of the tube is unimportant as long as enough serum for the assay is obtained. Tubes may be obtained from any of several manufacturers. After the clot forms, centrifuge and remove the serum for testing.

Note: The Reviewer suggests that red-top non-siliconized Vacutainer Tubes (Becton, Dickinson and Co., Rutherford, NJ 07070) are preferred and that, if siliconized tubes are used, they should be inverted to facilitate clotting.

The sera are stable in capped tubes for at least 4 h at room temperature, two days refrigerated, and several months when stored frozen.

Plasma may be used when collected in heparinized tubes, but ethylenediaminetetraacetate and fluoride anticoagulants will interfere with the assay.

Note: The Submitter does not use plasma because anticoagulants that contain sodium, potassium, or lithium may interfere with sodium and potassium assays on the same specimen.

Cerebrospinal fluid should be collected in three sterile tubes, with samples from the first or second tube used for chloride determination.

Procedure

1. Label small test tubes or cuvets to receive 75, 100, and 125 mmol/L chloride standards and at least two controls (normal, abnormal), as well as the number of serum samples to be assayed and a blank.

2. Pipet 1 mL of working solution into each cuvet or test tube. Using an SMI pipettor (Scientific Manufacturing Industries, Emeryville, CA 94608), add 10 µL of sample, standard, or blank to each. The working solution may be scaled up to 5 mL or down accordingly. If an automatic pipettor is used, allow the solution to run down the side of the tube or cuvet so that vigorous mixing will *not* occur.

Note: The degree of reproducibility will in part be governed by the pipettors. For dispensing the working solution, good precision can be obtained by using plunger-type pipettors with polypropylene tips, available from many manufacturers. This approach is simple for manual pipetting of multiple samples, but each laboratory should determine its own precision to be certain the pipettor used does not allow variation beyond a clinically acceptable range. For manual pipetting of the sample, the Submitter and the Reviewer strongly suggest using an SMI pipettor.

3. Mix the solution gently by inversion one or two times. Samples in a microcuvet should be mixed gently with the pipettor tip as a stirring rod. *Do not shake or stir vigorously.*

Note: Vigorous mixing causes formation of a coagulum, the significance of which is considered in the *Discussion*.

4. Incubate the mixtures for 5 to 10 min at a selected constant temperature between 25 and 37 °C and perform the assay at the incubation temperature.

Note: The Submitter has found that a 5-min incubation period is satisfactory when the incubation and assay are performed at 37 °C with an ABA-50 or a Perkin-Elmer spectrophotometer (Model 55; Perkin-Elmer Corp.) fitted with a 250-µL glass-jacketed flow cell. Evaluator G. N. recommended a 10-min incubation period when the incubation and assay are performed at ambient temperatures.

5. After adjusting the blank absorbance to zero, read the absorbance of each sample at a specific wavelength between 460 and 480 nm. The color is stable for at least 30 min.

Alternative procedure

Although not preferred, spectrophotometers that give precise measurements up to only 0.80 absorbance units (such as the Coleman Junior II) may be used by the following maneuver:

Follow steps 1 through 4 of the procedure above but prepare a 50 mmol/L standard. Substitute in step 5 the 50 mmol/L standard for the blank. Adjust the instrument to zero with the 50 mmol/L standard and read the absorbance of each sample at a specific wavelength between 460 and 480 nm.

Direct concentration readout

Follow steps 1 through 4 of the procedure above. Substitute in step 5 the 75 mmol/L standard for the blank. For the lower setting, adjust the instrument to read 75 with the zero adjustment control. Using the appropriate adjustment control, adjust the 100 mmol/L standard to read 100 for the upper setting. The 125 mmol/L standard should read between 123 and 127. Read the concentration of each sample.

Note: Evaluator G. N. indicates that when used with a Gilford 300 N spectrophotometer (Gilford Instruments, Oberlin, OH 44074), the direct concentration readout approach provides results more simply, accurately, and rapidly than reading absorbances and plotting the standard curve. The Submitter's experience indicates that the direct readout technique can provide results with an ABA-50 at a rate greater than 180 samples per hour.

Preparation of standard curve (when direct concentration readout is not available)

1. Plot the absorbance of the standards against their known concentrations on regular graph paper.
2. Draw the line that best fits the three points. All points should lie nearly on the line. Extend the line from 75 to 125 mmol/L. The line will not necessarily pass through zero.

Note: The concentration of mercuric nitrate is adjusted to provide linearity over a range of nearly 70 to 140 mmol/L. Below 70 mmol/L, the absorbance rapidly drops nonlinearly as the chloride concentration decreases, and above 140 mmol/L the absorbance plateaus as chloride concentration increases. Because the clinically important range for chloride is between about 75 and 125 mmol/L, this nonlinear behavior does not hinder the effective assay of chloride in serum or cerebrospinal fluid.

Quality Control

At least one sample of each 75, 100, and 125 mmol/L standard should be assayed with each batch, to ensure that linearity is intact over the entire physiological range for serum.

At least two controls (normal and abnormal) should be assayed with each batch. Each control should come within appropriately established limits.

Samples of 125 mmol/L or more (or 75 mmol/L or less) may not be within the linear range and must be reassayed with a diluted sample (or more sample). Samples exceeding 125 mmol/L can be diluted by adding one volume of water to two volumes of sample; those less than 75 mmol/L can be pipetted with a 15-µL SMI pipettor. Multiply or divide, respectively, the measured concentrations so obtained by 1.5 to obtain the actual concentrations.

Note: To check the accuracy of the chloride measurement, calculate the differences between the sodium value and the sum of the chloride and total CO_2 values, in mmol/L, for each patient. Normally, the difference is 7 to 14 mmol/L; most patients will have values within this range. Sera that do not conform should have the electrolytes rechecked unless a clinical reason for this discrepancy can be found. A more detailed discussion of this quality-control maneuver is given elsewhere (9, p 875).

The Reviewer cautions that the above arithmetic should not be attempted unless the CO_2 values are available in addition to the sodium and chloride values.

Results

Using this procedure with an ABA-50 in a direct concentration readout mode at 37 °C, and with the Abbott micro-dilutor for pipetting, the Submitter had the following results:

Normal range. The reference interval obtained by assaying 48 sera obtained from patients admitted for minor surgery who were without laboratory evidence of illness was 95–106 mmol/L, and agrees well with previously documented values (9, p 1209).

Precision and accuracy. Day-to-day variability over a period of one year, for controls with mean values of 105 and 111 mmol/L, gave CVs of 2.5 and 2%, respectively. Within-run variation showed a CV of less than 1% (8).

Correlation between this method and that of Schales and Schales (chemical titration) was excellent over a range of 82 to 120 mmol/L with the ABA-50 (8) and with a spectrophotometer (6).

Agreement of this chloride assay with the College of American Pathologists basic survey over the past three years is excellent, as shown in Table 1.

Note: Evaluator S. C. L. obtained an analytical recovery of between 95.7 and 100% for five experiments.

Evaluator G. N., using a Gilford 300 N spectrophotometer with direct concentration readout, found that for within-run assays the CV for a 125 mmol/L standard was 0.22%, and that the CV for control sera varied between 1.24 and 1.5%. These results are similar to those obtained by the Submitter using an ABA-50 with direct concentration readout. Also, Evaluator G. N. obtained a correlation coefficient of 0.973, with a regression line of $y = 0.996x - 0.986$, when 129 patients' sera ranging between 72 and 125 mmol/L were assayed by the present (y) and by a coulometric–amperometric (x) titration method. These results agree well with the correlation coefficient of 0.951 obtained by the Submitter (8) when data from 36 sera ranging between 82 and 120 mmol/L were compared with values obtained

Table 1. Agreement of the Chloride Manual Colorimetric Assay with the College of American Pathologists Survey

	Cl, mmol/L	
	Observed	Reference
	95	95
	101	101
	94	97
	96	101
	84	82
	85	84
	98	102
	101	106
	100	100
	102	105
	103	101
	104	105
Total	1163	1179
Mean	97	98
SD	6.6	7.8

The correlation coefficient between the observed (y) and the reference (x) values is $r = 0.95$, with the regression line $y = 1.12x - 10.67$. The observed data were obtained with an ABA-50 with a direct-concentration readout mode at 37 °C.

by a chemical titration method. Evaluator S. C. L. obtained a slightly higher within-run CV of between 1.34 and 2.13%, assaying control sera 20 times with a Gilford Stasar II spectrophotometer without direct concentration readout; he also obtained a higher CV for between-run precision for two concentrations of control sera (20 assays): 3.7% at 79 mmol/L and 3.2% at 116 mmol/L.

Evaluator R. F. D. obtained a much larger within-run CV of between 2.2 and 4.7% for control sera with a Coleman Jr. spectrophotometer, and a poorer correlation coefficient ($r = 0.845$) when results for 20 sera between 90 and 118 mmol/L were compared with results by a coulometric–amperometric method. The poorer correlation coefficient is consistent with the basic principle that results are only as good as the instrument used to obtain them: those obtained by an instrument with precision better than 0.001 absorbance units will be more accurate and reproducible than results obtained with an instrument with precision poorer than 0.01 absorbance units such as the Coleman Jr. Use of the less-precise spectrophotometer and the alternative procedure should be reserved only for "back-up."

It is difficult to compare the results obtained with the Gilfords and the ABA-50 because they were obtained with controls from different manufacturers and different types of pipettors. Still, the data suggest that the direct concentration readout mode will produce more precise results than will indirect results calculated from a standard curve after plotting absorbance vs concentration.

The Submitter recommends using a direct concentration readout mode, when available, and a spectrophotometric instrument that can monitor absorbances near 0.001 absorbance units.

The Reviewer notes that high-precision instruments with microcuvets similar to those used in these studies (300 N, Stasar II, and ABA-50) cost nearly $5000 and may not be available in many small laboratories. For such laboratories, the Submitter suggests using a Perkin-Elmer Model 35, a less-expensive digital spectrophotometer with a direct-concentration readout that is designed to replace the Coleman Jr.; its cuvets are also similar to those for the Coleman Jr. This would provide these laboratories a modern spectrophotometric system for performing chemistries and enzymes at a cost currently less than $1500.

Interfering Substances

Bilirubin, bovine serum albumin, and triglycerides in concentrations of less than 120 mg/L, 150 g/L, and 6.0 g/L, respectively, did not significantly alter the assay (8).

Note: Evaluator G. N. found that bilirubin up to 220 mg/L did not significantly alter the assay when performed with a Gilford 300 N spectrophotometer. Also, lipemic specimens assayed with the Gilford instrument gave nearly the same results as when assayed by coulometric–amperometric titration.

Halogens compete with chloride for displacement of mercuric ion, but because fluorides, bromides, and iodides normally contribute a combined total of less than 1.0 mmol/L of serum, their presence can be neglected (5, 6). Rarely, bromide poisoning or iodide treatment will require use of an alternative method. The effects of bromide or iodide are not additive, because of differential competition for free mercuric ion (10).

Discussion

The amount of mercuric nitrate added to the color reagent changes the sensitivity of the reagent and determines the range of concentrations over which linearity can be obtained (10). Although the concentration used here provides linearity between 70 and 140 mmol/L of chloride, samples that exceed 125 mmol/L should be diluted, thereby bringing the apparent sample concentration into the linear range. Concentrations less than 75 mmol/L should be brought within the linear range by increasing the amount of serum used.

Because chloride measurement in urine is rarely useful clinically, the detailed description of its measurement is omitted. Nevertheless, it can be assayed by using appropriate sample volumes to bring the wide range found in urine within the linear range of the serum method and its standards.

The wavelength at which maximum color is obtained is 460 nm, but a broad peak provides substantial absorbance between 460 and 480 nm.

At room temperature, the reagent provides high absorbance in the region of 0.8 to 1.4 absorbance units when a micro sample volume is used. This minimizes the effects of turbidity, bilirubin, and other interferences, thereby eliminating the need to run serum blanks with most samples. The absorbance at each concentration increases slightly with increased temperature of the assay.

The results in Table 1 indicate that the method agreed well with other methods of good quality over a period of three years. During this period, the results were always within acceptable limits as defined by the survey, a further indication of good quality.

Previously published papers described the formation of a precipitate when serum was added to the reagent used here (6, 7). To eliminate this effect, Hamilton (7) suggested the addition of perchloric acid and urea to the reagent. I have observed that a coagulum does occur if the tubes are mixed vigorously after adding the serum. As described elsewhere (6), this coagulum rises to the top, upon standing for several minutes, and then floats above the light beam, not interfering with the assay. In more than three years of performing this assay with an ABA-50, using these simpler and commercially prepared reagents, I have not noticed a coagulum or an interference when the sample is mixed gently as described. Two reasons for this lack of interference may be suggested: first, by manipulation of the concentration of mercuric nitrate, a greater dilution of sample can be used; second, the smaller total volumes (5 mL or less) applicable to modern instrumentation allows for gentle mixing to obtain a homogeneous distribution of constituents.

I greatly appreciate the valuable assistance of Kathy Levinson of North Shore Community College for helping to review the original manuscript.

References

1. Schales, O., Chloride. *Stand. Methods Clin. Chem.* **1**, 37–42 (1953).
2. Klein, B., Chloride. *Stand. Methods Clin. Chem.* **2**, 22–25 (1958).
3. Cotlove, E., Chloride. *Stand. Methods Clin. Chem.* **3**, 81–92 (1961).
4. Lustgarten, J. A., Wenk, R. E., Byrd, C., and Hall, B., Evaluation of an automated selective ion-electrolyte analyzer for measuring Na^+, K^+, and Cl^- in serum. *Clin. Chem.* **20**, 1217–1221 (1974).
5. Zall, D. M., Fisher, D., and Garner, M. Q., Photometric determination of chlorides in water. *Anal. Chem.* **28**, 1665–1670 (1956).
6. Schoenfeld, R. G., and Lewellen, C. J., A colorimetric method for determination of serum chloride. *Clin. Chem.* **10**, 533–539 (1964).
7. Hamilton, R. H., A direct photometric method for chloride in biological fluids, employing mercuric thiocyanate and perchloric acid. *Clin. Chem.* **12**, 1–17 (1966).
8. Levinson, S. S., Direct determination of serum chloride with a semiautomated discrete analyer. *Clin. Chem.* **22**, 273–274 (1976).
9. Tietz, N. W., Ed., *Fundamentals of Clinical Chemistry.* W. B. Saunders Co., Philadelphia, PA, 1970.
10. Levinson, S. S., and Rieder, S. V., The effect of free mercuric ion on apparent bromide concentrations when measured by the standard AutoAnalyzer method for chloride. *Clin. Chim. Acta* **52**, 249–251 (1974).

Chloride, Coulometric–Amperometric Method

Submitters: Albert A. Dietz and Epperson E. Bond, *Research Service, Veterans Administration Hospital, Hines, IL 60141*

Evaluators: Catherine Baker, Edward W. Bermes, Jr., and T. Oeser, *Loyola University Medical Center, Maywood, IL 60153*
Daniel K. Y. Chan, *Michigan Biomedical Laboratory, Flint, MI 48504*

Reviewer: Alan Mather, *Clinical Chemistry Division, Center for Environmental Health, Centers for Disease Control, Atlanta, GA 30333*

Introduction

Cotlove *(1)* reviewed many of the methods that have been used for the quantitative estimation of chloride. Most popular have been the titrametric procedures, in which the chloride is precipitated with a soluble silver or mercuric salt and the first excess of the heavy metal ion is detected colorimetrically or electrometrically. The electrometric method outlined here has received wide acceptance.

Principle

The coulometric–amperometric quantitation of chloride was first described by Cotlove et al. *(2)*. Silver is generated from a silver-wire anode to react with the chloride of a measured volume of serum or other source. The solution is stirred continuously so that the generated Ag^+ reacts immediately with the Cl^-. The first excess of silver ion is sensed by a second pair of silver electrodes, acting as a conductivity cell in an amperometric circuit. The sensing of the Ag^+ by the conductivity cell triggers a mechanism that automatically shuts off the current and records the time interval during which a constant current flowed. The moles of chloride in solution equals the coulombs multiplied by the Faraday constant (96 500 C/mol of chloride). The number of coulombs is proportional to the titration time multiplied by the amperes of constant coulombic current. In practice, the current is not measured, but the time required to titrate a standard chloride solution is compared with the time needed to titrate an unknown, after making proper correction for blanks. Use of the blanks corrects for traces of silver-combining substances in the reagents and for the amount of Ag^+ excess necessary to detect the endpoint.

As with any method involving the reaction of chloride with silver (or mercury) ion, the present method measures, in addition to chloride, other silver ion-combining substances such as bromide, iodide, thiocyanate, sulfide, or sulfhydryl ions, if present in the sample.

In the method described by Cotlove as an earlier Standard Method *(3)*, gelatin was used to suppress the current maxima *(4)* and prevent reduction of AgCl *(2)*. The gelatin solution must be prepared freshly daily. By substituting polyvinyl alcohol for gelatin, a stable reagent can be prepared in acid medium *(5)*, which prevents the precipitation of basic salts of silver. Nitric acid provides the necessary acidity and conductivity. Including acetic acid sharpens the endpoint by making the solution less polar and reducing the solubility of AgCl.

Materials and Methods

Reagents

1. *Chloride-free water.* De-ionized or distilled water, free of chloride, must be used for the preparation of all reagents and for washing the electrodes.

2. *Polyvinyl alcohol (PVA).* A soluble form of PVA must be used (available from the instrument manufacturers[1]). Elvanol 72–60 or 71–30 (DuPont Instruments, Wilmington, DE 19898) is also satisfactory.

3. *PVA acid reagent (5).*[2] Add 1.8 g of PVA to 100 mL of room-temperature water and stir while bringing the mixture to boiling. After cooling, add the solution to a mixture of 6.4 mL of concentrated nitric acid and 100 mL of glacial acetic acid in about 600 mL of water, and dilute to 1 L. Because the volume need not be exact, the dilution may be made in a graduated cylinder.

4. *Chloride standard* (160 mmol/L). Dissolve 9.352 g of dry NaCl in water and dilute to 1.0 L.

---
[1] In the United States, Cotlove Chloride Titrators are available from: American Instrument Co., Silver Spring, MD 20910, and Buchler Instruments, Fort Lee, NJ 07024. The Beckman Chloride/Carbon Dioxide Analyzer (Beckman Instruments, Inc., Fullerton, CA 92934) is also based on this principle.

[2] Because of its stability, the PVA acid reagent may be purchased (Buchler Instruments).

Note: Evaluators C. B., E. W. B., and T. O. and Reviewer A. M. find better correlation with other methods by using a 100 mmol/L chloride standard, which is nearer the concentration found in serum. This change especially improves correlation for the newer instruments with automatic readout and higher currents.

5. *Silver polish.* Any commercial paste silver polish is satisfactory.

Apparatus

The circuitry of the needed apparatus has been described by Cotlove et al. *(2).* Commercial instruments based on this circuitry are available.[1] Since the early description of the apparatus, modifications have been made to facilitate operations, including additions of automatic diluters, blank correctors, and digital readouts (giving results directly in millimoles of chloride per liter). The method we describe is for the simpler instruments, because the more automated procedures can easily be modified from this one.

Cleaning the electrodes. The generating electrodes *must* be clean: erratic results can frequently be traced to improperly cleaned electrodes. The initial procedure *(2)* required frequent polishing of the generating electrodes with silver polish, but using PVA in place of gelatin lessens the tarnish. In addition, AgCl may deposit on the electrodes to interfere with reproducibility. To avoid this, we have modified the suggestion of Toribara and Koval *(6)* so that the silver portions of the electrodes are washed after each titration—first with a solution of NH_4OH/water (1/3 by vol) to dissolve the AgCl, then with a solution of HNO_3/water (1/3 by vol)—and finally the whole electrode assembly is washed with water. It is convenient to have the cleaning reagents in titration vials and to rinse the electrodes by immersion. Whenever the anode is newly extended, or when the instrument has been idle for 24 h or longer, the electrodes should be polished with silver polish. The indicating electrodes need to be cleaned only if they become tarnished.

Note: Evaluator D. K. Y. C. suggested the above method for rinsing the electrodes with NH_4OH and HNO_3 and found the solutions could be used for at least 100 titrations.

Note: Reviewer A. M. finds that too much stress has been put on the details of washing the electrodes, and advocates washing only with water between each titration. Conditions probably differ in each laboratory, so that users will need to make their own decisions as to the requirement for washing, based on the reproducibility of replicate runs and the accuracy of control titrations. The Submitters have frequently been able to correct irregularities by proper cleaning of the generating electrodes.

Specimens

1. *Serum or plasma.* Chloride is readily determined in serum. Plasma from heparinized or oxalated blood is also satisfactory. Changes in the pH of blood at the time of collection and before serum is separated cause a slight but measurable shift of chloride between cells and serum *(7)*. For the most nearly accurate results, draw blood with minimum venous stasis and prevent exposure of the drawn blood to air (with loss of CO_2) before separating serum *(3)*.

2. *Urine, sweat, and gastric fluids.* These sources may be used directly without special preparation. Should the concentration of urine be very low, a larger sample may be used, with correspondingly less PVA acid reagent, but make sure that the mixture is acidic. Problems in the collection of sweat have been discussed by Ibbott *(8)*.

The presence of free sulfhydryl groups in the titration solution produces a negative initial amperometric current, and leads to erroneously high results, owing to the combination of these anions with silver ions. Accurate analyses of chloride in samples containing free sulfhydryl groups is possible, if these groups are first oxidized *(3)*. Avoid contact of reagent solutions or samples with rubber stoppers or tubing, because aqueous solutions, especially if alkaline, extract sulfhydryl groups from rubber. Use glass, polyethylene, or cork stoppers, or Parafilm covering, and plastic tubing (Tygon or vinyl). Avoid formation of free sulfhydryl groups caused by bacterial or mold growth in biological fluids. Titrate samples promptly, or store at freezer temperature, or store at refrigerator temperature with thymol crystals or other chloride-free chemical preservative.

Avoid contact of reagent solutions or samples with metal surfaces. Metal ions with more than one ionic valence (such as iron or copper) can produce a positive amperometric current *(3)*.

Procedure

The following procedure is for 100-μL samples (automatic pipetters are convenient). Use the same pipetter for measuring all samples: as long as the delivered volume of the sample, standard, and blank is constant, the exact volume of the sample is immaterial.

In describing the use of the instrument, we followed the instructions for the titrator manufactured by the American Instrument Co.[1]

Initial steps

1. Turn on power switch of the automatic titrator. Set Titration Rate Switch at high. Wait 15 s for warm-up, and stop stirrer by manual shut-off. (Subsequently, shut-off occurs automatically at the end of each titration.)

2. Manual shut-off. With the titration switch at position 2, move the adjustable (red) pointer of the meter-relay to coincide with the indicator (black)

pointer. (An audible click sounds as the relay is activated.) Move the adjustable pointer, with the indicator pointer attached, to about 20 µA.

Note: The conditions required for titration are as follows: *(a)* the total volume in the vial should be the same for a group of analyses: 4.0–4.5 mL; *(b)* the final concentration of reagents in the vial should be 0.1 mol/L strong acid (HNO_3) and 100 mL/L (1.75 mol/L) acetic acid, both within 10% of the concentrations noted; and *(c)* the preferable ranges of chloride content in the vial are 0.25 to 3 µmol at low titration rate (for which the net titration time is about 64 s/µmol), 2.5 to 5 µmol at medium (about 17 s/µmol). When the titration time is longer than 160 s at high titration rate [chloride content exceeding 40 µmol (400 mmol/L)], the large bulk of precipitate interferes with an accurate titration; terminate the titration by manual shut-off, and repeat with a smaller amount of sample *(3)*.

Note: Reviewer A. M. reports that the above times and volumes may not apply to all instruments; users should consult the instrument manual. He has encountered slippage in the stirrer drive, which gives poor reproducibility of results. On newer instruments with fast titration times, the major portion of the increased blank titration time (about 5 s) is used in transporting the excess Ag^+ from the generating to the sensing electrodes. This is enhanced if the volumes are larger than 4.5 mL.

Titration

1. Fill 20 × 40 mm vials with 4 mL of PVA acid solution, and add 100 µL of blank, standard, unknown, or quality-control sample to each.

Note: The solution must be acid during the titration; under normal conditions, it always will be with the PVA acid reagent prepared as given. If there should be any doubt of this acidity, add 35 mg of thymol blue to each liter of PVA acid. On dissolving, the thymol blue has a pink color that turns yellow in a few days; neither color interferes.

2. Position the vial on the apparatus.
3. Turn titration switch to position 1 and adjust.
4. Reset timer to zero.
5. Set adjustable pointer at 10 divisions above the stable reading of the indicating pointer (which is reached in 10–20 s).
6. Turn titration switch to position 2 (Titrate). Avoid any delay of more than 1 min between steps 2 and 5.
7. When timer stops automatically, record time to nearest 0.1 s.
8. Remove the vial and rinse the generating electrodes rapidly in succession with the NH_4OH and HNO_3 solutions and water. To prevent deterioration of the plastic materials, be careful to rinse only the silver portion of the electrode assembly with the NH_4OH and HNO_3 solution. The whole assembly should then be rinsed well with water.

The instrument is now ready for the next titration.

At the start of the day's work, titrate three or four blanks and three or four standards before the unknowns. The first titration may be low and should be disregarded. Titrate unknowns and quality-control standards. At the end of the assay, or periodically when titrating many samples, titrate two blanks and standards.

Note: Cotlove *(3)* recommended that gelatin always be present when titrating the blank and standard. He found that in its absence, the electrode became impaired and required repolishing. The Submitters found this precaution unnecessary when the electrodes were washed after each titration, as described above. Evaluator D. K. Y. C. found that aqueous standards showed shorter titration times unless three drops of gelatin was added to the vial; however, this resulted in falsely high values for the chloride concentrations of the unknowns unless corrected for. He suggested either using serum standards or increasing the concentration of PVA to 7 to 9 g/L.

The Submitters have since assayed aqueous standards with and without the addition of four drops of 0.62% (6.2 g/L) gelatin to standards and blanks. PVA concentrations were varied between 0.18 and 1 g/L. The titration time was always 2 to 3% longer when the gelatin was used with the standard. Thus, either gelatin should be added to aqueous standards, or a standard chloride-containing serum should be used.

Calculations

When the same volume of sample is used throughout, the chloride concentration is calculated from the equation:

$$\text{Chloride, mmol/L} = \frac{Cl^- \text{ of standard, mmol/L}}{\text{aver. net s (standard)}} \times \text{aver. net s (unknown)}$$

The "net s" is the titration time for the respective samples minus the time to titrate the blank. This subtraction is automatic on some instruments, in which the blank titration value is set into the instrument. When titrating many samples, a constant, K, can be calculated for the first two factors on the right-hand side of the equation. The more sophisticated instruments have a scaling circuit, which will automatically calculate the results in mmol/L and display them on a digital readout. The instructions in the operator's manual should be followed for these functions.

Reference Intervals *(3)*

The chloride concentration of serum in 31 ostensibly normal subjects averaged 104 mmol/L; the range was 101 to 109 mmol/L. Venous blood was collected from subjects before breakfast and was kept under oil until the serum was separated.

The chloride concentration of 15 normal cerebrospinal fluids averaged 124 mmol/L; the range was 122 to 128 mmol/L.

Discussions of the changes in the chloride composition in health and disease may be found in any text. A recent comprehensive review is that of Natelson and Natelson *(9)*.

Table 1. Precision of the Chloride Titration Method (n = 20 each)

Evaluator	Cl concn, μmol/L Mean	SD	CV,%
Within-day			
C.B., E.W.B., T.O.	105.5	0.61	0.58
D.K.Y.C.	100.8	0.53	0.53
Between-day			
C.B., E.W.B., T.O.	101.4	0.68	0.67
D.K.Y.C.[a]	98.8	1.26	1.27

[a] Based on 20 assays on each of 10 days.

Table 2. Comparison of Chloride Titration and Chloride–CO_2 Analyzer Methods (n = 10 each)

Evaluator	Chloride titration, μmol/L Mean	SD	Chloride–CO_2 Analyzer, μmol/L Mean	SD
C.B., E.W.B., T.O.	102	4.2	101[a]	3.4
D.K.Y.C.	99.6	0.52	99.8[b]	0.78

[a] IL-446; Instrumentation Laboratory, Inc., Lexington, MA 02173.
[b] Beckman Instruments, Inc.

Discussion

Linearity and Reproducibility of Automatic Titration Results (3)

There is a precise linear relationship, over a wide range, between titration time and chloride content. The standard deviations from this linear relationship were between 0.2 and 0.8 s (0.5–2 mmol/L) in titrations of 2.5 to 50 μmol of chloride (50–400 mmol/L in 100-μL samples). The exact value will depend on the instrument used. The titration time of the reagent blank sample is an accurate measure of the excess titration time of each sample, as determined from the time intercept of the linear relationship of gross seconds to amount of chloride.

Twenty replicate pipettings of a single normal serum were performed with the same 0.1-mL Lang–Levy micropipette with rinse-out delivery into 4 mL of acid reagent. The results at high titration rate showed a standard deviation (SD) of 0.1 s (0.25 mmol/L) and a coefficient of variation (CV) of 0.3%. This variability includes that of pipetting. The average difference between duplicate titrations of 40 different sera was 0.05 s, and of 76 different urines was 0.2 s (3).

The accuracy and percentage variation of results improved at the longer titration times, but closely approached an optimum at titration times of 40 to 60 s at the titration setting. This optimum will vary greatly with the instrument used.

In the experience of the Submitters, the CV of five to 20 within-day samples of standards or control sera was usually less than 2%. The between-day variation of a control serum assayed on 10 different days was 103.5 (SD 1.6) mmol/L, CV = 1.5%.

Analytical Recovery of Chloride Added to Serum and Urine (3)

Chloride added to dialyzed serum having an initial titration value of 0.56 mmol/L, to yield concentration increments of 25 to 400 mmol/L, was accurately recovered: the average analytical recovery was 100.0% (range, 99.7 to 100.2%). Chloride added to urine having an initial concentration of 1.7 mmol/L, to yield concentration increments of 2 to 80 mmol/L, was also accurately recovered: average recovery was 100.1% (range, 98.7 to 101.3%).

Evaluations

The Evaluators made both within- and between-day precision studies of the methods, and their results are shown in Table 1. They also compared the chloride titration method with the results of chloride–CO_2 analyzer procedures. The results of the two methods, given in Table 2, did not differ significantly. In addition, Evaluator D. K. Y. C. did recovery studies on five samples of serum that had been supplemented to give chloride concentrations of between 107 and 130 mmol/L; analytical recovery averaged 100.3% (range, 99.3 to 100.7%).

References

1. Cotlove, E., Determination of chloride in biological materials. *Methods Biochem. Anal.* 12, 277–391 (1964).
2. Cotlove, E., Trantham, H. V., and Bowman, R. L., An instrument and method for automatic, rapid, accurate, and sensitive titration of chloride in biologic samples. *J. Lab. Clin. Med.* 51, 461–468 (1958).
3. Cotlove, E., Chloride. *Stand. Methods Clin. Chem.* 3, 81–92 (1961).
4. Willard, H. H., Merritt, L. L., and Dean, J. A., *Instrumental Methods of Analysis*, 4th ed. Van Nostrand, Princeton, NJ, 1966, p 678.
5. Dietz, A. A., Chloride determination by the Cotlove method. *Chicago Clin. Chem.* 3, 43–44 (1970).
6. Toribara, T. Y., and Koval, L., Improving the precision of chloride analysis. Reprint 332, American Instrument Co., Silver Springs, MD.
7. Schales, O., Chloride. *Stand. Methods Clin. Chem.* 1, 37–42 (1953).
8. Ibbott, F. A., Chloride in sweat. *Stand. Methods Clin. Chem.* 5, 101–111 (1965).
9. Natelson, S., and Natelson, E. A., *Principles of Applied Clinical Chemistry I. Maintenance of Fluid and Electrolyte Balance*. Plenum Press, New York, NY, 1975.

Chloride, Direct Mercurimetric Titration (Provisional)[1]

Submitter: Alan Mather, *Clinical Chemistry Division, Center for Environmental Health, Centers for Disease Control, Atlanta, GA 30333*

Introduction

Although titrimetric analyses for the determination of chloride in serum have largely been superseded by more instrumented and automated procedures, they still have a place in laboratories where this analysis is performed on small or intermittent workloads. A single determination, for example, can be performed in a few minutes without an extensive setup of an analytical run.

Perhaps the best of these methods is that of Schales and Schales *(1)*, which was also presented in the first volume of this series *(2)*. When a deproteinized tungstate filtrate is carefully titrated, the results can be highly precise and accurate. In extensive comparisons with a more laborious argentimetric reference method *(3)* or with the chloride-titrating apparatus *(4)*, the mercurimetric titration showed zero bias with a large variety of human serum specimens (unpublished data).

A variant of this method, involving a direct titration of serum, was coincidentally described as an alternative procedure that was widely used in clinical laboratories *(5)*. The primary problem with direct serum titration arose from difficulties with determining the endpoint. Because the nitric acid necessary to provide a mildly acidic reaction medium was incorporated into the titrant, a false endpoint was observed in serum samples, arising from the excess of competing protein early in the titration; also, the shade and intensity of the endpoint are altered in the presence of protein.[2] Pigmented specimens may thus give trouble in the endpoint determination, and the 1–2 mmol/L increase in direct titrimetric results noted by the original authors is probably accounted for by the tendency to overtitrate. The first of these problems, and to some extent the second, can be ameliorated by incorporating part of the requisite nitric acid in a composite indicator in which the sample is diluted; the indicator concentration is increased here so that the first permanent tinge of color is used as the endpoint. Except for this modification, the direct procedure is essentially that described in *Standard Methods (2)*.

Materials and Methods

Equipment

1. *Sample delivery.* A 200-μL micropipette or, preferably, a positive displacement pipetter, such as the SMI Micro/Pettor no. 1058–E200 (Scientific Manufacturing Industries, Emeryville, CA 94608) or a similar model from other sources.

Note: The use of air-piston types, such as the Oxford or Eppendorf pipetters, shows a small but significant bias between the delivered volumes of aqueous standards and serum samples, which can amount to 3–5 μL at 200 μL. Each manufacturer recommends a different means of "correcting" this error, which only introduces compensating errors. The recommended positive-displacement types require care in operating and maintaining the pipetters, but are capable of excellent precision; the actual volume of sample delivered is not critical, because the same pipetter is used for standards and specimens. The same tip should be used for all; rinse the tip each time with the succeeding sample.

2. *Diluent dispenser.* A pipette or reagent dispenser, set for delivering 2-mL volumes of acid indicator. Precision and accuracy are not critical.

3. *Microburette.* A number of microburettes are available. Select one that is easily readable to 0.01

---

[1] Based on the method of Schales and Schales *(1, 2)*.

Editors' Note: This chapter is marked "Provisional," indicating that it may not have been completely evaluated, and has not met the arbitrary criteria of our reviewing process. On the other hand, the method is used daily in the Submitter's laboratory, and is not only clinically useful, but has met current standards of quality for many years.

[2] The authors also pointed out the effect of pH on the dye color and recommended a pH of 6 for the titration of urine samples.

mL (or 10 μL) per division and is convenient to fill. Probably the most suitable is a Koch-type "automatic" 2-mL burette with a titrant reservoir, three-way Teflon® stopcock, and glass tips with small orifices. Unfortunately, all of these features are no longer easily available in one burette. Kimble no. 13135F (Kimble Products Div., Owens-Illinois, Inc., Toledo, OH 43666) has a detachable rare-metal tip, which must be rinsed thoroughly between titrations, or else one may use a fine glass tip with a Luer syringe hub; model no. 13132F has two stopcocks and a fixed tip that must be inserted into the sample during titration and rinsed afterward. A. H. Thomas, Philadelphia, PA 19105, supplies a proper burette with interchangeable tips and a glass stopcock. The Micro-Metric syringe-type burette (Micro-Metric Instruments Co., Cleveland, OH 44122) with a 3-mL syringe is quite precise, but is not quite so convenient to refill as, e.g., Scientific Products B9430 plus syringes and tips. Larger burettes, such as the 5-mL size, are not recommended, owing to their considerably less precise readability.

The burette should be securely clamped (with removable clamps) to a support stand having a white ceramic base, and should be placed near a small fluorescent desk lamp for illumination.

Reagents

1. *Reagent water.* Distilled water or demineralized water from a high-efficiency mixed-bed resin de-ionizer.

Note: Reagent water for all electrolyte analyses should ideally be specially processed by redistillation or demineralization of central-supply distilled water, because the latter may not be systematically monitored or serviced and may occasionally show an unacceptable blank titration. If this is not possible, the blank titration becomes important for detecting chloride intrusion in the supply. Older two-stage demineralizers can easily introduce a significant chloride blank in the effluent if used intermittently; it is also important to ensure absence of resin particles by using a post-resin filter.

2. *NaCl standard solution, 100 mmol/L (5.845 g/L).* Use ACS reagent grade NaCl from a freshly opened bottle or oven-dried overnight at 110 °C. Dissolve 1.4613 g in distilled water in a 250-mL volumetric flask and dilute to the mark at 20 ± 2 °C. When stored in a screw-capped borosilicate bottle, this standard is stable at room temperature, but should be checked or replaced each three months. A convenient method is to prepare or procure a stock standard solution of exactly 1.000 mol/L (58.45 g/L) and prepare precise 10-fold volumetric dilutions (e.g., 25.0 mL diluted to 250 mL) as needed or whenever the calibration is suspect.

Note: Adequate NaCl standards or stock concentrates are available commercially, but the proper concentration should be checked: commercially available reagent sets for the Schales assay may include a standard at one-tenth the concentration (i.e., 10 mmol/L) required for the 200-μL aliquots.

3. *Indicator-diluent.* Stock indicator: Dissolve 5 mg of diphenylcarbazone in 20 mL of 95% (950 mL/L) ethanol and store in a dark bottle in the refrigerator for not more than one week.

Note: Preweighed 5-mg tablets of this reagent are available (no. 830–105; Sigma Chemical Co., St. Louis, MO 63178); bulk reagent is also available (J. T. Baker Chemical Co., Phillipsburg NJ 08865, no. 2-K617, 10 g; Eastman Organic Chemicals, Rochester, NY 14650, no. 4459; and others).

Working indicator-diluent: Prepare fresh solution each working day. Dilute 2.0 mL of stock indicator to 10.0 mL (or dilute 5 mL to 25 mL) with 25 mmol/L nitric acid.

4. *Nitric acid, 25 mmol/L.* Dilute about 1.5 mL of concd. HNO_3 to 1 L with reagent water. Concentration is not critical.

5. *Mercuric nitrate titrant, approximately 120 meq/L.* Dissolve about 2.3 g of ACS grade $Hg(NO_3)_2 \cdot H_2O$ in 1 L of distilled water. After solution and thorough mixing, filter the solution through a retentive filter paper, if needed. This solution should be adjusted by dilution so that between 1.6 and 1.7 mL of titrant is needed to titrate 200 μL of standard in 2 mL of diluent; this permits the titration of samples having chloride concentrations up to approximately 120 mmol/L with a 2.0-mL burette full of titrant.

Note: The original method specified, and some kits furnish, 10 mmol/L mercuric nitrate titrant. If the recommended 2-mL microburette is used, this requires refilling the burette to complete the titration of most samples. If purchased reagents of this concentration must be used (e.g., Sigma no. 830–2), it would be advisable to invest in a pipettor adjustable to about 170 μL volume (e.g., an adjustable SMI pipette 1500E, 100 μL–200 μL), to keep the titrant volume within 2 mL for samples containing chloride up to about 117 mmol/L.

Procedure

1. Deliver 2 mL of freshly diluted indicator-diluent into the requisite number of 25-mL Erlenmeyer flasks; include a blank (no sample) and one 100 mmol/L standard.

2. Pipet 200 μL (by micropipette) of each sample into the diluent, rinse the pipette twice with diluent, and empty the pipette completely.

Note: If a positive-displacement pipetter is used, deliver each sample against the inside of each empty flask and rinse down with 2 mL of indicator-diluent.

3. Fill the reservoir of the microburette with *well-mixed* mercuric nitrate titrant.

Note: The titrant may be left in the reservoir but, because the heavy mercuric salt tends to settle into a concentration gradient, it is necessary to mix the solution each day before use. Close the reservoir stopper, remove the burette from its clamp, and invert several times to mix the solution thoroughly. Rinse the burette several times and discard the rinses. Refill to the zero mark.

4. Titrate the blank flask containing 2 mL of indicator-diluent; the violet-blue endpoint should be reached with the smallest first drop (about 5 µL) of titrant. If more is required, the source of distilled water should be investigated.

5. Readjust the level in the burette to zero. While swirling the flask, titrate the standard flask to the *first permanent tinge* of blue color, delivering the titrant at a moderate, steady rate to approximately 1.5 mL. Proceed with dropwise additions approximately to the endpoint. Deliver half drops against the inside of the flask, and swirl the contents to wash them in.

Record the milliliters of standard titration, and compare this value with previous titrations, which are recorded on a separate quality-control chart. The titration should fall within 15 µL of the previous average titration. If not, prepare another standard and repeat the titration; if the concentration of the titrant has apparently changed (the two values are near each other and show a difference from previous values), use the average of the new standard titration values in the computations.

Note: Because the titrant is relatively stable and not light-sensitive, the standard need be titrated only once each day, if the burette is left filled for subsequent analyses by the same analyst throughout the day.

6. Readjust the level of titrant in the burette to zero, and titrate each sample to the first permanent shade change and increase in color. The intensity and shade of the endpoint for serum samples will be lighter and less blue than that of the standard, so do not attempt to match these colors. Record the titration of each sample.

7. Compute the concentrations of the specimens (mmol/L) by the equation:

$$\text{Concn of Cl}^- = \frac{\text{mL titration of sample}}{\text{mL titration of standard}} \times 100$$

8. To clean up, titrant may be left in the reservoir, but the burette should be drained, flushed with water, and allowed to drain; remove and dry the tip.

References

1. Schales, O., and Schales, S. S., A simple and accurate method for the determination of chloride in biological fluids. *J. Biol. Chem.* **140**, 879–884 (1941).
2. Schales, O., and Schales, S. S., Chloride. *Stand. Methods Clin. Chem.* **1**, 37–42 (1953).
3. van Slyke, D. D., and Hiller, H., Application of Sendroy's iodometric chloride titration to protein-containing fluids. *J. Biol. Chem.* **167**, 107–124 (1947).
4. Cotlove, E., Chloride. *Stand. Methods Clin. Chem.* **3**, 81–92 (1961).
5. Asper, S. P., Schales, O., and Schales, S. S., Importance of controlling pH in the Schales and Schales method of chloride determination. *J. Biol. Chem.* **168**, 779–780 (1947).

Introduction to Analysis of Cholesterol, Triglyceride, and Lipoprotein Cholesterol: Reference Values and Conditions for Analyses

Submitter: Gerald R. Cooper, *Clinical Chemistry Division, Center for Environmental Health, Centers for Disease Control, Public Health Service, U.S. Department of Health and Human Services, Atlanta, GA 30333*

Reviewers: Leila V. Edwards, *Clinical Chemistry Laboratory, Erie County Laboratory, Buffalo, NY 14218*

G. Russell Warnick, *Northwest Lipid Research Clinic, Harborview Medical Center, Seattle, WA 98104*

Introduction

The purpose of this chapter is to introduce briefly the methods of measuring total cholesterol selected for discussion in this volume; to discuss in general terms the common procedural factors that affect cholesterol measurement; and to provide reference information needed for interpreting cholesterol measurements.

Many types of cholesterol methods are used in clinical laboratories and in clinical chemistry research (1–6). To help the small laboratory choose a procedure, separate chapters have been devoted to each of three selected methods: an enzymic method, an iron–uranyl acetate method, and a Liebermann–Burchard reagent method, which is similar to the automated procedure adopted by the laboratories of the Lipid Research Clinics. Because variations in specimen collection and calibration procedures often result in lack of comparability of results, whatever method is used for cholesterol determination, these procedures are discussed briefly here. Interpretation of cholesterol results from the point of view of reference conditions and reference values that appear valid are also discussed. Reference ("normal") or observed population distribution values included in this chapter and applicable to serum specimens and to all three methods are based largely on the recently published Lipid Research Clinics findings (7–10).

Conditions for Analyses

Specimen Collection

The patient should have no medication or alcohol within 24 h before blood collection for lipid and lipoprotein analysis and should fast for 12 to 16 h (11). Ideally, the patient should be on a steady-state diet that does not cause gain or loss of weight.

Precautions should be taken to prevent excessive exchange of fluids from the blood into the tissues (12, 13) and the nonreproducible transfer of water from the cells into the plasma in the evacuated collection tube (11, 14). Have the patient sit or lie down for 10 min before taking blood, and apply a tourniquet for not more than 30 s. If plasma is to be separated from whole blood, dissolve the anticoagulant completely to avoid the problems associated with fibrin formation after separation. The anticoagulant can be dissolved more easily if the bottom of the tube is tapped to loosen the anticoagulant residue before the blood is drawn. Also, completely fill the tube. To obtain serum, allow the sample to coagulate about 30 min at room temperature before centrifugation.

Avoid cellular and bacterial contamination of specimens while collecting and processing the blood.

Fresh samples are preferred for all methods of cholesterol analysis because some specimens develop unknown interfering materials during storage (15). Specimens collected for measurement of low-density lipoprotein and high-density lipoprotein cholesterol should be stored at temperatures below −60 °C if there is a delay in analysis, and should not be thawed more than once.

Note (Reviewer G. R. W.): Lipoprotein separations should be made as soon as possible after blood collection. HDL cholesterol, for example, can change substantially within 24 h in specimens kept at 4 °C, but lipoproteins are stable for at least a year in specimens stored below −60 °C. Cholesterol in specimens or lipoprotein fractions is reasonably stable at 4 °C for approximately two weeks and at −20 °C for at least a year.

Calibration

When possible, calibration with a primary standard (pure analyte) is generally preferable. Alcohols are suitable solvents for preparing cholesterol standard solutions unless alcohol is incompatible with the chemical or enzymic reaction used. Aqueous cholesterol standards are difficult to prepare, even with solubilizers, and difficult to use because cholesterol is highly insoluble in water.

Purity of cholesterol used for preparing standards in the laboratory can be checked against the pure cholesterol reference preparations distributed by the National Bureau of Standards (16). Preparations of cholesterol that contain impurities or that deteriorate can be purified by the dibromide precipitation method (17).

Calibration with pure cholesterol is limited in methods in which the Liebermann–Burchard reagent is used without hydrolysis, because this reagent reacts differently with free and esterified cholesterol. Calibration of the enzymic method with pure cholesterol also is difficult because incomplete enzymic hydrolysis of cholesterol esters causes underestimation of cholesterol in specimens. Detergents that solubilize cholesterol in water inhibit the cholesterol oxidase, whereas the alternative solvent, alcohol, inhibits the action of cholesterol esterase. Calibration with a serum standard (secondary) has been found necessary and effective with the enzymic and Liebermann–Burchard reagent direct or extraction methods. Prerequisites for accurate calibration of a method with a single serum pool are: (a) accurate target value for the serum used in calibration, (b) linearity of absorbance readings with target values determined by the reference method for serum pools of different cholesterol concentrations, (c) comparability of values by the method in which the serum calibrator is used and by the reference method on recently collected ("fresh") individual specimens, and (d) comparability of target values of old and new serum calibrators in overlapping experiments.

Note (Reviewer L. V. E.): It is important to check each batch of reference material with overlapping assays not only for stated value, but also for assurance that behavior has not changed between old and new batches.

Manufacturers who produce serum pools for use as calibrators or controls should label these pools with target values assigned by the proposed cholesterol reference method recommended by the Cholesterol Reference Method Study Group of the American Association for Clinical Chemistry Standards Committee (18).

Observed Population Values

The interpretation of a single cholesterol measurement depends upon many physiological variables in the individual blood donor (19–21). For laboratorians, the main variables to consider are age and sex of the donor, types of lipoproteins, and plasma anticoagulant. Because lipoproteins in plasma become diluted when the anticoagulant draws water from the blood cells by osmosis, analysis of plasma gives systematically lower results than analysis of serum (22). Tables of expected values for a healthy or defined population with respect to age, sex, plasma or serum, and type of lipoprotein are thus needed for interpretation of cholesterol measurements.

Serum. The following tables on population lipid and lipoprotein cholesterol and triglyceride values are prepared for use with serum samples, serum specimens being the kind mainly used in the clinical chemistry laboratory. The data in the tables are given in grams per liter; factors are provided for converting values to mg/dL or mmol/L. Serum (s) values (g/L) were calculated from plasma (p) values of the Lipid Research Clinics Prevalence Study by the equations (22) for cholesterol (C) and triglyceride (TG):

$$C_s = 1.036 (C_p) - 0.0138$$
$$TG_s = 1.031 (TG_p) - 0.0037$$

The analytes for which population distributions are presented (Tables 1–4) were selected on the basis that a clinical chemistry laboratory report of values for total cholesterol, triglyceride, LDL cholesterol, and HDL cholesterol appears to satisfy the needs of most clinicians caring for patients with lipid abnormalities.

Plasma. The plasma lipid and lipoprotein results used to prepare Tables 1 through 4 were derived from prevalence studies on well-defined North American populations by 10 collaborating Lipid Research Clinics (7–10). These provide epidemiologic "hard" data on a large scale for a wide distribution of populations in various places across North America. All of the cholesterol values reported in the Lipid Research Clinics studies were standardized against the cholesterol values by the proposed, modified Abell–Levy–Brodie–Kendall cholesterol reference method established at the Centers for Disease Control (18) and by the proposed definitive isotope dilution–mass spectrometric method developed at the United States National Bureau of Standards (23).

Reference observed population. The laboratory data of the prevalence study consist of cholesterol and triglyceride measurements made on specimens collected during Visit 1, and of lipid and lipoprotein measurements made upon a 15% random sample of Visit 1 participants at a follow-up visit (Visit 2). Plasma data collected on 48 431 individuals tested on Visit 1 were used to prepare tables of observed population distributions of serum values for clinical chemistry laboratories for total cholesterol (Table 1) and total triglyceride (Table 2). Plasma data collected

Table 1. Distribution of Total Cholesterol—Observed Population Values[a]

Total cholesterol in serum, g/L[b]

Age, yr	Male (n = 24 485) Mean	Range	Female (n = 18 168) Mean	Range	Female, hormone user (n = 5891) Mean	Range
0–cord	0.65	0.37–0.96	0.65	0.37–0.96		
0–1	1.51	1.02–1.90	1.56	1.13–2.01		
2–3	1.59	1.13–1.99	1.60	1.17–2.04		
4–5	1.62	1.17–2.18	1.63	1.20–2.02		
6–7	1.63	1.24–2.04	1.67	1.30–2.10		
8–9	1.66	1.25–2.11	1.70	1.28–2.13		
10–11	1.65	1.29–2.10	1.66	1.30–2.09		
12–13	1.61	1.21–2.05	1.62	1.26–2.04		
14–15	1.55	1.16–2.01	1.60	1.23–2.05		
16–17	1.54	1.15–2.04	1.61	1.23–2.06	1.74	1.24–2.38
18–19	1.57	1.17–2.04	1.64	1.23–2.13		
20–24	1.72	1.27–2.24	1.69	1.25–2.22	1.84	1.34–2.43
25–29	1.87	1.36–2.51	1.76	1.31–2.29	1.89	1.45–2.43
30–34	1.98	1.42–2.62	1.80	1.33–2.37	1.94	1.43–2.53
35–39	2.07	1.50–2.78	1.89	1.44–2.49	2.00	1.50–2.57
40–44	2.13	1.55–2.76	2.00	1.51–2.60	2.05	1.56–2.66
45–49	2.18	1.62–2.85	2.09	1.56–2.73	2.15	1.59–2.85
50–54	2.19	1.62–2.86	2.24	1.66–2.94	2.24	1.72–2.88
55–59	2.20	1.60–2.85	2.38	1.77–3.10	2.24	1.69–2.92
60–64	1.69	1.62–2.85	2.38	1.77–3.06	2.31	1.77–2.94
65–69	2.19	1.62–2.82	2.40	1.76–3.13	2.30	1.78–2.88
70+	2.13	1.55–2.78	2.35	1.74–2.98	2.22	1.64–2.83

[a] Sources: 0–cord values (26); age groups 0–19 (10); age groups 20–70+ (7).
[b] Serum values were calculated by equations in ref. 22; mg/dL = g/L × 100; mmol/L = g/L × 2.586. Range values are 5th and 95th percentiles.

Table 2. Distribution of Triglyceride—Observed Population Values

Triglyceride in serum, g/L

Age, yr	Male (n = 24 485) Mean	Range	Female (n = 18 168) Mean	Range	Female, hormone user (n = 5891) Mean	Range
0–cord	0.36	0.10–0.98	0.36	0.31–0.98		
0–1	0.70	0.31–0.82	0.78	0.31–0.98		
2–3	0.55	0.30–0.84	0.66	0.32–1.03		
4–5	0.55	0.28–0.91	0.59	0.32–0.97		
6–7	0.56	0.31–1.05	0.60	0.33–1.07		
8–9	0.58	0.31–1.09	0.64	0.34–1.10		
10–11	0.59	0.30–1.09	0.72	0.37–1.24		
12–13	0.72	0.36–1.38	0.81	0.42–1.38		
14–15	0.76	0.36–1.38	0.79	0.40–1.37		
16–17	0.80	0.38–1.57	0.73	0.40–1.20	1.09	0.50–2.06
18–19	0.85	0.43–1.69	0.79	0.41–1.35		
20–24	1.03	0.45–2.07	0.74	0.37–1.35	1.08	0.56–1.81
25–29	1.19	0.47–2.56	0.77	0.38–1.49	1.13	0.58–1.97
30–34	1.32	0.51–2.74	0.81	0.40–1.55	1.19	0.59–2.12
35–39	1.49	0.55–3.31	0.88	0.41–1.81	1.30	0.57–2.48
40–44	1.55	0.56–3.30	1.01	0.46–1.97	1.33	0.59–2.44
45–49	1.56	0.59–3.37	1.08	0.47–2.20	1.34	0.54–2.68
50–54	1.56	0.59–3.30	1.18	0.53–2.40	1.34	0.58–2.55
55–59	1.45	0.59–2.94	1.29	0.56–2.70	1.31	0.64–2.45
60–64	1.46	0.59–3.00	1.31	0.57–2.46	1.30	0.58–2.47
65–69	1.41	0.58–2.75	1.35	0.61–2.50	1.34	0.62–2.41
70+	1.34	0.59–2.66	1.35	0.61–2.44	1.25	0.62–2.22

See footnotes to Table 1. For triglyceride, mmol/L = g/L × 1.1293.

on 7055 white individuals tested on Visit 2 were used to prepare serum distributions for LDL cholesterol (Table 3) and HDL cholesterol (Table 4). Serum triglyceride values (Visit 1) are included because they are needed to estimate very low-density lipoprotein (VLDL) and LDL cholesterol (24) and to interpret and understand lipid metabolic states (25).

Data for subjects older than 70 years are tabulated together because data published for 70–74, 75–79, and 80+ age groups show little difference (7, 8). Values for neonates (0–cord) originated from plasma values by Tsang et al. (26). Results on subjects younger than 20 years are reported in two-year intervals as published by Christensen et al. (10), but are avail-

Table 3. Distribution of LDL Cholesterol—Observed Population Values[a]

LDL cholesterol in serum, g/L[b]

Age, yr	Male (n = 3584) Mean	Range	Female (n = 2627) Mean	Range	Female, hormone user (n = 699) Mean	Range
0–cord	0.28	0.12–0.48	0.28	0.12–0.48		
5–9	0.95	0.64–1.32	1.02	0.69–1.44		
10–14	0.99	0.65–1.35	0.99	0.69–1.40		
15–19	0.96	0.63–1.33	0.97	0.61–1.39		
20–24	1.06	0.67–1.51	1.00	0.65–1.49	1.10	0.63–1.68
25–29	1.20	0.71–1.69	1.09	0.71–1.55	1.19	0.73–1.74
30–34	1.29	0.79–1.90	1.12	0.68–1.54	1.20	0.73–1.74
35–39	1.36	0.83–1.94	1.22	0.77–1.77	1.22	
40–44	1.40	0.89–1.91	1.28	0.78–1.79	1.29	
45–49	1.48	1.00–2.08	1.33	0.82–1.93	1.32	
50–54	1.46	0.91–2.03	1.50	0.92–2.21	1.24	
55–59	1.50	0.90–2.09	1.56	0.97–2.19	1.37	
60–64	1.50	0.85–2.16	1.60	1.02–2.41	1.41	
65–69	1.54	1.00–2.16	1.67	0.99–2.29	1.30	
70+	1.47	0.90–1.91	1.53	0.98–2.13		

[a] Sources: 0–cord values (26); all others (7).
[b] Serum values calculated by equations in ref. 22. Range values are 5th and 95th percentiles (omitted if sample size is insufficient). mmol/L = g/L × 2.586.

Table 4. Distribution of HDL Cholesterol—Observed Population Values

HDL cholesterol in serum, g/L

Age, yr	Male (n = 3563) Mean	Range	Female (n = 2593) Mean	Range	Female, hormone user (n = 601) Mean	Range
0–cord	0.29	0.05–0.50	0.29	0.05–0.50		
5–9	0.57	0.38–0.75	0.54	0.36–0.74		
10–14	0.56	0.37–0.75	0.53	0.37–0.71		
15–19	0.47	0.30–0.64	0.53	0.35–0.74		
20–24	0.46	0.30–0.64	0.53	0.35–0.74	0.56	0.34–0.79
25–29	0.46	0.31–0.64	0.57	0.37–0.83	0.57	0.35–0.83
30–34	0.47	0.28–0.64	0.56	0.38–0.76	0.59	
35–39	0.44	0.29–0.63	0.56	0.34–0.84	0.58	
40–44	0.45	0.27–0.68	0.58	0.33–0.89	0.62	
45–49	0.46	0.30–0.65	0.59	0.33–0.88	0.67	
50–54	0.45	0.28–0.64	0.61	0.37–0.91	0.69	
55–59	0.49	0.28–0.72	0.60	0.36–0.88	0.72	
60–64	0.53	0.30–0.75	0.63	0.36–0.93	0.72	
65–69	0.52	0.30–0.79	0.62	0.34–0.91	0.77	
70+	0.52	0.31–0.76	0.61	0.33–0.93		

See footnotes to Table 3.

able also in five-year intervals *(7)*. There is a relatively wide distribution of mean values in some age groups for different population groups, as indicated in the published range of mean values for the different Lipid Research Clinics *(7)*. Findings of subset populations of the Lipid Research Clinics *(27, 28)* and the Cooperative Lipoprotein Phenotyping Studies *(29, 30)* were used for comparison and discussion of tables. Data on users of hormones and on different ethnic groups are included to the extent available. Users of hormones are women who reported usage of gonadal hormone during the previous two weeks. Sex hormones taken before menopause were mainly oral contraceptive preparations, whereas estrogen replacement was predominant among postmenopausal users. The number of participants in the prevalence study and the standard deviation values about the mean in each group are available in the original Lipid Research Clinics publications. All values < 0.995 g/L are rounded to two significant figures; > 0.995, to three significant figures.

Total Cholesterol in Serum

Table 1 gives "normal" reference values for total cholesterol in serum. Examination of results on specimens from 24 485 white males shows a gradual increase in mean total cholesterol serum values from 1.51 g/L for the 0–1 age group to a range of 1.65–1.66 g/L for the 8–11 age group, and then a decrease to 1.54–1.57 g/L range for the 14–19 age group. The mean concentrations of cholesterol then gradually increase at a fairly constant rate from age 20 to age 45, reaching the range of 2.18–2.20 g/L by age group 45–69. There is a slight decrease in total cholesterol after age 70. The 95th percentile values, usually considered the upper limit of "normal" values, follow the same type of curve as the mean values. These are consistent with results of other investigators *(29, 30)*, and give a maximum value of 2.86 g/L for 50- to 54-year-old men.

For 18 168 white females not using sex hormones, total cholesterol reference serum values show a gradual increase in mean cholesterol from 1.56 g/L at age 0–1 to 1.70 g/L at age 8–9; values decrease to 1.60–1.62 g/L for age group 12–17. Thereafter, gradual increases occur from 1.64 g/L (age group 18–19) to 2.38–2.40 g/L (age group 55–69). After this, the mean cholesterol values decline slightly to 2.35 g/L. The 95th percentile values for the total cholesterol in females follow the same type of curve as the mean values. These 95th percentile values vary for age groups younger than 20: from 2.01 to 2.06 g/L for age groups 0–5 and 12–17, and from 2.09 to 2.13 for age groups 6–11 and 18–19. For women over 20, the 95th percentile values increase gradually to 3.06–3.13 g/L for age groups 55–69 and then decrease slightly to 2.98 g/L.

Hormone intake. White women taking hormones (n = 5891) show a 5 to 8% increase in total cholesterol concentrations in age groups 15–39 and a tendency to have lower cholesterol in age groups 55–70+, compared with women not taking hormones *(8)*. Thus, women not taking hormones have higher cholesterol concentrations in the older age groups.

Racial differences. Comparison of Lipid Research Clinics data on blacks and whites younger than 20 showed that total cholesterol values for whites were higher than for blacks *(7,10)*. Above 20 years of age, men of both races have similar total cholesterol values, whereas black women tend to have higher values than white women between the ages of 20 and 39. In another population, however, mean values of serum cholesterol have been observed to be higher in black children than in white children *(31)*. These results are consistent with findings that total cholesterol can vary with ethnic group and location of population *(29–33)*. Because environmental and lifestyle effects appear to exert a major influence on lipid values, and because findings for different ethnic groups in different locations are not consistent, only the Lipid Research Clinics results on white males and females are presented here as representative reference "normal" values for practical use in the clinical chemistry laboratory.

Cord values for total cholesterol, as well as triglyceride cord values shown in Table 2, are much less than the corresponding values for postneonate children *(26, 34)*.

Serum Triglyceride

Table 2 presents serum population distributions for triglyceride measurements on 24 485 white males and 18 168 white females *(7, 8)*. For fasting white males, mean triglyceride values are in the range 0.55–0.59 g/L for age group 2–11, increase gradually to range from 1.55 to 1.56 g/L for age group 40–54, and then decline slowly to 1.34 g/L for age group older than 70. The 95th percentile values increase regularly from range 0.82–0.84 g/L for age group 0–3 to range 3.30–3.37 g/L for age group 35–54 and then decline to 2.66 g/L for those over 70.

For fasting white females not using sex hormones, mean triglyceride values range from 0.59 to 0.66 g/L for age group 2–9, increasing gradually to 1.29–1.35 g/L for age group 55+. The 95th percentile values are 0.97–1.10 for age group 0–9, increasing to a range of 2.40–2.70 for age group 50+.

Hormone intake. For fasting white women on hormones, compared with non-hormone users, mean triglyceride concentrations were about 45% higher in age group 15–39 and declined gradually to only about 15% more in age group 50–54 *(7, 8)*. The 95th percentile triglyceride values were increased similarly. Thus in women taking hormones, triglyceride values are markedly and consistently higher at all ages than those for women not taking hormones (7, 8).

Racial differences. In the Lipid Research Clinics Prevalence Study, specimens collected from black males gave lower mean triglyceride values and lower 95th percentile values than those of white males at all ages (7, 10). In black females, mean triglyceride values were lower than those of white females only in women over 50, but 95th percentiles for black females were lower than for white females for all age groups. Because the 5th and 95th percentiles of triglyceride distributions are similar for both sexes of both races, the representative reference "normal" values in Table 2 appear suitable for use in the clinical chemistry laboratory for triglyceride analytical results for both blacks and whites.

LDL Cholesterol

Table 3 presents serum population distributions for LDL cholesterol ("normal" values) (7, 9). LDL cholesterol values tend to parallel the same type of curves observed with mean and 95th percentile values of total cholesterol. For white men (n = 3584) between the ages of 20 and 70, mean LDL values increase steeply from 1.06 g/L to a range of 1.46–1.54 g/L for the age group 45–70+. The LDL cholesterol curve of 95th percentile values with respect to age resembles that of mean values and reaches a plateau of 2.03–2.16 g/L for the age group 45–69. For white women (n = 2627), mean LDL cholesterol serum values increase steadily between ages 20 and 49, more slowly than for men; but beginning in age group 50–54, LDL cholesterol increases markedly to a maximum of 1.67 g/L for age group 65–69. The 95th percentile values for white women plateau at 2.21–2.41 g/L for age group 50–69. In women using exogenous sex hormones, LDL cholesterol is slightly higher in age group 20–34, similar in age group 35–49, and lower in age group 50–69, than mean values of women not using hormones. Thus LDL cholesterol values in women over 50 and not taking hormones are higher than in men and in women taking hormones.

HDL Cholesterol

Table 4 gives serum population distributions for HDL cholesterol (7, 9, 35, 36). Between age 5 and 14, mean HDL cholesterol values are similar in both sexes. In white males, mean HDL cholesterol concentrations decrease sharply after age 14 to a plateau range of 0.44–0.47 g/L for age group 15–54. For age group 55–70+, mean values of HDL cholesterol increase to a range 0.49–0.53 g/L, and 95th percentiles to a range of 0.72–0.79 g/L (7, 35). In white females, mean HDL cholesterol values range from 0.53 to 0.54 g/L for age group 5–24; 0.56–0.59 g/L for age group 25–49; and 0.60–0.63 g/L for age group 50–70+.

For women taking hormones, HDL cholesterol mean values show a linear increase from age 20 to 69, from 0.56 to 0.77 g/L. At all ages, women using exogenous hormones have higher HDL cholesterol values than women not taking hormones. Thus women not on hormones have markedly and consistently higher concentrations of HDL cholesterol than men, but lower concentrations than women taking hormones.

HDL cholesterol has been observed to be higher in blacks than whites, when compared on the basis of age, sex, and total serum cholesterol (30).

Interpreting Test Results

Comparison of a result of a measurement for total cholesterol, triglyceride, LDL cholesterol, or HDL cholesterol with the values listed in Tables 1 through 4 relates the result to those of a representative population with *average risk factors* for cardiovascular disease; these population distributions are not necessarily *ideal* or *healthy* values (32). Interpretation is complicated because the risk associated with a single total cholesterol result can be affected by many other risk factors that are mediated through complex biochemical reactions and by lipid metabolism, which is not well understood. Thus each total and lipoprotein cholesterol result must be interpreted in light of the presence and degree of nonlipid risk factors in addition to the lipid risk factors such as the distribution of classes of lipoproteins; genetic, dietary, and other environmental factors must also be considered (19–21, 25, 32). Reference values also must be considered from the point of view of what cholesterol quantities are essential for host immunity and metabolism of different organs.

Note (Reviewer G. R. W.): It is important to emphasize that the population distributions are from a population with a substantial incidence of cardiovascular disease. Therefore, even though a value below a 95th percentile for a particular age and sex may be interpreted as "normal," this does not necessarily imply that the value is "healthy." According to Stamler (32), adult total cholesterol values above a 40th percentile are associated with increased risk of cardiovascular disease. Therefore, only values below the 40th percentile could be interpreted as "healthy."

Note (Reviewer L. V. E.): The epidemiological data of the Lipid Research Clinics should be considered as "observed value distribution," whereas, on the basis of studies in other populations, the "ideal" total cholesterol values for adults of all ages appear to have a mean near 1.60 g/L. The observed data in the U.S. population are probably due in large part to the combined effects of diet, environment, and lifestyle superimposed on whatever genetic factors may dominate in our varied population.

This distribution of observed values from the Lipid Research Clinics studies, dealing primarily with American whites and blacks, does not offer much help to laboratories serving large population groups of different ethnic backgrounds, such as oriental and Hispanic patients. A good laboratory determines or confirms its own reference values for the population(s) it serves.

When compared with results of studies conducted more than 10 years ago, the Lipid Research Clinics studies (7) indicate that the mean cholesterol value among the population is decreasing and mean triglyceride is increasing (37). The values of total cholesterol in adults, however, are still quite high in population groups in North America, particularly if the "desired" or healthy value approximates the lower 5th percentile values of the distribution curve of measured "normal" values (32).

This study has been collaboratively supported by the Lipid Research Clinics Program, National Heart, Lung, and Blood Institute, National Institutes of Health, Bethesda, MD 20205.

References

1. Cooper, G. R., The World Health Organization–Center for Disease Control Lipid Standardization Program. In *Quality Control in Clinical Chemistry* (Transactions of the VI International Symposium, April 23–25, 1974), G. Anido, S. B. Rosalki, E. J. van Kampen, and M. Rubin, Eds., Walter de Gruyter, Berlin–New York, 1975, pp 95–109.
2. Zak, B., Cholesterol methodologies: A review. *Clin. Chem.* 23, 1201–1214 (1977).
3. Tonks, D. B., The estimation of cholesterol in serum: A classification and critical review of methods. *Clin. Biochem.* 1, 12–29 (1967).
4. Fasce, C. F., Jr., and Vanderlinde, R. E., Factors affecting the results of serum cholesterol determinations. *Clin. Chem.* 18, 901–908 (1972).
5. Boone, D. J., and Knouse, R. W., Proficiency Testing Summary Analysis, Clinical Chemistry Survey IV, November 13, 1979. Center for Disease Control, Atlanta, GA.
6. Haeckel, R., Sonntag, O., Kulpmann, W. R., and Feldmann, U., Comparison of nine methods for the determination of cholesterol. *J. Clin. Chem. Clin. Biochem.* 17, 553–563 (1979).
7. *The Lipid Research Clinics Population Studies Data Book*, 1, *Prevalence Study*. DHHS, PHS (NIH) National Heart, Lung, and Blood Institute, Publication No. 80-1527, Bethesda, MD 20014, 1980.
8. Lipid Research Clinics Program Epidemiology Committee, Plasma lipid distributions in selected North American populations: The Lipid Research Clinics Program Prevalence Study. *Circulation* 60, 427–439 (1979).
9. Heiss, G., Tamir, I., Davis, C. E., Tyroler, H. A., Rifkind, B. M., Schonfeld, G., Jacobs, D., and Frantz, I. D., Jr., Lipoprotein–cholesterol distributions in selected North American populations: The Lipid Research Clinics Program Prevalence Study. *Circulation* 61, 302–315 (1980).
10. Christensen, B., Glueck, C., Kwiterovich, P., Degrott, I., Chase, G., Heiss, G., Mowery, R., Tamir, I., and Rifkind, B., Plasma cholesterol and triglyceride distributions in 13,665 children and adolescents: The prevalence study of the Lipid Research Clinics Program. *Pediatr. Res.* 14, 194–202 (1980).
11. *Lipid Research Clinics Program: Manual of Laboratory Operations*, 1. *Lipid and Lipoprotein Analysis*. DHEW publication No. (NIH) 75-628, Government Printing Office, Washington, DC, 1974.
12. Page, I. H., and Moinuddin, M., The effect of venous occlusion on serum cholesterol and total protein concentration—a warning. *Circulation* 25, 651–652 (1962).
13. Tan, M. H., Wilmshurst, E. G., Gleason, R. E., and Soeldner, J. S., Effect of posture on serum lipids. *N. Engl. J. Med.* 289, 416–418 (1973).
14. Cooper, G. R., Roland, D. M., and Eavenson, E., Determination of total cholesterol with the AutoAnalyzer. *Automation in Analytical Chemistry, Technicon International Symposia*. Mediad, New York, NY, 1964.
15. Wood, P. D., Bachorik, P. S., Albers, J. J., Stewart, C. C., Winn, C., and Lippel, K., Effects of sample aging on total cholesterol values determined by the automated ferric chloride–sulfuric acid and Liebermann–Burchard procedures. *Clin. Chem.* 26, 592–597 (1980).
16. Office of Standard Reference Materials, Cholesterol SRM No. 911a. National Bureau of Standards, Washington, DC 20234.
17. Fieser, L., Cholesterol and companions. VII. Steroid bromides. *J. Am. Chem. Soc.* 75, 5421–5422 (1953).
18. Cooper, G. R., Akins, J. R., Biegeleisen, J., Hazlehurst, J., Myers, C. M., and Bayse, D. D., Comparison of selected cholesterol colorimetric and enzymatic reference methods. *Clin. Chem.* 24, 1046 (1978). Abstract.
19. Fredrickson, D. S., Levy, R. I., and Lees, R. S., Fat transport and lipoproteins—an integrated approach to mechanisms and disorders. *N. Engl. J. Med.* 276, 32–44, 94–103, 148–156, 215–224, 273–281 (1967).
20. Kannel, W. B., Dawber, T. R., Kagan, A., Revotskie, N., and Stokes, J., Factors of risk in the development of coronary heart disease: Six-year follow-up experience. *Ann. Intern. Med.* 55, 33–50 (1961).
21. Zilversmit, Z. B., Atherogenesis: A postprandial phenomenon. *Circulation* 60, 473–485 (1979).
22. Laboratory Methods Committee of the Lipid Research Clinics Program, Cholesterol and triglyceride concentrations in serum–plasma pairs. *Clin. Chem.* 23, 60–63 (1977).
23. Cohen, A., Hertz, H. S., Mandel, J., Paule, R. C., Schaffer, R., Sniegoski, L. T., Sun, T., Welch, M. J., and White, E., Total serum cholesterol by isotope dilution–mass spectrometry: A candidate definitive method. *Clin. Chem.* 26, 854–860 (1980).
24. Friedewald, W. T., Levy, R. I., and Fredrickson, D. S., Estimation of the concentration of low-density lipoprotein cholesterol in plasma, without the use of the ultracentrifuge. *Clin. Chem.* 18, 499–502 (1972).
25. Goldstein, J. L., Schrott, H. G., Hazzard, W. R., Bierman, E. L., and Motulsky, A. G., Hyperlipidemia in coronary heart disease. II. Genetic analysis of lipid levels in 176 families and delineation of a new inherited disorder, combined hyperlipidemia. *J. Clin. Invest.* 52, 1544–1568 (1973).
26. Tsang, R. C., Fallat, R. W., and Glueck, C. J., Cholesterol at birth and age 1: Comparison of normal and hypercholesterolemic neonates. *Pediatrics* 53, 458–470 (1974).
27. Hewitt, D., Jones, G. J. L., Godin, G. J., McComb, K., Breckenridge, W. C., Little, J. A., Steiner, G., Mishkell, J. A., Bailie, J. H., Martin, R. H., Gibson, E. S., Pendergrast, W. F., and Parliament, W. J., Normative standards of plasma cholesterol and triglyceride concentrations in Canadians of working age. *Can. Med. Assoc. J.* 117, 1020–1024 (1977).
28. Miskel, M. A., Neonatal plasma lipids as measured in cord blood. *Can. Med. Assoc. J.* 111, 775–780 (1974).
29. Castelli, W. P., Cooper, G. R., Doyle, J. T., Garcia-Palmieri, M., Gordon, T., Hames, C., Hulley, S. B., Kagan, A., Kuchmak, M., McGee, D., and Vici, W. J.,

Distribution of triglycerides and total, LDL, and HDL cholesterol in several populations: A cooperative lipoprotein-phenotyping study. *J. Chron. Dis.* 30, 147–169 (1977).
30. Tyroler, H. A., Hames, C. G., Krisham, I., Heyden, S., Cooper, G. R., and Cassel, J. C., Black–white differences in serum lipids and lipoproteins in Evans County. *Prev. Med.* 4, 541–549 (1975).
31. Frerichs, R. R., Srinivasan, S. R., Webber, L. S., and Berenson, G. S., Serum cholesterol and triglyceride levels in 3,446 children from a biracial community: The Bogalusa Heart Study. *Circulation* 54, 302–309 (1976).
32. Stamler, J., Lifestyles, major risk factors, proof and public policy. *Circulation* 58, 3–19 (1978).
33. Keys, A., Ed., Coronary heart disease in seven countries. *Circulation* 41 (Suppl. 1), 1–211 (1970). Monograph 29, American Heart Association, Dallas, TX.
34. Mishkel, M. A., Neonatal plasma lipids as measured in cord blood. *Can. Med. Assoc. J.* 111, 775–780 (1974).
35. Heiss, G., Davis, C. E., Tyroler, H. A., Tamir, I., and Barrett-Conner, E., HDL levels in 12,500 white men and women ages 5 to 96; the LRC population survey. 18th Conference on Cardiovascular Disease Epidemiology, Orlando, FL, March 1978.
36. Rifkind, B. M., Tamir, I., and Heiss, G., Preliminary high-density lipoprotein findings. The Lipid Research Clinics Program, In *High Density Lipoproteins and Atherosclerosis*, A. M. Gotto, Jr., N. E. Miller, and M. F. Oliver, Eds., Elsevier-North Holland Press, New York, NY, 1978, pp 109–120.
37. Beaglehole, R., LaRosa, J. C., Heiss, G., Davis, C. E., Williams, O. D., Tyroler, H. A., and Rifkind, B. M., Serum cholesterol, diet, and the decline in coronary heart disease mortality. *Prev. Med.* 8, 538–547 (1979).

Cholesterol, Enzymic Method[1]

Submitters: Gerald R. Cooper, Patricia H. Duncan, J. S. Hazlehurst, Dayton T. Miller, and D. D. Bayse, *U.S. Department of Health and Human Services, Public Health Service, Centers for Disease Control, Center for Environmental Health, Clinical Chemistry Division, Atlanta, GA 30333*

Evaluators: G. M. Widdowson, H. J. Mulvihill, and M. L. Kuehneman, *Institutes of Medical Sciences, San Francisco, CA 94115*
Ingeborg Kupke and S. Zeugner, *Department of Clinical Chemistry and Lipid Research, Medizinische Errichtungen der Universität Dusseldorf, 4000 Dusseldorf 1, F.R.G.*

Reviewer: James L. Driscoll, *Department of Pathology, Rhode Island Hospital, Providence, RI 02902*

Introduction

Primary and secondary lipoproteinemia disorders in patients are monitored in the clinical laboratory mainly by the determination of total cholesterol and the cholesterol content of lipoproteins *(1)*. Enzymic methods for determining cholesterol have been developed *(2–4)* and are now used frequently, as are the Liebermann–Burchard and the iron methods *(5)*. Use of the enzymic method is limited when either aqueous or alcoholic pure cholesterol standards are used, but it can be calibrated accurately with a homogeneous and stable serum pool. This method involves a direct measurement in serum, the enzymic analytical determination being calibrated by a serum labeled with a target value assigned by an accepted reference method; no blank corrections are made. The proposed procedure and findings, with use of a control serum pool for internal surveillance, are applicable also to substitution of commercial enzymic assay reagents in this method. Using the optimized method presented here, clinical chemists may prepare their own reagents or validate commercial reagents.

Principle

The total cholesterol in serum is primarily unesterified and esterified cholesterol in association with lipoproteins. The cholesteryl esters are freed from the lipoproteins and enzymically hydrolyzed by cholesterol ester hydrolase (EC 3.1.1.13) to form free cholesterol and fatty acids. The free cholesterol produced by cholesterol ester hydrolase and that already present in the sample is oxidized by another enzyme, cholesterol oxidase (EC 1.1.3.6), to form cholest-4-en-3-one and hydrogen peroxide (H_2O_2). The peroxide, in the presence of peroxidase (EC 1.11.1.7), oxidatively couples with phenol and 4-aminoantipyrine to produce a quinoneimine dye, with maximum absorbance measured at 500 nm. The absorbance readings at 500 nm are proportional to the concentration of total cholesterol in samples.

Materials and Equipment

Reagents[2] *(6)*

All chemicals are analytical reagent grade unless otherwise specified. Microbial cholesterol ester hydrolase (cat. no. 161772) and microbial cholesterol oxidase (cat. no. 126934) were obtained from Boehringer Mannheim Biochemicals, Indianapolis, IN 46250; 1,4-piperazinediethanesulfonic acid (PIPES)[3] and horseradish peroxidase (cat. no. P8375) were obtained from Sigma Chemical Co., St. Louis, MO 63178. 4-Aminoantipyrine and sodium cholate were purchased from ICN Pharmaceuticals, Inc., Plainview, NY 11803; phenol from J. T. Baker Chemical Co., Phillipsburg, NJ 08865; and Triton X-100 from Rohm and Haas, Philadelphia, PA 19105. The other reagents were from various commercial sources.

---

[1] This study has been collaboratively supported by the Lipid Research Clinics Program, National Heart, Lung, and Blood Institute, NIH, Bethesda, MD, and was part of an investigation carried out at CDC under the sponsorship of the Cholesterol Reference Method Study Group of the AACC Standards Committee. The members of the Study Group are: J. G. Batsakis, G. N. Bowers, Jr., W. C. Breckenridge, M. D. Chilcote, B. E. Copeland, L. V. Edwards, C. S. Furfine, B. Gerson, H. F. Martin, R. Schaffer, and G. R. Cooper (Chairman).

[2] The use of trade names is for identification only and does not constitute endorsement by the Public Health Service, the U.S. Department of Health and Human Services, or the AACC.

[3] Nonstandard abbreviations: PIPES, 1,4-piperazinediethanesulfonic acid; CDC, Centers for Disease Control; ALBK, Abell, Levy, Brodie, and Kendall (cholesterol method); HDL, LDL, high- and low-density lipoprotein, respectively; HPTLC, high-performance thin-layer chromatography.

Cholesterol oxidase was also purchased from Beckman Instruments, Carlsbad, CA 92008. Pure cholesterol was obtained from Sigma (99+%; CH-S; no. 74C7440) and the U.S. National Bureau of Standards (Standard Reference Material 911a).

Enzymes were stored at 4 °C according to manufacturers' directions.

Solutions

1. PIPES *buffer, 50 mmol/L, pH 6.9.* To a 1000 mL-beaker containing approximately 200 mL of distilled water, add 15.2 g of PIPES. This solution will be cloudy. While stirring with a magnetic mixer, slowly add about 6 mL of 50% w/v NaOH (500 g/L, 12.5 mol/L) until mixture becomes clear. Add approximately 600 mL of distilled water and adjust the pH to 6.9 with a pH meter. Quantitatively transfer the solution to a 1000-mL volumetric flask, and cool to ambient temperature. When the solution has equilibrated at room temperature, adjust to 1000-mL volume with distilled water. It is best to filter the solution through a 0.45-μm (av. pore size) disposable plastic filter before use.

Note: Evaluator I. K. found that filtration was not necessary with her reagents.

2. *Stock mixed reagent.* The stock mixed reagent contains no cholesterol ester hydrolase or cholesterol oxidase. The amount of stock mixed reagent prepared depends upon the number of samples to be analyzed and is best prepared freshly daily. This stock reagent, if prepared without peroxidase, has been stored successfully one month at 4 °C.

a. After weighing, place the dry ingredients in a 100-mL graduated cylinder. The concentration of the constituents in the stock mixed reagent for a 50-mL volume is shown in Table 1.

Table 1. Constituents in Stock Mixed Reagent

Constituents	Concn.	Amount for 50 mL
4-Aminoantipyrine	2 mmol/L = 0.4066 mg/mL	20.3 mg
Na cholate	6 mmol/L = 2.5800 mg/mL	129.2 mg
KCl	200 mmol/L = 14.9200 mg/mL	746 mg
Peroxidase	50 U/test = 25 U/mL	1250 U
Triton X-100	2 mL/L	0.1 mL

After being mixed with the phenol reagent, the final concentration will be one-half that of the stock reagent.

b. Dissolve with shaking and adjust to 50-mL volume with 50 mmol/L PIPES buffer. This volume is sufficient for 24 test determinations. If 24 serum blank determinations are necessary also, then double the quantities (100 mL).

3. *Stock phenol reagent, 80 mmol/L.* Phenol stored over Drierite in a desiccator is used to prepare the stock reagent. Carefully but quickly weigh 0.38 g of phenol crystals into a 50-mL glass beaker. With careful swirling, dissolve the phenol in the beaker in about 10 mL of the PIPES buffer. Transfer the corrosive solution to a 100-mL graduated cylinder, rinsing the beaker several times with buffer. Bring the phenol reagent to 50-mL volume with the PIPES buffer. If serum blanks are to be run, prepare twice this quantity (100 mL). This reagent has been stored successfully as long as a month at 4 °C in a tightly closed glass container.

4. *Working blank reagent.* This reagent is prepared only if serum blank runs are required, i.e., when it is observed that interferences, such as high amounts of bilirubin or dense turbidity, are present in the specimens to be analyzed. Mix 50 mL of the stock mixed reagent and 50 mL of the stock phenol reagent to form the working blank reagent.

5. *Working test reagent.* Mix 50 mL of the stock mixed reagent and 50 mL of the stock phenol reagent. Add 25 U each of cholesterol oxidase and cholesterol esterase, either from a concentrated solution of the enzymes or weighed amounts of the solid enzymes.

Alternative procedure: Prepare the stock mixed reagent without peroxidase and the stock phenol reagent in quantities sufficient for one month (store at 4 °C). Prepare fresh working test reagent daily by adding peroxidase, cholesterol oxidase, and cholesterol esterase to the mixture of stock reagents.

Note: Evaluator I. K. found it advantageous to prepare the stock reagent without peroxidase and to store it at 4 °C.

Quality-Control Materials

1. *Serum reference pools used in this study.* Serum pools of various cholesterol concentrations were prepared by adding a concentrate of human lipoproteins to a human serum pool (7). These pools were analyzed by the Centers for Disease Control (CDC) Reference Laboratory, with a modified method of Abell, Levy, Brodie, and Kendall (ALBK) (8, 9). The pools were assigned target values by analyses over a period of two to six months to assure with 95% confidence that each assigned target value differed by less than 1.0% from the expected target value with that method.

2. *Serum pool for calibration of the enzymic method.* A homogeneous serum pool labeled with a target value by a defined cholesterol reference method should be acquired from a commercial source. A lyophilized material can be used.

Note: Reviewer J. L. D. states that regardless of the accuracy of the analytical method used to standardize serum calibrators, secondary standards are always a potential source of error. This precaution should be stressed in detail.

Note: Evaluator G. M. W. states that the absolute accuracy of this selected method depends to a major degree on the accuracy of the assigned value of the serum calibrator used. He does not know of any such commercially available material with a target value established by the ALBK reference method. He comments that general availability of a serum calibrator such as the one supplied to us by CDC would quiet the above concern.

3. *Serum pool for internal quality surveillance.* A homogeneous serum pool made from clear serum, dispensed into 1-mL aliquots, and stored at −20 °C, can be used for internal surveillance. Duplicates from a single vial permit internal monitoring and a measurement of within-vial variation. A commercial, lyophilized, homogeneous serum pool lot can also be used for this purpose.

Equipment

Micropipettes with disposable tips (such as Oxford, Eppendorf, or SMI), capable of dispensing 50 to 100 µL, may be used for manual sampling. The spectrophotometer must be capable of measuring absorbance linearly at 500 nm. A water bath or heating block is needed for temperature control at 37 °C.

Assay of Enzymes for Specific Activity

New lots of purchased enzyme preparations should be assayed for specific activity. The methods described in the *Appendix* to this chapter were found dependable and can be used as alternatives to the manufacturers' suggested assay methods. Enzyme preparations can lose specific activity over extended periods of time.

Sample Collection

Collect blood from a vein, using minimum obstruction from a tourniquet to prevent excessive extravasation flow of fluid water from veins into tissues. Do not apply the tourniquet longer than 30 s; release it as soon as the blood starts flowing into the collecting tube. To control the effect of posture, standardize the length of time a person has been sitting or lying down before the collection of blood. A good rule: Have the patient sit for 10 min before collecting blood from the vein. Whenever possible, blood should be collected after 12–16 h of fasting, to minimize turbidity.

Note: Evaluator I. K. notes that in pediatrics small sample volumes such as serum from capillary blood have to be analyzed. Therefore, differences in cholesterol concentrations of capillary and venous blood samples must be taken into consideration. Her studies have shown that venous and capillary blood serum cannot be used interchangeably for cholesterol determination (*Clin. Chim. Acta* 97: 279–283, 1979).

Procedure

At a preselected time interval add 50 µL of sample to 4.0 mL of working test reagent, in 13 × 100 mm test tubes. Mix gently by inversion and place in a 37 °C water bath for 12 min. Read the absorbance (A) in a spectrophotometer at 500 nm at the prescribed time interval. The volumes of sample and reagent can be altered proportionately, if desired; for example, using a 25-µL sample and 2.0 mL of working test reagent would halve the cost of the reagents.

Note: Evaluator I. K. found that Beer's Law applies up to about 5.5 g of cholesterol per liter; however, because of the high sensitivity of the CDC method, this cholesterol concentration yields an absorbance of approximately 1.0. For the manual procedure, dilute the sample when the cholesterol concentration exceeds 4.0 g/L.

1. Turn on the spectrophotometer. Set the wavelength to 500 nm.
2. List the samples to be analyzed numerically on a suitable data recording form. For example, for 20 single unknowns, or 10 unknowns in duplicate, use a control pool in duplicate, a single serum calibrator specimen, and a water sample. Follow the protocol in Table 2 in positioning samples and reference materials in a run.

Table 2. Position of Samples and Reference Materials in an Analysis

Tube position no.	Sample
1	Water
2	Serum calibrator
3	Quality control pool
4–14	Unknowns (singles or duplicates)
15	Quality control pool
16–24	Unknowns (singles or duplicates)

3. Prepare a rack of 13 × 100 mm screw-capped test tubes, numbered according to the data recording form.
4. Sample with either manual or sampler-diluter technique at a preselected time interval:

Manual: Zero or read the spectrophotometer against water. Include the constant reagent blank absorbance (about 0.020) in the test absorbance (A) values. Next, dispense 4.0 mL of working test reagent into the rack of tubes. Add 50 µL of sample at 30-s or other selected time intervals. Use pipettes that have a precision (coefficient of variation, CV) better than 0.5%. The reagent and samples should be at room temperature before pipetting.

Sampler-diluter: At set intervals (30 s or other selected intervals), dispense 50 µL of sample and 4 mL of working test reagent with a precise sampler-diluter instrument. The instrument used should have a precision of dispensing better than 0.5% CV.

5. Cap the tubes, mix gently by inversion, and keep each tube in a 37 °C water bath for 12 min.

6. Ten seconds before 12 min have elapsed, remove the first sample from the water bath so that the absorbance can be read in the spectrophotometer at exactly 12 min. Continue to read the samples at the prescribed time interval, and record the data on the data record sheet.

7. Record the data in the test absorbance column of the data record sheet.

8. Calculate test results as follows:

Test result, g/L = (A test/A calibrator) × assigned value of calibrator, g/L

In this procedure, the absorbance of a serum reagent blank (where the reagent contains no cholesterol enzymes) is determined for each sample only when the samples are suspected to contain a high concentration of an interfering substance (such as icteric, turbid, or hemolyzed samples).

The time interval to be used between sampling depends on the pipette and spectrophotometer used and the number of analysts available. Sufficient time must be allowed for sampling and for reading and recording the data. Reading of absorbance after 12 min allows an analyst to perform 24 readings on samples and reference materials at 30-s intervals as one run. Multiple 24-reading runs, of course, can be carried out during a day if needed.

Calibration

It is difficult to calibrate the enzymic determination of cholesterol with aqueous or organic solutions of cholesterol, because of problems of solubility of

Table 3. Comparison of Experimental Results of the Enzymic Cholesterol Method, Corrected (Corr) and Uncorrected (Unc) for Blanks, with Results of the CDC Cholesterol Reference Method for 10 Serum Pools

Pools, g/L[a]	Absorbance Mean	SD	CV, %	Calibrator, 1.62 g/L Mean	SD	Calibrator, 2.55 g/L Mean	SD	CDC reference method (n = 20), g/L Mean	SD
1.33									
Corr	0.256	0.008	3.1	1.32	0.03	1.34	0.04	1.34	0.02
Unc	0.267	0.007	2.6	1.32	0.03	1.34	0.03		
1.62									
Corr	0.315	0.010	3.4	1.63	0.04	1.64	0.04	1.63	0.02
Unc	0.326	0.011	3.4	1.62	0.04	1.64	0.04		
1.83									
Corr	0.354	0.006	1.7	1.81	0.02	1.84	0.03	1.84	0.02
Unc	0.363	0.006	1.7	1.83	0.03	1.83	0.02		
2.38									
Corr	0.457	0.012	2.4	2.36	0.05	2.38	0.05	2.39	0.04
Unc	0.476	0.010	2.1	2.38	0.05	2.41	0.05		
2.47									
Corr	0.465	0.012	2.6	2.41	0.04	2.43	0.04	2.46	0.02
Unc	0.481	0.011	2.0	2.42	0.03	2.45	0.03		
2.55									
Corr	0.488	0.008	1.6	2.52	0.03	2.54	0.03	2.54	0.04
Unc	0.502	0.007	1.4	2.51	0.02	2.55	0.02		
2.89									
Corr	0.557	0.009	1.6	2.88	0.05	2.91	0.05	2.88	0.03
Unc	0.571	0.009	1.6	2.84	0.04	2.88	0.04		
3.01									
Corr	0.576	0.010	1.7	2.96	0.04	2.99	0.04	3.02	0.06
Unc	0.591	0.008	1.4	2.96	0.04	3.00	0.03		
3.41									
Corr	0.658	0.012	1.8	3.41	0.06	3.44	0.06	3.41	0.04
Unc	0.672	0.011	1.6	3.37	0.04	3.41	0.04		
3.50									
Corr	0.666	0.012	1.8	3.45	0.08	3.48	0.08	3.51	0.04
Unc	0.684	0.010	1.5	3.44	0.07	3.50	0.07		

[a] Previously assigned reference values.

cholesterol and differences in reaction rates in various solvent media *(10)*. In addition, uncertainty exists about the extent of hydrolysis of different cholesterol esters *(10, 11)*, the effect of interfering substances *(10, 12)*, and the complexity of the color-developing chemical reaction *(10)*. Calibration with a correctly labeled serum pool, however, appears to be the best way to achieve the highest level of accuracy with the enzymic method.

Experimental studies have been devised to prove the suitability of serum calibration and to determine whether test results should be corrected for serum blanks or should be used as uncorrected values to calculate results. The results of studies on 10 serum pools with previously assigned reference values are given in Table 3. For the enzymic method, absorbance data on duplicates per run are tabulated in rows to show test data corrected and uncorrected for serum blanks. Results were closely comparable, with or without correction. One set of data shown in Table 3 was calculated by comparison with a serum calibrator of lower total cholesterol (1.62 g/L), and another set with a serum calibrator of higher concentration of total cholesterol (2.55 g/L).

Means and SDs of 20 parallel results on each pool with the CDC cholesterol reference method are also included in the last two columns of Table 3. The CV of the CDC cholesterol reference method varied from 0.8 to 2.0% (depending upon cholesterol concentration in each pool) for 20 analyses on each pool.

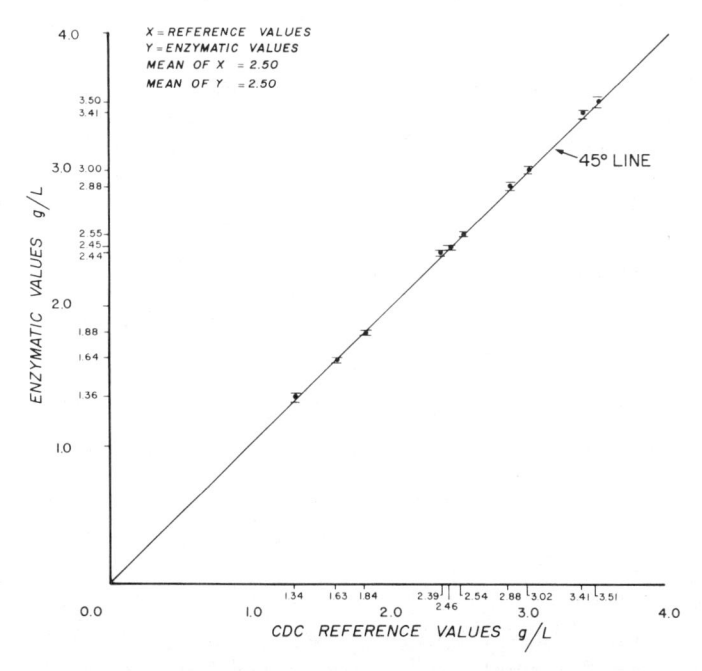

Fig. 1. Comparison of cholesterol values of 10 serum pools analyzed by the enzymic method with a serum calibrator (2.55 g/L) and uncorrected for blanks, and by the CDC reference method

A 45° line is drawn on the plot for use as a line of reference

A plot of the data of Table 3 (Figure 1) confirmed agreement between the values of the serum-calibrated enzymic and the CDC reference methods.

Results of calibration with the two serum pools were similar, verifying the feasibility of using serum pools of different concentrations as calibrators. Thus, on the basis of feasibility, less cost, and decreased time of analysis, it is advantageous to use an enzymic method based on serum calibration of absorbance data uncorrected for serum blanks.

Results

The enzymic procedure in which absorbance data were uncorrected for serum blanks was evaluated by performing duplicate analyses within the same run on 75 individual serum specimens from working adults. Two serum pools previously labeled as having 2.47 and 3.50 g of cholesterol per liter were also analyzed 14 times during these analyses over a period of three weeks. The results obtained with the enzymic procedure and the use of a serum calibrator (2.47 g/L) are presented in Table 4.

Table 4. Experimental Results of Enzymic Cholesterol Method, Corrected (Corr) and Uncorrected (Unc) for Blanks and Calibrated against a Serum Pool (2.47 g/L), Compared with Results of CDC Cholesterol Method

	Enzymic method				CDC reference method mean, g/L
	Absorbance			Mean cholesterol, g/L	
Sample	Mean	SD	CV, %[a]		
Individual human sera (n = 75)					
Corr	0.428	0.095	22	2.29	2.24
Unc	0.450	0.100	22	2.27	
Quality control pool (n = 14)[b]					
Corr	0.647	0.009	1.4	3.49	3.51
Unc	0.684	0.004	0.6	3.47	

[a] CV (coefficient of variation) = (SD/mean) × 100. The SD and CV of 75 individual sera are large because they are representing both physiologic and analytical variation among specimens.
[b] Reference value, 3.50 g/L.

Cholesterol in the 75 individual sera was also determined concurrently with the CDC cholesterol reference method, to serve as a point of reference. The results obtained by the enzymic method calibrated with the 2.47 g/L serum pool are about 1% higher than those by the CDC cholesterol reference method. The results of the quality-control serum pool (3.50 g/L), though, are about 1% lower by the enzymic method than by the CDC reference method. Plots of the 75 individual results revealed close agreement between the results of the enzymic method, either uncorrected or corrected for blanks, and the results of the CDC reference method (Figures 2 and 3).

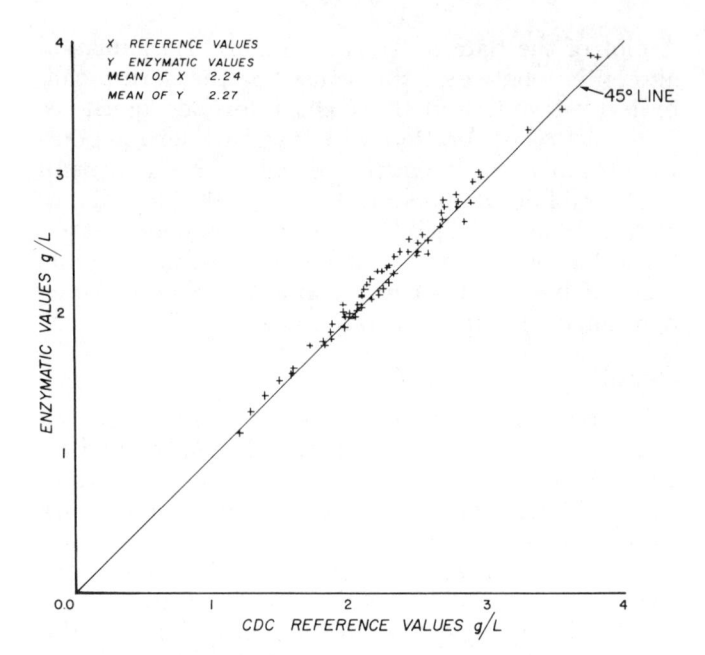

Fig. 2. Comparison of cholesterol values of 75 individual serum samples analyzed by the enzymic method with a serum calibrator (2.47 g/L) and uncorrected for blanks, and by the CDC reference method
A 45° line is drawn on the plot for use as a line of reference

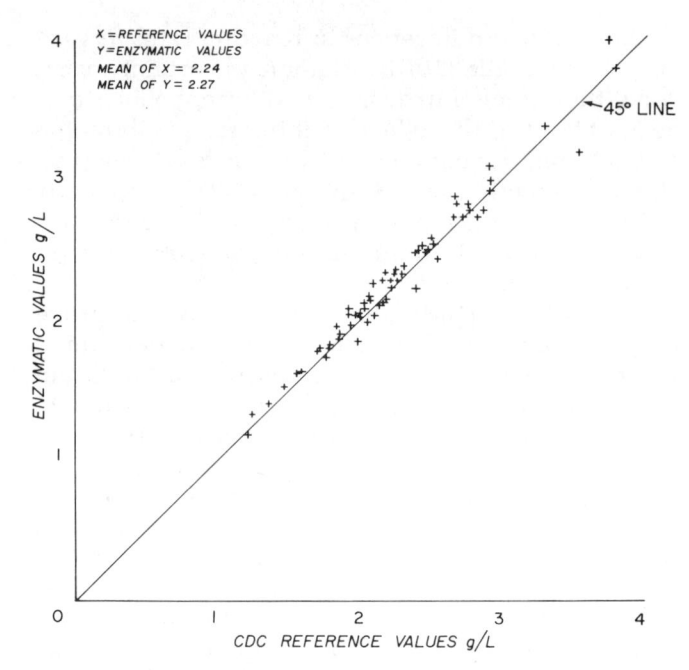

Fig. 3. Comparison of cholesterol values of 75 individual serum samples analyzed by the enzymic method with a serum calibrator (2.47 g/L) and corrected for blanks, and by the CDC reference method
A 45° line is drawn on the plot for use as a line of reference

Results of the analyses on the 75 individual sera were also calculated on the basis of calibration with aqueous cholesterol standards solubilized with Triton X-100 (150 mL/L) and corrected for serum/reagent blanks. The results, calibrated against cholesterol standards, averaged about 5.5% lower than the results of the CDC cholesterol reference method.

Reference Values

The enzymic determination of cholesterol is used primarily to determine total and lipoprotein cholesterol in serum or plasma. The lipoprotein cholesterol values of serum are often determined from total cholesterol, total triglyceride, and high-density lipoprotein (HDL) cholesterol values (13). Values in the physiologically critical range of atherosclerosis risk, which require accurate and precise measurement, are HDL cholesterol values below "normal" limits, low-density lipoprotein (LDL) cholesterol values above "normal" limits, and total cholesterol values outside "normal" limits. Epidemiologists emphasize that each total and lipoprotein cholesterol value has a certain risk of atherosclerosis associated with it.

In the United States, the Lipid Research Clinics Prevalence Study, involving standardized lipid laboratories, has recently published data on distribution of plasma values of lipids and lipoproteins (14). The results of these studies were corrected for serum–plasma differences and used to develop a set of tables of normal values for total cholesterol, total triglyceride, and lipoprotein cholesterol that can be used as a point of reference for serum cholesterol determination. These tables are presented in the preceding chapter (pages 159–160). The proposed reference values tabulated with respect to age and sex are applicable to the enzymic method proposed here, because this method is calibrated against a serum pool labeled by a modified proposed reference cholesterol method (9), which is in agreement with the proposed isotope dilution/mass spectrometer definitive method (15).

Note: Evaluator I. K. emphasizes that environmental factors appreciably influence reference values; her laboratory's recent studies on Japanese children living in Dusseldorf strongly suggest that environmental influences predominate over ethnic factors. German (n = 335) and Japanese (n = 80) children showed mean total cholesterol values of 1.79 (SD 0.28) and 1.65 (SD 0.22) g/L, respectively. Their unpublished comparison of population samples living in various European cities reveal a decline, from north to south, in total cholesterol values.

Discussion

The simplicity, sensitivity, and precision of the enzymic determination of cholesterol have made it popular in many clinical laboratories (16, 17). The uncertainty associated with enzymic hydrolysis, blanks, and color development (10) appears to be largely controlled by calibration with an aliquot from a serum pool with a known target value. This permits easy application to automation as well as to a direct manual analytical method.

The analytical capabilities of the enzymic method appear suitable for use in the clinical laboratory. The results indicate an attainable CV of between 1 and 2% on the same serum specimens. The use of a serum calibrator labeled with a target value by a cholesterol reference method provided values on 75 individual specimens that agreed within 2% with those determined concurrently by the CDC reference method. The sensitivity and specificity permit utilizing a direct reaction; the higher cost of enzyme preparations is thus partly offset by the time saved in the analysis.

The enzymic analytical system has been adopted by many "reagent kit" systems. Like the method proposed in this chapter, competent kit systems calibrated against a serum calibrator with a target value accurately assigned by an accepted reference method should be capable of successful application in the clinical laboratory. The enzymic method can meet the prerequisites for calibration with a serum calibrator because it shows linearity and proportionality of absorbance with "true" values of cholesterol in serum pools of various concentrations and in individual serum specimens.

In the enzymic procedure, errors may be introduced from interferences in either serum or reagents (10, 12). Cholesterol concentration must not exceed its solubility in any standard or serum matrix. Turbidity can decrease in the sample as the reaction proceeds because the reagent contains a solubilizer. Bilirubin and ascorbic acid appear to exert negative interfering effects (11, 12, 18). Bilirubin exerts a doubly interfering effect, reacting with hydrogen peroxide to lose what absorbance it had when nonoxidized. Various interfering substances, such as sterols other than cholesterol (3, 17), are believed to exist naturally in serum. Chemical reducing agents usually found in clinical samples do not show obvious interference effects in the enzymic method (3). These findings have caused indecision about whether it is better to calibrate with or without correction for blanks. The effects of these sources of error, therefore, appear to be compensated for, or controlled reasonably well, when the procedure is calibrated against serum of known total cholesterol concentration without correction for serum blanks.

The enzymic determination of cholesterol also provides a simple way to perform unesterified cholesterol analyses. Unesterified cholesterol determinations can be calibrated against an aqueous solution of cholesterol solubilized in Triton X-100, 150 mL/L. This calibration appears suitable, giving a linear response with concentration for standard solutions of pure cholesterol up to 4.0 g/L. Measurements of unesterified cholesterol by the enzymic method with no esterase in the reagent and with Triton X-100 aqueous standards for calibration, and measurements of total cholesterol with use of esterase and a serum for calibration, provide an estimate of the relationship of unesterified cholesterol to cholesterol ester. These measurements are useful in detecting and evaluating certain metabolic disorders and diseases.

Reports of Evaluators

Evaluators G. M. W., H. J. M., and M. L. K. followed the instructions for the method exactly, except that the final absorbance readings were at 37 °C in the thermo-cuvet, and the sample and reagent volumes were adjusted to suit the equipment and amounts of reagents available. They used 750 μL of reagent, 20 μL of sample, and a further 750 μL of reagent added to each tube with a Micromedic diluter (final reagent volume was therefore 1520 μL).

The stock mixed reagent was filtered after preparation through a Millipore filter (RAWPO 1300; 1.2-μm pore size).

No serum blanks were run in this study.

Assays were performed on nine different days. Each run consisted of a serum calibrator in duplicate, three serum controls of differing cholesterol concentrations, and 19 test samples. The results were calculated by comparison with the mean absorbance value of the pair of serum calibrators and the value of 299 mg/dL assigned to the calibrator by the ALBK reference method at CDC.

Results: Within-run precision was determined by running 10 consecutive aliquots of a serum pool: $\bar{x} = 246.47$ mg/dL, SD = 0.998 mg/dL, CV = 0.40%. Day-to-day precision was determined by running on each of the nine days three serum controls having different concentrations:

Control	n	$\bar{x}$	Target value[a]	SD	CV, %
MQ-3	8	55.79	—	0.810	1.45
Q-7	9	192.28	188	1.901	0.99
Q-10	8	269.89	269	3.714	1.38

[a] Supplied by CDC.

Replication of serum calibrator absorbance readings for nine days was $\bar{x} = 0.653$, SD = 0.006, and CV = 0.9%.

For method comparison, 136 serum samples were analyzed in eight different runs by the enzymic method (*y* values) and a modified AutoAnalyzer II (Technicon Instruments, Tarrytown, NY 10591) method (*x* values) described in the *Manual of Laboratory Operations, Lipid Research Clinics Program*, 1 [May 1974, DHEW Publication No. (NIH) 75–628]. The values obtained by this latter method were corrected with the *same* serum calibrator (Sercal 6) used to calibrate the enzymic method. The results of the comparison are summarized: $\bar{x} = 237.9$ mg/dL, $\bar{y} = 249.7$ mg/dL, SD = 5.079, correlation coefficient = 0.993, paired $t = 26.99$ (significant at $p = 0.001$), regression equation $y = 1.019\,x + 7.292$, SD of y about regression line = 5.037 mg/dL.

These evaluators had no difficulty following the instructions, preparing the reagents, or making the analytical runs.

Within-day and day-to-day precision was excellent. Comparison of the results for the 136 sera analyzed by the enzymic method and the Lipid Research Clinics AutoAnalyzer II method showed a constant difference of 7.3 mg/dL and a proportional difference of 1.9% (as shown by the regression equation). These differences, though probably of little consequence in a clinical setting, are larger than expected because both methods were calibrated with the same calibrator, which had an ALBK assigned target value. In view of this difference, it is interesting (and a little disturbing) to note that the mean value of 269.9 obtained for pool Q-10 is almost exactly the same as the CDC target value of 269 and that the mean value of 192.3 mg/dL obtained for pool Q-7 is also close to the CDC target value of 188 mg/dL. One might speculate that the candidate method handles the CDC pools and (or) calibrator in a slightly different manner than it does patients' sera.

Evaluator I. K. reduced the volumes used as follows: 12.5 µL of serum (syringe) and 1.0 mL of working test reagent (dispenser) were mixed. Absorbance was measured in semimicrocuvets (1-cm diameter). In practice, 10 µL of serum plus 1.0 mL of working test reagent is convenient. Performing micro-analysis as described saves 75% of cost of reagents.

She compared this method with the "Biochemica Test Combination Cholesterol" (no. 15 738, Boehringer/Mannheim) katalase method, which involves use of sample blanks and "high-performance" thin-layer chromatography (HPTLC) (*J. Chromatogr.* **146**: 262–271, 1978). All methods were calibrated against "Precisit."

Results: Within-run precision from results of four series (n = 10) of high- and low-concentration cholesterol pool samples was as follows:

Method	CV, % High pool	Low pool
CDC	0.73	1.15
Katalase	1.22	1.02

Precision from day to day on pools was determined by analyses performed on six pools, two of which had target values set by CDC and two by the distributors. Two other pools were prepared in her laboratory from human serum. Six samples from each pool were analyzed within a period of 10 days by the three methods.

Pool (and target value)	Katalase	HPTLC	CDC
CDC-reference sample No.60 (1.67 g/L)	1.63 ± 0.05 (3.0)	1.63 ± 0.04 (2.2)	1.60 ± 0.05 (2.9)
CDC-reference sample No. 52 (3.12 g/L)	2.93 ± 0.03 (0.9)	2.99 ± 0.10 (3.5)	2.88 ± 0.08 (2.8)
Boehringer Mannheim Precilip (1.23 ± 0.07 g/L)	1.23 ± 0.04 (2.9)	1.20 ± 0.02 (1.8)	1.19 ± 0.02 (1.9)
Dade Lipid-Trol (2.85 ± 0.06 g/L)	2.90 ± 0.05 (1.6)	—	2.88 ± 0.05 (1.9)
Pool a	1.58 ± 0.04 (2.4)	1.61 ± 0.05 (3.2)	1.52 ± 0.04 (2.6)
Pool b	3.31 ± 0.04 (1.3)	3.12 + 0.09 (2.8)	3.18 ± 0.08 (2.3)

Mean ± SD, cholesterol, g/L (and CV, %)

To compare methods on individual serum samples, 39 different "normal" serum samples were analyzed. Data were examined by paired *t*-test statistics and linear regression analysis. These samples were nonlipemic, nonicteric, and nonhemolytic. Data obtained by the CDC-method were *not* corrected for the sample blank. Comparison of CDC method values *(y)* against katalase values *(x)* and against HPTLC values *(z)* gave the following equations:

$$y = 0.96x + 0.04; \quad y = 0.97z - 0.02.$$

The mean values obtained by the CDC method were about 2.2% lower than those of the katalase method and about 3.5% lower than those of the HPTLC method. The agreement between these methods is plausible because the HPTLC procedure is calibrated against the katalase method.

Thirty-three different serum samples (lipemic, hemolytic, or icteric) were also analyzed by the three methods. Studies on hyperlipemic sera indicate that it is reasonable to neglect a sample blank for analysis of turbid serum samples. A special experiment carried out to observe the effect of a hyperlipemic serum (triacylglycerols, 5.80 g/L) on the serum blank and on results of analysis confirmed that the reagents and the volumes used minimize the effect of hyperlipemia.

Studies on hemolytic sera showed that hemolysis interferes to a certain extent with the CDC method. According to Evaluator I. K., moderately hemolytic samples can be analyzed *without* a sample blank; extremely hemolytic samples should be excluded from analysis because of interference by cellular sterols.

Analyses on icteric sera showed that bilirubin interferes with the CDC method. Her experience is that icteric sera with bilirubin <50 mg/L can be analyzed without a sample blank, but with more highly icteric samples correction with sample blanks is not possible. The following table summarizes the results on "normal" sera and sera with potentially interfering substances:

	Mean ± SD, cholesterol, g/L			
Method	Nonlipemic (n = 39)	Lipemic[a] (n = 19)	Hemolytic[b] (n = 5)	Icteric[c] (n = 9)
Katalase	2.25 ± 0.95	2.89 ± 1.16	1.74 ± 0.43	1.09 ± 0.33
HPTLC	2.28 ± 0.93	2.90 ± 1.13	1.83 ± 0.41	1.10 ± 0.31
CDC	2.20 ± 0.91	2.82 ± 1.15	1.64 ± 0.45	0.99 ± 0.37

[a] Lipemic serum samples (triglycerides = 3.84 ± 2.05 g/L) from obese, diabetic and atherosclerotic patients.
[b] Hemolytic samples (hemoglobin = 0.45 ± 0.33 g/L) from sick children.
[c] Icteric samples from sick children, and extremely icteric samples from neonates with neonatal hyperbilirubinemia. Mean bilirubin (both groups combined) was 10.2 ± 3.5 mg/dL.

Evaluator I. K. comments that the equipment needed for this method is available in any laboratory; the reagents can be easily prepared; the procedure is simple; handling is easy; and the method is easily adapted to microanalysis. Except for some restrictions (icteric sera), sample blanks can be neglected, and the precision and accuracy of the CDC method compare well with those of the katalase and HPTLC methods.

As a result of practical experience with the CDC method, Evaluator I. K. recommends this method as highly suitable for routine analysis.

References

1. Fredrickson, D. S., Levy, R. I., and Lees, R. S., Fat transport and lipoproteins—an integrated approach to mechanisms and disorders. *N. Engl. J. Med.* **276,** 32–44, 94–103, 148–156, 215–224, 273–281 (1967).
2. Richmond, W., Preparation and properties of a cholesterol oxidase from *Nocardia* sp. and its applications to the enzymic assay of total cholesterol in serum. *Clin. Chem.* **19,** 1350–1356 (1973).
3. Allain, C. C., Poon, L. S., Chan, C. S. G., Richmond, W., and Fu, P. C., Enzymatic determination of total serum cholesterol. *Clin. Chem.* **20,** 470–475 (1974).
4. Roschlau, P., Bernt, E., and Gruber, W., Enzymatische Bestimmung des Gesamt-Cholesterins in Serum. *J. Clin. Chem. Clin. Biochem.* **12,** 403–407 (1974).
5. Boone, D. J., and Knouse, R. W., Proficiency Testing Summary Analysis, Clinical Chemistry Survey II, May 15, 1979. Center for Disease Control, Atlanta, GA 30333.
6. Duncan, P. H., and Cooper, G. R., A report on the development of a candidate reference method for cholesterol. *Clin. Chem.* **22,** 1193 (1976). Abstract.
7. Williams, J. H., Taylor, L., Kuchmak, M., and Witter, R. F., Preparation of hypercholesterolemic and/or hypertriglyceridemic sera for lipid determinations. *Clin. Chim. Acta* **28,** 247–253 (1970).
8. Abell, L. L., Levy, B. B., Brodie, B. B., and Kendall, F. E., A simplified method for the estimation of total cholesterol in serum. *J. Biol. Chem.* **195,** 357–366 (1952).
9. Cooper, G. R., Akins, J. R., Biegeleisen, J., Hazlehurst, J. S., Myers, C. M., and Bayse, D. D., Comparison of selected cholesterol colorimetric and enzymatic reference methods. *Clin. Chem.* **24,** 1046 (1978). Abstract.
10. Cooper, G. R., Ullman, M. D., Hazlehurst, J., Miller, D. T., and Bayse, D. D., Chemical characteristics and sources of error of an enzymatic cholesterol method. *Clin. Chem.* **25,** 1074 (1979). Abstract.
11. Deacon, A. C., and Dawson, P. J. G., Enzymic assay of total cholesterol involving chemical or enzymic hydrolysis—a comparison of methods. *Clin. Chem.* **25,** 976–984 (1979).
12. Pesce, M. A., and Bodourian, S. H., Interference with the enzymatic measurement of cholesterol in serum by use of five reagent kits. *Clin. Chem.* **23,** 757–760 (1977).
13. Friedewald, W. T., Levy, R. I., and Fredrickson, D. S., Estimation of the concentration of low-density lipoprotein cholesterol in plasma, without the use of the ultracentrifuge. *Clin. Chem.* **18,** 499–502 (1972).
14. The Lipid Research Clinics Population Studies Data Book, **1,** Prevalence Study. DHHS, PHS (NIH), National Heart, Lung, and Blood Institute, Publication No. 80–1527, Bethesda, MD 20014, 1980.
15. Cohen, A., Hertz, H. S., Mandel, J., Paule, R. C., Schaffer, R., Sniegoski, L. T., Sun, T., Welch, M. J., and White, E. V., Total serum cholesterol by isotope dilution–mass spectrometry: A candidate definitive method. *Clin. Chem.* **26,** 854–860 (1980).
16. Slicker, K. A., Enzyme-linked assays for cholesterol. *CRC Lab. Sci.* **8,** 193–212 (1977).
17. Witte, D. L., Brown, L. F., and Feld, R. D., Enzymatic analysis of serum cholesterol and triglycerides: A brief review. *Lab. Med.* **9,** 39–44 (1978).
18. Witte, D. L., Brown, L. F., and Feld, R. D., Effects of bilirubin on detection of hydrogen peroxide by use of peroxidase. *Clin. Chem.* **24,** 1778–1782 (1978).

Appendix: Specific Activity Assays of Enzymes

Cholesterol Esterase (CEH)

Label value: 20 U/mg

Stock (in vial): 5 mg in 2.5 mL = 2 mg in 1 mL = 40 U in 1 mL

Dilution for assay:

1. Dilute 0.1 mL (0.2 mg) to 1 mL = 10-fold dilution = 4 U/mL.

2. Dilute 0.1 mL (0.02 mg) to 1 mL = 100-fold final dilution = 0.4 U/mL.

Substrate: Preferably serum, or aqueous cholesterol oleate solution > 2.5 g/L.

Reaction:

1. 3.0 mL of mixed working reagent
   0.1 mL of substrate
   0.1 mL of ChO stock (2.5 U)
   <u>0.1 mL of POD stock (100 U)</u>
   3.3 mL, total volume

2. Incubate at 37 °C for 5 min.

3. Add 0.05 mL of CEH (100-fold dilution). Note zero time with stopwatch.

4. After a lag phase of about 3 min, measure the change in absorbance at 500 nm for ≃ 6 min with a spectrophotometer at 37 °C.

5. Calculate ΔA/min from linear part of the curve.

6. Calculate activity:

$$U/mL = \frac{\Delta A}{5.33} \times \frac{TV}{SV} \times \text{dilution}$$

TV = total volume of reaction mixture
SV = sample volume.

Example: $U/mL = \frac{0.032}{5.33} \times \frac{3.35}{0.05} \times 100 = 40.2$

$$U/mg = \frac{40.2 \, U/mL}{2 \, mg/mL} = 20.1 \, U/mg$$

Horseradish Peroxidase (POD)

Label value: 1000 U/mg

Stock (in vial): 10 mg. Dissolve to prepare stock solution. Dilute to 10 mL, or 1 mg/mL = 1000 U/mL.

Dilution for assay:
1. Dilute 0.1 mL (0.1 mg) to 10 mL = 100-fold dilution = 10 U/mL.
2. Dilute 0.1 mL (0.001 mg) to 10 mL = 10 000-fold final dilution = 0.1 U/mL.

Substrate: Alcoholic or aqueous cholesterol solution > 3.00 g/L.

Reaction:
1. 3.0 mL of working reagent
   0.1 mL of substrate
   0.1 mL of ChO stock (2.5 U)
   3.2 mL, total volume
2. Incubate at 37 °C for 5 min.
3. Add 0.1 mL of POD (10 000-fold dilution). Note zero time with stopwatch.
4. Measure the change in absorbance at 500 nm for ≃ 6 min with spectrophotometer at 37 °C.
5. Calculate $\Delta A/$min.
6. Calculate activity: $U/mL = \frac{\Delta A}{5.33} \times \frac{TV}{SV} \times \text{dilution}$

Example: $U/mL = \frac{0.25}{5.33} \times \frac{3.3}{0.1} \times 10\,000 = 1548$

$$U/mg = \frac{1548 \, U/mL}{1 \, mg/mL} = 1548 \, U/mg$$

Cholesterol Oxidase (ChO)

Label value: 25 U/mg

Stock (in vial): 1 mg in 1 mL = 25 U/mL

Dilution for assay:
1. Dilute 0.1 mL (0.1 mg) to 1 mL = 10-fold dilution = 2.5 U/mL.
2. Dilute 0.1 mL (0.01 mg) to 1 mL = 100-fold final dilution = 0.25 U/mL.

Substrate: Alcoholic or aqueous cholesterol solution > 3.00 g/L.

Reaction:
1. 3.0 mL of working reagent
   0.1 mL of substrate
   0.1 mL of POD stock (≃ 100 U)
   3.2 mL, total vol
2. Incubate at 37 °C for 5 min.
3. Add 0.1 mL of ChO (100-fold dilution). Note zero time with stopwatch.
4. Measure the change in absorbance at 500 nm for ≃ 6 min with spectrophotometer at 37 °C.
5. Calculate $\Delta A/$min.
6. Calculate activity: $U/mL = \frac{\Delta A}{5.33} \times \frac{TV}{SV} \times \text{dilution}$

Example: $U/mL = \frac{0.040}{5.33} \times \frac{3.3}{0.1} \times 100 = 24.8$

$$U/mg = \frac{24.8 \, U/mL}{1 \, mg/mL} = 24.8 \, U/mg$$

Notes on enzyme assays

1. Absorbance change per minute should be in range of 0.025 to 0.045.
2. If reaction is too fast, decrease the volume of the enzyme to be assayed or increase the dilution. If reaction is too slow, do the opposite.
3. Test tubes or cuvets should be mixed by gentle inversion only.
4. Units (U) are defined as follows:

One unit of cholesterol esterase (U) is that amount which hydrolyzes 1 μmol of cholesterol ester per minute at 37 °C, pH 6.9.

One unit of cholesterol oxidase (U) is that amount which liberates 1 μmol of H_2O_2 per minute at 37 °C with cholesterol as the substrate, or that amount which oxidizes 1 μmol of cholesterol to cholest-4-en-3-one per minute at 37 °C, pH 6.9.

One unit of peroxidase (U) is that amount which decomposes 1 μmol of peroxide per minute at 37 °C, pH 6.9.

Cholesterol, Liebermann–Burchard Method with 2-Propanol Extraction

Submitters: G. Russell Warnick and Chien Yu, *Northwest Lipid Research Clinic, Harborview Medical Center, Seattle, WA 98104, and Department of Medicine, University of Washington, Seattle, WA*

Evaluators: Samuel Y. Chu, *Department of Laboratory Medicine, Ottawa General Hospital, Ottawa KIN 5C8, Canada*
Paul C. Fu and Michael M. Lubran, *Department of Pathology, Harbor General Hospital, University of California at Los Angeles, Torrance, CA 90509*
Betty J. Lash, *Animal Reference Laboratories, Inc., Houston, TX 77086*

Reviewer: James L. Driscoll, *Department of Pathology, Rhode Island Hospital, Providence, RI 02901*

Introduction

The reaction of cholesterol with sulfuric acid in acetic anhydride to form a colored product was initially described by Liebermann and provided the basis for an assay reported by Burchard in 1890 (1). Subsequently, various cholesterol methods involving the Liebermann–Burchard (L-B) reaction have been described (reviewed in ref. 2 and 3). These methods can be categorized into three groups: (a) direct—serum added directly to color reagent; (b) extraction—cholesterol extracted into an organic solvent before addition to color reagent; (c) hydrolysis—esterified cholesterol hydrolyzed before solvent extraction and color reaction.

The direct methods are simple and convenient but are subject to interference by compounds normally present in serum, for example, bilirubin and proteins. Extraction of cholesterol with an organic solvent before the color reaction eliminates much of this interference. Hydrolysis of the cholesterol in serum present as esterified cholesterol (approximately 70%) is advantageous because the esterified cholesterol produces more color than free cholesterol, leading to overestimation of total cholesterol. However, use of a hydrolysis step with, for example, the Abell–Kendall method (4), which is accepted as the reference procedure for cholesterol (5), is generally considered too tedious for routine analysis. An acceptable method for routine use must give reasonable accuracy and yet not be excessively difficult or time consuming.

Principle

This L-B method has been modified from the semiautomated version used by the Lipid Research Clinics (6), which has proven to be quite accurate and precise (7, 8) in extensive population studies (9, 10). Although the procedure is somewhat more complicated than the direct methods, the demonstrated accuracy and precision make it worthy of consideration. Cholesterol is extracted from serum samples into 2-propanol, which eliminates interfering substances. In the Lipid Research Clinics method, extracts are mixed directly with L-B reagent. In this manual version, however, the color reaction is substantially accelerated by the heat from the exothermic reaction of 2-propanol with sulfuric acid and can not be controlled sufficiently to allow reproducible measurement. Therefore, measured aliquots of the extracts must be evaporated before adding the color reagent. After a timed incubation, absorbance is measured at 630 nm. As an alternative to a hydrolysis step, a secondary serum standard can be used for calibration. This approach to calibration gives acceptable accuracy, provided the serum calibrator has the usual proportion of free to esterified cholesterol and an accurate target value.

Materials and Methods

Reagents

All chemicals are of reagent grade unless otherwise specified.

1. *Reference serum:* Serum pools in lyophilized or liquid form are commercially available. We recommend a pool with cholesterol value between 2 and 3 g/L. Because accuracy is dependent upon a reliable reference value, the target value should preferably have been established by the Abell–Kendall reference procedure (4, 5).

Note: For example, a liquid pool containing bovine lipoproteins is available (Cholesterol Calibrator, Beckman Microbics, Carlsbad, CA 92008). The label values for cholesterol on this material have been established by an Abell–Kendall assay *(4)*. We expect that commercial, human-based pools with accurate reference method target values in the 2–3 g/L range will soon be available.

2. *Quality-control serum:* A serum pool with cholesterol between 1 and 2 g/L established by the reference procedure *(4, 5)* is recommended.

Note: Lipid Fraction Control available from Hyland Div., Travenol Laboratories, Deerfield, IL 60015, meets these specifications.

3. *NaCl, 0.15 mol/L:* Dissolve 8.77 g of NaCl in de-ionized water in a volumetric flask and dilute to 1.0 L.

4. *L-B reagent:* In a fume hood, transfer 600 mL of acetic anhydride to a 2.0-L Erlenmeyer flask containing a magnetic stir bar. Place the flask in wet ice in an ice bucket on a stir plate. Add 300 mL of glacial acetic acid with slow stirring and allow the contents of the flask to cool. When ice crystals appear, begin adding 30 mL of concd. sulfuric acid in 5-mL portions at 5-min intervals. Stir contents thoroughly for 10 min. Transfer the reagent to an amber bottle, stopper tightly, and store at 4 °C. Before use, allow the reagent to equilibrate to room temperature. The reagent is stable at 4 °C for four weeks.

Note: This stability is confirmed by Evaluator S. Y. C. Evaluators P. C. F. and M. M. L. reported the L-B reagent was stable for four weeks at room temperature.

Specimen Collection

Blood is collected generally from the antecubital vein of a subject in the sitting position. After blood flow starts, remove the tourniquet used to distend the vein. Allow blood to stand 30 min at room temperature for clot formation. Store the serum, obtained after centrifugation, at 4 °C until analysis.

Procedure

1. *Extraction:* Label 16 × 150 mm screw-cap tubes for blank, reference serum (duplicate), quality-control serum, and patients' specimens. Add 9.5 mL of 2-propanol to each tube. Add 0.5 mL of 0.15 mol/L NaCl to the tube labeled blank. Similarly, add 0.5 mL of reference serum, quality-control serum, and patients' specimens to the appropriate tubes. To produce a finely granulated protein precipitate, samples must be injected forcefully directly into the alcohol and not against the side of the tube. Otherwise, cholesterol extraction may not be complete. Immediately after adding the sample, cap the tube and mix for 10 s with a vortex-type mixer. Allow the tubes to stand for 10 min, vortex-mix each for 10 s, and centrifuge (1500 × g, 10 min) to sediment the precipitates.

Note: Manual pipettes, e.g., Pipetman (West Coast Scientific, Oakland, CA 94618), capable of reproducibly dispensing 0.5 mL are suitable for measuring and dispensing the samples. Evaluator P. C. F. recommends the use of a repeating dispenser, for example, the Repipet (Lab Industries, Berkeley, CA 94710), to dispense the alcohol. Improved precision can be obtained by using a mechanical sampler-dilutor, e.g., Micromedic (Micromedic Systems, Inc., Philadelphia, PA 19105), for adding sample and alcohol to extraction tubes.

2. *Solvent evaporation:* Transfer 3.0 mL of each extract to a labeled culture tube. Evaporate the extracts in a fume hood by bringing the tubes to 60 °C (heating block or water bath). A gentle stream of nitrogen gas or dry air sweeping into the tubes speeds evaporation.

Note: Evaluator S. Y. C. indicates that evaporation without a stream of nitrogen gas requires 2–2.5 h at 60 °C and can be decreased to 1 h by heating at 70 °C. In our experience, evaporation requires approximately 20 min at 60 °C with a stream of nitrogen gas.

3. *Color development:* Cool tubes to room temperature. At timed intervals add 6.0 mL of L-B reagent to each tube with thorough mixing. The time interval selected should be based on the time required for measuring absorbance after incubation.

4. *Absorbance reading:* After 30 min of incubation at room temperature, adjust the spectrophotometer to zero at 630 nm with the reagent blank. Read the absorbance of each tube in sequence, maintaining the same time interval established in the previous step. This assures that each tube is incubated for exactly 30 min.

Calculations

Calculate the cholesterol concentration of each unknown and quality-control serum in relation to the reference serum value as follows:

(Absorbance of unknown/mean absorbance of duplicate reference serum specimens) × target value of reference serum = concentration of unknown. For example:

absorbance of reference serum, 0.376 and 0.374
target value of reference serum, 2.99 g/L
absorbance of unknown sample, 0.290

Calculation:

$0.290/[(0.376 + 0.374)/2] \times 2.99$ g/L = 2.31 g/L

Results

We determined the analytical recovery of the method by adding aliquots of a plasma sample and increasing amounts of cholesterol to 2-propanol. The total cholesterol concentration was determined as described previously. Cholesterol recovery ranged from 98 to 101% for cholesterol additions of 0.4, 0.8, 1.2, 1.6, and 2 mg, which are equivalent to cholesterol concentrations of 0.8, 1.6, 2.4, 3.2, and 4.0 g/L, respectively.

Note: Evaluator S. Y. C. observed recovery from 95 to 108% with a mean recovery of 102%.

Table 1 presents the mean and SD for three quality-control pools analyzed by the manual method 14 times over one month compared with results by the semi-automated AutoAnalyzer II (Technicon Instruments, Tarrytown, NY 10591) method for the same period. Also indicated are target values for the pools by the Abell–Kendall reference method (4). Results by the manual procedure are within 40 mg/L of results by the automated method and within 30 mg/L of target values. Precision by the manual method was acceptable, with CVs being 1.1% for the high-value Q6 pool, 2.3% for the normal Q5 pool, and 5.3% for the low MQ2 pool.

Within-run precision was assessed by analyzing a plasma pool 20 times in one analytical run. A CV of 1.0% was obtained for a pool having a mean cholesterol concentration of 2850 mg/L.

To determine comparability of results for patient's samples by the manual procedure with those by the semi-automated AutoAnalyzer II cholesterol method, we analyzed by both methods plasma samples and lipoprotein fractions with a wide range of cholesterol concentrations (Figure 1). The manual procedure (y) was highly correlated with the semi-automated method (x), with a correlation coefficient of 0.995. The relationship between the two methods for 218 samples was: $y = 0.959\,x + 36.8$ mg/L. On a subset of 40 high-density-lipoprotein fractions with low cholesterol concentrations, the agreement was similar, with correlation coefficient 0.992, slope 0.957, and y-intercept 16.4 mg/L.

Precision results obtained by the method evaluators are presented in Table 2. Reported CVs for within-run precision were 2.5 to 8.27%. For between-run precision, CVs of 4.7 to 10.5% were obtained.

Table 1. Analytical Accuracy and Precision for Quality Control Pools

Cholesterol, mean (SD), mg/L

Pool	Reference values[a]	AutoAnalyzer II[b]	Manual[b]
Q6	2760	2748 (25.7)	2779 (31.0)
Q5	1830	1850 (21.6)	1847 (43.4)
MQ2	690	678 (9.2)	716 (38.2)

[a] Established by Abell–Kendall method (4) at the Clinical Chemistry Standardization Section, Centers for Disease Control, Atlanta, GA 30333.
[b] Fourteen analytical runs over a one-month period.

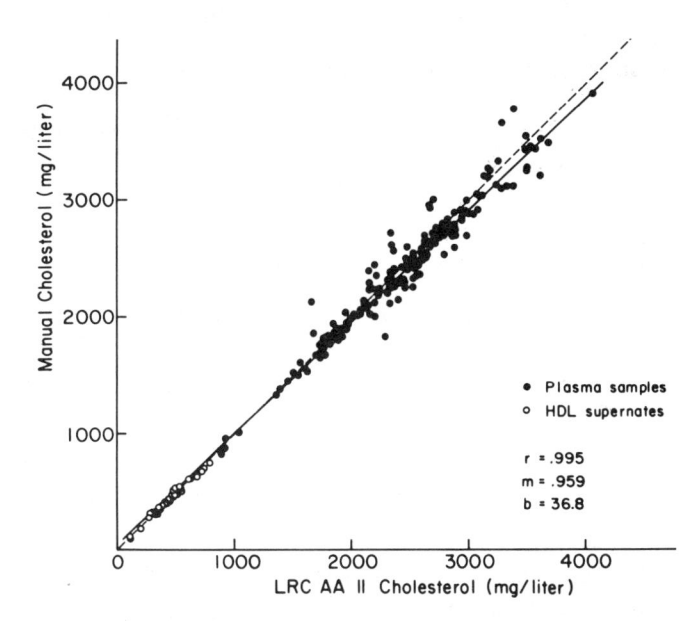

Fig. 1. Results of manual cholesterol procedure vs Lipid Research Clinics semi-automated procedure on 218 plasma samples, 40 of which (indicated by *open circles*) were high-density lipoprotein fractions obtained by heparin–Mn^{2+} precipitation (6)

The correlation coefficient (r), slope (m) and y-intercept (b) are indicated. The *solid line* indicates the actual agreement between the two methods. The *dashed line* indicated $y = x$ (the line of perfect agreement)

Evaluator S. Y. C. compared the manual cholesterol method with an unspecified enzymic procedure, which gave for the 2800 mg/L pool in Table 2 a mean value of 2820 mg/L, within-run SD of 16.89 mg/L, and CV of 0.67%. The agreement reported between the enzymic procedure (x) and the manual L-B method (y) on 20 samples was: $y = 0.9635\,x + 103$ mg/L, correlation coefficient, 0.9703.

Evaluators P. C. F. and M. M. L. compared the manual method (x) with an enzymic procedure (y) for the Vickers M-300 (Vickers Instruments, Inc., Woburn, MA 01801). The reported agreement for 21 specimens was: $y = 1.1539\,x - 142.2$ mg/L, correlation coefficient, 0.9294.

Discussion

The method as described here provides a linear response of absorbance with cholesterol concentration to 10 g/L. Color response for samples with very low cholesterol values can be increased by simply increasing the volume of extract evaporated. To calculate cholesterol concentration, correct the value obtained by using the inverse proportion of the volume change. For example, if one evaporates 6.0 mL rather than the usual 3.0 mL of extract, the cholesterol value calculated from the absorbance must be divided by 2.

The extraction step in this procedure minimizes interference from bilirubin, proteins, and hemoglo-

Table 2. Precision Results Reported by Evaluators

Evaluator	No. of samples	Mean mg/L	SD	CV, %
Within-day				
S. Y. C.	—	2800	69.15	2.5
	—	2130	66.0	3.1
P. C. F. and M. M. L.	5	1544	128	8.27
	5	2387	163	6.82
B. L.	20	1001	82.8	8.3
Between-day				
S. Y. C.	—	2690	127	4.7
	—	2100	102.6	4.9
P. C. F. and M. M. L.	5	2330	223	9.57
	5	1370	124	9.04
B. L.	20	1016	106.5	10.5

bin. However, extensive hemolysis in samples should be avoided, because the fluid released from the erythrocytes dilutes cholesterol in the serum.

An important factor in obtaining adequate precision by the manual L-B procedure is the complete extraction of cholesterol from the protein. One must inject the sample forcefully into the alcohol to obtain a uniform dispersion of the protein precipitate. If the sample is added to the alcohol in droplets or injected against the side of the extraction tube, lipid extraction may be incomplete.

The method described is calibrated with a secondary standard. Overall accuracy, therefore, depends on the reliability of the target value assigned to the reference serum. Reference pools are commercially available; however, it is advisable to select a pool with a target value assigned by an accurate reference method such as the Abell–Kendall method *(4, 5)*. Alternatively, a reference pool can be prepared in-house by pooling samples of appropriate cholesterol concentration. The homogenous pool is aliquoted into vials that are then sealed and frozen. A target value is subsequently established by an appropriate method *(4)*.

This L-B reagent method, when calibration is accurate, gives results comparable with those obtained by the Lipid Research Clinics method. Population distributions by this method are described in the introductory chapter to the cholesterol methods (this volume).

References

1. Burchard, H., Beitrage zur Kenntnis des Cholesterins. *Chem. Zentralbl.* **61**, 25 (1890).
2. Tonks, D. B., The estimation of cholesterol in serum: A classification and critical review of methods. *Clin. Biochem.* **1**, 12–29 (1967).
3. Zak, B., Cholesterol methodologies: A review. *Clin. Chem.* **23**, 1201–1214 (1977).
4. Abell, L. L., Levy, B. B., Brodie, B. B., and Kendall, F. E., A simplified method for the estimation of total cholesterol in serum and demonstration of its specificity. *J. Biol. Chem.* **195**, 357–366 (1952).
5. Cooper, G. R., Akins, J. R., Biegeleisen, J., Hazelhurst, J., Myers, C. M., and Bayse, D. D., Comparison of selected cholesterol colorimetric and enzymatic reference methods. *Clin. Chem.* **24**, 1046 (1978). Abstract.
6. *Manual of Laboratory Operations: Lipid Research Clinics Program,* **1**, *Lipid and Lipoprotein Analysis,* DHEW Publication No. (NIH) 75–628, May, 1974.
7. Bachorik, P. S., Wood, P. D. S., Williams, J., Kuchmak, M., Ahmed, S., Lippel, K., and Albers, J. J., Automated determination of total plasma cholesterol: A serum calibration technique. *Clin. Chim. Acta* **96**, 145–153 (1979).
8. Lippel, K., Ahmed, S., Albers, J. J., Bachorik, P., Cooper, G. R., Helms, R., and Williams, J., Analytical performance and comparability of the determination of cholesterol by 12 Lipid Research Clinics. *Clin. Chem.* **23**, 1744–1752 (1977).
9. The Lipid Research Clinics Program Epidemiology Committee. Plasma lipid distributions in selected North American populations: The Lipid Research Clinics Program Prevalence Study. *Circulation* **60**, 427–439 (1979).
10. Heiss, G., Tamir, I., Davis, C. E., Tyroler, H. A., Rifkind, B. M., Schonfeld, G., Jacobs, D., and Frantz, I. D., Jr., Lipoprotein-cholesterol distributions in selected North American populations: The Lipid Research Clinics Program Prevalence Study. *Circulation* **61**, 302–315 (1980).

Cholesterol, Parekh–Jung Method

Submitters: Leila Edwards[1] and Mildred Suszka, *Erie County Laboratory, and [1]Departments of Biochemistry, Pathology and Medical Technology, State University of New York at Buffalo, Buffalo, NY 14215*

Evaluators: Geert J. M. Boerma and Pieter J. Noordeloos, *Departments of Clinical Chemistry and Chemical Pathology, University Hospital and Erasmus University, Rotterdam, The Netherlands*

Reviewer: Charles S. Furfine, *Division of Medical Devices, Food and Drug Administration, Silver Spring, MD 20910*

Introduction

Cholesterol is probably the steroid most frequently determined in human body fluids; its popularity is attributed to its apparent role in causing disease of blood vessels. Analysis of cholesterol has often been inaccurate and imprecise and at times very cumbersome. There are more than 200 methods for determining cholesterol in serum, perhaps an indication of the difficulties in developing a reliable assay. The simple method presented here is based on precipitation of proteins and associated substances with uranyl acetate, followed by colorimetry of the reaction product of cholesterol with a $FeSO_4/H_2SO_4$ mixture. The method can be used to determine total cholesterol in serum, other body fluids, tissue homogenates, or their extracts.

Materials and Methods

Apparatus

Centrifuge, capable of centrifuging the tubes in the test at $700 \times g$ for 10 min.

Rotator or other mechanical mixer. Any rotator capable of fully inverting the tubes slowly (30 times per minute) or any vortex-type mixer. Mixing by strong agitation of handheld tubes is possible but discouraged for safety reasons.

Spectrophotometer. Any instrument capable of linear response in the absorbance range from 0.0 to 0.4. The Submitters use a Bausch and Lomb 100 (Bausch & Lomb, Inc., Rochester, NY 14602).

Micropipette, 50 µL or 25 µL. Full-displacement ones (such as Scientific Manufacturing Industries, Berkeley, CA, or Labindustries, Berkeley, CA 94710) are preferred, but other micropipettes such as Eppendorf (Brinkmann Instruments, Westbury, NY 11590) or Oxford (Oxford Labs., Foster City, CA 94404) or equivalents may be used.

Repetitive pipettor, 10 mL, 5 mL. Labindustries, Oxford, or equivalent.

Drying tube.

Timer capable of measuring in seconds.

Screw-capped culture tubes, 13×100 mm and (or) 16×125 or 20×150 mm, with caps lined with Teflon.

Culture tubes, 13×100 mm.

Glass wool.

Acid-resistant connecting tube.

Gloves, latex, surgeon-type.

Reagents

Note: As a general rule, only "reagent" or ACS-grade reagents should be used. Reagents below are from J. T. Baker Chemical Co., Phillipsburg, NJ 08865, unless specified otherwise.

1. Ferric chloride, iron (III) chloride hexahydrate.
2. Acetic acid, glacial.
3. Uranyl acetate dihydrate, $(CH_3CO_2)_2UO_2 \cdot 2H_2O$, powder.
4. Ferrous sulfate, anhydrous (Fisher Scientific Co., Pittsburgh, PA 15219).
5. Sulfuric acid, concd.
6. Silver nitrate, 1.0 mol/L solution (optional) (Fisher Scientific Co.).
7. Drierite—$CaSO_4$, anhydrous. Available from most suppliers of laboratory reagents.
8. Cholesterol, chromatographically pure standard. (J. T. Baker, "Ultrex" grade; Nu-Chek Prep., Inc., Elysian, MN 56028; Sigma Chemical Co., St. Louis, MO 63178; or Supelco Inc., Bellefonte, PA 16823, are sources of material 99+% pure by several chromatographic tests.)
9. Isopropanol or ethanol, reagent grade, aldehyde-free.
10. De-ionized or distilled water.
11. Ammonium hydroxide, concd.

Preparations prior to assay

1. Drying tube. Loosely fill the bulb of the drying

tube with clean glass wool, fill its body with Drierite and plug the top centimeter or so with more glass wool. To prevent the absorption of moisture by the reagents, this system should be connected to the air intake of the repetitive pipettor containing the reagent with sulfuric acid.

Note: Instead of using drying tubes, the Evaluators closed the dispensers and cleaned the dispenser part after the dispensing step was completed. The dispenser they used was a Seripettor (Labora Mannheim GmbH, 6800 Mannheim, F. R. G.).

2. Extension for ferrous sulfate repetitive pipettor. It is convenient, although not absolutely necessary, to extend the tip of the repetitive pipette containing the ferrous sulfate–sulfuric acid reagent with a thin Teflon tube so that the tip can reach the bottom of the test tube to be used in the reaction. It is possible to use a long, very narrow glass tube, but it is less convenient to work with a glass tip.

Connect the delivery tip of the repetitive pipette to a length of Teflon tube (1.5 mm i.d.) that reaches the bottom of the color-development tube by use of a short piece of acid-resistant tubing such as 0.09 Acidflex, then strengthen the connection by inserting a stronger piece of plastic tube over the connection.

When pipetting, test to make sure air bubbles are not introduced through the connection.

Preparation of Combined Reagents

Ferric acetate/uranyl acetate reagent

Work under a hood or in a well-ventilated place.

1. Weigh 100 mg of uranyl acetate dihydrate and place it in a 1-L volumetric flask.
2. Add 300 mL of glacial acetic acid and mix until solid is completely dissolved. Protect solution from light. (Placing it in a brown paper bag is satisfactory.)
3. Using mortar and pestle, grind lumps of ferric chloride hexahydrate to a powder.
4. Weigh 0.5 g of the powder and place it in a 50-mL centrifuge tube.
5. Dissolve the ferric chloride in water by adding 10 mL of de-ionized water at a time, while mixing, for a total of 40 mL of water added. All the solid should be in solution.
6. Add 3 mL of concd. ammonium hydroxide to the tube, mix thoroughly, then centrifuge to separate the sediment.
7. Aspirate and discard the supernate.
8. Wash the precipitate six times by completely resuspending it in 40 mL of de-ionized water each time, centrifuging, and aspirating the supernate. After six washings the supernate should be chloride-free, giving a negative test with silver nitrate. The chloride test consists of adding 2 to 3 drops of 1 mol/L silver nitrate solution to approximately 1.5 mL of supernate. If an interfering amount of chloride is present, a fine, white precipitate will form. If detectable amounts of chloride are present, continue washing the precipitate until it is chloride-free.

9. Aspirate as much supernate as possible and quantitatively transfer the residue into the volumetric flask containing the dissolved uranyl acetate, rinsing with glacial acetic acid to complete the transfer.
10. Mix until all solid is dissolved.
11. Cool to room temperature. Fill to the 1-L mark with glacial acetic acid, mix well, and store in a brown bottle covered with Parafilm.

Sulfuric acid/ferrous sulfate color reagent

Prepare this under a hood or in a very well-ventilated place.

1. Place 0.1 g of anhydrous ferrous sulfate in a graduated 1-L Erlenmeyer flask or a volumetric flask.
2. Add 100 mL of glacial acetic acid and mix with a magnetic stirrer.
3. Add 100 mL of concd. sulfuric acid while swirling the flask. **Be careful.** Cover with Parafilm.
4. Cool to room temperature and fill to the 1-L mark with sulfuric acid.
5. Store in a brown bottle covered securely with an acid-resistant stopper or with several layers of Parafilm. *DO NOT* use corks or rubber stoppers.

Note: If the atmospheric conditions in your laboratory are not carefully controlled, take precautions so that water is not absorbed into this reagent, either during preparation or storage.

Preparation of Standards

Store the dry standard over the desiccant, protected from humidity and light. The Submitters store this and other materials at reduced pressure in a desiccator in a freezer ($-20\ °C$), although storage at low temperature is not essential. If the standard is stored at low temperature, bring the desiccator to room temperature before opening it.

1. Dry the necessary amount of cholesterol in an oven at 70 °C for 2 h.
2. Cool in a desiccator protected from light.
3. Weigh 100-, 200-, 300-, or 400-mg portions of dried cholesterol into separate 100-mL volumetric flasks.
4. Dissolve cholesterol in alcohol (either isopropanol or ethanol) kept at 25 °C in a water bath.

Note: Either solvent is satisfactory for preparing the standard curve. Be consistent and always use the same alcohol. Ethanol, if used, should be free from aldehydes.

5. Bring all flasks to 25 °C in a bath and fill each flask to the mark with the alcohol kept at the same temperature.
6. Transfer aliquots of standards to be used into culture tubes tightly covered with Teflon-lined caps.

Store in a refrigerator (2 to 10 °C) until the standard is used.

Note: The Submitters prefer to prepare each standard individually but it is possible to prepare a concentrated standard from which all the others are prepared by dilution. However, an error in the preparation of the stock standard will not be readily apparent from your curve, and the entire assay will be in error. If it is desired to prepare the standards by serial dilution, one might attempt to overcome weighing and dilution errors by weighing two portions of cholesterol, each sufficient to prepare a concentrated stock solution. Then one stock may be used to prepare half the working standards and the other stock to prepare the remaining working standard solutions. The calibration curve must be linear within the range of the 50 to 400 mg/L standards if all dilutions are correct.

Procedure

1. Appropriately label a 20 × 150 mm (or 16 × 125 mm) culture tube covered with Teflon-lined screw caps for each standard (in duplicate), quality-control material, and sample.

2. Using a repetitive pipetting device, add 10 mL of the ferric acetate/uranyl acetate reagent to each tube. Just before adding the sample, wet the walls of each tube with the reagent by stirring the tube contents with a vortex-type mixer for 2 to 3 s. This facilitates complete suspension of the sample in the medium and allows complete extraction.

3. Into the appropriate tube add 50 μL of standard (brought to room temperature), control, or sample. Cap the tube, and vortex-mix thoroughly (25 s) after addition of each sample.

Note: If you do not use a micropipette with a complete displacement mechanism, make allowances for the difference in viscosity between the biological samples and the standards. One way to do this, for pipettors with conical plastic tips, is to deliver the sample to the first stop, wait for 5 s for the sample on the top wall of the tip to drain, then complete the delivery.

4. Let stand at room temperature for 10 min.
5. Centrifuge for 10 min at 700 × g.
6. Using 3.0 mL volumetric pipettes, place 3.0 mL of supernate in clean tubes labeled to correspond to those used in step 1.

Note: If necessary, one may stop here. It is possible to complete the test the next day, provided the stored tubes are tightly capped.

7. Using a repetitive pipette with an extended tip, add 2.0 mL of the $FeSO_4/H_2SO_4$ reagent. To deliver the color reagent, place the tip of the pipette at the bottom of the tube so that mixing of the two resulting layers is minimized. Place the tightly capped tube on a rotator set at 30 rpm and allow to mix for 20 s. Alternatively, mix with a vortex-type mixer until a few seconds after striations disappear (at least 20 s). By using the rotator, a tube may be processed completely every 30 s to keep a convenient pace.

Note: The Evaluators did not use an extended tip for this step. They simply added the reagent along the side of the test tube held at an angle of approximately 60°. The tube was then tightly closed and vortex-mixed within 3 s. In the Submitters' laboratory the technologists tried it this way when originally setting up the method but were reluctant to continue using it because the tubes, especially the 13 × 100 mm size, became very hot. To avoid the possibility of having warm to hot sulfuric acid leak out of the tube onto one's hands or clothes, we used rotator mixing. If tubes become too hot with mixing and cannot be held, mixing may be incomplete and the color thus developed is less stable. When using the larger size tubes (20 × 150 mm), make sure that the reagent added drains to the lower part of the tube where it will be uniformly mixed with the extract.

8. After mixing, cool the tubes in a rack, allowing empty spaces all around each tube to facilitate uniform cooling.

9. Read the absorbance of each sample and standard with a spectrophotometer set at 560 nm 20 min after mixing with the color reagent.

10. Plot the absorbance of each standard vs its concentration on linear graph paper. Determine the concentrations of samples with this standard curve.

Discussion

The method presented here *(1, 2)* has been presented as a Selected Method before *(3)*, modified by changing the proportion of sample to reagent and the wavelength at which the color is read. The data in these previous publications remain valid.

This method may be used to determine cholesterol in serum, serum extracts, and pleural fluid, and has been used to assay tissue cholesterol *(4)*. It can also be used to assess the concentration of cholesterol in serum calibrators to be used in enzymic methods. The Submitters prefer to use the proportion of ferric acetate/uranyl acetate reagent to serum originally proposed by Parekh and Jung because the absorbances observed with patients' samples generally fall within the limits of our standard curve and in a region of linear response in our instrument. For fluids or tissue of low cholesterol content, use 5 mL of the reagent for 50 μL of test material. It is possible to scale down the method for serum by using 25 μL of serum and 5 mL of reagent and work in the same absorbance range. This saves reagent cost.

Note: The Evaluators consider a sample/reagent ratio of 50 μL/5 mL as preferable because of the optimum absorbance range of 0.300–0.500 obtained with the most common cholesterol concentration in patients' sera.

Table 1. Linearity Study

Sample Standard, mg/L	Day 1 A	Day 1 Cholesterol, mg/L[b]	Day 2 A	Day 2 Cholesterol, mg/L	Day 3 A	Day 3 Cholesterol, mg/L
250	0.021	310	0.018	240	0.016	250
500	0.037	520	0.038	510	0.035	500
1000	0.074	1000	0.076	1010	0.072	990
2000	0.150	2000	0.151	2000	0.148	2000
3000	0.227	3000	0.226	2980	0.224	3010
4000	0.304	4010	0.304	4010	0.298	3990
Patients A,B, or C[a]	0.108	1450	0.086	1140	0.057	790
Diluted 1/1	0.051	700 × 2 = 1400	0.041	550 × 2 = 1100	0.026	380 × 2 = 760
Diluted 1/3	0.025	360 × 4 = 1440	0.022	300 × 4 = 1200	0.012	200 × 4 = 800

[a]The values obtained for these three patients on the original day of evaluation were 1410, 1170, and 770 mg/L, respectively.
[b]Concentrations obtained from linear graph of standards. To convert cholesterol concentration from mg/L to mmol/L, multiply by 0.002586.

Uranyl acetate is a "source material," so a radioisotope license is not needed to buy it in small amounts. Using 5 mL of the ferric acetate/uranyl acetate reagent per test tube, one can perform 2000 tests per gram of material.

A surgical mask and gloves should be used when handling the solid material, to avoid contamination by inhalation or contact with the skin.

Do not pipet by mouth. Carry out all reactions under a hood to avoid harm from the fumes of acetic and sulfuric acids; to avoid acid burns, handle all containers containing acid reagents (from bottles to cuvets) only while wearing thick rubber gloves.

The method presented in reference 2 is said not to give satisfactory results with pediatric samples. These usually exhibit concentrations of 450 mg/L or higher. The method used by the Submitters yields linear results with the lowest concentrations we have measured (250 mg/L). To illustrate this point we present data obtained on three different sera, each assayed on a different day (Table 1). Each dilution was prepared separately, and the regular proportion of uranyl acetate reagent to sample was used (10 mL to 50 μL).

Increased values of bilirubin interfere in the reaction. Before analysis, dilute the sample so that the bilirubin is 100 mg/L or less. Work done in the Submitters' laboratory (5), in cooperation with the Centers for Disease Control, and as part of the work of the Committee for the Standardization of Cholesterol shows that bilirubin at 100 mg/L produces an interference, when compared with a definitive method, of about 3% of expected value (unpublished results). The bias shown by other methods for cholesterol determination, even in the absence of bilirubin, is greater than this.

Strongly hemolyzed samples should not be assayed by this method. Hemoglobin at 2500 mg/L shows approximately 10% interference (increase) in the Submitters' laboratory and 6.7% interference at 2720 mg/L in a study made with the Centers for Disease Control.

Liquid serum control pools are best kept frozen in small aliquots sufficient to work from one tube per week. They begin to show deterioration in three months. This deterioration is related in an unknown manner to the color reaction because the same materials assayed by other methods show stable values (6). Lyophilized materials appear stable for approximately one year.

To avoid the problem of mismatched cuvets, use a single cuvet to read all the samples. As an example we use the following routine: rinse the cuvet (square, 10 mm i.d. × 10 mm × 35 mm, 3-mL capacity) with the sample, fill it, read the absorbance, discard the sample, touch the edge of cuvet against the beaker to help drain it, blot the mouth of the cuvet, rinse it with the next sample, drain, fill it with new sample, read, and repeat the cycle as necessary. This approach allows the reading of one sample every 30 s without undue haste. At the end of the run, rinse the cuvet thoroughly with distilled water, dry it, and put it away for the next run. If a cuvet longer than 35 mm is used, such as the round one, then there must be a cuvet available for each sample because there is not enough sample or time to rinse this cuvet thoroughly. Using the Submitters' approach and the 35-mm cuvet, there is no carryover. The coefficient of variation for the test is about 2%.

The Evaluators used a flow-through cuvet to read samples with a Vitatron colorimeter with a 2-mL flow-through cuvet. The flow-through instrument available to the Submitters has a smaller cuvet, and when the viscous sample was pulled through the system, air bubbles formed, which interfered in the measurement of the color formed.

References

1. Parekh, A. C., and Jung, D. H., Cholesterol determination with ferric acetate–uranium acetate and sulfuric acid–ferrous sulfate reagents. *Anal. Chem.* **42**, 1423–1427 (1970).
2. Jung, D. H., and Parekh, A. C., A new color reaction for cholesterol assay. *Clin. Chim. Acta* **35**, 73–78 (1971).
3. Jung, D. H., Biggs, H. G., and Moorehead, W. R., Colorimetry of serum cholesterol with use of ferric acetate/uranyl acetate and ferrous sulfate/sulfuric acid reagents. *Sel. Methods Clin. Chem.* **8**, 51–62 (1977).
4. Parekh, A. C., and Creno, R. J., Submicrogram assay of serum and adrenal cholesterol by a simple, direct, non-extraction procedure. *Biochem. Med.* **13**, 241–250 (1975).
5. Slickers, K. A., Edwards, L., Daly, J., and Ertingshausen, G., Automated reaction rate method for the direct determination of total serum cholesterol by use of the "CentrifiChem" parallel fast analyzer. *Clin. Chem.* **19**, 937–941 (1973).
6. Wood, P. D., Bachorik, P. S., Albers, J. J., Stewart, C. C., Winn, C., and Lippel, K., Effects of sample aging on total cholesterol values determined by the automated ferric chloride–sulfuric acid and Liebermann–Burchard procedures. *Clin. Chem.* **26**, 592–597 (1980).

Creatine Kinase in Serum

Submitters: John A. Lott and John W. Heinz, *Division of Clinical Chemistry, Department of Pathology, The Ohio State University, Columbus, OH 43210*

Evaluators: A. H. J. Maas, *Department of Cardiology, University Hospital, Utrecht, Netherlands*

C. van der Heyden, *Laboratory of Clinical Chemistry, Wilhelmina Childrens Hospital, Utrecht, Netherlands*

Thomas E. Burgess and Lainie Blum, *Laboratory Procedures, Inc., King of Prussia, PA 19406*

Reviewer: Robert C. Elser, *Department of Pathology, York Hospital, York, PA 17405*

Introduction

Creatine kinase (CK; ATP: creatine N-phosphotransferase, EC 2.7.3.2)[1] in serum was recognized in 1959 as having diagnostic value by Ebashi et al. (1), who observed abnormal serum CK activities in patients with muscular dystrophy. Shortly thereafter, Dreyfus et al. (2) reported that serum CK is increased after acute myocardial infarction. CK has become an important chemical determination in the diagnosis of cardiac and skeletomuscular disorders, and is increased after muscle trauma, intramuscular injections, and exercise, and in other conditions.

CK plays an important role in the energy-storing mechanism of tissue by catalyzing the reversible reaction between creatine and ATP to form creatine phosphate and ADP. CK is distributed in various organs; the highest activities are found (in decreasing order) in skeletal muscle, heart, and brain. Considerably lower activities are present in the urinary bladder, stomach, ileum, colon, and uterus (3, 4). The CK content of liver, erythrocytes, and kidney is less than 1% of the amount found in skeletal muscle.

After "uncomplicated" acute myocardial infarction, serum CK usually increases within 6 h, reaches a peak in 24 h, and returns to normal within 36 h.

CK activities have been reported as above normal in 94 to 100% of patients with acute myocardial infarction (5). The CK-MB isoenzyme of CK is highly sensitive and specific for injury to the myocardium, and its estimation is of great value in the diagnosis of myocardial infarction. Increased serum CK has also been seen in angina pectoris, congestive heart failure, myocarditis, and tachycardia (5).

Abnormal CK activities have been described in all forms of muscular dystrophy as well as polymyositis, dermatomyositis, and other myopathies (6, 7). Many nonaffected carriers of muscular dystrophy have abnormal serum CK activities, which provides a method for identifying such carriers. CK is also increased in 75% of individuals after a bout of acute alcohol intoxication (8). Surgery-induced muscle injury may result in marked increases in serum CK, and may be related to the type of anesthetic used and the duration of the surgery (9, 10). Maximum values are reached 24 to 48 h after surgery; they may remain abnormal for four to five days (9).

Some of the other causes of increased CK in serum are drug overdoses and poisoning, trauma to muscle or brain, hypothermia, hyperthermia, Reye's syndrome, infectious diseases, and hypothyroidism (5).

Caution must be exercised in making a diagnosis of heart or muscle disease when the serum CK activity is abnormal. Intramuscular injections of a variety of drugs have been shown to increase CK (8, 11–13). Moreover, CK activity can increase to above normal after mild to strenuous exercise and remain abnormal for several days (8, 14).

Methods have been described and reviewed (15) that involve both the forward reaction (i.e., creatine to creatine phosphate), and the reverse reaction, including colorimetric endpoint methods and continuous monitoring methods in the ultraviolet. The reverse reaction as presented here is superior to the forward reaction for several reasons: (a) it has a sixfold greater rate; (b) activation and restoration of enzyme activity by sulfhydryl compounds are possible in serum samples that have been stored; (c) there is less interference from side reactions; and (d) there is no need for a serum blank. Nearly all the current methods and the one described here are based on the reaction scheme proposed by Oliver (16) and

---

[1] Nonstandard abbreviations used: ADP, adenosine 5-diphosphate; AMP, adenosine 5-monophosphate; ATP, adenosine 5-triphosphate; CK, creatine kinase; CK-MM, CK-MB, and CK-BB, isoenzymes of CK; DPP, diadenosinepentaphosphate; EDTA, ethylenediaminetetraacetate.

Rosalki (17), with some modification of the conditions.

Principle

Creatine kinase is measured by the following coupled reactions:

$$\text{Creatine phosphate} + \text{ADP} \underset{\text{Mg}^{2+}}{\overset{\text{CK}}{\rightleftharpoons}} \text{creatine} + \text{ATP}$$

$$\text{ATP} + \text{D-glucose} \overset{\text{hexokinase}}{\rightleftharpoons} \text{D-glucose 6-phosphate} + \text{ADP}$$

$$\text{D-Glucose 6-phosphate} + \text{NADP}^+ \overset{\text{glucose-6-phosphate dehydrogenase}}{\rightleftharpoons} \text{D-glucono-}\delta\text{-lactone-6-phosphate} + \text{NADPH (indicator reation)}$$

The complete reagent we use here contains creatine phosphate, ADP, AMP, diadenosinepentaphosphate (DPP), glucose, NADP$^+$, N-acetylcysteine, hexokinase (EC 2.7.1.1), glucose-6-phosphate dehydrogenase (EC 1.1.1.49), magnesium acetate, ethylenediaminetetraacetate (EDTA), and imidazole buffer. Serum is added to the reconstituted complete reagent to start the reaction. After a 5-min preincubation to allow the lag phase to go to completion, the change in absorbance at 340 nm is monitored at 37 °C. The CK activity is determined from the linear, zero-order portion of the curve of absorbance vs time, and the results are expressed in IUB units per liter (U/L).

Materials and Methods

Reagents

1. *Complete reagent.* The method proposed by Szasz et al. (18) and adopted with very minor changes by the Scandinavian Society for Clinical Chemistry and Clinical Physiology (19) is used here, but with EDTA added to the reagent (20). The reagent and buffer are available from Bio-Dynamics/*bmc*, Indianapolis, IN 46250, as "CK-NAC (U.V.)" in three sizes: 2.5 mL for single tests, and 13 and 27 mL for multiple tests. The reagent is reconstituted in buffer before use; once prepared, it is stable at room temperature for one day, and for two days at 4 °C. Do not try to store the reconstituted reagent longer; prepare an amount appropriate to the number of samples on hand.

The final reaction conditions, after 50 µL of serum has been added to 2.5 mL of reagent, are: ADP 2.0 mmol/L, DPP 10.0 µmol/L, AMP 5.0 mmol/L, creatine phosphate 30.0 mmol/L, EDTA 2 mmol/L, glucose (from reagent) 20 mmol/L, glucose-6-phosphate dehydrogenase 1500 U/L, hexokinase 2500 U/L, imidazole buffer 100 mmol/L, pH 6.6 at 37 °C, magnesium acetate 10 mmol/L, NADP$^+$ 2.0 mmol/L, N-acetylcysteine 20 mmol/L, and volume fraction of serum (0.05 in 2.55) 0.020.

The preparation of this reagent from the individual components is clearly beyond the capability and desire of most laboratories.

The buffer solution should be clear and colorless and contain no undissolved particulate matter nor any growth of mold or bacteria. The dry reagent should be easily shakable in its flask, and free of moisture and large unbreakable lumps; it should be white to slightly cream colored; a yellow reagent is unsatisfactory. The Submitters have observed reagent that had absolutely no activity; it had been shipped during hot weather and presumably the reagent enzymes (hexokinase, glucose-6-phosphate dehydrogenase, or both) were destroyed.

Also make the following checks: (a) If water is substituted for serum, the starting absorbance should be less than 0.400 A, and the total change in absorbance *after* the lag phase (see *Procedure*) and during the 5-min measuring phase should be less than 0.005 A. (b) Check the pH with a pH meter at 37 °C. (c) Estimate the activity of the reagent enzymes from the length of the lag phase by analyzing a sample with a normal CK activity; the time for the reaction rate to become linear with time should be no more than 2 to 3 min. Longer times will generally mean there has been loss of one or both of the enzymes, and the reagent cannot be used. Glucose-6-phosphate dehydrogenase is the least stable component of the reagent (21). (d) Check the ability of the reagent to produce previously obtained or expected values for control samples (see *Procedure*).

2. *Dichromate blanking solution.* If, with turbid samples, the absorbance at the start of the reaction exceeds 0.8 A, use the instrument's absorbance offset. Or, use the blanking solution described here: Dissolve 50 mg of K$_2$Cr$_2$O$_7$ in 1 L of H$_2$SO$_4$ (5 mmol/L) to give an absorbance offset of about 0.5 A when setting the instrument to zero. A solution of 75 mg of K$_2$Cr$_2$O$_7$ per liter of 5 mmol/L H$_2$SO$_4$ will give an offset of about 0.75 A. Do not use more highly concentrated solutions of K$_2$Cr$_2$O$_7$ because they require an extra-wide slit opening; as a result, the spectrophotometer will no longer produce the expected values for the molar absorptivity of NADPH and will cause falsely low CK results.

Spectrophotometry

Accurate spectrophotometry is a prerequisite for the determination of serum CK activities. There are no acceptable CK standards, and controls cannot be used as standards; therefore, standardization relies on the ability of the instrument to reproduce, among other things, the widely used molar absorptivity of NADPH: 6220 L · mol^{-1} · cm^{-1} (22). Evaluation of the spectrophotometer is described in the chapter on lactate dehydrogenase in serum, in this volume.

Collection and Handling of Specimens

Collection. Blood should be collected in plain evacuated tubes or syringes, centrifuged promptly after the blood clots, and the serum taken off for analysis. Hemolyzed serum specimens with as much as 320 mg of hemoglobin per liter (trace hemolysis) can be tolerated because the CK value in such samples will be increased only slightly or remain unchanged *(23).* Erythrocytes contain very little CK; however, they contain adenylate kinase (EC 2.7.4.3), which must be inhibited because of its reaction with ADP:

$$2\ ADP \xrightleftharpoons{\text{adenylate kinase}} ATP + AMP$$

The ATP produced can then participate in the reactions described earlier, which leads to a falsely increased value for CK. With hemoglobin as great as 320 mg/L, inhibitors of adenylate kinase (i.e., AMP and DPP) are present in adequate concentration. With greater hemolysis, some increase in the observed CK is possible, owing to incomplete inhibition of adenylate kinase. EDTA is present in the reagent to stabilize *N*-acetylcysteine and to prolong the life of the reagent after it is reconstituted. It also increases the measured CK activity by 10 to 20% over that measured with the same reagent without EDTA *(21).* EDTA binds calcium and removes the inhibiting effect of calcium on CK *(20, 21, 24).* Blood should *not* be collected in tubes containing EDTA, because the EDTA can have an adverse effect on CK stability during storage *(21);* anticoagulants such as oxalate and citrate cannot be used because they inhibit CK *(21).* Heparin, however, reportedly does not affect CK in plasma *(17).* Turbid and icteric samples can be analyzed if the starting absorbance is not too high, and if a suitable dichromate solution can be used as a blank to offset the absorbance so as to be within the range of the spectrophotometer.

Storage. Serum must be stored in a tightly sealed container in the dark at 4 or −20 °C. After storage as plain serum at 4 °C for seven days or at −20 °C for one month, full CK activity is restored during the assay by *N*-acetylcysteine. Do not add sulfhydryl agents to serum before storage because they cause an irreversible loss of CK activity *(24).*

Note: Very lipemic samples can be cleared of some of their lipids by extracting the sample with water-saturated ether. Add 2 mL of distilled water to 10 mL of diethyl ether, shake well, and allow the phases to separate. Add three volumes of the upper ether layer to one volume of serum, mix thoroughly on a vortex-type mixer, centrifuge, and discard the upper layer. Analyze the lower serum layer.

Procedure

1. Evaluate the liquid and dry reagent for appearance (see above). Reconstitute the dry reagent with the buffer–glucose solution according to the package insert. Use the amount appropriate for the number of samples and controls to be analyzed.

2. Pipet 2.50 mL of the reagent, prewarmed to 37 °C, and 50 µL of serum or control, or proportionately smaller volumes of each, into a cuvet or suitable container that can be incubated at 37 °C in a water bath or other thermostated device. If there is insufficient sample, use 25 µL of serum added to 25 µL of 150 mmol/L saline instead, and note this on the worksheet. Set a timer to ring at 5 min after the serum is added to the reagent.

Note: The manufacturer of the reagent recommends reading after a 1-min incubation time. However, this is inadequate to fully activate CK in sera that have been stored; we recommend a 5-min incubation before taking readings.

3. For turbid or pigmented samples, adjust the spectrophotometer to zero absorbance at 340 nm with water or one of the dichromate blank solutions. Extract very lipemic samples with water-saturated ether before analysis (see above, *Collection*).

4. After the 5-min incubation in the thermostated container, place the reaction mixture into a 1-cm cuvet thermostated at 37 °C and take absorbance readings immediately and at 1-min intervals for 5 min, to give six readings.

5. If the change in absorbance per minute (ΔA/min) exceeds 0.180, repeat the analysis on a sample diluted with 150 mmol/L saline, and note this on the worksheet.

Quality Control

Quality-control serum must be stable for at least 24 h after reconstitution; even longer stability is desirable to keep costs down. The source should be human material, with 70 to 90% CK-MM and the rest CK-MB. The controls should mimic patients' samples as closely as possible. Because the K_m values of isolated CK-MM, CK-MB, and CK-BB are quite different *(25),* the use of controls prepared to contain largely CK-MB or CK-BB is inappropriate. Cloudy controls are undesirable; they have a high starting absorbance, which leads to poor precision, and are unrepresentative of most patients' samples, which are clear.

Analyze at least two controls with each group of samples, one at the upper reference limit and another exceeding the upper limit by two- to threefold. The upper reference limit is a "decision" point for diagnosis and treatment; the precision should be greatest here. The high-activity control will be most useful for detecting loss of linearity from a deteriorated reagent. How long the reaction rate is constant depends greatly on the NADP$^+$ concentration, particularly for samples with high activity *(18).* Do not use controls with activities exceeding the range of linearity because the additional dilution step adds potential uncertainty.

Note: Rosalki has described a simple procedure for preparing controls in the laboratory from tissue obtained at autopsy *(26)*.

For recording quality-control data, Levey–Jennings charts, which allow long-term trends in the assay to be easily seen, are recommended *(27)*. Ideally, the same control material should be used over a period of many months. Warning limits (i.e., ± 2 SD) and action limits (i.e., ± 3 SD), should be established by the laboratory during periods of "good" performance, when the assay has good stability. Limits provided by the manufacturer of the control material can be of some help, but must not be used uncritically because they are often much broader than the laboratory-established limits. CVs for CK in the Submitters' laboratory have been about 3 to 5% during periods of good performance.

Calculations

Average the five ΔA/min values. If the ΔA/min values are decreasing or erratic, repeat the analysis with a diluted specimen and investigate the source of the problem. CK activity (U/L) at 37 °C in 1-cm pathlength cuvets is calculated by:

$$U/L = \frac{\Delta A (\text{avg})}{\text{min}} \times \frac{(10^6 \,\mu\text{mol})}{(\text{mol})} \times \frac{(\text{mol} \cdot \text{cm})}{(6220 \text{ L})}$$
$$\times \frac{\text{total volume}}{\text{sample volume}}$$
$$= \frac{\Delta A (\text{avg})}{\text{min}} \times \frac{10^6}{6220} \times \frac{2.55}{0.05}$$
$$= \frac{\Delta A (\text{avg})}{\text{min}} \times 8199.$$

Discussion

Optimum Conditions

At least eight reports in the literature and technical reports available from reagent manufacturers discuss the optimization of CK methods by Oliver's approach. Morin *(28)* and Szasz et al. *(18)* have summarized the various reaction conditions. There is no agreement on the ideal method. In the reports, the work of Szasz et al. *(18, 21, 29)* stands out for its thoroughness and clarity; moreover, their results have been confirmed *(19)*. The method described here follows the work of Szasz, and EDTA is included in the reagent *(20, 21)*.

Imidazole is recommended as the buffer because its pK of 6.8 at 37 °C is very close to the desired assay pH of 6.6, and because it is available commercially in high purity. "Bis-tris" propane {1,3-bis[tris (hydroxymethyl)methylamino]propane} *(28)* is suitable but costs 12 times as much ($/mol) as imidazole.

A lag phase, or time for reaction to reach maximum velocity, is inherent in a coupled-reaction system. The lag phase is directly dependent on the activities of hexokinase and glucose-6-phosphate dehydrogenase *(18)*. With Szasz's method, the lag phase is 5 min or less for reactions started with serum specimens having CK activities within the linear range of the method.

The range of linearity of the method extends to 1500 U/L, *not* to the 2050 U/L given in the manufacturer's literature. The limiting factor for linearity appears to be the NADPH formed during the assay, which inhibits the indicator reaction.

Choice of temperature

The reaction can be carried out at 30 to 37 °C; The Submitters recommend 37 °C because the rates are faster and less sensitive to small temperature fluctuations *(30)*.

Choice of thiol compound

Many authors have shown that compounds containing sulfhydryl groups must be present in the assay mixture to reactivate serum CK. Szasz examined 27 such compounds for suitability *(31)*, eliminating 19 on the basis of either very high cost or instability in solution. The remaining eight, *N*-acetylcysteine, glutathione, *S*-(2-aminoethyl)isothiuronium bromide, dithiothreitol, dithioerythritol, thioglycolic acid, mercaptoethanol, and cysteine were essentially the same in activating ability. *N*-Acetylcysteine and mercaptoethanol were the most stable in solution at both 4 and 25 °C. Mercaptoethanol was rejected, however, because of its offensive odor and because it cannot be lyophilized.

Morin *(28)* concluded that thioglycerol, mercaptoethanol, dithiothreitol, and dithioerythritol were superior to *N*-acetylcysteine and certain other thiols in their ability to reactivate CK in a serum-initiated assay; mercaptoethanol and thioglycerol were found to be superior to other thiols in their ability to maintain CK activities during storage. Dithioerythritol and dithiothreitol are expensive and impractical; at concentrations exceeding 5 mmol/L, they gel serum *(31)*. Mercaptoethanol and thioglycerol are liquids and cannot be used in dry, single-reagent kits, but could be used in a reconstituting fluid *(28)*.

Whether there is a "best" activator remains to be proved. *N*-Acetylcysteine appears to be a good choice for a dry reagent, whereas mercaptoethanol and thioglycerol may be superior in a two-part reagent. Glutathione, originally proposed by the German Society for Clinical Chemistry *(32)*, must be avoided because glutathione reductase (EC 1.6.4.2) from erythrocytes can react with oxidized glutathione, consume NADPH, and thereby give falsely low CK results *(33)*. Mercaptoethanol and *N*-acetylcysteine showed a loss of 4 and 8%, respectively, after storage in solution for 43 h at 25 °C. At 4 °C, they showed no loss after two days, and 16 and 14% loss, respectively, after seven days *(31)*. EDTA enhances the stability of *N*-acetylcysteine; the reagent containing EDTA is stable for at least 12 h at 25 °C and for 48 h at 4 °C *(21)*. Longer storage of the reconstituted reagent is not recommended.

Inhibition of Adenylate Kinase

Adenylate kinase is found in skeletal muscle, brain, liver, heart, kidney, small intestine, spleen, erythrocytes, platelets, and lung. It is normally present in serum and is markedly increased in hemolyzed samples *(29)*. If it is not inhibited, falsely high CK values will be obtained. AMP, used as the inhibitor in most methods, is reasonably satisfactory at a concentration of 5 mmol/L to inhibit normal activities of adenylate kinase, but is inadequate in the presence of moderate hemolysis *(18, 23)*. AMP does not completely inhibit adenylate kinase from liver *(29)*; moreover, it also inhibits CK by 5 to 10%. Szasz et al. *(29)* recommended that 10 µmol of DPP per liter can be used in addition to 5 mmol of AMP per liter, because DPP is a potent inhibitor of erythrocyte adenylate kinase. However, DPP also does not completely inhibit this enzyme from liver.

Rosano et al. *(34)* found that F^- inhibits adenylate kinase but not CK; they recommend that 25 mmol of F^- per liter be used as the inhibitor. Unfortunately, this concentration of F^- will precipitate MgF_2 if the complete reagent contains both Mg^{2+} and F^-. It is possible to use F^- if the Mg^{2+} and F^- are combined at the point where the rate is measured, but this may also create a problem of turbidity *(19)*. Finally, Meiattini et al. *(35)* observed that AMP (2 mmol/L) and F^- (6 mmol/L) gave the best inhibition of human liver, muscle, heart, erythrocyte, and platelet adenylate kinase; the least inhibition of CK; and no precipitation of MgF_2. The issue of the best inhibitor of adenylate kinase is not completely resolved.

Inhibition of CK

Besides AMP and Ca^{2+}, several other inhibitors of CK have been identified. Warren *(36)* observed that uric acid is a potent CK inhibitor, but the inhibition is completely reversed by thiol-containing compounds. Cystine has recently been identified as a CK inhibitor *(37)*. It oxidizes the SH groups on the enzyme, and the addition of thiol-containing reducing agents reverses the inhibition.

Stability of CK in Serum

CK is *not* stable in serum; however, full CK activity can be restored by *N*-acetylcysteine during the 5-min incubation phase. Full activity is restored for serum stored without additives for 24 h at 25 °C, seven days at 4 °C, or one month at −20 °C *(32)*. Because CK undergoes irreversible loss when exposed to light *(38)*, samples should be stored in the dark. Alkaline phosphatase does not hydrolyze glucose 6-phosphate under the conditions of the assay, and therefore does not interfere *(31)*.

Reference Values

The Submitters examined sera from volunteer adult blood donors by this method. Blood was collected in red-top Vacutainer Tubes (Becton, Dickinson and Co., Rutherford, NJ 07070), centrifuged within 1 h, stored at 4 °C, and analyzed within 48 h of collection. Our findings for 84 men were: mean 178 (SD 102) U/L, central 95th percentile 80–460 U/L, median 159 U/L, mode 127 U/L, range of observed values 63–460 U/L. The Submitters observed 37 values (44%) exceeding 160 U/L, the upper reference limit given by the manufacturer. Clearly our population was different from theirs.

For 34 women we found the following values: mean 87 (SD 43) U/L, central 95th percentile 48–159 U/L, median 95 U/L, mode 95 U/L, range of observed values 16–206 U/L. The Submitters observed six values (18%) that exceeded the upper limit (130 U/L) given by the manufacturer.

It is recommended that users of this method confirm these reference ranges at their location with persons in good health.

Between-day results for lyophilized enzyme controls (Hyland, Costa Mesa, CA 92626) were: Control A, mean = 168 (SD 8.5) U/L, CV = 5.1% (n = 15); Control C, mean = 1045 (SD 35) U/L, CV = 3.3% (n = 14).

References

1. Ebashi, S., Toyokura, Y., Momoi, H., and Sugita, H., High creatine phosphokinase activity of sera in patients with progressive muscular dystrophy. *J. Biochem.* **46**, 103–104 (1959).
2. Dreyfus, J.-C., Schapira, G., Resnais, J., and Scebat, L., La créatine-kinase serique dans le diagnostic de l'infarctus myocardique. *Rev. Fr. Etud. Clin. Biol.* **5**, 386–387 (1960).
3. Dawson, D. M., and Fine, I. H., Creatine kinase in human tissue. *Arch. Neurol.* **16**, 175–180 (1967).
4. Tsung, S. W., Creatine kinase isoenzyme patterns in human tissue obtained at surgery. *Clin. Chem.* **22**, 173–175 (1976).
5. Lott, J. A., and Stang, J. M., Serum enzymes and isoenzymes in the diagnosis and differential diagnosis of myocardial ischemia and necrosis. *Clin. Chem.* **26**, 1241–1250 (1980).
6. Okinaka, S., Kumagai, J., Ebashi, S., Sugita, H., Momoi, H., Toyokura, Y., and Fujie, Y., Serum creatine phosphokinase. *Arch. Neurol.* **4**, 520–525 (1961).
7. Tyler, F. H., Muscular dystrophies. In *Metabolic Basis of Inherited Disease*, 3rd ed., J. B. Stanbury, J. B. Wyngaarden, and D. S. Fredrickson, Eds. McGraw-Hill, New York, NY, 1978, pp 1204–1217.
8. Nevins, M. A., Saran, M., Bright, M., and Lyon, L. J., Pitfalls in interpreting serum creatine phosphokinase activity. *J. Am. Med. Assoc.* **224**, 1382–1387 (1973).

9. Dixon, S. H., Jr., Fuchs, J. C. A., and Ebert, P. A., Changes in serum creatine phosphokinase activity following thoracic, cardiac, and abdominal operations. *Arch. Surg.* **103**, 66–68 (1971).
10. Phornphutkul, K. S., Anuras, S., Koff, R. S., Seeff, L. P., Mahler, D. L., and Zimmerman, H. J., Causes of increased plasma creatine kinase activity after surgery. *Clin. Chem.* **20**, 340–342 (1974).
11. Meltzer, H. Y., Mrozak, S., and Boyer, M., Effect of intramuscular injections on serum creatine phosphokinase activity. *Am. J. Med. Sci.* **259**, 42–48 (1970).
12. Knirsch, A. K., and Gralla, E. J., Abnormal serum transaminase levels after parenteral ampicillin and carbenicillin administration. *N. Engl. J. Med.* **282**, 1081–1082 (1970).
13. Meltzer, H., Intramuscular chlorpromazine and creatine kinase: Acute psychosis or local muscle trauma? *Science* **164**, 727–728 (1969).
14. Kaman, R. L., Goheen, B., Patton, R., and Raven, R., The effect of near maximum exercise on serum enzymes: The exercise profile versus the cardiac profile. *Clin. Chim. Acta* **81**, 145–152 (1977).
15. Batsakis, J. G., Creatine kinase. In *Second International Symposium on Clinical Enzymology*, N. W. Tietz, A. Weinstock, and D. O. Rodgerson, Eds. American Association for Clinical Chemistry, Washington, DC, 1976, pp 185–213.
16. Oliver, I. T., A spectrophotometric method for the determination of creatine phosphokinase and myokinase. *Biochem. J.* **61**, 116–122 (1955).
17. Rosalki, S. B., An improved procedure for serum creatine phosphokinase determination. *J. Lab. Clin. Med.* **69**, 696–705 (1967).
18. Szasz, G., Gruber, W., and Bernt, E., Creatine kinase in serum: 1. Determination of optimum reaction conditions. *Clin. Chem.* **22**, 650–656 (1976).
19. Committee on Enzymes, Scandinavian Society for Clinical Chemistry and Clinical Physiology, recommended method for the determination of creatine kinase in blood. *Scand. J. Clin. Lab. Invest.* **36**, 711–723 (1976).
20. Gruber, W., Inhibition of creatine kinase activity by Ca^{2+} and reversing effect of ethylenediaminetetraacetate. *Clin. Chem.* **24**, 177–178 (1978).
21. Szasz, G., Waldenstrom, J., and Gruber, W., Creatine kinase in serum: 6. Inhibition by endogenous polyvalent cations, and effect of chelators on the activity and stability of some assay components. *Clin. Chem.* **25**, 446–452 (1979).
22. Horecker, B. L., and Kornberg, A., The extinction coefficients of the reduced band of pyridine nucleotides. *J. Biol. Chem.* **175**, 385–390 (1948).
23. Frank, J. J., Bermes, E. W., Bickel, M. J., and Watkins, B. F., Effect of in vitro hemolysis on chemical values for serum. *Clin. Chem.* **24**, 1966–1970 (1978).
24. Rollo, J. L., Davis, J. E., Ladenson, J. H., McDonald, J. M., and Bruns, D. E., Effects of β-mercaptoethanol and chelating agents on the stability and activation of creatine kinase in serum. *Clin. Chim. Acta* **87**, 189–198 (1978).
25. Szasz, G., and Gruber, W., Creatine kinase in serum: 4. Differences in substrate affinity among the isoenzymes. *Clin. Chem.* **24**, 245–249 (1978).
26. Rosalki, S. B., Preparation of stable creatine kinase isoenzyme controls (MM, MB, BB). *Clin. Chem.* **22**, 1753–1754 (1976).
27. Grannis, G. F., Gruemer, H.-D., Lott, J. A., Edison, J. A., and McCabe, W. C., Proficiency evaluation of clinical chemistry laboratories. *Clin. Chem.* **18**, 222–236 (1972).
28. Morin, L. G., Creatine kinase: Re-examination of optimum reaction conditions. *Clin. Chem.* **23**, 1569–1575 (1977).
29. Szasz, G., Gerhardt, W., Gruber, W., and Bernt, E., Creatine kinase in serum: 2. Interference of adenylate kinase with the assay. *Clin. Chem.* **22**, 1806–1811 (1976).
30. Nealon, D. A., and Henderson, A. R., The apparent Arrhenius relationships of the human creatine kinase isoenzymes using the Oliver–Rosalki assay. *Clin. Chim. Acta* **66**, 131–136 (1976).
31. Szasz, G., Laboratory measurement of creatine kinase activity. In *Second International Symposium on Clinical Enzymology* (see ref. *15*), pp 143–179.
32. Recommendation of the German Society for Clinical Chemistry, Standardization of methods for the estimation of enzyme activities in biological fluids. *J. Clin. Chem. Clin. Biochem.* **10**, 281–291 (1972).
33. Klotzsch, S. G., Glutathione reductase as a possible source of error in common laboratory tests. *Clin. Chem.* **22**, 1236–1239 (1976).
34. Rosano, T. G., Clayson, K. J., and Strandjord, P. E., Evaluation of adenosine 5'-monophosphate and fluoride as adenylate kinase inhibitors in the creatine kinase assay. *Clin. Chem.* **22**, 1078–1083 (1976).
35. Meiattini, F., Giannini, G., and Tarli, P., Adenylate kinase inhibition by adenosine 5'-monophosphate and fluoride in the determination of creatine kinase activity. *Clin. Chem.* **24**, 498–501 (1978).
36. Warren, W. A., Identification of a creatine kinase inhibitor in human serum. *Clin. Biochem.* **8**, 247–253 (1975).
37. Jacobs, H. K., Phillipp, K. H., Sundmark, V., and Weselake, R. J., Isolation and identification of a new potent inhibitor of creatine kinase from human serum. *Clin. Chim. Acta* **85**, 299–309 (1978).
38. Thomson, W. H. S., An investigation of physical factors influencing the behaviour *in vitro* of serum creatine phosphokinase and other enzymes. *Clin. Chim. Acta* **23**, 105–120 (1969).

Creatine Kinase Isoenzymes in Serum

Submitter: John A. Lott, *Division of Clinical Chemistry, Department of Pathology, The Ohio State University, Columbus, OH 43210*

Evaluators: E. Christis Farrell, Jr., *Ohio Valley Hospital, Steubenville, OH 43952*

D. J. Moffa, *Biomedical Reference Laboratory, Inc., Fairmont, WV 26654*

Jannie Woo, *Department of Pathology and Laboratory Medicine, University of Texas Health Science Center, Houston, TX 77025*

Reviewer: Robert C. Elser, *Department of Pathology, York Hospital, York, PA 17405*

Introduction

Creatine kinase (CK)[1] exists in serum as a dimer with a relative molecular mass (M_r) of approximately 80 000. The monomers are B and M, and the three isoenzymes observed in serum are called CK-BB or CK-1, CK-MB or CK-2, and CK-MM or CK-3. The dual terminology reflects the primary tissues of origin and the order of electrophoretic migration, respectively. CK-BB is found in highest activity in brain tissue, CK-MM in skeletal muscle, and the hybrid CK-MB in heart. At pH 8.5 on agarose or cellulose acetate, CK-1 migrates the furthest toward the anode, CK-2 is intermediate, and CK-3, having the least negative charge, remains near the point of application. CK-MM is always present in serum, and in small, healthy, sedentary individuals, its activity can be very low. When analyzed by electrophoresis, serum CK appears to be all CK-MM except for the traces of CK-MB seen in a few individuals after strenuous exercise. CK-BB is not reported in sedentary or physically active healthy persons.

Injury to the myocardium, especially acute myocardial infarction, will usually result in an increased CK-MB in serum. Tests for CK-MB and other enzymes and the temporal sequence of increased serum activities are of great value in the diagnosis of acute myocardial infarction.

CK-MB activity in serum has been found in cases of myocarditis, and in a few patients with congestive heart failure, angina pectoris, and arrhythmias. Patients with Duchenne's muscular dystrophy can have greatly increased serum CK-MB activities. Other conditions showing an increased serum CK-MB in some patients are polymyositis/dermatomyositis, myositis, Reye's syndrome, malignant hyperpyrexia, drug overdose, alcoholic delirium tremens, and carbon monoxide poisoning. Extremely stressful exercise and trauma, including trauma to the thorax, usually result in an increased CK-MB in serum. Injury to the myocardium is not a prerequisite for an increased CK-MB, but must be considered in the differential diagnosis. The value of CK-MB in the diagnosis of myocardial injury has been reviewed *(1)*.

CK-BB appears to be an incidental finding in a variety of conditions; in prostatic cancer, however, it may have diagnostic significance.

Principle

A 1.6-μL sample of serum is applied to an agarose gel, and the gel is electrophoresed at 90 V for 20 min. The gel is overlaid with substrate containing the reagents listed below. The reaction proceeds as described by Rosalki *(2)*:

$$\text{Creatine phosphate} + \text{ADP} \xrightleftharpoons[\text{Mg}^{2+}]{\text{CK}} \text{creatine} + \text{ATP}$$

$$\text{ATP} + \text{glucose} \xrightleftharpoons{\text{hexokinase}} \text{glucose 6-phosphate} + \text{ADP}$$

$$\text{Glucose 6-phosphate} + \text{NADP}^+ \xrightleftharpoons{\text{G-6-PDH}}$$
$$\text{D-glucono-}\delta\text{-lactone-6-phosphate} + \text{NADPH}$$

After contact with the substrate, the gels are incubated at 37 °C for 20 min and then dried at 60 °C for 20 min. The fluorescent bands (from NADPH) are observed under ultraviolet light and are given a qualitative interpretation or are scanned with a fluorometric densitometer and expressed as percent of total fluorescence.

Materials and Methods

Reagents

Complete reagent. The reagent recommended (no. 470114; Corning ACI Electrophoresis System;

---

[1] Nonstandard abbreviations used: CK, creatine kinase (EC 2.7.3.2); CK-MM, CK-MB, and CK-BB, isoenzymes of CK; G-6-PDH, glucose-6-phosphate dehydrogenase (EC 1.1.1.49); LD, lactate dehydrogenase (EC 1.1.1.27); LD-1, LD-2, and LD-5, isoenzymes of LD; MES, 2-(*N*-morpholino)ethanesulfonic acid; MOPS, 3-(*N*-morpholino)propanesulfonic acid.

Corning Medical and Scientific/Corning Glass Works, Medfield, MA 02052) with agarose gels has the following composition, after the dry reagent is reconstituted with the 2-(N-morpholino)ethanesulfonic acid (MES)/sucrose solution: ADP 6.0 mmol/L, AMP 18 mmol/L, creatine phosphate 189 mmol/L, glucose 63 mmol/L, G-6-PDH (from *Leuconostoc mesenteroides*) 3000 U/L, glutathione 84 mmol/L, hexokinase (EC 2.7.1.1, from yeast) 9000 U/L, magnesium acetate 33 mmol/L, MES buffer 50 mmol/L (pH 6.20 at 37 °C), NADP$^+$ 6.0 mmol/L, and sucrose 584 mmol/L. AMP inhibits adenylate kinase (EC 2.7.4.3), and glutathione is the thiol activator of CK.

The preparation of this reagent is clearly beyond the desire of the majority of laboratories, and therefore a commercial source is recommended. The reagent should dissolve completely, and it should have no more than a faint yellow color. Do not use the reagent if the pH differs by more than 0.1 pH unit from 6.20 at 37 °C. Store the unreconstituted substrate at 2–8 °C. Higher temperatures may cause loss of hexokinase or G-6-PDH activities, and no bands will be seen after electrophoresis. The Submitter has observed that an unsatisfactory reagent may also cause a high background fluorescence.

Electrophoresis buffer, 3-(N-morpholino)propanesulfonic acid (MOPS), 50 mmol/L, pH 7.8 at 25 °C. Dissolve 1.046 g of MOPS, $C_7H_{15}NO_4S$, M_r 209.3 (e.g., cat. no. M1254; Sigma Chemical Co., St. Louis, MO 63178) in about 900 mL of distilled water. Using a pH meter to monitor, add 1 mol/L NaOH slowly, mixing well, and bring the pH to 7.80. Adjust the volume to 1 L and mix well. The buffer is stable for at least six months at 25 °C. Discard if any growth appears. The buffer is also available as a concentrate from Corning (no. 470015).

Agarose electrophoresis gels. The recommended gels are available from Corning (no. 470104) and consist of 10 g of agarose and 50 g of sucrose per liter of 2-amino-2-methyl-1-propanol buffer, 30 mmol/L, pH 8.6. Store the gels at 2–8 °C.

Apparatus

1. Electrophoresis cell and power supply (Corning no. 470136). The power supply is a fixed voltage unit (90 V ± 5%) designed for use with the Corning cell. With the banana plugs of the cell pointed away from you, the right-hand chamber is the cathode (negative).

Note: Evaluator J. W. successfully used a Beckman (Beckman Instruments Co., Fullerton, CA 92634) power supply and a Gelman (Gelman Instrument Co., Ann Arbor, MI 48106) cell for the electrophoresis.

2. Microsyringe, 10 µL (Hamilton Co., Reno, NV 89510), and disposable tips (Corning no. 470154) used for sample application.

3. Incubating and drying ovens. Use two Corning no. 712 ovens, or the equivalent, with one set at 37 (± 0.2) °C for incubating the gels and one set at 60 (± 1) °C for drying the gels. The ovens should have a temperature control that cycles the heat on and off to maintain temperatures within the above limits.

Note: Evaluator J. W. used a Thelco (Scientific Products, Div. American Hospital Supply Corp., McGaw Park, IL 60085) oven successfully.

Collection and Handling of Specimens

Collection. Collect blood in plain evacuated tubes or in syringes; centrifuge promptly after the blood clots, and take off the serum. Anticoagulants should not be present. Hemolyzed samples with as much as 4000 mg of hemoglobin per liter (grossly hemolyzed) can still be analyzed by this method. There is no detectable CK in erythrocytes; however, adenylate kinase appears as a band cathodal to CK-MM in samples having about 4000 mg of hemoglobin per liter. Increased serum bilirubin and lipemia do not interfere with the analysis.

Storage. Store the serum in tightly sealed containers in the dark at 4 or −20 °C. Full CK activity is restored during the assay by the sulfhydryl agent, glutathione, after storage for seven days at 4 or −20 °C. Do not add sulfhydryl agents, e.g., mercaptoethanol, to serum before storage because they can cause an irreversible loss of activity (3). The Submitter stored sera at 4 °C and found no change in the percentage of CK-MB in samples stored as long as 11 days (see Table 1).

Table 1. Stability of CK-MB in 19 Patients' Sera Stored at 4 °C

\	\	\	\	Days stored	\	\	\	\	\
0	1	2	3	4	5	6	8	10	11
\	\	\	\	Percent of total CK activity	\	\	\	\	\
22								23	24
24								24	24
6		4	5	4					
6		5	7	6					
7		8	6	7					
9		7		9					
19			17		16				
10			11		10				
12			14		15				
10			9		8				
3			4		4				
14			15		12				
3	4			4					
11						14	11	11	
5						7	6	5	
12						13	10	13	
6						6	4	5	
8			10	8		8			
6	5		5						

Sera assayed with the Corning ACI Electrophoresis System.

Procedure

Follow the procedure below in the listed order. These instructions are for the Corning electrophoresis cell and power supply. Evaluator J. W. used other equipment successfully (see *Apparatus*).

1. Fill each electrophoresis chamber at the base of the cell with 95 mL of fresh MOPS buffer. Level the buffer in the cell bases by tilting the cell. Wipe any excess buffer from the center partition of the cell to the power supply. Do *not* reuse the buffer.

2. Place a piece of Corning STA-MOIST filter paper (no. 470158) moistened with distilled water into an incubator tray and place the tray into the incubation chamber of the 37 °C oven. Allow at least 30 min for the oven and the incubator tray to come to 37 °C.

3. Remove the agarose film from its container. Handle it by the edges only and peel back the thin plastic cover in one continuous motion. Do not use the gel if any irregularities are present in the surface. Lay the gel on a clean, flat surface, gel side up and with the wells on the left side, arranged top to bottom. Do not allow the film to be exposed to air more than a total of 5 min, including the time the samples are applied. Films dry out quickly, so work with only one film at a time.

4. Place a plastic disposable tip on the needle of the microsyringe. Fill the tip with 0.8 µL of sample and wipe off the tip with lint-free tissue. Apply the sample to the gel, into the well only; do not allow serum to spread out beyond the well. Allow the sample to soak into the gel and then apply a second 0.8-µL sample. Dilute with physiological saline any sera with total CK activities exceeding 500 U/L so that they are 500 U/L or less. 500 U/L is fivefold the upper reference limit in the Submitter's laboratory. Also analyze the sample undiluted.

Note: The Submitter has observed that the CK-MB and CK-BB bands disappear when sera with greatly increased CK activities are diluted 10- to 100-fold. Run at least one control on the eight-position plate.

5. Insert the loaded film into the cassette holder, gel side up, and match the anodal (right) side to that of the cover. The edge grippers of the cover should support the film snugly in position.

6. Activate the power supply by placing the cassette cover on the cell. Set a timer and electrophorese the film for 20 min. For other power supplies, adjust to 90 V.

Note: Evaluator J. W. used 30 min for electrophoresis with good results.

7. During the time of electrophoresis, remove the substrate set from the refrigerator and allow the MES buffer to come to room temperature. Add 1 mL of buffer to the lyophilized reagent, allow it to dissolve completely (about 20 min), and mix well. The reconstituted reagent is stable for 8 h at room temperature.

8. After electrophoresis, open the cell, slowly drain off excess buffer (taking care not to wet the area of migration), and remove the gel from the holder.

9. Place the gel on a *flat* surface with the cathodal (negative) edge, i.e., the edge closest to the wells or numbers on the gel, toward you. Blot any residual buffer from the edges of the gel with a lint-free tissue. Do not blot the gel itself.

10. An even application of the substrate is very important for reproducible results. Careful technique on a *flat* surface is absolutely essential. Place a 5-mL serological pipette lengthwise along the cathodal edge of the gel. Use a pipette or dropper and dispense the 1 mL of substrate evenly onto the gel surface along the edge of the pipette facing away from you.

11. Slowly push, without pressing down or rolling, the serological pipette across the agarose film to the anodic edge of the film, but do not push the pipette off the gel. Without lifting the pipette, slowly pull back to the cathodic edge; again, do not pull the pipette off the gel. With a single, smooth motion, slowly push the pipette across the film to the anodic edge once more, pushing the pipette with the excess substrate off the gel; this makes a total of three passes across the film.

12. With lint-free tissue remove excess substrate from the edges and place the gel, gel side up, into an incubator tray containing a STA-MOIST filter paper. Incubate at 37 °C for exactly 20 min.

13. Remove the gel from the incubator tray, dry any moisture on the back of the film with a tissue, and then dry in an oven for 20 min at 60 °C. Do not dry at a hotter temperature, because excessive background fluorescence will result.

Quantitation

Some laboratories may prefer to give a qualitative interpretation of the results by simply observing the gels under an ultraviolet light (excitation wavelength 365 nm). What is sought primarily is the presence or absence of CK-MB and CK-BB. Qualitative interpretations could be "absent," "trace," "1+," "2+," etc. for CK-MB, and a comment if CK-BB is seen. The difficulty with this kind of interpretation is the consistency of evaluation. Thus "trace" may also be reported as "1+" and "2+" from day to day, depending on the observer. Comparing the fluorescence intensity of samples with a control may give better reproducibility of the qualitative remarks. For many laboratories, however, a qualitative interpretation can be quite satisfactory, provided the clinicians clearly understand what the laboratory is doing.

A fluorometric densitometer permits quantitation of the bands. The Corning 720 densitometer can be used successfully for this procedure, and the user is referred to the instrument manual for operating instructions.

Note: Evaluator J. W. successfully used a Gelman Model ACD-15 fluorometric densitometer to scan the gels.

When a scanning densitometer with automatic gain adjustment is used, scan gels for samples having low total CK activity carefully—background and spurious fluorescence may indicate "bands" where nothing is present. All gels should be looked at under ultraviolet light, to check for excessive background and satisfactory separation, and to rule out artifactual CK-MB and CK-BB bands.

In the Corning agarose system, CK-BB migrates anodic to albumin. Albumin will appear as a greenish-yellow fluorescence; in contrast, the CK isoenzymes will have a blue fluorescence. If there is doubt whether a band is truly a CK isoenzyme or an artifact, electrophorese the sample as usual but do not overlay with substrate, and simply dry the strip. Observe the strip under an ultraviolet light for CK-BB or other bands (4). It is also possible to obtain substrate that does not contain creatine phosphate (CPK n-1 Automation Pack; Worthington Diagnostics, Freehold, NJ 07728), to determine whether unusual bands between CK-MM and CK-MB or elsewhere are actually adenylate kinase, e.g., of erythrocytic or liver origin, and not CK bands. This is recommended in cases where unusual or suspicious bands are seen. Adenylate kinase catalyzes the reaction:

$$2\ ADP \xrightleftharpoons{\text{adenylate kinase}} AMP + ATP$$

The ATP will then react in the Rosalki scheme of reactions shown earlier, and it will appear that "CK" bands are present.

Quality Control

On each gel apply a control having CK-MM, CK-MB, and CK-BB activity. The control's activity should not exceed 500 U/L. Ideally, the control should give distinct but not intense bands for the three isoenzymes. Rosalki (5) has described a technique for preparing controls from tissue obtained at autopsy, which can reduce assay costs. The Submitter has used the Corning CK isoenzyme control containing all three isoenzymes (no. 470020) with satisfactory results. The control is reconstituted with 5 mL rather than 1 mL of diluent, to decrease the intensity of the bands so that they more closely resemble the bands from patients. The control can be divided into 0.2-mL aliquots and stored at −20 °C for at least four weeks without loss of activity. Table 2 shows precision of this method as determined with lyophilized control sera.

Table 2. Between-Day Precision of the Method with Lyophilized Control Sera

	Submitter[a]	Evaluator D.J.M.[b]
CK-MM, %	22.9	89.1
SD, %	2.9	10.2
CV, %	12.7	11.4
CK-MB, %	35.1	10.9
SD, %	2.3	2.8
CV, %	6.6	25.7
CK-BB, %	41.9	—
SD, %	3.7	—
CV, %	8.8	—

[a] Corning Isoenzyme Control, no. 47002; n = 22.
[b] n = 20.

If a fluorometric densitometer is available, record the percentage of each of the three isoenzymes on Levey–Jennings charts (6), to facilitate detecting trends and shifts. The allowable limits of ± 2 SD and ± 3 SD should be established by the laboratory during periods of good performance.

Discussion

CK Content of Tissues

The distribution of CK isoenzymes in human tissue obtained at surgery (7) and in baboon tissue (8) has been described. The tissue with highest CK activity is skeletal muscle (2500 U/g), and lower amounts of CK (U/g) are present in heart (380), brain (157), colon (137), ileum (161), and stomach (122) (7). Skeletal muscle CK is nearly all CK-MM, about 20% of the CK in heart is CK-MB (1), and nearly all the CK in brain, colon, ileum, and stomach is CK-BB. CK is present in trace amounts in liver and kidney (7, 8).

There is a small amount (<1%) of CK-MB in freshly obtained normal skeletal muscle (7, 8). The isoenzyme composition of normal and dystrophic muscle is different (9, 10). In Duchenne dystrophy, the CK composition of muscle resembles fetal muscle, in which CK-MM and CK-MB are present. This may explain the frequent observation of increased serum CK-MB in patients with Duchenne dystrophy, where no cardiac involvement is present (11–13).

Whether CK-MB is found in normal muscle may also depend on the freshness of the tissue and the sensitivity of the assay. Fresh normal skeletal muscle has more CK-MB than skeletal muscle obtained at autopsy (14–16). Yasmineh et al. (10) found 0.43% (range, ±0.18%) of total CK as CK-MB in normal skeletal muscle, which could have easily been missed with less-sensitive assay methods.

CK-MB in Myocardial Injury

The presence of CK-MB in serum is a very sensitive indicator of injury to the myocardium because of the very high CK-MB activity in myocardium and its presence in only trace amounts elsewhere in the body. This isoenzyme is a very sensitive and specific test for the diagnosis of acute myocardial infarction in those patients for whom there is a high suspicion of infarction (1). Measuring CK-MB is the best enzyme test currently available to document injury to the myocardium.

Of great importance to the diagnosis of acute myocardial infarction is the time sequence of enzyme abnormalities. CK-MB typically appears within 2 to 12 h after the onset of symptoms and before LD-1 exceeds LD-2. CK-MB usually persists for 24 to 72 h, whereas LD-1 is greater than LD-2 in some patients for four to five days after an "uncomplicated" infarction (1). Sampling every 12 h for 48 h is appropriate for patients with symptoms suggestive of infarction.

In the noncardiac surgery patient, CK-MM and LD-5 may be increased because of muscle trauma; CK-MB is usually normal (17–19). A postoperative increase in CK-MB in this group is suggestive of intra- or postoperative acute myocardial infarction. CK-MB is much less specific for acute myocardial infarction in patients having cardiac surgery, because manipulation of the heart can also release CK-MB into the blood (19). Increased CK-MB was seen postoperatively in 51 of 100 patients undergoing coronary bypass surgery (20). Twenty-six of these 51 showed CK-MB for only 12 h postoperatively and were diagnosed as having had no infarction. Of the 25 remaining, 17 had an electrocardiogram that was positive for acute myocardial infarction, but only five of these were diagnosed as having had infarction (20). After cardiac surgery, the absence or rapid disappearance of CK-MB is useful in ruling out acute myocardial infarction.

Increased Serum CK-MB Not Due to Acute Myocardial Infarction

The presence of CK-MB in serum does not mean that acute myocardial infarction has occurred (21). CK-MB has been seen in patients with coronary insufficiency and in patients after cardiopulmonary resuscitation (22–24). CK-MB has also been seen in some patients with tachyarrhythmias (26, 27), atrial fibrillation (24–27), cardioversion (19, 25), cardiac catheterization (28), angina pectoris (23, 24), and congestive heart failure (23, 24). In other reports, CK-MB was not seen after cardioversion (29), or cardiac catheterization (18, 30), or in angina pectoris (29).

Serum CK-MB is increased in most cases of muscular dystrophy (12, 13, 31), especially in the commonest form, Duchenne dystrophy. Increased CK-MB activity has been seen in a variety of myopathies such as dermatomyositis (32), polymyositis (32–34), viral myositis (32), and myoglobulinuria (28). Serum CK-MB has also been increased in pulmonary embolism (23, 24), malignant hyperpyrexia (35), hypothyroidism (36, 37), and carbon monoxide poisoning (38). By contrast, CK-MB has been reported not to appear in blood in pulmonary embolism (29), convulsions (27), trauma (39), hypothyroidism (40), exercise (41), and injections (39).

Strenuous exercise can increase serum CK-MB activity. One of 13 varsity swimmers had 40 U of CK-MB and 1000 U of CK per liter (10-fold normal) during the course of active physical training. This athlete had a normal electrocardiogram, no chest pain, and no other symptoms. Four others showed increased total CK and traces of CK-MB (1). The CK-MB may have come from skeletal muscle and was measurable because of the high total CK activity.

Significance of CK-BB

CK-BB is found infrequently in the serum of patients. It has been observed in acute brain trauma, cerebrovascular trauma, and cerebral infection, and after cerebrovascular accidents (42, 43). CK-BB has also been observed in 22% of patients receiving intensive care (44), in half of patients with metastatic prostatic cancer (45, 46), in renal disease (47–49), and in 1% of a large miscellaneous group (50, 51). CK-BB reported in some cases of renal disease may have been an electrophoretic artifact (4).

CK-BB is the least stable of the three isoenzymes and disappears rapidly from blood. CK-BB also appears to be transformed in vivo to CK-MB and then to CK-MM (52, 53), which, if true, may explain why CK-BB is observed so infrequently.

Isoenzyme variants of CK have been observed infrequently in serum. Mitochondrial CK, seen by the Submitter in a few patients, has not been observed in the peripheral blood of myocardial infarction patients (54); it migrates cathodal to CK-MM. There are also reports of an infrequently seen band in serum migrating between CK-MM and CK-MB (55–58), which may have been adenylate kinase in some cases (14). Whether these bands have any clinical importance remains to be proved. CK-MM and CK-MB can be further divided into subbands under special circumstances (59).

Methods Available for Determination of CK Isoenzymes

The five approaches for determining CK isoenzymes are electrophoresis, ion-exchange chromatography, immunological inhibition, selective activation, and radioimmunoassay. Each method can be evaluated in terms of four criteria: *(a) resolution*, or the ability to measure each fraction without cross-contamination from other fractions; *(b) reliability*, or the reproducibility of the results from day to day; *(c) sensitivity*, or the ability to detect minor fractions; and *(d) ease* of use in the laboratory (60).

1. Electrophoresis. CK isoenzymes have different charges at pH 8.6 and are thus easily separated by electrophoresis *(28, 61).* Satisfactory separations and fluorometric quantitation can be carried out on both cellulose acetate and agarose. Colorimetric staining with tetrazolium salts can be carried out with commercially available kits (e.g., Sigma no. 715GM); however, the bands are fainter than with fluorometric detection. A disadvantage of fluorometric procedures is the occasional appearance of background fluorescence, which can interfere with quantitation. Resolution with electrophoretic methods is excellent, but reproducibility is only fair; it is a semiquantitative technique at best. The sensitivity of the method is high, but less than with column chromatographic methods *(27).* Electrophoresis requires skill in applying the sample and substrate, and it probably cannot be done by all technologists. Nonetheless, electrophoresis is the method of choice for small and large laboratories. The overriding advantage is that *it is possible to see the bands:* there is no confusion as to whether CK-MB or CK-BB is present or not.

2. Ion-exchange chromatography. This is a popular alternative technique for either batch fractionation *(62, 63)* or column elution *(16, 34, 64),* the latter being available commercially in kit form. Various anion-exchange resins have been used. In batch methods, resin is mixed with diluted serum, and CK-MB and CK-BB are selectively adsorbed to the resin. In column methods, a serum sample is applied to the top of the column and is eluted with NaCl solutions of increasing concentrations. The eluate is fractionated and analyzed for CK activity. Ion-exchange fractionation has the possibility for high resolution, reproducibility, and very good sensitivity. Commercially available column methods, however, have compromised the intent of the original methods in column size, volume of eluent, number of samples analyzed, and the CK method itself *(60, 65).* Decreasing the volume of eluent can lead to cross contamination of the isoenzymes in the collected samples, whereas increasing the volume of eluent necessitates the analysis of many fractions per serum sample, each fraction having low CK activity. With a given column, there is no way to tell whether a particular sample has been adequately fractionated or whether there has been "bleed-over" of CK-MM into the CK-MB region. On balance, column methods are not as desirable as electrophoresis for the routine laboratory.

3. Immunological inhibition. In these procedures, total CK activity is measured, then the M subunit is inhibited with anti-M antibody, and the remaining activity is measured again *(66–69).* This method does not distinguish between CK-MB and CK-BB. In the presence of low activities of CK-MB, extremely stable spectrophotometry is necessary to measure the small changes in absorbance, making this technique questionable in view of the quality of many common laboratory spectrophotometers *(68).*

4. Selective activation. As described by Rao et al. *(70),* this method relies on the differential activation of CK isoenzymes. Glutathione reportedly does not activate CK-MB, but dithiothreitol does. CK activity is measured with glutathione alone and then with both thiols, and the difference of these two (large) numbers is reported to be CK-MB. The method was found not to correlate at all with electrophoretic *(71)* or column *(72)* methods for CK-MB, and should not be used.

5. Radioimmunoassay. Methods are becoming available for measuring either CK-MM or CK-BB or both *(73–75).* They are still largely in the research stage and are too time-consuming and costly for present use in the routine laboratory.

Reference Values

Sera from 38 volunteer adult blood donors were examined by this electrophoretic method. The blood was collected in red-top Vacutainer Tubes (Becton, Dickinson and Co., Rutherford, NJ 07070), centrifuged at $1500 \times g$ for 15 min within 1 h after collection, and analyzed within 48 h of collection. Storage was at 4 °C.

In 36 samples, the Submitter observed only a CK-MM band at the origin. The two remaining samples had 4% CK-MB and less than 1% CK-MB. CK-BB was not seen in any of the samples. The least amount of CK-MB that could be detected was about 3 U/L.

Electrophoretic reference values for CK-MB in the literature range from "absent" or "undetectable" to "trace" *(29, 39, 68, 76–82).* Results from patients have been used to establish discrimination values for CK-MB, to distinguish between persons with and without acute myocardial infarction. Patients with acute myocardial infarction will have a CK-MB of >3% *(23),* >4% *(26, 83),* >5% *(27),* >3 U/L *(69),* or >5 U/L *(29, 84).* The variability in the literature not only reflects different methodologies, but more importantly indicates that there is no clear demarcation, in terms of CK-MB, between patients with or without acute myocardial infarction.

References

1. Lott, J. A., and Stang, J. M., Serum enzymes and isoenzymes in the diagnosis and differential diagnosis of myocardial ischemia and necrosis. *Clin. Chem.* 26, 1241–1250 (1980).
2. Rosalki, S. B., An improved procedure for serum creatine phosphokinase determination. *J. Lab. Clin. Med.* 69, 696–705 (1967).
3. Rollo, J. L., Davis, J. E., Ladenson, J. H., McDonald, J. M., and Bruns, D. E., Effects of β-mercaptoethanol and chelating agents on the stability and activation of creatine kinase in serum. *Clin. Chim. Acta* 87, 189–198 (1978).
4. Coolen, R. B., Herbstman, R., and Hermann, P., Spuri-

ous brain creatine kinase in serum from patients with renal disease. *Clin. Chem.* **24**, 1636–1638 (1978).
5. Rosalki, S. B., Preparation of stable creatine kinase isoenzyme controls (MM, MB, BB). *Clin Chem.* **22**, 1753–1754 (1976).
6. Grannis, G. F., Gruemer, H.-D., Lott, J. A., Edison, J. A., and McCabe, W. C., Proficiency evaluation of clinical chemistry laboratories. *Clin. Chem.* **18**, 222–236 (1972).
7. Tsung, S. W., Creatine kinase isoenzymes patterns in human tissue obtained at surgery. *Clin. Chem.* **22**, 173–175 (1976).
8. Yasmineh, W. G., Pyle, R. B., Hanson, N. Q., and Hultman, B. K., Creatine kinase isoenzymes in baboon tissues and organs. *Clin. Chem.* **22**, 63–66 (1976).
9. Kuby, S. A., Keutel, H. J., Okabe, K., Jacobs, H. K., Ziter, F., Gerber, D., and Tyler, F. H., Isolation of human ATP-creatine transphosphorylases (creatine phosphokinases) from tissues of patients with Duchenne muscular dystrophy. *J. Biol. Chem.* **252**, 8382–8390 (1977).
10. Yasmineh, W. G., Ibrahim, G. A., Abbasnezhad, M., and Awad, E. A., Isoenzyme distribution of creatine kinase and lactate dehydrogenase in serum and skeletal muscle in Duchenne muscular dystrophy, collagen disease, and other muscular disorders. *Clin. Chem.* **24**, 1985–1989 (1978).
11. Somer, H., Dubowitz, V., and Donner, M., Creatine kinase isoenzymes in neuromuscular diseases. *J. Neurol. Sci.* **29**, 129–136 (1976).
12. Silverman, L. M., Mendell, J. R., Sahenk, Z., and Fontana, M. B., Significance of creatine kinase isoenzymes in Duchenne dystrophy. *Neurology* **26**, 561–564 (1976).
13. Cao, A., deVirgiliis, S., Lippi, C., and Coppa, G., Serum and muscle creatine kinase isoenzymes and serum aspartate aminotransferase isoenzymes in progressive muscular dystrophy. *Enzyme* **12**, 49–62 (1971).
14. Dawson, D. M., and Fine, I. H., Creatine kinase in human tissue. *Arch. Neurol. (Chicago)* **16**, 175–180 (1967).
15. Smith, A. S., Separation of tissue and serum creatine kinase isoenzymes on polyacrylamide gel slabs. *Clin. Chim. Acta* **39**, 351–359 (1972).
16. Mercer, D. W., Separation of tissue and serum creatine kinase isoenzymes by ion-exchange column chromatography. *Clin. Chem.* **20**, 36–40 (1974).
17. Mauney, F. M., Ebert, P. A., and Sabiston, D. C., Postoperative myocardial infarction: A study of predisposing factors, diagnosis, and mortality in a high risk group of surgical patients. *Ann. Surg.* **172**, 497–503 (1970).
18. Roberts, R., Gowda, K. S., Ludbrook, P. A., and Sobel, B. E., Specificity of elevated serum MB creatine phosphokinase activity in the diagnosis of acute myocardial infarction. *Am. J. Cardiol.* **36**, 433–437 (1975).
19. Tonkin, A. M., Lester, R. M., Guthrow, C. E., Roe, C. R., Hackel, D. B., and Wagner, G. S., Persistence of MB isoenzyme of creatine phosphokinase in the serum after minor iatrogenic cardiac trauma. Absence of postmortem evidence of myocardial infarction. *Circulation* **51**, 627–631 (1975).
20. Dixon, S. H., Limbird, L. S., Roe, C. R., Wagner, G. S., Oldham, H. N., and Sabiston, D. C., Recognition of postoperative acute myocardial infarction. *Circulation* **47**, Suppl. III, 137–140 (1973).
21. Guzy, P. M., Creatine phosphokinase-MB (CPK-MB) and the diagnosis of myocardial infarction. *West. J. Med.* **127**, 455–460 (1977).
22. Roe, C. R., Diagnosis of myocardial infarction by serum isoenzyme analysis. *Ann. Clin. Lab. Sci.* **7**, 201–209 (1977).
23. Galen, R. S., Reiffel, J. A., and Gambino, S. R., Diagnosis of acute myocardial infarction: Relative efficiency of serum enzyme and isoenzyme measurements. *J. Am. Med. Assoc.* **232**, 145–147 (1975).
24. Galen, R. S., and Gambino, S. R., Creatine kinase isoenzyme MB and heart disease. *Clin. Chem.* **21**, 1848–1849 (1975).
25. Ehsani, A., Gordon, A. E., and Sobel, B. E., Effect of electrical countershock on serum creatine phosphokinase (CPK) isoenzyme activity. *Am. J. Cardiol.* **37**, 12–18 (1976).
26. Varat, M. A., and Mercer, D. W., Cardiac specific creatine phosphokinase isoenzyme in the diagnosis of acute myocardial infarction. *Circulation* **51**, 855–859 (1975).
27. Mercer, D. W., and Varat, M. A., Detection of cardiac-specific creatine kinase isoenzyme in sera with normal or slightly increased total creatine kinase activity. *Clin. Chem.* **21**, 1088–1092 (1975).
28. Galen, R. S., and Gambino, S. R., Isoenzymes of CPK and LDH in myocardial infarction and certain other diseases. *Patholbiol. Ann.* **5**, 283–315 (1975).
29. Konttinen, A., and Somer, H., Specificity of serum creatine kinase isoenzymes in diagnosis of acute myocardial infarction. *Br. Med. J.* **i**, 386–389 (1973).
30. Roberts, R., Ludbrook, P. A., Weiss, E. S., and Sobel, B. E., Serum CPK isoenzymes after cardiac catheterization. *Br. Heart J.* **37**, 1144–1149 (1975).
31. Takahashi, K., Shutta, K., Matsuo, B., Takai, T., Takao, H., and Imura, H., Serum creatine kinase isoenzymes in Duchenne muscular dystrophy. *Clin. Chim. Acta* **75**, 422–435 (1977).
32. Brownlow, K., and Elevitch, F. R., Serum creatine phosphokinase isoenzyme (CPK 2) in myositis, a report of six cases. *J. Am. Med. Assoc.* **230**, 1141–1144 (1974).
33. Goto, I., Nagamine, M., and Katsuki, S., Creatine phosphokinase isoenzymes in muscles. *Arch. Neurol. (Chicago)* **20**, 422–429 (1969).
34. Yasmineh, W. G., and Hanson, N. A., Electrophoresis on cellulose acetate and chromatography on DEAE-Sephadex A-50 compared in the estimation of creatine kinase isoenzymes. *Clin. Chem.* **21**, 381–386 (1975).
35. Zsigmond, E. K., and Starkweather, W. H., Abnormal serum and muscle creatine phosphokinase (CPK) isoenzyme pattern in a family with malignant hyperthermia. *Anaesthetist* **22**, 16–22 (1973).
36. Griffiths, P. D., Serum enzymes in diseases of the thyroid gland. *J. Clin. Pathol.* **18**, 660–663 (1965).
37. Doran, G. R., and Wilkinson, J. H., The origin of the elevated activities of creatine kinase and other enzymes in the sera of patients with myxoedema. *Clin. Chim. Acta* **62**, 203–211 (1975).
38. Anido, V., Conn, R. B., Mengoli, H. F., and Anido, G., Diagnostic efficacy of myocardial creatine phosphokinase using polyacrylamide disc gel electrophoresis. *Am. J. Clin. Pathol.* **61**, 599–605 (1974).
39. Klein, M. S., Shell, W. E., and Sobel, B. E., Serum creatine phosphokinase (CPK) isoenzymes after intramuscular injections, surgery and myocardial infarction: Experimental and clinical studies. *Cardiovasc. Res.* **7**, 412–418 (1973).
40. Goto, I., Serum creatine phosphokinase isozymes in hypothyroidism, convulsion, myocardial infarction and other diseases. *Clin. Chim. Acta* **52**, 27–30 (1974).
41. Kaman, R. L., Goheen, B., Patton, R., and Raven, R., The effects of near maximum exercise on serum enzymes: The exercise profile versus the cardiac profile. *Clin. Chim. Acta* **81**, 145–152 (1977).

42. Kaste, M., Somer, H., and Konttinen, A., Brain-type creatine kinase isoenzymes. Occurrence in serum in acute cerebral disorders. *Arch. Neurol. (Chicago)* **34**, 142–144 (1977).
43. Somer, H., Kaste, M., Troupp, H., and Konttinen, A., Brain creatine kinase in blood after acute brain injury. *J. Neurol. Neurosurg. Psychiatry* **38**, 572–576 (1975).
44. Mercer, D. W., Frequent appearance of creatine kinase isoenzyme BB in sera of critical-care patients. *Clin. Chem.* **23**, 611–612 (1977).
45. Feld, R., and Witte, D., Presence of creatine kinase BB isoenzyme in some patients with prostatic carcinoma. *Clin. Chem.* **23**, 1930–1932 (1977).
46. Feld, R. D., Van Steirteghem, A. C., Zweig, M. H., Weimar, G. W., Narayana, A. S., and Witte, D. L., The presence of creatine kinase BB isoenzyme in patients with prostatic cancer. *Clin. Chim. Acta* **100**, 267–273 (1980).
47. Chuga, D. J., and Bachner, P., Creatine kinase isoenzyme BB in the serum of renal-disease patients, distinct from an albumin-like artifact. *Clin. Chem.* **24**, 1286 (1978).
48. Pascual, C., Segura, R. M., and Schwartz, S., Situations that can lead to increased creatine kinase isoenzyme BB activity in serum. *Clin. Chem.* **24**, 729–730 (1978).
49. Wesely, S. A., Byrnes, A., Alter, S., Solangi, K. B., and Goodman, A. I., Presence of creatine phosphokinase brain band in the serum of chronic renal disease patients. *Clin. Nephrol.* **8**, 345–348 (1977).
50. Lamar, W., Woodward, L., and Statland, B. E., Clinical implications of creatine kinase-BB isoenzyme. *N. Engl. J. Med.* **299**, 834–835 (1978).
51. Lang, H., Würzburg, U., Neumeier, D., Knedel, M., Prellwitz, W., Kattermann, R., Schlebusch, H., and Schürmann, J., Creatinkinase—Isoenzyme: Idiopathisches Auftreten von Creatinkinase-BB-Aktivitäten im Serum von Patienten. *Klin. Wochenschr.* **56**, 641–646 (1978).
52. Cho, H. W., Meltzer, H. Y., Joung, J. I., and Goode, D., Effect of incubation in human plasma on electrophoretic mobility of brain-type creatine phosphokinase. *Clin. Chim. Acta* **73**, 257–265 (1976).
53. Frotscher, U., Dominik, B., Richter, R., Zschaege, B., Schulte-Lippern, M., Jenett, G., Messerschmidt, W., Schmidtmann, W., and Wilbrandt, R., Die Instabilität der Kreatin-Phosphokinase-Isoenzyme im Serum. *Klin. Wochenschr.* **51**, 801–805 (1973).
54. Somer, J., Uotila, A., Konttinen, A., and Saris, N.-E., Creatine kinase activity and its isoenzyme pattern in heart mitochondria. *Clin. Chim. Acta* **53**, 369–372 (1974).
55. Sax, S. M., Moore, J. J., Giegel, J. L., and Welsch, M., Atypical increase in serum creatine kinase activity in hospital patients. *Clin. Chem.* **22**, 87–91 (1976).
56. Velletri, K., Griffiths, W. C., and Diamond, I., Abnormal electrophoretic mobility of a creatine kinase MM isoenzyme. *Clin. Chem.* **21**, 1837–1838 (1975).
57. Lott, J. A., and Heinz, J. W., A stable, unusual isoenzyme of creatine kinase observed in a case of acute anoxia. *Clin. Chem.* **24**, 1047 (1978). Abstract.
58. Leroux, M., Jacobs, H. K., Rabkin, S. W., and Desjardins, P. R., Measurement of creatine kinase in human sera using a DEAE-cellulose mini column method. *Clin. Chim. Acta* **80**, 253–264 (1977).
59. Wevers, R. A., Wolters, R. J., and Soons, J. B. J., Isoelectric focusing and hybridization experiments on creatine kinase. *Clin. Chim. Acta* **78**, 271–276 (1977).
60. Morin, L. G., Evaluation of current methods for creatine kinase isoenzyme fractionation. *Clin. Chem.* **23**, 205–210 (1977).
61. Elevitch, F. R., *Fluorometric Techniques in Clinical Chemistry*, Little Brown Co., Boston, MA, 1973, pp 244–249.
62. Henry, P. D., Roberts, R. R., and Sobel, B. E., Rapid separation of plasma creatine kinase isoenzymes by batch adsorption on glass beads. *Clin. Chem.* **21**, 844–849 (1975).
63. Morin, L. G., Improved separation of creatine kinase cardiac isoenzymes in serum by batch fractionation. *Clin. Chem.* **22**, 92–97 (1976).
64. Nealon, D. A., and Henderson, A. R., Separation of creatine kinase isoenzymes in serum by ion-exchange column chromatography. *Clin. Chem.* **21**, 392–397 (1975).
65. Lum, G., and Levy, A., Chromatographic and electrophoretic separation of creatine kinase isoenzymes compared. *Clin. Chem.* **21**, 1601–1604 (1975).
66. Griffiths, J., and Handschuh, G., Creatine kinase isoenzyme MB in myocardial infarction: Methods compared. *Clin. Chem.* **23**, 567–570 (1977).
67. Jockers-Wretou, E., and Pfleiderer, G., Quantitation of creatine kinase isoenzymes in human tissue and sera by an immunological method. *Clin. Chim. Acta* **58**, 223–232 (1975).
68. Obzansky, D., and Lott, J. A., Clinical evaluation of an immunoinhibition procedure for creatine kinase-MB. *Clin. Chem.* **26**, 150–152 (1980).
69. Würzburg, A., Hennrich, N., Lang, H., Prellwitz, W., Neumeier, D., and Knedel, M., Bedeutung der Aktivität von Creatinkinase MB im Serum unter Verwendung inhibierende Antikörper. *Klin. Wochenschr.* **54**, 357–360 (1976).
70. Rao, P. S., Lukes, J. J., Ayres, S. M., and Mueller, H., New manual and automated method for determining activity of creatine kinase isoenzyme MB, by use of dithiothreitol: Clinical applications. *Clin. Chem.* **21**, 1612–1618 (1975).
71. Vacca, G., Creatine kinase MB isoenzyme: A comparison of the electrophoretic method with the selective activating method and the immunological method. *Clin. Chim. Acta* **75**, 175–176 (1977).
72. Balkcom, R. M., Sheehan, M., Leipper, K., Hultman, B. K., Yasmineh, W. G., and Mushenheim, F., Evaluation of the chemical activation procedure (Rao's method) for the measurement of the MB isoenzyme of creatine kinase. *Clin. Chem.* **22**, 929–933 (1976).
73. Roberts, R., Sobel, B. E., and Parker, C. W., Radioimmunoassay for creatine kinase isoenzymes. *Science* **194**, 885–857 (1976).
74. Fang, V. S., Cho, H. W., and Meltzer, H. Y., Radioimmunoassay for MM and BB isoenzymes of creatine kinase substantiated by clinical application. *Clin. Chem.* **23**, 1898–1902 (1977).
75. Van Steirteghem, A. C., Zweig, M. H., Schechter, A. N., and Berzofsky, J. A., Radioimmunoassay of creatine kinase isoenzymes in human serum: Isoenzyme MM. *Clin. Chem.* **24**, 414–421 (1978).
76. Roberts, R., and Sobel, B. E., Creatine kinase isoenzymes in the assessment of heart disease. *Am. Heart J.* **95**, 521–528 (1978).
77. Roe, C. R., Limbird, L. E., Wagner, G. S., and Nerenberg, S. T., Combined isoenzyme analysis in the diagnosis of myocardial injury: Application of electrophoretic methods for the detection and quantitation of the creatine phosphokinase MB isoenzyme. *J. Lab. Clin. Med.* **80**, 577–590 (1972).
78. Konttinen, A., and Somer, H., Determination of serum creatine kinase isoenzymes in myocardial infarction. *Am. J. Cardiol.* **29**, 817–820 (1972).
79. Roark, S. F., Wagner, G. S., Izlar, H. L., and Roe,

C. R., Diagnosis of acute myocardial infarction in a community hospital. Significance of CPK-MB determination. *Circulation* **53**, 965–969 (1976).
80. Blomberg, D. J., Kimber, W. D., and Burke, M. D., Creatine kinase isoenzymes. Predictive value in the early diagnosis of acute myocardial infarction. *Am. J. Med.* **59**, 464–469 (1975).
81. Irvin, R. G., Cobb, F. R., and Roe, C. R., Acute myocardial infarction and MB creatine phosphokinase. *Arch. Intern. Med.* **140**, 329–334 (1980).
82. Gann, D., Cabello, B., DiBella, J., Rywlin, A., and Samet, P., Optimal enzyme test combination for diagnosis of acute myocardial infarction. *South. Med. J.* **71**, 1459–1462 (1978).
83. Marmor, A., Grenadir, E., Keidar, S., Edward, S., and Palant, A., The MB fraction of creatine phosphokinase. An indicator of myocardial involvement in acute pericarditis. *Arch. Intern. Med.* **139**, 819–820 (1979).
84. Shell, W. E., Kligerman, M., Rorke, M. P., and Burnham, M., Sensitivity and specificity of MB creatine kinase activity determined with column chromatography. *Am. J. Cardiol.* **44**, 67–75 (1979).

Creatinine in Serum and Urine with Fuller's Earth

Submitter: Mariet Iosefsohn, *Children's Hospital National Medical Center, Washington, DC 20010*

Evaluators: Manik L. Das, *Sigma Chemical Co., St. Louis, MO 63178*

Iris M. Osberg, *Pediatric Microchemistry Laboratory, University of Colorado Medical Center, Denver, CO 80262*

Richard Pavlovec, Harry S. Kim, and Sheshadri Narayanan, *Department of Pathology, New York Medical College, Metropolitan Hospital Center, New York, NY 10029*

Peter D. Spare, *McKellar General Hospital, Thunder Bay, Ontario, Canada P7E 1G6*

Reviewer: Thomas R. Koch, *Clinical Chemistry, University of Maryland Hospital, Baltimore, MD 21201*

Introduction

Creatinine, a waste product derived from creatine, is removed from the plasma by glomerular filtration and excreted in the urine without any significant reabsorption by kidney tubules. The determination of creatinine has one advantage over the blood urea nitrogen measurement for the diagnosis, prognosis, and monitoring of renal function; namely, it is not affected by a high-protein diet as is urea nitrogen. Because the excretion of creatinine in any individual (adult) is relatively constant, its concentration in a 24-h urine specimen may be used as a check on the completeness of the urine collection.

This method for determining creatinine was developed by Jaffé in 1886 *(1)*. Creatinine reacts with alkaline picrate to form an amber-yellow solution, the intensity of which is measured spectrophotometrically. Because many pseudocreatinine substances such as ascorbic acid, pyruvate, and glucose also react with alkaline picrate, the Jaffé method has been improved by incorporating deproteinization of serum *(2)* and a specific adsorption of creatinine on Fuller's earth (an aluminum silicate), Lloyd's reagent *(3, 4)*, or ion-exchange resin *(5, 6)*. (Also, see the alternative creatinine method presented in this volume.) The ever-increasing workload in the clinical chemistry laboratory had led to many modifications of methods for determining creatinine. The continuous-flow system (AutoAnalyzer; Technicon Instruments Corp., Tarrytown, NY 10591), introduced commercially in 1957, was the first mechanical instrument for creatinine analysis. This method removed proteins but not noncreatinine chromogens, and therefore served as a "screening" method.

"Kinetic" methods of analysis have become popular in routine chemical analysis, because they presumably offer the advantages of blanking out serum and reagent background absorbances; if the initial reaction rate is used, they also are much faster than conventional endpoint methods. A linear relationship exists between the initial change of absorbance and the creatinine concentration, but most other Jaffé-positive components of serum react with significantly slower reaction velocities. Thus, interference by most of the known Jaffé-positive serum constituents, except for keto acids, is minimized by selection of the optimum time interval during which the reaction rate is measured *(7)*. Kinetic methods that do not require deproteinization of serum and reduce or eliminate interferences have been described and applied via centrifugal analyzers such as CentrifiChem (Union Carbide Research Institute, Tarrytown, NY 10591) *(8)*, GEMSAEC (Electro-Nucleonics, Inc., Fairfield, NJ 07006), and more recently the Beckman Creatinine Analyzer (Beckman Instruments, Inc., Fullerton, CA 92634) *(9)*. It should be emphatically noted that the kinetic methods will, however, give *falsely high* values for patients with high concentrations of keto acids *(10)*.

Although the method described here is more time-consuming than those mentioned above, it is a manual method of choice. It is more accurate, minimizes or eliminates interferences, and is applicable for use in a low-volume clinical laboratory with a minimum amount of equipment. Because it provides "true" creatinine values, it is more reliable than kinetic methods for determining creatinine clearances and for measuring creatinine in patients with ketoacidosis.

Principle

The serum sample is deproteinized with tungstic acid and the supernatant fluid is mixed with an acidified Fuller's earth suspension. Creatinine is adsorbed on Fuller's earth and separated from the other non-specific chromogens, which remain in solution. Creatinine is eluted in alkaline picrate solution, and the absorbance of the yellow-orange color formed is determined spectrophotometrically at 520 nm *(11)*.

One usually selects the wavelength at the peak of maximum absorbance, to achieve maximum sensitivity; however, it may be desirable to choose another wavelength to minimize interfering substances. The color developed in the alkaline picrate procedure for creatinine produces a relatively flat peak in the visible region at approximately 485 nm, but the reagent blank itself absorbs light strongly at wavelengths less than 500 nm. Using the 520 nm wavelength minimizes the contribution of the blank reading, and improves precision and accuracy *(12)*.

Materials and Methods

Reagents

1. *Sodium tungstate* ($Na_2WO_4 \cdot 2H_2O$). Mallinckrodt (Mallinckrodt Inc., St. Louis, MO 63147) or Fisher (Fisher Scientific Headquarters, Pittsburg, PA 15219) analytical reagent, or any other brand meeting ACS specifications, 0.303 mol/L. Dissolve 100 g of sodium tungstate in about 600 mL of distilled water, and dilute to 1000 mL with distilled water. Allow to stand for several days and decant or filter if any precipitate has formed. The reagent, kept in a glass or polyethylene bottle, is stable at room temperature (approx. 25 °C) for six months.

2. *Sulfuric acid* (H_2SO_4). Mallinckrodt, Fisher, or any other brand meeting ACS specifications, 0.33 mol/L. Add slowly 18.7 mL of concd. H_2SO_4 to a flask containing approximately 500 mL of distilled water. Cool, mix, and dilute to 1000 mL with distilled water. The reagent is stable at 25 °C for at least six months.

3. *Fuller's earth suspension (Lloyd's reagent)*. Fuller's earth powder (FX 470; Matheson, Coleman and Bell Manufacturing Chemists, Norwood, OH 45212). Suspend 100 g of Fuller's earth in 1000 mL of distilled water. Fuller's earth suspension, after having been mixed very well, can be aliquoted according to the volume used in the procedure and kept refrigerated in test tubes covered with Parafilm.

Note: Evaluator P. D. S. observed that some batches of Fuller's earth do not adsorb creatinine.

4. *Oxalic acid, anhydrous* ($H_2C_2O_4$) *(Fisher), about 1.1 mol/L solution*. Add 10.3 g of oxalic acid and bring to 100 mL with distilled water to form a saturated solution (100 g of aqueous solution saturated at 25 °C contains 9.81 g of oxalic acid). Filter the solution and store the filtrate in a glass or plastic bottle at 25 °C (room temperature); stable for six months.

5. *Picric acid* $[(NO_2)_3C_6H_2OH]$ *(Fisher), 0.040 mol/L*. Dissolve 9.16 g of anydrous or 10.17 g of reagent grade picric acid (containing 100 to 120 mg of water per gram) in distilled water (may be heated to facilitate dissolving), and dilute to 1000 mL. This solution is stable at room temperature for three months in a dark container.

6. *Sodium hydroxide, 0.75 mol/L*. Dissolve 30 g of NaOH, ACS grade, in 800 mL of distilled water, allow to cool, and dilute to 1000 mL. When stored in a polyethylene bottle, the reagent is stable for six months at room temperature.

7. *Creatinine stock standard* ($C_4H_7N_3O$), *100 mg/dL (8.84 mmol/L)*. Preferably use National Bureau of Standards Standard Reference Material No. 914. Merck No. 5208 (purity 99%) or Fisher certified reagent may also be used. Dissolve 100 mg of anhydrous creatinine in 100 mL of 0.1 mol/L HCl. The solution is stable for two months when stored in a refrigerator (4 °C).

8. *Creatinine working standards*. Dilute 1.5 and 3 mL of stock standard solution to 100 mL each with distilled water to obtain 15 and 30 mg/L working standards, respectively. The working standards are stable for one week when stored in a refrigerator. They are used in the assay with each run to check accuracy and to calculate the unknowns.

Apparatus

This method requires the following equipment: 12 × 75 mm glass or plastic test tubes; microliter pipettes; volumetric and serological pipettes; 1.5-mL spectrophotometer cells with 1-cm lightpath; Parafilm; test tube mixer, or any available vibrator; liquid aspirator or a source of reduced pressure; centrifuge, 2500–3000 rpm; any spectrophotometer that measures absorbance between 500 and 560 nm, such as a Coleman 55 (Perkin-Elmer, Norwalk, CT 06856), Gilford 300 N (Gilford Instrument Laboratories, Oberlin, OH 44074), or Beckman Model 25 (Beckman Instruments, Inc.); and an analytical balance. Use any good commercial assayed control containing predetermined creatinine values, or prepare a control from a serum pool.

Collection and Handling of Specimens

Transfer freshly drawn blood (venipuncture or heel stick) to a tube without anticoagulant and permit it to clot spontaneously at room temperature *(12)*. Separate the serum from the clot by centrifugation within 1 h and store in a stoppered test tube or a glass or plastic container at 25 °C. The serum can be stored for as long as 96 h without alteration of the creatinine concentration (13). Urine should be collected in a clean container without preservative and stored at 4 °C until analysis. When a creatinine clearance test is requested, measure the volume of the 24-h urine and store an appropriate aliquot at 4 °C until the test can be performed (12). Dilute urine samples 10- or 20-fold with distilled water before analysis. Deproteinization and adsorption onto Fuller's earth can often be omitted for urine specimens. However, Sadilek has reported interferences in urine creatinine from alcaptonuric patients (14), and those very rare urine specimens should be analyzed without omitting any steps.

Procedure

1. Into labeled test tubes pipet the following:

	Blank	Standard	Control	Unknown
Distilled water	0.50 mL	0.30 mL	0.30 mL	0.30 mL
Standard	—	0.20 mL	—	—
Control	—	—	0.20 mL	—
Unknown	—	—	—	0.20 mL

Cover the tubes with Parafilm and mix by inverting the tubes or using a vibrator.

2. Into all tubes add 0.10 mL of sodium tungstate solution, mix, and add 0.20 mL of the sulfuric acid. Mix, and let stand for 5 min at room temperature.

3. Centrifuge for 5 min at 2500 rpm.

4. Transfer 0.30 mL of the supernate into 10 × 75 mm labeled test tubes.

5. Add to each tube 0.20 mL of distilled water, 0.05 mL of oxalic acid solution, and 0.10 mL of the Fuller's earth suspension.

6. Allow to stand for 10 min and mix repeatedly with a vibrator or vortex-type mixer.

7. Centrifuge for 5 min at 2500 rpm.

8. Aspirate as much of the supernate as possible and discard.

9. To the precipitate in each tube add: 0.30 mL of distilled water, 0.10 mL of picric acid solution, and 0.10 mL of the sodium hydroxide.

10. Allow to stand for 10 min and mix repeatedly.

11. Centrifuge at 2500 rpm for 10 min; after an additional 10 min read the absorbances of the supernates for the standard, control, and unknowns at 520 nm on the spectrophotometer, setting the blank to read zero absorbance. The volumes for serum, urine, and reagents can be doubled or multiplied by a higher factor (see fourth note below).

Note: Evaluators I. M. O. and P. D. S. suggest that decanting the supernatant fluid instead of aspirating it (step 8, above) is faster and more efficient, decreasing the possibility of disturbing or aspirating some of the Lloyd's reagent.

Note: Evaluator I. M. O. suggests doubling the volume of supernate used in step 4, and subsequently doubling all the other volumes in steps 5–9. It is easy to obtain 0.6 mL of supernate if Haden's modification (15) of the Folin–Wu method is used. Haden reverses the addition of sodium tungstate and sulfuric acid, adding the acid first. The advantages of this modification are that the reaction is immediate and that a larger volume of supernate is obtained.

Note: Evaluator P. D. S. recommends using a combination of the reagents in step 5, above. He mixed 15 mL of distilled water and 5 mL of saturated oxalic acid solution with 10 mL of Fuller's earth suspension and added 0.3 mL of this mixture to 0.3 mL of supernatant fluid. The slight difference in total volume did not influence the results.

Note: Evaluators R. P., H. S. K., and S. N. used sixfold greater volumes for all components. Instead of 0.2 mL of serum, they used 1.2 mL, and the final volume was 3.0 mL, instead of 0.5 mL. For the precision studies they used a normal (Ortho Diagnostics Inc., Raritan, NJ 08869) and an abnormal (Monitrol; DADE Division, American Hospital Supply Corp., Miami, FL 33152) control serum. The within-day and day-to-day variations were established on paired data run through 10 consecutive days; the within-day CV was 2.3%, and the day-to-day CV was 3.3%. When 20 samples were assayed for creatinine by the above method and compared with the SMA 12/60 values (based on the Jaffé reaction, but without Lloyd's reagent), the regression coefficient was 0.9993.

Calculations

Calculate results from absorbances (A) of unknowns and standards, as follows:

Creatinine, mg/L = (A unk/A std)
$\quad$ × concentration of the standard, mg/L

Discussion

Linearity. The standard curve for creatinine determined with Lloyd's reagent is linear to 120 mg/L (9). Assay of creatinine exceeding 120 mg/L should be repeated with a diluted specimen; include the dilution factor in the calculation.

Precision. The within-run precision derived by 20 analyses of a serum with a mean value of 10.7 mg/L gave an SD of 0.44 mg/L and a CV of 4.08%. Precision and accuracy obtained by the Evaluators are presented in Tables 1 and 2.

Table 1. Recovery Studies

Evaluator	Creatinine, mg/dL Expected	Obtained	Recovery, %
I. M. O.	0.63	0.7	111
	5.00	5.0	100
	10.0	10.2	102
R. P., H. S. K., and S. M.	2.00	1.98	97
	2.25	2.16	96
P. D. S.	1.00	0.98	99
	2.0	2.05	102.5
	5.0	4.96	98.3
	10.0	9.64	96.4
M. L. D.	1.50	1.40	93
	2.00	1.93	97

Table 2. Precision

Evaluator	Within-day (n = 20 each) x̄, mg/dL	SD, mg/dL	CV, %	Day-to-day (n = 20 each) x̄, mg/dL	SD, mg/dL	CV, %
I. M. O.	0.77	0.047	6.1	0.81	0.107	13.2
	0.86	0.068	7.9			
	10.75	0.051	0.5	10.32	0.375	3.6
	10.41	0.186	1.8			
R. P., H. S. K., and S. N.	—	—	2.3	—	—	3.3
M. L. D.	1.065	0.075	7.0	1.100	0.086	7.8
	6.985	0.203	2.9	7.180	0.182	2.5
P. D. S.	10.14	0.39	3.8	14.15	0.67	4.7

Method comparison. Results by the manual Jaffé technique with Lloyd's reagent (y) were compared with those obtained with the Beckman Creatinine Analyzer (x), a semiautomated instrument (9) for the determination of creatinine in serum and urine. The regression equation, calculated by least-squares regression for 114 serum specimens, was: $y = 0.878x + 1.78$ mg/L ($r = 0.994$, $S_{y/x} = 0.3301$).

Note: Evaluator P. D. S. compared the manual micromethod involving Lloyd's reagent with an ion-exchange reference method of Rockerbie and Rasmussen (16) and the ABA-100 kinetic method (Abbott Bichromatic Analyzer; Abbott Diagnostics, N. Chicago, IL 60064). There was no significant difference among the results obtained with the three methods. Evaluator P. D. S. observes that the Lloyd's reagent method has the advantages of being easier to perform because it is less complex than the ion-exchange method and does not require the automated equipment necessary for kinetic measurements.

Table 3 presents data obtained by the Evaluators on several different methods and instruments.

Interferences. Grossly lipemic serum, hemoglobin (up to 2.0 g/L), bilirubin (up to 250 mg/L), and protein (up to 150 mg/L) do not interfere with this creatinine method. Acetoacetate and acetone, each exceeding 1.0 g/L, increase the creatinine value (8, 17–20). Acetoacetate, 1 g/L, increased a creatinine of 9 mg/L (79 μmol/L) to 36 mg/L (317 μmol/L), and 250 mg of acetoacetate per liter increased this creatinine to 16 mg/L (141 μmol/L); 1 g of acetone per liter increased the apparent creatinine from 9 mg/L (79 μmol/L) to 13 mg/L (114 μmol/L) (9).

Normal values determined by some of the Evaluators are listed in Table 4. Reference intervals for creatinine reported in the literature (4, 12) are:

Serum		4–12 mg/L
Urine		
Premature infants	2–12 wk	8.3–19.9 mg/kg per 24 h
Full-term infants	1–7 wk	10.0–15.5
Children	2–5 yr	10.0–12.0
	6–11 yr	6.4–21.9
	12–16 yr, M	12.2–29.4
	12–16 yr, F	20.7–28.2

Table 3. Comparison between This Method (y) and Other Methods (x)

Evaluator	Comparison method (x)
I. M. O.	GEMENI centrifugal analyzer: $y = 1.018 x + 0.062$ ($r = 0.998$)
R. P., H. S. M., and S. N.	Technicon SMA 12/60: $t = 2.09$ (no significant difference between means) $F = 2.12$ (no significant difference in variances) $r = 0.9993$
M. L. D.	Heinegard and Tiderstrom (21): $y = 1.020 x + 0.037$ ($r = 0.997$)
P. D. S.	Rockerbie and Rasmussen (16): $y = 1.02 x + 0.2$ ($r = 0.999$)

Table 4. Normal Values in Healthy Subjects Determined by This Method or Supplied from the Literature by the Evaluators

Evaluators	Sample	Creatinine content
R. P., H. S. K., and S. N.	Adults, serum (mg/L)	4–12
	Adults, urine (g/24 h)	
	Men	1.0–1.8
	Women	0.7–1.5
	Children, urine (mg/kg per 24 h)	15–25
I. M. O.	Children, serum (mg/L) (22)[a]	
	3 d	2–9
	1 yr	2–5
	2–3 yr	3–6
	4–7 yr	2–7
	8–10 yr	3–8
	11–12 yr	3–9
	13–17 yr	3–11
	18–20 yr	3–11
	Children, urine (mg/kg per 24 h)(23)[a]	
	Premature	8.1–15.0
	Full-term	10.4–19.7
	1½–7 yr	10.4–15.0
	7–15 yr	
	M	5.2–41.0
	F	11.5–29.1

[a] References.

References

1. Jaffé, M., Ueber den Niederschlag, welchen Pikrinsaure in normalem Harn erzeugt und über eine neue Reaction des Kreatinins. *Hoppe-Seyler's Z. Physiol. Chem.* **10**, 391–400 (1886).
2. Folin, O., and Wu, H., A system of blood analysis. *J. Biol. Chem.* **38**, 81–110 (1919).
3. Owen, J. A., Iggo, B., Scandrett, F. J., and Stewart, C. P., The determination of creatinine in plasma or serum, and in urine; a critical evaluation. *Biochem. J.* **58**, 426–437 (1954).
4. Knoll, E., and Stamm, D., The specific determination of serum creatinine. *J. Clin. Chem. Clin. Biochem.* **8**, 582–587 (1970).
5. Teger-Nilsson, A. C., Serum creatinine determination using an ion-exchange resin. *Scand. J. Clin. Lab. Invest.* **13**, 326–331 (1961).
6. Stoten, A., A micromethod for creatinine using resin to remove interfering substances. *J. Med. Lab. Technol.* **25**, 240–247 (1968).
7. Mark, H. B., and Rechnitz, G. A., *Kinetics in Analytical Chemistry*, Interscience Publishers, New York, NY, 1968, p 67.
8. Fabiny, D. L., and Ertingshausen, G., Automated reaction-rate method for determination of serum creatinine with the CentrifiChem. *Clin. Chem.* **17**, 696–700 (1971).
9. Hicks, J. M., Iosefsohn, M., and Lewis, A. S., Evaluation of the Beckman Creatinine Analyzer. *Clin. Chem.* **25**, 1005–1008 (1979).
10. Lustergarten, J. A., and Wenk, R. E., Simple, rapid, kinetic method for serum creatinine measurement. *Clin. Chem.* **18**, 1419–1422 (1972).
11. Mattenheimer, H., *Micromethods for the Clinical and Biochemical Laboratory*, 2d ed., Ann Arbor Science Publishers, Inc., Ann Arbor, MI, 1971, p 69.
12. Tietz, N. W., *Fundamentals of Clinical Chemistry*, 2nd ed., W. B. Saunders Co., Philadelphia, PA, 1976, pp 47–51, 113.
13. Kirberger, E., and Keller, H., Errors in the determination of creatinine caused by storage. *J. Clin. Chem. Clin. Biochem.* **11**, 205–208 (1973).
14. Sadilek, U., Creatinine determination in the urine of alcaptonuric patients. *Clin. Chim. Acta* **12**, 436–439 (1965).
15. Haden, R. L., A modification of the Folin–Wu method for making protein-free blood filtrates. *J. Biol. Chem.* **56**, 469–471 (1923).
16. Rockerbie, R. A., and Rasmussen, K. L., Rapid determination of serum creatinine by an ion-exchange technique. *Clin. Chim. Acta* **15**, 475–479 (1967).
17. Koenst, M. H., and Freier, E. F., An evaluation of two methods for the determination of true creatinine. *Am. J. Med. Technol.* **37**, 473–479 (1971).
18. Narayanan, S., and Appleton, H. D., Specificity of accepted procedures for urine creatinine. *Clin. Chem.* **18**, 270–274 (1972).
19. Mitchell, R. J., Improved method for specific determination of creatinine in serum and urine. *Clin. Chem.* **19**, 408–410 (1973).
20. Osberg, I. M., and Hammond, K. B., A solution to the problem of bilirubin interference with the kinetic Jaffé method for serum creatinine. *Clin. Chem.* **24**, 1196–1197 (1978).
21. Heinegard, D., and Tiderstrom, G., Determination of serum creatinine by a direct colorimetric method. *Clin. Chim. Acta* **43**, 305 (1973).
22. Schwartz, G. J., Haycock, G. B., Chin, B., and Spitzer, A., Plasma creatinine and urea concentration in children: Normal values for age and sex. *J. Pediatr.* **88**, 828 (1976).
23. Novak, L. P., Age and sex differences in body density and creatinine excretion of high school children. *Ann. N.Y. Acad. Sci.* **110**, 545 (1963).

Creatinine in Serum and Urine, Column-Chromatographic Method

Submitter: John E. Sherwin, *Valley Children's Hospital and Guidance Clinic Fresno, CA 93703* (formerly at *Department of Pediatrics, Michael Reese Hospital, Chicago, IL 60616*)

Evaluators: Willard A. Stacer, *Chemistry Department, Harper Hospital, Detroit, MI 48201*
Don S. Miyada, *Department of Pathology, University of California Medical Center, Orange, CA 92668*
Orlando R. Flores, Lilla Sun, and Mary Phillips, *Beckman Instruments, Inc., Fullerton, CA 92634*

Reviewer: Thomas R. Koch, *Clinical Chemistry, University of Maryland Hospital, Baltimore, MD 21201*

Introduction

Creatinine is a waste product derived from creatine and excreted by the kidneys. Methods for its analysis and its usefulness in health monitoring are discussed in the introduction to the preceding chapter.

This spectrophotometric procedure for determining "true" creatinine is a modification of the procedure of Mitchell (1). Creatinine is isolated by column chromatography. Quantitative recovery of creatinine in the elution step of the chromatography permits the use of as little as 50 µL of serum with no loss in accuracy and precision. The procedure may be used for measuring either serum or urinary creatinine.

Techniques currently used to assay creatinine in biological fluids include Lloyd's reagents with alkaline picrate (2), alkaline picrate (3, 4), enzymic analysis with creatininase (5), kinetic analysis with alkaline picrate (6), and "high-pressure" liquid chromatography (7). These methods are difficult to perform, expensive, insufficiently quantitative, and less sensitive than this procedure. The manual procedure for creatinine described here is sensitive, relatively simple to perform, accurate, and precise, and can be performed with a micro sample.

Principle

Creatinine is reacted with alkaline sodium picrate to form a colored complex that absorbs strongly at 505 nm (8), as follows:

To separate creatinine from other substances reacting with picric acid, creatinine in serum or urine samples is purified by chromatography on a cation-exchange resin, then reacted with alkaline picrate; its concentration is calculated from comparison with standard creatinine samples treated in the same manner as the unknowns.

Materials and Methods

Reagents

Analytical grade reagents and distilled de-ionized water are used for all solutions.

1. *Sodium hydroxide solution, 2.5 mol/L.* Dissolve 100 g of sodium hydroxide in 500 mL of water. *Cool under running tap water.* Dilute to a final volume of 1 L. Store at room temperature.

2. *Adsorption buffer, pH 3.0.* Dissolve 7.73 g of citric acid and 2.84 g of disodium phosphate (Na_2HPO_4) in 500 mL of water and dilute to 1 L with water. If necessary, adjust to pH 3 with a 10% (100 g/L) sodium hydroxide solution. This buffer is stable for about one month if refrigerated.

3. *Elution buffer, 0.5 mol/L, pH 12.2.* Dissolve 20.0 g of sodium hydroxide and 71.0 g of disodium phosphate (Na_2HPO_4) in 500 mL of water and dilute to 1 L with water. This solution is stable for one month if kept refrigerated.

Note: Evaluator W. A. S. reports that the elution buffer can precipitate upon refrigeration. He eliminates this problem by adding 0.5 mL of pentachlorophenol solution [2 g/L in 95% (950 mL/L) ethanol] to 1 L of buffer. This buffer can be stored at room temperature.

4. *Saturated picric acid solution.* Suspend 1.4 g of picric acid in 100 mL of water. Stir overnight. This solution is stable for one month at room temperature.

5. *Creatinine standard solution, 1.0 g/L.* Dissolve 100.0 mg of creatinine (certified grade; Fisher Scientific Co., Fairlawn, NJ 07410) in 50 mL of 0.2 mol/L hydrochloric acid (1.7 mL of concd. HCl diluted to 1 L with water). Dilute this solution to a final volume of 100 mL of water, cover with 0.5 mL of toluene, and store tightly capped in the refrigerator. This solution is stable for one month. Prepare fresh standards daily by diluting this solution with appropriate amounts of 0.1 mol/L HCl and storing them in the refrigerator (Table 1).

Table 1. Preparation of Creatinine Standard Solutions from the 1 g/L Stock Solution and 0.1 mol/L HCl

Volumes to mix, mL		Creatinine, mg/L
Standard	HCl	
0	50.0	0
0.5	49.5	10
1.0	49.0	20
2.0	48.0	40
4.0	46.0	80
6.0	44.0	120

Equipment

1. *Ion-exchange columns.* Plastic columns, 0.5 × 110 cm (Kontes, Vineland, NJ 08360), are used.

Note: Evaluator W. A. S. uses the Bio-Rad columns (Bio-Rad Laboratories, Richmond, CA 94804) that are used for catecholamine measurement.

Cation-exchange resin, Ag 50 W × 2, 100–200 mesh (Bio-Rad Laboratories), is suspended in water, drawn into a graduated pipette, and allowed to settle. Add 0.5 mL of settled resin to each column. These columns should be prepared with fresh resin every three months.

Note: Evaluators D. S. M., O. R. F., L. S., and M. P. observe absorbance decreases as great as 10% as the column resin ages. Therefore, they caution that only columns with resin of the same age be used in each analysis.

2. *Spectrophotometer.* An 8-nm bandpass spectrophotometer (Model 102 Hitachi; Nissei Sangyo Inc., Sunnyvale, CA 94086) was used for all analyses. A narrow bandpass spectrophotometer is essential for these analyses.

Procedure

1. Prepare the resin in the column by washing it with 2.0 mL of 2.5 mol/L sodium hydroxide.
2. Rinse the column with 2.0 mL of H_2O and then with 4.0 mL of adsorption buffer.
3. Add 0.1 mL of serum, standards, or controls to 5.0-mL aliquots of adsorption buffer. Urine samples should be diluted 50-fold (1.0 mL to 50 mL total; 1 + 49) with H_2O before adding 0.1 mL to 5.0 mL of adsorption buffer. These diluted samples are added to the columns and allowed to drain through the resin. Each resin is then washed with 4.0 mL of H_2O. Discard these washings.
4. Add 3.0 mL of elution buffer to each column and carefully collect all of the eluate in a test tube.
5. Add 0.5 mL of saturated picric acid solution to each test tube containing the column eluate, and mix thoroughly.
6. Incubate these solutions at room temperature for 30 min.
7. After this incubation, determine the absorbance of each sample at 505 nm, using a spectrophotometer adjusted to zero with H_2O.
8. Graphically determine the creatinine concentration of the sample by plotting absorbance of the standards (after subtracting the absorbance of the zero standard) vs their concentration. Alternatively, because the standard curve is linear, analyze only one standard and calculate the results by the proportional relationship between the absorbance of the standard and its concentration. Calculate urinary creatinine similarly but multiply the result by 50 (dilution factor).

Collection of Blood Samples

Use either serum or plasma for this analysis. Draw 5 mL of whole blood and allow it to clot 20 min before centrifugation. If pediatric or geriatric microsamples are desired, collect the blood sample by skin puncture into Natelson blood-collecting tubes or any similar tube; centrifuge the sample in the same tube and collect the serum or plasma by cutting the tube at the cell/plasma interface.

Results

Figure 1 shows the excellent linearity and reproducibility of the creatinine standard curve up to at least 100 mg/L. The absorbance of the 120 mg/L standard is 0.249 ± 0.0066 at 505 nm. Figure 2 compares the linearity and reproducibility of the standard curve for sample volumes of 100, 50, and 25 µL. Although these curves are linear and have correlation coefficients near 1.0, the absorbance of the highest standard decreases from about 0.250 with 100 µL to only 0.090 with 25 µL; thus, samples in the normal range will exhibit absorbance of less than 0.010 if 25-µL sample volumes are used. This decreasing sensitivity is reflected in the observed CV of normal creatinine samples when the smaller sample volumes are used (Table 2).

Table 3 compares the analytical results of two creatinine-specific methods with those of the chromato-

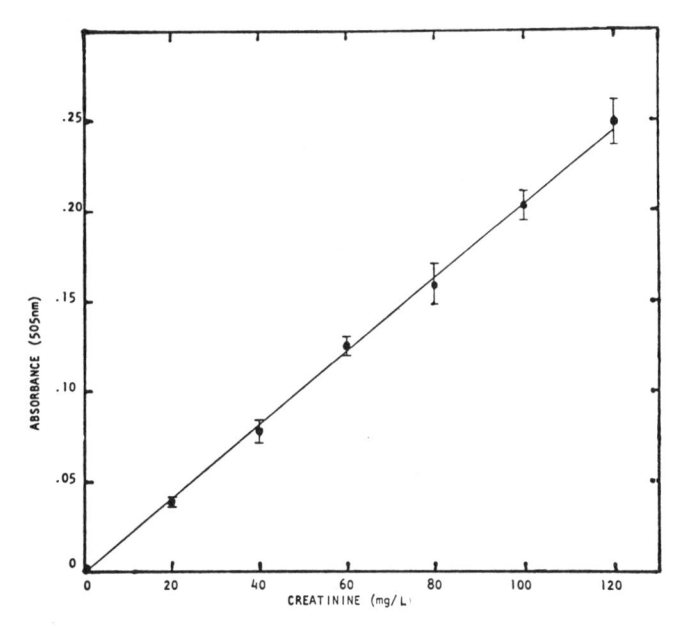

Fig. 1. Standard curve for determination of creatinine in serum or urine

Bars indicate 2 SD from the mean. Linear regression analysis yields $y = -0.0017 + 0.0206x$, with a correlation coefficient (r) of 0.9995 and a covariance of 0.0031

graphic procedure. The only statistically significant differences occur at high creatinine values and do not represent clinically significant differences. The average analytical recovery of standard creatinine samples after chromatography is more than 98% (Table 4).

Table 2. Effect of Sample Volume on Creatinine Analysis

Sample vol, μL	Creatinine, mg/L Mean	SD	CV,%
100	9.8	0.046	4.8
50	10.4	0.167	16.0
25	11.6	0.204	17.6

Table 3. Comparison of Creatinine Results from Three Creatinine-Specific Methods

	Creatinine, mg/L[a]		
	Sample A	Sample B	Sample C
Chromatographic purification analysis	10.2 (0.94)	43.1 (1.6)	6.28 (0.15)[b]
Alkaline picrate kinetic analysis[c]	10.0 (0.45)	29.6 (8.7)	5.97 (0.13)
Acid fading fraction analysis[d]	9.8 (0.9)	42.6 (2.1)	5.97 (0.20)

[a] Mean (and SD); n = 25.
[b] Values are statistically different ($p < 0.001$) from those obtained by the other two methods for the same sample.
[c] Done with the Beckman Creatinine Analyzer and reagents (Beckman Instruments, Inc., Fullerton, CA 92634).
[d] Done with the Rapid Stat Creatinine procedure (Pierce Chemical Co., Rockford, IL 61105).

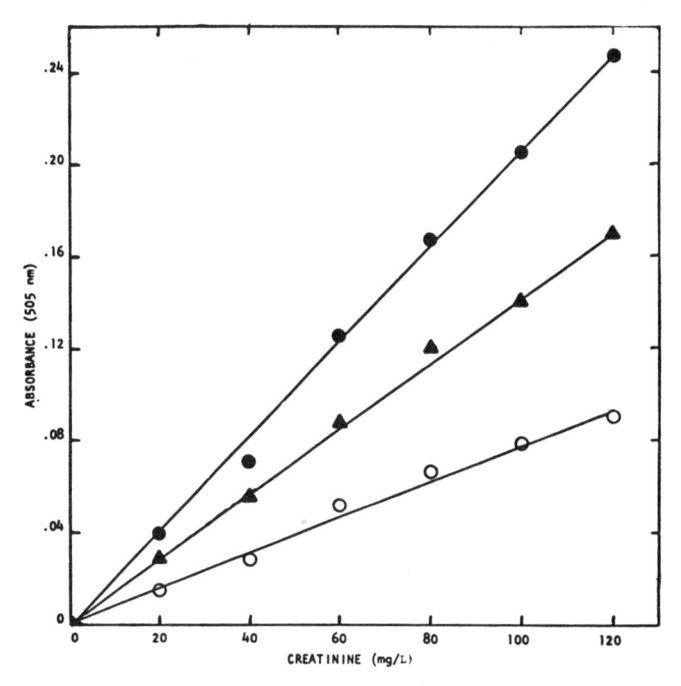

Fig. 2. Comparison of creatinine standard curves based on use of 100-μL (●), 50-μL (▲), and 25-μL (○) samples

Linear regression analysis yields $y = -0.0030 + 0.0209x$ ($r = 0.9984$) for the 100-μL sample and $y = 0.0009 + 0.0007x$ ($r = 0.9954$) for the 25-μL sample

Table 4. Analytical Recovery of Creatinine after Chromatography

Creatinine concn, mg/L	Mean absorbance (and SD) Untreated	Chromatographed	% recovery
20	0.0298 (0.0222)	0.0395 (0.0008)	99.3
40	0.0793 (0.0016)	0.0782 (0.0033)	98.6
60	0.1292 (0.0033)	0.1250 (0.0030)	96.8
80	0.1582 (0.0017)	0.1588 (0.0059)	100.4
100	0.2078 (0.0051)	0.2027 (0.0040)	97.6
120	0.2478 (0.0050)	0.2485 (0.0066)	100.3
		Average recovery	98.8

Interferences due to hemolysis are removed by this procedure (Table 5). Severe hemolysis (2.0 g/L) results in an increase of measured creatinine by only 2.0 mg/L in the normal range. Bilirubin concentrations of less than 150 mg/L result in only a slightly increased creatinine result (Table 5). Interferences due to acetoacetate and ascorbic acid are negligible in clinically encountered concentrations and only slightly affect the creatinine value in high concentrations (Table 5).

Analysis of serum from 25 apparently normal adults gave a mean creatinine value of 7.35 (SD 1.77) mg/L. These sera were chosen randomly from a group of outpatient specimens being analyzed as part of a routine annual physical. The distribution of these samples with regard to sex and race is unknown. The normal range for creatinine determined by this procedure is effectively the same as the nor-

Table 5. Effect of Interfering Substances on Creatinine Results with and without Chromatographic Purification

	Creatinine, mg/L[a]	
Interfering substance	Untreated	Chromatographed
None	26.7 (1.98)	9.9 (0.77)
Acetoacetate, 2.5 g/L	42.8 (3.49)	10.8 (1.81)
Bilirubin, 150 mg/L	43.6 (1.35)	11.0 (1.49)
Hemoglobin, 2.0 g/L	40.1 (4.58)	12.0 (1.33)
Ascorbic acid, 2.5 g/L	4.60 (3.71)	10.7 (1.16)

[a] All samples contained 10 mg of creatinine per liter and were analyzed in triplicate. Data represent the mean (and SD) of three separate experiments.

mal range determined by use of other specific methods *(4, 9)*.

Creatinine clearance was calculated for 15 adults with various degrees of renal dysfunction, with use of the Boyd surface-area formula *(10)*. Comparison of the values derived from the ion-exchange chromatographic data with those by a fading fraction method and a kinetic method (see Table 3) demonstrated a significant correlation (Figure 3). However, by linear regression analysis, the relationship between the fading fraction method and the ion-exchange chromatographic method has a slope that is more nearly 1.0 and an intercept more nearly zero than the relationship of the chromatographic method with the kinetic method. Statistical analysis with the "paired *t*-test" indicated no significant difference between the fading fraction method and the chromatographic method ($t = 0.26$, $df = 15$). However, the kinetic procedure and the chromatographic procedure give significantly different results

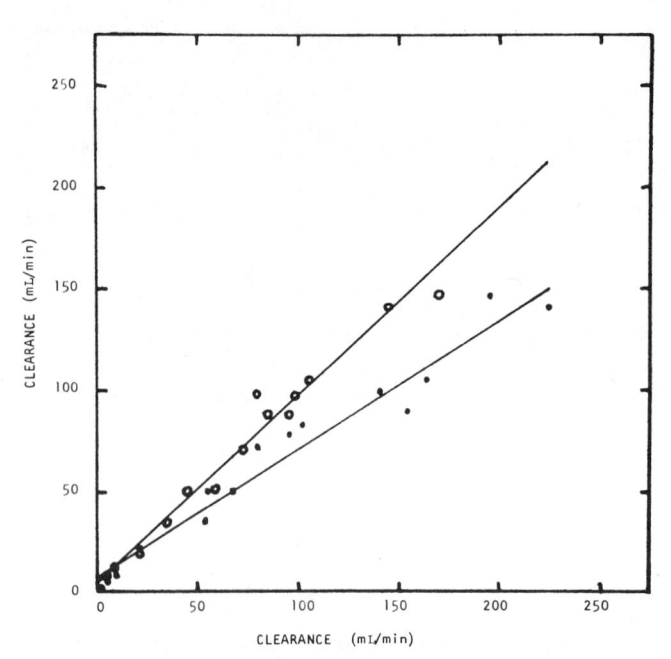

Fig. 3. Comparison of the ion-exchange chromatographic procedure (*x*-axis) with an acid fading fraction method (○) and a kinetic alkaline picrate procedure (●) in the measurement of creatinine clearance (n = 15)
Linear regression analysis yields $y = 4.1367 + 0.9316x$ ($r = 0.9864$) for the acid fading fraction comparison and $y = 7.7759 + 0.6236x$ ($r = 0.9761$) for the kinetic method comparison

($p < 0.005$). This difference, although not specifically investigated, is probably due to the accumulation of interfering substances (see Table 5), as a result of altered metabolism in patients with significant renal disease.

Table 6. Summary of Evaluators' Data

		Precision						
		Within-day				Between-day		
Evaluator	N	$\bar{x}$, mg/L	SD, mg/L	CV, %	N	$\bar{x}$, mg/L	SD, mg/L	CV, %
D. S. M.	20	50.4	1.85	3.66	20	50.4	3.37	6.7
W. A. S.					20	47.5	2.15	4.53

		Method comparison		
	Comparison method	Regression line		r
D. S. M.	Trace-3	$y = 1.07x - 0.04$		0.994
	Beckman Creatine Analyzer-2	$y = 0.992x - 0.06$		0.996
	Technicon SMAC	$y = 0.915x + 0.04$		0.992
W. A. S.	Technicon AutoAnalyzer-II	$\bar{x}$, mg/L	47.5	53.0
		SD, mg/L	2.15	1.3
		CV, %	4.53	2.5

		Normal values	
		From the literature	From Harper Hospital
W. A. S.	Men	6.0–12.0 mg/L	6.0–13.0 mg/L
	Women	5.0–10.0 mg/L	

Discussion

This ion-exchange chromatographic method is specific for creatinine and is easily performed in any small laboratory. A good-quality spectrophotometer with a spectral bandpass of less than 20 nm at 505 nm is required to ensure linearity of the standard curve. Evaluators D. S. M., O. R. F., L. S., and M. P. report linearity to 150 mg/L with a Coleman II spectrophotometer.

The analytical precision of this method compares well with other creatinine-specific methods. Evaluators D. S. M., O. R. F., L. S., and M. P. report within-day precision of 3.7% and a between-day precision of 6.7%. Evaluator W. A. S. observed a between-day precision of 4.5% at a creatinine concentration of 47.5 mg/L and a precision of 7.8% at a creatinine concentration of 12.1 mg/L.

Results by this chromatographic method compare favorably with those of other creatinine-specific methods (Table 3). Evaluators D. S. M., O. R. F., L. S., and M. P. confirm this correlation (Table 6).

The small ion-exchange columns used in this procedure have rapid flow rates, and the analysis is easily performed in less than 1 h. In routine practice, 25 columns can be assayed together without difficulty. The requirement of only 0.1 mL of serum or less for this analysis makes it attractive for use in geriatric and pediatric laboratories.

The excellent technical assistance of Mrs. Margie Webb is gratefully acknowledged.

References

1. Mitchell, R. J., Improved method for specific determination of creatinine in serum and urine. *Clin. Chem.* **19**, 408–410 (1973).
2. Owen, J. A., Iggo, B., Scandrett, F. J., and Stewart, C. P., The determination of creatinine in plasma or serum, and in urine; a critical evaluation. *Biochem. J.* **58**, 426–437 (1954).
3. Folin, O., On the determination of creatinine and creatine in urine. *J. Biol. Chem.* **17**, 469–473 (1914).
4. Slot, C., Plasma creatinine determination, a new and specific Jaffé reaction method. *Scan. J. Clin. Lab. Invest.* **17**, 381–387 (1965).
5. Moss, G. A., Bondar, R. J. L., and Buzzelli, D. M., Kinetic enzymatic method for determining serum creatinine. *Clin. Chem.* **21**, 1422–1426 (1975).
6. Larsen, L., Creatinine assay by a reaction-kinetic principle. *Clin. Chim. Acta* **41**, 209–217 (1972).
7. Soldin, S. J., and Hill, J. G., Micromethod for determination of creatinine in biological fluids by high-performance liquid chromatography. *Clin. Chem.* **24**, 747–750 (1978).
8. Vasiliades, J., Reaction of alkaline sodium picrate with creatinine: I. Kinetics and mechanism of formation of the non-creatinine picric acid complex. *Clin. Chem.* **22**, 1664–1671 (1976).
9. Knoll, E., and Stamm, D., The specific determination of serum creatinine. *J. Clin. Chem. Clin. Biochem.* **8**, 582–587 (1970).
10. Boyd, E., *Growth of the Surface Area of the Human Body*, University of Minnesota Press, Minneapolis, MN, 1935, p 132.

Drug Screening

Submitter: R. Thomas Chamberlain, *Laboratory Service, Veterans Administration Medical Center, and University of Tennessee Center for the Health Sciences, Memphis, TN 38104*

Evaluators: John P. Aitchison, *Department of Clinical Pathology, University of Oregon Health Sciences Center, Portland, OR 97201*
Francis T. Fox, *Laboratory Service, Veterans Administration Medical Center, New Orleans, LA 70146*
Max E. McIntosh and Joyce A. McIntyre, *Laboratory Service, Veterans Administration Medical Center, Sepulveda, CA 91343*

Reviewer: Peter I. Jatlow, *Department of Laboratory Medicine, Yale University, School of Medicine, New Haven, CT 06504*

Introduction

The number of drug-screening laboratories has rapidly increased in the last decade (1). Because of problems with "turnaround" times and sample handling, hospitals outside the larger medical complexes have found it necessary to establish their own reliable drug-screening programs. Laboratories with even limited capabilities for toxicology testing can contribute significantly to hospital services (2). The two methods for screening described here are different though complimentary as to the types of drugs detected, analytical speed, cost, and flexibility. However, not all drugs that may be commonly encountered are detected by these two techniques. Drug screening for medical emergencies in the small laboratory should also be capable of assessing problems involving acetaminophen, alcohol, and salicylate, which are addressed in separate chapters of this book.

In the first method, amberlite XAD-2 (Rohm and Haas Co., Philadelphia, PA 19105) nonionic resin is used for urinary drug extraction. This method has been successfully used in urine-screening programs (3–5); indeed, the use of thin-layer chromatography (TLC) for detection of large numbers of drugs in urine has been part of the toxicologists' armamentarium for several years (6,7).

Enzyme-multiplied immunoassay (EMIT™; Syva Co., Palo Alto, CA 94304), used in the second method, is a newly developed technique (8). This method is limited by the fact that only drugs or groups of drugs can be determined for which special reagents have been developed; however, its capabilities continue to grow. This system, originally designed for assaying drugs of abuse in urine, has since expanded into therapeutic drug monitoring (not covered in this chapter).

Drug screening is usually performed initially to rule out what drugs (or groups of drugs) are not in the urine and then to identify tentatively the presence of certain other drugs. We emphasize that this is tentative identification—an unequivocal finding requires at least one separate confirmatory technique. (The small-volume clinical laboratory may have limited usefulness for the more advanced confirmatory techniques.) However, for medical emergencies, a drug-screening coupled with a good patient history can quickly rule in or out certain drugs or drug groups. Turnaround time of results can be a very important factor in a drug crisis.

Other clinical uses of drug screens include employee tests, drug abuse counseling, and patient compliance with medication instructions. Therefore, drug screening in community hospitals can be a very important service even though the results may not be unequivocally confirmed. However, an answer of "none detected" can sometimes provide useful information, e.g., to the clinician treating a suspected overdose.

Many factors should be considered before determining which method would be most suited for a community hospital. Costs (1980) for reagents and equipment will vary from less than $500 for TLC to more than $7000 for enzyme immunoassay (EIA). However, turnaround time approaches 2 h for TLC but can be less than 15 min for EIA. Turnaround time is less of a factor when a laboratory is processing 15 to 20 specimens at a time. Costs for two TLC plates and their development and spraying—sufficient for analysis of 15 specimens and controls—will be less than $5. The same number of specimens analyzed for five drugs (or drug groups) by EIA will cost more than $35. Of course, other factors such as sensitivity and specificity should be considered before one selects a method. TLC may be more sensitive for barbiturates, but EIA is usually more sensitive for methadone, cocaine, and opiates. TLC analysis is not limited by availability of reagents, and the number of drugs detectable is limited only by the experience and competence of the technician. Elaborate comparisons as to costs, technician time, startup, turnaround time, equipment, sensitivity, and specificity have been made *(9)*. EIA in some cases can be used to confirm TLC results, but such data must be used with extreme caution because some of the EIA reagents give less-specific results than does the TLC technique.

THIN-LAYER CHROMATOGRAPHY

Principle

The porous polymer (XAD-2 amberlite nonionic) resin used for the extraction process adsorbs drugs from the buffered urine specimen placed in contact with it. Washing the resin column with the proper organic solvent elutes the drugs *(10, 11)*. The eluate is passed through a phase-separating filter to remove the aqueous (urine) fraction, which is collected in a beaker. After evaporation, the concentrated drugs in the dried residue are ready for TLC.

The dried residue is suspended in a small amount of organic solvent. An aliquot of this suspension is then applied to a glass plate on which a thin layer of silica gel has been coated and fixed. The plate is then placed in a closed chamber containing specific solvents. The solvent travels up the adsorbent layer by capillary action. As the solvent moves up the plate, various compounds (drugs, metabolites, etc.) in the sample are carried by the solvent at different rates. The relative solubilities of the components of the sample in the solvent mixture, the degree of activation of the thin-layer plate, the polarity of the solvent, and the polarities of the solutes of interest all affect the rates at which the different compounds move. After the solvent has traveled a measured distance, the plate is removed from the container, dried, and sprayed with various chemicals for color development. Drugs are identified by comparing the relative distance traveled by each and by comparing their colors with those of standards. The colors develop because certain functional groups (i.e., primary amines) in the drug molecule react with a given reagent. Though several drugs give similar color reactions, they may be differentiated and identified because their spots on the plate will be in different locations *(12, 13)*.

Materials and Methods

Reagents

1. *Sodium carbonate–bicarbonate buffer*. Add 45 g of $NaHCO_3$ to 400 mL of water and mix to dissolve. Add 65 g of Na_2CO_3 to 180 mL of water and dissolve. Allow both solutions to settle; then mix 135 mL of the second solution (saturated sodium carbonate) with 365 mL of the sodium bicarbonate solution. Check the pH (9.5 ± 0.1) with a pH meter and adjust with either solution.

2. *Elution solvent, 1,2-dichloroethane/ethyl acetate (2/3 by vol)*. Mix two parts of 1,2-dichloroethane (AR grade, cat. no. 4966; Mallinckrodt, Inc., St. Louis, MO 63147) with three parts of ethyl acetate (AR grade, Mallinckrodt cat. no. 4992) in a mixing cylinder. Prepare freshly every other day.

3. *HCl, 0.1 mol/L in methanol*. Dilute 1.0 mL of concd (12 mol/L) HCl with 119 mL of methanol.

4. *Resin columns (XAD-2)*, nonionic resin in plastic columns (prepacked columns are available from Brinkmann Instruments, Inc., Westbury, NY 11590). Keep the columns sealed at both ends and refrigerated, so that the resin remains moist before use.

5. *Thin-layer plates, glass plates (20 × 20 cm) coated with 250-mm thick silica gel G-60* (E. Merck, E. M. Laboratories, Inc.; cat. no. 5763). Depending on the humidity of the laboratory, it may be necessary to keep the plates dry in a drying chamber (e.g., Tupperware E-60 storage container).

6. *Development tank solutions*. Prepare freshly daily using clean, stoppered, 100-mL graduated cylinders. Allow tanks to equilibrate at least 15 min before use.

   a. Alkaloid I. Mix 90 mL of ethyl acetate, 10 mL of methanol, and 5 mL of concd. ammonium hydroxide.

   b. Alkaloid II. Mix 90 mL of ethyl acetate and 10 mL of methanol.

   c. Barbiturate tank. Mix 90 mL of chloroform and 10 mL of acetone.

7. *Spray reagents*.

   a. Potassium permanganate. Dissolve 0.1 g of potassium permanganate (laboratory grade, cat. no. 77645; Fisher Scientific Co., Fair Lawn, NJ 07410) in 100 mL of distilled water. Stable for at least two months.

   b. Mercuric sulfate. Suspend, while stirring, 5 g of mercuric oxide (AR grade, Mallinckrodt cat. no. 1426) in 100 mL of water. Slowly and cautiously add 20 mL of sulfuric acid, while the solution is mixing. Cool thoroughly, then dilute to 250 mL with distilled water. Stable for at least two months.

   c. Ninhydrin. Add enough ninhydrin (cat. no. 2495; Eastman Organic Chemicals, Rochester, NY 14650) to a small amount of acetone to produce a straw-colored solution. Prepare solution freshly daily. Clean the sprayer with dry acetone after use each day.

   d. Iodoplatinate. First prepare the following solutions:

   —Platinum chloride. Dissolve 1 g of chloroplatinic acid (Mallinckrodt cat. no. 6680) in 20 mL of distilled water.

—Potassium iodide. Dissolve 9 g of potassium iodide (USP, Mallinckrodt cat. no. 1112) in 90 mL of distilled water.

—HCl, 2 mol/L. Dilute concd. hydrochloric acid sixfold.

Mix 10 mL of the platinum chloride solution, 90 mL of the potassium iodide solution, and 100 mL of distilled water to make the iodoplatinate stock. Just before use, combine equal parts of iodoplatinate stock and 2 mol/L hydrochloric acid for working solution. Stable for at least two months.

8. *Urine controls.* Prepare a high-concentration control of 3 mg each of amphetamine, methamphetamine, codeine, methadone, morphine, phenobarbital, propoxyphene, and secobarbital per liter of urine. Prepare a low-concentration control of the same drugs at a concentration of 1 mg/L, except morphine should have a final concentration of 0.5 mg/L. To prepare stock standards of the above drugs plus other drugs to be identified, dissolve 20 mg of the drug in 20 mL of reagent grade methanol to give a concentration of 1 mg/mL. Stock standards are stable for a minimum of six months. These stock standards can also be applied (5–10 µL) directly onto the TLC plate for comparison purposes. Prepare high-concentration controls by adding 3 mL of each appropriate stock drug standard to 1 L of previously assayed "negative" urine. Low-concentration urine controls are prepared by adding 1 mL of each appropriate stock drug standard (0.5 mL for morphine) to 1 L of "negative" urine.

Apparatus

1. *Columns.* Prepacked plastic columns (Figure 1) containing nonionic XAD-2 adsorbent resin and attachable sample reservoirs (Brinkmann Instruments: prefilled columns, cat. no. 3500050–0; sample reservoirs, cat. no. 3500010–1; empty columns, cat. no. 3500700–8).

2. *Filter cartridges.* Plastic cartridges containing phase-separating filters (Brinkmann, cat. no. 3500030–5).

3. *Evaporation beakers.* Conical evaporation tubes (Brinkmann, cat. no. 3500420–3), 30 mL. (These tubes must be scrupulously cleaned before reuse.)

Note: If 30-mL beakers are used, the evaporator below can be replaced by a water bath, as described in *Procedure.*

4. *Racks for columns and evaporation tubes.* (Brinkmann: stainless steel column rack for 27 columns, cat. no. 3500440–8; tube racks for 27 tubes, cat. no. 3500450–8).

5. *Evaporator.* Concentrator heating block (Brinkmann, cat. no SC-27). Shallow heating baths can also be used, with a hairdryer blowing across the tops of the tubes or beakers placed in the bath.

6. *Blower.* Hairdryer (gun type), 1000 W.

7. *Capillary tubes.* Glass capillaries capable of holding at least 20 µL (Scientific Products, Div. of Am. Hosp. Supply Corp., McGaw Park, IL 60085; cat. no. B4195–2).

8. *Developing tanks.* Glass tanks (4 × 12 × 10 in.; 10 × 30 × 25.4 cm) with glass lids (Applied Science Labs., Inc., State College, PA 16801; cat. no. 17103).

9. *Cabinet with ultraviolet light source.* (Applied Science Labs., cat. no. 17005).

10. *Spray bottles.* TLC reagent glass spray bottles, 125 mL (Applied Science Labs., cat. no. 17017).

11. *Centrifuge tubes.* Polypropylene, 50 mL (Falcon 2070).

Specimen

A freshly collected random (untimed) urine specimen (30 mL) is most suitable. Preservatives should not be used. Refrigerate the specimen if it is not analyzed on the day of collection. Freezing is recommended if the specimens are going to be held for more than seven days before analysis. Specimen containers must be leakproof.

Procedure

Caution: Because of toxicity, flammability, and the irritating effects of some of the reagents, the evaporation, TLC development, and spraying of the plates should be done in an efficient, explosion-proof hood.

Urine preparation

1. Place 50-mL plastic centrifuge tubes in racks and number them consecutively, 1–1 through 1–15, 2–1 through 2–15, placing tubes marked H-C and L-C (for high- and low-concentration urine controls) between tubes 9 and 10 of each series.

2. Number all urines received for screening as above and pour 20–25 mL of each urine into the appropriate tube. Thaw the control urines and pour a 20-mL aliquot of high control and of low control into the tubes marked H-C and L-C, respectively.

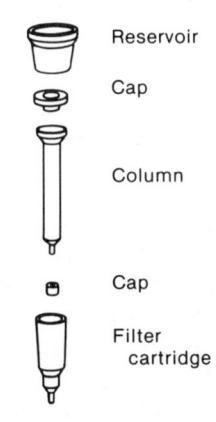

Fig. 1. Elution column

3. Add approximately 1 mL of sodium carbonate–bicarbonate buffer to all urines and controls to bring the pH to approximately 9. Use pH paper to check this (pHydrion paper, pH 1–11, Scientific Products).

4. Centrifuge all urines at 1500 rpm for 3–4 min to settle any sediment. Urines are now ready for resin extraction.

Column preparation and extraction

1. Remove the necessary number of packed columns from the refrigerator and attach a sample reservoir to each one.

2. Place the columns in a column rack and pour about 20 mL of distilled, de-ionized water onto each column. Discard the eluent.

3. When the water has passed through the columns, begin to pour the supernates of the buffered, centrifuged urine specimens through the columns consecutively, beginning with 1–1, processing up to 27 urines at one time. Allow 4 to 10 min for the urines to pass through the columns. Discard the eluate.

4. When the urines have drained through the columns, remove the cotton plug from the top of each column and discard.

5. Assemble phase-separating filters in holders, making sure rings are seated firmly. Wet each filter with a few drops of eluting solvent and place the filters on the stainless steel rack under the columns. Place the evaporator tube rack under the filter cartridges. Make sure the evaporator tubes are appropriately labeled to correspond to the urines being extracted. Add three 5-mL aliquots of eluting solvent to each column, allowing each aliquot to drain through (approximately 5 min) before the next is added. (At this stage, as well as when the urine is in contact with the resin, contact time plays an important part in the recovery rate of the drugs.)

6. After the third aliquot of eluting solvent has been added and allowed to drain, blow air through all the columns by using an unpacked plastic column attached to a dry air source. (The outlet end of the unpacked column is connected to the air source by tubing, and the reservoir end is placed over the top of the eluting column.) After all the solvent has been forced through the column and collected in the evaporator tubes, discard the columns or save them for repacking. Using a Pasteur pipette, gently bubble dry, clean air into any filter holders that are still full. (This will dislodge any aqueous solution covering the filter and allow the organic solvent to filter through into the concentrator beakers.) Between specimens rinse the pipette in a small beaker containing eluting solvent, to prevent any transfer of drugs.

7. After all urines have filtered completely, remove the filter holders and carefully examine all beakers to see if any need refiltering.

Note: Do not dry extracts containing any visible aqueous phase, because this interferes with the TLC process.

8. Sometime during the extraction process, turn on the concentrator and allow it to warm up 10–15 min. Place a beaker of water in the concentrator, put a thermometer into the beaker, and measure the water temperature at the end of the 15 min. Record this temperature (it should be 56–62 °C) and the oven temperature on a daily work sheet.

9. Place the rack of eluted, filtered extracts into the concentrator; close the top; turn on the water aspirator and air flow connected to the concentrator; and allow the samples to evaporate. When evaporation is about halfway complete (i.e., 20–30 min), add one to two drops of 0.1 mol/L HCl in methanol to each beaker, to prevent loss of volatile drugs. An alternative method for drying the extracts is to use a hotplate to heat flat pans of water in which the evaporator tube racks can be placed; hairdryers can be positioned to blow air over the tops of the tubes.

10. When all beakers are dry, remove the rack from the concentrator and allow it to cool. The dried residues are now ready to be applied onto a thin-layer plate.

Plate preparation

1. Use two 20 × 20 cm silica gel G-60 TLC plates for each 15 extracts of patients' urine and two extracted urine controls. Mark one plate "A" for alkaloid development and detection, and the other "B" for the barbiturates.

2. Draw a light pencil line across both plates, approximately 1.5 cm above the bottom edge (samples will be applied along this line). Make small pencil marks on either outside edge of the alkaloid plate at 10 cm and 15 cm above the origin. The barbiturate plate should be marked at 18 cm above the origin.

Note: A plastic TLC spotting guide template can be used (Applied Science Labs., cat. no. 17197).

Sample application

Add approximately 50 µL of a methanol to one beaker at a time. Mix well with a vortex-type mixer, then take up all remaining methanol with two short capillary pipettes and spot an equal amount on the "A" and "B" plates in the appropriate position. While using a hairdryer (or other source of warm, dry air) to evaporate the methanol, take care to spot aliquots evenly, keeping the spots small (less than 1 cm in diameter). Normally, 17 extracts (15 patients' samples and two controls) may be spotted on each plate. In addition, any desired standards may be spotted in the remaining two "channels" of the plate (10 µg of methamphetamine is routinely spotted on the alkaloid plate and 10 µg of phenytoin on the barbiturate plate). Dry each plate thoroughly with a hairdryer.

Plate development

1. Place the alkaloid plate in the development tank marked Alkaloid I and allow solvent to travel to the 10-cm mark (about 17–18 min). Remove the plate, dry until ammonia fumes are not apparent (minimum of 5 min with hairdryer), and place the plate in the Alkaloid II tank, allowing the solvent front to reach the 15-cm mark (about 30 min). Remove the plate, mark the solvent line across the width of the plate lightly with a pencil, and dry thoroughly with warm air (use a hairdryer) for at least 10 min.

2. Place the B plate in the barbiturate tank; remove when the solvent line has reached the 18-cm mark. Dry 2–3 min with the hairdryer.

Spraying sequence and identification of spots

1. Alkaloid-plate sequence:

   a. Place plate in drying oven (75–80 °C) for at least 10 min.

   Note: Evaluator F. T. F. suggests using 125 °C to allow for better removal of the ammonia from the silica.

   b. Remove and immediately spray forcefully the bottom half of the warm plate with a heavy application of ninhydrin. Place the plate under an ultraviolet light source (340–350 nm) for 4 to 5 min, marking quinine and other fluorescent spots of interest.

   c. Remove the plate and identify all primary amines (pink spots), particularly amphetamine.

   d. Place the plate in the oven for an additional 5 to 10 min. Remove, mark, and record location

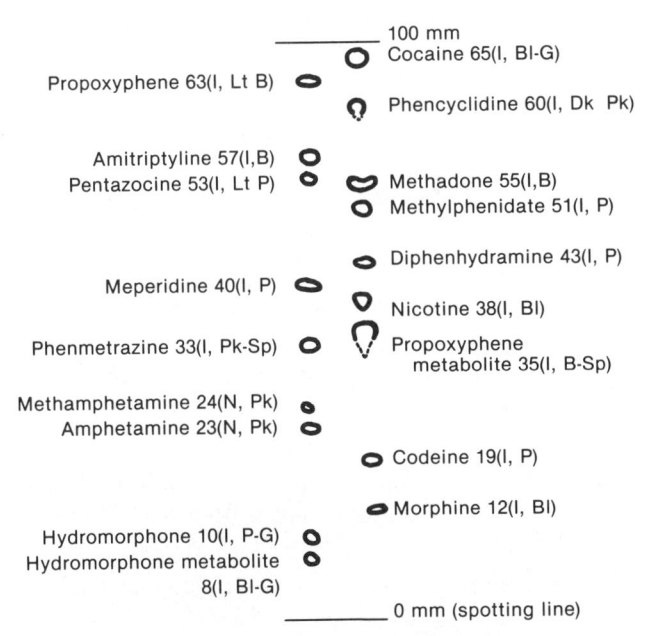

```
                              ——————— 100 mm
                              O  Cocaine 65(I, Bl-G)
Propoxyphene 63(I, Lt B)  O
                          Q  Phencyclidine 60(I, Dk Pk)
Amitriptyline 57(I,B)   O
Pentazocine 53(I, Lt P) O   Methadone 55(I,B)
                          O  Methylphenidate 51(I, P)
                          o  Diphenhydramine 43(I, P)
Meperidine 40(I, P)   O
                          Q  Nicotine 38(I, Bl)
Phenmetrazine 33(I, Pk-Sp) O  Propoxyphene
                                metabolite 35(I, B-Sp)
Methamphetamine 24(N, Pk)  o
Amphetamine 23(N, Pk)      O
                          O  Codeine 19(I, P)
                          o  Morphine 12(I, Bl)
Hydromorphone 10(I, P-G) O
Hydromorphone metabolite O
    8(I, Bl-G)
                              ——————— 0 mm (spotting line)
```

Fig. 2. Relative location of organic bases on a twice-developed (Alkaloid I and Alkaloid II) TLC plate

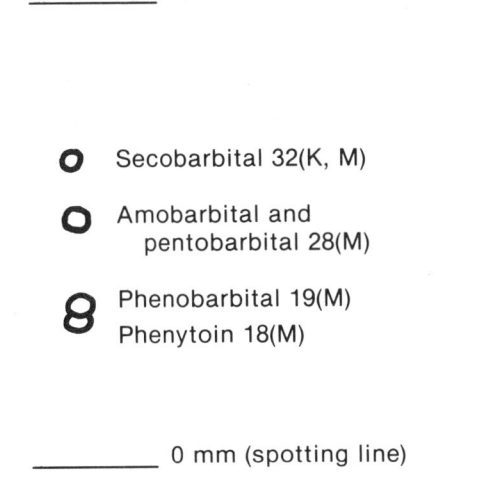

```
                   ——————— 100 mm

O  Secobarbital 32(K, M)

O  Amobarbital and
     pentobarbital 28(M)

8  Phenobarbital 19(M)
   Phenytoin 18(M)

                   ——————— 0 mm (spotting line)
```

Fig. 3. Relative location of barbiturates and phenytoin on a barbiturate TLC plate

of any secondary amines, especially methamphetamine (lavender spot).

   e. Spray the entire plate with iodoplatinate, marking and recording location of all tertiary amines. Spots in the control urine will be (in ascending order) morphine, codeine, methadone, and propoxyphene (see Figure 2).

2. Barbiturate-plate spraying sequence:

   a. Spray the entire plate with potassium permanganate until secobarbital appears as a yellow spot on a pink background. Mark lightly with a pencil any other well-formed white spots.

   b. Allow the plate to fade. Spray with mercuric sulfate until any barbiturates appear as white spots on a gray background. Mark and record the spots before the plate fades (see Figure 3).

3. Spots can be located and recorded as millimeters moved up the plate or as R_f (distance the spot travels divided by the distance the solvent travels). R_f values may be smaller than usual if high concentrations of a drug are present.

Discussion

This procedure for screening for drugs in urine is inexpensive and rapid for ruling out the presence of several drugs. Confirmatory procedures, good histories, or reference laboratory backup should be used before spots on the thin-layer plates are considered *unequivocal* evidence for drug identifications. The minimum detectable concentrations of certain drugs successfully identified by this technique are indicated in Table 1. These sensitivities should routinely be checked, however, because slight changes in columns, plates, sprays, or technique may alter the amount detectable.

Table 1. Routine Sensitivity Limits of Drug in Urine

	mg/L		mg/L
Amitriptyline	1	Methamphetamine	1
Amobarbital	1	Methylphenidate	2
Amphetamine	1	Morphine (free)	0.5
Cocaine	1	Pentazocine	2
Codeine	1	Pentobarbital	1
Diphenhydramine	2	Phencyclidine	2
Fluphenazine	2	Phenmetrazine	3
Hydromorphone	1	Phenobarbital	1
Meperidine	2	Propoxyphene	2
Methadone	2	Secobarbital	1

Sensitivity of some of these drugs may be enhanced by the appearance of metabolites, such as methadone and propoxyphene metabolites.

Because of the plethora of drugs available and the number that may appear in the urine, along with their possible metabolites, interferences should always be considered. A *conservative* approach to identification of spots on the TLC plate should be the rule! Therefore, be cautious and do not try to identify more drugs than you have had experience with.

For example, there are possible interferences among the drugs shown in Figures 1 and 2; i.e., pentazocine and methadone appear very close in R_f and color on the alkaloid plate, and phenytoin appears very close to phenobarbital on the barbiturate plate. As one gains experience with this method of drug identification, samples can be screened for more drugs. However, results by this method still should not be considered unequivocal and should be considered as screening only. This procedure is most useful in determining when a drug is below detectable limits.

Alternative extraction procedures involve use of a Jetube (cat. no. 2420; Harlan Associates, Pittsburgh, PA 15209) or Tox Elut (cat. no. 3120; Analytichem International, Lawndale, CA 90260), mixed-resin bead columns that have given results at least comparable with the XAD-2 method for those drugs checked. The sensitivity for the detection of methadone has been increased to 1 µg/mL of urine. Lantz and Eisenberg report that recoveries by the use of these tubes are better than with the XAD-2 extraction procedure *(14)*.

A major difficulty for the analyst is to determine which drugs can be detected by any given procedure. For instance, the procedure as presented here will not detect methaqualone metabolites, which is usually the only way to determine whether methaqualone has been ingested. Moreover, in reporting results the laboratory must be explicit as to what drugs are reported as "negative" or "none detected." These descriptive terms have meaning only when used in the context of the sensitivity limits of the method for a particular drug (e.g., Table 1). For example, where phenobarbital is less than 1 µg/mL, the report should indicate "none detected."

ENZYME IMMUNOASSAY
Principle

This immunoassay procedure is based on the ability of similar antigens to compete for specific sites on an antibody molecule *(15)*. The antibodies used in the EIA procedure have been developed so that they bind specific drugs or drug groups. The assay mixture contains a covalently bound drug–enzyme complex that can compete for the antibody site with the drug of interest, which might be found in the urine. If a particular drug is not present in the urine, this drug–enzyme complex will be bound by the large antibody molecule and physically hindered from reacting with the substrate. Conversely, if the drug is present, the drug–enzyme complex will be unbound and therefore capable of catalyzing a subsequent reaction *(16)*. This competitive-binding process and enzymic reaction is illustrated diagramatically, where D is the drug to be assayed, Ab is the antibody for drug D, and D-ENZ is the drug–enzyme complex.

$$D + Ab + D\text{-}ENZ \longrightarrow Ab\text{-}D + D\text{-}ENZ$$
$$\text{Substrate} \xrightarrow{D\text{-}ENZ} \text{product}$$

The substrate in this case is a bacterial suspension of *Mycococcus luteus,* which is lysed by the enzyme lysozyme. The reaction is monitored kinetically by measuring the decrease in turbidity of the cell suspension with a spectrophotometer set at 436 nm. The decrease in turbidity of the cell suspension is proportional to the amount of drug present.

Currently available reagents for detecting drugs in urine include antibodies and drug–enzyme complexes for amphetamine/methamphetamine, barbiturates, benzodiazepines (such as oxazepam), cocaine (such as benzoylecgonine), methadone, opiates, and propoxyphene. Phencyclidine is also available but involves different reagents.

Materials and Methods
Reagents

1. *Buffer.* Tris(hydroxymethyl)methylamine maleate (Tris maleate), 25 mmol/L, pH 6.0. Add buffer vial to 1 L of distilled water.

2. *Bacterial suspension.* Add 75 mg of the lyophilized bacteria *(M. luteus)* to 100 mL of Tris maleate buffer.

Note: Tris buffer concentrate and the freeze-dried bacterial substrate must be obtained from Syva Co. as cat. no. 3A039.

3. *Amphetamine assay reagents* (Syva cat. no. 3C019).

4. *Barbiturate assay reagents* (Syva cat. no. 3D019).

5. *Benzodiazepine metabolite assay reagents* (Syva cat. no. 3F019).
6. *Cocaine metabolite assay reagents* (Syva cat. no. 3H019).
7. *Methadone assay reagents* (Syva cat. no. 3E019).
8. *Opiate assay reagents* (Syva cat. no. 3B019).
9. *Propoxyphene assay reagents* (Syva cat. no. 3G019).

Note: Each of the above assay reagents contains Reagent A (antibody) and Reagent B (enzyme–drug) complex.

10. *Calibrator set.* Three urine standard calibrators containing the following drug concentrations (µg/mL):

Drug	Negative	Low	Medium
Amphetamine	0	1.0	5.0
Benzoylecgonine	0	1.0	5.0
Methadone	0	0.3	5.0
Morphine	0	0.3	5.0
Oxazepam	0	0.5	5.0
Propoxyphene	0	1.0	5.0
Secobarbital	0	1.0	5.0

11. *Hydrochloric acid, 0.1 mol/L.* Add 1 mL of concd. (12 mol/L) HCl to 120 mL of distilled water.
12. *Sodium hydroxide, 0.1 mol/L.* Add 4 g of NaOH to 100 mL of distilled water.

Apparatus

1. *Pipettor-dilutor.* This device should be capable of picking up a 50-µL sample and dispensing the sample with 250 µL of buffer (Syva cat. no. 5B129).
2. *Digital spectrophotometer with thermally regulated flow cell (37° C, 20 µL).* A timer–printer accessory should be integrally connected and capable of a 10-s delay time, measuring absorbance at 10 and 40 s after the reagents are mixed.

Note: Some EMIT chemistries have been successfully performed with the Gilford Stasar or 300 series, Beckman 24/25, Abbott ABA-100, LKB 8600, Perkin-Elmer KA-150, and the centrifugal analyzers (IL MCA-III). Manufacturers of the above instruments should state the capability of their particular instrument.

3. *Dispenser.* This should be capable of delivering 200 µL of bacterial suspension (Syva cat. no. 3A021).
4. *Disposable plastic cups, 2.0-mL capacity* (Syva cat. no. 3A031 or similar).
5. *Lucite cup rack.* This is useful for holding continuous-flow analyzer-type cups (Syva cat. no. 3A041), but any suitable rack will suffice.

Specimens

A freshly collected random (untimed) urine specimen is most suitable. Do not use preservatives. Refrigerate the specimen if it is not analyzed on the day of collection. Freezing is recommended if the specimen is to be kept more than seven days before analysis.

Procedure

The procedure presented is for the manual system used with the Gilford Stasar or Beckman 35 spectrophotometer:

Note: Automated systems adapted for the EMIT chemistries will vary according to the manufacturer.

1. Turn on the spectrophotometer and set at 3-s sample time, absorbance mode, 37 °C temperature, and 436 nm wavelength.
2. Set timer–printer to measure from a high absorbance to low at 10 and 40 s after the reaction begins. Select a setting for reading change in absorbance (ΔA).
3. Place enough sample cups into the rack, and put 200-µL aliquots of freshly mixed bacterial suspension into each cup.

Note: The bacterial suspension settles rapidly and must be mixed before dispensing.

4. Using the pipettor-dilutor, draw up 50 µL of the low-concentration calibrator, wipe the pipettor tip carefully, and deliver the automatically diluted calibrator sample directly into the center of a sample cup, wiping the tip again when delivery is complete.
5. Similarly, add 50 µL of Reagent A of the desired test reagent, along with 250 µL of buffer, to the sample cup containing calibrator and bacterial suspension.

Note: EIA reagents should be close to room temperature when used. Cold reagents may result in low absorbance readings.

6. Adjust the spectrophotometer to zero with distilled water by turning the zero control knob, and adjust the temperature meter if necessary by means of the null knob. After purging the flow cell to empty it (press the stop in back of the aspirator all the way back with a firm push), draw up and deliver a 50-µL aliquot of Reagent B along with 250 µL of buffer to the sample cup. Without delay, aspirate the contents of the cup into the flow cell by depressing the stop only halfway back.
7. The timer–printer will now read absorbance of the calibrator at 10 and 40 s after introduction of the mixture into the flow cell. The number printed at the end of this time is the low-calibrator value (in EMIT units) and is used for the lower cutoff value for determining a "none detected" specimen or a "positive" specimen.
8. Repeat the steps above for all specimens to be tested, making sure to empty the flow cell thoroughly before introducing the next sample. When the presence of another drug must be tested, flush

the flow cell with water and, if necessary, readjust the instrument to zero.

9. When all tests have been performed, clean the flow cell as follows before the end of the work day:

   a. Flush with water and adjust the zero knob so that the digital meter reads 2.000.

   b. Draw up some dilute HCl (0.1 mol/L) into the flow cell and allow it to remain for a few minutes.

   c. Flush the flow cell with water again.

   d. Draw up some dilute NaOH (0.1 mol/L) into the flow cell and allow it to remain for a few minutes.

   e. Flush the cell again with distilled water and then leave the cell full of water.

   Note: The flow cell should not be left dry or without liquid for any period of time.

Standardization and Quality Control

Run one or more calibrator urines (Syva) containing a known amount of drug (e.g., barbiturates, opiates) each time a specimen or batch is analyzed. This allows for the determination of a cutoff point—the change in absorbance value below which a sample will be reported as "none detected." Because some enzyme activity will be observed in all specimens, include a negative calibrator in each run. The values of the calibrators should remain fairly consistent in day-to-day use. Record all results.

Discussion

Using enzyme-"tagged" antigens in an immunoassay procedure is a rapid way to rule out the presence of certain drugs or drug groups in the urine. The cost of equipment for this technique is low compared with costs for radioimmunoassay procedures *(17)*. As with the TLC procedure, unequivocal answers are possible only when confirmatory techniques are used. The use of EIA with TLC increases the reliability of the results reported, if one is appropriately careful interpreting results.

Some of the reagents are not specific for a particular drug assayed. The opiate assay, as its name implies, will detect opiates other than morphine (codeine, hydromorphone) as well as the opiate antagonists nalorphine and naloxone (in high concentrations, 120 µg/mL) *(18)*. Therefore, a positive opiate result does not give conclusive evidence that morphine is present. This type of nonspecificity can be used advantageously when this assay is used as a screen; that is, negative results would rule out opiates as a group for further testing.

The barbiturate assay will detect many of the barbiturates in the urine and is not known to cross react with nonbarbiturate compounds; however, butalbital has been found not to react in some cases (Chamberlain and Stafford, unpublished results, 1975). The assays for the cocaine metabolites, methadone and propoxyphene, have been relatively specific and are very useful in detemining the presence or absence of these compounds. The benzodiazepine metabolite assay picks up only those benzodiazepines metabolized to or occurring as oxazepam in the urine (chlordiazepoxide, diazepam, and oxazepam). The amphetamine assay is the least specific of the assays, known to react with phentermine, phenmetrazine, ephedrine, and others. A confirmatory test such as a secondary extraction procedure or gas chromatography *(19, 20)* should be made when the amphetamine assay is positive, before even a tentative result is reported.

On rare occasions, a strongly positive reaction may occur with all of the drug essays for a particular specimen (i.e., ΔA exceeds 200), because of the presence of lysozyme in the urine. If this occurs, proceed as follows:

1. Fill the sample cup with buffer only.
2. Fill a cup with the bacterial suspension.
3. Add the diluted urine to the cup containing the bacteria as done in the assay procedure steps 4 through 8.
4. Draw buffer from the buffer-only cup twice and add this 500 µL of buffer to a third specimen cup instead of using any of Reagent A or Reagent B as called for in the procedure.
5. Introduce this into the flow cell and check to see if a high reading is obtained. When lysozyme is *not* present, a reading of only 5–10 EMIT units will be seen. If ΔA is greater than 200, *none* of the results should be considered valid.

As indicated in the TLC procedure, when results are reported, the drugs reported as "positive" or as "none detected" should be explicit as to sensitivity of the procedure. If the results are not confirmed, this also should be reported. Results from analytical toxicology laboratories must be reported conservatively, to promote credibility and to prevent legal problems. So that the analyst can gain confidence and experience, initial results on patients' samples from the laboratory should be confirmed by a more specific alternative procedure, such as by using a reference laboratory to compare its results with the analyst's results.

Again, we recommend that the procedures described herein be used only for screening purposes. Possible exceptions to this rule are the use of the EIA procedure for cocaine, methadone, and propoxyphene to confirm TLC-"positive" results for these drugs.

The Submitter acknowledges the invaluable assistance of Mrs. Linda White in typing, retyping, and proofreading this manuscript.

General References

Bastos, M. L., Kananen, G. E., Monforte, J. R., and Sunshine, I., TLC of basic organic drugs. In *Methodology for*

Analytical Toxicology, I. Sunshine, Ed., CRC Press, Inc., Cleveland, OH, 1975, pp 434–442.

Clarke, E. C. G., *Isolation and Identification of Drugs*. Pharmaceutical Press, London, 1969, especially vol. **1**, pp 43–58, 637–655, and vol. **2**, pp 1122–1127.

Forney, R., Jr., Modified Davidow TLC drug screening procedure. In *Methodology in Analytical Toxicology (op. cit.)*, pp 443–466.

Mulé, S. J., Sunshine, I., Baude, M., and Willette, R. E., Eds., *Immunoassays for Drugs Subject to Abuse*. CRC Press, Cleveland, OH, 1974.

Sunshine, I., Ed., *Handbook of Analytical Toxicology*. CRC Press, Inc., Cleveland, OH, 1969, pp 2–123, 772–795.

References

1. Finkle, B. S., Forensic toxicology of drug abuse: A status report. *Anal. Chem.* **44**, 18A–26A (1972).
2. Jatlow, P. I., Analytical toxicology in the clinical laboratory—an overview. *Lab. Med.* **6**, 10–14 (1975)
3. Fujimoto, J. M., and Wang, R. I. H., A method of identifying narcotic analgesics in human urine after therapeutic doses. *Toxicol. Appl. Pharmacol.* **16**, 186–193 (1970).
4. Mulé, S. J., Bastos, M. L., Jukofsky, D., and Saffer, E., Routine identification of drugs of abuse in human urine II. Development and applications of the XAD-2 resin column method. *J. Chromatogr.* **63**, 89–301 (1971).
5. Weisman, N., Lowe, M. L., Beattie, J. M., and Demetriou, J. A., Screening method for detection of drugs of abuse in human urine. *Clin. Chem.* **17**, 875–881 (1971).
6. Sunshine, I., Use of thin layer chromatography in the diagnosis of poisoning. *Am. J. Clin. Pathol.* **40**, 576–582 (1963).
7. Davidow, B., Petri, N., and Quame, B., A thin layer chromatographic screening procedure for detecting drugs of abuse. *Am. J. Clin. Pathol.* **50**, 714–719 (1968).
8. Rubenstein, K. E., Schneider, R. S., and Ullman, E. F., "Homogeneous" enzyme immunoassay. A new immunochemical technique. *Biochem. Biophys. Res. Comm.* **47**, 846–851 (1972).
9. Catlin, D. H., A guide to urine testing for drugs of abuse. Monograph Series B,2. Executive Office of the President. Special Action Office for Drugs of Abuse Program. Washington, DC, 1973.
10. Miller, W. L., Kullberg, M. P., Banning, M. E., Brown, L. D., and Doctor, B. P., Studies on the quantitative extraction of morphine from urine using nonionic XAD-2 resin. *Biochem. Med.* **7**, 145–158 (1973).
11. Kullberg, M. P., Miller, W. L., McGowan, F. J., and Doctor, B. P., Studies on the single extraction of amphetamine and phenobarbital from urine using XAD-2 resin. *Biochem. Med.* **7**, 323–335 (1973).
12. Mulé, S. J., Routine identification of drugs of abuse in human urine 1. Application of fluorometry, thin layer and gas liquid chromatography. *J. Chromatogr.* **55**, 255-266 (1971).
13. Kaistha, K. K., and Jaffe, J. H., TLC techniques for identification of narcotics, barbiturates, and CNS stimulants in a drug abuse screening program. *J. Chromatogr. Sci.* **61**, 679–689 (1972).
14. Lantz, R. K., and Eisenberg, R. B., Evaluation of the JET technique for extracting drugs from urine. *Clin. Chem.* **24**, 821–824 (1978).
15. Yalow, R. S., and Berson, S. A., Assay of plasma insulin in human subjects by immunological methods. *Nature* **184**, 1648–1651 (1959).
16. Bastiani, R. J., Phillips, R. C., Schneider, R. S., and Ullman, E. F., Homogeneous immunochemical drug assays. *Am. J. Med. Technol.* **39**, 211–216 (1963).
17. Brattin, W. V., and Sunshine, I., Immunological assays for drugs in biological samples. *Am. J. Med. Technol.* **39**, 223–230 (1973).
18. Schneider, R. S., Lundquist, P., Tong-in-Wong, E., Rubinstein, K. E., and Ullman, E. F., Homogeneous enzyme immunoassays for opiates in urine. *Clin. Chem.* **19**, 821–825 (1973).
19. McIntyre, J. A., England, L., McIntosh, M. E., and Ling, W., An improved system for confirmatory testing of drugs of abuse in urine. 31st Annual Meeting, Am. Acad. Forensic Sci., Atlanta, GA, Feb. 12–17, 1979.
20. Chamberlain, R. T., and McNatt, B., Confirmation and identification of alkaloidal drugs in biological tissues by an easily modifiable solvent extraction and gas chromatographic procedure. *Clin. Chem.* **20**, 901 (1974). Abstract.

Electrophoresis of Serum Proteins on Cellulose Acetate

Submitters: Michael A. Pesce and Genevieve C. Covolo, *Special Chemistry Service, Columbia Presbyterian Medical Center, New York, NY 10032*

Evaluators: Fram R. Dalal, *Clinical Laboratories, Health Sciences Center, Temple University School of Medicine, Philadelphia, PA 19140*
Robert B. Foy, *Department of Laboratories, Edward W. Sparrow Hospital Association, Lansing, MI 48909*
Carl R. Jolliff, *Clinical Laboratory Section, Lincoln Clinic, P.C., Lincoln, NB 68501*
F. Y. Leung, *Department of Clinical Biochemistry, University Hospital, London, Ontario, Canada N6A 5A5*

Reviewers: Basil T. Doumas, *Department of Pathology, The Medical College of Wisconsin, Milwaukee, WI 53226*
G. V. Freile-Gagliardo, *Laboratorio Biochuimico Clinico, Guayaquil, Ecuador*

Introduction

Electrophoresis is defined as migration of a charged particle in an electric field. The word electrophoresis is derived from the Greek words *elektron*, meaning amber, and *phore*, meaning bearer. Electrophoresis can be performed in the absence (moving-boundary method) or presence of a stabilizing medium. The moving-boundary method, introduced in 1892 by Picton and Linder *(1)* and successfully used by Tiselius *(2)* in 1937 to separate serum proteins, was cumbersome and time consuming and was eventually replaced by zone electrophoresis. Zone electrophoresis is the migration of charged particles supported on an inert homogeneous medium. Filter paper *(3)* was initially used as a support medium but was eventually replaced by cellulose acetate *(4)*.

Proteins in serum are normally separated by electrophoresis on cellulose acetate membranes into five fractions: albumin, alpha$_1$-, alpha$_2$-, beta-, and gamma-globulins. Albumin migrates the fastest followed by alpha$_1$-, alpha$_2$-, beta-, and gamma-globulins. In certain pathological conditions, the gamma-globulin component can split into gamma$_1$ and gamma$_2$ fractions.

Protein electrophoresis is useful in detecting: *(a)* clinical disorders involving abnormal production of immunoglobulins, *(b)* disturbances in the metabolism of albumin, *(c)* nephrosis, *(d)* alpha$_1$-antitrypsin deficiency, *(e)* malnutrition, and *(f)* liver damage. A comprehensive classification of disorders involving serum proteins has been reported *(5)*.

Principle

Protein molecules at pHs other than their isoelectric points are charged particles and, when placed in an electric field, will migrate toward one of the electrodes, depending on the charge of the particle, the strength of the electric field, the pH of the buffer, and the medium used to support the particle.

In protein electrophoresis, serum is placed on a cellulose acetate support medium, buffered at pH 8.6. Because of the amphoteric nature of the proteins, at an alkaline pH they will be negatively charged particles. A direct current potential is applied for a specified length of time and the negatively charged proteins will migrate toward the positively charged electrode (anode). Because the proteins migrate at different rates, they are separated into the five fractions mentioned above, the isoelectric point of albumin being about 4.8 and of gamma-globulins about 7.4. At alkaline pH, albumin will have a greater number of negative charges than the gamma-globulins and therefore will migrate faster towards the anode. Isoelectric points of the other serum proteins lie between albumin and gamma-globulin, so these proteins will migrate at intermediate rates. As can be seen by noting the application point, the slowest gamma-globulins drift by endosmosis towards the cathode.

After proteins are separated by electrophoresis, they are stained with a Ponceau-S dye/trichloroacetic acid solution and quantitated by densitometry. The protein fractions are reported in grams per volume and as a percentage of the total serum protein.

Materials and Methods

Reagents

1. *Barbital buffer B-2, 90 mmol/L, ionic strength 0.075* (no. 320024; Beckman Instruments, Inc., Fullerton, CA 92634). Each tube contains 2.76 g of diethylbarbituric acid and 15.40 g of sodium diethylbarbiturate. Dissolve the contents of one tube in 1000 mL of distilled water, heating, if necessary, to dissolve the contents. Cool to room temperature before using. Store at 4 °C to retard bacterial growth. This solution is stable for two weeks at 4 °C.

2. *Fixative-dye solution* (no. 324340, Beckman). Dilute the contents of one tube of fixative-dye solution to 250 mL with distilled water. This solution contains, per liter, 2 g of Ponceau-S dye, 30 g of trichloroacetic acid, and 30 g of sulfosalicylic acid, and is stable at room temperature for one week.

3. *Glacial acetic acid, reagent grade ACS* (Fisher Scientific, Springfield, NJ 07081).

4. *Destaining solution, 5% acetic acid (50 mL/L)*. Dilute 150 mL of glacial acetic acid to 3 L with distilled water.

5. *Dehydration solution, 95% ethanol (950 mL/L)* (IMC Chemical Group, Inc., Newark, NJ 07105).

Note: Evaluator F. Y. L. suggests that absolute methanol (ACS grade) can replace 95% ethanol.

6. *Clearing solution.* Add 50 mL of glacial acetic acid to 150 mL of 95% ethanol. Prepare a fresh solution daily.

Note: Evaluator F. Y. L. prepared a clearing solution of 15% glacial acetic acid/methanol (i.e., 15/85 by vol).

7. *Normal serum control* (Versatol no. 30101; General Diagnostics, Div. of Warner Lambert, Morris Plains, NJ 07950). Reconstitute one vial of control serum with 5 mL of distilled water. This control is stable for five days at 4 °C.

Apparatus

Note: The Submitters use the Beckman apparatus.

1. *Microzone electrophoresis cell, model R-101.* The fluid-leveling siphon of earlier models may be removed.
2. *Duostat power supply* (no. 320800).
3. *Single sample applicator* (no. 324399).
4. *Nine trays and covers* (no. 327426).
5. *Two glass drying plates* (no. 324323).
6. *Squeegee* (no. 324352).
7. *Cellulose acetate blotters* (no. 324326).
8. *Plastic envelopes* (no. 326189).
9. *Parafilm* (Dixie-Marathon, Div. of American Can Co., Greenwich, CT 06830).
10. *Cellulose acetate membrane* (no. 655420). These membranes are supplied in dry form and measure 145 mm long × 57 mm wide. Eight samples can be measured with each membrane. Store membranes in a cool, dry area.
11. *Air-convected oven* (Model 1304; Hotpack Corp., Philadelphia, PA 19406).
12. *Densitometer* (Model 730; Corning-Clifford, Medfield, MA 02052).

Collection and Handling of Serum

Collect blood either by venipuncture or by finger or heel stick and allow it to clot long enough to separate serum. Do not use anticoagulants, because plasma fibrinogen will contaminate the sample, causing an extra peak in the electropherogram between the beta- and gamma-globulin regions. After the blood has clotted, centrifuge it, and remove the serum. Serum can be stored at 4 °C in a stoppered container for three days without producing any change in the protein patterns.

Procedure

Note: If possible, read the Beckman instruction manual to become familiar with the equipment and the techniques used. The procedure presented here is a modification of the Beckman procedure.

Preparation of the cell

1. Place the microzone cell with the electrode terminal pins facing the operator. Remove the upper lid, the cell cover, and the bridge assembly.
2. Pour 132 mL of barbital buffer into each side of the cell, or pour 265 mL of buffer into the whole cell and tilt it, to level the buffer on each side. Wipe the buffer off the top of the cell partition. Check before each use that the buffer level is between the two Fluid Level lines. Place fresh buffer in the cell after every four electrophoresis determinations.
3. Replace the cell cover and upper lid.

Arrangement of the trays

1. Pour 100 mL of barbital buffer into a single pre-buffering tray. If one membrane is to be used, prepare four trays. Prepare a second identical set of four trays if two membranes are to be used. Fill the four trays for one set as follows.
2. Pour 125 mL of fixative-dye solution into the fixative-dye tray.
3. Pour 100 mL of the destaining solution into one tray.
4. Pour 100 mL of the dehydration solution into one tray.
5. Place a clean glass plate into the clearing tray, and pour in 100 mL of the clearing solution.
6. Keep all trays covered except while using them. The trays containing the barbital buffer, the destaining, the dehydration, and the clearing solutions must be emptied and cleaned daily. The fixative-dye tray must be filled with fresh solution after being used

to stain the protein fractions on 10 membranes, or after five days of use.

Note: Evaluator C. R. J. changes the fixative-dye solution after staining for five membranes.

Equilibration of the cellulose acetate membrane

1. With tweezers (to avoid contamination from fingers) remove the white cellulose acetate membrane from its box, grasping the membrane by one end. Carefully place it on top of the buffer in the pre-buffering tray, letting the far end of the membrane drag across one end of the tray and fall on the surface of the buffer before letting the end held by the tweezers drop, to avoid trapping air bubbles. The membrane must be floated on top of the buffer for a few seconds until completely moistened from below. A membrane immediately immersed in the buffer should not be used because air bubbles might be trapped inside the porous structure of the membrane and cause irregular separations.

2. After the membrane has been saturated with buffer, completely submerge it in the pre-buffering solution by agitating the tray. Allow it to remain submerged for 2 min.

3. Using tweezers, remove the membrane from the buffer tray. Place the membrane, with reference hole to the lower left, between two blotters and gently blot by passing the edge of your palm twice over the top blotter. Only excess buffer should be removed from the membrane; too much blotting will dry the membrane, resulting in irregular protein patterns. To identify the membranes (as 1, 2, etc.), number them by penciling dots near the reference hole.

Installing the membrane in the cell

1. Lift the hinged membrane guides wide open on the bridge assembly so that they lift away from the operator.

2. With tweezers, remove the blotted membrane and place it on the bridge. Line up the holes in the membrane with the teeth on the bridge, keeping the reference hole at the lower left. Carefully press down the membrane with tweezers and close the hinged guides against the ends of the membrane to secure it in place. Do not touch the membrane with your hands.

3. Remove the lid and cell cover from the cell and place the bridge assembly containing the membrane into the cell, with the ends of the membrane hanging free in the buffer. The reference hole on the membrane should be on the lower left side; this corresponds to sample number 1 position.

4. Replace the cell cover and upper lid, and allow the membrane to equilibrate for 2 min before applying the samples. During this time a small amount of buffer from the chamber will diffuse up into the membrane. The 2-min interval is critical to the proper uptake of the samples.

5. While the membrane is equilbrating, place a small drop of each serum sample on a piece of Parafilm. Each sample drop should be covered with a small plastic cover (labeled with the sample number) to minimize evaporation.

Note: To cover the serum, the Submitters use the disposable caps for Technicon AutoAnalyzer sample cups. These disposable caps can be obtained from Scientific Products, Edison, NJ 08817 (no. B2716-1).

Dispensing the sample

Note: Inspect the applicator and its tip before using. The dual ribbons across the tip must be flat and parallel. Any distortion in the ribbons can result in irregular-shaped protein patterns. A dirty applicator tip can result in the appearance of an irregular-shaped band at the point of application.

1. Press the white button on the applicator to extend the applicator tip. Prewet the tip with distilled water and remove excess water by touching the tip to absorbent tissue. Pick up a small quantity of sample by moving the tip slowly across the top surface of the sample drop without breaking the surface.

2. Retract the tip by pressing the red button. Remove the upper lid from the cell. Line up the applicator so that the samples are applied in the center of the membrane. Allow 5 to 7 s for complete transfer of sample to the membrane. Depress the red button to retract the tip and carefully lift the applicator from the cell cover. The applied specimen should look like a distinct yellowish bar with a slight dumbbell shape at the ends. If it does not, the sample has not been properly applied.

3. After the sample has been applied to the membrane, extend the applicator tip and clean it with a strong downward stream of water. Dry by carefully touching the tip to absorbent tissue. Repeat steps 1 and 2 for each sample and control serum that is placed on the membrane. After the last sample has been applied to the membrane, replace the lid. The applicator tip is fragile and should not be touched with the fingers.

Electrophoresis

1. If two membranes are to be electrophoresed together (two cells), complete one preparation, from pre-buffering to application, before starting the other. Then set up the second membrane quickly so as to minimize sample diffusion in the first membrane.

Note: The sample will diffuse because of the porous nature of the membrane.

Note: Beckman now offers a Multizone Electrophoresis Cell (no. 655275), which accommodates three cellulose acetate membranes. Apply samples by use of a multi-sample applicator kit (no. 655540).

2. Place the cell in front of the power supply so that the electrode terminal pins of the cell face the operator. Connect the cell(s) to the power supply by means of the interconnecting cables (which simply connect with one Duostat plug at one end and with one cell at the other end).

Note: To make the right-hand electrode of the cell the anode, use the interconnecting cable plugged into the red power-supply plug. Alternatively, to make the left-hand electrode the anode, plug the other interconnecting cable into the black power supply plug.

When only one membrane is being electrophoresed at a time, any unused connection with no interconnecting cable must be grounded by being plugged into the terminals at the back of the power supply.

3. Set the power source. Set the output selection switch to constant voltage, 0 to 300 V. Set the meter range switch to voltmeter, 0 to 500 V. Turn the output adjust control to the right until the pilot light comes on behind the control switch. As the power source warms up, adjust the output control until the meter reads 260 V. Start timing the run. The optimum migration distance, 16 to 18 mm, requires an electrophoresis time of 22 min.

4. Record the milliamperage. If the proper buffer has been used, the initial current should be between 2.9 and 3.5 mA. "Creeping" of crystallized buffer at the corners of the center partition may cause excessively high currents and may heat up the buffer solution. Use of silicone grease (Dow Corning Corp., Midland, MI 48640) at the corners next to this partition and over it at its ends will prevent this salt formation. At the end of the run the current should be between 4.1 and 5.0 mA, if all has progressed normally.

5. At the end of the timed run turn off the power supply, remove the Duostat cable plug from the cell, and remove the upper lid and cover from the cell.

6. During the 22-min electrophoresis period, determine the total protein values of the samples (see chapter on *Total Protein*, this volume, for measurement of total protein by refractometry and by the biuret procedure).

Note: The Submitters use a refractometer from American Optical Corp., Buffalo, NY 14211. The protein/nitrogen ratio used with this instrument is 6.54. The total protein values are adjusted to a protein/nitrogen ratio of 6.10 by use of a table, which must be calculated by each laboratory. In 1969 when our laboratory used the refractometer to measure total protein, it was thought that a protein/nitrogen ratio of 6.10 was accurate. Total protein values so obtained and values obtained by the biuret procedure correlated excellently.

Post-electrophoresis processing

1. To dye the protein fractions, use tweezers to place the membrane into the tray containing the fixative-dye solution. Allow the membrane to remain submerged in the staining solution for 7 to 10 min.

2. Using tweezers, remove the membrane from the dye and place it into the destaining solution. Agitate for 2 min. Empty the tray, leaving the membrane in the tray, and add fresh destaining solution. Repeat the above procedure until the destaining solution is colorless and the membrane is clear of color except at the protein fractions.

3. Using tweezers, remove the membrane from the destaining solution and drain off excess liquid against the sides of the tray. Place the membrane into the tray containing 95% ethanol. Gently agitate the tray for 1 min. The alcohol in the tray should be changed daily.

4. Using tweezers, remove the membrane from the dehydration tray and drain off excess liquid against the sides of the tray. Place the membrane into the clearing solution over the glass plate. Agitate gently for 1 min. With the membrane resting on the glass plate, grasp the membrane and glass plate at one end and remove it from the clearing solution (use rubber gloves or tweezers).

Tip the plate to allow the excess liquid to drain off. Hold the membrane and plate at a slight angle with one end resting on a pad of absorbent tissue. Gently draw the squeegee evenly over the full length of the membrane and down into the tissue, causing the membrane to adhere uniformly to the glass, with no pools of liquid remaining on it. It is important to remove excess clearing solution from the membrane; otherwise, the protein patterns on the membrane will be distorted. Work rapidly because the membrane begins to soften at this point.

5. Use an air-convection oven (75–80 °C, for 5 min) to dry the membranes, to prevent their softening and shrinking. Remove the glass plate with the dry membrane from the oven and allow it to cool to room temperature.

6. If the undistorted membrane adheres evenly to the glass plate, scan it with a densitometer while still on the glass plate, membrane-side down (nearest to the light beam). The membrane can also be scanned in the plastic envelope in which it is to be stored. To remove the membrane from the glass plate, place a length of pressure-sensitive tape half on and half off one end of the membrane. As the tape is lifted, the membrane comes off the glass and can be slipped into the plastic envelope while the other end still adheres to the glass plate. Trim the ends to fit the envelope after peeling the remaining end off the plate.

Note: Evaluator C. R. J. does not recommend scanning the membranes while they are on the glass plate, but prefers placing the membranes in plastic envelopes and scanning them.

Scanning the membranes

The Submitters scan the membranes with the Corning-Clifford Scanner densitometer. Scan according to the manufacturer's directions, first setting the dial to total protein. The Versatol control should have been applied in position 1 on each membrane; scan it first. Values are read as grams of protein per deciliter and simultaneously as percent of total protein. The densitometer is calibrated by scanning a test pattern supplied by the manufacturer. The linearity of the densitometer is checked from the peak heights of the test pattern. Recently, Beckman introduced a normal (no. 667660) and an abnormal (no. 667490) liquid serum control, which can be used in place of the Versatol control.

Note: In the Submitters' laboratory the scan is traced on a sheet of paper printed on one side with the normal values and on the other side with a model of an electrophoresis scan of a sample from a normal adult.

Discussion

The destaining solution used is a 5% aqueous acetic acid solution (50 mL of acetic acid per liter of water). Beckman recommends that the destaining solution of 5% acetic acid be prepared in specially denatured alcohol. We obtained similar electrophoresis patterns with either solvent.

The clearing solution used in our procedure is an acetic acid/95% ethanol solution. A 30% solution of cyclohexanone in denatured alcohol (300 mL/L of alcohol) has been suggested by Beckman as a substitute for the acetic acid/alcohol clearing solution. We found this did not improve electrophoresis patterns; therefore, because of its harmful vapor and undesirable order, cyclohexanone need not be used in the clearing solution.

The cellulose acetate membranes used are relatively new, introduced by Beckman in 1978. These membranes provide a clear background, in contrast with the older Beckman membranes, which had a cloudy background. It is essential to dry these membranes for 5 min at 75 °C in an oven with forced air. If dried at 90 °C (or at 75 °C in an oven without air currents), the edges of the membranes will soften and shrink.

Note: Evaluator F. R. D. dries the membranes in an oven (non-air-convection) at 100 °C for 5 min.

If the total protein concentration in the serum exceeds 15 g/dL, the protein patterns may be distorted. The serum should be diluted with an equal volume of saline solution (9 g/L NaCl) before electrophoresis.

Note: Evaluator C. R. J. suggests that serum should be diluted if total protein exceeds 9 g/dL.

Reference Intervals

The reference values for protein electrophoresis are shown in Table 1. These results were obtained over eight years from adult patients in Presbyterian Hospital, with Beckman's original cellulose acetate membranes. Because the protein electrophoresis patterns obtained with the newer and the original membranes showed no significant differences, the values listed in Table 1 are also applicable to results

Table 1. Reference Intervals for Electrophoresis of Serum Proteins from Healthy Adults

	g/L	(g/dL)
Total protein	66–82	(6.6–8.2)
Albumin	35–48	(3.5–4.8)
Alpha$_1$-globulin	2–4	(0.2–0.4)
Alpha$_2$-globulin	5–9	(0.5–0.9)
Beta-globulin	6–12	(0.6–1.2)
Gamma-globulin	7–20	(0.7–2.0)

with the newer membranes. A comprehensive listing of the reference (normal) values for newborns and children has been reported (6). The reference values for total protein are dependent on posture and are higher in samples taken from patients who are not lying down.

Note: The effect of posture on the reference intervals for total protein and the protein fractions in serum has been determined by Evaluator B. T. D. as follows:

	g/L	
	Recumbent	Ambulatory
Albumin	> 34	> 38
Alpha$_1$-globulin	1–3	1–3
Alpha$_2$-globulin	4–8	4–8
Beta-globulin	5–10	6–12
Gamma-globulin	6–15	5–16
Total protein	58–77	65–83

Precision

The precision of the method was determined with use of a normal serum and a serum with a large monoclonal component in the gamma-globulin region of the electropherogram. For each serum, 24 replicate determinations were performed. To determine the day-to-day precision of the method, we used the Versatol normal control serum for 20 days. The results are shown in Table 2.

Clinical Significance

Figure 1 shows where the albumin, alpha$_1$-, alpha$_2$-, beta-, and gamma-globulin fractions are found in the electropherogram; lists the proteins most likely to be found in these fractions; and shows

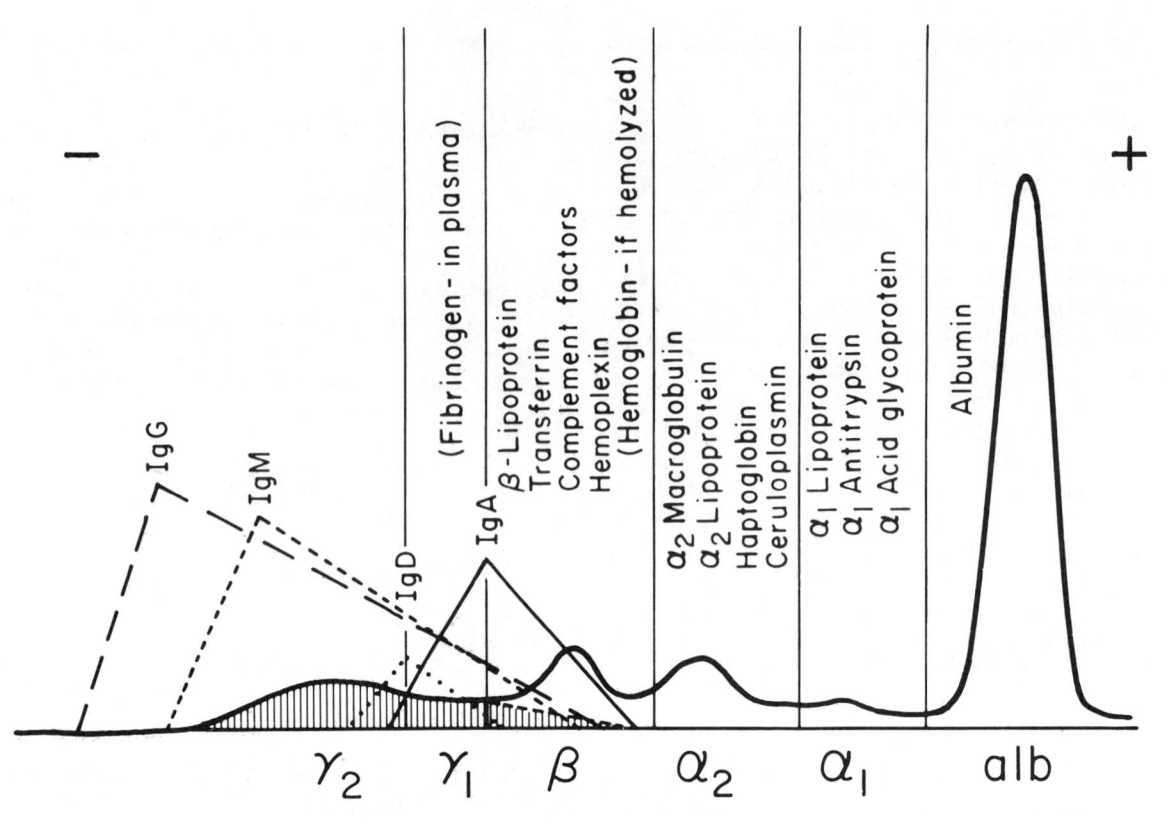

Fig. 1. Normal mobility distribution of proteins in serum

Table 2. Precision of the Method

	Albumin	Alpha₁-globulin	Alpha₂-globulin	Beta-globulin	Gamma-globulin
Within-run					
Normal serum					
Mean, g/L	42.4	2.8	5.0	6.1	12.7
SD, g/L	0.41	0.16	0.15	0.60	0.28
CV, %	0.97	5.72	3.03	9.79	2.22
Serum with large IgG peak					
Mean, g/L	34.7	2.2	5.4	4.3	80.5
SD, g/L	1.66	0.20	0.33	0.31	2.11
CV, %	4.79	9.00	6.06	7.29	2.62
Day-to-day					
Versatrol normal control					
Mean, g/L	38.3	2.4	6.7	6.7	11.3
SD, g/L	1.05	0.46	0.42	0.34	0.96
CV, %	2.74	19.17	6.29	5.12	8.52

containing the immunoglobulin fraction IgG. In Figure 3 the IgG component is minimal and located in the slow gamma₂-globulin region. All the protein fractions, including the gamma-globulin fraction, show normal values; only by careful inspection of the electropherogram can the presence of an abnormal immunoglobulin be detected. In Figure 4 the IgG component is large and located in the mid gamma₂-globulin region. The IgG peak is larger than the albumin peak, in part because of a decreased albumin with a slight increase in alpha₂-globulin. In Figure 5 the IgG component is medium-sized and located in the beta-globulin region; all other protein fractions are normal. These patterns indicate the range in the electropherogram where the IgG peak

the actual mobility distribution of the monoclonal immunoglobulins IgG, IgM, IgA, and IgD observed in the Submitters' laboratory from August 1974 to November 1978. The immunoglobulins were determined by immunoelectrophoresis. Figure 2 shows the normal electrophoretic pattern for an adult. Figures 3 through 11 show electropherograms with monoclonal gammopathies.

Figures 3 through 5 show electrophoretic patterns

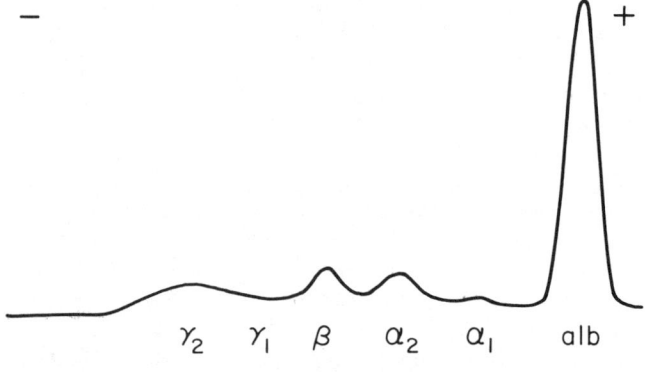

Fig. 2. Normal electrophoresis pattern for an adult

is most likely to occur and also shows the various sizes and shapes that can be obtained for IgG.

Figures 6 through 8 show electropherograms for IgM. In Figure 6 the IgM component is small and located in the slow gamma$_2$-globulin region. The electrophoretic protein values are normal and the immunoglobulin is detectable only by careful examination of the electropherogram. Figure 7 shows an medium-sized IgM component, located in the mid gamma-globulin region; albumin is decreased and alpha$_2$-globulin increased. Figure 8 shows an IgM located in the fast gamma-globulin region, between the gamma$_1$-globulin and the beta-globulin. These patterns indicate the range in the electropherogram where IgM usually occurs and also shows the various sizes and shapes most likely to be obtained for IgM.

Figure 9 shows an electrophoretic pattern for IgA; here, the IgA is located in the fast gamma-globulin

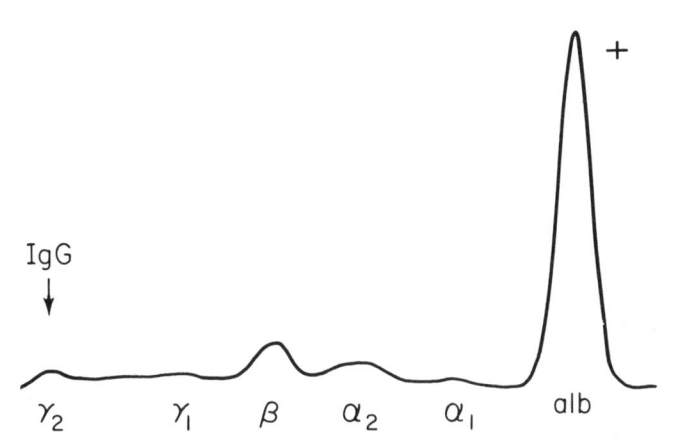

Fig. 3. Electrophoresis pattern with a small monoclonal IgG peak in the slow gamma-globulin region

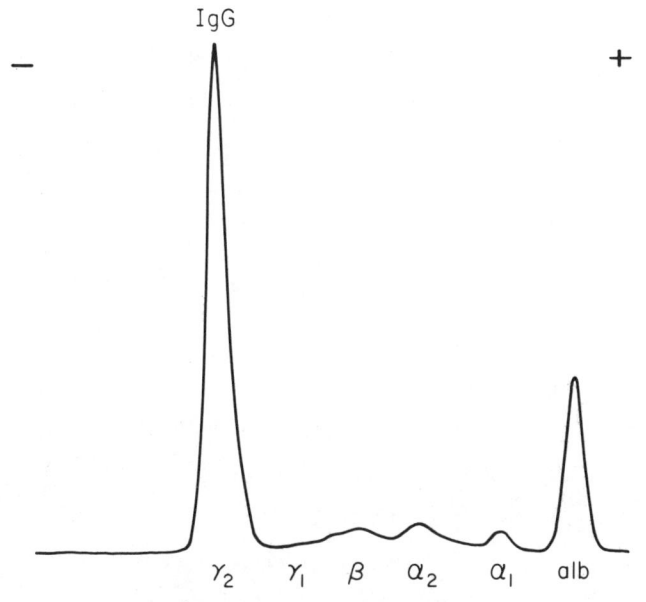

Fig. 4. Electrophoresis pattern with a large monoclonal IgG peak in the mid gamma-globulin region

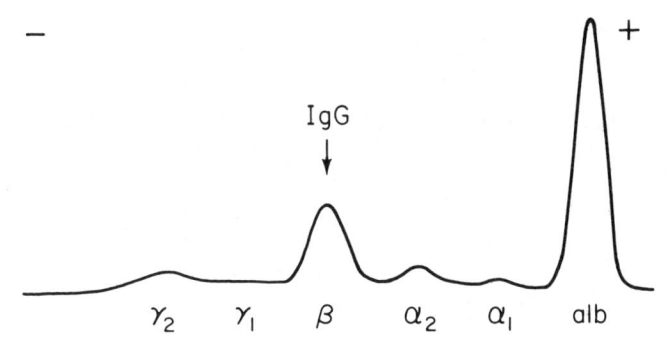

Fig. 5. Electrophoresis pattern with a medium-sized monoclonal IgG peak in the beta-globulin region

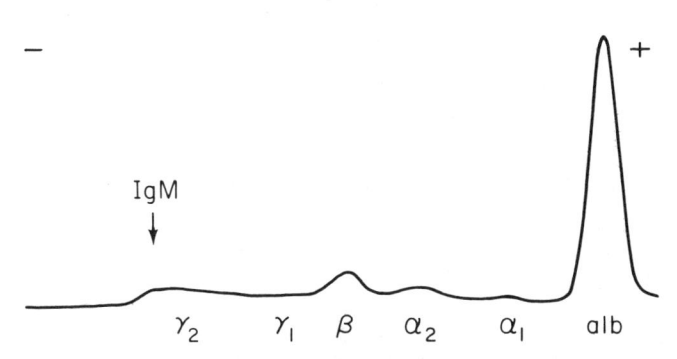

Fig. 6. Electrophoresis pattern with a small monoclonal IgM peak in the slow gamma-globulin region

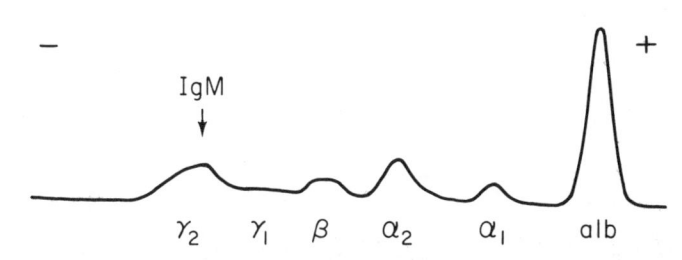

Fig. 7. Electrophoresis pattern with a medium-sized monoclonal IgM peak in the mid gamma-globulin region

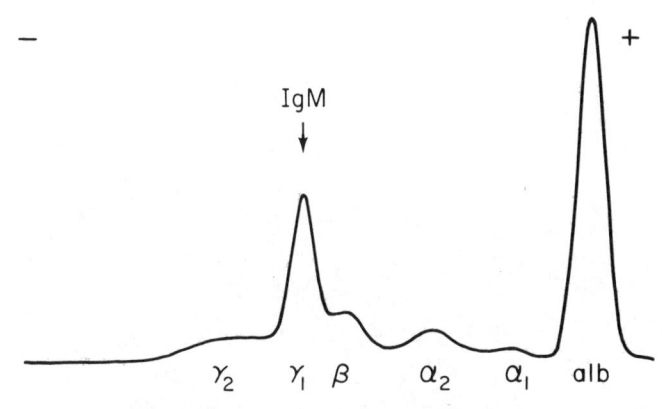

Fig. 8. Electrophoresis pattern with a large monoclonal IgM peak in the fast gamma-globulin region

region. Our observed mobility range for IgA peaks is fast gamma (gamma₁)-globulin through the beta-globulins.

Figures 10 and 11 show two electrophoretic patterns containing IgD. In Figure 10 the IgD is located in the mid gamma₂-globulin region. In Figure 11 the IgD is located in the fast gamma-globulin region between the gamma₁-globulin and beta-globulin region; alpha₁- and alpha₂-globulins are slightly decreased. IgD is usually not observed in serum. We have electrophoresed more than 20 000 sera, but have observed IgD in only four patients.

Samples that contain large monoclonal peaks of either IgG or IgA are usually indicative of multiple myeloma. However, 25% of patients with multiple myeloma do not show an abnormal immunoglobulin peak in the electropherogram (7). These patients are usually identified by the presence of Bence Jones protein in their urine. Samples that contain a large monoclonal IgM component usually indicate Waldenstrom's macroglobulinemia (8). Patients with small immunoglobulin peaks should be followed up to determine whether there is an increase in the abnormal immunoglobulin. Some individuals can have monoclonal gammopathies with no associated disease (9, 10).

Figure 12 shows a biclonal electrophoretic pattern, with one component in the gamma-globulin region and one in the beta-globulin region. These peaks were identified as IgG and IgA, respectively. This pattern also shows a decrease in albumin. Figure 13 shows a triclonal electrophoretic pattern. The electropherogram contains three immunoglobulins; IgG, located in the slow gamma-globulin region; IgA, in the mid gamma-globulin region; and IgM, in the fast or gamma₁-globulin region. Biclonal and triclonal patterns are not very common.

Figure 14 shows the electropherogram of a plasma

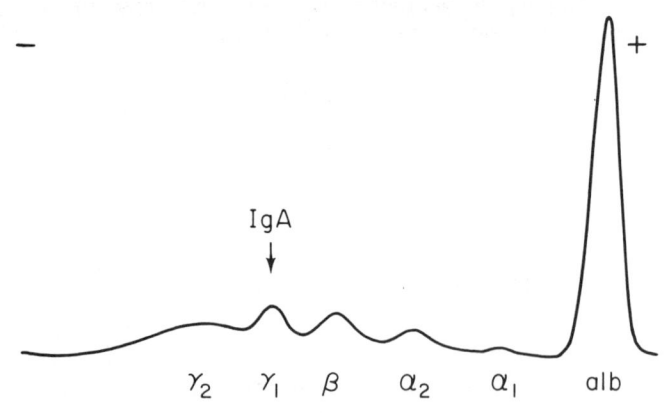

Fig. 9. Electrophoresis pattern with a monoclonal IgA peak in the fast gamma-globulin region

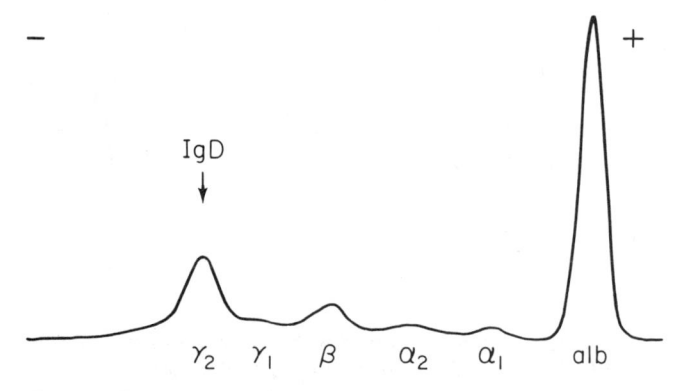

Fig. 10. Electrophoresis pattern for a monoclonal IgD peak in the mid gamma-globulin region

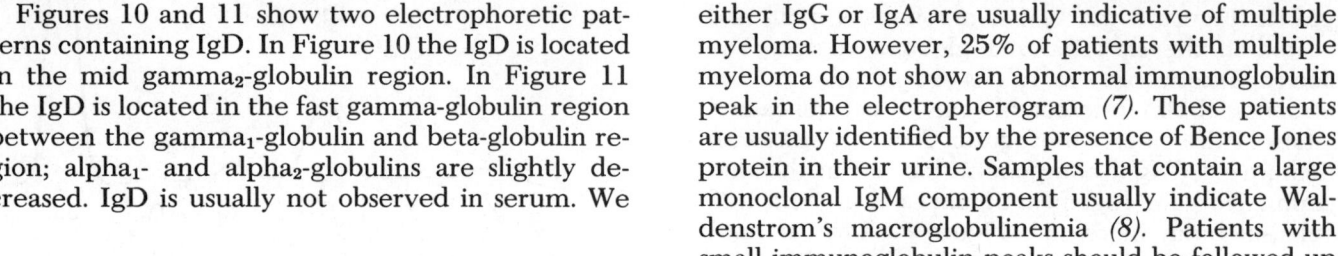

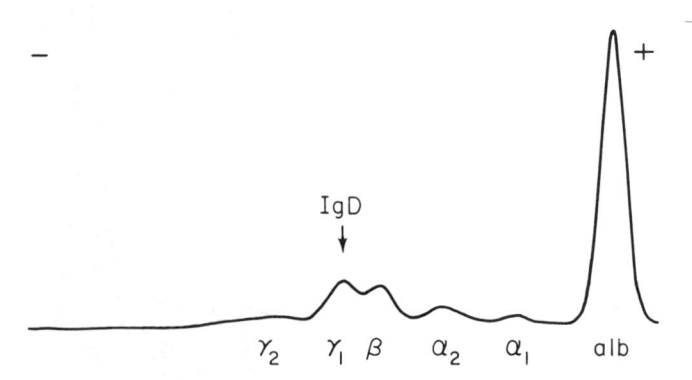

Fig. 11. Electrophoresis pattern for a monoclonal IgD peak in the fast gamma-globulin region

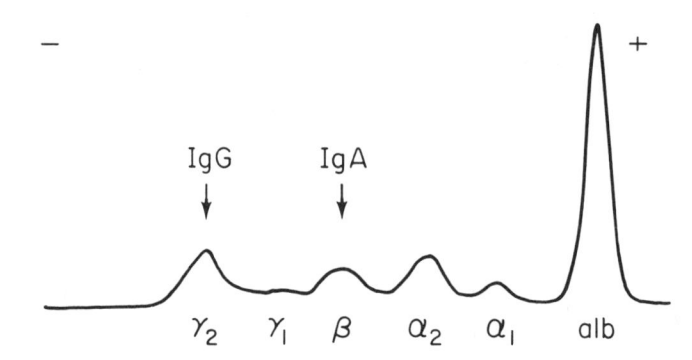

Fig. 12. Electrophoresis pattern showing biclonal gammopathy

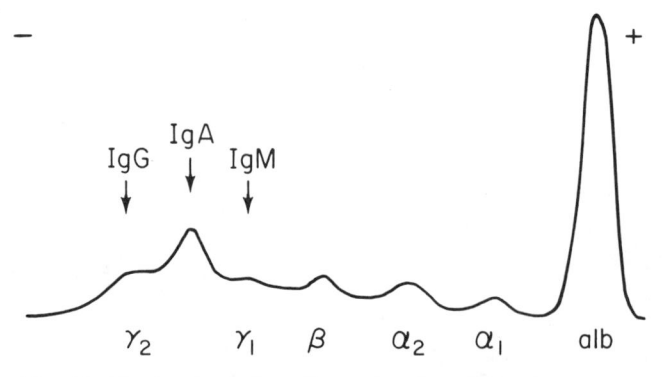

Fig. 13. Electrophoresis pattern showing triclonal gammopathy

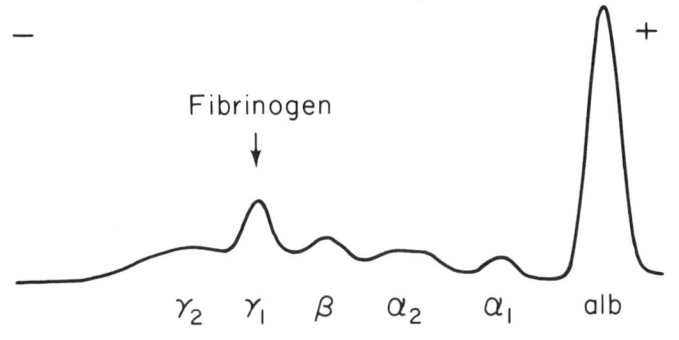

Fig. 14. Electrophoresis pattern of a plasma sample

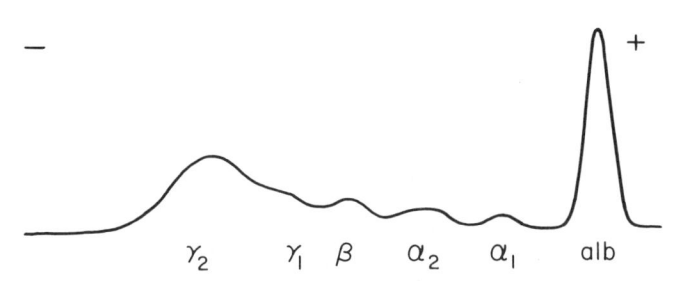

Fig. 15. Electrophoresis pattern showing liver disease

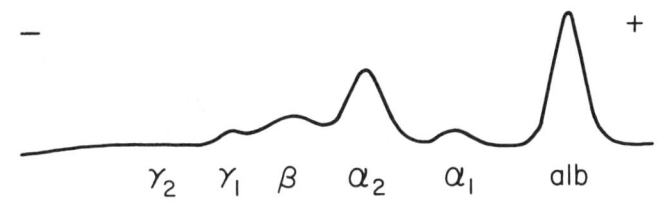

Fig. 16. Electrophoresis pattern showing nephrosis

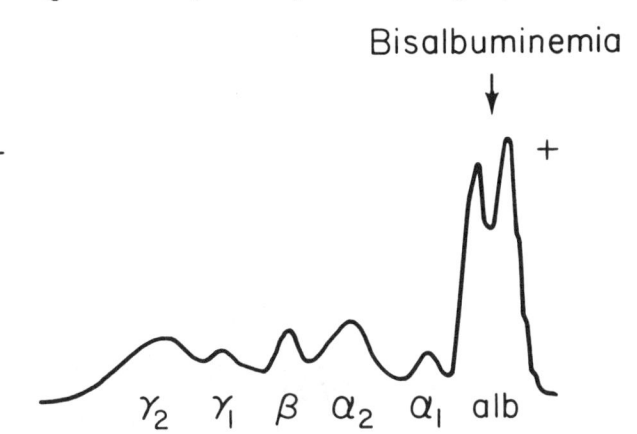

Fig. 17. Electrophoresis pattern showing bisalbuminemia

sample. In this pattern fibrinogen is seen as a peak in the gamma$_1$-globulin region and can either be mistaken for an immunoglobulin or conceal an immunoglobulin peak. A hemolyzed specimen may also produce an abnormal peak that can be mistaken for an immunoglobulin. Therefore, only nonhemolyzed serum samples should be used for protein electrophoresis.

Figure 15 shows the electropherogram of a patient with liver disease. This pattern shows an increased and diffuse gamma-globulin fraction and a decrease in the albumin fraction. Figure 16 shows an electropherogram of a patient with nephrotic syndrome; albumin and gamma-globulin fractions are decreased and the alpha$_2$-globulin fraction increased. Figure 17 shows an electropherogram of a patient with bisalbuminemia. There are two albumin peaks of about equal intensity, a fast and a slow albumin. Bisalbuminemia is a rare genetic disease with apparently no clinical symptoms. In more than 100 000 sera analyzed at Presbyterian Hospital, only two cases of bisalbuminemia have been detected. Detection of bisalbuminemia in the North American Indian population, however, is not uncommon *(11)*.

In conclusion, electrophoresis of serum proteins provides a means of assessing disturbances in the metabolism of albumin and globulins. Protein electrophoresis is particularly useful in detecting individuals with abnormal immunoglobulins. However, many patients will have very small immunoglobulin peaks in their electropherograms, and abnormalities will be detected only if the electrophoresis patterns are carefully inspected by the laboratory personnel. The physician sees only the electrophoresis scan; laboratory personnel can correlate a suspicious abnormality in the scan with a corresponding discrete band in the membrane.

The Submitters thank Dr. E. F. Osserman, Columbia University, College of Physicians and Surgeons, for identifying the specific abnormal immunoglobulins, and David Rubinstein and Ann-Marie Krause, Columbia-Presbyterian Medical Center, for technical assistance.

References

1. Picton, H., and Linder, S. E., XI, Solutions and pseudosolution. Part 1. *J. Chem. Soc. Trans.* **61**, 148–172 (1892).
2. Tiselius, A., A new apparatus for electrophoresic analysis of colloidal mixtures. *Trans. Faraday Soc.* **33**, 524–531 (1937).
3. Durrum, E. L., Microelectrophoretic and microionophoretic technique. *J. Am. Chem. Soc.* **72**, 2943–2948 (1950).
4. Kohn, J., A cellulose acetate supporting medium for zone electrophoresis. *Clin. Chim. Acta* **2**, 297–303 (1957).
5. Joliff, C. R., The clinical value of protein analysis. In *Beckman Electrophoresis Information*, **4**, Beckman Instruments, Fullerton, CA, 1977, pp 1–18.
6. *Pediatric Clinical Chemistry*, 2nd ed., S. Meites, Ed., American Association for Clinical Chemistry, Washington, DC, 1981, p 381.
7. Osserman, E. F., Plasma cell dyscrasias. In *Cecil–Loeb Textbook of Medicine*, P. B. Beeson and W. McDermott, Eds., W. B. Saunders Co. Philadelphia, PA, 1971, pp 1574–1589.
8. Bhoopalam, N., Lee, B. M., Yakulis, V. J., and Heller,

P., IgM heavy chain fragments in Waldenström's macrogloblinemia. *Arch. Intern. Med.* **128**, 437–440 (1971).
9. Axelsson, U., and Hällén, J., A population study on monoclonal gammopathy. *Acta Med. Scand.* **191**, 111–113 (1972).
10. Kyle, R. A., Monoclonal gammopathy of undetermined significance. Natural history of 241 cases. *Am. J. Med.* **64**, 814–826 (1978).
11. Frohlich, J., Kozier, J., Campbell, D. J., Curnow, J. V., and Tárnoky, A. L., Bisalbuminemia. A new molecular variant, albumin Vancouver. *Clin. Chem.* **24**, 1912–1914 (1978).

General Reference

Gebott, M. J., and Peck, J. M., Beckman Microzone Electrophoresis Manual, Beckman Instruments, Inc., Fullerton, CA, 1978, Chapters 1 and 2.

ADDENDUM: ELECTROPHORESIS OF SERUM PROTEINS (PROVISIONAL)[1]

Submitter: Richard L. Imblum, *Laboratory Service, Veterans Administration Medical Center, Salisbury, NC 28144*

Materials and Methods[2]

Reagents

1. *Electra HR 750 Buffer* (no. 5805). Dissolve the contents of one packet in 750 mL of distilled water. Each packet contains 5.75 g of Tris [tris(hydroxymethyl)aminomethane], 2.45 g of barbituric acid, and 9.80 g of sodium barbital. The final pH should be 8.8 ± 0.2. Store at 2–8 °C; stable for two weeks. Discard after each day of use in the electrophoresis chamber.

2. *Cellulose acetate plate* (Titan III, 2-3/8 × 3 in., no. 3023). The plates consist of a layer of cellulose acetate on a rigid Mylar plastic support. Each plate holds up to eight samples.

3. *Ponceau S stain* (no. 5525), 5 g/L Ponceau S in 100 g/L sulfosalicylic acid. Stable unopened for two years at 20–25 °C; opened, for up to six months.

4. *Glacial acetic acid*, reagent ACS grade (Fisher Scientific Co., Raleigh, NC 27604).

5. *Destaining solution, acetic acid 5%* (50 mL/L). Dilute 50 mL of glacial acetic to 1000 mL.

6. *Methanol, certified ACS* (Fisher Scientific Co.).

7. *Clear-Aid, ethylene glycol* (no. 5005).

8. *Clearing solution.* Add 25 mL of glacial acetic acid to 71 mL of methanol, then add 4 mL of Clear-Aid and mix. Prepare freshly each day of use.

9. *Quality control serum.* Several commercial control sera are suitable. Serum used in a daily quality-control program is most convenient.

Apparatus

1. *Sample applicator* (Super Z applicator no. 4084).
2. *Super Z Sample Well Plate* (no. 4085).
3. *Super Z Aligning Base* (no. 4086).
4. *Electrophoresis chamber* (Zip Zone Chamber, no. 1283).
5. *Disposable paper wicks* (no. 5081).
6. *Titan Power Supply,* 110 V (no. 1500) or 220 V (no. 1501).

Note: Helena Laboratories also has a less expensive and less versatile power supply available, but Submitter has no experience with it.

Note: Helena Laboratories has a Zip Zone applicator, sample well plate, and aligning base similar to items 1, 2, and 3 above at about one half the cost.

7. *Microdispenser and tubes* (no. 6008).
8. *Blotters* (no. 5034).
9. *Staining Kit* (no. 5113). Set of seven plastic containers and carrying racks for soaking and staining cellulose acetate plates.
10. *Forced air oven:* 110 V (no. 5116) or 220 V (no. 5117).
11. *Electrophoresis densitometer:* Quick-Scan Flur-Vis, 110 V (no. 1115) or 220 V (no. 1116).

Note: If a densitometer with fluorescence capabilities is not used, a less-expensive, less-versatile scanner (the Quick-Scan Jr.) is available from Helena Laboratories.

12. *Electrophoresis computer (optional):* Quick Quant II, 115 V (no. 1011) or 220 V (no. 1012). The Quick Quant II calculates the area under the curve and prints out for each of the six protein fractions concentration (g/dL) and percent of total protein.

Procedure

The following procedure is that recommended by Helena Laboratories for using their equipment and reagents. Because Ponceau S does not bind equally to all protein fractions, follow the time schedule closely to obtain reproducible results. Avoid touching the cellulose acetate surface.

1. Presoak the required number of cellulose acetate plates (each plate holds eight samples) by placing the plates in the carrying rack and slowly lowering the rack into HR 750 buffer until the plates are completely immersed. Soak for about 15 min.

2. During the soaking period pour 50 mL of HR 750 buffer into the two outer compartments of the electrophoresis chambers. Wet two paper wicks and drape one over each support bridge of the chamber. Put on the chamber cover.

3. Using the microdispenser and starting with the

---

[1] Cellulose acetate system of Helena Laboratories, Beaumont, TX 77704.
[2] Unless otherwise noted, the items listed by catalog number are obtained from Helena Laboratories.

first sample, fill the sample wells with 5 µL of sample. Place the sample applicator over the wells of the sample well plate and load the applicator by depressing the tips into the sample wells several times. Depress the sample onto a blotter pad and reload the applicator. Proceed rapidly through steps 4, 5, and 6.

4. Place a drop of water or buffer in the middle of the aligning base. Remove the soaking cellulose acetate plate, blot it, and place it on the aligning base, cellulose acetate side up with the bottom end of the strip lined up on a mark 0.25 in. above the line underscoring "Helena Laboratories." This will give an off-center sample application. Cut a small piece off the upper-left corner of the cellulose acetate plate, just enough to distinguish it from the other corners. This corner will be on the same side as the first sample and serves as a reference to distinguish samples.

5. Depress the sample applicator tips into the sample wells again and move the applicator to the aligning base. Depress the sample applicator and hold down for 10 s to transfer the sample to the cellulose acetate plate.

6. Place the plate cellulose acetate side down in the electrophoresis chamber so that it bridges the two buffer compartments and rests on the wet paper wicks; place the clipped corner toward the anode and the off-center samples toward the cathode. Replace the chamber cover.

7. Plug the color-coded wires of the chamber into the terminals on the power supply, turn timer to 15 min, and adjust the voltage to 180 V. The power is automatically cut off at the end of 15 min and the timer bell rings.

8. During the electrophoresis run, pour about 50 mL of Ponceau S stain solution into a plastic container (see staining kit), about 50 mL of 5% acetic acid into each of three plastic containers (washes), about 50 mL of methanol into one container, and 100 mL of clearing solution into another container. The above amounts are for use on one plate.

9. At the end of the run remove the cellulose acetate plate from the chamber, blot it gently, and stain it for 6 min in Ponceau S. If more than one plate is being stained, use the carrying rack to carry out the entire procedure.

10. Remove the plate from the stain with forceps, drain excess stain, and destain in the three successive washes of 5% acetic acid.

11. Place the plate in methanol for 2 min.

12. Blot the plate and place in the clearing solution for 5 min.

Note: The 5% acetic acid washes, the methanol, and the clearing solution should be discarded after each day of use.

13. Air-dry the plate for about 2 min on a blotter pad, cellulose acetate side up, then place in a 50–60 °C oven until the plate is clear. The plate is now ready to scan.

14. Scan the plate, using a 525 nm filter and following the manufacturer's instructions for the densitometer.

Note: The printed circuit board electrical connections of the densitometer should be cleaned annually to minimize noise in the pen tracings.

Quality Control

Include a quality-control serum on each cellulose acetate plate. Establish the 95% confidence limits for each of the six serum protein fractions.

Calculation

If the densitometer is equipped with a computer, the results are printed out as g/dL and as percent of total.

If the densitometer is not equipped with a computer, the relative amount of each fraction must be determined. Add up the total integration marks and the integration marks in each fraction. Divide the number for each fraction by the total. Multiply each of these numbers by the total protein concentration to obtain g/dL for each fraction and by 100 to obtain percent of total.

Note: The total protein can be determined by the biuret or refractive index methods.

Estrogens (Total Pregnancy) in Urine

Submitters: Edward J. Rosenthal and Howard C. Leifheit, *Stanbio Laboratory, Inc., San Antonio, TX, 78202*
Paige K. Besch, *Department of Obstetrics and Gynecology, Baylor College of Medicine, Houston, TX 77030*

Evaluators: Pennell C. Painter, *University Pathologists, University of Tennessee Memorial Hospital, Knoxville, TN 37920*
Norman Huang, *Reproductive Research Laboratory, St. Luke's Episcopal Hospital, Houston, TX 77030*

Reviewer: Alan Broughton, *Nichols Institute, San Pedro, CA 90731*

Introduction

Quantitative urinary assay of estriol as a means of monitoring high-risk pregnancies has been well established in patients presenting with complications such as toxemia, hypertension, diabetes, premature labor, and missed abortions (1). Excretion of this steroid increases during the second and third trimesters of pregnancy, indicating the well-being of the fetoplacental unit. Any rapid or sustained decrease in estriol excretion in late pregnancy, or its failure to increase steadily, is taken as evidence of fetal distress. Therefore, results of such serial determinations can aid the physician in deciding whether to permit a pregnancy to continue or to remove the fetus from unfavorable surroundings.

In late pregnancy, estriol conjugates (glucuronides and sulfates) constitute about 90% of total urinary estrogens, the remaining conjugates being those of estrone and estradiol-17β. Therefore, their collective assay as "total pregnancy estrogens" (TPE),[1] also termed "total urinary placental estrogens," can provide a useful clinical estimate of estriol excretion in urine from pregnant women (2).

We stress that a single determination is of limited value. Serial measurements are necessary to indicate the *trend* of excretion. Because an abrupt decrease reflects fetal jeopardy, the assay method must be both rapid and reliable (1). Moreover, the *same* method must be used throughout the monitoring period (3); it is misleading to compare results of the "salting out" method presented here with results by gas chromatography or by techniques involving acid or enzymic hydrolysis.

The relatively large amounts of estriol conjugates present in urine during the second half of pregnancy have been assayed by several techniques. The colorimetric methods reported by Brown et al. (4, 5), modifying the Kober reaction (6), are particularly suitable for the clinical laboratory.

Methods for the estimation of TPE have routinely involved an initial acid-hydrolysis step to convert these steroids from their water-soluble conjugates of glucuronic and sulfuric acids to the free form. However, loss of estrogens and formation of artifacts during this step adversely influence the recovery of estrogens, mainly estriol. Although nonspecific background absorbance can be eliminated by using the Allen correction, variations in reported TPE concentrations must indicate the change in estriol excretion and not its destruction.

The TPE determination we present is a modification of the method reported in 1968 by Rourke et al. (7), evaluated later by Crooke et al. (8) and used successfully at the Baylor College of Medicine, Houston, TX. This method, which omits hydrolysis, is the most rapid and accurate for obtaining data to assess fetal well-being. Recent reports reveal the advantage of using urine estrogen/creatinine (E/C) ratios from a series of single-voided specimens, instead of complete 24-h collections, as an index of fetal well-being in screening normal and high-risk pregnancies (2, 9–12).

Note: The Reviewer states that the requesting physician should decide whether 2-h or 24-h urine collections are to be assayed, and that the laboratory should schedule this "stat" procedure for same-day reporting.

When 24-h urine collections are used, assuming an average daily urine creatinine excretion for females of 0.8–1.8 g (13), results of creatinine assays can reflect errors in collection and prevent faulty interpretation (14). However, use of the E/C ratio concept, with single specimens (usually 2-h collections), precludes lengthy and careful collection, and provides the clinician with prompter, more objective data.

Principle

A small aliquot of a 24-h urine specimen is acidified, saturated with sodium chloride, diluted with

---

[1] Nonstandard abbreviations used: TPE, total pregnancy estrogens; E/C, ratio of total estrogen to creatinine; TCA, trichloroacetic acid.

an equal volume of water, and extracted with ethyl acetate. Part of the organic solvent layer, containing the estrogens as sulfate and glucuronide conjugates, is evaporated to dryness. The Kober color reaction is performed on the dry residue by using sulfuric acid containing hydroquinone, the latter compound preventing estrogen oxidation and facilitating formation of the Kober color complex. The pink chromogens are extracted into trichloroacetic acid, back-extracted into chloroform, and the absorbance values read spectrophotometrically at three wavelengths (505, 535, and 565 nm). The Allen correction (15) is applied to correct background absorbance values at 535 nm, and the concentration of TPE in the unknown specimens is calculated by comparison with estriol standards similarly treated.

Materials and Methods

Reagents

1. *Ethyl acetate,* ACS reagent (Mallinckrodt, St. Louis, MO 63147).

2. *Color reagent,* 14.3 mol/L sulfuric acid solution containing hydroquinone and oxalic acid, 182 and 7.9 mmol/L, respectively. Add 2.0 g of hydroquinone (Photo-Purified crystals) and 0.1 g of oxalic acid (dihydrate crystals, ACS reagent, both from Mallinckrodt) to a 100-mL volumetric flask containing 15 mL of de-ionized or distilled water. Swirl to partly dissolve, place flask on a magnetic stir plate, and *slowly* add 79 mL of ACS reagent concd. sulfuric acid; *be careful!* Mix until completely dissolved, cool to room temperature, and dilute to 100 mL with water. Stopper the flask and mix well by inversion. At this point the temperature of the solution will increase, so *be careful.* This solution is stable for one year when stored at 4–8 °C. Protect from light and contamination with organic material and avoid further dilution with water. *The solution is corrosive!* Do not pipet by mouth.

3. *Trichloroacetic acid (TCA) solution, 2 mol/L.* Dissolve 32.7 g of TCA (AR grade, Mallinckrodt) in water and dilute to 100 mL. Stored in a glass bottle at room temperature, it is stable for one year.

4. *Chloroform.* ACS reagent.

5. *Estriol stock standard, 100 mg/dL.* Dissolve 100 mg of dry 1,3,5-estratriene-3,16α,17β-triol (Biochemical Laboratories, Redondo Beach, CA 90278) in ethanol and dilute to 100 mL. Store tightly capped at 4–8 °C; protect from light and contamination. Stable for one year.

6. *Estriol working standard, 1 mg/dL.* Dilute 1.0 mL of estriol stock standard (brought first to room temperature) to 100 mL with ethanol. Store tightly capped at 4–8 °C; protect from light and contamination.

7. *Extraction tubes, 15 × 100 mm.* These tubes may be of flint glass, equipped with stoppers or caps resistant to ethyl acetate (e.g., polypropylene). To each add approximately 2.5 g of ACS grade dry sodium chloride. Store tightly sealed at room temperature. Stable indefinitely.

8. *Color-development tubes, 15 × 100 mm.* These must be Pyrex or borosilicate glass (to withstand thermal shock), with stoppers or caps resistant to sulfuric acid and chloroform. To each add 20 mg of hydroquinone and 0.5 mg of oxalic acid (same quality and suggested source as in color reagent). Store at room temperature, tightly sealed, and protected from light. Discard if contents assume a purple color.

Note: These are conveniently prepared by dissolving the correct amount of hydroquinone and oxalic acid for the number of tubes required in diethyl ether/ethanol (1/1 by vol), pipetting the appropriate volume of this solution into each tube, and allowing the solvent to completely evaporate (under a fume hood at room temperature). Using contrasting colored caps, such as red for "extraction tube" and white for "color tube," facilitates tube identification.

9. *Hydrochloric acid, concd.* ACS grade.
10. *Ethanol (ethyl alcohol),* USP, 95% (950 mL/L).

Apparatus

1. *Vacuum aspirator* (e.g., faucet-type fitted with disposable Pasteur pipette).
2. *Vortex-type mixer*
3. *Nitrogen source and gas manifold,* for evaporating steroid extracts.

Note: Compressed air may be used when properly filtered to remove oil, water, and particulate matter. A Gas Purifier (model no. 450) with Filter Cartridge (model no. 451) from Matheson Gas Products, LaPorte, TX 77571 is a suitable filter.

Collection and Handling of Specimens

The sensitivity of this method makes it applicable *only* to urines from women past the 20th week of pregnancy, and preferably in the third trimester. Collect a 24-h urine specimen by discarding the first voiding on the morning of day 1 and collecting all subsequent voidings up to and including the first voiding of day 2. Measure and record total urine volume (milliliters).

Many laboratories assess the accuracy of 24-h collection by determining the amount of creatinine in the specimen submitted. Normal creatinine excretion for females is 1.3 ± 0.5 g/24 h (13); results within these limits indicate a true 24-h collection, except in cases of renal disease and diabetes. Assay the urine as soon as possible after collection. No preservative is necessary if the specimen is kept under refrigeration after (and preferably during) collection; however, use of boric acid (5–10 g) is permissible. Do not use formaldehyde as a perservative.

In applying the E/C ratio to a single-voided specimen, use a 2-h volume. Have the patient empty her bladder; discard the urine. Two hours later collect a complete voiding, and assay both total estrogens and creatinine.

Procedure

Salting out and extraction

1. To an extraction tube add 2.0 mL of a well-mixed and measured 24-h urine and 2 mL of distilled water.

2. Add two drops of concd. hydrochloric acid and mix well.

Note: The pH *must* be 2–3; if not, add more acid dropwise until the proper pH is reached.

3. Add 5.0 mL of ethyl acetate, cap the tube, and shake it vigorously for 30 s.

4. Centrifuge 2 to 3 min, or until solvent (upper) layer is clear.

5. Transfer 1.0 mL of solvent layer to a color development tube marked U (unknown).

Note: For urine in which total estrogen concentration is somewhat low, use 2 or 3 mL of solvent layer (extract). The volume of aliquot used must, of course, be considered in the final calculations. Refer to *Procedural Notes No. 2.*

6. Set up standards as follows: (a) Bring the working standard to room temperature and mix by inversion. (b) Transfer 0.5 mL (5 µg) and 1.0 mL (10 µg) of estriol working standard to each of two color-development tubes; label these tubes S_5 and S_{10}, respectively.

7. Carefully evaporate the contents of all tubes to *complete dryness* at about 40 °C under a gentle stream of nitrogen gas. Refer to *Procedural Notes No. 4.*

Note: Well-filtered compressed air may be used instead (see *Apparatus* above). Evaluator P. C. P. found that the tubes could be stored overnight at 5 °C after this evaporation-to-dryness step. No significant difference in results was evident if this stopping point was used.

Color development

8. Add 2.0 mL of color reagent to each color-development tube, cap, and mix well. See *Procedural Notes No. 9.*

9. Place all tubes, uncapped, in boiling water bath for 20 min. Mix thoroughly twice during the first 10 min.

10. Cool all tubes in a cold water bath, add 1.0 mL of distilled water to each, and mix well.

11. Replace all tubes in a boiling water bath for 10 min.

12. Cool all tubes to room temperature in a water bath.

Color extraction

13. Add 1.0 mL of TCA solution to each tube, cap, shake vigorously for 15 s, and place in an ice bath for 2 to 3 min.

14. To each tube add 3.0 mL of chloroform, cap, and again shake vigorously for 15 s.

15. Centrifuge until solvent is clear, then aspirate and discard all of the aqueous (upper) layer.

16. Carefully transfer the lower, colored layer to cuvets and read absorbance (A) of each vs chloroform at 505, 535, and 565 nm.

Note: Specimens having A_{535} values exceeding 0.800 should be diluted in the cuvet with chloroform and reread at the three wavelengths. Remember to use the appropriate dilution factor in the final calculations.

Procedural notes

1. This method is designed for use *only with urine specimens from women past the 20th week of pregnancy*, and preferably in the third trimester. It is unreliable at very low estrogen concentrations and is *not* intended for estrogen determination in blood or in the urine of nonpregnant women.

Note: Evaluator N. H. reports the procedure to be straightforward, and easy to set up and perform, requiring about one week of training time for a technician. He further states that one technician can perform 30 assays, including quality-control specimens, in a single day. The Reviewer warns against mouth-pipetting of organic solvents and acids.

2. If the estrogen concentration in a given urine specimen is suspected of being low, assay more than 1.0 mL of the ethyl acetate layer (step 5). The final result must then be divided by the volume used. For example, if 2.0 mL of ethyl acetate layer was used, the final value must be divided by 2.

3. To use 19 × 150 mm cuvets, increase the volume of chloroform to 6 mL (step 14). This will not affect calculations, as long as specimen and standard tubes are treated alike.

4. Ethyl acetate extracts and standards *must* be evaporated to complete dryness (step 7). To maximize recoveries, rinse the sides of the color tubes with ethyl acetate and *completely* redry.

5. Both stock and working standards should be tightly capped and refrigerated to prevent loss by evaporation. Allow a small volume of stock standard to come to room temperature before preparing the working standard. This working standard, when stored as directed, is stable for one year. Bring aliquots of the working standard to room temperature before pipetting.

6. To enhance stability of all reagents, store as directed and *avoid pipetting directly from the reagent container.* Pour slightly more than the volume needed into an appropriate flask or tube, pipet from that vessel, and discard the remaining reagent. Do

not return an unused portion to the reagent container.

Note: Evaluator N. H. states that, in his experience, all reagents are very stable, the color reagent being stable for more than a year if stored refrigerated in a brown bottle. Contents of the dry color-development tubes also are stable for at least a year at room temperature when protected from light.

7. Evaporate the extracts carefully and completely, using a *gentle* stream of nitrogen (or well-filtered air). Too strong a stream will blow particles from the tube (step 7).

8. For both the specimen extraction (step 3) and color extraction (steps 13 and 14), mix vigorously. Any emulsion formed may be broken by centrifugation.

9. When color reagent is added, contents of color-development tubes will not dissolve until heated. Mixing must be complete to ensure solution during heating period (step 9). Because of the relative insolubility of the dry residue in diluted color reagent, two successive boiling steps are required.

10. After shaking with TCA solution (step 13), minimize evaporation by cooling the reaction mixture thoroughly before adding chloroform (step 14). Carefully transfer the lower, colored (chloroform) layer, excluding all traces of aqueous phase from the cuvets (step 16).

Quality Control

Most large laboratories store frozen aliquots of urine from pregnant women for use in their quality-control program *(3);* however, preparations for this purpose are commercially available. Products such as Ortho Control Urine I (Ortho Diagnostics, Raritan, NJ 08869) and UriChem™ Urine Chemistry Control, Level I (Fisher Diagnostics, Orangeburg, NY 10962) are satisfactory.

Calculations

1. To obtain corrected absorbance (CA) for each unknown (CAu) and each standard (CAs), apply the Allen correction:

$$CA = A_{535} - [(A_{505} + A_{565})/2]$$

2. Calculate results as follows:

Urine TPE (μg/tube) = (CAu/CAs) × [s]
Urine TPE (mg/24 h) = urine TPE (μg/tube) × 0.001 × 2.5 × 24-h urine vol (mL)

[s] represents concentration of the standard (either 5 or 10 μg), for which the corrected absorbance (CAs) is closest to that of the unknown (CAu). The factor 2.5 is derived from the use of 1 mL of the 5-mL extraction solvent, the latter containing the total estrogens in 2 mL of urine sample; hence, 1 × 5/2 = 2.5.

Example A: The following absorbance data were determined from assay of a 24-h urine (1400 mL) from a woman in the 37th week of pregnancy, with a Bausch and Lomb Spectronic 70 spectrophotometer and 12 × 75-mm cuvets:

Tube	A_{505}	A_{535}	A_{565}
Unknown, U	0.140	0.262	0.090
Standard (5 μg), S_5	0.118	0.229	0.040
Standard (10 μg), S_{10}	0.212	0.448	0.050

Applying the calculations yields:

$$CAu = 0.262 - [(0.140 + 0.090)/2]$$
$$= 0.262 - 0.115 = 0.147$$
$$CAs(5) = 0.229 - [(0.118 + 0.040)/2]$$
$$= 0.229 - 0.079 = 0.150$$
$$CAs(10) = 0.448 - [(0.212 + 0.050)/2]$$
$$= 0.448 - 0.131 = 0.317$$

Urine TPE (μg/tube) = (0.147/0.150) × 5 = 4.9
Urine TPE (mg/24 h) = 4.9 × 0.001 × 2.5 × 1400
$$= 17.2$$

Alternatively, results may be expressed as the E/C ratio for a given urine volume (e.g., *either* a 24-h excretion *or* a random voiding).

Note: In both instances the total estrogen concentration is expressed in milligrams per a given volume and creatinine in grams *per the same volume.*

Example B: In the preceding illustration, where the 24-h total estrogen was 17.2 mg, total creatinine content for the 24-h (1400 mL) urine volume was 1.6 g. The E/C ratio would then be calculated as follows:

E/C = TPE (mg/24 h)/creatinine (g/24 h)
$$= 17.2/1.6 = 10.8$$

Example C: For a 2-h urine sample from the same patient, where volume is irrelevant for single-voided specimens, the urine TPE (μg/tube) was calculated to be 5.1. For this purpose TPE is conveniently expressed as milligrams per 100 mL (mg/dL), and creatinine as grams per 100 mL (g/dL). Therefore:

Urine TPE (mg/dL) = 5.1 × 0.001 × 2.5 × 100
$$= 1.27$$

In the same urine specimen, creatinine was 125 mg/dL (or 0.125 g/dL for use in the calculation of the E/C ratio):

$$E/C = 1.27/0.125 = 10.2$$

Note: Again the E/C ratio is merely a simple expression of milligrams of estrogen divided by grams of creatinine, with both being per the same volume. Use of "100 mL" is for convenience in calculation, as well as making volume measurement of the random urine sample unnecessary.

Discussion

Precision

Replicate analyses were performed on urine samples from two pregnant women; one had a mean

TPE value of 6.2 mg/L (n = 38), the other 15.0 mg/L (n = 49). Standard deviations calculated were 0.48 and 0.57 mg/L, respectively, and CVs were 7.7 and 3.8%. Results of precision studies conducted by Evaluators P. C. P. and N. H. are summarized in Table 1.

Table 1. Precision Studies

Evaluator	Within-day $\bar{x}$(SD), mg/L	CV, %	Day-to-day $\bar{x}$(SD), mg/L	CV, %
P. C. P.	16.4 (0.30)	1.85	16.5 (0.42)	2.54
N. H.	14.77(0.70)	4.73	14.84(0.71)	4.78
	45.80(0.88)	1.92	44.36(2.09)	4.71

Note: Evaluator P. C. P reports the precision of this method to be somewhat better (CV = 2.54% for TPE of 16.5 mg/L) than that found by the Submitters or by Rourke et al. *(7)*. He presumes his precision data are related to the use of more reproducible and accurate pipettors, as would be found in larger clinical laboratories.

Correlation

Twenty-six urine samples from pregnant women, having TPE values in the range of 4 to 35 mg/24 h, were assayed by the salting-out method presented here (y) and by a procedure involving acid–heat hydrolysis (x). Statistical analysis of results yielded a correlation coefficient (r) of 0.967, with the regression equation being $y = 0.918x + 0.723$.

Note: Evaluator N. H. assayed 88 urines from pregnant women, comparing the method presented as available in the Stanbio TPE Kit (y) and the same procedure performed with in-house reagents from the Reproductive Research Laboratory (x). By his results, $y = 0.93x + 0.1$, $r = 0.985$.

Recovery

Results of analytical recovery studies conducted by Evaluator N. H. are shown in Table 2.

Table 2. Analytical Recovery Studies

Estrogen added	TPE found	Recovery, %
mg		
0	15.45	—
5.55	21.35	106.3
11.10	26.21	97.0
16.65	32.45	102.1
22.20	39.17	106.8
27.75	42.93	99.0

Expected Values

An example of the value of E/C ratios applied to serial determinations on 24-h urine collections is taken from the 1969 report of Cummings et al. *(14)* and is presented in Table 3. Obviously, had not the total volume and the creatinine value for the third sample been known, TPE results (19.2 mg/24 h) could have been misinterpreted as indicating fetoplacental distress rather than a collection error. On the contrary, the consistent increase in the E/C ratio from week 36 to week 40 reflected fetal well-being.

Table 3. Estrogen/Creatinine (E/C) Ratios on Serial 24-h Urines from a Pregnant Woman

Gestation, weeks	Total vol, mL	Total creatinine, g/24 h	Total pregnancy estrogen, mg/24 h	E/C
36	1140	1.08	23.5	22
37	1640	1.53	36.7	24
38	480	0.66	19.2	29
39	1600	1.35	51.7	38
40	1480	1.24	55.9	45

In their 1975 report on random urine E/C ratios as a practical and reliable index of fetal welfare, Aubry et al. *(12)* demonstrated the close parallel between the E/C ratios of 24-h urine samples and those of paired random urine collection. Table 4 summarizes their data on 182 such collections from 20 subjects with "normal outcome pregnancies."

Table 4. E/C Ratios in 24-h Urine and Paired Random Urine Samples Compared

Weeks of gestation	Mean E/C ratios 24-h urine	Paired random urine
21	5	4
23	6	5
25	8	7
27	10	10
29	11	11
31	14	15
33	15	16
35	16	17
37	23	23
39	26	25

In both of the above examples, estrogen values were determined as total pregnancy estrogens (TPE) by the rapid method of Rourke et al. *(7)*. Creatinine values were estimated by the classical Jaffé (alkaline picrate) reaction.

Interfering Substances

Various carbohydrates such as glucose, high concentrations of urinary protein, and various drugs can interfere with the isolation or quantitation of urinary TPE by the usual procedures. However, the salting-out technique used here avoids interference by glucose and other carbohydrates, which are retained in the urinary phase during extraction. Protein also remains in the urinary phase, but excess amounts may cause a troublesome emulsion; this difficulty can be avoided by diluting the urine twofold with 50 g/L TCA, mixing well, centrifuging, and extracting 2 mL of the supernate. For subsequent assay, 2 mL of the ethyl acetate extract, instead of 1 mL, must be used.

Several drugs reportly interfere with this determination *(16)*. For example, the diuretic hydrochlorothiazide directly interferes chemically with the formation of the Kober chromogen in this method, resulting in lower apparent values *(17)*. Mandelamine (methenamine mandelate), because it breaks down to formaldehyde under acid conditions in the bladder or in the collecting vessel, has been reported to destroy urinary estriol *(18)*. Among other medications implicated are hexamine, ampicillin, phenazopyridine, prochloroperazine, and the tetracyclines. Some investigators *(1, 19)* advise withholding meprobamate, phenolphthalein, cascara, senna, stilbesterol, and cortisone for 24 h before and during collection.

Reference Intervals

After the establishment of an initial value for either TPE concentration or E/C ratio for a given pregnant woman's urine, the significance of results lies in the *trend* of subsequent values as gestation progresses. Therefore, aside from the two typical sets of values cited, no attempt is made to present TPE concentrations or E/C ratios as a "normal" or "usual" range.

Note: Evaluator P. C. P. states that in his hospital, when a decision is made to monitor total urinary estrogens of a patient, two 24-h urine specimens are collected and assayed to establish a baseline at about the 32nd or 33rd week. Then two samples are collected per week in the 35th and 36th weeks, and daily samples from the 37th week. Daily samples may be requested sooner if clinical indications suggest impending problems. To chart the urinary estrogen trend, total estrogens are plotted on the *y*-axis against the gestational age on the *x*-axis. Prompt obstetrical management is suggested if on any day there is a decrease to 50% or less of the mean TPE concentration of the preceding three days. The Reviewer states that although the method presented is "tried and true" and suitable for the small clinical laboratory, his method of choice for monitoring fetal distress is the radioimmunoassay of plasma unconjugated estriol.

We thank Charles E. Miles, past President, Stanbio Laboratory, Inc., for his encouragement and interest in the writing of this chapter. We further wish to acknowledge the help and patience of Linda Winston and Chris Harris in typing the manuscript.

References

1. Chattoraj, S. C., Endocrine function. In *Fundamentals of Clinical Chemistry*, 2nd ed., N. W. Tietz, Ed. W. B. Saunders, Philadelphia, PA, 1976, pp 776–781.
2. Grannis, G. F., and Dickey, R. P., Simplified procedure for determination of estrogen in pregnancy urine. *Clin. Chem.* **16**, 97–102 (1970).
3. Besch, P. K., Buttram, V. C., Jr., and Besch, N. F., Stanbio and "Tekit" methods for placental estrogens compared. *Clin. Chem.* **19**, 434 (1973). Letter.
4. Brown, J. B., A chemical method for the determination of oestriol, oestrone and oestradiol in human urine. *Biochem. J.* **60**, 185–193 (1955).
5. Brown, J. B., Macnaughtan, C., Smith, M. A., and Smyth, B., Further observations on the Kober colour and Ittrich fluorescence reactions in the measurement of oestriol, oestrone and oestradiol. *J. Endocrinol.* **40**, 175–188 (1968).
6. Kober, S., Eine kolorimetrische Bestimmung des Brunsthormons (Menformon). *Biochem. Z.* **239**, 209–212 (1931).
7. Rourke, J. E., Marshall, L. D., and Shelley, T. F., A simple rapid assay of estrogens in pregnancy. *Am. J. Obstet. Gynecol.* **100**, 331–335 (1968).
8. Crooke, N., Madeira, L. B., Haller, W. S., and Besch, P. K., Second generation steroid assays II: Total placental estrogens. *J. Am. Med. Technol.* **35**, 161–163 (1973).
9. Shelley, T. F., Cummings, R. V., Rourke, J. E., and Marshall, L. D., Estrogen–creatinine ratios: Clinical application and significance. *Obstet. Gynecol.* **35**, 184–190 (1970).
10. Osofsky, H. J., Nesbitt, R. E. L., Jr., and Hagen, J. H., High-risk obstetrics IV. Estrogen/creatinine ratios in routine urine samples as a method of screening a high-risk obstetrics population. *Am. J. Obstet. Gynecol.* **106**, 692–698 (1970).
11. Dickey, R. P., Grannis, G. F., Hanson, F. W., Schumacher, A., and Ma, S., Use of the estrogen/creatinine ratio and the "estrogen index" for screening of normal and "high-risk pregnancy. *Am. J. Obstet. Gynecol.* **113**, 880–886 (1972).
12. Aubry, R. H., Rourke, J. E., Cuenca, V. G., and Marshall, L. D., The random urine estrogen/creatinine ratio: A practical and reliable index of fetal welfare. *Obstet. Gynecol.* **46**, 64–68 (1975).
13. Faulkner, W. R., and King, J. W., Renal function. In *Fundamentals of Clinical Chemistry* (see ref. *1*), p 998.
14. Cummings, R. V., Rourke, J. E., and Shelley, T. F., Serial rapid assay of total urinary estrogens in pregnancy. *Am. J. Obstet. Gynecol.* **104**, 1047–1052 (1969).
15. Allen, W. M., Jr., A simple method for analyzing complicated absorption curves of use in the colorimetric determination of urinary steroids. *J. Clin. Endocrinol.* **10**, 71–83 (1950).
16. Young, D. S., Pestaner, L. C., and Gibberman, V., Effects of drugs on clinical laboratory tests. *Clin. Chem.* **21**, 294D (1975). Special Issue.
17. Acosta, A. A., Madeira, L. B., Besch, P. K., and Buttram, V. C., Jr., Additional observations on hydrochlorothiazide interference with measurement of total urinary placental estrogens. *Clin. Chem.* **19**, 261–263 (1973).
18. Eraz, J., and Hausknecht, R., Diminished urinary estriol due to mandelamine administration during pregnancy. *Am. J. Obstet. Gynecol.* **104**, 924–925 (1969).
19. Hobrick, R., and Metcalf-Gibson, A., Urinary estrogens. *Stand. Methods Clin. Chem.* **4**, 65–83 (1963).

Glucose, Direct Hexokinase Method

Submitter: Jane W. Neese, *Nutritional Biochemistry Branch, Clinical Chemistry Division, Center for Environmental Health, Centers for Disease Control, Atlanta, GA 30333*

Evaluators: Seymour Bakerman, Prabhaker G. Khazanie, and Joseph W. Litten, *Department of Pathology and Laboratory Medicine, School of Medicine, East Carolina University, Greenville, NC 27834*

Thorne J. Butler, *Southern Nevada Memorial Hospital, Las Vegas, NE 89102*

Reviewer: Basil T. Doumas, *Department of Pathology, Medical College of Wisconsin, Milwaukee, WI 53226*

Introduction

This coupled-enzyme method for the determination of glucose in serum and plasma involves hexokinase (EC 2.7.1.1; ATP:D-hexose-6-phosphotransferase) and glucose-6-phosphate dehydrogenase (EC 1.1.1.49; D-glucose-6-phosphate:NAD(P)$^+$ oxidoreductase). It is an adaptation of a candidate glucose reference method *(1)*. In lieu of deproteinization, a specimen blank is used to correct for interfering substances that absorb at 340 nm, the wavelength used for quantitation. The direct, specimen-blanking approach was evaluated and extensively compared with the candidate reference method during the development and validation of the latter procedure *(2)*. Better specificity and precision are obtained with deproteinization; however, the performance characteristics of direct specimen-blanking are more than adequate for clinical use. This method is especially suitable for small laboratories, because of the simplicity and adaptability of the procedure and the small volume of specimen required.

Principle

A series of chemical reactions is used, involving two enzymes and two coenzymes. The catalytic phosphorylation of glucose by adenosine triphosphate (ATP), reaction 1, is an exergonic reaction that is essentially irreversible *(3)*.

1. Phosphorylation (catalyzed by hexokinase)

$$\text{D-Glucose} + (\text{ATP·Mg})^{2-} \xrightarrow{\text{hexokinase}} \text{D-glucose 6-phosphate} + (\text{ADP·Mg})^{-} + \text{H}^{+}$$

Specificity derives from the second reaction, the catalytic dehydrogenation step, which also provides the basis for the assay: for each mole of glucose present in the reaction mixture, 1 mol of nicotinamide adenine dinucleotide (NAD$^+$) is reduced *(4)*, and the absorbance of the reduced dinucleotide at 340 nm is measured spectrophotometrically.

2. Dehydrogenation (catalyzed by glucose-6-phosphate dehydrogenase, G6PDH)

$$\text{D-glucose 6-phosphate} + \text{NADP}^+ \text{ or NAD}^+ \xrightarrow{\text{G6PDH}} \text{6-phosphoglucono-}\delta\text{-lactone} + \text{NADPH or NADH} + \text{H}^+$$

The unstable gluconolactone produced undergoes spontaneous hydrolysis *(5)*, and the reaction sequence goes virtually to completion *(6)*, within about 10 min.

3. Hydrolysis

$$\text{6-phosphoglucono-}\delta\text{-lactone} + \text{H}_2\text{O} \rightarrow \text{6-phosphogluconic acid}$$

Materials and Methods[1]

Reagents

Estimates on the stability of reagents are based on the use of high-purity chemicals. Water used in preparation and for dilutions must be of high quality, preferably de-ionized/distilled or reagent grade *(7)*.

1. *Benzoic acid diluent, 1.0 g/L.* Dissolve 2.0 g of ACS-grade benzoic acid in 2.0 L of hot water in a large flask. Mix well, cover, and set aside to cool to room temperature.

---

[1] Use of trade names is for identification only and does not constitute endorsement by the Public Health Service, the U.S. Department of Health and Human Services, or the AACC.

2. *Glucose stock standard, 10.00 g/L (0.555 mol/L)*. Use National Bureau of Standards Standard Reference Material no. 917 or ACS-grade anhydrous D-glucose (dextrose) that has been stored at reduced pressure in a desiccator over anhydrous calcium sulfate with cobalt chloride added as an indicator. Weigh and transfer 10.000 g of glucose to a 1-L Class A volumetric flask, using the benzoic acid diluent for the quantitative transfer. Add diluent to approximately one-half full, mix well to dissolve the glucose, then make to volume with benzoic acid diluent.

Invert the stoppered flask several times, dispense about 200 mL into each of five labeled 250-mL (8-oz.) plastic or borosilicate screw-cap bottles, and tighten the caps securely. Store frozen at −20 °C. With care to avoid evaporation, this stock standard can be used for one year.

3. *Glucose working standards*. Thaw a bottle of glucose stock standard and allow it and the benzoic acid diluent to come to room temperature, but not warmer than 30 °C. Using Class A volumetric pipettes and 100-mL volumetric flasks, prepare the following dilutions:

Stock standard, mL	Final glucose concn, after dilution to 100 mL with benzoic acid diluent	
	mg/dL	mmol/L
5.00	50	2.78
10.00	100	5.55
15.00	150	8.33
20.00	200	11.10
30.00	300	16.65
40.00	400	22.20
60.00	600	33.30

Mix each dilution well and transfer to a labeled, clean, dry 120-mL (4-oz.) screw-cap plastic or borosilicate bottle. In an eighth bottle is benzoic acid diluent for reagent blanks (zero concentration standard). Store working standards at 4 °C. With care to prevent microbial contamination, mold growth, and evaporation, these working standards will remain stable for several months. Never sample directly from the container; mix the refrigerated standard, pour an aliquot into another container or tube, and return the bottle of standard to the refrigerator.

Note: All eight standards are not required for every run. A minimum of three to four acceptable points is necessary, plus the zero concentration standard, which must always be included. Standard concentrations of 0, 50, 150, and 300 mg/dL are adequate to cover normal fasting and tolerance testing ranges. If specimens from newborns will also be analyzed, a 25 mg/dL glucose working standard should be added. For a population containing a large number of diabetics, standard concentrations should be selected to extend the upper range—for example, 0, 50, 100, 200, and 400 mg/dL. Specimens with glucose concentrations greater than 400 mg/dL should be either diluted with water and repeated or re-analyzed with a smaller sample size. The 600 mg/dL standard is used to assess the adequacy of the enzyme reagent and for quality control, but is not used in calculating results for unknowns.

4. *Tris buffer, (0.1 mol/L, pH 7.5 at 25 °C)*. Prepare stock Tris base and Tris·HCl as follows. For the Tris base, weigh 6.06 g of reagent grade tris(hydroxymethyl)aminomethane (Tris), dissolve and dilute with water to 500 mL in a volumetric flask, and mix well. To make stock Tris·HCl, weigh 31.52 g of reagent grade Tris hydrochloride, then dissolve and dilute to volume with water in a 2-L volumetric flask. Mix 400 mL of stock Tris base and 1600 mL of stock Tris·HCl in a large beaker and measure the pH. If necessary, adjust the pH to 7.5 by adding additional Tris base (if too acid) or Tris·HCl (if the pH is greater than 7.5). After adjusting the pH, transfer the buffer into a screw-cap bottle and refrigerate. About half of this buffer is used to make the enzyme reagent; the remaining half is dispensed in aliquots for preparing specimen blanks.

5. *Enzyme reagent*. Weigh and measure the following and add to a 1-L volumetric flask, using Tris buffer for washing and diluting to volume: 0.92 g of magnesium acetate tetrahydrate (reagent grade); 0.551 g of adenosine 5′-triphosphate (ATP, disodium salt, trihydrate); 750 U of hexokinase; 750 U of glucose 6-phosphate dehydrogenase; and 0.633 g of β-nicotinamide adenine dinucleotide, oxidized form (β-NAD$^+$, dihydrate) *or* 0.765 g of nicotinamide adenine dinucleotide phosphate, oxidized form (NADP$^+$, sodium salt, hydrate).

Note: If the source of the glucose 6-phosphate dehydrogenase is bacterial *(Leuconostoc mesenteroides)*, then NAD$^+$ *must* be used as the cofactor. If the dehydrogenase is from yeast, then NADP$^+$ *must* be used. The enzyme from *L. mesenteroides* is recommended, being slightly less susceptible to certain interferences *(2)*.

Mix well and immediately perform the following steps to test for enzyme reagent adequacy:

(a) Pour about 15 mL of reagent into a container and refrigerate the remainder.

(b) Prepare three spectrophotometric cuvets, matched at 340 nm with water. Fill the first cuvet with water and the second with enzyme reagent (reagent blank). Place the cuvet holder containing the three cuvets in the spectrophotometer.

(c) Dispense 6 mL of enzyme reagent into a test tube and add 50 µL of 600 mg/dL glucose standard that has come to room temperature; mix well. Start a timer immediately and transfer this solution to the third cuvet.

(d) Record the absorbances of the reagent blank and the glucose test solution against the water at 340 nm after 6, 8, and 30 min. With respect to timing, the reading of the glucose test solution at 8 min is most important. The criteria for acceptability are:

- Absorbance of the reagent blank must be less than 0.1 at 30 min.
- The difference between the absorbance readings of the reagent blank at 6 and at 30 min must be 0.01 or less.

- Absorbance of the 600 mg/dL glucose test solution must be between 1.6 and 1.8 at 30 min.
- The difference between the absorbance readings of the glucose test solution at 8 and at 30 min must be 0.01 or less.

If the enzyme reagent meets all four criteria for adequacy, dispense it on the day it is prepared in aliquots sufficient for one run or one day's use, on the basis of 6 mL for each test. Store the aliquots in clean, dry, heavy borosilicate glass screw-cap bottles at −20 °C. Enzyme reagent prepared and stored in this manner is stable for approximately six months; therefore, large amounts can be prepared and stored.

Aliquot and freeze the remaining Tris buffer in similar amounts for each day's use in preparing specimen blanks.

Note: The most common reason for the enzyme reagent to fail this adequacy test is insufficient enzyme, caused by deterioration on storage, overestimation of activity by the manufacturer, or the use of temperatures greater than 25 °C for the enzyme assays. An insufficient amount of either of the two enzymes required will be detected by the last two criteria above. Rather than discarding unacceptable reagent, one may add additional enzymes and repeat the test. Unacceptable results for the first two criteria listed are usually due to impurities in the NAD^+ or $NADP^+$, which cannot be remedied. This test for enzyme reagent adequacy also monitors spectrophotometer performance; excessive drift and nonlinearity of the spectrophotometer can result in rejection of an adequate enzyme reagent.

Note: The highest purity enzymes and coenzymes obtainable should be used in this reagent.[2] The increased cost is justified by the increased stability of the resulting reagent. Coenzyme preparations of 98% purity or better (formula weight basis) are readily available. For the enzymes, either a suspension in ammonium sulfate solution or a lyophilized powder is acceptable. Assay values provided by the manufacturer may be used, if they are given in useful units, such as activity per unit volume or weight. Preferably, these values will have been determined at 25 °C by the Darrow–Colowick (hexokinase) and the Olive–Levy (glucose-6-phosphate dehydrogenase) methods (8, 9).

Note: Evaluator T. J. B. experienced difficulty with the solubility of a hexokinase preparation. To circumvent this, he dissolved both the hexokinase and the glucose-6-phosphate dehydrogenase in 20 mL of 500 mL/L glycerol before addition to 1 L of buffer.

Apparatus

1. *Spectrophotometer.* A high-quality, stable spectrophotometer with good spectral resolution at 340 nm (half bandwidth of 10 nm or less) and linearity through 2.0 absorbance is required. The cuvet pathlength should be 1.0 cm. Performance should be pretested with solutions of NADH, acid dichromate, or other suitable spectrophotometric standards (10).

Note: The 6-mL test volume is sufficient for use with a flow-through cuvet or a single cuvet (with plane-parallel faces) with rinsing with the next sample between readings. Control of temperature in the cuvet compartment (at 25 °C) is recommended, but not essential.

2. *Volumetric measurements.* Class A volumetric pipettes and flasks must be used for preparing glucose standard solutions. Because these standards are taken through the entire procedure, accuracy of other volumetric measurements is less critical than precision. Volumetric Class A pipettes (6-mL) or various dispensing devices may be used to measure buffer and enzyme reagent. For measuring specimens, take care to minimize the effect of the different viscosities of serum and plasma and aqueous standards. This can potentially introduce considerable error, particularly with microvolumes, if pipettes with large-bore tips are used. For this reason, Lang–Levy, Kirk, or Lambda-type pipettes are recommended. Automatic pipettors can also be used.

3. *Water baths.* With this equilibrium/endpoint method, the actual incubation temperature used in less important than keeping the temperature constant, because the absorptivity of reduced pyridine nucleotides varies with temperature. If room temperature is fairly constant, then a water bath is not necessary. Otherwise, a pan of water is usually adequate to maintain temperature during a small- or medium-size run. If room temperature is warmer than 30 °C, then a 25 °C water bath should be used in preparing standards (to bring the stock and working standards to a constant temperature before making to volume) and for equilibrating standards and specimens before sampling, to minimize volumetric errors and evaporation.

Collection and Handling of Specimens

Important considerations for serum and plasma specimens for glucose analysis include preventing hemolysis; minimizing the time in which the serum or plasma remains in contact with the clot, both before and after centrifugation; keeping the sample sterile; storing at low temperature; preventing evaporation; and preventing cellular or chemical contamination. Repetitive sampling from the same container risks potential deterioration of the specimen from repeated thawing, evaporation, or contamination. Sodium fluoride, 2 mg/mL of blood, is the recommended preservative for collecting blood and must be combined with an anticoagulant such as disodium ethylenediaminetetraacetate (EDTA), 1 mg/mL of blood. This mixture is available in conventional evacuated collection tubes (such as from Becton Dickinson Co., Rutherford, NJ 07070). Other anticoagulants that can be used are heparin, oxalate, and citrate. For addition to "harvested" serum or

---

[2] Sources for the enzymes and coenzymes include: Boehringer-Mannheim Corp., New York, NY 10017; Calbiochem, La Jolla, CA 92037; Microbics Division of Beckman Instrument Co., Carlsbad, CA 92008; Sigma Chemical Co., St. Louis, MO 63178; and Worthington Biochemicals Corp., Freehold, NY 07728.

plasma, sodium fluoride, 1 mg/mL of specimen, can be used as a preservative. For storage longer than 48 h before analysis, even preserved specimens should be frozen in containers that are securely closed and have a low permeability to water vapor.

Note: Pooled serum or plasma used for preparing control materials should be filtered through a series of membranes (Millipore Corp., Bedford, MA 01730) to remove tissue cells and microorganisms. Be careful to prevent recontamination during dispensing. Store liquid control materials frozen.

Procedure

1. Remove the appropriate glucose standards and one bottle each of Tris buffer and enzyme reagent from the refrigerator or freezer. Mix the standards well, pour aliquots into labeled tubes, and return the remainder to the refrigerator.

2. Allow reagents, specimens, standards, and controls to come to room temperature.

Note: In a 25 °C water bath, small volumes of previously refrigerated standard or specimens will equilibrate to temperature in about 15–20 min; frozen materials will require from 30 to 45 min.

3. Label two tubes for each sample (standards, specimens, and controls). One tube will be for the "blank" and the other for the "test."

4. Dispense 6 mL of Tris buffer into each of the "blank" tubes. Dispense 6 mL of enzyme reagent into each of the "test" tubes.

5. Measure 50 µL (0.05 mL) of the first sample. After sample uptake, wipe the outside of the pipette with a clean lint-free tissue, using a downward motion and taking care not to touch the tip.

6. Deliver the sample into the "blank" tube, mix the tube contents, and set it aside.

7. Measure another 50 µL of the first sample as in Step 5. Deliver this into the "test" tube.

8. Proceed with the next sample, first dispensing 50 µL of sample into the "blank" tube, then 50 µL into the "test" tube.

Note: If a sampler-dilutor is used, the reagent volume, either Tris buffer or enzyme reagent, may be used to wash the sample into the receiving tube. All "blanks" can be sampled and dispensed, followed by the "tests." Timing is critical only for the "tests," but the set of "tests" should be sampled as soon after the set of "blanks" as possible, in case the specimens should become microbiologically contaminated during the first sampling.

9. Adjust the spectrophotometer to zero at 340 nm with water. Using matched cuvets, record the absorbance of the "test" vs the absorbance of its "blank" for each sample. Absorbance must be measured within 10 to 30 min after the sample is added to the enzyme reagent. Between samples, drain the two cuvets, rinse them once or twice with water, then rinse once with the next sample, and discard. Read all "blanks" in one cuvet and all "tests" in the other.

Note: Instead of readjusting the spectrophotometer to zero with each "blank" before reading the absorbance of the respective "test," the absorbance of the "blank" and "test" can be read vs water and the absorbance of the "blank" subtracted from the absorbance of the "test."

Several options can be used for obtaining the minimum 10 min and maximum 30 min timing. One or two samples can be processed at a time and read at 10 min. For a small number of samples, all samples can be processed and absorbance measurements begun within 30 min after adding enzyme reagent to the first sample. Batches of samples can be processed for reading at any designated time between 10 and 30 min. The same incubation time should always be used.

Note: Evaluators J. W. L., P. G. K., and S. B. found that strongly lipemic samples have high blank readings and must be either diluted or cleared before analysis.

Calculations

After spectrophotometry has been completed, plot a standard curve on graph paper. The concentrations of the standards are recorded on the abscissa (x-axis) and absorbance on the ordinate (y-axis). Connect the points to form the standard curve, which should approximate a straight line, as shown in Figure 1.

Results for unknown specimens can be obtained in one of three ways:

1. Read the concentration for the corresponding absorbance from the standard curve.

2. Subtract the absorbance of the zero concentration standard (reagent blank) from the absorbance of each standard and unknown to obtain corrected absorbances (Corr A). Select the standard whose absorbance most nearly matches that of the sample and apply the formula:

$$\text{Concn}_{sample} = \frac{\text{Concn}_{standard}}{\text{Corr } A_{standard}} \times \text{Corr } A_{sample}$$

3. Calculate the least-squares estimate of the slope (b) and intercept (a) of the calibration curve, where each individual uncorrected absorbance value is taken as the dependent variable (y) and its concentration as the independent variable (x). Use these statistics to obtain a multiplication factor (F) and a correction factor (C) for calculating the concentrations of samples as follows: $F = 1/b$, and $C = a/b$.

$$\text{Concn of sample} = (A_{sample})(F) - C$$

Example: For the standard curve shown in Figure 1, the calculated slope was 0.0026527 with an intercept of 0.04776.

$$F = 1/0.0026527 = 376.97$$
$$C = 0.04776/0.0026527 = 18.00$$

The uncorrected absorbance in the 600 mg/dL standard in this run was 1.640. The calculated concentration for this standard was therefore:

$$(1.640)(376.97) - 18.00 = 600.2 \text{ mg/dL}$$

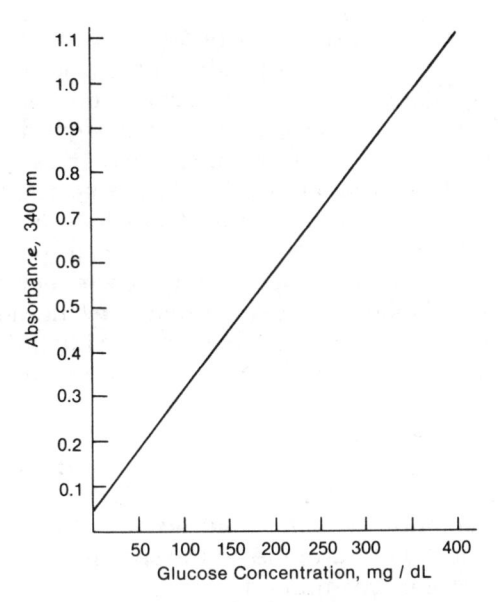

Fig. 1. Full-range standard curve
Zero concentration = reagent blank

Note: The 600 mg/dL glucose standard is one of the most sensitive indicators of enzyme reagent deterioration. In addition to its use in validation of the enzyme reagent, this standard should be included as an unknown in every run. Its absorbance readings should be accumulated and inspected to detect slow degradation, and an absorbance of 1.5 should be the lowest acceptable reading. The concentration should also be calculated and plotted, as for a regular quality-control specimen. Once the method is in operation, new standards and new enzyme reagent should never be introduced in the same run.

Note: Analyze specimen-simulating control samples routinely. As with calibration standards, the number of samples and the concentrations used will depend upon the patient population. There should be a control near each medical decision point. For general use, concentrations near 40, 110, and 180 mg/dL glucose are recommended. A laboratory serving a diabetic clinic should add a control in the 250–350 mg/dL range.

Micro adaptation

To conserve specimen or enzyme reagent, the volumes used in this procedure can be halved, i.e., 25 µL of sample added to 3 mL of Tris buffer or enzyme reagent. This will require use of a microvolume cuvet having a 1.0-cm pathlength, rigidly mounted inside the spectrophotometer during the determination; load and rinse with an appropriate pipetting system.

"Stat" analysis

The simplicity and consistency of this method is such that it can be readily adapted for emergency use.

1. Thaw one container each of Tris buffer and enzyme reagent, equilibrate them to room temperature, and dispense in 6-mL amounts into two sets of test tubes. Cap the tubes tightly and label the two sets as Tris buffer or enzyme reagent. Place the tubes in the refrigerator immediately after dispensing and store them at 4 °C until needed. A less desirable alternative, if care is taken to minimize the time at which the enzyme reagent is at room temperature, is to dispense reagents remaining after a regular run for later use. In either case, enzyme reagent once thawed should not be refrozen, nor used after 24 h.

2. At the beginning of the work shift, turn on the spectrophotometer to "warm up." Pour aliquots of the working standards into labeled tubes and return the standards and aliquots to the refrigerator, except for the zero and 600 mg/dL aliquots. Place the latter and two tubes of enzyme reagent in a 25 °C water bath. After the standards and reagents have equilibrated to temperature and the spectrophotometer has stabilized, test the adequacy of the enzyme reagent as described earlier.

3. While the patient's specimen is being centrifuged, perform the following tasks: Place aliquots of appropriate working standards and controls and one tube each for each sample of Tris buffer and enzyme reagent in the water bath. Check the spectrophotometer and adjust to zero at 340 nm with water.

4. Perform the glucose analysis according to steps 5 through 9 of the regular procedure.

5. Calculate the glucose concentration of unknowns and controls by comparison with the nearest standard:

$$\text{Concn}_{\text{sample}} = \frac{\text{Concn}_{\text{standard}}}{\text{Corr } A_{\text{standard}}} \times \text{Corr } A_{\text{sample}}$$

Calculations can also be performed by using the "F factor" and "C factor" generated in the previous run, provided concentrations calculated with these factors for standards and controls analyzed with the "stat" specimens meet quality-control criteria.

Use of Commercially Prepared Enzyme Reagent

A large number of diagnostic kits are commercially available in which these principles are used for measuring glucose. The lyophilized reagent can usually be purchased in many sizes. Although the composition may be similar to that of this method, the concentrations of the reactants will vary. These commercial reagents can be used, if they are pretested according to the adequacy test procedures discussed previously and meet the four criteria for acceptability. Most directions call for reconstituting the enzyme reagent just before use; therefore, it may be advisable to order different sizes from the same batch or lot, for example, a large size for routine use and a smaller size for emergency analyses. Because a buffer solution for preparing specimen blanks is not generally supplied, the user will have to prepare this separately. It should be of the same composition, concentration, and pH as the buffer used in the commercial reagent.

Discussion

The performance characteristics of automated, micro versions of this procedure, with both bacterial and yeast dehydrogenase, were tested and compared with four other methods (2): the hexokinase glucose reference method, which involves specimen deproteinization with barium hydroxide and zinc sulfate, glucose-6-phosphate dehydrogenase from *L. mesenteroides*, and NAD$^+$ as the coenzyme; a similar deproteinization procedure with yeast dehydrogenase and NADP$^+$; the glucose oxidase method of Fales (11), which in addition involves a Somogyi filtrate; and an adaptation of the AutoAnalyzer ferricyanide procedure.[3] In this study, the enzymes used for the four hexokinase procedures were assayed to assure exact concentrations, and additional adequacy tests were performed on the enzyme reagents to detect contaminating enzymes.

All four hexokinase methods gave results that were linear with concentration through 600 mg/dL, based on full-range standard curves analyzed on 19 days with three different sets of enzyme reagents. On six control pools analyzed in these same runs, the hexokinase methods involving deproteinization were most precise: both had total standard deviations (1 SD) less than 2.0 mg/dL at all concentrations tested, which ranged from 41 to 209 mg/dL. The specimen-blanked direct approach used in this procedure was slightly less precise at all concentrations. Still, the SD was less than 1.8 mg/dL for the five pools with a glucose concentration less than 135 mg/dL, and did not exceed 2.5 mg/dL at 209 mg/dL.

Analytical recovery of glucose was tested for amounts equivalent to 40, 87.5, and 150 mg of glucose per 100 mL, added to a bovine serum pool with an endogeneous glucose concentration of 47 mg/dL. The recovery obtained with the specimen-blanked, direct procedures ranged from 99.4 to 101.6%. The average for yeast enzyme with NADP$^+$ was 100.5%.

Precision, recovery, and accuracy data of the Evaluators are shown in Table 1.

Nineteen substances that could possibly cause interference were evaluated individually in aqueous standard and bovine serum. These included other substrates for the primary enzymes or their contaminants, glucose-containing disaccharides, and common serum constituents or drug metabolites previously reported to affect glucose methods. None of these potential interferences when tested in aqueous glucose standards had an effect on the hexokinase methods. In serum, of the five substances with a determinable effect on any of the hexokinase techniques, three (glucose 1-phosphate, fructose 6-phosphate, and maltose) apparently resulted from conversion to glucose during preparation of the serum interference pools, because they were detected by all six glucose methods. The interference by maltose appeared to be an artifact in the bovine serum base. On assay, it demonstrated a maltase-like activity not found in four fresh human serum pools. Reduced glutathione, which had a slight effect on the specimen-blanked, direct hexokinase method involving yeast dehydrogenase and NADP$^+$, did not affect any of the other hexokinase procedures. It is reportedly co-precipitated with Somogyi deproteinization. Glutathione reductase, an erythrocyte enzyme that can also be a contaminant of the reagent enzymes, preferentially utilizes NADP$^+$ as the pyridine nucleotide, which could account for the lack of influence of glutathione on the specimen-blanked, direct hexokinase method with NAD$^+$. All four of these interferences with hexokinase methods were minor, requiring at least 10-fold the normal physiological concentrations for detection.

The fifth substance, fructose, was recovered by the ferricyanide method from both serum and aqueous glucose solution, but of the five enzymic procedures, it affected only the blanked direct hexokinase methods in serum. Phosphoglucose isomerase, which converts fructose phosphate (resulting from hexokinase reaction with fructose) into glucose 6-phosphate, occurs naturally in human serum and can be a contaminant of both enzymes used in the assay. Although only small amounts of fructose are usually present in normal fasting sera, it is the potentially most serious interference in this procedure. After healthy persons ingest 2 g of sucrose (table sugar) per kilogram of body weight, serum fructose increases to 8 to 10 mg/dL within an hour and persists for 2 h. In fructose-tolerance testing and in individuals with essential fructosuria or hereditary fructose intolerance, serum fructose concentrations can reach 100 mg/dL. Use of fructose as a sugar substitute in dieting is also becoming popular. Because glucose is the preferred substrate of hexokinase, reaction with fructose mainly occurs after glucose in the sample has been depleted (Figure 2); therefore, one way to minimize possible fructose interference is to take absorbance readings as soon as possible after the sample is added to enzyme reagent, for example, using 10 min rather than 30 min as the incubation time. Obviously, solutions administered during glucose tolerance testing should not contain fructose or fructose-containing saccharides.

To test for idiosyncratic, specimen-related bias, analyses were performed on individual fresh human plasma specimens from a group of apparently healthy subjects and from an appropriate clinical population. Blood specimens from 64 "normals" and 92 diabetic patients (being controlled with diet) were

---

[3] Technicon N-2B Methodology Bulletin, Technicon Instruments Corp., Tarrytown, NY 10591.

Table 1. Statistical Evaluation Data

		Within-day				Day-to-day		
Evaluator	n	Mean, mg/dL	SD, mg/dL	CV, %	n	Mean, mg/dL	SD, mg/dL	CV, %
Precision								
J. W. L., P. G. K.,	20	69.4	1.60	2.31	20	69.6	2.24	3.21
and S. B.	20	90.2	1.77	1.96	20	90.8	2.31	2.54
	20	309.1	5.82	1.88	20	315.1	7.72	2.45
T. J. B.					20	84.3	2.52	3.0
					20	252.8	3.36	3.0

		Glucose, mg/dL		
	Sample	Calculated	Assayed	Recovery, %
Recovery				
J. W. L., P. G. K.,	Serum		85.0	
and S. B.	+19.6 mg/dL[a]	104.6	101.7	97.2
	+47.6 mg/dL[a]	132.6	128.7	97.1
	+90.9 mg/dL[a]	175.9	166.0	94.4
T. J. B.	Standards	50.0	49.7	99.4
		100.0	100.8	100.8
		200.0	201.4	100.7
		300.0	299.5	99.8
		400.0	400.8	100.2

	Sample	Comparison method[b]	y-intercept	Slope	r
Comparison					
J. W. L., P. G. K., and S. B.	97 patients' specimens, with values from 56 to 424 mg/dL	Glucose oxidase	0.190	1.01	0.998
T. J. B.	120 patients' specimens, with values from 25 to 500 mg/dL	Glucose oxidase (oxygen uptake)			0.984

[a] Glucose concentration added to serum.
[b] Both with the Beckman Astra 8.

collected the morning after an overnight fast, into evacuated collection tubes containing EDTA and NaF. Overly turbid or hemolyzed plasma specimens were rejected. The excellent agreement between the specimen-blanked, direct hexokinase method and the candidate glucose reference method (NAD⁺ supernate) is shown in Figure 3. With dehydrogenase from either bacteria or yeast, only 10 of the 156 specimens gave results by the direct procedures that differed by more than 10 mg/dL from those obtained with the reference method. Five of these were higher and five were lower; nine of these ten were from diabetics. Extremes were −18.5 to +18.8 mg/dL. With respect to the other glucose methods tested, the estimated mean differences between the specimen-blanked, direct hexokinase procedure and the glucose oxidase method indicated results by glucose oxidase were 1 to 3 mg/dL higher. Ferricyanide also yielded higher results than by this method, by about 3 to 5 mg/dL.

The specimen blanks were also examined to determine the magnitude of the sample blank correction for materials absorbing at 340 nm in human plasma (Table 2). The average correction is equivalent to 7 mg of glucose per 100 mL, but errors as great as 20 to 30 mg/dL could result if correction for specimen background is not made. With grossly hemolyzed, turbid, or icteric specimens, which were excluded from this study, blank correction might be higher.

On the plasma specimens collected from the 64 fasting "normal" donors, the range of glucose con-

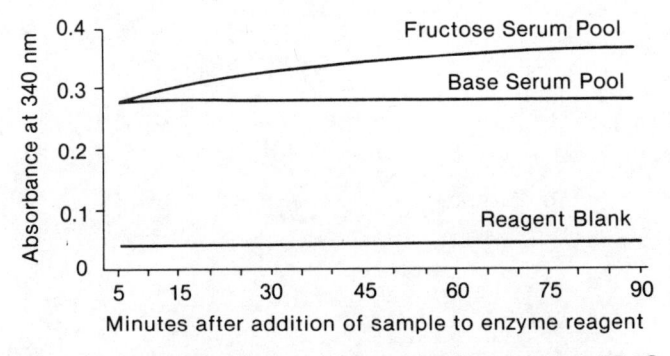

Fig. 2. Fructose-modifying activity of bovine serum at 25 °C

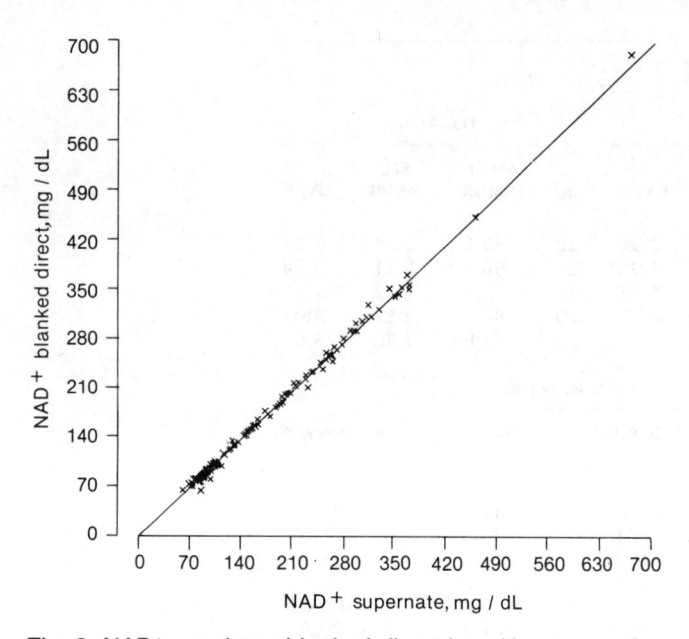

Fig. 3. NAD⁺ specimen-blanked direct hexokinase vs reference hexokinase (NAD⁺ supernate)

No. of pairs = 152 (values obtained on diluted specimens were excluded from calculations); estimate of slope of least-squares line = 1.005; estimate of intercept of least-squares line = −0.081; correlation coefficient = 0.9988

centration was 70 to 105 mg/dL. The donors were 33 males and 31 females, ages 8 to 62 years.

References

1. Proposed performance standard for in vitro diagnostic devices used in the quantitative measurement of glucose in serum or plasma. Food and Drug Administration Medical Device Standards Publication Technical Report, May 1, 1980. USDHHS, PHS, FDA, Bureau of Medical Devices, Silver Spring, MD 20910.
2. Neese, J. W., Duncan, P., Bayse, D., Robinson, M., Cooper, T., and Stewart, C., Development and evaluation of a hexokinase/glucose-6-phosphate dehydrogenase procedure for use as a national glucose reference method. HEW Publication No. (CDC) 77–8830, Center for Disease Control, Atlanta, GA 30333, 1976. (Available on request to the first author.)
3. McGlothlin, C. D., and Jordan, J., Enzymatic enthalpimetry, a new approach to clinical analysis: Glucose determination by hexokinase-catalyzed phosphorylation. *Anal. Chem.* **47**, 787–790 (1975).
4. Warburg, O., Christian, W., and Griese, A., Hydrogen-transferring coenzyme, its composition and mode of action. *Biochem Z.* **282**, 157–205 (1935).
5. Horecker, B. L., and Smyrniotis, P. Z., Reversibility of glucose 6-phosphate oxidation. *Biochim. Biophys. Acta* **12**, 98–102 (1953).
6. McComb, R. B., Bond, L. W., Burnett, R. W., Keech, R. C., and Bowers, G. N., Jr., Determination of the molar absorptivity of NADH. *Clin. Chem.* **22**, 141–150 (1976).
7. NCCLS Publication No. TSC-3, Reagent water specification and test methods for water use in the clinical laboratory. National Committee for Clinical Laboratory Standards, 771 East Lancaster Ave., Villanova, PA 19085.
8. Darrow, R. A., and Colowick, S. P., Hexokinase from baker's yeast. *Methods Enzymol.* **5**, 226–235 (1962).
9. Olive, C., and Levy, H. R., The preparation and some properties of crystalline glucose 6-phosphate dehydrogenase from *Leuconostoc mesenteroides*. *Biochemistry* **6**, 730–736 (1967).
10. Rand, R. N., Practical spectrophotometric standards. *Clin. Chem.* **15**, 839–863 (1969).
11. Fales, F. W., Glucose (enzymatic). *Stand. Methods Clin. Chem.* **4**, 102–112 (1963).

Table 2. Magnitude of Sample Blank Corrections for Materials Absorbing at 340 nm in Human Plasma

	"Normal" population	Diabetic patients	Total
No.	64	92	156
Mean absorbance	0.0164	0.0166	0.0165
Approx. glucose equivalent, mg/dL	7	7	7
Absorbance range	0.007 to 0.047	0.008 to 0.068	0.007 to 0.068
Approx. glucose equivalent, mg/dL	2 to 20	3 to 30	2 to 30

Glucose, o-Toluidine Method

Submitters: Richard B. Passey, Ronald L. Gillum, and Mary Louise Baron, *Department of Pathology, University of Oklahoma Health Sciences Center, Oklahoma City, OK 73125*

Evaluators: Don S. Miyada, A. Akiyama, and C. Ying, *University of California at Irvine, Medical Center, Orange, CA 92668*
O. R. Flores, *Beckman Instrument Co., Fullerton, CA 92634*
Paul F. A. Richter, *Department of Pathology and Nuclear Medicine, Presbyterian–University of Pennsylvania Medical Center, Philadelphia, PA 19104*
Angelo Burlina, *Instituti Ospitalieri, Verona, Italy*
Callum G. Fraser and Craig R. Hearne, *Flinders Medical Centre, Bradford Park, South Australia*

Reviewer: Edward A. Sasse, *Department of Pathology, The Medical College of Wisconsin, Milwaukee, WI 53226*

Introduction

Reid *(1)*, in 1896, published a method for the estimation of sugar in blood, requiring "about 50 cc of blood"; since then, many procedures for glucose have been introduced *(2)*. The three predominant surviving methods are glucose oxidase, hexokinase, and o-toluidine. The o-toluidine method for the determination of glucose is well suited to manual *(3–7)* as well as automated *(7–9)* applications.[1] The method described here *(6)* involves the use of o-toluidine in dilute acetic acid, with thiourea as a stabilizer, and boric acid as an accelerator. The presence of a clarifying agent ("cremophor EL") prevents interference from turbidity.

Principle

o-Toluidine condenses with an aldose or ketose to form an intermediate that equilibrates with its Schiff base *(4)* and possibly forms other complex chromogens *(12)*.

These reactions occur in acid medium with added heat. This method displays sufficient specificity because glucose is the only aldose or ketose found in serum or plasma in significant quantity *(13)*. The final color is measured spectrophotometrically and its intensity is linearly related to glucose concentration.

Materials and Methods

Reagents

To prepare the o-toluidine reagent, add 33 mmol (2.5 g) of thiourea (Eastman Kodak Co., Rochester, NY 14650) to 500 mL of glacial acetic acid (J. T. Baker Chemical Co., Phillipsburg, NJ 08865) and 350 mL of saturated (25 °C) boric acid (J. T. Baker) in de-ionized water.

Note: Evaluators C. G. F. and C. R. H. found that the saturated boric acid solution, when prepared during the working day, became supersaturated during the cooler night. The solution was refiltered before use.

Completely dissolve the thiourea and boric acid before adding 150 mL of o-toluidine (Matheson Scientific, Elk Grove Village, IL 60007), and 50 mL of 10 g/L "cremophor EL" (Sigma Chemical Co., St. Louis, MO 63178). Stored in a brown bottle at 4–8 °C, this reagent is stable at least 90 days.

Note: Reviewer E. A. S. observed *(8)* that the sequence of addition of the components of the o-toluidine reagent is important in enhancing the reagent's stability; i.e., thiourea, glacial acetic acid, followed by o-toluidine, and water. This may not be a factor for the Ceriotti reagent *(6)* we describe.

Note: Evaluators D. S. M., A. A., C. Y., and O. R. F. demonstrated no reagent instability over a two-month period.

Do not use the o-toluidine reagent if it is discolored (brown). All chemicals are analytical grade. Prepare

---

[1] Glucose methods previously published as *Selected Methods* or under the older *Standard Methods* are: *(a)* evaluation and comparison of 10 glucose methods and the reference method recommended in the proposed product class standard (1974) *(7)*; *(b)* the determination of glucose by the o-toluidine method (filtrate and direct procedure) *(5)*; *(c)* ultramicro glucose (enzymic) *(10)*; and *(d)* glucose (enzymic) *(11)*.

standards by weighing D-glucose (Standard Reference Material 917, National Bureau of Standards, Washington, DC 20234) into 1.0 g/L (8.19 mmol/L) benzoic acid (J. T. Baker) in de-ionized water.

Apparatus

The spectrophotometer used was a Coleman Jr. II with 20-nm bandpass (Coleman Instruments, Oak Brook, IL 60521) and round cuvets of 12-mm outside diameter by 75-mm length. The pipettes were 50.0 ± 0.3 µL (± 2 SD) (Scientific Manufacturing Industries, Emeryville, CA 94608). Heating was performed in a dry block heater (Model 2090 with Model 2073 Anodized Aluminum Block; Lab Line Instruments, Inc., Melrose Park, IL 60160).

Collection and Handling of Specimens

For best results, collect specimens so that they contain at least 4.3 g/L (0.10 mol/L) sodium fluoride anticoagulant.

Note: Evaluators C. G. F. and C. R. H. found that heparinized plasma gave identical values with plasma in oxalate/fluoride, provided the assays were performed less than 30 min after the heparinized specimens were collected.

If promptly removed from the erythrocytes, serum gives excellent results. Centrifuge specimens at 1000 × g for 8–10 min. If the assay will be delayed more than 12 h, freeze the serum (−20 °C).

In all cases, protect the specimen from evaporation by tightly capping the tube. Before analysis, mix the specimen thoroughly, and take an aliquot for the assay.

Procedure

1. Add 3.0 mL of *o*-toluidine reagent to clean 16 × 100 mm test tubes marked blank, standard, or unknown (up to 10 unknowns, specifically identified).

Note: Evaluators C. G. F. and C. R. H. reported that it was difficult to perform more than 10 assays (plus standard, blank, and quality-control specimens) in a single batch.

2. Add 50 µL of water to the blank tube, 50 µL of standard 100, 150, or 200 mg/dL (5.6, 8.3, or 11.1 mmol/L) glucose in 1 g/L benzoic acid, and 50 µL of each unknown to the appropriate tubes.

Note: Evaluators D. S. M., A. A., C. Y., and O. R. F. found no significant differences when reagents and samples were added in the reverse order. With a vortex-type mixer, or by inversion, mix each tube thoroughly after covering it with Parafilm (American Can Co., Greenwich, CT 05830).

3. Quickly place all tubes in the dry block heater set at 90 ± 1 °C. A hot water bath may also be used.

Note: Evaluators D. M. S., A. A., C. Y., and O. R. F. demonstrated poor precision when they used a dry block heater; they suggest that a boiling water bath gives more consistent results.

Note: Evaluators C. G. F. and C. R. H. found that if a boiling water bath is used for heating, the bath must be boiling vigorously before inserting the tubes, to obtain good color development.

Place a clean marble (solid glass sphere greater than 17-mm diameter) on the top of each tube to reduce evaporation while preventing the pressure from increasing to dangerous levels. The "cremophor EL" precipitates in the hot solution but redissolves as the solution cools to produce a clear final solution.

4. After 10.0 min, remove all tubes in the same order as they were put into the heating block, and place them for 5 min in an ice-water bath (as deep as the top of the reaction mixture).

Note: Reviewer E. A. S. observed that the Submitters specify a sample-to-reagent ratio different from the Ceriotti reference *(6)* and a 10-min heating time, whereas Ceriotti recommended 15 min. From the excellent results reported here, these changes probably have no adverse effects, despite the fact that Ceriotti showed significant loss of color within a short time after only 5 min of heating.

5. Read the absorbance of each specimen at 630 nm with the spectrophotometer set to 0.000 absorbance with the blank. Each specimen must be free of turbidity or the absorbance value will be erroneous.

Quality Control and Standardization

One or more of the samples in each run should be a control specimen.

Note: Reviewer E. A. S. recommends using two control samples (normal and abnormal), especially because the method uses only one standard. Control specimens with mean glucose values greater than 150 mg/dL (8.3 mmol/L) should give a ±2 SD range of 5 mg/dL (0.3 mmol/L) from the mean. The expected *(7)* limits of linearity are 10–400 mg/dL (0.6–22.2 mmol/L). For greatest accuracy, specimens with values less than 50 mg/dL (2.8 mmol/L) or more than 300 mg/dL (16.7 mmol/L) should be reassayed along with a standard that has a glucose value near the specimen's value.

Note: Evaluators D. S. M., A. A., C. Y., and O. R. F. report linearity to 300 mg/dL (16.7 mmol/L). Evaluators C. G. F. and C. R. H. showed linearity to 500 mg/dL (27.8 mmol/L), using a Gilford 300 T-1 spectrophotometer. (Also see discussion of linearity under *Analytical Performance*.)

Calculations

Calculate results from absorbance *(A)* at 630 nm as follows:

$$\text{glucose concn} = (A_{\text{unknown}} / A_{\text{standard}}) \times \text{concn of standard}$$

Report values to the nearest 1.0 mg/dL *(6)*.

Table 1. Precision of the Method

	Within-run			Day-to-day		
Evaluators	N	Mean (SD), mg/dL	CV, %	N	Mean (SD), mg/dL	CV, %
D. M. S., A. A., C. Y., and O. R. F.	20	101 (<2.0)	2.0			
C. G. F. and C. R. H.	20	70 (1.0)	1.4	20	68 (1.0)	1.4
	20	206 (4.0)	1.9	20	212 (5.0)	2.3
	20	219 (7.0)	3.2	20	203 (6.0)	2.9
A. B.	24	129 (3.0)	2.3		87 (3.0)	3.4
					188 (5.0)	2.6
	21	253 (4.0)	1.5		353 (13.0)	3.6

Discussion

The reaction of o-toluidine with hexoses and ketoses is one of a generalized condensation of aldehydes and ketones with aromatic amines. Among other aromatic amines that have been used are 2,6-dimethylanaline *(3)*; 2,5-dimethylaniline *(3)*; p-aminosalicylic acid *(14)*; p-aminohippuric acid *(14)*; m-aminophenol *(3, 15)*; aniline *(16)*; benzidine *(16)*; xylidine *(16)*; p-bromoaniline *(16)*; and o-aminodiphenyl *(17)*. The reaction with o-toluidine is widely accepted because of its specificity, availability, purity of the reagent used, and reported lack of carcinogenic effects *(18)*.

Note: Evaluators C. G. F. and C. R. H. stress that care must be exercised in handling glacial acetic acid, which can cause chemical burns. o-Toluidine, an aniline derivative, may cause poisoning and hematuria if ingested, inhaled, or absorbed through the skin *(19)*.

Note: Evaluator P. F. A. R. suggests that the procedure be run under a hood because of the noxious fumes from glacial acetic acid.

Many modifications of the reaction conditions have been proposed. Among them are:

1. Changing the concentrations of o-toluidine to 60 mL/L *(3–5, 9)*, 150 mL/L *(6, 20)*, 167 mL/L *(8)*, 5–90 mL/L *(12)*, or 50 mL/L *(20)*.
2. Decreasing the concentration of acetic acid with water *(3, 6, 12, 20 21)*.
3. Substituting other solvents and (or) proton sources for acetic acid *(9, 12, 21)*.
4. Using a protein-free filtrate *(4, 20)*.
5. Using clarifying agents *(6, 8)*.
6. Changing the heating time *(5, 6, 12, 20, 21)*.
7. Adding reaction accelerators and enhancers: thiourea *(5, 6, 8, 9, 12, 16, 20)*, lactic acid *(6, 12)*, boric acid *(6, 12, 21)*, and citric acid *(9, 12)*.

The spectral characteristics of the complex colored products have been documented *(3, 4, 9, 12)*.

Analytical Performance

The method described here gives excellent analytical performance.

1. Within-run precision was measured by the Submitters for six different human serum pools *(7)*. The means for 20 assays on each of the pools ranged from 77 to 202 mg/dL (4.3–11.2 mmol/L) with a corresponding SD of 2 to 3 mg/dL (0.1–0.2 mmol/L). The day-to-day precision was measured on five commercial lyophilized sera *(7)*, each reconstituted daily. The means for 27 to 33 assays on each pool ranged from 76 to 205 mg/dL (4.2–11.4 mmol/L) with a corresponding SD of 2 to 3 mg/dL (0–0.2 mmol/L).

The precision obtained by the Evaluators is shown in Table 1.

2. Linearity was determined with glucose standards of the following concentrations in 1 g/L benzoic acid in water: 12.5, 25, 50, 100, 150, 200, 250, 300, 400, and 500 mg/dL (0.7, 1.4, 2.8, 5.6, 8.3, 11.1, 13.9, 16.7, 22.2, and 27.8 mmol/L). Excellent linear correlation was obtained between glucose concentration and spectrophometric absorbance.

3. The results of recovery studies carried out by the Evaluators, using the method described here, are shown in Table 2.

4. Comparison of the manual direct o-toluidine with the Beckman glucose oxidase/oxygen-rate method was excellent *(7)*. The bias between these two methods, when 228 specimens [glucose range

Table 2. Analytical Recovery Studies

	Glucose, mg/dL		
Evaluators	Added[a]	Found	Recovery, %
C. G. F. and C. R. H.	80	77	96.3
	160	152	95.6
	240	234	97.9
	320	310	96.5
	400	397	99.3
A. B.	50	53	107
	100	102	102
	150	155	104
	200	202	101
	300	282	94

[a] Glucose standards added to a serum pool.

3.4–19.6 mmol/L (61–353 mg/dL)] were assayed by both methods, was 0.1 mmol/L (2.0 mg/dL). These two methods (glucose oxidase/oxygen-rate = x, o-toluidine = y) correlated over the above-mentioned range as follows: slope 0.987, intercept 3.0 mg/dL (0.2 mmol/L), and standard deviation of y about x ($S_{y \cdot x}$ or standard error of the estimate) 5.0 mg/dL (0.3 mmol/L).

Note: The instructions for including the clarifying agent "cremophor EL" were erroneously omitted from the first draft of this chapter submitted to the Evaluators. Consequently, each Evaluator noted and commented on a large, positive, constant bias when this method was compared with other methods. In Table 3, we have documented similar bias and ascribe this bias, as do Evaluators D. S. M., A. A., C. Y. and O. R. F., C. G. F. and C. R. H., to turbidity in the final reaction mixture. This turbidity is eliminated by the use of the clarifying agent. Without the addition of the clarifying agent the positive bias renders this method unacceptable.

5. The analysis of the 95% confidence estimates for random, systematic, and total errors according to the method of Westgard et al. *(22)* showed little systematic error *(7)* in the manual direct o-toluidine method when clarifying agent is added to the reagent. The glucose oxidase/oxygen rate method was used as the comparison method. The systematic error is between 1.0 and 4.0 mg/dL (0.1–0.2 mmol/L) on specimens whose mean glucose is 50–300 mg/dL (2.8–16.8 mmol/L). The random component of error *(7)* is acceptable, being between 3.0 and 8.0 mg/dL (0.2–0.4 mmol/L) for specimens whose means are between 50 and 300 mg/dL (9.8–16.7 mmol/L). The 95% confidence estimates for total error *(7)* were very good, total error being between 5.0 and 9.0 mg/dL (0.3–0.5 mmol/L) for the same range.

Note: Evaluators C. G. F. and C. R. H. found "general agreement" with assay values by the proposed method

Table 3. Effects of a Clarifying Agent and Protein Precipitation on the Manual o-Toluidine Method; Correlation with Other Methods

Evaluators and Submitters	Comparison method	Slope[a]	Intercept, mmol/L	r
Assays by o-toluidine method without clarifying agent or protein precipitation				
D. S. M., A. A., C. Y., and O. R. F.	Hexokinase/Beckman Trace 1[b]	1.00	0.9	0.984
	Glucose oxidase/oxygen rate (Beckman)[b]	1.09	0.5	0.988
	Hexokinase/Gilford 3500[c]	1.02	0.6	0.984
	Hexokinase/Technicon SMAC[d]	1.02	0.4	0.958
A. B.	Glucose oxidase	1.10	0.9	0.990
C. G. F. and C. R. H.	Glucose oxidase/oxygen rate (Beckman)[b]	0.93[e]	1.2	0.992
		0.90[f]	1.5	0.993
	Glucose oxidase/Trinder AutoAnalyzer II (Technicon)[d]	0.83	2.2	0.980
R. B. P., R. L. G. and M. L. B.	Hexokinase manual *(7)*	0.99	1.2	0.996
	Glucose oxidase/oxygen rate (Beckman)[b]	0.99	1.2	0.995
Assays by o-toluidine method after protein precipitation[g]				
C. G. F. and C. R. H.	Glucose oxidase/oxygen rate (Beckman)[b]	0.92[e]	0.1	0.994
		0.91[f]	0.3	0.987
Assays by o-toluidine method with clarifying agent ("cremophor EL") added				
R. B. P., R. L. G., and M. L. B.	Glucose oxidase/oxygen rate (Beckman)[b]	0.99	0.2	0.998

[a] y = o-toluidine method, x = comparison method.
[b] Beckman Instruments Inc., Fullerton, CA 92634.
[c] Gilford Instruments, Inc., Oberlin, OH 44074.
[d] Technicon Instruments, Inc., Tarrytown, NY 10591.
[e] First evaluation by these authors.
[f] Second evaluation by these authors.
[g] The protein-free filtrate modification was performed as follows: Add 100 µL of sample to 1.0 mL of 3% (30 g/L) trichloracetic acid; centrifuge; add 500 µL of supernate to 3.0 mL of o-toluidine (without "cremophor EL"), then follow the selected method exactly.

with international consensus values for multi-method assays of eight lyophilized serum pools.

6. Substances causing interference greater than 2.0 mg/dL (0.1 mmol/L) as apparent glucose per millimole of added substance (when added to serum pools in relatively high concentrations) are shown in Table 4.

Table 4. Substances Causing Significant Interference[a] When Added to Serum Pools (7)

Substance	Apparent glucose, mg/dL (mmol/L)[b]
D(−)-Fructose	3.0 (0.17)
D(+)-Mannose	11.5 (0.64)
D(+)-Galactose	11.0 (0.61)
Na salicylate	2.0 (0.11)
Creatinine	2.7 (0.15)
Uric acid	2.2 (0.12)
Streptomycin	5.8 (0.32)
Methyldopa	4.1 (0.23)
Levodopa	13.9 (0.77)
Dextran 40	94 (5.2)[c]
Bilirubin	51.0 (2.86)
Lipid[d]	9.9 (0.55)[e]

[a] Greater than 2.0 mg/dL (0.1 mmol/L) as apparent glucose per millimole of substance added. A complete listing of substances tested and their concentrations is given in ref. 7. Twenty other substances tested (7) showed no significant interference.
[b] Per millimole of substance added.
[c] Per liter of Dextran 40 added.
[d] 10.2 mmol/L triglyceride as triolein (9000 mg/L) and 4.4 mmol/L cholesterol (1700 mg/L). The effect of lipid is apparent only in grossly turbid specimens.
[e] Per millimole of triglyceride as triolein.
Note: Reviewer E. A. S. observed that mannose, at the same concentration as glucose, produces about 1.6-fold the color of glucose with o-toluidine methods not involving boric acid or "cremophor EL."

Reference Intervals

The normal adult range for glucose as determined by the o-toluidine method has been reported by a number of investigators (Table 5).

Table 5. Reference Intervals

Subjects	mg/dL	mmol/L	Ref.
1330 patients	70–120	3.9–6.7	4
32 selected adults	70–122	3.9–6.8	5
300 specimens	65–105	3.6–5.8	6
550 specimens	60–105	3.3–5.8	20
228 patients	65–110	3.6–6.1	7

References

1. Reid, E. W., A method for the estimation of sugar in blood. *J. Physiol.* **20**, 316–321 (1896).
2. Cooper, G. R., Methods for determining the amount of glucose in blood. *CRC Crit. Rev. Clin. Lab. Sci.* **4**, 101–145 (1973).
3. Hultman, E., Rapid specific method for determination of aldosaccharides in body fluids. *Nature* **183**, 108–109 (1959).
4. Dubowski, K. M., An o-toluidine method for body-fluid glucose determination. *Clin. Chem.* **8**, 215–235 (1962).
5. Cooper, G. R., and McDaniel, V., The determination of glucose by the *ortho*-toluidine method (filtrate and direct procedure). *Stand. Methods Clin. Chem.* **6**, 159–170 (1970).
6. Ceriotti, G., Blood glucose determination without deproteinization, with use of o-toluidine in dilute acetic acid. *Clin. Chem.* **17**, 440–441 (1971).
7. Passey, R. B., Gillum, R. L., Fuller, J. B., Urry, F. M., and Giles, M. L., Evaluation and comparison of 10 glucose methods and the reference method recommended in the proposed product class standard (1974). *Clin. Chem.* **23**, 131–139 (1977); *Sel. Methods Clin. Chem.* **8**, 9–19 (1977).
8. Moorehead, W. R., and Sasse, E. A., An automated micromethod for determination of serum glucose, with an improved o-toluidine reagent. *Clin. Chem.* **16**, 285–290 (1970).
9. Snegoski, M. C., and Freier, E. F., An automated o-toluidine glucose procedure without acetic acid. *Am. J. Med. Technol.* **39**, 140–150 (1973).
10. Meites, S., Ultramicro glucose (enzymatic). *Stand. Methods Clin. Chem.* **5**, 113–120 (1965).
11. Fales, F. W., Glucose (enzymatic). *Stand. Methods Clin. Chem.* **4**, 101–112 (1963).
12. Yee, H. Y., and Goodwin, J. F., Evaluation of some factors influencing the o-toluidine reaction with glucose. *Anal. Chem.* **45**, 2162–2165 (1973).
13. Pileggi, V. J., and Szustkiewicz, C. P., Carbohydrates. In *Clinical Chemistry Principles and Techniques*, 2nd ed., R. J. Henry, D. C. Cannon, and J. W. Winkleman, Eds., Harper and Row, Hagerstown, MD, 1974, p 1268.
14. Ek, J., and Hultman, E., A new method for determining aldosaccharides. *Scand. J. Clin. Lab. Invest.* **9**, 315–316 (1957).
15. Ek, J., and Hultman, E., Determination of glucose and laevulose in body fluids. *Nature* **181**, 780–781 (1958).
16. Roe, J. H., and Rice, E. W., A photometric method for the determination of free pentoses in animal tissues. *J. Biol. Chem.* **173**, 507–512 (1948).
17. Timell, T. E., Glaudemans, C. P. J., and Currie, A. L., Spectrophotometric method for determination of sugars. *Anal. Chem.* **28**, 1916–1920 (1956).
18. Indriksons, A., Hazards of o-toluidine. *Clin. Chem.* **21**, 1345 (1975).
19. *The Merck Index*, Merck and Co., Rahway, NJ, 1968, pp 6, 85, 1059.
20. Abraham, C. V., A modified o-toluidine reagent for glucose analysis. *Clin. Chim. Acta* **70**, 209–211 (1976).
21. Fasce, C. F., Jr., Rej, R., and Pignataro, A. J., A study of the direct o-toluidine blood glucose determination. *Clin. Chim. Acta* **43**, 105–111 (1973).
22. Westgard, J. O., Carey, R. N., and Wold, S., Criteria for judging precision and accuracy in method development and evaluation. *Clin. Chem.* **20**, 825–833 (1974).

Glucose-6-Phosphate Dehydrogenase Activity in Erythrocytes

Submitters: N. Vasudeva Paniker, *Nogales Medical Laboratory, Nogales, AZ 85621*
James S. Roloff, Andy Burton, and John N. Lukens, *Department of Pediatrics, Children's Hospital, Vanderbilt University Medical Center, Nashville, TN 37232*

Evaluators: V. Thomas Alexander, *Department of Pathology, University Hospital, Augusta, GA 30902*
Angelo Burlina, *Laboratorio di Chimica Clinica e di Ematologia, Instituti Ospitalieri de Verona, Verona, Italy*

Reviewer: Carl A. Burtis, *Chemical Technology Division, Oak Ridge National Laboratory, Oak Ridge, TN 37830*

Introduction

Glucose-6-phosphate dehydrogenase (G6PD) (EC 1.1.1.49; D-Glucose 6-phosphate:NADP$^+$ 1-oxidoreductase) is the initial enzyme in the hexose monophosphate pathway of glucose metabolism.[1] It catalyzes the oxidation of glucose 6-phosphate (G6P) to 6-phosphogluconic acid (6-PGA) and simultaneously reduces one molecule of nicotinamide adenine dinucleotide phosphate (NADP$^+$) to NADPH. An inherited deficiency of this enzyme in human erythrocytes may lead to hemolytic anemia, especially when the cells are challenged in vivo with stressful conditions such as exposure to certain oxidant drugs or infectious agents. Investigation of G6PD in different populations has revealed many genetic variants. The normal enzyme is designated G6PD B. About 22% of black males have a variant (G6PD A) that, though having a different electrophoretic mobility, has normal activity and is without clinical significance. Over 140 variants with deficient activity have been identified. The most common of these are G6PD A- (present in approximately 10% of black males), G6PD Mediterranean (occurring primarily in Caucasians), and G6PD Canton (prevalent among oriental groups).

Quantitative determination of erythrocyte G6PD is often performed on blood samples that have been previously subjected to inexpensive screening procedures; however, several of the screening procedures such as the brilliant cresyl blue reduction test *(1)* and the ascorbate–cyanide screening test *(2)* are not always reliable. The most dependable screening procedure depends upon the fluorescence (under longwave ultraviolet light) of NADPH formed by the reduction of NADP$^+$ in the G6PD reaction. A similar principle is utilized in the quantitative determination of enzyme activity. The method described here has been adapted from the methods of Löhr and Waller *(3)* and Beutler *(4)*.

Principle

The oxidation of one molecule of G6P to 6-PGA by G6PD is accompanied by the reduction of one molecule of NADP$^+$ to NADPH. The rate of reduction of NADP$^+$ to NADPH may be monitored spectrophotometrically because the NADPH has a millimolar absorptivity of 6.22 L · μmol^{-1} · cm^{-1} at 340 nm. Another enzyme present in erythrocytes, 6-phosphogluconate dehydrogenase (6PGD; EC 1.1.1.44) generates a second molecule of NADPH by oxidizing 6-PGA to ribulose 5-phosphate with the release of one molecule of carbon dioxide (CO$_2$).

$$\text{G6P} \xrightarrow[\text{NADP}^+ \quad \text{NADPH+H}^+]{\text{G6PD}} \text{6-PGA} \xrightarrow[\text{NADP}^+ \quad \text{NADPH+H}^+]{\text{6PGD} \quad \text{CO}_2} \text{ribulose 5-phosphate}$$

To differentiate between these two enzyme activities, an aliquot of the erythrocyte hemolysate is mixed with saturating concentrations of 6-PGA and NADP$^+$, while an additional aliquot is added to a mixture of G6P, 6-PGA, and NADP$^+$. The amount of NADPH formed in the former reaction (6PGD activity) is subtracted from the amount of NADPH formed in the latter reaction (G6PD activity plus 6PGD activity) to obtain the amount of NADPH formed by G6PD alone. The activity of G6PD is cal-

---

[1] Nonstandard abbreviations used: G6PD, glucose-6-phosphate dehydrogenase; G6P, glucose 6-phosphate; 6-PGA, 6-phosphogluconic acid; 6-PGD, 6-phosphogluconate dehydrogenase (decarboxylating), EC 1.1.1.44; EDTA, ethylenediaminetetraacetate; Tris, tris(hydroxymethyl)aminomethane; Hb, hemoglobin.

culated from the quantity of NADPH produced per minute.

Materials and Methods

Reagents

Note: Sigma Chemical Co., St. Louis, MO 63178, was the source of all reagents, unless otherwise noted. Evaluator A. B. also used reagents from Boehringer Mannheim, F.R.G.

1. *Tris·HCl buffer, 0.1 mol/L, pH. 7.5.* Dissolve 12.1 g of tris(hydroxymethyl)aminomethane (Tris) in approximately 750 mL of water. Using a pH meter, adjust the pH of the solution to about 8.5 by the addition of concd. HCl while the solution is stirred. Carefully adjust the pH to 7.5 with 2 mol/L HCl. Dilute the solution to 1 L. The solution is stable for at least four to six months at 4 °C if no mold forms.

2. *EDTA, 0.05 mol/L.* Dissolve 1.68 g of anhydrous[2] disodium salt of ethylenediaminetetraacetic acid (EDTA) in 90 mL of water. Dilute to 100 mL with water. The solution is stable indefinitely at 4 °C.

3. *$NADP^+$, 5 mmol/L.* Dissolve 382.7 mg of anhydrous[2] $NADP^+$, monosodium salt, in 90 mL of water. Dilute to 100 mL. Dispense the solution into 10-mL aliquots[3] and store at 4 °C. The reagent is stable for at least four to six months if no mold forms.

4. *Glucose 6-phosphate, 6 mmol/L.* Dissolve 169.3 mg of anhydrous[2] glucose 6-phosphate, monosodium salt (G6P), in 100 mL of water. Dispense solution into 4-mL aliquots[3] and freeze at −20 °C. Thaw the required amount for daily work. The reagent is stable for at least four to eight months at −20 °C.

5. *6-Phosphogluconic acid.* Dissolve 205.3 mg of anhydrous[2] 6-phosphogluconic acid, trisodium salt (6-PGA), in 100 mL of water. Dispense solution into 8-mL aliquots[3] and freeze at −20 °C. Thaw the required amount for daily work. The reagent is stable for at least four to eight months at −20 °C.

6. *Saponin, 5 g/L.* Dissolve 500 mg of saponin in 100 mL of water. The reagent is stable for at least four to six months at 4 °C if no mold forms.

7. *Buffered isotonic saline, pH 7.4.* (a) Sodium chloride, 85 g/L, 90 mL; (b) K_2HPO_4, 1 mol/L, 7.1 mL; and (c) KH_2PO_4, 1 mol/L, 2.9 mL. These reagents can be stored at 4 °C for several months if no mold appears. Mix a, b, and c and dilute to approximately 900 mL with water. Adjust the pH to 7.4, using NaOH (1 mol/L) or HCl (1 mol/L) and a pH meter. Dilute to 1 L with water. Recheck pH and adjust to 7.4 if necessary. The reagent is stable at 4 °C for four to six weeks if no mold forms.

8. *Drabkin's reagent* (Stable Cyanmethemoglobin Reagent Dry Pack, Hycel No. 116C; Hycel, Inc., Houston, TX 77036). Reconstitute according to the manufacturer's directions.

Note: The reagents for determining hemoglobin by the cyanmethemoglobin method may also be obtained from other commercial sources. The AACC does not endorse specific manufacturers.

9. *Hemoglobin standards* (Hycel No. 117, Hycel Cyanmethemoglobin Standards). Plot a standard curve according to manufacturer's directions.

10. *Hemolyzing solution.* Add 2 mL of Reagent 1 and 1 mL of Reagent 6 to 15 mL of water. Keep cold on ice. Prepare fresh solution each day.

11. *Reaction mixture.* Prepare fresh reaction mixture each day: 5 parts of Reagent 1, 1 part of Reagent 2, and 1 part of Reagent 3. Readjust pH to 7.5 if necessary.

Apparatus

1. Recording spectrophotometer with temperature-controlled sample chamber (Model 2400S; Gilford Instrument Laboratories, Inc., Oberlin, OH 44074).

Note: Instead of a recording spectrophotometer, a good-quality spectrophotometer suitable for taking readings at 340 nm may be used. Evaluator V. T. A. used a Gilford Stasar III with satisfactory results.

2. Centrifuge (Model J-21; Beckman Instruments, Inc., Fullerton, CA 97634).
3. pH meter.
4. Micropipettes.
5. Test tubes.
6. Three 3-mL cuvets with 1-cm light path.

Procedure

1. Centrifuge about 2 mL of anticoagulated blood (EDTA, heparin, or sodium citrate may be used as anticoagulant) at $1000 \times g$ for 10 min.[4] Carefully aspirate the plasma and the top layer of leukocytes ("buffy coat"), taking care to remove few or no erythrocytes during this process.

Note: For improved separation of erythrocytes from contaminating leukocytes and platelets, Evaluator A. B. suggests a column separation technique (5) recommended by the ICSH Expert Panel on Red Cell Enzymes (6). However, as a general rule, this refinement is unnecessary for the clinical assay of G6PD.

Mix the packed erythrocytes in the tube with about 5 volumes of cold buffered isotonic saline (Reagent

---

[2] Because these reagents vary in water content and purity, those factors must be included in calculating the equivalent weight of the particular material used. Reagent grade chemicals with a purity of 98% or above may be used. Those from the Sigma Chemical Co. have been satisfactory.

[3] This is sufficient reagent for assaying 12 blood samples.

[4] Packed erythrocytes obtained from several heparinized microhematocrit tubes may be used also.

7). Centrifuge and remove the supernate as before. Wash the cells two more times with buffered isotonic saline.

2. Prepare hemolysate by adding 1 volume of the washed packed erythrocytes to 9 volumes of cold hemolyzing solution (Reagent 4). Allow to stand in ice for 15 min and centrifuge at 4500 × g for 10 min. Transfer the supernatant hemolysate to another test tube, taking care not to disturb the sedimented stroma.

Note: Evaluator A. B. prefers to use a hemolysate prepared with 9 volumes of a "stabilizing solution" made of 2-mercaptoethanol, 0.7 µmol/L, and EDTA, 2.7 mmol/L (pH 7.0), plus 1 volume of a washed erythrocyte 50% suspension. This combined solution is then alternately frozen and thawed. Again, this is as recommended by the ICSH Expert Panel on Red Cell Enzymes (6). For ease of technique, here we have dispensed with the use of 2-mercaptoethanol and the additional manipulation of the erythrocytes, which is acceptable if the assay is performed within 2 h of preparation of the hemolysate.

Determine the hemoglobin content of the hemolysate as follows: add 0.1 mL of the hemolysate to 5 mL of Drabkin's reagent (Reagent 8), mix, allow to stand at room temperature for 10 min, and determine the absorbance at 540 nm against a Drabkin's reagent blank. Read the hemoglobin content from a calibration curve prepared from standard solutions of hemoglobin according to manufacturer's specifications. Adjust the hemoglobin concentration of the hemolysate to 1.0 ± 0.3 g/dL by diluting with the hemolyzing solution. Keep the hemolysate in an ice bath. Determine the G6PD activity within 2 h or preparation.

3. Adjust the spectrophotometer to 340 nm and the temperature-controlled sample chamber to 37 °C. Set the recorder expansion to 1.0 if a recording spectrophotometer is used.

4. Arrange 3-mL cuvets (light path, 1 cm) in the cuvet carrier as follows:

Cuvet	Water	Reaction mixture	Hemolysate
1	0.9	2.0	0.1
2	0.6	2.0	0.1
3	0.3	2.0	0.1

Additions, mL

Incubate the samples at 37 °C for 10 min in the sample chamber of the spectrophotometer (or a suitable temperature-controlled block).

5. Add 0.3 mL of G6P (Reagent 4) to cuvet 3 and 0.3 mL of 6-PGA (Reagent 5) to cuvets 2 and 3. Quickly remove the cuvet carrier from the sample chamber, cover the cuvets, mix the contents well, and replace into the sample chamber. Without delay, adjust the positions of the recorder pen and record the increase in absorbance of cuvets 2 and 3 for 10 min, using cuvet 1 as a reference blank. A recorder speed of 1 in./5 min (0.5 cm/min) is satisfactory. If a spectrophotometer suitable for taking readings at 340 nm is used instead of a recording spectrophotometer, measure the absorbance changes of cuvets 2 and 3 between 0 and 10 min.

Calculation

Erythrocyte G6PD activity is usually expressed in IUB units (U) per gram of hemoglobin. One unit is defined as the amount of G6PD that oxidizes 1 µmol of G6P to 6-GPA per minute at 37 °C. From the linear portion of the spectrophotometric recording, calculate the change in absorbance per minute for cuvets 2 and 3. The absorbance change per minute ($\Delta A'$/min) of cuvet 2 is subtracted from that of cuvet 3 to obtain ΔA/min (the absorbance resulting from NADPH generated by G6PD in cuvet 3). The activity of G6PD is calculated from the equation:

$$\text{G6PD acty, U/g Hb} = \frac{(\Delta A/\text{min}) \times 3}{6.22 \times 0.1} \times \frac{100}{\text{Hb concn, g/dL}}$$

where 3 is the volume (mL) of the cuvet, 6.22 is the absorbance of a 1 mmol/L solution of NADPH in a 1-cm light path, 0.1 is the volume (mL) of the hemolysate added to the cuvet, and 100 is milliliters per deciliter. The equation may be simplified to

$$\text{G6PD acty, U/g Hb} = \frac{(\Delta A/\text{min})}{\text{Hb concn}} \times 482.3$$

Results

The G6PD activity was 6.54 ± 0.79 U/g Hb (mean ± 1 SD) at 37 °C, in blood from 25 healthy adult volunteers. There is no difference in G6PD activity between normal men and women or between normal adults and children.

Note: Evaluator V. T. A. tested the procedure on 20 different normal pediatric samples. All had values greater than 6.0 U/g Hb. In addition, 10 samples recognized as deficient by a screening procedure were tested and found to have values less than 5.0 U/g Hb (0.0 to 5.0). Evaluator A. B., using this procedure, obtained a reference interval of 6.03 ± 2.08 U/g Hb on 20 samples.

Note: Reproducibility is good. Evaluator A. B. determined a CV of 4.5% within a single day on 20 samples, and a day-to-day CV of 5.0% on 20 samples.

Discussion

G6PD deficiency is the most common metabolic abnormality of the human erythrocyte. Under unique circumstances it may produce severe to life-threatening hemolysis. The defect is worldwide in distribution but is found most commonly among equatorial populations. It is genetically transmitted as an X-linked recessive trait, so that it usually is seen in males only. The heterozygous female shows no evidence of the abnormality; on the average, half of her erythrocytes contain normal amounts of the enzyme (the Lyon hypothesis) (7). Nevertheless, the heterozygous state is detected with some screening procedures, and will demonstrate intermediate values of G6PD activity when the enzyme is quantitatively assayed. Because G6PD is a cellular age-dependent enzyme, the mean age of the erythrocyte population will affect measured activity. If there is a relative increase in the number of young cells (e.g., reticulocytosis in *any* hemolytic anemia), the value for G6PD activity may be misleadingly high. Consequently, the hematologic status of the patient must be taken into account in interpreting results of G6PD assays. If the patient has experienced a recent acute hemolytic event, so that the erythrocyte population has been modified temporarily, it is advisable to repeat the assay when the patient has returned to the premorbid state.

This assay was developed to be usable in any small laboratory. Although it does not produce values that are the same as those established for G6PD by the ISCH Expert Panel on Red Cell Enzymes (6), the ease of performance and use of readily available, non-noxious reagents should provide an acceptable rationale for its use.

> *Note:* A factor may be calculated, if so desired, to convert the results obtained by this method to that obtained by the method of the Expert Panel. To do this, we multiply our results by the factor 1.3.

The pH of the reaction mixture (Reagent 11) is critical in this assay. After the reaction mixture is prepared, the pH must be checked and adjusted to 7.5. The acidity of the NADP$^+$ solution produces a decrease in pH. It has been shown by Löhr and Waller (3) that the pH optimum for this enzyme is between 7.4 and 8.6. The measurements are made at pH 7.5 because not only is this close to physiological pH, but also results are highly reproducible.

The hemolysate must be kept in an ice-water bath and the assay performed within 2 h of preparation, because G6PD is quite unstable in the hemolysate. If the assay cannot be completed on the day the specimen is collected, the whole blood should be stored at 4 °C. G6PD is stable for at least 20 days in whole blood kept at 4 °C (4).

The hemoglobin concentration of the hemolysate should be 1.0 ± 0.3 g/dL. At higher hemoglobin concentrations the values of enzyme activity are lower, possibly as a result of spectral interference by hemoglobin.

References

1. Motulsky, A. G., and Kampbell-Kraut, J. M., Population genetics of glucose 6-phosphate dehydrogenase deficiency of the red cell. In *Proceedings of the Conference on Genetic Polymorphism and Geographic Variations in Disease,* B. S. Blumberg, Ed., Grune and Stratton, New York, NY, 1961, p 159.
2. Jacob, H. S., and Jandl, J. H., A simple visual screening test for glucose 6-phosphate dehydrogenase deficiency employing ascorbate and cyanide. *N. Engl. J. Med.* **274,** 1162–1167 (1966).
3. Löhr, G. W., and Waller, H. D., Glucose 6-phosphate dehydrogenase. In *Methods of Enzymatic Analysis,* **2,** 2nd ed., H. U. Bergmeyer, Ed., Academic Press, New York, NY, 1974, pp 636–643.
4. Beutler, E., *Red Cell Metabolism: A Manual of Biochemical Methods,* 2nd ed., Grune and Stratton, New York, NY, 1975, pp 8–11, 24–26, and 66–69.
5. Beutler, E., West, C., and Blume, K.-G., The removal of leukocytes and platelets from whole blood. *J. Lab. Clin. Med.* **88,** 328–333 (1976).
6. Beutler, E., Blume, K.-G., Kaplan, J. C., Löhr, G. W., Ramot, B., and Valentine, W. N., International Committee for Standardization in Hematology: Recommended methods for red-cell enzyme analysis, *Br. J. Haematol.* **35,** 331–340 (1977).
7. Beutler, E., Glucose-6-phosphate dehydrogenase deficiency. In *The Metabolic Basis of Inherited Disease,* 4th ed., J. B. Stanbury, J. B. Wyngaarden, and D. S. Frederickson, Eds., McGraw-Hill Book Co., New York, NY, 1978, pp 1430–1451.

Hematocrit (Packed Cell Volume)[1]

Submitter: Eugene W. Rice, *Department of Health and Human Services, Public Health Service, Food and Drug Administration, Silver Spring, MD 20910*

Evaluators: John A. Koepke, *Department of Pathology, Duke University Medical Center, Durham, NC 27710*

O. W. van Assendelft, *Hematology Division, Center for Infectious Diseases, Centers for Disease Control, Atlanta, GA 30333*

Reviewer: Bruce Lee Evatt, *Hematology Division, Center for Infectious Diseases, Centers for Disease Control, Atlanta, GA 30333*

Introduction

The determination of the hematocrit (Hct) value or ratio, or packed (red) cell volume (PCV), measures the relative volume occupied by erythrocytes in samples of capillary or venous blood, after the sample has been centrifuged, under specified conditions. In conformity with the use of SI in clinical laboratory measurements, the Hct should be expressed as a fraction, for example, 0.41, rather than as a percentage, 41%.

Other less practical methods for determining the relative volume occupied by the cellular components of blood include the indicator-dilution technic, measurement of the interference of electrical conductivity of the supporting medium where cells are present, and ultrasound.

In dilution methods a known volume of blood is added to a known volume of Evans blue dye (T-1824) or to a solution containing radioactive tracer material (^{51}Cr-labeled ethylenediaminetetraacetate) that distributes only in plasma. By the dilution of the dye or isotope in the plasma, one can calculate the ratio of plasma volume to solution volume, and, hence, the "plasma-crit" of the blood.

Because erythrocytes are a poorer conductor of electricity than plasma, conductivity has been utilized to measure the Hct of noncentrifuged blood. Such measurements yield relative, not absolute, values and are influenced by the shape and orientation of erythrocytes in plasma or diluting medium, by other components, and by variability in instrument calibration.

When ultrasound was used to estimate Hct in the range 0.03–0.55, a correlation coefficient of 0.96 was noted *(1)*.

Although the terms "packed cell volume" or "hematocrit value or ratio" are not strictly correct when noncentrifugal methods are used, it is acceptable by common usage, especially when it is part of data derived from interrelated analyses with automated hematology instrument systems.

Principle

The determination of Hct involves centrifuging an oxygenated sample of anticoagulated venous or capillary blood in either a "macro" or, preferably, a "micro" hematocrit tube. Centrifugation is continued until packing of cells is complete, i.e., until additional centrifugation, under the same conditions, leaves the length of the cell column unchanged. The Hct is expressed as the length of the packed erythrocyte column relative to the total length of the column.

Collection and Handling of Specimens

Collect blood samples from a freely bleeding capillary puncture, e.g., finger, earlobe, heel (in infants), or from a venous sample. Avoid exerting excess pressure to obtain blood from a capillary puncture, because dilution of the blood with tissue fluids may cause errors. Collect venous samples with minimum stasis into the customary amount of solid anticoagulant (see *Reagents*). For further instructions the phlebotomist should study the proposed laboratory standards published by the National Committee for Clinical Laboratory Standards (NCCLS) on blood-specimen collection with evacuated tubes (ASH-1), by venipuncture (PSH-3), and by skin puncture (PSH-4). A proposed Selected Method on skin-puncture and blood-collecting techniques for infants has

---

[1] Based on the methods prepared on behalf of the World Health Organization by van Assendelft, Coster, and Crosland-Taylor for the International Committee for Standardization in Hematology. This WHO publication also forms part of the basis for the National Committee for Clinical Laboratory Standards Tentative Standard TSH-7, Standard procedure for the determination of packed cell volume by the microhematocrit method *(1)*.

been published (2). There is also an NCCLS standard procedure for the storage and transport of diagnostic hematological, biochemical, and serological body fluid specimens (PSH-5) (3). Centrifugation should be done within 6 h of collecting the blood sample. Upon storage for 24 h at room temperature, Hct increases in many samples.

MACROMETHOD

Materials and Methods

Reagents

Anticoagulant-containing blood-collection tubes. Various solid anticoagulants that will not change the size or shape of the blood cells (ammonium oxalate/potassium oxalate, K_2 or K_3 ethylenediaminetetraacetate, heparin) are satisfactory. Tubes may be prepared or purchased. It is important to fill the tubes with the correct quantity of blood in relation to the amount of the anticoagulant. The blood-collection tubes should be large enough that there is an airspace left over when the proper amount of blood is added. This enables proper mixing and permits oxygenation of the blood sample. The Hct of oxygenated blood averages about 1% less than that of deoxygenated blood, caused by loss of carbon dioxide upon aeration.

Apparatus

1. *"Wintrobe" hematocrit tube.* A "Wintrobe" hematocrit tube is a thick-walled glass tube of uniform bore (internal diameter, 3.0 mm) with a flat, sealed base and graduated in millimeters from the base for a length of 100 mm. About 0.6 mL of blood is required.

2. *Centrifuge.* A table-top or floor laboratory centrifuge with an arm length of 15 cm or more (as measured from the axis of rotation to the base of the tube holders) is satisfactory. The centrifuge must be capable of sustaining a force corresponding to about $2300 \times g$ at the base of the tube holders.

Procedure

1. Fill a clean, dry Wintrobe hematocrit tube from the bottom upwards with a well-mixed oxygenated blood sample, using a Pasteur pipette or syringe equipped with a long needle and taking care to avoid air bubbles. After removing the pipette or needle, the top of the column of blood should be at the 100-mm mark.

Note: Because slight evaporation may occur during centrifugation, it is inaccurate to wait until after centrifugation to estimate the original length of the blood column.

2. Centrifuge the blood-filled Wintrobe tubes for 30 min at $2000-2300 \times g$ (Table 1).

Table 1. Revolutions per Minute Required to Attain Approximately 2000 and $2300 \times g$ at Various Centrifuge Arm Lengths (Radii) from the Axis (rpm = $291\sqrt{g/cm}$)

Radius, cm	$2000 \times g$	$2300 \times g$
	rpm	
15	3400	3600
20	2900	3100
25	2600	2800

Note: In some few cases, especially if the sample is polycythemic, the packing of the cells may not be complete after this period, because the gravitational force applied to the top of the erythrocyte column is only part of that applied to the base, especially in short-armed centrifuges.

3. Remove the tubes from the centrifuge, read the length of the erythrocyte column from the graduated millimeter scale on the tube, and express as a fraction of the original length of the column of blood (100 mm) to give the Hct. Exclude, as much as possible, the platelet layer adjacent to the plasma and the leukocyte layer. This "buffy coat" is about 1 mm for a leukocyte count of $10.0 \times 10^9/L$. Read column marking the same way each time, to avoid reading errors that are due to parallax.

Note: Clean the tubes immediately after use by submerging them inverted in water. Force water through them by pushing a long steel needle into the bottom to displace the column of blood. Rinse the tubes finally with distilled or de-ionized water, and dry the emptied tubes in a hot-air oven. After repeated usage Wintrobe tubes become coated with a film of blood proteins, which can be removed by soaking them in dilute hydrochloric acid (about 0.1 mol/L) or detergent solutions for 24 h, and then cleaning and drying as described above. Disposable Wintrobe hematocrit tubes are available commercially.

MICROMETHOD

Materials and Methods

Reagents

1. *Anticoagulant-containing capillary blood-collection tubes.* If capillary blood is used, commercial tubes are available for a micro-Hct method that requires not more than 50 µL of blood. The internal surface of these tubes is coated with anticoagulant. Two international units of heparin per tube is suitable. These disposable soft-glass capillary tubes are 7.5-cm long with a uniform bore of 1.555 ± 0.085 mm, have a wall thickness of 0.2 to 0.25 mm, and are not graduated. To avoid errors, it is recommended that one end of the capillary tubes have a colored band; use one color if the tube contains no anticoagulant, for sampling from specimens col-

lected by venipuncture, and a different color if it contains a dried film of anticoagulant, usually heparin, in which case the tube may be filled directly with capillary blood by direct skin puncture. Also available (Arthur H. Thomas Co., Philadelphia, PA 19105) are anticoagulant-containing capillary blood-collection tubes, 32 mm long, used with the Drummond centrifuge in the Strumia-Sherwood method.

2. *Micro-capillary centrifuge.* The centrifuge is designed to hold several capillary tubes (usually 24) in numbered slot positions. It should have a radius greater than 8 cm and be able to sustain a constant force corresponding to $10\,000-15\,000 \times g$ at its outer edge for periods of 5 min, without causing the temperature to go above 45 °C. The centrifuge temperature increases by 6–7 °C above the room temperature after a 5-min run. Allow an interval of a few minutes between runs so the centrifuge can cool down.

Procedure

1. Fill a capillary tube with capillary blood or with a well-mixed oxygenated blood sample for two-thirds to three-fourths of its total length; seal the dry unfilled end by forcing a special "sealing compound" into it in the upright perpendicular position. Pay particular attention to obtaining a flat seal.

2. Place the filled, sealed tubes in the centrifuge, with the sealed end toward the outer edge, record the position of each tube, and centrifuge for 3–5 min.

3. Remove the tubes one at a time, and read the Hct value by means of a micro-hematocrit reader device for determining the ratio of the length of the erythrocyte column to that of the entire column (erythrocytes plus plasma). Alternatively, the ratio can be calculated from measuring the columns by placing the tube against arithmetic graph paper or a metric ruler, but these procedures are more laborious. After obtaining the Hct, discard the tube.

Discussion

The results will always include some trapped leukocytes, platelets, and particularly plasma. Trapped plasma will be increased in patients with polycythemia or certain other conditions, such as iron-deficiency anemia or sickle-cell anemia. More is trapped also when centrifuges with short arms are used or if the tubes are centrifuged at a low rather than high centrifugal force.

Note: When reporting the Hct for publication, state the gravitational force and the time used for centrifugation, to facilitate reproducibility of results.

In routine practice the simpler micromethod is strongly recommended; it has largely replaced the macromethod because of the advantages of the shorter time and the smaller quantities of blood needed. The micromethod involves proportions of about 1–2% trapped plasma, and duplicate tests should not differ by more than 0.02. On the average, the Hct by the micromethod is about 0.01 to 0.02 lower (more complete cell packing) than that measured by the macromethod.

Note: Quality-control procedures should include use of commercial hematology quality-control products and a regular check of centrifuge performance. The rpm may be ascertained with a calibrated tachometer or a suitable stroboscope, or by determining the minimum time after which no further packing of cells occurs (use a normal and a polycythemic sample).

Evaluator J. A. K., using 60 duplicates, found microhematocrit maximum within-day variability of 0.005 hematocrit units. Comparisons with the Coulter S apparatus (Coulter Electronics, Hialeah, FL 33010) varied an average of 0.01 hematocrit units. Day-to-day precision studies showed no changes during a five-day experiment.

The Wintrobe macrohematocrit and the Coulter S hematocrit (not corrected for trapped plasma) were run in parallel with the proposed method for 20 samples. As expected, the Wintrobe method yielded results 0.015–0.035 hematocrit units (3–5%) higher, because of the tighter packing possible with the microhematocrit method. Two large micro vs macro discrepancies (0.36 vs 0.40, and 0.32 vs 0.39) were unexplained, but were assumed to be related to differences in centrifugal force applied.

For additional relevant information about hematocrit, consult references *1, 4,* and *5.*

Reference Values

There is a large variation in the Hct of blood of normal healthy individuals. Different values are quoted in various texts, but the following are commonly accepted for normal adults at sea level: men, 0.47 (2 SD 0.07); women, 0.42 (2 SD 0.05). Each laboratory should determine the range of normal values valid for its locale and population.

Note: Evaluator J. A. K. reported the following Hct ranges from the University of Iowa School of Medicine:

	M	F
Newborn, full-term	0.45–0.65	0.45–0.65
1 month old, full-term	0.33–0.55	0.33–0.55
3 months old, full-term	0.31–0.41	0.31–0.41
6 months–11 years old	0.35–0.44	0.35–0.41
12–18 years old	0.37–0.48	0.34–0.44
Adults	0.40–0.52	0.35–0.47

Like hemoglobin, Hct shows diurnal variations and fluctuates with exercise and with body posture (see comments in chapter on *Hemoglobin*).

The measurement of Hct is a simple and reliable means for detecting the presence or absence of anemia or polycythemia. The interpretation of Hct (and hemoglobin) may be distorted when there is fluctuation in plasma volume, as in splenomegaly and in pregnancy.

The values for erythrocyte count, hemoglobin concentration, and Hct are used to obtain Wintrobe's cell indices, which define the size and hemoglobin content of erythrocytes as follows:

1. Mean cell volume (MCV) in femtoliters = [Hct/no. of erythrocytes $(10^{12}/L)] \times 10^3$. Normal erythrocytes have an average mean cell volume of 90 (range 76–96) fL (1 fL = 10^{-15} L). Low results indicate that the erythrocytes are, on an average, microcytic; similarly, a value above 95 indicates that the cells are macrocytic. The mean cell volume is usually higher than normal in newborns and infants.

2. Mean cell hemoglobin (MCH) in picograms = [hemoglobin (g/dL)/no. of erythrocytes $(10^{12}/L)] \times 10$. The results gives the average content of hemoglobin per erythrocyte. The accepted mean cell hemoglobin value is 30 pg, range 27–32 pg (1 pg = 10^{-12} g). Hypochromic erythrocytes (< 27 pg) occur typically in iron-deficiency and pernicious-type anemias. Hyperchromic cells (> 32 pg) are found in newborns and young infants (1–2 months old), in uncomplicated macrocytic anemias, in some cases of spherocytosis, and in patients with extreme dehydration.

3. Mean cell hemoglobin concentrations (MCHC) = (hemoglobin, g/dL)/Hct. The normal mean cell hemoglobin concentration is 30–35 g/dL, and low values indicate hypochromic cells with a low hemoglobin concentration.

In summary, these indices, in conjunction with the appearance of the erythrocytes on fixed stained smears, yield an accurate picture of erythrocyte morphology. According to the mean cell volume, erythrocytes may be classified as normal (normocytic), small (microcytic), or large (macrocytic). According to the mean cell hemoglobin concentration, erythrocytes may be normal (normochromic) or deficient (hypochromic). A higher than normal hemoglobin concentration does not occur except in some cases of hereditary spherocytosis and in patients with extreme dehydration. The mean cell hemoglobin value expresses only the mean weight of hemoglobin per erythrocyte.

References

1. NCCLS Tentative Standard TSH-7. Standard procedure for the determination of packed cell volume by the microhematocrit method. National Committee for Clinical Laboratory Standards, 771 E. Lancaster Ave., Villanova, PA 19085, 1979.
2. Meites, S., and Levitt, M. J., Skin-puncture and blood-collecting techniques for infants. *Clin. Chem.* **25**, 183–189 (1979).
3. Monaghan, J. C., New lab standards cover blood collection, storage, and transport. *Med. Lab. Observer*, May, 105–108 (1978).
4. Chanarin, I., Critical appraisal of the PCV. In *Quality Control in Hematology*, S. M. Lewis and J. F. Coster, Eds., Academic Press, New York, NY, 1975, pp 103–110.
5. Miale, J. B., *Laboratory Medicine Hematology*, 5th ed., The C. V. Mosby Co., St. Louis, MO, 1977, pp 422–428, 443–448.

Hemoglobin[1]

Submitter: Eugene W. Rice, *Department of Health and Human Services, Public Health Service, Food and Drug Administration, Silver Spring, MD 20910*

Evaluators: John A. Koepke, *Department of Pathology, Duke University Medical Center, Durham, NC 27710*

O. W. van Assendelft, *Hematology Division, Center for Infectious Diseases, Centers for Disease Control, Atlanta, GA 30333*

Reviewer: Bruce Lee Evatt, *Hematology Division, Center for Infectious Diseases, Centers for Disease Control, Atlanta, GA 30333*

Introduction

The quantitation of blood hemoglobin, (B)Hb, maintains a prominent role in multiphasic biochemical screening and clinical medicine *(3)*.[2] The chemical composition of the adult normal Hb molecule has been completely elucidated *(4)*. On the basis of having two alpha-chains, two beta-chains, and four heme-groups, it is calculated to have a molecular weight (relative molecular mass) of 64 458 (anhydrous). The iron(II) concentration is 3.47 mg/g (mass fraction 0.00347).

Makarem *(5)* has discussed briefly the more contemporary technics of routine clinical hemoglobinometry. In 1967 the International Committee for Standardization in Hematology (ICSH) adopted the photometric determination of the most stable Hb derivative known, namely, cyanmethemoglobin (hemiglobincyanide, HiCN), as the recommended method *(6)*. If any other method is used (e.g., photometric determination of oxyhemoglobin, iron determination, gasometric oxygen methods), an appropriate mathematical factor should be used to make values obtained comparable with those by the HiCN method. The ICSH recommendation was reviewed, revised, and republished in 1978 *(7)*. In 1968, the World Health Organization established the international HiCN reference solution as the WHO International HiCN Reference Preparation. Since then, the international HiCN reference solution has also been accepted by the College of American Pathologists, The American Society of Hematology, and the National Committee for Clinical Laboratory Standards (NCCLS).

Detailed instructions and discussions regarding routine HiCN spectrophotometry have been widely published (for example, *1, 3–5, 8, 9*). This chapter reviews and updates such procedural accounts, particularly as they are applicable to manual (in contrast to automated) hemoglobinometry in the small-volume clinical laboratory.

Principle

The determination of (B)Hb as HiCN involves diluting a specimen suitably (e.g., 1 to 251, or 251-fold) with a nonionic detergent containing buffered ferricyanide/cyanide reagent diluent. The reagent rapidly converts all of the Hb derivatives in normal human blood (hemoglobin, oxyhemoglobin, carboxyhemoglobin, hemiglobin), except sulfhemoglobin, into HiCN. The clear and stable solutions of HiCN have an absorption spectrum with a broad and relatively flat maximum around 540 nm. Absorbance measurements obtained on these solutions with photoelectric colorimeters or spectrophotometers typically follow the Lambert–Beer Law over a wide concentration range. Photometric instruments are calibrated by means of "calibrator" HiCN solutions (see below).

Materials and Methods

Reagents

1. *Ferricyanide/cyanide reagent,* modified van Kampen and Zijlstra *(1)*. In a 1-L volumetric flask (Class A), dissolve 200 mg of reagent grade potassium ferricyanide, $K_3Fe(CN)_6$; 50 mg of potassium cyanide, KCN (**poison!**); 120 mg of potassium dihydrogen phosphate, KH_2PO_4; 4.0 g of sodium chloride, NaCl; and 0.5 mL of Sterox S. E. (Harleco Co., Philadelphia, PA) in about 800 mL of distilled or de-ionized water, and dilute to the mark with water. The reagent must be a clear, pale yellow, and the pH must be between 7.2 and 7.5 (at 25 °C; check with

---

[1] Based on the method of van Kampen and Zijlstra *(1)*, modified by Matsubara et al. *(2)*. Portions of this chapter are also based on NCCLS Tentative Standard TSH-15, Reference Procedure for the Qualitative Determination of Hemoglobin in Blood. National Committee for Clinical Laboratory Standards, 771 E. Lancaster Ave., Villanova, PA 19085, 1979.

[2] Abbreviations used: (B)Hb, blood hemoglobin; HiCN, hemiglobincyanide; HbCO, carboxyhemoglobin; ICSH, International Committee for Standardization in Hematology; NCCLS, National Committee for Clinical Laboratory Standards.

a pH meter). Store the reagent in a well-stoppered brown borosilicate (Pyrex) glass bottle at room temperature. It is stable for several months.

Notes:

1. Sterox S. E. is a nonionic detergent added to enhance hemolysis and to prevent turbidity by precipitation of plasma proteins. The following materials have also been reported to yield satisfactory results: Triton X-100; Berol L-048 (a Swedish product); and Nonidet P40, 1 mL/L (Shell International Chemical Co., The Hague, The Netherlands). The NaCl is added to prevent development of turbidity in blood samples containing increased amounts of plasma proteins, as in cases of multiple myeloma. Rarely, HiCN solutions from patients having leukemia (with extreme leukocytosis) or multiple myeloma (with marked hypergammaglobulinemia) may be turbid and must be clarified by centrifuging for 15 min before measuring the absorbance of the clear supernatant solution. Also, turbidity can be prevented by increasing the concentration of phosphate buffer in the van Kampen–Zijlstra reagent to 33 mmol/L *(2)*. Alternatively, a turbid HiCN solution may be cleared for spectrophotometry by filtration with membrane filters, mean pore size 0.45 μm.

2. When reagent is stored in polyethylene bottles, some cyanide is lost, so that Hi is formed but not HiCN. As a result, erroneously low values for (B)Hb are reported.

3. Ferricyanide/cyanide reagents used for determining (B)Hb become colorless on freezing because the ferricyanide is reduced by the cyanide ion in an alkaline medium. Oxidation of the samples to Hi does not occur with such colorless reagents. The resulting solution will contain HbO_2 and give results averaging about 2.5% higher than actual amounts present. Under unusual circumstances such as field investigations in cold environments, where freezing may be unavoidable, Rice and LaPara *(10)* have shown that reagents prepared with aqueous ethylene glycol (50/50 by vol) instead of with water alone may be used, in which case there is less than a 1% error. Subsequent studies *(11)* have shown that the reaction that occurs on freezing can also be prevented by the addition of ethanol, methanol, ethylene glycol, or glycerol.

2. *Calibrator HiCN solutions.* HiCN solutions conforming, in general, to international specifications as designated by the ICSH *(6)* are commercially available. They are dispensed as sterile solutions and contain 55–85 mg of HiCN per 100 mL. The solutions are stable for two years or longer. Calibrator HiCN solutions may be used to check the reliability of instruments and scale readings, to prepare a calibration graph, or to serve as a reference value for the calculation of (B)Hb concentration when comparing analytical techniques.

Note: The purity of HiCN calibrator solutions may be checked by judging the shape of the absorption curve from 450 to 800 nm *(7)*. Solutions of HiCN may also be evaluated from the ratio $\epsilon_{540\,nm}/\epsilon_{504\,nm} = A_{540\,nm}/A_{504\,nm}$. The theoretical value is $11.0/6.83 = 1.61$. The range acceptable by ICSH requirements is 1.59–1.63. Solutions of HiCN do not absorb light from 720 to 1000 nm, and absorbance at 750 nm should be less than 0.002 per 1-cm lightpath. If an HiCN solution does not meet this requirement, it is contaminated or turbid, and is not acceptable as a calibrator solution.

3. *Hematology quality-control materials.* Numerous Hb quality-control products—assayed stabilized suspensions of erythrocytes, or Hb solutions—are commercially available. Such material may be used for quality control, but *not* for instrument calibration.

Apparatus

1. *Photometers.* Various photometers may be used to measure the absorbance of HiCN solutions at or around 540 nm. Numerous direct-reading photoelectric "hemoglobinometers" have a scale calibrated directly in Hb, g/dL blood. Check this calibration regularly with calibrator HiCN solutions.

Photoelectric colorimeters should have an interference filter giving a narrow half-intensity bandwidth (less than 20 nm) with a nominal wavelength of 540 nm. Calibrate the instrument regularly with calibrator HiCN solutions.

Spectrophotometers for measuring HiCN solutions should be calibrated for wavelength accuracy and absorbance. Measure absorbance of samples at 540 nm vs a reagent blank. Obtain the equivalent value of calibrator HiCN solutions, expressed as (B)Hb concentration in g/dL, by multiplying the labeled assay value (mg/dL) by the sample dilution factor (e.g., 251) and dividing by 1000 (mg/g). For example, a 60.0 mg/dL standard solution of HiCN is equivalent to a (B)Hb of $(60.0 \times 251)/1000$, or 15.1 g/dL.

2. *Volumetric ware.* Use a 5.0-mL volumetric pipette with an accuracy within 0.5% to measure the reagent, and 0.02-mL (20-μL) Salhi-type pipettes, or disposable capillary-type pipettes, having an accuracy within 0.5% to transfer blood samples. If very many hemoglobin determinations are performed regularly, use an automatic pipetting device of the same accuracy.

For best results the accuracy of each new pipette should be verified by either weighing the amount of water it delivers when filled to the mark, by reading the absorbance of the amount of dye solution each delivers when filled to the mark, or by making replicate determinations of HiCN of a blood sample of accurately known Hb content.

Collection and Handling of Specimens

Collect blood samples from a freely bleeding skin puncture, e.g., finger, earlobe, heel (in infants), or from a venous sample. Avoid exerting excess pressure to obtain a sufficient amount of blood from a capillary puncture, because an error may result from diluting the blood with tissue fluids. Collect venous samples with minimum stasis into the customary amount of any solid anticoagulant (ammonium oxalate/potassium oxalate, K_2 or K_3 ethylenediaminetetraacetate, heparin). For further instructions the phlebotomist should study the proposed laboratory standards published by the NCCLS on blood specimen collection with evacuated tubes (ASH-1), by venipuncture (ASA-3), and by skin puncture (TSH-4). A proposed Selected Method on skin-puncture and blood-collecting techniques for infants has been published (12). There is also an NCCLS standard procedure for the storage and transport of diagnostic hematological, biochemical, and serological body fluid specimens (TSH-5) (13). (B)Hb is stable in anticoagulated samples for at least two to three days at room temperature and for at least five to six days at 4 °C if not microbially contaminated.

Procedure

1. Transfer to photometrically matched test tubes or cuvets 5.0 mL of reagent, using a volumetric transfer pipette.

2. Carefully add to the reagent 0.02 mL (20 µL) of well-mixed blood or quality-control samples with Sahli pipettes or disposable capillary pipettes. With a Salhi pipette, expel the contents by gently blowing; then partly withdraw the pipette from the liquid and rinse three times from the top portion of the solution. When using a capillary pipette, carefully drop the blood-filled pipette into the reagent. (Not all types of cuvets are suitable for this technique.)

3. Mix thoroughly, preferably with a vortex-type mechanical mixer, and allow 3–5 min for complete conversion of the Hb pigments to HiCN. If a capillary pipette was left in the test tube or cuvet (step 2), remove before continuing to step 4.

4. Measure the absorbance of the HiCN solutions in the cuvets with a previously calibrated photoelectric "hemoglobinometer," colorimeter, or spectrophotometer, using the reagent as a blank at 540 nm. HiCN solutions are stable for about two days if kept well-stoppered in a refrigerator or at room temperature in the dark.

Notes:

1. The rate of conversion of carboxyhemoglobin (carbon monoxide hemoglobin, HbCO) to HiCN with ferricyanide/cyanide reagents is much slower than the rate of HbO_2. If this slower reaction time (about 30 min) is not considered when determining total Hb in blood containing 10% or more HbCO (as with heavy smokers or in carbon monoxide poisoning), the results may be erroneously low. The correct total Hb concentrations can be obtained readily for blood containing HbCO by simply heating the diluted (251-fold) specimen in a 56 °C water bath for 3–5 min with continuous gentle mixing, then cooling to room temperature before determining the absorbance at 540 nm (14).

2. As noted previously, sterile HiCN solutions sealed in amber glass ampules are stable at 4 °C for several years. However, unsterilized solutions are photometrically stable for only a day or so before absorbances at 540 nm may begin to change, owing to microbial contamination.

3. Although the conventional procedure requires only 20 µL of blood, as described, it may be expedient occasionally to use a still smaller volume (15). HiCN solutions have two absorption peaks in the spectrum, the well-known broad and relatively flat peak at about 540 nm, and another sharp peak at 421 nm (Soret band). The absorptivity of HiCN at the latter peak is about 11-fold greater than that at 540 nm, and accordingly forms a basis for a convenient ultramicro modification of the standard procedure: Add 2 µL of blood to 5.0 mL of reagent/diluent to attain a 2500-fold dilution, and determine the absorbance against a reagent blank [not water, because $K_3Fe(CN)_6$ absorbs significantly at 420 mm] at 421 nm. This modification yields a mean coefficient of variation (CV) of 1.5–2.0%. (Investigations reported in reference 14 were conducted with a spectrophotometer having a 10-nm half-intensity bandwidth, and an incorrect wavelength of 418 nm was used as the Soret band wavelength peak.)

Calculation

Direct-reading "hemoglobinometers" have a scale calibrated directly for Hb in g/dL blood. Check the calibration regularly with a calibrator HiCN solution. When using a colorimeter or spectrophotometer, calculate the (B)Hb concentration by the usual formula: $(A_{unk}/A_{std}) \times$ concn of standard. Alternatively (but potentially less accurately), calculate the (B)Hb concentration from a previously prepared calibration graph or table, relating measured absorbance to hemoglobin concentration, or by using an absorbance multiplicand factor. Check the graph, table, or factor regularly with a calibrator HiCN solution.

Note: If the measurement at 540 mm is obtained with 1.000-cm cuvets, calculate (B)Hb by multiplying the absorbance at 540 nm by a factor of 36.8, for a sample diluted 251-fold. The factor 36.8 is derived as follows: (B)Hb is calculated in mmol/L from absorbance at 540 nm by using the equation: $(B)Hb = (A \times F)/(11.0 \times 1.000)$, where F is the dilution factor (251), 11.0 is one-fourth of the millimolar absorptivity of HiCN at 540 nm, and 1.000 is the lightpath in cm. Taking the relative molecular mass of one-fourth of the hemoglobin molecule to be 16 114.5, (B)Hb in g/dL is therefore: $[(A \times 251)/(11.0 \times 1.000)] \times 16\,114.5 \times 10^{-4} = A \times 36.8$, where 10^{-4} is used to convert mg/L to g/dL.

Note: Despite strong efforts by numerous scientific bodies to implement the SI system in clinical chemistry throughout the world, (B)Hb concentration is currently reported by various expressions: g/dL, mmol/L, g/L. The International Federation of Clinical Chemistry Expert Panel on Quantities and Units has recommended since 1972 that Hb be reported as B-hemoglobin (Fe), substance concentration, expressed in mmol/L. The International Committee for Standardization in Hematology recommends "g/dL," and the Editorial Board of this book recommends "SI." The mathematical relationship for converting the traditional and SI units is as follows:

hemoglobin (Fe), g/dL × 0.6205 = mmol/L
hemoglobin (Fe), mmol/L × 1.612 = g/dL

Discussion

Data from various laboratory proficiency testing programs have documented a considerable improvement in hemoglobinometry since 1965. Manual Hb methodology has been greatly aided by the standardization of technique and the development of stable HiCN reference solutions. With manual procedures of the types described here, CVs of about 3.5% may be attained, compared with about 1.5% obtained with automated instruments *(16, 17)*.

Note: Evaluator J. A. K., using 60 duplicates from polycythemic, normal, and anemic patients, found the following within-day precision: 52 pairs agreed within 0.2 g/dL, two within 0.3 g/dL, three within 0.4 g/dL, two within 0.5 g/dL, and one within 0.6 g/dL. In his five-day day-to-day precision study, involving three duplicate samples from polycythemic, normal, and anemic patients, the mean difference between duplicates was 0.14 g/dL. Likewise, he evaluated the accuracy of the proposed method by comparing the values for 20 polycythemic, normal, and anemic patients obtained by a manual cyanmethemoglobin method with "Coulter S" results. Excellent correlations were obtained ($r = 0.99$).

Reference Intervals

There is a large variation of Hb concentration in the blood of healthy individuals. Hb is at its greatest concentration at birth, varying from 16 to 24 g/dL and gradually decreasing to a mean of about 12.5 g/dL within two months. The "normal" range for children is about 12–14 g/dL in both sexes up to the time of puberty. After puberty, the (B)Hb concentration in boys continues to increase until age 16 or 17. The maximum Hb in women tends to reach a lower plateau than that for men. During pregnancy, the Hb concentration generally decreases, rising again after delivery of the baby. These lower hemoglobin values in pregnancy are due only partly to iron depletion and more significantly to dilution (increased total blood volume).

Reference intervals generally acceptable for clinical purposes are (g/dL): men under 65, 14.5–17; women, 13–15. In normal adults between 65 and 80 years of age, the range of (B)Hb decreases to 10–15 g/dL. These values are for persons at sea level; Hb concentration increases in individuals living in higher altitudes. (B)Hb has a diurnal variation of 0.2–0.5 g/dL. Posture is also a variable. (B)Hb concentration is greater in a standing subject than in a person lying down.

Note: Evaluator J. A. K. reported the following range of values from the University of Iowa School of Medicine (all values in g/dL).

	M	F
Newborn full-term infants	15.0–24.0	15.0–24.0
1 month old, full-term	10.0–17.0	10.0–17.0
3 months old, full-term	10.0–14.0	10.0–14.0
6 months old, full-term	11.0–14.0	11.0–14.0
1 year old	11.0–15.0	11.0–15.0
6–11 years old	12.0–15.0	12.0–15.0
12–18 years old	12.8–17.0	12.0–15.0
Adults	13.3–17.7	12.0–15.5

(B)Hb concentration less than 5 g/dL (3.1 mmol/L) represents a "critical laboratory value," indicating an immediate threat to the patient's life through heart failure and anoxemia. The laboratory should notify responsible clinicians so that action can be taken without delay.

References

1. van Kampen, E. J., and Zijlstra, W. G., Determination of hemoglobin and its derivatives. *Adv. Clin. Chem.* **8**, 141–187 (1965).
2. Matsubara, T., Okuzono, H., and Senba, U., A modification of van Kampen–Zijlstra's reagent for the haemiglobincyanide method. *Clin. Chim. Acta* **93**, 163–164 (1979).
3. Davidson, I., and Nelson, D. A., The blood. Chapter 4 in *Todd–Sanford Clinical Diagnosis by Laboratory Methods*, 15th ed., I. Davidson and D. A. Nelson, Eds., W. B. Saunders Co., Philadelphia, PA, 1974.
4. Fairbanks, V. F., Hemoglobin, hemoglobin derivatives, and myoglobin. In *Fundamentals of Clinical Chemistry*, 2nd ed., N. W. Tietz, Ed., W. B. Saunders Co., Philadelphia, PA, 1976, pp 401–454.
5. Makarem, A., Hemoglobins, myoglobins and haptoglobins. In *Clinical Chemistry, Principles and Technics*, 2nd ed., R. J. Henry, D. C. Cannon, and J. W. Winkelman, Eds., Harper and Row, Hagerstown, MD, 1976, pp 1125–1138.
6. ICSH recommendations for haemoglobinometry in human blood. *Br. J. Haematol.* **13** (Suppl.), 71 (1967).
7. ICSH recommendations for reference method for haemoglobinometry in human blood (ICSH Standard EP 6/2: 1977), and Specifications for international haemiglobincyanide reference preparation (ICSH Standard EP 6/3: 1977). *J. Clin. Pathol.* **31**, 139 (1978).
8. Hainline, A., Jr., Hemoglobin. *Stand. Methods Clin. Chem.* **2**, 49–60 (1958).
9. Izak, G., and Lewis, S. M., Eds., *Modern Concepts in Hematology*, Academic Press, Inc., New York, NY, 1972.
10. Rice, E. W., and LaPara, C. Z., Reliability of hemoglobin values obtained with ferricyanide–cyanide reagents subjected to freezing temperature. *Clin. Chem.* **11**, 531–532 (1965).
11. Zweins, J. H., Frankena, H., and Zijlstra, W. G., Decomposition on freezing of reagents used in the ICSH-recommended method for the determination of total haemoglobin in blood; its nature, cause and prevention. *Clin. Chim. Acta* **91**, 337–352 (1979).
12. Meites, S., and Levitt, M. J., Skin-puncture and blood-collecting techniques for infants. *Clin. Chem.* **25**, 183–189 (1979).
13. Monaghan, J. C., New lab standards cover blood collection, storage, and transport. *Med. Lab. Observer* (May), 105–108 (1978).
14. Rice, E. W., Rapid determination of total hemoglobin as hemiglobicyanide in blood containing carboxyhemoglobin. *Clin. Chim. Acta* **18**, 89–91 (1967).
15. Rice, E. W., Spectrophotometric ultramicrodetermination of hemoglobin as hemiglobincyanide, oxyhemoglobin, or hemiglobinazide at the Soret band maxima (418, 412, and 415 nm). *Clin. Chem.* **14**, 841 (1968).
16. Koepke, J. A., The calibration of automated instruments for accuracy in hemoglobinometry. *Am. J. Clin. Pathol.* **68**, 180–184 (1977).
17. Gilmer, P. R., Williams, L. J., Koepke, J. A., and Bull, B. S., Calibration methods for automated hematology instruments. *Am. J. Clin. Pathol.* **68**, 185–190 (1977).

Iron and Total Iron-Binding Capacity[1]

Submitters: Thomas J. Giovanniello[2] and Joseph Pecci, *Laboratory Service, Veterans Administration Medical Center, Boston MA 02130*
Philip J. Garry, *University of New Mexico, School of Medicine, Albuquerque, NM 01939*

Evaluator: Manik L. Das, *Sigma Chemical Co., St. Louis, MO 63178*

Reviewer: Richard J. Carter, *Technical Services, I. E. duPont de Nemours & Co., Wilmington, DE 19898*

Introduction

The determination of serum iron and total iron-binding capacity has grown increasingly popular in the clinical chemistry laboratory, particularly for pediatric patients. For this reason, a microtechnique is obviously necessary.

This method described here is simple, requires minimum equipment, and is adaptable to small clinical chemistry laboratories.

Principle

The method presented is that of Barkan and Walker (1), as modified by Peters et al. (2). The sample is treated with hydrochloric, thioglycolic, and trichloroacetic acids. After separation of the precipitated proteins by centrifugation, the chromogen ferrozine is added to form a colored iron complex, which is measured spectrophotometrically. The Submitters have introduced this chromogen (3) to replace the bathophenanthroline used by Peters et al.

The total iron-binding capacity is determined by adding excess ferric ammonium citrate to the sample. Excess iron, not bound by the serum, is removed by addition of a small amount of buffered ion-exchange resin (Amberlite IRA 410). This sample is diluted and centrifuged, and an aliquot of the supernate is analyzed for iron content to give the total iron-binding capacity.

Serum Iron

Specimens

Use iron-free equipment to collect blood specimens, and perform the analysis with iron-free glass or plastic-ware.

Note: The Submitters find Falcon 2054 disposable plastic tubes (12 × 75 mm) with snap caps acceptable (available from American Scientific Products, Div. of American Hospital Supply Corp., McGaw Park, IL 60085).

Because samples with hemoglobin concentrations exceeding 50 mg/dL can contribute as much as 15 μg of iron per deciliter, do not process specimens showing visible hemolysis: Either analyze the sample immediately after its separation from the clot or, if storage is required, freeze the serum specimen at −20 °C (no more than two weeks) in an iron-free container with a secure closure.

Reagents

1. *Iron-free water,* with a resistivity of at least 1 MΩ · cm (mega-ohm · cm).

2. *Trichloroacetic acid* (G. Frederick Smith, Columbus, OH; iron- and copper-free, cat. no. 390).

3. *Thioglycolic acid ($HSCH_2COOH$, mercaptoacetic acid),* purity 98% (Sigma Chemical Co., St. Louis, MO 63178; cat. no. T-3758).

4. *Hydrochloric acid, 6 mol/L* (G. Frederick Smith; iron- and copper-free, cat. no. 504), available as 6 mol/L.

5. *Sodium acetate, 2 mol/L solution* (G. Frederick Smith; iron- and copper-free), available as 2 mol/L.

6. *Ferrozine* (Hach Chemical Co., Ames, IA).

7. *Iron stock standard (1 g/L).* Prepare from reagent-grade iron wire. Accurately weigh 1 g of iron wire and dissolve it in a minimum amount of 6 mol/L HCl. Dilute to 1 L with iron-free water. Store in an iron-free glass-stoppered Pyrex bottle. Stable up to six months at room temperature.

8. *Protein-precipitating reagent.* Add sufficient iron-free water to dissolve 100 g of trichloroacetic acid in a 1-L volumetric flask. Add 30 mL of thioglycolic acid to the same flask and 334 mL of 6 mol/

---

[1] Some of the material used is taken from data produced by the Iron Study Group, a Working Group of The AACC Committee on Standards, 1979: R. J. Carter, J. L. Barnes, G. N. Bowers, Jr., R. Schaffer, R. Cooper, E. Berman, P. J. Garry, and T. J. Giovanniello.

[2] Present address: Milton Medical Laboratory, Inc., Milton, MA 02187.

L HCl, using a 500-mL graduated cylinder. Mix and dilute with iron-free water. Transfer the resulting solution to an iron-free amber-colored bottle and store at room temperature. Stable up to three months.

9. *Chromogen solution.* Sodium acetate, 2 mol/L, containing 100 mg of ferrozine per liter.

10. *Working standards,* to calculate unknowns selectively and to check linearity. Prepare three working standards of between 50 and 400 µg/dL in 0.1 mol/L HCl. Transfer 1.0 and 3.0 mL of iron stock standard to individual 100-mL volumetric flasks and dilute to 100 mL with 0.1 mol/L HCl (100 and 300 µg/dL standards). Transfer another 1.0-mL aliquot of stock iron standard to a 200-mL volumetric flask and dilute to the mark with 0.1 mol/L HCl (50 µg/dL standard).

Procedure

1. To disposable 12 × 75 mm tubes, add the following: 0.20 mL of serum, controls, standards, or blank (HCl, 0.1 mol/L), followed by 0.20 mL of protein-precipitating reagent. Cap and vortex-mix for 10 s. Let stand at room temperature for 30 min. Use normal and abnormal controls and three standards with each run. Air-displacement pipettes with disposable plastic tips are suitable for the transfers.

2. Centrifuge at 1000 × g for 10 min.

3. Remove a 0.20-mL aliquot of the clear supernatant fluid from each tube and transfer it into correspondingly labeled disposable spectrophotometer microcuvets. Use any type of automatic pipette that has a disposable tip. A new tip should be used for each serum and supernate transfer.

4. Add 0.20 mL of de-ionized water to each cuvet.

5. Add 0.20 mL of chromogen solution and stopper tightly with cuvet cap.

6. Mix by inverting five times and allow the cuvet to stand for 30 min at room temperature.

7. Measure the absorbance of the unknowns, standards, and control at 562 nm within 30–120 min after adding the chromogen solution. Clean, dry, semimicro disposable spectrophotometer cuvets (no. 3166; Evergreen Scientific, Los Angeles, CA 90058) are acceptable.

Notes:

1. Prepare fresh working standards monthly and store them in clean, tightly capped bottles.

2. Serum or heparinized plasma may be used for determination of both serum iron and total iron-binding capacity. Citrated, oxalated, or ethylenediaminetetraacetate (EDTA)-treated plasma samples may not be used for determination of total iron-binding capacity, but are permitted for measurement of serum iron.

3. Plastic tubes, caps, and pipette tips can be checked for iron contamination by performing the serum-iron determination on a representative number of tubes or caps that would simulate reagent blanks. Pipette tips can be added to another set of tubes and the procedure carried through as a reagent blank. When read against distilled de-ionized water, the absorbance should not exceed that of the reagent blank, 0.015.

4. Iron-free reagents may be obtained from sources other than those listed.

Total Iron-Binding Capacity (TIBC)

Specimens

Samples obtained for the determination of TIBC must be free of oxalate or citrate. Such anticoagulants will bind with the resin, releasing calcium ions and causing the specimen to clot. *Heparinized* plasma may be used.

Reagents

The reagents used for determining serum iron are also used here.

1. *Ferric ammonium citrate (Fe, 50 mg/L).* In a 50-mL centrifuge tube, dissolve 60 mg of $FeCl_3 \cdot 6H_2O$ in 20 mL of water. Precipitate $Fe(OH)_3$ by adding 0.20 mL of concd. NH_4OH. After mixing, centrifuge and discard the supernatant liquid. Add 1 mL of water and 70 mg of citric acid ($C_6H_8O_7 \cdot H_2O$), and heat gently until the solution is clear. Add about 25 mL of water and adjust the pH to 6.5–7.0 with dilute NH_4OH. Quantitatively transfer the resulting solution to a 250-mL volumetric flask, and dilute to volume. When refrigerated, the solution is stable for at least six months, as indicated by a light green color; discard if the solution turns yellow. The amount of iron contained in 20 µL of the ferric ammonium citrate solution used in step 2 of the procedure should be sufficient to exceed the iron-binding capacity of the sample.

2. *Barbital buffer, pH 7.5,* containing NaCl (0.11 mol/L) and barbital (diethylbarbituric acid, 44 mol/L).

3. *Hydrochloric acid, concd.*

4. *Amberlite, IRA 410 resin* (Rohm and Haas, Philadelphia, PA 19105). Suspend the resin overnight in HCl (about 3 mol/L, made by carefully adding one part of concd. HCl to three parts of distilled water), to saturate it with chloride ion. Wash the resin several times with distilled water, and remove the water each time by decantation; i.e., allow the resin to settle, then pour off the supernatant fluid. Suspend the resin slurry in barbital buffer and adjust the pH to 7.5 with NaOH (0.1 mol/L). Dry overnight at 95 °C and store in a desiccator.

Procedure

1. To 12 × 75 mm Falcon disposable tubes, add 0.20 mL of unknowns, controls, or blank (0.1 mol/L HCl). (Use air-displacement pipettes throughout.)

2. Add 20 µL of ferric ammonium citrate solution. Centrifuge the tubes briefly to spin down any adher-

ing sample or reagent. Let stand for about 10 min at room temperature.

3. Add about 50 mg of dry resin to each tube and shake about 20 times during a period of 5 to 10 min. The amount of resin should be small, yet sufficient to remove all unbound or excess iron. About 50 mg of the dry resin is adequate and provides sufficient supernate for final iron analyses. Use a nickel–stainless steel micro-spatula (American Scientific Products, cat. no. 1580; or Fisher Scientific Co., Pittsburgh, PA 15219, cat. no. 21-401-25A) to add the resin. Two level scoopfuls will provide sufficient resin.

4. To all tubes, add 0.20 mL of buffer and shake occasionally during 5 to 10 min.

> *Note:* If serum iron and TIBC determinations are done simultaneously, the absorbance value of the 300 µg/dL iron standard can be used for TIBC calculations; otherwise, include an iron standard of 300 µg/dL in this part of the procedure.

5. Centrifuge for 5 min at approx. 500 × g.
6. Transfer 0.20 mL of supernate to a properly labeled plastic test tube.
7. To the tubes containing the supernates add 0.20 mL of protein-precipitating solution. Cap and vortex-mix for 10 s, then allow tubes to stand for 30 min at room temperature.
8. Centrifuge at 1000 × g for 10 min.
9. Remove 0.20 mL of an optically clear aliquot of the supernatant fluid and transfer it into correspondingly labeled disposable spectrophotometer microcuvets.
10. Add 0.20 mL of de-ionized water to each cuvet.
11. Add 0.20 mL of chromogen solution and stopper tightly with a cuvet cap.
12. Mix by inverting five times and allow the cuvets to stand for 30 min at room temperature.
13. Measure the absorbance at 562 nm within 30 to 120 min after adding the chromogen solution. Set the spectrophotometer absorbance to zero for the reagent blank and then read the absorbance of the unknowns, standards, and controls. The reagent blank should have an absorbance no greater than 0.015.

Calculations

Serum Iron

$(A_x/A_s) \times C_s =$ Fe concn, µg/dL

$A_x =$ absorbance of samples (unknowns or controls)
$A_s =$ absorbance of standard
$C_s =$ concentration of standard, µg/dL

Example: sample reading 0.250, standard reading 0.340, concentration of standard 300 µg/dL:

$(0.250/0.340) \times 300 = 221$ µg/dL, serum iron

Assay three standards with each batch to validate the linearity of the procedure.

The absorbance of the iron–ferrozine complex is linear to an iron concentration of 400 µg/dL. Samples with more iron than 400 µg/dL should be diluted and reanalyzed.

Total Iron-Binding Capacity

Samples for TIBC are unlikely to require dilution because the sample is diluted in Steps 2 and 4 by a factor of 2.1. Consequently, a dilution factor must be used to correct for the sample dilution made by the addition of the ferric chloride reagent and buffer (0.20 mL of sample, 0.02 mL of reagent, and 0.20 mL of buffer).

$$(A_x/A_s) \times C_s \times f = \text{TIBC, µg/dL}$$

$A_x =$ absorbance of sample (unknowns or controls)
$A_s =$ absorbance of standard
$C_s =$ concentration of standard, µg/dL
$f =$ dilution factor, 2.1

Example: sample reading 0.168, standard reading 0.340, concentration of standard, 300 µg/dL:

$(0.168/0.340) \times 300 \times 2.1 = 311$ µg/dL, TIBC

Discussion

This method (x) was compared with the automated procedure of the Technicon AutoAnalyzer II (y) *(4)*. The results show good agreement. Serum iron: $y = 0.98x + 0.42$ ($r = 0.99$, n = 34, SEM = 4.8 µg/dL); TIBC: $y = 0.94x + 16.9$ ($r = 0.94$, n = 29, SEM = 13.1 µg/dL).

Tables 1 through 3 present the precision, recovery, and additional intercomparison studies carried out by Evaluator M. L. D.

Table 1. Precision Studies (Evaluator M. L. D.)

Sample	Mean (SD), µg/dL	CV, %
Serum iron, within-day, 20 replicates		
Normal	93.05 (4.15)	4.4
Above normal	287.00 (9.60)	3.3
TIBC, within-day, 20 replicates		
Normal	265.45 (13.53)	5.1
Normal	381.60 (18.13)	4.8
Serum iron, day-to-day (20 days)		
Normal	90.20 (4.94)	5.5
Above normal	281.25 (10.14)	3.6
TIBC, day-to-day (20 days)		
Normal	254.90 (19.46)	7.6
Normal	368.15 (20.22)	5.5

Table 2. Recovery Studies (Evaluator M. L. D.)

Test specimen[a]	Fe concn, µg/dL			
	Estimated	Added	Recovered[b]	% recovered[c]
SP + IS (1:1)	235	100.0	96.5	97.0
SP + IS (1:2)	232	133.1	139.7	105.0
SP + IS (1:3)	225	150.0	155.8	104.0
SP + IS (2:1)	248	66.7	63.3	95.0
SP + IS (3:1)	254	50.0	46.3	93.0

[a] Test specimens with added iron were prepared from a serum pool, SP (277 µg/dL) and an iron standard, IS (200 µg/dL). Iron was then assayed in the test specimens.
[b] Measured iron less estimated iron.
[c] (Recovered/added) × 100.

Table 3. Correlation Studies (Evaluator M. L. D.)[a]

Method	Serum iron, µg/dL[b]		TIBC, µg/dL[c]	
	Range	Mean	Range	Mean
This method	46–246	104.8	250–480	358.7
Modified Goodwin		96.9		316.6

[a] 10 serum specimens were assayed for iron and TIBC by this procedure and by the modified procedure of Goodwin et al. (5) (see Technical Bulletin No. 565, Sigma Chemical Co.).
[b] Slope = 0.975; intercept = −5.253; correlation coefficient = 0.995.
[c] Slope = 0.871; intercept = 4.227; correlation coefficient = 0.980.

Serum iron and TIBC were determined for 38 apparently healthy men and 39 apparently healthy women (who were not using oral contraceptives).

Using a nonparametric percentile technique *(6, 7)*, the Submitters established the following reference ranges at the 95% limit:

	M	F
Serum iron, µg/dL	70–168	38–153
TIBC, µg/dL	264–394	293–478

Each laboratory should establish its own reference range for serum iron and TIBC by this method.

References

1. Barkan, G., and Walker, B. S., Determination of serum iron and pseudohemoglobin with *o*-phenanthroline. *J. Biol. Chem.* **135**, 37–42 (1940).
2. Peters, T., Giovanniello, T. J., Apt, L., and Ross, J. G., A simple method for the determination of serum iron. *J. Lab. Clin. Med.* **48**, 280–288 (1956).
3. Giovanniello, T. J., and Pecci, J., Measurement of serum iron and total iron-binding capacities: Manual and automated techniques. *Stand. Methods Clin. Chem.* **7**, 127–141 (1972).
4. Technicon AutoAnalyzer Methodology for Serum Iron Method File No. SF4-0025FL4, Technicon Instruments Corp., Tarrytown, NY, 1974.
5. Goodwin, J. F., Murphy, B., and Guillemette, M., Direct measurement of serum iron and binding capacity. *Clin. Chem.* **12**, 47–57 (1966).
6. Herrera, L., The precision of percentiles in establishing normal limits in medicine. *J. Lab. Clin. Med.* **52**, 34–42 (1958).
7. Martin, H. F., Gudzinowicz, B. J., and Fanger, J., *Normal Values in Clinical Chemistry*, Marcel Dekker, New York, NY, 1975, pp 214–216.

Lactate Dehydrogenase in Serum

Submitters: John A. Lott and Kathie Turner, *Division of Clinical Chemistry, Department of Pathology, The Ohio State University, Columbus, OH 43210*

Evaluators: Quentin C. Belles, *G. N. Wilcox Memorial Hospital, Lihue, Kauai, HI 96766*
June D. Boyett, *St. Vincent Hospital and Medical Center, Portland, OR 97225*
A. R. Henderson, *University Hospital, London, Ontario*
Robert L. Lynch, *Medical College of Virginia, Richmond, VA 23298*
P. A. Govinda Malya, *Medical Research Department, I.C.I. United States Inc., Wilmington, DE 19899*

Reviewer: Steven N. Buhl, *Technicon Instruments Corp., Tarrytown, NY 10591*

Introduction

Lactate dehydrogenase (LD; EC 1.1.1.27, L-lactate:NAD$^+$ oxidoreductase) is distributed very widely in the body and is invariably found in the cell cytoplasm in very high activities in the heart, liver, kidney, and skeletal muscle and in lesser amounts in lung, smooth muscle, erythrocytes, brain, and pancreas. Because activities of LD are much higher in tissues than in plasma, injury to tissue, with accompanying leakage of the cytoplasm into the peripheral blood, can increase the serum LD dramatically.

Above-normal LD activities in serum are seen in several hematologic, neoplastic, cardiac, hepatic, skeletomuscular, and renal diseases. Very high serum LD activities have been seen in megaloblastic anemia, pernicious anemia, extensive carcinomatosis, viral hepatitis, shock, hypoxia, and extreme hyperthermia. Somewhat lower activities occur after myocardial or pulmonary infarction, leukemia, and hemolytic anemia. Moderately increased activities occur in cirrhosis, obstructive jaundice, and various renal, skeletomuscular, and neoplastic diseases. The very many conditions in which high serum LD activities occur make an abnormal value for LD a nonspecific finding and therefore of limited value. The test continues to be popular with clinicians, because a normal value can be helpful in ruling out a variety of diseases. Assays for LD isoenzymes are also used because the isoenzymes are more organ specific. The clinical enzymology of LD and LD isoenzymes has been reviewed by Zimmerman and Henry (1).

Principle

LD catalyzes the reduction of pyruvate to lactate, accompanied by the oxidation of NADH to NAD$^+$ according to the following stoichiometric equation:

$$CH_3-\underset{\text{pyruvate}}{\overset{\overset{O}{\|}}{C}}-COOH + H^+ + NADH \underset{\longleftarrow}{\overset{LD}{\longrightarrow}}$$

$$CH_3-\underset{\text{lactate}}{\overset{\overset{OH}{|}}{C}}-COOH + NAD^+$$

The above reaction is used in the method described here. Serum is added to a solution of NADH in buffer, after which the mixture is incubated for 10 min to destroy endogenous oxo-acids. The reaction is initiated by adding pyruvate, and the decrease in absorbance at 340 nm is monitored at 37 °C with a spectrophotometer having a bandpass of 10 nm or less. The LD activity is determined from the linear, zero-order portion of the curve of absorbance vs time, and the results are expressed in IUB units per liter (U/L).

Materials and Methods

Reagents

Imidazole buffer, pH 6.90 at 37 °C, 100 mmol/L. Dissolve 6.80 g of imidazole, $C_3H_4N_2$, rel. molecular mass (M_r) 68.08 (e.g., cat. no. I-0125; Sigma Chemical Co., St. Louis, MO 63178) in approximately 940 mL of distilled water. Calibrate a suitable pH meter at 25 °C and pH 6.87 with National Bureau of Standards (NBS) or equivalent buffer containing, per liter, 25 mmol each of KH_2PO_4 and Na_2HPO_4. Add 45 mL of 1.0 mol/L HCl to the imidazole, mix well, and use the pH meter to adjust to pH 7.10 at 25 °C with additional 1.0 mol/L HCl (about 50 mL of HCl solution will be required). Bring buffer to 1 L, mix well, and store in a Pyrex container. Note that the pH of the prepared buffer decreases to 6.90 at 37 °C. The buffer is stable for at least one month at 4 °C. If any mold or other growth appears, discard.

NADH/buffer solution (NADH 166 µmol/L). Dissolve 13.0 mg of Na₂NADH · 4H₂O (e.g., Sigma no. N-8129), M_r 781, in enough imidazole buffer to make 100 mL of solution. Preweighed vials of NADH are available (Sigma no. 340–101 to 125) and can be used to avoid the potentially inaccurate weighing of very small quantities of NADH:

Preweighed NADH, mg	Buffer, mL	No. of determinations
1	7.7	2
2	15.4	5
5	38.4	13
10	76.8	27
25	192	70

The absorbance of the NADH solution in buffer should be between 0.96 and 1.16 A. Because the NADH has a variable water content, it may be necessary to adjust the weight of NADH used or the volumes of buffer taken to obtain an absorbance in the above range. The NADH solution is stable for 8 h at room temperature. NADH forms a potent LD inhibitor in solution, even at 4 °C, so prepare the solution freshly each day.

Sodium pyruvate/buffer solution (sodium pyruvate 22.50 mmol/L). Dissolve 247.6 mg of sodium pyruvate, NaC₃H₃O₃, M_r 110.06 (e.g., Sigma no. P-2256), in enough imidazole buffer to make 100 mL of solution. This is stable for at least one month at 4 °C. Discard if any growth appears.

Dichromate blank solution (50 or 75 mg/L). Use a dichromate blank with turbid samples and with instruments that do not have an absorbance offset. Dissolve 50 mg of K₂Cr₂O₇ in 1 L of 5 mmol/L H₂SO₄ to give an offset of about 0.5 A. Use 75 mg of K₂Cr₂O₇ per liter if an offset of 0.75 A is desired for use with more turbid solutions. K₂Cr₂O₇ solutions are stable indefinitely if kept from evaporating.

The final concentrations of reagents after the reaction has been initiated with pyruvate are: imidazole buffer, 100 mmol/L; NADH, 0.15 mmol/L; and sodium pyruvate, 1.5 mmol/L. The sample fraction is 0.1 in 3.0, or 0.033.

Apparatus

Precise continuous-monitoring methods for LD require a spectrophotometer having a bandpass of 10 nm or less and in proper adjustment. At a 10-nm bandpass, the absorbance readings for NADH will be 1.5% less than the absorbance at 1-nm bandpass; 10 nm is acceptable, but instruments with wider bandpass are not recommended (2). There are no *accurate* LD standards; therefore, the accuracy of the procedure is dependent, in part, on the ability of the instrument to produce the accepted molar absorptivity of NADH. The accuracy depends on wavelength and absorbance calibration, the linearity of the photomultiplier response, the absence of stray light, the adequacy of the thermostating, the accuracy of the measurement of time (either by the instrument or by a strip-chart recorder), the dimensions of the cuvets, and the calibration of any sample dilutor and (or) pipettes used. Inaccuracies in cuvet path length, sample volume, or reagent volume; the presence of stray energy; and instrument noise greatly reduce the accuracy and precision (3).

Wavelength accuracy can be checked if the instrument has a D_2 lamp that emits a prominent line at 486 nm (4, 5). Move the wavelength dial slowly; the dial reading at the point of maximum deflection of the galvanometer or pen must be very close to 486 nm. A less satisfactory approach is to use holmium oxide or didymium glass filters, or solutions of K₂Cr₂O₇ (5, 6). Absorbance accuracy at 440, 465, 546, 590, and 635 nm is checked easily with NBS Standard Reference Material (SRM) no. 930 glass filters, and absorbance accuracy at 350 nm can be checked with solutions of NBS K₂Cr₂O₇. Superficial moisture can be removed from K₂Cr₂O₇ without decomposition by drying at 140 °C. After cooling it in a desiccator, prepare 50.0 and 100.0 mg/L solutions in 5 mmol/L H₂SO₄. The absorbance values of these two solutions at 350 nm and 25 °C should be within 1 to 2% of 0.535 and 1.073, respectively (5). The linearity of the absorbance response can also be checked with these two solutions, with a 150.0 mg/L K₂Cr₂O₇ solution, and with mixtures of these. The absorbance accuracy should be satisfactory to at least 1.5 A, the starting absorbance for turbid samples.

Stray light is easily detected with 10 g/L solutions of NaI; the absorbance should exceed 3 at 240 nm.

Note: Evaluator R. L. L. recommended using acetone to check for stray light in instruments with a tungsten source. Because acetone has an absorbance of about 0.9 in 1-cm cells at 330 nm and about 7 at 320 nm, it therefore could be used to check for stray light at 320 nm (6, 7). Stray light will lead to underestimations of enzyme activities, particularly when the starting absorbances are high.

The temperature control of the solution in the cuvet should be within 0.1 °C of 37 °C. Calibrated thermometers are available from the NBS as SRM no. 934 (8); recently, the gallium melting-point standard became available from the NBS as SRM no. 1968 (9) and can be used to calibrate thermometers and temperature probes. In instruments where it is not possible to put a probe into the cuvet, an indirect colorimetric method can be used to estimate the temperature (10).

A standard cuvet is available from the NBS as SRM no. 932 and can be used with K₂Cr₂O₇ solutions to make comparisons with laboratory cuvets and flow-through cuvets. Sample and reagent volumes dispensed from manual or automatic pipettes can be checked by using distilled water and weighing. Dilutors can be checked by diluting 10- to 50-fold a 1.00 g/L solution of K$_2$Cr$_2$O$_7$ in 5 mmol/L H$_2$SO$_4$, measuring the absorbance of the dilution at 350 nm, and then comparing the result with the expected absorbance. The instrument drift of the absorbance of K$_2$Cr$_2$O$_7$ solutions should be less than 0.002 A/min.

Collection and Handling of Samples

Collection. Blood should be collected in plain evacuated tubes or syringes, centrifuged, and the serum taken off the clot within 30 min of clotting. When serum remains in contact with the clot, serum LD activity increases, owing to leakage of LD from the clotted erythrocytes *(11)*. Even faintly visible hemolysis renders a sample unusable, because the LD activity of erythrocytes is 100- to 150-fold that normally found in serum *(12)*. Icteric and lipemic samples can be analyzed, provided *(a)* the starting absorbance is not beyond the limits of the instrument, i.e., usually less than 1.5 A, and *(b)* a suitable dichromate blank can be used. Very lipemic samples can be cleared of some turbidity by extracting the sample with water-saturated ether as follows: Add 2 mL of distilled water to 10 mL of diethyl ether, shake well, and allow the phases to separate. Add three volumes of the upper (ether) layer to one volume of serum. Mix thoroughly, centrifuge at 1500 rpm, discard the upper layer, and analyze the lower (serum) layer. Plasma containing heparin or EDTA can be used and will give the same results as serum, provided the platelets have been removed by centrifugation at 1300 × g or more for 10 min *(13)*. Oxalated plasma cannot be used because oxalate preferentially inhibits LD-1 *(14)*.

Storage. LD in serum at normal and above-normal activities decreases when the serum is stored at 25, 4, or −20 °C. The Submitters prepared two pools from fresh sera—one with normal activity and one with above-normal activity—and analyzed them in duplicate after storage at the times and temperatures given in Table 1. Table 2 lists the results from an Evaluator. LD-4 and LD-5 are the least stable of the isoenzymes of LD during storage, and LD-1 is the most stable *(15)*. LD-5 is inactivated rapidly at 37 °C if leukocytes remain in the sample *(14)*. Serum samples should be analyzed as quickly as possible after the sample is drawn and stored only at room temperature if analysis is delayed.

Procedure

1. Pipet 2.70 mL of the NADH/buffer solution and 0.10 mL of serum into a cuvet or suitable container that can be incubated at 37 °C. Extract very lipemic samples with water-saturated ether before analysis (see *Collection* section). If there is insufficient serum, use 0.05 mL of serum added to 0.05 mL of imidazole buffer, and note this on the worksheet. Mix, then incubate at 37 °C for 10 to 30 min.

Table 1. Stability of LD on Storage

Storage time, h	Percentage of original activity after storage at		
	25 °C	4 °C	−20 °C
	Low-activity pool[a]		
0	100	—	—
24	98	95	94
48	94	90	89
	High-activity pool[b]		
0	100	—	—
24	99	96	96
48	99	91	91

[a] Value near midpoint of reference range.
[b] 2.1-fold upper limit of reference range.

Table 2. Stability of LD on Storage (Evaluator P. A. G. M.)

Storage time, h	Mean activity[a] after storage at		
	25 °C	4 °C	−80 °C
0	100	—	—
24	99	94	102
48	99	89	103
72	100	87	107

[a] Percent of original mean activity of 26 fresh serum samples.

Times between 10 and 30 min are satisfactory if evaporation is prevented.

2. Adjust the spectrophotometer to zero absorbance at 340 nm with water or with a dichromate blank solution in case of a turbid or pigmented sample. Check that the cuvet is thermostated at 37 °C.

3. Add 0.20 mL of pyruvate/buffer solution, prewarmed to 37 °C, to the reaction mixture, mix rapidly, then read absorbance *at once* and at 30-s intervals for 3 min (seven readings, six intervals).

4. If the ΔA/30 s exceeds 0.1 A, repeat the determination on serum diluted with imidazole buffer and note this on the worksheet. To be acceptable, individual ΔA/30 s values must be within 10% of the mean ΔA/30 s value. If the starting absorbance vs water is less than 0.75, too much NADH has been consumed in the pre-incubation period and the analysis should be repeated on a diluted sample. Turbid samples may have a starting absorbance exceeding 0.75 A in the presence of suboptimal amounts of NADH.

Quality Control

Controls should be stable after reconstitution for at least 24 h, and should mimic patients' samples if at all possible. To resemble most patients' samples, controls should be clear; cloudy controls cause a high starting absorbance and poor precision *(16)*.

The activities of controls should be at the upper reference limit and at two- to threefold the upper limit. The upper limit is a "decision" area, where the method should be as precise as possible. Controls at above-normal activities will detect a loss of linearity more readily. The presence and development of inhibitor in the NADH *(17)* will be most apparent with above-normal controls.

At least two controls should be run with each group of samples to monitor precision. Controls with activities above the limit of linearity of the method, about 1000 U/L, should not be used; they would require additional dilution before use, a procedure not done with most specimens. Controls are not standards and cannot be used as such. Levey–Jennings charts are the most satisfactory for recording data and monitoring long-term trends *(18)*. It is very desirable to use the same lot of control material for one year or more to reveal small shifts in the values with time.

The allowable limits of the assay should be developed in the laboratory. Manufacturers' limits can be a guide, but are usually broader than what the laboratory can do in periods of *good performance*. The laboratory's ± 2 SD limits should be used as the warning limits, and values outside of ± 3 SD should be the action limits for the procedure.

Calculations

Average the six $\Delta A/30$ s values, noting the above criteria for the possible rejection of the data. Activity (U/L) at 37 °C in 1-cm cuvets is calculated as follows:

$$U/L = [\Delta A/30 \text{ s (avg)}] \times 2 \times \frac{10^6 \ \mu\text{mol}}{\text{mol}}$$
$$\times \frac{1 \text{ cm} \cdot \text{mol}}{6220 \text{ L}} \times \frac{\text{total vol}}{\text{sample vol}}$$

$$U/L = [\Delta A/30 \text{ s (avg)}] \times 2 \times (10^6/6220) \times (3.0/0.1)$$

$$U/L = [\Delta A/30 \text{ s (avg)}] \times 9646$$

For example, if the $\Delta A/30$ s values are 0.080, 0.079, 0.081, 0.075, 0.077, and 0.080, the activity in 0.1 mL of serum is 759 U/L.

Discussion

A procedure for LD in serum was described in *Standard Methods of Clinical Chemistry* in 1963 *(12)*. The pyruvate-to-lactate reaction in a pH 7.45 Tris buffer [tris(hydroxymethyl)aminomethane] at 37.5 °C was used with 0.9 mmol of pyruvate and 0.25 mmol of NADH per liter in the final reaction mixture. These conditions are not optimum for the reaction.

A value of 6220 L · cm^{-1} · mol^{-1} has been used here for the molar absorptivity for NADH, because of the widespread use of this value *(19)*. A molar absorptivity of 6317 at 25 °C *(20)* is probably a more reliable value; better yet would be a value determined in imidazole buffer at 37 °C and pH 6.90.

NADH is an unstable reagent; it develops a potent LD inhibitor in the presence of moisture and in solution *(21, 22)*. Even lyophilized NADH in well-closed containers stored at 4 °C develops the inhibitor. A fraction containing the inhibitor can be isolated from some lots of NADH *(17)*. We have found that lyophilized NADH obtained in smaller amounts (e.g., 50 mg or less per bottle) appears to be more stable than the same material in larger quantities per bottle. The manufacturer may be able to dry smaller quantities more completely during packaging.

An off-white color or the presence of lumps indicates decomposition and the NADH should not be used. NADH completely free of LD inhibitor probably does not exist *(21, 22)*, but two simple procedures can be used to check NADH for inhibitor *(23)*. First, check the residual absorbance at 340 nm of NADH that has been completely oxidized. Prepare a solution in the imidazole buffer to contain, per liter, 150 μmol of NADH and 1.2 mmol of pyruvate. After measuring the absorbance vs a blank (buffer plus pyruvate), add 30 U of LD (e.g., Sigma no. L-2500) to *both* the sample of NADH and pyruvate in buffer and the blank solution. Allow the reaction to go to completion at 37 °C for 30 to 45 min; the residual absorbance at 340 nm should be less than 1% of the starting absorbance. As the inhibitor forms, the residual absorbance increases *(23)*.

For the second procedure, use a solution of NADH like the above, but without added LD or pyruvate, and measure the ratio of the absorbance at 260 nm to 340 nm vs a buffer blank; an acceptable NADH will have an absorbance ratio of 2.3 or less *(23)*. The 260/340 nm absorbance ratio of the isolated inhibitor is about 9, so the formation of inhibitor in the NADH will be reflected by a steady increase in the absorbance ratio.

In the determination of LD activity, both the pyruvate-to-lactate and lactate-to-pyruvate reactions have been used. The pyruvate-to-lactate reaction is preferred on theoretical grounds because the equilibrium constant (2.7×10^{11}) is so large *(24)*. Also, the amount of costly NADH required in the pyruvate-to-lactate reaction is only 3% of the amount of NAD$^+$ used in the lactate-to-pyruvate method *(24)*. The pH optima for the pyruvate-to-lactate reaction for LD-1, 3, and 5 are all in the range of pH 7.1 to 7.4, whereas for the lactate-to-pyruvate reaction the range is broader, from pH 8.3 to 8.9 *(25)*. Because the rates for a given amount of enzyme are approximately threefold faster for the pyruvate-to-lactate reaction than for the reverse, smaller samples and shorter observation periods can be used and still maintain accurate spectrophotometry. The pyruvate-to-lactate reaction has the same rate regardless of whether serum, pyruvate, or NADH is used to initiate the reaction. The activity of LD observed in the lactate-to-pyruvate reaction is depen-

Table 3. Reference Values

Source	Population	n	Observed range, U/L[a]
Submitters	Adult volunteer blood donors, men & women	56	211–421
Evaluator J. D. B.	Adult volunteers	59	218–458
Evaluator P. A. G. M.	Adult employees	69	289–518

[a] Reference ranges for central 95% of population were determined at 30 °C and converted to 37 °C by dividing values by 0.57 (28).

Table 4. Between-Day Precision of the Method with Lyophilized Control Sera

		LD, U/L		
Source	n	Mean	SD	CV, %
Submitters	13	172	6.2	3.6
	13	315	7.3	2.3
	13	632	13.7	2.2
Evaluators				
A. R. H.	200	252	20.6	8.2
	200	558	36.7	6.6
R. L. L.	20	226	10.1	4.5
	21	362	12.8	3.5
	20	835	24.8	3.0
J. D. B.	24	196	7.8	4.0
	16	521	30.0	5.8
	16	721	41.0	5.7
P. A. G. M.	17	175	13.1	7.5
	18	376	29.5	7.8
	22	582	45.2	7.7

dent on the initiator used (26). A disadvantage of the pyruvate-to-lactate reaction is its more rapid loss of linearity compared with the lactate-to-pyruvate direction (27); for accurate measurements, the rate must be measured at once after the reaction is initiated (28, 29). Another serious problem is the potent LD inhibitor in some lots of NADH; the quality of the NADH must be evaluated before a given lot is used (17). A reference-grade NADH may be available soon from the NBS for comparison purposes. NAD^+ usually does not contain inhibitors, but exceptions have been reported (30). Also, NAD^+ undergoes appreciable decomposition at the pH of the lactate-to-pyruvate assay (24).

The concentration of pyruvate used here (1.5 mmol/L) is necessarily a compromise, the optimum concentrations of pyruvate for LD-1 and LD-5 being 0.7 and 2.5 mmol/L, respectively (25). The Scandinavian Committee on Enzymes has recommended 1.2 mmol/L (31), and the German Society for Clinical Chemistry 0.6 mmol/L (32). The latter optimum was for normal sera, where LD-1 and LD-2 predominate. Imidazole is recommended as buffer at pH 6.90 because at 37 °C the pH equals the pK_a (28). Per mole, imidazole costs one-third as much as Tris, which has been used in many LD procedures.

Reference Values and Precision Data

The Submitters analyzed sera from 56 volunteer adult blood donors. Blood was collected in red-top Vacutainer Tubes (Becton Dickinson and Co., Rutherford, NJ 07070), centrifuged at 1500 × g for 15 min within 1 h after collection, and analyzed on the day of collection. Our findings and the findings of two of the Evaluators are in Table 3. Precision data are given in Table 4. Laboratories using the method described here should check the reference range for LD by assaying samples from a healthy population of individuals who are reasonably well-matched in age and sex with the patients who will be examined.

References

1. Zimmerman, H. J., and Henry, B. J., Clinical enzymology. In *Clinical Diagnosis and Management by Laboratory Methods*, 1, 16th ed., J. B. Henry, Ed. W. B. Saunders Co., Philadelphia, PA, 1979, pp 347–384.
2. Surles, T., and Erickson, J. O., Absorbance measurements at various spectral bandwidths. *Clin. Chem.* 20, 1243–1244 (1974).
3. Maclin, E., Rohlfing, D., and Ansour, M., Relationship between variables in instrument performance and results of kinetic enzyme assays—a system view. *Clin. Chem.* 19, 832–837 (1973).
4. Rand, R. N., Practical spectrophotometric standards. *Clin. Chem.* 15, 839–863 (1969).
5. Lott, J. A., Practical problems in clinical enzymology. *C.R.C. Crit. Rev. Clin. Lab. Sci.* 8, 277–301 (1977).
6. Frings, C. S., and Broussard, L. A., Calibration and monitoring of spectrometers and spectrophotometers. *Clin. Chem.* 25, 1013–1017 (1979).
7. Slavin, W., Stray light in ultraviolet, visible, and near-infrared spectrophotometry. *Anal. Chem.* 35, 561–566 (1963).
8. Mangum, B. W., Standard Reference Materials 933 and 934: The National Bureau of Standards' precision thermometers for the clinical laboratory. *Clin. Chem.* 20, 670–672 (1974).
9. Bowers, G. N., Jr., and Inman, S. R., The gallium melting-point standard: Its application and evaluation for temperature measurement in the clinical chemistry laboratory. *Clin. Chem.* 23, 733–737 (1977).
10. Bowie, L., Esters, F., Bolin, J., and Gochman, N., Development of an aqueous temperature-indicating technique and its application to clinical laboratory instrumentation. *Clin. Chem.* 22, 449–455 (1976).
11. Ghosh, B. P., and Mitra, A. K., Methods of estimation of lactic dehydrogenase (LDH) activity in serum. *Med. Exp.* 8, 28–34 (1963), cited in *Chem. Abstr.* 59, 11811C.
12. Bowers, G. N., Jr., Lactic dehydrogenase. *Stand. Methods Clin. Chem.* 4, 163–172 (1963).
13. Rothwell, D. J., Jendrzejczak, B., Becker, M., and Doumas, B. T., Lactate dehydrogenase activities in serum and plasma. *Clin. Chem.* 22, 1024–1026 (1976).
14. Qureshi, A. R., and Wilkinson, J. H., Inactivation *in vitro* of lactate dehydrogenase-5 in blood. *Ann. Clin. Biochem.* 14, 48–52 (1977).
15. Kreutzer, H. H., and Fennis, W. H. S., Lactic dehy-

drogenase isoenzymes in blood serum after storage at different temperatures. *Clin. Chim. Acta* **9,** 64–68 (1964).
16. Maclin, E., A systems analysis of GEMSAEC precision used as a kinetic enzyme analyzer. *Clin. Chem.* **17,** 707–714 (1971).
17. Berry, A. J., Lott, J. A., and Grannis, G. F., NADH preparations as they affect reliability of serum lactate dehydrogenase determinations. *Clin. Chem.* **19,** 1255–1258 (1973).
18. Grannis, G. F., Gruemer, H.-D., Lott, J. A., Edison, J. A., and McCabe, W. C., Proficiency evaluation of clinical chemistry laboratories. *Clin. Chem.* **18,** 222–236 (1972).
19. Horecker, B. L., and Kornberg, A., The extinction coefficients of the reduced band of pyridine nucleotides. *J. Biol. Chem.* **175,** 385–390 (1948).
20. McComb, R. B., Bond, L. W., Burnett, R. W., Keech, R. C., and Bowers, G. N., Jr., Determination of the molar absorptivity of NADH. *Clin. Chem.* **22,** 141–150 (1976).
21. McComb, R. B., and Gay, R. J., A comparison of reduced-NAD preparations from four commercial sources. *Clin. Chem.* **14,** 754–763 (1968).
22. Strandjord, P. E., and Clayson, K. J., The control of inhibitory impurities in reduced nicotinamide adenine dinucleotide in lactic dehydrogenase assays. *J. Lab. Clin. Med.* **67,** 144–153 (1966).
23. Gerhardt, W., Kofoed, B., Westlund, L., and Pavlu, B., Quality control of NADH: Evaluation of method for detection of inhibitors and specifications for NADH quality. *Scand. J. Clin. Lab. Invest.* **33,** Suppl. *139,* 1–51 (1974).
24. Howell, B. F., McCune, S., and Schaffer, R., Lactate-to-pyruvate or pyruvate-to-lactate assay for lactate dehydrogenase: A re-examination. *Clin. Chem.* **25,** 269–272 (1979).
25. Gay, R. J., McComb, R. B., and Bowers, G. N., Jr., Optimum reaction conditions for human lactate dehydrogenase isoenzymes as they affect total lactate dehydrogenase activity. *Clin. Chem.* **14,** 740–753 (1968).
26. Buhl, S. N., Jackson, K. Y., Lubinski, R., and Vanderlinde, R. E., Effect of reaction initiator on human lactate dehydrogenase assay. *Clin. Chem.* **22,** 1098–1099 (1976).
27. Buhl, S. N., and Jackson, K. Y., Optimal conditions and comparison of lactate dehydrogenase catalysis of lactate-to-pyruvate and pyruvate-to-lactate reactions in human serum at 25, 30, and 37 °C. *Clin. Chem.* **24,** 828–831 (1978).
28. Buhl, S. N., Jackson, K. Y., and Graffunder, B., Optimal reaction conditions for assaying human lactate dehydrogenase pyruvate-to-lactate at 25, 30, and 37 °C. *Clin. Chem.* **24,** 261–266 (1978).
29. Sims, G. M., and Schoen, R. L., A quantitation of lactate dehydrogenase nonlinearity. *Clin. Chim. Acta* **69,** 317–322 (1976).
30. Babson, A. L., and Arndt, E. G., Lactic dehydrogenase inhibitors in NAD. *Clin. Chem.* **16,** 254–255 (1970).
31. Committee on Enzymes, Scandinavian Society for Clinical Chemistry and Clinical Physiology, Recommended methods for the determination of four enzymes in blood. *Scand. J. Clin. Lab. Invest.* **33,** 291–306 (1974).
32. Recommendations of the German Society for Clinical Chemistry, Standardization of methods for the estimation of enzyme activities in biological fluids. *J. Clin. Chem. Clin. Biochem.* **10,** 281–291 (1972).

Magnesium in Biological Fluids (Provisional)[1]

Submitters: Eugene S. Baginski and Slawa S. Marie, *Chemistry Laboratory, Doctors Hospital, Detroit, MI 48207*
Raymond E. Karcher, *Department of Clinical Pathology, William Beaumont Hospital, Royal Oak, MI 48072*
Bennie Zak, *Department of Pathology, Wayne State University School of Medicine, Detroit Receiving Hospital and University Health Center, Detroit, MI 48201*

Evaluator: Karen Saniel-Banrey, *The Children's Hospital, Columbus, OH 43205*

Introduction

Although the role of magnesium in human physiology is well-established, the monitoring of magnesium in biological fluids on a routine basis along with other electrolytes has not as yet become a reality. Perhaps this is because conflicting reports indicate that magnesium determination in blood serum may not represent a reliable index of total body magnesium deficit *(1–3)*. Also some clinicians apparently believe that magnesium determination requires sophisticated instrumentation that is not available in most clinical laboratories. However, the monitoring of serum magnesium under various clinical conditions is considered generally useful *(4–7)*, particularly in digitalized patients *(8–10)*, in those treated with diuretics *(2, 11, 12)*, in diabetic patients *(13–15)*, and in many other clinical circumstances associated with body deficit, some of which should be of great concern to the clinician *(16, 17)*.

To make serum magnesium determinations available to more clinical laboratories, a simple, rapid, and reliable method is needed that could serve as an alternative to atomic absorption spectrometry (generally considered to be the method of choice). The technique described here appears to meet these criteria. In addition, deproteinization is not needed, as many as 10 specimens can be processed for measurement within 10 min, and the method is applicable to pediatric patients because as little as 10 μL of specimen can be used.

[1] *Editors' Note:* This chapter is marked "Provisional," indicating that it may not have been completely evaluated, and has not met the arbitrary criteria of our reviewing process. On the other hand, the method is used daily in the Submitter's laboratory, and is not only clinically useful, but has met current standards of quality for many years.

Principle

Magnesium forms a colored complex with sulfonated Magon in a strongly basic medium. The absorbance maximum of the complex is at 548 nm. Calcium interference is circumvented by incorporating [ethylenebis(oxyethylenenitrilo)]tetraacetic acid (EGTA) in the system, and the interferences by potentially interfering trace metals are masked by cyanide. The final mixture also contains about 100 mL/L dimethyl sulfoxide, which eliminates the effect of proteins. The latter phenomenon is a matrix effect responsible for a difference in slopes between calibration curves obtained with aqueous standards and standards prepared with protein. The solvent also appears to repress somewhat the absorbance of the blank.

Materials and Methods

Chemicals

1. *Magon sodium sulfonate (Magon)*, 1-azo-2-hydroxy-3(2,4-dimethylcarboxanilido)-naphthalene-1'-(2-hydroxy)benzene-5-sulfonate sodium salt (ICI Pharmaceuticals, Inc., Life Sciences Group, Plainview, NY 11803).
2. *Dimethyl sulfoxide (DMSO)*, reagent grade.
3. *EGTA*.
4. *Potassium cyanide*, reagent grade (**Danger! extremely toxic**).
5. *Potassium hydroxide*, reagent grade pellets.
6. *Magnesium sulfate heptahydrate* ($MgSO_4 \cdot 7H_2O$).

Reagents

1. *Magnesium standard, 2.0 meq/L.* Dissolve 246.48 mg of $MgSO_4 \cdot 7H_2O$ in water in a 1-L volumetric flask. Add 1.0 mL of concd. HCl and dilute to 1 L with metal-free water. Alternatively, the magnesium stock standard (Mg, 1 g/L) can be purchased and a working standard prepared by diluting 24.3 mL of the stock to 1 L with metal-free water.
2. *Dye.* Transfer 200 mL of DMSO into a 500-mL glass beaker. Place a magnetic stirring bar inside and, while stirring the solvent, add 100 mg of Magon. Continue stirring until the dye completely dissolves. The dye dissolves without difficulty.

Transfer 800 mL of metal-free water into a separate 2-L glass beaker. Place a magnetic stirring bar inside and, while stirring, add 100 mg of EGTA. Add two pellets of potassium hydroxide to help the EGTA dissolve. When all dissolves, pour the Magon solution prepared above into the beaker containing EGTA, mix well, and transfer into a polyethylene bottle. Stable indefinitely.

3. *Base.* Dissolve 16.0 g of potassium hydroxide pellets and 0.5 g of potassium cyanide in metal-free water and dilute to 1 L with the water. Store in a polyethylene bottle. Stable indefinitely.

4. *Reagent for sample blank.* Transfer 800 mL of metal-free water into a 2-L glass beaker. Place a magnetic stirring bar inside and, while stirring, add 100 mg of EGTA. Add two pellets of potassium hydroxide to help the EGTA dissolve. When all dissolves, add 200 mL of DMSO, cool to room temperature, and transfer into a polyethylene bottle. Stable indefinitely.

Apparatus

Perform the test in 4.0-mL polystyrene cups (Technicon AutoAnalyzer ™) with polyethylene caps. Polyethylene or polystyrene test tubes can also be used. It is convenient and speedy to deliver the reagents from dispensers equipped with syringes having plungers covered with Teflon. Regular glass plungers tend to "freeze" when exposed to alkali and should not be used. If an automated dilutor is used, deliver the dye with it.

Specimen Requirement

Use serum, heparinized plasma, cerebrospinal fluid, or acidified 24-h urine. Avoid hemolyzed specimens.

Procedure

1. Pipet a 20-µL sample of the specimen into one 4-mL polystyrene cup and 20 µL of magnesium standard into another.

2. Add 1.0 mL of dye reagent to both cups and to a third one to be used as a reagent blank.

3. Add 1.0 mL of base reagent to each of the three cups, cap them, and mix by inversion.

4. After about 5 min determine the absorbance of each sample (A_S) and standard (A_{ST}) against the reagent blank with a spectrophotometer at 548 nm.

5. When turbid or chylous specimens are encountered, proceed as follows: Pipet a 20-µL aliquot of specimen into each of two cups. Process one aliquot as described above and record the absorbance (A_S). Pipet 1.0 mL of the reagent for the sample blank into the cup containing the other aliquot and into an additional cup designated for the reagent blank. Add 1.0 mL of base reagent to both cups and mix. After about 5 min determine the absorbance of the sample blank (A_{SB}) against the reagent blank at 548 nm. Use the difference in absorbances between the sample and the sample blank ($A_S - A_{SB}$) for calculations.

Calculations

Calculate results as follows:

$$(A_S/A_{ST}) \times 2.0 = Mg, meq/L$$

or $[(A_S - A_{SB})/A_{ST}] \times 2.0 = Mg, meq/L$

To convert meq/L to mg/dL multiply by 1.22; to convert meq/L to mmol/L, multiply by 0.5.

Reference Intervals

The range of magnesium values in healthy individuals is as follows *(18):* newborns, 1.5–2.3 meq/L (0.75–1.15 mmol/L); children, 1.4–1.9 (0.70–0.95); adults, 1.3–2.5 (0.65–1.25). The Submitters found the following range: 1.6–2.3 meq/L (0.8–1.1 mmol/L).

Discussion

Method Evaluation

Note: Evaluator K. S.-B. states that the method was easily set up and performed, and that reagents were stable for the duration of the testing period of three months.

Precision, analytical recovery, and correlation of results by this method with those by atomic absorption were assessed; color stability and citrate interference were also studied. In addition, conditions were established for automating the procedure for use with the Abbott VP ™ analyzer (see Table 1).

Table 1. Abbott VP Test Parameters

Test Name	MAG
Temperature	25 °C
Filter Value	550/650 nm
Units of Measurement	meq/L
Dilution Ratio Setting	1:51
Rev Time	2
Aux Disp Station	N/A
Air Mix?	N/A
Aux Disp Volume	N/A
FRR?	Y
Test Type	End Point
Reaction Direction	Up
Revolutions	3
High Standard	2
Low Standard	1
Assay Factor	N/A
Substrate Depletion	N/A
Max Absorbance Limit	1.0
Reagent Degradation	N/A

Abbott Diagnostic Instruments, South Pasadena, CA 91030.

Within-day precision was determined by assaying 20 replicates of each of three different specimens with mean Mg concentrations of 0.78, 3.37, and 4.96 mg/dL. The CVs observed for these specimens were 4.86, 0.95, and 1.25%, respectively. Day-to-day precision was similarly measured over 20 days on eight specimens with mean Mg values ranging from 0.74 to 3.87 mg/dL. The average CV was 5.11%, with CVs for individual specimens ranging from 3.4 to 6.4%. The analytical recovery study included nine different concentrations of Mg, from 1.98 to 6.70 mg/dL, and yielded recoveries ranging from 93 to 100% (mean, 98%).

Method comparisons between atomic absorption and the proposed method performed both manually and with the Abbott VP produced regression and correlation statistics tabulated in Table 2. The regression data show good correlation between the Magon methods and atomic absorption and between the manual and automated procedures. Mean values in all comparisons differed by 0.11 meq/L or less. The range of results was 0.4 to 2.3 meq/L.

Table 2. Method Comparisons

	Study[a]		
	1	2	3
n	20	16	16
Slope	0.929	0.890	0.978
Intercept	+0.224	+0.221	−0.002
r	0.972	0.969	0.988
$\bar{x}$	1.62	1.52	1.61
$\bar{y}$	1.73	1.58	1.58

[a] 1: Manual Magon (y) vs atomic absorption (x).
2: Abbott VP Magon (y) vs atomic absorption (x).
3: Abbott VP Magon (y) vs manual Magon (x).

Five samples studied over a period of 120 min indicated that the produced color was stable for at least 60 min. Absorbance changes measured at 15-min intervals relative to the 5-min reading showed randomly scattered differences for each specimen, ranging from +0.36% to −2.18%. No consistent change was evident in the values measured for any specimen, and differences observed in the specimen readings were essentially equivalent to those in the reagent blank. The average change at 120 min for the five specimens was −1.71%; that of the blank was −1.50%.

In the final study the influence of citrate on the method was investigated. Two controls with Mg concentrations of 2.3 and 4.0 meq/L were assayed in triplicate in the presence of citrate concentrations ranging from 0.0 to 0.80 g/dL. No consistent trend was apparent in the results obtained for the two controls, except that in both cases results were decreased by an average of 7.5% in the presence of 0.2 g of citrate per deciliter. Both higher and lower citrate concentrations deviated from the original baseline values by +0.34% to −6.69%.

The method described here was first published in abstract form (19). Because of space limitations in the abstract, not all findings could be reported, including the study concerning the effect of citrate. That study had been prompted by a report indicating citrate interference (20) in Mann's method (21), but no such interference was found in our studies. However, upon receiving the Evaluator's comments concerning the interference by citrate, we reinvestigated the problem. Using quantities of citrate equivalent to and larger than those used by the Evaluator, we found no significant difference in values for magnesium determined with or without added citrate, and conclude that the presence of citrate does not seem to interfere in the method described here. This does not preclude the fact that citrate binds magnesium, a problem well recognized in patients transfused with citrated blood. Under these circumstances determination of plasma magnesium apparently reflects total magnesium concentration; although from the analyst's point of view the magnesium may be within the normal range, it is chelated and may not function for the moment to maintain normal physiology.

Because erythrocytes contain much larger concentration of magnesium than does plasma, the use of hemolyzed specimens should be avoided. Icteric specimens can be used because the absorbance spectrum maxima for bilirubin and the Magon–Mg complex are sufficiently distant that the spectra do not overlap at 548 nm.

Moderate lipemia can be tolerated because the presence of DMSO in the system has a solubilizing effect on lipids and that, along with the large dilution of specimen (100-fold), minimizes the effect. However, if a very lipemic specimen is encountered, a sample blank correction should be carried out as described in the procedure. Such a correction should be considered as only approximate.

The absorbance of the reagent blank varies with the purity of the dye but usually does not exceed 0.9 (determined with a Beckman Acta III spectrophotometer) with water as a reference. However, the high blank does not seem to affect the sensitivity, which is sufficient to allow use of as little as 10 µL of serum in 1.0 mL of the reagents. The color is stable for at least 1 h, and both reagents seem to be stable for at least one year.

Physiological and Clinical Considerations

Magnesium is the second most abundant intracellular cation. Approximately one-half of the body magnesium is present in bone; the rest is distributed in soft tissues and blood cells, and a small amount is present in blood plasma. The latter appears to serve as a pool for the ion, which can be supplemented from reserves in bone if needed to maintain equilibrium with magnesium in tissues. The variation of plasma magnesium stays within relatively narrow limits under normal conditions, although it is not clear how the equilibrium is controlled. There is a considerable disagreement among investigators as to the possible control mechanism, but the parathyroid gland may be involved to some extent *(22–25)*, perhaps in a fashion analogous to the metabolism of calcium. However, one basic difference is that hypomagnesemia may often decrease the release of parathormone *(23, 25)*, causing secondary hypocalcemia; apparently, adequate supplies of magnesium are needed to maintain the activity of both vitamin D and the hormone in bone resorption *(26)*.

Because only about up to one-third of the total magnesium is actually absorbed *(27)* and the kidney loss of the ion is often increased under various conditions, including excessive diuresis caused by diabetes *(6, 13, 15)*, alcoholism *(28–30)*, or diuretics *(6, 28)*, it is not surprising to encounter a body magnesium deficit frequently. However, there is a difference of opinion as to how the deficit should be best ascertained. Obviously, the measure of magnesium in serum should be the method of choice, but some argue that serum magnesium does not accurately reflect total body deficit *(1–3)*, and that skeletal muscle magnesium should be determined instead *(2, 31, 32)*. However, changes in muscle magnesium reflect changes in body potassium and therefore are not valid indicators of magnesium deficiency *(33)*. When experimental human subjects or laboratory animals were placed on a magnesium-free diet, hypomagnesemia developed first *(34–37)*, followed by either a decrease in muscle magnesium *(36)* or by no change *(34, 37)*. Others found that muscle biopsy specimens from hypomagnesemic patients contained low concentrations of magnesium *(38)*. Serum magnesium appears to correlate well with bone magnesium *(33)*, the latter serving as a reservoir for plasma magnesium *(37)*, and magnesium deficiency usually although not invariably results in hypomagnesemia *(39)*. Magnesium, like potassium, may undergo a rapid shift between intra- and extracellular compartments under several clinical conditions, particularly those involving blood pH changes *(14, 40, 41)*, and it appears that serum magnesium determinations would be more meaningful if they were carried out more often along with determinations of other electrolytes.

Hypomagnesemia has been reported to be due to the following:

diabetes	hypoparathyroidism
alcoholism	malabsorption
pregnancy	hyperalimentation
oral contraceptives	myocardial infarction
some diuretics	congestive heart failure
hyperthyroidism	liver cirrhosis

Hypermagnesemia may be encountered in renal failure, excessive ingestion of antacids containing magnesium, and excessive injection of magnesium solutions.

References

1. Flink, E. B., McCollister, R., Prasad, A. S., Melby, J. C., and Doe, R. P., Evidences for clinical and magnesium deficiency. *Arch. Intern. Med.* **47**, 956–968 (1957).
2. Lim, P., and Jacob, E., Magnesium deficiency in patients on long-term diuretic therapy for heart failure. *Br. Med. J.* iii, 620–622 (1972).
3. Kaya, G., and Özsoylu, S., Serum magnesium levels in children with cirrhosis. *Acta Paediatr. Scand.* **61**, 442–444 (1972).
4. Seller, R. H., The role of magnesium in digitalis toxicity. *Am. Heart J.* **82**, 551–556 (1971).
5. Beller, G. A., Hood, W. B., Smith, T. W., Abelmann, W. H., and Wacker, W. E. C., Correlation of serum magnesium levels and cardiac digitalis intoxication. *Am. J. Cardiol.* **33**, 225–229 (1974).
6. Havill, J. H., Hypermagnesaemia caused by parenteral nutrition in patients with renal failure. *Anaesth. Intens. Care* **3**, 154–157 (1975).
7. Abraham, A. S., Eylath, U., Weinstein, M., and Czaczkes, E., Serum magnesium levels in patients with acute myocardial infarction. *N. Engl. J. Med.* **296**, 862–863 (1977).
8. Szekely, P., and Wynne, N. A., The effects of magnesium on cardiac arrhythmias caused by digitalis. *Clin. Sci.* **10**, 241–253 (1951).
9. Iseri, L. T., Freed, J., and Bures, A. R., Magnesium deficiency and cardiac disorders. *Am. J. Med.* **58**, 837–846 (1975).
10. Singh, R. B., and Jha, V. K., Serum magnesium concentrations in pulmonary heart disease. *Indian J. Chest Dis.* **17**, 15–19 (1975).
11. Smith, W. O., Kyriakopoulos, A. A., and Hammarsten, J. F., Magnesium depletion induced by various diuretics. *Okla. State Med. Assoc.* **55**, 248–250 (1962).
12. Singh, R. B., Srivastava, D. K., Dube, S. S., Dube, K. P., and Vaish, S. K., Hypomagnesaemia and digoxin therapy. *J. Assoc. Physicians India* **22**, 427–430 (1974).
13. Haury, V. G., and Cantarow, A., Variations of serum magnesium in 52 normal and 440 pathologic patients. *J. Lab. Clin. Med.* **27**, 616–622 (1942).
14. Martin, H. E., Mehl, J., and Wertman, M., Clinical studies of magnesium metabolism. *Med. Clin. North Am.* **36**, 1157–1171 (1952).
15. Jackson, C. E., and Meier, D. W., Routine serum magnesium analysis. Correlation with clinical state in 5,100 patients. *Ann. Intern. Med.* **69**, 743–748 (1968).
16. Caddell, J. L., Magnesium deprivation in sudden unexpected infant death. *Lancet* ii, 258–262 (1972).
17. Turlapaty, P. D. M. V., and Altura, B. M., Magnesium deficiency produces spasms of coronary arteries: Relationship to etiology of sudden death ischemic heart disease. *Science* **28**, 198–200 (1980).
18. Meites, S., Ed., *Pediatric Clinical Chemistry*, 2nd ed., Am. Assoc. Clin. Chem., Washington, DC, 1981, p 317.
19. Baginski, E. S., Marie, S. S., and Zak, B., Direct microdetermination of magnesium in biological fluids. *Clin. Chem.* **25**, 1139 (1979). Abstract.
20. Caddell, J. L., and Damrongsak, D., Interference from citrate using Mann's dye method for magnesium determination. *Clin. Chim. Acta* **44**, 273–276 (1973).
21. Mann, C. K., and Yoe, J. H., Spectrophotometric determination of magnesium with sodium 1-azo-2-hydroxy-3-(2,4-dimethylcarboxanilido)-naphthalene-1'-(2-hydroxybenzene-5-sulfonate). *Anal. Chem.* **28**, 202–205 (1956).
22. Heaton, F. W., and Pyrah, L. N., Magnesium metabo-

lism in patients with parathyroid disorders. *Clin. Sci.* **25,** 475–485 (1963).
23. Muldowney, F. P., McKenna, T. J., Kyle, L. H., Freaney, R., and Swan, M., Parathormone-like effect of magnesium replenishment in steatorrhea. *N. Engl. J. Med.* **282,** 61–68 (1970).
24. Targovnik, J. H., Rodman, J. S., and Sherwood, L. M., Regulation of parathyroid hormone secretion in vitro: Quantitative aspects of calcium and magnesium ion control. *Endocrinology* **88,** 1477–1482 (1971).
25. Anast, C. S., Mohs, J. M., Kaplan, S. L., and Burns, T. W., Evidence for parathyroid failure in magnesium deficiency. *Science* **177,** 606–608 (1972).
26. Medalle, R., Waterhouse, C., and Hahn, T. J., Vitamin D resistance in magnesium deficiency. *Am. J. Clin. Nutr.* **29,** 854–858 (1976).
27. Seelig, M. S., The requirement of magnesium by the normal adult. *Am. J. Clin. Nutr.* **14,** 342–390 (1964).
28. Martin, H. E., Clinical magnesium deficiency. *Ann. N.Y. Acad. Sci.* **162,** 891–900 (1969).
29. Zieve, L., Influence of magnesium deficiency on the utilization of thiamine. *Ann. N.Y. Acad. Sci.* **162,** 732–743 (1969).
30. Heaton, F. W., Pyrah, L. N., Beresford, C. C., Bryson, R. W., and Martin, D. F., Hypomagnesaemia in chronic alcoholism. *Lancet* **ii,** 802–805 (1962).
31. Lim, P., Chir, B., Dong, S., and Khoo, O. T., Intracellular magnesium depletion in chronic renal failure. *N. Engl. J. Med.* **280,** 981–984 (1969).
32. Lim, P., Jacob, E., Dong, S., and Khoo, O. T., Values for tissue magnesium as a guide in detecting magnesium deficiency. *J. Clin. Pathol.* **22,** 417–421 (1969).
33. Alfrey, A. C., Miller, N. L., and Butkus, D., Evaluation of body magnesium stores. *J. Lab. Clin. Med.* **84,** 153–162 (1974).
34. Dunn, M. J., and Walser, M., Magnesium depletion in normal man. *Metabolism* **15,** 884–905 (1966).
35. Shils, M. E., Experimental human magnesium depletion. *Medicine* **48,** 61–85 (1969).
36. Chutkow, J. G., Lability of skeletal muscle magnesium in vivo. *Mayo Clin. Proc.* **49,** 448–453 (1974).
37. Aikawa, J. K., and David, A. P., ^{28}Mg studies in magnesium-deficient animals. *Ann. N.Y. Acad. Sci.* **162,** 744–757 (1969).
38. Stendig-Lindberg, G., Bergström, J., and Hultman, E., Hypomagnesaemia and muscle electrolytes and metabolites. *Acta Med. Scand.* **201,** 273–280 (1977).
39. Wallach, S., and Dimich, A., Radiomagnesium turnover studies in hypomagnesemic states. *Ann. N.Y. Acad. Sci.* **162,** 963–972 (1969).
40. Martin, H. E., and Wertman, M., Serum potassium, magnesium, and calcium levels in diabetic acidosis. *J. Clin. Invest.* **26,** 217–228 (1947).
41. Nabarro, J. D. N., Spencer, A. G., and Stowers, J. M., Metabolic studies in severe diabetic ketosis. *Q. J. Med.* **21,** 225–248 (1952).

Magnesium, Titan Yellow Method

Submitters: W. William Spencer and Samuel J. Gentile, *St. Elizabeth Medical Center, Dayton, OH 45408*

Evaluators: Michael McAneny, *Chemistry Laboratory, Kadlec Hospital, Richland, WA 99352*
Marjorie Uhl, *Tule Road, Bolton, ME 01740*
Robert M. Johnson, *Ann Arbor, MI 48103*

Reviewer: George N. Bowers, Jr., *Clinical Chemistry Laboratory, Hartford Hospital, Hartford, CT 06115*

Introduction

Magnesium is one of the most abundant cations in intracellular fluid. Usually magnesium concentrations in the healthy individual are quite constant. Magnesium, like calcium, is absorbed in the upper intestine, but apparently does not require a factor such as vitamin D for absorption to occur.

Decreased serum magnesium values (hypomagnesemia) have been found in alcoholic cirrhosis, malabsorption syndrome, acute pancreatitis, chronic glomerulonephritis, and excessive loss of magnesium in the urine [1].

Increased serum magnesium values (hypermagnesemia) have been reported in cases of dehydration, severe diabetic acidosis, Addison's disease, and chronic and acute renal failure [1].

Several of the first methods for the determination of magnesium involved the precipitation of magnesium either as magnesium ammonium phosphate [2] or as molybdivanadate [3]. These methods were relatively accurate but were both cumbersome and time-consuming. Flame photometry [4] and atomic absorption [5] yield excellent results, but both require instrumentation probably not available to the small-volume clinical chemistry laboratory. Colorimetric procedures such as that of Orange and Rhein [6] have been reported, but in general may give erratic results. Complexometric methods involving Eriochrome, murexide, and Magon are rarely used because of the difficulties involved [1].

In this chapter the Titan Yellow procedure [8] is presented. This method was chosen for its simplicity, precision, accuracy, and amount of equipment required.

Principle

A trichloroacetic acid filtrate of serum is treated with the dye Titan Yellow, 2,2'-(1-triazene-1,3-diyldi-4,1-phenylene)bis[6-methyl-7-benzothiazolesulfonic acid] disodium salt, in alkaline solution. The red lake that forms is thought to be dye adsorbed onto the surface of the collodial particles of magnesium hydroxide, which are kept in solution with the aid of polyvinyl alcohol. (The sensitivity of the method is increased twofold by the presence of the polyvinyl alcohol.) The chemical reaction is not fully understood [1]. The color intensity, which is proportional to the magnesium concentration, is measured spectrophotometrically at 540 mm.

Materials and Methods

Reagents

1. *Trichloroacetic acid, 306 mmol/L.* Dissolve 50.0 g of trichloroacetic acid in de-ionized water and dilute to 1 L with de-ionized water. Store the solution in a brown bottle. Titrate to this molarity with a standard NaOH solution.

Note: The original procedure [8] recommended a glass-stoppered bottle, but this is unnecessary; a polyethylene-lined cap is acceptable.

2. *Sodium hydroxide, 2.5 mol/L.* Dilute 100 mL of a 12.5 mol/L solution of sodium hydroxide to 500 mL with boiled and cooled de-ionized water. The molarity should be checked by titration with a standard acid solution. Store in a brown polyethylene bottle securely stoppered with a polyethylene-lined stopper.

3. *Polyvinyl alcohol, 1 g/L.* Suspend 1.0 g of polyvinyl alcohol (99% hydrolyzed; Matheson, Coleman and Bell, Norwood, OH 45212) in 40–50 mL of 95% ethanol (950 mL/L) and pour the mixture into 500–600 mL of distilled water. Keep the water swirling with a magnetic stirrer while adding the polyvinyl alcohol suspension. Warm the solution on a hot plate until clear; this may take as long as an hour, depending on the temperature of the hot plate. Allow the solution to cool to room temperature and dilute to

1 L with de-ionized water. Polyvinyl alcohol is very difficult to dissolve directly in water without first mixing it with 95% ethanol.

4. *Titan Yellow stock solution.* Dissolve 75 mg of Titan Yellow (Clayton Yellow, Thiazole Yellow) in the 1 g/L polyvinyl alcohol solution and dilute to 100 mL with the same solution. Store in a brown bottle at room temperature. This solution is stable for at least two months; filter it if it is not clear.

5. *Titan Yellow working solution.* Dilute one part of the stock solution with nine parts of the polyvinyl alcohol solution. This solution is stable for a week if stored in a brown bottle at room temperature.

6. *Standard magnesium solution.* For the 1 mmol/L (2.0 meq/L) standard, dissolve 44.6 mg of magnesium iodate, $Mg(IO_3)_2 \cdot H_2O$, in 100 mL of de-ionized water. For the 3 mmol/L (6.0 meq/L) standard, dissolve 133.8 mg of magnesium iodate in 100 mL of de-ionized water. An alternative and preferable standard is magnesium gluconate from the National Bureau of Standards (Standard Reference Material No. 929).

Apparatus

Graduated cylinders, volumetric flasks, pipettes, flasks, and beakers for preparing standards, stock, and working solutions. A narrow bandwidth spectrophotometer is needed. The Turner Model 350 (Turner Associates, Palo Alto, CA 94043) with a bandwidth of approximately 9 mm is satisfactory. In a prior publication (8) it was reported that narrow-band spectrophotometers gave results that followed Beer's Law, while relatively broad-band instruments showed some deviation from a linearity at higher concentrations.

Note: Evaluator R. M. J. found that the test was not linear over the range 2–4 meq/L when either a Coleman Jr. II (bandpass 20 mm; Coleman Systems, Irvine, CA 92715) or a Gilford 300 N (bandpass 8 mm; Gilford Instrument Labs., Oberlin, OH 44074) was used.

Collection and Handling of Specimens

Use either serum or plasma collected with sodium or lithium heparin. Do not use other anticoagulants such as EDTA (ethylenediaminetetraacetate), citrates, or oxalate. Separate the serum from the cells within 30 min after collection. Because the magnesium concentration is three- to fivefold greater in cells than in blood, avoid hemolysis or contamination with tissue fluid during collection. Freeze the serum if the determination cannot be done within 12 h. Bring the frozen serum to room temperature and mix well before analyzing it. Serum concentrations in newborns are essentially the same as values reported for adults (1). Magnesium concentrations in plasma and serum do not change appreciably throughout the day (1).

Procedure

1. Place 1 mL of the unknown serum into a 15 × 150 mm test tube and add 5 mL of the trichloroacetic acid solution.

2. Mix contents of tubes gently but thoroughly, let stand for 5 min, and centrifuge at moderate speed (approximately 1000 rpm) for 10 min.

3. Transfer 3 mL of the clear supernatant fluid to a suitable cuvet.

4. Prepare a standard set by pipetting 0.5 mL of each of the two magnesium standards into separate cuvets, followed by 2.5 mL of the trichloroacetic acid solution.

5. Prepare a reagent blank by adding 0.5 mL of de-ionized water to 2.5 mL of the trichloroacetic acid solution.

6. Add to all cuvets 2 mL of Titan Yellow working solution and 1 mL of 2.5 mol/L sodium hydroxide solution. Mix thoroughly and read absorbance (A) after 5 min, but not later than 30 min, at a wavelength of 540 nm with the reagent blank set to zero absorbance.

Quality Control and Standardization

Use an assayed serum control, in both the normal and abnormal ranges, with each batch of samples. The value for each control should be within the manufacturer's stated accepted range.

Use a calibration curve or calculate the concentration from the following equation:

Magnesium concn = (A unknown/A standard)
$$\times \text{concn of standard}$$

The 1 mmol/L (2 meq/L) standard is usually used unless the concentration of the unknown is very high.

Discussion

The specific performance characteristics of the method are as follows:

Within-run precision (n = 20 each)

Mean, meq/L	SD, meq/L	CV, %
4.22	0.05	1.2
2.74	0.04	1.6 (0.7)
1.51	0.05	3.0
1.34	0.02	1.5 (1.1)

Note: Evaluator R. M. J. found that within-run precision was significantly better for the Gilford 300 N than for the Coleman Jr. II, presumably because of the former's better stability and digital read-out. CVs for the Gilford instrument are in parentheses above. Evaluator M. U. reported an SD of 0.03 meq/L and a CV of 1.9% for a sample having 1.73 meq/L.

Day-to-day precision (n = 20 each)

Mean, meq/L	SD, meq/L	CV, %
4.37	0.29	6.7
2.87	0.14	4.8
1.90	0.11	5.5
1.32	0.05	4.0

The results for the 4.37 and 1.90 meq/L samples were obtained by the Submitters; results for the 2.87 and 1.32 meq/L samples were reported by Evaluator R. M. J.

Analytical recovery, assessed for 10 separate samples, ranged from 90 to 102%.

Note: Evaluator M. McA. found recoveries of 96–107% and Evaluator R. M. J. found recoveries of 102–114%.

Normal values (reference intervals) have been reported by Tietz *(1)* as 0.7–1.2 mmol/L (1.4–2.4 meq/L). Basinski *(7)* reported a mean value of 0.9 mmol/L (1.8 meq/L) with an SD of 0.13 mmol/L; this would give a reference interval of 0.65–1.15 mmol/L (1.3–2.3 meq/L).

Note: Evaluator R. M. J. determined the reference interval to be 1.4–2.3 meq/L in normal volunteers.

The Submitters were able to reduce the sample size from 1 mL to 0.5 mL without sacrificing accuracy. When sample sizes smaller than this were used, accuracy and precision were greatly reduced.

Note: Evaluator M. U. found the 0.5-mL sample size to be suitable.

References

1. Tietz, N. W., *Fundamentals of Clinical Chemistry*, 2nd ed., W. B. Saunders Co., Philadelphia, PA, 1976, pp 917–921.
2. Kramer, B., and Tisdall, F. F., A simple technique for the determination of calcium and magnesium in small amounts of serum. *J. Biol. Chem.* **47**, 475–481 (1921).
3. Simonsen, D. G., Westover, L. M., and Wertman, M., The determination of serum magnesium by the molybdivanadate method for phosphate. *J. Biol. Chem.* **169**, 39–47 (1947).
4. Wacker, W. E. C., and Vallee, B. L., A study of magnesium metabolism in acute renal failure employing a multichannel flame spectrometer. *N. Engl. J. Med.* **257**, 1254–1262 (1957).
5. Gorfien, P. C., and Kramer, B., Quantitative determination of magnesium in biological fluids: A comparative study. *Fed. Proc.* **22**, 457 (1963).
6. Orange, M., and Rhein, H. C., Microestimation of magnesium in body fluids. *J. Biol. Chem.* **189**, 379–386 (1951).
7. Basinski, D. H., Magnesium (Titan Yellow). *Stand. Methods Clin. Chem.* **5**, 137–142 (1965).

Note: For a pertinent review, see Wacker, W. E. C., and Vallee, B. L., Magnesium metabolism. *N. Engl. J. Med.* **259**, 431–438, 475–482 (1958).

Osmolality

Submitters: Craig C. Foreback and Robert C. King, *Department of Pathology, Henry Ford Hospital, Detroit, MI 48202*

Evaluators: Gerald E. Clement and John M. Waud, *Allentown and Sacred Heart Hospital Center, Allentown, PA 18105*

Rita Powers, *St. John's Hospital, Springfield, IL 62702*

Sharad Patel, *Augustana Hospital and Health Care Center, Chicago, IL 60614*

Karen Oates, *American Medical Laboratories, Inc., Fairfax, VA 22030*

Montgomery C. Hart, *Perinatal Laboratory, St. Joseph's Hospital and Medical Center, Phoenix, AZ 85013*

Reviewers: Harry F. Weisberg, *Department of Pathology and Laboratory Medicine, Mount Sinai Medical Center, Milwaukee, WI 53201*

Roy B. Johnson, Jr., *Scripps Clinic and Research Foundation, La Jolla, CA 92037*

John Vasiliades, *Department of Pathology, University of Michigan, Ann Arbor, MI 48109*

Introduction

Osmotic pressure is one of the primary factors regulating the homeostatic equilibrium between cytoplasm and extracellular fluid. Refractive index and specific gravity are often considered to provide information that is interchangeable with osmotic pressure, but there are some differences between them. Specific gravity depends on the total weight of dissolved material in a unit volume. Refractive index of a solution is a function of concentration, but the change in refractive index differs according to the optical properties of the solutes.

Osmotic pressure is a measure of the effect of all the particles in solution and does not differentiate between molecules and ions, nor vary according to the size of the particles. Refractive index and specific gravity only approximate osmotic pressure in biological fluids. The term osmolality has been conceived for biological applications and is expressed in terms of the molality of the dissolved solutes.

Osmolality can be calculated from measurements of several other properties, and formulas used to calculate serum osmolality have been reviewed *(1)*. However, calculated osmolalities of serum are usually lower than the determined osmolality, particularly when hyperproteinemia or hyperlipidemia is present. Because determination of osmolality is a relatively easy and inexpensive procedure to set up and perform, measurement is preferred to calculation.

Theory

If a semipermeable membrane is interposed between a solution and its pure solvent, or between two solutions differing in solute concentration, the solvent molecules, usually water, will diffuse through the membrane faster than the solute molecules. If, for example, a sucrose solution is enclosed within the membrane, water will diffuse across the membrane into the sucrose solution. This movement of water is called osmosis and results in the buildup of pressure within the membrane-enclosed area. The amount of pressure that must be imposed on a solution to prevent passage of solvent into it through a semipermeable membrane is defined as the osmotic pressure.

Osmotic pressure is one of the colligative properties of solutions. Other colligative properties—boiling point, freezing point, and vapor pressure—are related to one another and are mathematically interconvertible. Thus, if a solute is dissolved in a solvent such as water, the following events occur:

- Osmotic pressure is increased
- Vapor pressure is lowered
- Boiling point is raised
- Freezing point is lowered

All are related to the number of particles dissolved, independent of their nature, per mass of solvent.

The term *osmolality* is used to express the reactive osmotic pressure of a solution in terms of mass of solute per mass of solvent.

In the ideal case, 1 mol of a nondissociated solute dissolved in 1 kg of water is a 1 molal[1] solution (1

---
[1] In contrast to a molar solution, in which 1 mol of solute is dissolved in a final volume of 1 L.

osmolal solution). The vapor pressure of this solution is 0.3 mmHg (40 Pa) lower than the vapor pressure of pure water. This solution boils at 0.52 °C higher and freezes 1.86 °C lower than pure water. The osmotic pressure is increased to 22.4 atmospheres.

Because these changes are all interrelated, osmotic pressure can be measured indirectly (osmometry) by determining one of the other parameters. In clinical laboratories, freezing-point depression is most frequently used for measuring osmolality. However, the use of vapor-pressure depression has become popular in the last few years, and will be discussed later.

Sample Collection and Specimen Handling

Collect blood by venipuncture, with a minimum of stasis. Evacuated tubes containing a gel for serum separation are acceptable for obtaining specimens for determination of osmolality. Separate the serum by centrifugation soon after collection. Specimens collected in tubes that do not have a gel separating device should be centrifuged twice to lessen the possible presence of particulate matter. Heparinized plasma is also satisfactory, but plasma obtained with other anticoagulants (e.g., ethylenediaminetetraacetate, citrate) is not. Hemolysis does not interfere.

Collect urine in clean, dry, capped containers without preservatives; centrifuge to remove gross particulate matter.

Complete the analysis of serum and urine as soon as possible after specimen collection. If there is a delay in the analysis, refrigerate the capped specimens to avoid a change in the original solute concentration either by evaporation of H_2O or by the decomposition or combination of solutes. Before analysis, refrigerated covered specimens *must* be warmed to room temperature (see precautions, under *Procedure*).

FREEZING POINT OSMOMETRY

At present, osmometers based on freezing-point thermodynamics are manufactured by three U.S. companies: The Osmette (Precision Systems, Sudbury, MA 01776); the Fiske (Fiske Instruments, Uxbridge, MA 01569); and Advanced Digimatic 3D and 3DII (Advanced Instruments, Needham Heights, MA 02194). The instrument used by the Submitters is the Digimatic 3DII.

Principle

The freezing-point osmometer measures the freezing point of a solution by super-cooling it several degrees below its freezing point, then vibrating it internally. This starts the formation of ice crystals, which rapidly liberate heat of fusion as the temperature of the sample is raised to its freezing point and then held at equilibrium while the temperature is measured. The measurement cycle is shown in Figure 1. The procedure for most instruments is essentially similar with only minor differences between them. The exact procedure for one instrument is given below.

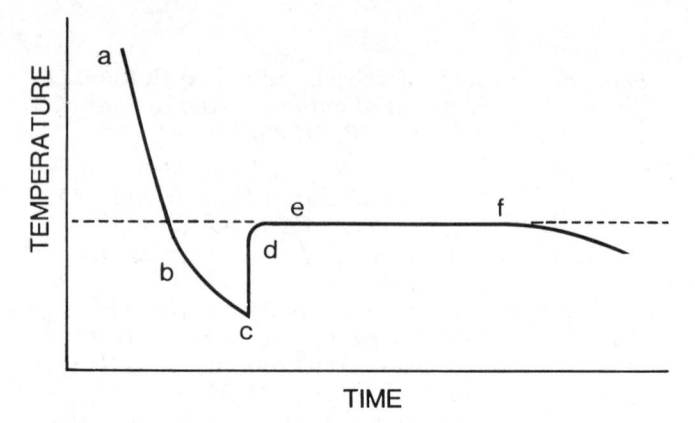

Fig. 1. Uniform freezing curve
a, fast cool; *b*, slow cool; *c*, freeze; *d*, heat of fusion release; *e*, plateau (sample freezing point); *f*, readout

Materials and Methods

Apparatus

1. Freezing point osmometer.
2. Sample tubes.
3. Sample tube rack.

Reagents

1. *Standards, 100 and 500 milli-osmol/kg (mmol/kg) H_2O.* Dissolve 3.094 g and 15.93 g of NaCl (previously dried overnight at 100 °C) in each of two 1-kg quantities of de-ionized H_2O (these standards may be obtained from Advanced Instruments as product no. 3LA010 and 3LA050).

Note: Evaluator R. P. prefers to use standards of 14 and 900 mmol/kg (these may be obtained from Scientific Products, McGaw Park, IL 60085). Reviewer R. B. J. recommends that, in view of the high quality of standards now commerically available, it is unwise for the small laboratory to prepare them.

2. *Bath liquid, ethylene glycol in water (1/2 by vol).* Prepare by measuring 333 mL of analytical grade ethylene glycol (product no. L715; J. T. Baker Chemical Co., Phillipsburg, NJ 08865) into a 1-L beaker. Add 667 mL of de-ionized water.

Procedure

1. Pipet 250 μL of serum or urine into a sample tube. The exact sample size is not critical, but should be kept uniform for best accuracy and precision.

2. Place the sample tube in the elevator well, and lower the head by pressing the Head Control button.

Note: On the model 3D, the operator must manually lower the head by pushing it down from the back.

3. At this point the instrument automatically completes the cooling and freezing cycle. When the test is completed, the head will rise; the test result is then locked into the digital display where it remains until the head is lowered for the next test. Once thawed, the sample may be used for measuring other electrolytes.

Calibration

Once each day or on every work shift, depending on the frequency of use, run standards to check the calibration. Standards of 100 or 500 mmol/kg of H_2O are recommended. If the standards read 100 ± 2 mmol and 500 ± 4 mmol, respectively, the instrument is calibrated properly. If the instrument is not calibrated correctly, follow the procedures on the Calibration section of the Operator's Manual. After the calibration check, measure at least one reference serum daily. The values should fall within the precision limits established by the laboratory. Day-to-day CVs should be less than 3%.

Note: Evaluator R. P. recommends running Advanced Instruments' reference solution (no. 3LA029) daily. The Submitters recommend determining two concentrations (normal and high) of an unassayed control serum at the beginning of each 8-h shift.

Routine Daily Maintenance

1. Measure and record the temperature of the cooling bath liquid. The recommended temperature is −6 °C.
2. Check the volume level of the bath liquid. For other periodic maintenance and troubleshooting, see the appropriate sections of the Operator's Manual.

Precautions

The probe, sample tube, and the sample must all be at the same temperature. If the probe or the sample tube is much warmer than the sample, heat will flow into the sample, causing a distorted freezing curve *(2)*.

Discussion

Precision and Accuracy of the Method

Evaluator S. P. reports a day-to-day CV of less than 1% with the Osmette A; recovery studies averaged 99.5% for three serum specimens and 99.6% for three urine specimens. Evaluator K. O., using the Advanced Osmometer, reports a day-to-day CV less than 2% with 100 mmol/kg and less than 0.5% with 500 mmol/kg. Submitter C. C. F. observed CVs of 0.6 and 0.4% over a nine-month period for normal and high abnormal control sera, respectively.

Clinical Significance

Although the osmolality of serum or urine alone is of limited clinical significance, many causes of hyper- and hypo-osmolal plasma are known. In uncontrolled diabetes mellitus, serum osmolality is often increased because of excess glucose and the accumulation of ketone bodies. In some cases diabetic coma is caused by the hyperosmolality rather than the glucose itself. Serum osmolality increases in dehydration, e.g., from fluid intake restriction, vomiting, or diarrhea. In chronic renal failure the concentrating ability of the tubules is diminished, as shown by the urine osmolality. Inappropriate high production of antidiuretic hormone (ADH), or of an ADH-like substance, may be associated with poorly differentiated bronchiogenic carcinoma, mediastinal tumors, myxedema, porphyria, expanding intracranial lesions, or craniocerebral trauma. In this condition, water is retained by the kidney, causing decreased serum osmolality, increased urinary osmolality, and decreased urinary output. Diabetes insipidus results from ADH deficiency. Urinary output is markedly increased, even when fluids are restricted; urinary osmolality is decreased, while serum osmolality is increased.

The diuretic action of alcohol has led to the assumption that the acute or chronic alcoholic patient is dehydrated. However, Beard and Knott *(3)* have shown that alcohol increases urine formation only as long as blood alcohol concentration is increasing. Thereafter, water is retained, accompanied by a decrease in urinary sodium and potassium; therefore, serum osmolality decreases in the absence of vomiting, malnutrition, or diarrhea. For this reason, assessment of serum and urine osmolality is extremely important. Measurement of serum osmolality in cases of suspected alcohol abuse can help avoid the pitfalls of the alcohol dehydrogenase procedure for alcohol assays when isopropanol is involved *(4)*. (See also *Discussion* section under *Vapor Pressure Osmometry*.)

Free-Water Clearance

Acute renal failure is a significant mortality factor in the postoperative patient or in the patient surviving an episode of shock or other extreme stress. The concentrative capacity of the kidney is its most sensitive and biologically important function, which is best assessed by measuring urinary and plasma osmolality and determining the ratio of the two. A value of 1.8 *(5)* or greater indicates adequate renal function. However, this ratio may be misleading in the presence of oliguria. Thus the early prediction of acute renal failure is best accomplished by measuring osmolal clearance (C_{osmol}), which involves simultaneously quantitating plasma and urinary osmolality and urinary flow rate.

Free-water clearance (C_{H_2O}) is calculated from absolute urine flow, and is a measure of renal concentrative capacity that takes into account the rate of urinary output.

To determine C_{H_2O}, measure osmolality of a timed urine specimen (preferably 1 h), and of a plasma specimen obtained during the period of the urine collection. Calculation is as follows:

$$C_{osmol} = \frac{\text{urinary osmolality}}{\text{plasma osmolality}} \times \text{rate of urine excretion}$$

$$C_{H_2O} = \text{urine volume} - C_{osmol}$$

Note: C_{H_2O} is reported in mL/h. Therefore, C_{osmol} must be divided by the exact time in hours if the collection is not exactly 1 h.

Reference Intervals

Reference intervals for serum osmolality range from 280 to 290 mmol/kg of H_2O; for random (untimed) urine samples, values range from 280 to 310 for serum, and from 300 to 1400 for urine.

Reference intervals for free-water clearance taken from the literature are -100 to -25 mL/h *(9)*. Values that approach a positive number indicate early onset of acute renal failure.

VAPOR PRESSURE OSMOMETRY

Principle

A thermocouple hygrometer measures the decrease in dew point temperature of serum or urine in a closed chamber. Although called a vapor pressure osmometer, it actually determines water potential as a measure of relative vapor pressure and therefore osmolality. The instrument measures the temperature at which water vapor condenses to liquid, whereas a true vapor pressure osmometer is based on evaporation of the solvent. Only one U.S. company manufactures a vapor pressure osmometer (Model 5100; Wescor, Inc., Logan, UT 84321).

Procedure

The operation of the analyzer is simple and straightforward, requiring careful and consistent sampling techniques as practiced in any microanalytical procedure. No special sample containers are needed. The specimen (10 µL of heparinized whole blood, serum, or urine) is absorbed or pipetted onto a disc of ordinary filter paper supplied by the manufacturer. The disc (filter paper) is inserted into the sample chamber by means of the sample slide. This activates the programmed sequence *(10)*: "(1) equilibration of the sample and nulling of the ambient temperature. (2) cooling the thermocouple sensor to a temperature below the dew point. (3) convergence of the thermocouple temperature to the dew point, and finally, (4) program completion with the result stored by a sample-hold circuit." The final reading is directly proportional to the measured temperature differential, which, in turn, is a function of vapor pressure of the test solution.

Operation

Leave the instrument ON at all times. Rotate the chamber-sealing knob counterclockwise (one-quarter turn) and withdraw the sample slide from the instrument. Measure the osmolality as follows:

1. Pick up a paper disc by grasping the edge with thumb forceps. A grip on the edge makes it easier to transfer the disc from the forceps to the sample holder and allows full immersion of the disc.

2. Immerse the disc completely in the solution to be tested, allowing the liquid to fully saturate the disc. Do not immerse the forcep tips, as this will pick up excess solution. For dilute solutions, such as NaCl calibrating standards, allow at least 2 s for absorption. Give more viscous liquids such as serum additional time, to ensure complete saturation of paper. Saturation requires 10–20 µL of sample. Alternatively, use an automatic dispenser to pipet 10 µL of sample or standard onto the disc.

Note: Evaluator M. C. H. obtains the best precision by using the method of saturation.

3. Promptly place the saturated disc in the concave depression in the sample holder. The wet disc should adhere uniformly to the bottom of the depression and have a glistening appearance over all of its upper surface, but without a convex meniscus. Avoid touching the wet disc on the outer surface of the sample holder, because this contaminates the thermocouple chamber when the sample is inserted. Should this occur accidentally, wipe the sample holder clean, and repeat steps 1, 2, and 3.

4. Gently push the sample slide into the instrument until it stops. The In Process indicator will light.

5. Seal the chamber by rotating the chamber-sealing knob clockwise. Important: Rotate the knob to a firm positive stop.

6. Approximately 2 min after the sample is inserted, the In Process indicator will go out, and a beep tone will sound. The number on the panel meter represents osmolality of the specimen, in mmol/kg of water. The instrument holds this final reading until a new specimen is inserted. A slight drift in the reading may occur, but will not exceed 1 mmol in 5 min.

7. Loosen the chamber-sealing knob by rotating counterclockwise at least one-quarter turn and with-

draw the sample slide. Using a lint-free tissue (such as Kimwipe), carefully remove the wet disc and any traces of residual liquid from the sample holder. If any visible residue remains on the holder surface, the tissue may be moistened with water to facilitate cleanup. The sample holder should appear bright, shiny, and dry before the next specimen is inserted.

Note: When the unit is idle, leave the sample slide inserted with the sample holder clean and dry. It is not necessary to seal the chamber. The instrument will complete a measurement cycle and will indicate a full-scale reading (infinite osmolality). This reading remains after the In Process indicator goes out. *The maximum reading of the Model 5100 (digital) is 1.999. A reading greater than this is signified by a 1 in the "thousand" position followed by three blanks.*

Note: A complete program cycle is initiated whenever the sample is inserted into the instrument. It is not necessary to wait for the In Process indicator to go out before starting a new process, but if the indicator is still on when the slide is withdrawn, allow 5 s for the timing circuits to reset before reinserting the slide.

Balancing

The function of the balance adjustment is to balance heat flow to and from the thermocouple junction. Unless there are substantial temperature fluctuations in the room, or a thermocouple head has been changed, the balance control should not ordinarily have to be readjusted. The instrument must be thermally equilibrated with its surroundings to make a proper adjustment. If balancing is necessary, see the appropriate section of the Operator's Manual.

Calibration

The instrument is usually calibrated with a 290 mmol/kg standard. This is accomplished by running the standard as a sample, then adjusting the Calibrate Knob if the reading is not in the range of 288–292 mmol/kg. If the instrument cannot be calibrated satisfactorily, it may be necessary to clean the thermocouple or check the balance adjustment.

The response of the instrument is linear with osmolality in the range of 200–2000 mmol/kg. If the osmolality of a patient's sample is less than 200 mmol/kg, run a 100 mmol/kg standard after calibration with the 290 mmol/kg standard, and set the readout to 100 by using the Compensate Knob.

Standards obtained from Wescor may give improper calibration (11). Therefore, it is advisable for a laboratory to prepare its own standards or obtain them from a second source such as Instrumentation Laboratories, Scientific Products, or Advanced Instruments for comparison.

After calibration, run a 500 mmol/kg standard to check linearity; the reading should be in the range of 485–515 mmol/kg. Also, analyze serum-based quality control material on a daily basis. Day-to-day CV should be less than 3%.

Note: Evaluator M. C. H. reports within-day CVs of 2–13%, depending on the operator.

Maintenance

Routine maintenance consists only of periodic cleaning of the thermocouple. The frequency depends on the number of samples processed.

Note: In the Submitters' laboratory, where large numbers of serum and urine specimens are processed on an around-the-clock basis, twice-weekly cleaning was required. In a less intense environment (St. John's Hospital in Detroit) ordinarily bimonthly cleaning is satisfactory (R. R. Calam, personal communication). Thermocouple cleaning is best accomplished with a Wescor accessory OM-230 thermocouple head cleaner. The instructions for its use are included with the assembly.

Reference Intervals

In general, osmolality measured with the Wescor Osmometer correlates well with values for normal sera measured with freezing-point osmometers; thus, reference intervals are as previously stated.

Discussion

As noted by Evaluator M. C. H., the Wescor is somewhat difficult to standardize and operate, whereas the freezing-point osmometers are simple to operate.

Note: Editor S. M. reports little difficulty standardizing and operating two Wescor instruments. In his institution more than 15 technologists operate them on three shifts in a pediatric hospital laboratory.

Volatile substances such as ethanol, isopropanol, and acetone will not be accounted for when using the Wescor Osmometer (see Table 1). Weisberg (1) has also observed that the contribution of volatiles to serum osmolality is not measured by the Wescor,

Table 1. Increased Osmolality Caused by Four Volatiles Added to Serum, as Measured by a Freezing-Point and Dew-Point Osmometer (12)[a]

Substance	Concn, mg/L	Expected increase, mmol/kg	Measured increase, mmol/kg Freezing point	Measured increase, mmol/kg Dew point
Ethanol	3500	80	80	2
Isopropanol	3400	60	61	4
Methanol	800	27	26	3
Acetone	550	10	11	5

[a] All determinations done in duplicate. The variation within either instrument was 2 milli-osmol/kg. Osmolality was underestimated by the dew-point osmometry at ethanol concentrations of 1000–3000 mg/L.

probably because of its principle of operation. A primary advantage of the Wescor is the small sample-size requirement, which is extremely important for pediatric measurements. This point is somewhat negated, however, because the sample used with the freezing-point apparatus can be recovered for most other measurements typically performed on micro specimens.

References

1. Weisberg, H. F., Osmolality—calculated, "delta," and more formulas. *Clin. Chem.* **21**, 1182–1185 (1975). Letter.
2. Osmometer Manual for Model 3D and 3DII. Advanced Instruments Inc., Needham Heights, MA, 1976.
3. Beard, J. O., and Knott, D. H., The effect of alcohol on fluid and electrolyte metabolism. Chapter 11 in *The Biology of Alcoholism*, **1**, B. Kissin and H. Begleiter, Eds., Plenum Press, New York, NY, 1971, pp 353–376.
4. Vasiliades, J., Pollock, J., and Robinson, C. A., Pitfalls of the alcohol dehydrogenase procedure for the emergency assay of alcohol: A case study of isopropanol overdose. *Clin. Chem.* **24**, 383–385 (1978).
5. Faulkner, W. R., and King, J. W. In *Fundamentals of Clinical Chemistry*, 2nd ed., N. W. Tietz, Ed., W. B. Saunders Co., Philadelphia, PA, 1976, p 1008.
6. Freier, E. F., *ibid.*, Chapter 3, p 153.
7. Johnson, R. B., and Hoch, H., Osmolality of serum and urine. *Stand. Methods Clin. Chem.* **5**, 159–168 (1965).
8. Weisman, N., and Pileggi, V. J., Inorganic ions. In *Clinical Chemistry: Principles and Technics*, 2nd ed., R. J. Henry, D. C. Cannon, and J. W. Winkleman, Eds., Harper and Row, Hagerstown, MD, 1974, pp 736–738.
9. Brown, R. S., Lechy, M. S., and Farber, E. R., Free water clearance. ASCP Check Sample Program, Advanced Clinical Chemistry No. ACC-18, 1976.
10. Osmometer Manual, Wescor Inc., Logan, UT, 1975.
11. Nielson, K. T., and Bresler, E. H., Reference standards for osmometry. *N. Engl. J. Med.* **299**, 493–494 (1978). Letter.
12. Rocco, R. M., Volatiles and osmometry. *Clin. Chem.* **22**, 399 (1978). Letter.

pH and Blood Gas Analysis

Submitters: Harvey L. Kincaid, *Department of Clinical Laboratories, Children's Hospital National Medical Center, and George Washington University School of Medicine, Washington, DC 20010* [former address: *Department of Pathology, University of Kentucky, College of Medicine, Albert B. Chandler Medical Center, Lexington, KY 40536*]

Richard A. Kaufman, *Clinical Chemistry and Quality Control, General Diagnostics, Div. of Warner-Lambert Co., Morris Plains, NJ 07950*

Evaluators: R. F. Moran, *Technical Consultation Center, Corning Medical, Corning Glass Works, Medfield, MA 02052*

Ronald J. Byrnes, Susan Myerow, Wayne Lambert, and Richard Brown, *Evaluation Laboratory, Instrumentation Laboratory, Inc., Lexington, MA 02173*

Reviewers: A. H. J. Maas, *Cardiology Laboratory and Cardiovascular Surgery, University Hospital, Utrecht, The Netherlands*

Robert W. Burnett, *Clinical Chemistry Laboratory, Hartford Hospital, Hartford, CT 06115*

Introduction

Direct measurements of pH, p_{CO_2}, and p_{O_2} are requested to determine the degree of blood oxygenation and to evaluate a patient's acid–base status. This information helps the physician manage acid–base imbalances in a comatose or semicomatose individual suffering from compromised respiratory function, renal failure, or shock from a variety of causes. Although plasma water represents less than 10% of the total body water, measurements on blood samples will reflect changes in a patient's overall acid–base status and gas-exchange processes. This chapter delineates the principles and procedures involved in pH and blood gas measurements, the collection and handling precautions required, and quality-control procedures. Advances in technology necessitate an updating of these topics covered in earlier volumes of this series *(1, 2)*; nonetheless, the reader is encouraged to review these chapters, if available, for a further appreciation of these measurements.

Principle

pH

Early pH measurements were made with a hydrogen electrode as described by Böttger *(3)*. In 1903, Höber used this electrode to measure the pH of ox blood *(4)*. The modern glass electrode dates from 1906, when it was demonstrated that the potential across a thin glass membrane was proportional to the hydrogen ion concentration *(5)*. In 1925, Kerridge was able to show that the glass electrode was useful for determining human blood pH *(6)*. Shortly thereafter, a micro capillary glass electrode was developed for blood pH measurements in a temperature-controlled system *(7)*. Further refinements in the 1950s by Sanz *(8)* and Astrup *(9)* led to the Sanz electrode, used in today's blood gas instruments.

The glass capillary membrane of the modern pH electrode consists of a special formulation of SiO_2 and certain oxides, making it especially sensitive to hydrogen ions. On one side of the glass membrane is the solution or blood sample to be measured. On the other side, in a sealed tube containing an aqueous solution of constant pH, is a silver–silver chloride indicator electrode. The most commonly used reference electrode for these measurements is the saturated calomel (Hg, Hg_2Cl_2) electrode. A saturated KCl solution (salt bridge) connects the two electrodes and functions to reduce the liquid junction (diffusion) potential. A difference in hydrogen ion activity across the glass membrane causes a change in the potential between the indicator and reference electrodes. This change in potential is measured with a voltmeter sensitive to 0.1 mV.

The scale used for expressing hydrogen ion concentration (activity) was developed by Sörensen, who wanted a simplified notation covering a wide range of concentrations. A logarithmic scale, "*power hydrogen*," and the abbreviation, pH, came into use *(10)*: a hydrogen ion concentration of 0.0001 mol/L thus became −4. For convenience, the minus sign was dropped and the formal definition of pH became $-\log[H^+]$. However, the logarithmic scale obscures the relationship between hydrogen ion concentration and other physiological ions.[1]

p_{CO_2}

Three methods are utilized in the determination of p_{CO_2}. In the *indirect method* p_{CO_2} is calculated from a measured total CO_2 and pH, by means of the Henderson–Hasselbalch equation.

$$p_{CO_2} \text{ (mmHg)} = \frac{\text{total } CO_2}{0.0303 \left[\text{antilog}(pH - pK) + 1 \right]}$$

In the *interpolation method*, the blood pH is first measured and then two aliquots of the sample are equilibrated with gases differing in p_{CO_2}. The pH is determined after each equilibration, and the sample's p_{CO_2} is interpolated from a plot of pH vs log p_{CO_2}.

Today, blood gas instruments measure p_{CO_2} *directly* with an electrode that was first described by Stow et al. *(12)* and further developed by Severinghaus and Bradley *(13)*. The Stow–Severinghaus electrode is a pH combination electrode immersed in a bicarbonate solution and placed behind a gas-permeable membrane. An equilibrium is established between the sample or calibrating gas and a thin film of bicarbonate solution. The electrode measures decrease in pH, which is proportional to the p_{CO_2}. Membrane materials in use today allow gas diffusion to occur while remaining impermeable to charged ions (H^+ and HCO_3^-). As a result, the sample pH does not influence the p_{CO_2} measurement.

p_{O_2}

One approach to the measurement of blood p_{O_2} involved equilibration of a blood sample with a small bubble of air, followed by analysis of the gas composition of the bubble *(14)*. Later, polarography was attempted, but the platinum electrode was easily fouled by contact with blood. In the middle 1950s, Clark developed an electrode for the routine amperometric measurement of blood p_{O_2} *(15)*. By covering the platinum tip of the electrode with a cellophane membrane, he prevented its becoming "poisoned" by contact with blood. The modern Clark electrode consists of a thin platinum cathode (10–25 μm in diameter) polarized to −0.65 V, and combined with a silver–silver chloride reference anode. In the measurement, an amount of oxygen proportional to the p_{O_2} of the sample will diffuse across the membrane and be reduced at the cathode. The flow of electrons (four electrons per mole of O_2) is electronically processed through a fixed resistor, causing a change in voltage. This voltage change is measured with a sensitive voltmeter, the same device used in both the pH and p_{CO_2} measurements. The cellophane used in Clark's original electrode has been replaced by more durable materials such as polypropylene, polyethylene, and Mylar. It is important that the membrane is selective, so that reducible ions are not allowed to cross it and contribute to the current flow.

Materials and Methods

Specimen Collection and Handling

Choice of arterial vs venous blood

Venous blood may serve quite well in many instances in the assessment of acid–base balance when only pH and p_{CO_2} are needed *(16, 17)*. Only small differences in these properties are noted when venous and arterial blood samples are compared; inexperience in performing arterial punctures or other restrictions may require the use of venous blood. On occasion, however, nonrespiratory disease states can lead to arterio-venous differences in p_{CO_2} and pH *(18)*. Arterial blood thus is considered the primary reference material for blood gas studies *(19)*, especially in cases that involve an evaluation of pulmonary function. For a p_{O_2} value to be clinically useful, arterial blood is required *(20)*; consequently, blood gas determinations are most often performed on arterial blood. For pediatric patients, arterialized capillary blood is used.

General blood-collection procedures

Anaerobic collection and handling are most important to ensure the integrity of the sample. The presence of bubbles or brief exposure to the atmosphere can significantly alter all blood gas parameters. Calculations with aqueous solutions show that a ratio of 1/10 (air/sample) results in large errors in both p_{CO_2} and p_{O_2} *(21)*. Blood p_{O_2} will show only small changes at values near the ambient p_{O_2}. The blood p_{CO_2}, however, decreases markedly, ambient p_{CO_2} being less than 1 mmHg.

To keep the specimen anaerobic, a tight seal must be formed between the barrel and plunger of the syringe used for collection. Plastic syringes are usually quite good in this respect, but some are unacceptable because the plastic is either permeable to the gases or absorbs them from the blood *(22)*. Mineral oil or lubricants should not be used as sealants because of the high solubility of CO_2 and O_2 in these substances *(23)*. A tightly fitting glass syringe is preferred. Vacutainer Tubes (Becton Dickinson and Co., Rutherford, NJ 07070) can be used, provided they are completely filled *(24)*. A heparin solution (1000 int. units/mL) is used to coat the inside walls of the syringe, to prevent coagulation and to fill dead spaces in the needle and syringe. Other anticoagulants such

---

[1] Because the hydrogen ion concentration of blood covers only a narrow range, the use of a logarithmic scale is not required. These considerations are reflected in the use of nmol/L in preference to pH as the Système International (SI) unit for expressing hydrogen ion concentration *(11)*.

as ethylenediaminetetraacetate (EDTA), citrates, and oxalates are not satisfactory because they alter the blood pH (19). In contrast, studies show that heparin in a concentration of 1 mg/mL has almost no effect on blood pH (25).

Collection is performed without stasis. If a tourniquet is used, draw the sample quickly while the tourniquet is in place. As soon as the blood is collected, seal the tip of the needle by inserting it into a rubber stopper; crimping the needle does not always give an airtight seal. The syringe is immersed in ice water for preservation of the blood gas values, and the specimen is analyzed without delay. The rate of decay of p_{O_2} in storage depends on the initial p_{O_2} value of the blood, the number of leukocytes and reticulocytes, and the temperature (26, 27), but does not depend on the number of erythrocytes or the degree of hemolysis (27, 28). At 37 °C, metabolism increases the p_{CO_2} of collected blood by approximately 5 mmHg/h (29). Cooling the blood to 2 °C reduces this rate to 0.5 mmHg/h (29). Similarly, blood pH decreases by 0.04–0.08 pH units/h at 37 °C but only by 0.004–0.008 pH units/h at 2 °C (29).

Note: 1 mmHg ≈ 133 Pa.

Special procedures for capillary samples

Newer instrumentation allows for microsampling, which is especially important for neonatal care. Capillary blood is arterialized by warming the skin with a towel soaked in hot running tap water (39–44 °C); this gives a p_{O_2} comparable with that of arterial blood (30, 31). The heel is the site for collection with newborns (31). Clean the area with a swab soaked in isopropanol (70%)[2] and wipe dry with a sterile gauze to prevent hemolysis and dilution. Make the puncture so that blood flows freely and rapidly, discarding the first drop. Because squeezing may release tissue juices, avoid "milking" the puncture, to prevent dilution and alteration of the blood gas values (20). Take the sample from well below the surface of the drop clinging to the skin, which serves as an anaerobic barrier. Fill heparinized Natelson collecting pipettes (75 and 200 µL) completely, seal with wax or caps without heating, and rapidly roll them between the thumb and forefinger for thorough mixing. A metal "flea" inserted before sealing the capillary can be used to mix the sample by passing a magnet rapidly along the length of the tube. Glass capillary tubes are preferred to plastic because of the lack of data on the gas permeability of the plastics used in these products.[3]

---

[2] Isopropanol/water (70/30, by vol, 70%) is preferred (31). Disposable pads soaked in 70% isopropanol are available from several manufacturers (see reference 31).

[3] See reference 31 for further details on the technique of collecting capillary blood samples, especially from newborns.

Reagents and Procedures

1. *pH electrode buffers, pH 7.38 and 6.84 (37 °C).* Two buffers are needed to calibrate the pH channel. One has a pH near that of the internal reference buffer in the glass electrode and sets the balance point. The other (usually at a pH within the physiological range) is used to calibrate the slope adjustment.

In 1969, the National Bureau of Standards (NBS) defined an aqueous phosphate buffer with a pH of 7.386 at 37 °C as a primary standard for the measurement of blood pH (32). To prepare, dissolve 1.179 g of reagent grade potassium dihydrogen phosphate (KH_2PO_4) and 4.302 g of reagent grade disodium hydrogen phosphate (Na_2HPO_4) in distilled water and dilute to 1000 mL.

The second buffer is an NBS formulation with a pH of 6.840 at 37 °C. To prepare, dissolve 3.388 g of KH_2PO_4 and 3.533 g of Na_2HPO_4 in distilled water and dilute to 1000 mL (32).

These buffers are also commercially available [Corning Medical, Corning Glass Works, Medfield, MA 02051; Instrumentation Laboratory, Lexington, MA 02173; The London Company (Radiometer), Cleveland, OH 44145].

Phosphate buffers near neutral pH are susceptible to bacterial and fungal contamination. Addition of a few thymol crystals (reagent grade) and refrigeration (4 °C) of freshly opened bottles increase shelf-life up to at least two months (1). Buffers stored in tightly sealed containers have an extended shelf-life but should be discarded at the first sign of precipitate, growth, or turbidity. An alternative is the use of sealed ampules containing sufficient buffer for daily routine calibration. Calibration buffers obtained from instrument manufacturers (see above) contain a germicide that prolongs shelf-life without the need of refrigeration.

2. p_{CO_2} *electrolyte.* This is an aqueous solution of sodium bicarbonate and a chloride salt. Optimum bicarbonate concentration for stability, responsiveness, and linearity is between 5 and 10 mmol/L (33). Addition of chloride and silver ions stabilizes the potential of the reference electrode.

Electrolyte formulations for p_{CO_2} electrodes are as follows: Corning—1.0 mol/L KCl, 5 mmol/L $NaHCO_3$, saturated with AgCl; Instrumentation Laboratory—100 mmol/L NaCl, 20 mmol/L $NaHCO_3$, half-cell stabilizer (stabilizer formulation is proprietary information); The London Co. (Radiometer)—20 mmol/L NaCl, 7 mmol/L $NaHCO_3$.

3. p_{O_2} *electrolyte.* This solution contains a chloride salt to stabilize the potential of the AgCl reference electrode and provides a continuous medium for oxygen diffusion to the cathode. A neutral or alkaline buffer is included to maintain the pH and thus control the reduction potential of oxygen.

To prepare, mix equal volumes of 100 mmol/L KCl and 500 mmol/L NaHCO$_3$ (1). Electrolyte formulations for p_{O_2} electrodes are as follows: Corning—356 mmol/L Na$_2$HPO$_4$, 183 mmol/L KH$_2$PO$_4$, 134 mmol/L KCl; Instrumentation Laboratory—100 mmol/L KCl, alkaline phosphate buffer (proprietary information); The London Co. (Radiometer)—298 mmol/L Na$_2$HPO$_4$ · 2H$_2$O, 192 mmol/L KH$_2$PO$_4$, 140 mmol/L KCl.

4. *KCl (saturated or 4 mol/L)*. This solution maintains the reference electrode at a constant potential and forms a liquid junction to complete the electrical circuit between the reference and glass electrodes. To prepare, add 4 mol of KCl to 1000 mL of distilled water.

5. *Calibrating gases*. Calibrating gases are obtained in almost any desired composition of O$_2$, CO$_2$, and N$_2$. Analyses accompanying each cylinder should be accurate to the nearest 0.05%. The gas mixtures most commonly used (available commercially, as listed above) are:

Gas I: CO$_2$/O$_2$/N$_2$, 5/12/83 or 5/20/75 (by vol)
Gas II: CO$_2$/N$_2$, 10/90 (by vol)

6. *Saline rinse, 0.85 g of NaCl in 1000 mL of distilled water*. Use to remove the blood sample after analysis.

7. *Reagent grade water*. Reagent grade distilled or de-ionized water is used to prepare all the solutions.

8. *Flush solutions*. These are used with automated instruments to remove the sample and residual protein from the tubing network and electrode assemblies. Instrumentation Laboratory: concentrate contains 3.8 g of polyethylene glycol P-150 octylphenyl ether per liter plus NaCl; dilute to use. Corning: concentrate contains 2 g of isooctylphenoxy polyethoxyethanol per liter plus a mold inhibitor; dilute to use. The London Co. (Radiometer): saline.

Note: Evaluator R. F. M. cautions that interchanging flush solutions may lead to protein precipitation.

9. *Sodium hypochlorite, 0.1% (1 g/L)*. Dissolve 100 mg of sodium hypochlorite in 100 mL of water, or dilute a commercially prepared bleach solution [approximately 5% (50 g/L) hypochlorite] 50-fold with water. Regular use of this solution helps clean thoroughly the internal parts of the instrument.

Calibration Procedure

The following is a generalized procedure for blood gas instrumentation. Some modifications will be required in the use of particular instruments and are covered in their respective manuals.

1. Turn instrument on and allow 14–20 min to reach thermal equilibrium.

Note: Evaluator R. F. M. recommends that the instrument be left on continuously to obtain best results.

2. Place the instrument in the Calibrate mode.

3. Calculate p_{CO_2} and p_{O_2} values for calibration Gases I and II (see below).

4. Perform a rapid 30-s purge of the system with Gas I; then use a slower flow rate (1–2 bubbles/s) to equilibrate the gas electrodes (3–5 min).

5. Thoroughly rinse the pH-measuring electrode and then fill it with pH 7.386 buffer. Obtain a reading of 7.386 with the pH Cal or Balance control.

6. Adjust the p_{CO_2} and p_{O_2} with the Cal (Balance) control to the partial pressures calculated for Gas I.

7. Rapidly purge the system with Gas II; then follow with a slower flow rate (1–2 bubbles/s) to equilibrate the gas electrodes (3–5 min).

8. Thoroughly rinse the pH-measuring electrode and then fill it with pH 6.840 buffer. Obtain a reading of 6.840 with the pH Slope control.

9. Adjust the p_{CO_2} reading with the p_{CO_2} Slope control to the calculated value for Gas II and adjust the p_{O_2} Zero to 000.0.

10. Re-equilibrate the gas electrodes with Gas I and adjust the Cal (Balance) control to the calculated p_{O_2} and p_{CO_2} values.

pH electrode

Buffers chosen for pH calibration are based on those of the National Bureau of Standards and have an accuracy of ± 0.005 pH (see *Reagents*). Accurate blood pH measurements depend on the instrument's capacity to maintain a constant temperature (37 ± 0.1 °C) and rapidly reach thermal equilibrium with the sample. Adjusting the pH Cal (Balance) to 7.386 does not significantly alter the slope of the electronic calibration line. The slope control moves the line's slope and is used only when the measuring electrode is filled with pH 6.840 buffer. After the slope is adjusted, the measuring electrode is filled with pH 7.386 buffer and the Cal (Balance) control fine-tuned to 7.386, if necessary.

Note: The pH should fluctuate by no more than 0.01 pH units of the final reading. If readings are erratic or replicates do not agree (± 0.005), a problem in the pH measuring system is indicated (see *Troubleshooting the blood gas electrodes*). Recommendations for ascertaining the accuracy and exactness of a calibration are presented under *Quality Control*.

Gas electrodes

Gases used to calibrate the p_{O_2} and the p_{CO_2} channels should be accurate to within ± 0.05% (see *Reagents*). They are obtained either from instrument manufacturers or from a local supplier. Verification of the composition usually requires the assistance of an analytical laboratory and specialized methods. Knowing the percentages of CO$_2$ and O$_2$ in the calibration gases, the ambient barometric pressure (BP, mmHg), and the vapor pressure of water at 37 °C

(47 mmHg), the partial pressure of each component gas (p_i) is calculated from the following formula:

$$p_i = (\text{percentage of gas i}/100)(BP - 47)$$

In the absence of a laboratory barometer, get the barometric pressure for the local altitude from a nearby weather bureau; do not correct to sea level.

Note: Because the gas electrodes have a finite linear range, calibration gases are chosen to bracket the expected partial pressures of arterial blood samples. For example, a p_{CO_2} of 40 mmHg is normal for arterial blood. Usually, the p_{CO_2} channel is calibrated with 5% and 10% CO_2 (50 and 100 mL/L), having partial pressures of 35 and 70 mmHg, respectively. The normal p_{O_2} of arterial blood is between 85 and 100 mmHg. Calibration of the p_{O_2} channel involves use of an oxygen-free solution or gas and either 20% or 12% O_2 (200 or 120 mL/L), having a nominal p_i of 140 or 84 mmHg, respectively.

Bubbles trapped under the gas membranes will cause erratic electrode response. It is good practice at the beginning of each calibration procedure to remove the O_2 electrode and gently tap it with the membrane end pointing down. This dislodges bubbles that may have been clinging to the membrane surface. Smaller bubbles are removed by rubbing the thumb over the membrane. For the CO_2 electrode, pull the electrode body a short distance backward from the membrane and tap the outer jacket gently. If the electrodes are removed one at a time and replaced immediately, thermal equilibration is quickly re-established and readings are usually stable again within 5 min.

Note: Evaluator R. F. M. indicates that this procedure is necessary with systems such as IL 113 and 213; Radiometer BMS 3; and Corning 160. Newer systems do not require a check for bubbles except after changing electrolytes.

A rapid purge of the system with calibration gas helps to eliminate any contamination with room air. A flow rate of 1–2 bubbles/s will minimize drying of the membranes and evaporation of the CO_2 electrolyte, and will equilibrate the electrodes within 3 to 5 min.

As with the pH electrode, adjustment of the p_{CO_2} Cal (Balance) control moves the calibration line without changing its slope. The O_2 Cal (Balance) control is different, however, in that its adjustment makes a change in the slope of the calibration line. The Zero O_2 control is used to establish a zero-point for the p_{O_2} calibration line.

The one-point calibration procedure

The pH and the p_{CO_2} are potentiometric measurements, and electrode drift creates an anomalous line parallel to the true calibration line. In contrast, drift for the p_{O_2} channel changes the slope of the calibration line. The one-point calibration procedure (steps 5 and 10, above) corrects for this drift and is usually performed before each patient's sample is analyzed. A two-point calibration procedure is generally performed every 8 h to keep the instrument in complete readiness.

Sample Analysis

Note: After collection, place the blood samples in an ice water bath without delay and analyze them as quickly as possible, to minimize changes in values resulting from cellular metabolism. The p_{O_2} deteriorates the most rapidly of the three parameters, in samples having a high p_{O_2}. Delays in analysis will affect results according to the temperature of the blood sample (Table 1) (20).

Table 1. Effect of Temperature on Blood Gas Values 10 Min after Collection[a]

	37 °C	4 °C
pH	−0.01	−0.001
p_{CO_2}	+1 mmHg	+0.1 mmHg
p_{O_2}	−1 mL/L	−0.1 mL/L

[a] Approximate changes after the sample is drawn into the syringe. At 37 °C, the blood remains at body temperature in the syringe; at 4 °C, the sample is properly iced immediately after being drawn.

Inspect specimens drawn in syringes for bubbles. If large bubbles are present, expel them without shaking before the sample is mixed and analyzed. Just before analysis, roll the syringe or evacuated collection tube rapidly between the palms of the hands, to mix contents thoroughly. Roll capillary tubes between the thumb and the forefinger; for tubes containing metal fleas, pass a magnet along the length of the capillary.

Note: Evaluator A. H. J. M. uses a metal (gold) platelet in the syringe to ensure thorough mixing of the blood.

Once the sample is introduced into the instrument, examine the measuring chamber and pH capillary, if possible. The pH measurement requires an unbroken column of blood extending from the glass electrode to the KCl junction. Bubbles in the measuring chamber indicate contamination with either room air or calibration gases and may require reintroducing the sample. Occasionally, blood clots are introduced into the instrument, which interferes with sample removal; this is caused by insufficient anticoagulant or mixing and will require repeat collection.

Capillary samples require some special techniques. After mixing, score the ends of the capillary with a file and break them off, or remove the endcaps. Be careful to avoid letting the blood contact air. Most instruments now have specialized adapters for the introduction of capillary samples, to minimize exposure to air.

With manual instruments, stable readings are usually obtained within 1–2 min and the data recorded. Other instruments will display results after a fixed time interval, during which stable readings are obtained.

Routine Maintenance and Troubleshooting

Routine maintenance

Use the following general checklist at the beginning of each 8-h laboratory shift to keep blood gas instrumentation in good working order.

Note: The operator should follow the manufacturer's recommended procedures.

1. Check the pH calibration buffers for sediments or turbidity, which may indicate microbial contamination. The shelf life of laboratory-prepared buffers is increased by adding a few crystals of thymol and refrigerating them after they are opened (1).

Note: Evaluator R. F. M. cautions that thymol added to commercially manufactured buffers could adversely react with bacteriostats already present.

2. Check the salt bridge for a continuous liquid junction between the reference and pH electrodes (i.e., make certain there are no bubbles in the conductivity path). A saturated KCl solution is indicated by the presence of a few KCl crystals, but make sure that the crystals do not entrap bubbles.

3. Check that the vent hole of the pH reference electrode is open.

4. Check that regulators on the gas tanks are set at 3–5 psig (about 160–260 mmHg or 21–35 kPa, gauge).

5. Calculate the gas calibration values for the day, based on the current barometric pressure.

6. Check the level of water in the bubble chambers.

7. Purge the gas lines with the calibration gases at the beginning of each two-point calibration procedure.

8. Check the p_{O_2} and p_{CO_2} electrodes for (a) intact membranes with integral "membrane check" circuit, (b) presence of bubbles under the membranes, (c) sufficient electrolyte, (d) proper seating in their respective ports, and (e) secure electrical connection to the instrument.

9. Check that the instrument temperature is stable at 37 °C.

10. Check for plugs and leaks by placing the instrument in the sample mode and aspirating either water or a blood sample through the entire system.

After each blood sample is analyzed, rinse the entire system by alternate aspirations of air and saline or by the manufacturer's recommended procedure. Introduction of bubbles into the rinse cycle helps remove residual protein from the membrane surfaces and the inner walls of the tubing. Low p_{O_2} recovery is often resolved by leaving a weak solution of hypochlorite (0.1%; 1 g/L) in the system for 3–5 min. Detergents with and without proteolytic enzymes are also useful in reducing protein buildup. Solutions containing enzymes are most effective when left in contact with the membrane surfaces for several hours or overnight, if possible. How often these cleaning procedures should be used depends on the workload of the laboratory. Inadequate cleaning leads to unstable pH and p_{CO_2} readings and low p_{O_2} recoveries. If these problems are not resolved by a thorough cleaning of the system, then a troubleshooting procedure is indicated (see below).

Electrode maintenance

The following are general guidelines for electrode maintenance. Detailed procedures for specific instruments are found in the manufacturer's manuals.

pH electrode system:

1. Replenish the KCl bridge solution with KCl crystals or fresh 4 mol/L KCl solution as necessary.

2. Remove salt deposits around the reference electrode, the salt bridge, and the junction, to prevent leakage of current.

3. To maintain its responsiveness, periodically soak the glass electrode in 1 mol/L KOH for 5 min.

Note: The pH electrode glass is easily scratched, resulting in a loss of responsiveness.

4. Before putting a new glass electrode into use, hydrate it by soaking it in 0.1 mol/L HCl overnight.

5. Keep spare glass electrodes hydrated by storing them in a neutral buffer.

p_{CO_2} electrode:

1. Replace the CO_2 electrolyte approximately every two weeks. Use fresh electrolyte to rinse out the electrode before refilling it.

2. Periodically soak the combination electrode tip in 1 mol/L KOH for 5 min to maintain its responsiveness. Avoid scratching the pH electrode glass.

3. Install a new p_{CO_2} membrane and spacer approximately every two weeks, taking care not to entrap bubbles under the membrane.

4. Install the new p_{CO_2} membrane free of wrinkles, check it for leaks, and wipe the assembly dry.

5. Thoroughly dry the port; insert the electrode assembly with its vent hole unobstructed and facing up.

6. After replacing the electrolyte, remove the electrode after 30 min and tap it to dislodge bubbles from the membrane.

7. Store spare p_{CO_2} electrodes in CO_2 electrolyte.

p_{O_2} electrode:

1. Polish the electrode tip approximately every two weeks, using a gentle circular motion with emery cloth (supplied by the manufacturer) as an abrasive.
2. Install a new p_{O_2} membrane approximately every two weeks, taking care not to entrap bubbles under the membrane.

Note: Air in the electrode is not a problem as long as it does not contact the membrane or the cathode tip.

3. Mount the membrane free of wrinkles, fill the electrode with fresh electrolyte, wipe the electrode dry, and check the membrane for leaks. Rinse the inside of the electrode with fresh electrolyte before refilling it with fresh electrolyte. Use a syringe with a small-gauge needle.
4. Thoroughly dry the port before re-inserting the electrode.
5. Remove the electrode after 30 min and tap it to dislodge bubbles from the membrane.
6. Store spare p_{O_2} electrodes in O_2 electrolyte.

Troubleshooting the blood gas electrodes

Detailed troubleshooting procedures regarding mechanical and electronic problems are included in the manuals supplied by the instrument manufacturers and are not covered here.

Failure of one of the electrodes to respond usually indicates that there is a problem in electrical continuity (i.e., the electrode is broken, a bubble or plug is present, or a connector is loose or broken). Problems associated with calibration are often traced to either contamination or a mixup (e.g., switching) of the usual buffers and gases used to calibrate. Imprecision and drift problems are usually associated with insufficient gas flow, bubbles present on the electrode surface, unstable temperature, protein or bacterial contamination, improper electrical grounding, and holes in the membranes. The ability to characterize a problem as belonging to one of these three categories will help resolve the difficulty quickly. Application of the quality-control procedures discussed below identify not only the obvious problem of the failure of an electrode to respond, but also the more subtle problem of a loss of accuracy and precision in the measurements.

Quality Control

Specimen integrity

Quality control begins with a properly collected specimen (see *Specimen Collection and Handling*). On receipt in the laboratory, the specimen is matched with its request slip by patient's name and hospital number, and this information is entered in the Blood Gas Log (Figure 1). Time of specimen receipt is especially important in cases where treatment of acid–base disturbances is monitored with serial samples. Some advocate making a correction of blood gas values whenever the patient's body temperature deviates from the temperature of the measurements (37 °C). A nomogram *(29)* or table *(35)* can be used for making these corrections. Others argue against correcting, because reference values for patient's temperatures other than 37 °C are not available. Currently, there is no clear consensus on this subject.

Tonometry

Besides ensuring specimen integrity, the quality-control program checks instrument performance for proper calibration and function. Control samples should not be given preferential treatment—introduce them into the work flow in the same manner as a patient's sample. Ideally, the matrix of the control will resemble closely or be identical to that of the patient's sample.

These criteria are conveniently met in the tonometered blood sample, which provides a check for the p_{CO_2} and p_{O_2} channels. An aliquot of blood is equilibrated with a gas of known composition and then analyzed. The observed p_{CO_2} and p_{O_2} values are compared with their expected values, based on the percentage composition of the equilibrating gas. A recovery in the range of 100 ± 3% (see Table 3) is expected for an instrument that is correctly calibrated and functioning properly. For effective quality control, use a gas with 7% O_2 and 7% CO_2 (70 mL/L each), which gives partial pressures approximately midway between the usual low and high calibration points. Additional gas mixtures are useful in checking the linear response of the instrument *(36)*.

The following is a general procedure for preparing tonometered samples:

1. Fill the hydration chamber to the designated level with water.
2. Check gas connections and purge the system with equilibration gas (300 mL/min for 10 min).
3. Calculate the p_{O_2} and p_{CO_2} of the equilibration gas, taking into account the vapor pressure of water and the current barometric pressure.
4. Check that the tonometer temperature (37 ± 0.1 °C) is in agreement with the blood gas instrument.
5. Place approximately 3 mL of heparinized whole blood in the equilibration vessel and mechanically mix it for 10 min while maintaining the gas flow rate at approximately 300 mL/min. (Larger volumes will require longer equilibration periods.)
6. Remove an aliquot of blood after "washing" the sampling syringe with equilibration gas. (To "wash" the syringe with equilibration gas, insert the empty

BLOOD GAS LOG

Name / Sample Hosp. No.	RM	Time	Date	pH	pCO_2	pO_2	Bar (mmHg)	Temp.	Tech / Comments
QC Specimen Level I	Lot	AM PM							Expected
									Observed
QC Specimen Level II	Lot	AM PM							Expected
									Observed
QC Specimen Level III	Lot	AM PM							Expected
									Observed
Tonometered Sample		AM PM							Expected
									Observed
		AM PM							
		AM PM							
		AM PM							

Fig. 1. The blood gas log provides a record of the specimens received for analysis, patient information, and the results obtained, and can also be used to record quality-control results

syringe into the tonometer chamber, draw gas into the syringe, and expel it several times just before sampling the tonometered blood.)

7. Perform the analysis without delay and evaluate the results.

Tonometered blood most often is used to check recovery of p_{O_2} and p_{CO_2} values. Recently, a tonometry system involving aqueous buffers was introduced (Dynex; Analytic Products, Inc., Belmont, CA 94002). A modified Henderson–Hasselbalch equation is used to calculate pH, thus providing control of the pH channel as well. Tonometry of aqueous sodium bicarbonate also provides a check for all three parameters (21). Use of outdated packed erythrocytes reconstituted with saline and sodium bicarbonate for blood gas quality control has the possible advantage of presenting essentially the same matrix as the patient's sample (37).

Quality-control materials

Commercial control products with established values for pH, p_{CO_2}, and p_{O_2} are available in sealed ampules. There is some controversy over the effectiveness of aqueous controls vs protein-based controls in the detection of errors (36, 38). Both types of controls are available in three ranges with a stated mean ±3 SD for each parameter (ContrIL, Instrumentation Laboratory; G. A. S., General Diagnostics, Div. of Warner-Lambert Co., Morris Plains, NJ 07950; Quantra, DADE Div., American Hospital Supply Corp., Miami, FL 33152). A lyophilized serum control (Versatol Acid-Base; General Diagnostics) is also available, but with values for only pH and p_{CO_2}, however. These products are used routinely in conjunction with tonometered samples to assess the analytical quality of the results. A series of controls is usually run after each two-point calibration procedure and perhaps once again during each 8-h shift.

Reasonableness of results

The critical examination of the blood gas values before reporting results is a valuable addition to any quality-control program. Values for pH that are outside the range 7.1 to 7.6 should be followed up by investigating the patient's status. An arterial blood p_{O_2} value becomes suspect if it exceeds 100 mmHg and the patient is not receiving O_2 therapy; a value below 50 mmHg may indicate venous blood contamination of the sample and requires investigation. Resolve questions concerning p_{CO_2} measurements by performing a total CO_2 assay on the separated plasma, if conditions permit; the assayed value should agree within 2 mmol/L of the calculated total CO_2.

Table 2. Accuracy and Precision in Blood Gas Measurement

Sample	Expected or calculated values			Observed values (mean ± SD)		
	pH	p_{CO_2} (mmHg)	p_{O_2} (mmHg)	pH	p_{CO_2} (mmHg)	p_{O_2} (mmHg)
Tonometered aqueous bi-	7.35	86	149	7.35 ± 0.014	81 ± 5	144 ± 3
carbonate solutions[a]	7.21	122	187	7.20 ± 0.015	110 ± 5	170 ± 6
Tonometered whole blood[b]	NA	NA	NA	7.360 ± 0.007	46.5 ± 2.0	51.4 ± 1.95
				7.420 ± 0.010	33.9 ± 0.77	137.5 ± 2.1
				7.500 ± 0.013	20.7 ± 0.58	75.0 ± 2.8
Tonometered hemolyzed	NA	70.0 ± 0.5	39.1 ± 0.3	7.197 ± 0.009	68.0 ± 1.4	40.3 ± 1.2
blood product (out-dated blood)[c]	NA	39.2 ± 0.3	140 ± 1.0	7.388 ± 0.010	38.5 ± 0.8	137 ± 2.4
Tonometered whole blood[d]	NA	15	NA	NA	16 ± 1.0	NA
	NA	30	50	NA	31 ± 1.0	52 ± 1.8
	NA	50	150	NA	51 ± 1.5	143 ± 5.0
	NA	80	250	NA	77 ± 2.4	224 ± 8.0

NA, not available.
[a] Data obtained over a six-month period (21, 50).
[b] Data obtained over a 30-day period (51).
[c] Data obtained over a 38-week period (37).
[d] Data obtained over an 89-day period (36).

Results and Interpretation

Precision and Accuracy

Absolute whole blood standards for pH, p_{CO_2}, and p_{O_2} are not available for determining the accuracy of these measurements. Relative accuracy and precision of pH measurements are verified with the use of precision buffers accurate to ±0.005 pH units. Interlaboratory comparison of blood pH measurements, however, indicate discrepancies as great as 0.06 pH units (38). Such differences may be due to the liquid junction, the solution used in the salt bridge, temperature differences between salt bridges, and the effects of erythrocytes on liquid-junction potentials (21, 39). Thus, when determining whole blood or plasma pH, the measured value will be relative to a given instrument calibrated with a given set of buffers.

Accuracy and precision data for the p_{CO_2} and p_{O_2} measurements are determined with either aqueous bicarbonate or whole blood samples and are presented in Table 2.

Note: Accuracy and precision data provided by the Evaluators are summarized in Table 3.

Reference Values

Reference values are obtained from a well-specified population (age, sex, race, geography, nutritional and health status, etc.) and add to the predictive value of diagnostic testing (40). Reference values for blood gas parameters are incomplete; however, selected data from the literature are given in Table 4. "Normal" values are usually established from a subset of a healthy population and a range defined by the mean ±2 SD (95% confidence limits). If the distribution obtained is not gaussian, then this approach is invalid. Even with a gaussian distribution, the 95% confidence limits mean that on a statistical basis, 5% (or 1 out of 20) in the "healthy" population being defined will have an abnormal value. In the same way, some values in the "normal range" may indeed be abnormal (41).

Note: Evaluator R. J. B. has supplied data on "normal" values, which are included in Table 4.

Classification of Acid–Base Disorders[4]

The practical determination of acid–base status involves use of the blood bicarbonate–carbonic acid buffer as an index of the body's total acid–base systems, which are in equilibrium with each other and with the bicarbonate–carbonic buffer as well. The relationship between pH and these buffer systems is given by the Henderson–Hasselbalch equation:

$$pH = 6.1 + \log([HCO_3^-]/[H_2CO_3])$$

where $[H_2CO_3]$ is equal to 0.03 (p_{CO_2})[5] and 6.1 is the pK of carbonic acid in blood. Acid–base homeostasis is accomplished chiefly through the collaborative actions of the lungs and kidneys, which maintain the blood H_2CO_3 and HCO_3^- concentrations at 1.25 and 25 mmol/L, respectively (43). A normal bicarbonate buffer ratio is thus 20/1, giving a plasma pH of 7.40.[6]

---

[4] Refer to other sources (20, 43–45) for a more thorough treatment of acid–base balance than space permits here.
[5] The solubility coefficient for CO_2 gas in normal plasma is 0.0306 ± 0.0006 mmol/L per mmHg.
[6] Log 20 = 1.3; substitution into the Henderson–Hasselbalch equation gives a normal pH of 7.40.

Table 3. Accuracy and Precision in Blood Gas Measurement

	Expected or calculated values				Observed values (mean ± SD)		
		mmHg				mmHg	
Sample	pH	p_{CO_2}	p_{O_2}	N	pH	p_{CO_2}	p_{O_2}
Tonometered aqueous bicarbonate solutions[a]	7.17			72	7.17 ± 0.003		
	7.38			72	7.38 ± 0.003		
	7.54			72	7.54 ± 0.003		
Commercial aqueous control product[a]	7.09			15	7.10 ± 0.004		
	7.41			15	7.41 ± 0.004		
	7.61			15	7.61 ± 0.004		
Commercial aqueous control product[b]	7.17–7.21	62–72	65–75	336	7.193 ± 0.004	67.3 ± 1.34	68.4 ± 2.87
	7.36–7.40	41–47	99–109	300	7.378 ± 0.003	45.0 ± 0.88	104.4 ± 1.74
	7.63–7.67	17–21	147–157	444	7.640 ± 0.005	19.8 ± 0.50	153.4 ± 1.53
Tonometered whole blood[a]		21.4	85.6	80		21.4 ± 0.3	86.7 ± 0.6
		35.6	378	80		35.5 ± 0.3	361 ± 3.1
		49.9	49.9	50		49.8 ± 0.3	50.8 ± 0.4

[a] Evalulator R.F.M., within-day precision.
[b] Evaluator R.J.B., day-to-day precision.

It is vital that the pH is maintained within a narrow range, to keep the body's enzymic processes intact. A decrease in blood pH to less than 7.35 ($[HCO_3^-]$/$[H_2CO_3]$ ratio less than 20/1) results in an *acidemia*. An increase in blood pH to above 7.45 ($[HCO_3^-]$/$[H_2CO_3]$ ratio greater than 20/1) results in an *alkalemia*. The pathophysiological causes of these pH changes are termed acidosis and alkalosis, respectively.

Normal acid–base status is the result of a balance between a respiratory component and a metabolic component. The Henderson–Hasselbalch equation may be rewritten to include the organ systems involved (44):

$$pH = pK + \log(\text{renal function}/\text{pulmonary function})$$

Compromised renal or pulmonary function leads to a change in the bicarbonate buffer ratio, with a concurrent change in pH away from 7.40. A compensation process corrects acid–base imbalances when they occur. Both the respiratory and renal systems will function to compensate for any shifts, bringing the pH back to normal through a change in the component that is the reciprocal of the primary cause of the imbalance.

Alterations in acid–base homeostasis are usually categorized into the four types discussed briefly below and summarized in Table 5.

Table 4. Reference Ranges for Blood Gas Values[a]

Age	Blood	pH	mmHg p_{CO_2}	p_{O_2}	O_2 satn.	Base excess	$[HCO_3]$, mmol/L	Total CO_2
Newborn	art. cap.[b]	7.33–7.49	27–40	60–70		−10 to −2	17–24[c]	
	arterial			65–76				
2 months– 2 years	arterial	7.34–7.46	26–41			−6.6 to +0.2	16–24[c]	
Child and adult	arterial	7.35–7.45	33–46	80–105[d]	95–99	−4 to +3	18–29	20–24
	venous	7.32–7.42	40–50	25–47	60–85	−3.3 to +2.2[c] (capillary blood)	14–16[c]	21–28
Adult[e]	venous	7.27–7.42	40–65	14–45				

[a] According to age group (20, 42, 46, 47).
[b] art. cap., arterialized capillary blood.
[c] Values calculated from pH and p_{CO_2}.
[d] Normal arterial p_{CO_2} decreases with age because of alveolar degeneration associated with the aging process (49) to the extent that p_{O_2} values of 70 and 60 mmHg are considered normal for 70- and 80-year-old individuals, respectively (20).
[e] Data supplied by Evaluator R.J.B.

Table 5. Summary of Changes in Blood Gas Values Associated with Alterations in Acid–Base Homeostasis[a]

	pH	p_{CO_2}	$[HCO_3^-]_p$[b]	Base excess
Reference values[c]	7.35–7.45	35–45	22–28	0 ± 2.3
Acidosis (nonrespiratory)				
Initial (acute)	↓7.35	N	↓22	↓
Compensated	N	↓35	↓	↓
Acidosis (respiratory)				
Initial (acute)	↓7.35	↑45	N	N
Compensated	N	↑	↑↑28	↑
Alkalosis (nonrespiratory)				
Initial (acute)	↑7.45	N	↑28	↑
Compensated	N	N or ↑	↑↑	↑
Alkalosis (respiratory)				
Initial (acute)	↑7.45	↓35	N or ↓	N
Compensated	N	↓	↓↓22	↓

[a] Compiled from several sources (20, 45, 46). N denotes normal. The direction and relative magnitude of alterations are indicated by arrows.
[b] Actual plasma $[HCO_3^-]$.
[c] Adult arterial blood gas values (20, 47).

Respiratory acidosis (↑p_{CO_2})

In addition to an increased p_{CO_2}, the total CO_2 is increased and the pH is decreased. The H_2CO_3 is increased, resulting in a $[HCO_3^-]/[H_2CO_3]$ ratio less than 20. The causes of respiratory acidosis include obstructive lung disease, oversedation, central nervous system lesions, disorders of respiratory muscles, inhaled CO_2, and mechanical asphyxia. Renal compensation leads to a retention of HCO_3^-, re-establishing the bicarbonate buffer toward a ratio of 20/1 and a pH near 7.40.

Respiratory alkalosis (↓p_{CO_2})

This condition, the opposite of respiratory acidosis, is characterized by an increased pH and a mildly decreased p_{CO_2} and total CO_2. The H_2CO_3 concentration is decreased, causing the bicarbonate buffer ratio to exceed 20. Causes of respiratory alkalosis include hypoxia, nervousness and anxiety, salicylate poisoning (early stages), and pulmonary embolus. Renal compensation results in an increased excretion of HCO_3^- until the buffer ratio approaches 20/1 with a pH near 7.40.

Metabolic acidosis (↓HCO_3^-)

A significantly lower pH than is found with respiratory acidosis accompanies severe metabolic acidosis. A lower total CO_2 and often an increased anion gap are also associated with metabolic acidosis. Conditions resulting in metabolic acidosis include diabetic ketoacidosis, renal failure, diarrhea, salicylate intoxication, and ammonium chloride ingestion. Compensation occurs through the lungs to decrease the p_{CO_2} and to bring the buffer ratio and pH to near normal.

Metabolic alkalosis (↑HCO_3^-)

Metabolic alkalosis is characterized by an increased HCO_3^-, pH, and total CO_2. Causes include loss of acid (vomiting, gastric lavage), corticotropin-secreting tumors, Cushing's syndrome, excessive diuretic intake, and excessive alkali administration. Compensation is accomplished by hypoventilation, which increases the p_{CO_2} (and H_2CO_3) and returns the bicarbonate buffer ratio to near normal. The pH approaches 7.40 in the fully compensated state.

References

1. Gambino, S. R., pH and p_{CO_2}. *Stand. Methods Clin. Chem.* **5**, 169–198 (1965).
2. Gambino, S. R., Oxygen partial pressure (p_{O_2}) electrode method. *Stand. Methods Clin. Chem.* **6**, 171–182 (1970).
3. Böttger, W., Die Anwendung des Electrometers als Indikator beim Titrieren von Sauren und Basen. *Z. Phys. Chem.* **24**, 253 (1897).
4. Höber, R., Über die Hydroxylionen des Blutes. *Arch. Ges. Physiol.* **99**, 572–593 (1903).
5. Cremer, M., Über die Ursache der electromotorischen Eigenschaften der Gewebe, zugleich ein Beitrag zur Lehre von den polyphasischen Electrolytketten. *Z. Biol.* **47**, 562 (1906).
6. Kerridge, P. T., The use of the glass electrode in biochemistry. *Biochem. J.* **19**, 611–617 (1925).
7. Stadie, W. C., O'Brien, H., and Laug, E. P., Determination of the pH of serum at 38 °C with the glass electrode and an improved electron tube pontentiometer. *J. Biol. Chem.* **91**, 243–269 (1931).
8. Sanz, M. C., Ultramicro methods and standardization of equipment. *Clin. Chem.* **3**, 406–419 (1957).
9. Siggaard-Andersen, O., Engel, K., Jørgensen, K., and Astrup, P., A micro method for determination of pH, carbon dioxide tension, base excess and standard bicarbonate in capillary blood. *Scand. J. Clin. Lab. Invest.* **12**, 172–176 (1960).
10. Sörensen, S. P. L., Enzymstudien. II. Über die Messung und die Bedeutung der Wasserstoffionenkonzentration bei enzymatischen Prozessen. *Biochem. Z.* **21**, 131–304 (1909).
11. IUPAC and IFCC, Quantities and units in clinical chemistry. Recommendation 1973. *Pure Appl. Chem.* **37**, 519–546 (1974).
12. Stow, R. W., Baer, R. F., and Randall, B. F., Rapid measurement of the tension of carbon dioxide in blood. *Arch. Phys. Med. Rehabil.* **38**, 646–650 (1957).
13. Severinghaus, J. W., and Bradley, A. F., Electrodes for blood p_{O_2} and p_{CO_2} determination. *J. Appl. Physiol.* **13**, 515–520 (1958).
14. Riley, R. L., Prommel, D. D., and Franke, R. E., A direct method of determination of oxygen and carbon dioxide in blood. *J. Biol. Chem.* **161**, 621–633 (1945).
15. Clark, L. C., Jr., Monitor and control of blood and tissue oxygen tensions. *Trans. Am. Soc. Artif. Intern. Organs* **2**, 41–57 (1956).
16. Gambino, S. R., Comparisons of pH in human arterial, venous, and capillary blood. *Am. J. Clin. Pathol.* **32**, 298–300 (1959).
17. Brooks, D., and Wynn, V., Use of venous blood for pH and carbon-dioxide studies, especially in respiratory failure and during anesthesia. *Lancet* **i**, 227–230, (1959).
18. Austin, W. H., Arterial and venous samples in acid–base balance. In *Blood pH, Gases, and Electrolytes*, R. A. Durst, Ed., National Bureau of Standards Special

Publication 450, U.S. Government Printing Office, Washington, DC 20402, 1977, pp 167–170.
19. Gambino, S. R., Astrup, P., Bates, R. G., Campbell, E. J. M., Chinard, F. P., Nahas, G. G., Siggaard-Andersen, O., and Winters, R., Report of the *ad hoc* committee on acid–base methodology. *Am. J. Clin. Pathol.* **46**, 376–381 (1966).
20. Shapiro, B. A., Harrison, R. A., and Walton, J. R., *Clinical Application of Blood Gases*, 2nd ed., Year Book Medical Publ. Inc., Chicago, IL, 1977, pp 79–160.
21. Noonan, D. C., and Burnett, R. W., Quality-control system for blood pH and gas measurements, with use of a tonometered bicarbonate–chloride solution and duplicate samples of the whole blood. *Clin. Chem.* **20**, 660–665 (1974).
22. Scott, P. V., Horton, J. N., and Mapleson, W. W., Leakage of oxygen from blood and water samples stored in plastic and glass syringes. *Br. Med. J.* **iii**, 512–516 (1971).
23. Gambino, S. R., Mineral oil and carbon dioxide. *Am. J. Clin. Pathol.* **35**, 268–269 (1961).
24. Fleisher, M., and Schwartz, M. K., Use of evacuated collection tubes for routine determination of arterial blood gases and pH. *Clin. Chem.* **17**, 610–613 (1971).
25. Siggaard-Andersen, O., Sampling and storing of blood for determination of acid–base status. *Scand. J. Clin. Lab. Invest.* **13**, 196–204 (1961).
26. Jalavisto, E., Oxygen consumption of blood and plasma and the percentage of reticulocytes. *Acta Physiol. Scand.* **46**, 244–251 (1959).
27. Lenfant, C., and Aucutt, C., Oxygen uptake and change in carbon dioxide tension in human blood stored at 37 °C. *J. Appl. Physiol.* **20**, 503–508 (1965).
28. Harrop, G. A., The oxygen consumption of human erythrocytes. *Arch. Intern. Med.* **23**, 745–752 (1919).
29. Kelman, G. R., and Nunn, J. F., Nomograms for correction of blood p_{O_2}, p_{CO_2}, pH, and base excess for time and temperature. *J. Appl. Physiol.* **21**, 1484–1490 (1966).
30. Ladenson, J. H., Non-analytical sources of laboratory error in pH and blood gas analysis. In *Blood pH, Gases, and Electrolytes* (see ref. *18*), pp 175–190.
31. Meites, S., and Levitt, M. J., Skin-puncture and blood-collecting techniques for infants. *Clin. Chem.* **25**, 183–189 (1979).
32. Bates, R. G., Revised standard values for pH measurements from 0 to 95 °C. *J. Res. Natl. Bur. Stand. Sect. A*, **66a**, 179–184 (1962).
33. Severinghaus, J. W., and Bradley, A. F., *Blood Gas Electrodes*, Radiometer, Copenhagen NV, Denmark, 1971, p 42.
34. Scholander, P. F., Analyzer for accurate estimation of respiratory gases in one-half cubic centimeter samples. *J. Biol. Chem.* **167**, 235–250 (1947).
35. Burnett, R. W., and Noonan, D. C., Calculations and correction factors used in determination of blood pH and blood gases. *Clin. Chem.* **20**, 1499–1506 (1974).
36. Leary, E. T., Delaney, C. J., and Kenny, M. A., Use of equilibrated blood for internal blood-gas quality control. *Clin. Chem.* **23**, 493–503 (1977).
37. Steiner, M. C., Shapiro, B. A., Kavanaugh, J., Walton, J. R., and Johnson, W., A stable blood product for pH–blood-gas quality control. *Clin. Chem.* **24**, 793–795 (1978).
38. Weisbrot, I. M., Kambli, V. B., and Gorton, L., An evaluation of clinical laboratory performance of pH–blood gas analyses using whole blood tonometer specimens. *Am. J. Clin. Pathol.* **61**, 923–935 (1974).
39. Burnett, R. W., Problems associated with the definition of measured and calculated quantities in blood pH and gas analysis. In *Blood pH, Gases, and Electrolytes* (see ref. *18*), pp 163–165.
40. Galen, R. S., and Gambino, S. R., *Beyond Normality: The Predictive Value and Efficiency of Medical Diagnoses*, John Wiley & Sons, New York, NY 1975, pp 2–7.
41. Bermes, E. W., Erviti, V., and Forman, D. T., Statistics, normal values, and quality control. In *Fundamentals of Clinical Chemistry*, 2nd ed., N. W. Tietz, Ed., W. B. Saunders Co., Philadelphia PA, 1976, pp 60–102.
42. Ibbott, F. A., LaGranga, T. S., Gin, J. B., and Inkpen, J. A., Blood pH, CO_2, and O_2. In *Clinical Chemistry; Principles and Technics*, 2nd ed., R. J. Henry, D. C. Cannon, and J. W. Winkleman, Eds., Harper & Row, New York, NY, 1974, pp 760–785.
43. Tietz, N. W., and Siggaard-Andersen, O., Acid–base and electrolyte balance. In *Fundamentals of Clinical Chemistry* (see ref. *41*), pp 945–974.
44. Tuller, M. A., *Acid–Base Homeostasis and Its Disorders*, Medical Examination Publ. Co., Flushing, NY, 1971.
45. Zilva, J. F., and Pannall, P. R., *Clinical Chemistry in Diagnosis and Treatment*, 2nd ed., Year Book Medical Publishers Inc., Chicago, IL, 1975, pp 76–108.
46. Fleisher, W. R., and Gambino, S. R., *Blood pH, pO_2 and Oxygen Saturation*, American Society of Clinical Pathologists Council on Continuing Education, Chicago, IL, 1972.
47. *Pediatric Clinical Chemistry*, 2nd ed., S. Meites, Ed., American Association for Clinical Chemistry, Washington, DC, 1981, pp 47–53.
48. Fiereck, E. A., Appendix. In *Fundamentals of Clinical Chemistry* (see ref. *41*), p 1220.
49. Davies, C. T. M., The oxygen-transporting system in relation to age. *Clin. Sci.* **42**, 1–13 (1972).
50. Noonan, D. C., Quality control and standards. In *Blood pH, Gases, and Electrolytes* (see ref. *18*), pp 275–278.
51. Malenfant, A. L., and Faller, K. D., Development of reference methods. In *Blood pH, Gases, and Electrolytes* (see ref. *18*), pp 279–283.

Phenylalanine, Fluorometric Method

Submitter: Paul M. Tocci, *Genetic Services Laboratory, University of Miami School of Medicine, Miami, FL 33101*

Evaluators: Mary H. Cheng, *Clinical Laboratories, Childrens Hospital of Los Angeles, Los Angeles, CA 90027*
Rui-Guan Chen, *Department of Pediatrics, Shanghai Second Medical College, Shanghai, China*
Edwin W. Naylor, *Department of Pediatrics,* and Adam P. Orfanos, *Biochemical Genetics Laboratory, State University of New York at Buffalo, Buffalo, NY 14214*
G. H. Thomas, *Genetics Laboratory, Johns Hopkins University School of Medicine, Baltimore, MD 21205*

Reviewer: F. W. Spierto, *Clinical Chemistry Division, Center for Environmental Health, Centers for Disease Control, Atlanta, GA 30333*

Introduction

Phenylalanine in blood is most frequently determined by the fluorometric procedure of McCaman and Robins (1) or one of its various modifications, including automated versions (2–10), and by the Guthrie bacterial inhibition assay (11). Although the Guthrie test does not have the accuracy of chemical methods, it is sufficiently sensitive to be used as a semiquantitative assay and is the method used in many states to detect newborns with phenylketonuria (PKU).

Note: Evaluators E. W. N. and A. P. O. state that, when used as a screening test, the Guthrie bacterial inhibition assay can be described as a semiquantitative assay. They further state that the bacterial response to phenylalanine in this assay is extremely sensitive. If appropriate blood-spot standards are used, a standard curve can be drawn, and, after accurate measurement of the growth zones, the precise phenylalanine concentration can be determined. The Evaluators are using the semiquantitative Guthrie test for monitoring nearly 50 cases of PKU. Standard values of 2, 4, or 6 mg/dL are considered adequate for weekly monitoring of the diet. In several states, the laboratories screening newborns for PKU are also involved in routine monitoring, using the same Guthrie inhibition assay. Furthermore, this inhibition assay is used in the majority of treatment centers outside the United States. Specimen collection for the Guthrie test is blood dried on filter paper. This type of specimen is being used for the fluorometric assay as well (12–14).

An excellent review by Faulkner of the history of the development of phenylalanine determinations appeared in volume 5 of the *Selected Methods* series (15). The method presented in that paper is almost identical with the one reported here for liquid specimens, except that smaller volumes of specimen are used here and L-glycyl-L-leucine instead of L-leucyl-L-alanine.

Principle

The fluorometric method of McCaman and Robins (1) measures a fluorescent compound produced when ninhydrin reacts with phenylalanine and copper in the presence of a dipeptide (usually L-leucyl-L-alanine). The method presented here incorporates several changes previously described (2, 6), especially the use of the dipeptide L-glycyl-L-leucine in place of L-leucyl-L-alanine (3). L-Glycyl-L-leucine is three- to fourfold less expensive than L-leucyl-L-alanine but is somewhat less sensitive. However, substitution with L-glycyl-L-leucine gives satisfactory results for monitoring concentrations of phenylalanine in blood.

The elution of phenylalanine from blood-soaked filter paper started with the adaptation of the McCaman and Robins method (1) to continuous-flow analysis by Hill et al. (12) in 1965. In that assay, a dialyzer was incorporated into the system to remove protein. In 1966, Bourdillon and Vanderlinde (14) used steamed blood specimens to obviate the use of a dialyzer, which can be quite troublesome. Searle et al. converted the autoclaving method to a manual procedure in 1967 (13), a method that works quite well and is still used in many centers. Elution of the blood spots with trichloroacetic acid (TCA) inhibits the development of fluorescence and, until recently, was not used. Wu et al. reported in 1979 an aqueous elution technique, followed by TCA precipitation of proteins at 0 °C and high-speed centrifugation, which considerably reduced the inhibition of fluorescence (16).

The methanol-elution technique of Andriaenssens et al. (17) is the method currently used by the Submitter and is presented here. Methanol precipitates the proteins onto the paper fibers and simultaneously dissolves the free phenylalanine, yielding a clear, protein-free filtrate containing 95% of the phenylalanine present in whole blood. The blood-collection technique described here permits the specimens to be drawn easily by parents and mailed to the laboratory.

Other methods used for the determination of phenylalanine in biological fluids include: paper chromatography *(18, 19)*, enzymic conversion of phenylalanine to products measured spectrophotometrically *(20, 21)*, and ion-exchange chromatography *(22)*. Compared with the fluorometric procedure, these methods are more difficult to perform, are not quantitative, are more expensive, or are not as sensitive.

Materials and Methods

Reagents

Use reagents that are ACS quality or better (e.g., fluorescent grade). Use water that is glass-distilled and (or) de-ionized.

1. *Succinate buffer, 0.3 mol/L (pH 5.8 ± 0.1)*. Dissolve 14.85 g of disodium succinate dihydrate ($Na_2C_4H_4 \cdot 2H_2O$) in 240 mL of water; adjust to correct pH with 1 mol/L HCl (about 4 mL), using a pH meter, and dilute to 250 mL. Overruns can be back-titrated with 1 mol/L NaOH. Alternatively, dissolve 8.85 g of succinic acid in 230 mL of water; titrate to pH 5.8 ± 0.1 with saturated (18 mol/L) NaOH, and dilute to 250 mL. The pH is critical because sensitivity is decreased at pH values below 5.7 or above pH 5.9. Store in a refrigerator. The succinate buffer is subject to deterioration, presumably as a result of bacterial action. At 4 to 10 °C it may be stable for only a few weeks but is stable indefinitely when frozen at −20 °C. A degraded buffer is apparent when standards fluoresce less than the blank.

> *Note:* Evaluators E. W. N. and A. P. O. stress the fact that the pH of 5.8 ± 0.1 is critical; a decrease by even 0.1 pH unit decreases the sensitivity, and increasing the pH decreases specificity.

2. *Ninhydrin, 20 mol/L*. Dissolve 534 mg in water and dilute to 100 mL. Store in a brown bottle at room temperature.

3. *Peptide reagent, L-glycyl-L-leucine* (Sigma Chemical Co., St. Louis, MO 63178). Dissolve 100 mg of L-glycyl-L-leucine in water and dilute to 100 mL. Store this reagent frozen; it may deteriorate in a few weeks if stored at 4 to 10 °C. It is convenient to freeze the reagent in quantities of the volume needed for a day's run (40 samples or standards require 1 mL). Discard any unused thawed reagent. The symptoms of degraded peptide solution are the same as with degraded succinate buffer, i.e., loss of fluorescence.

4. *Copper reagent*. Dissolve the following compounds separately in about 300-mL portions of water each: sodium carbonate, anhydrous (Na_2CO_3), 1.6 g; potassium sodium tartrate ($KNaC_4H_4O_6 \cdot 4H_2O$), 0.100 g; copper sulfate ($CuSO_4 \cdot 5H_2O$), 0.060 g. Mix together in the order listed, stirring after each addition. Dilute the combined mixture to 1 L. Store at room temperature. The reagent is stable for several weeks. Discard if it becomes cloudy.

5. *TCA, 0.6 mol/L*. Dissolve 98 g of TCA in water and dilute to 1 L.

6. *TCA, 0.3 mol/L*. Dilute 500 mL of the 0.6 mol/L acid with an equal volume of water.

7. *Methanol, absolute, reagent grade*.

8. *Preparation of standards for liquid specimens*. For the stock standard, dissolve 100 mg of L-phenylalanine in 0.3 mol/L TCA and dilute to 100 mL with 0.3 mol/L TCA. Store in refrigerator. For working standards, dilute 1.00, 2.00, and 3.00 mL of the stock standard separately to 100 mL with 0.3 mol/L TCA to provide standards of 1, 2, and 3 mg/dL. Because serum will be diluted with an equal volume of 0.6 mol/L TCA, these standards are equivalent to serum concentrations of 2, 4, and 6 mg of phenylalanine per 100 mL. Store in a refrigerator.

> *Note:* Evaluators M. H. C. and R. G. C. use an additional 4 mg/dL standard.

9. *Preparation of standards on filter paper*. Blood filter-paper disc standards with phenylalanine concentrations of 2, 4, 6, 8, and 10 mg/dL may be prepared from out-dated transfusion blood. Pre-analyze the phenylalanine concentration of this blood, using the procedure for liquid specimens. Then add the appropriate amount of stock phenylalanine solution [1 g/dL of phenylalanine in physiologic (8.5 g/L) saline]. Table 1 shows the quantities of phenylalanine stock solution added to the required amount of transfusion blood, assuming that the blood contains 1 mg of phenylalanine per 100 mL. Allow aliquots of each standard to fall dropwise onto the filter paper (no. 903; Schleicher & Schuell, Inc., Keene, NH 03431),

Table 1. Quantity of Phenylalanine Solution (1 g/dL) Required to Prepare Standards from Blood

Concn of standard, mg/dL	Vol, mL	
	Phenylalanine solution	Whole blood[a]
2	0.1	99.9
4	0.3	99.7
6	0.5	99.5
8	0.7	99.3
10	0.9	99.1

[a] Assumed to contain 1 mg of phenylalanine per deciliter.

with each drop on a different area of the paper and each standard on separate papers. Allow spots to air-dry overnight.

10. *Working reagent mixture.* Just before each test series, mix the succinate buffer, the ninhydrin solution, and the L-glycyl-L-leucine solution in proportions of 10/4/2 (by vol); a total of 0.4 mL of mixture is required for each tube to be run.

Collection and Handling of Specimens

Preparation of protein-free serum or plasma. Place equal volumes of 0.6 mol/L TCA and serum or plasma in micropolyethylene (500-μL) centrifuge tubes (as little as 25 μL of each is sufficient), mix thoroughly, and allow to stand for 10 min. Centrifuge at moderate speed for 10 min. The supernatant fluid is used for the analysis.

Preparation of blood on filter paper. For newborns the specimen is usually collected by heel or finger stick. Allow the blood to fall dropwise onto the filter paper; it should soak completely through the filter paper and form a circle at least 11 mm (½ in.) or more in diameter. Collect at least two, and preferably three, such spots. Be very careful to avoid double-spotting blood on the same place because this will lead to erroneously high results. Allow to dry at room temperature, protected from dust. If blood is drawn by venipuncture, the sample may be spotted directly from the syringe or a dropper for more convenient transport to a distant laboratory by mail.

With a paper punch (¼-in. hole diameter), cut discs from the centers of at least two spots. Place each disc in a separate 10 × 75 mm cuvet, then add 0.2 mL of absolute methanol. Methanol precipitates the protein on the disc and elutes 95% of the phenylalanine. Shake for 15 min and carefully remove the disc, allowing it to drain down the side of the tube during removal. The entire eluate is used for analysis.

Note: The discs can be stored at this stage in 75 × 10 mm Pyrex test tubes protected from light and moisture; they will be stable for several weeks. When ready for analysis, add 0.35 mL of distilled, de-ionized water to each tube. Place the tubes in a rack and shake on a mechanical shaker for 15 min.

Alternatively, the discs may be autoclaved to eliminate the necessity of using TCA, as described next for the liquid specimen. Arrange the discs on the bottom of a glass container so they do not touch each other. Cover the container with filter paper so that condensed steam does not fall on the discs when the autoclave is opened. Autoclave the discs for 3 min at 15 lb. of pressure.

Analytical Procedure for Liquid Specimens

1. Add 0.4 mL of reagent mixture to 10 × 75 mm cuvets.

Note: Evaluators M. H. C. and R. G. C. use 13 × 100 mm test tubes for easier mixing.

2. Add 25 μL of reagent blank (TCA, 0.3 mol/L), working standards, and protein-free plasma samples in duplicate.
3. Cover the cuvets or tubes with Parafilm and mix thoroughly by inversion.
4. Incubate samples in a rack in a 60 °C water bath for 2 h.
5. Cool in an ice bath and add 2.5 mL of copper reagent. Mix thoroughly by inversion and let stand at room temperature (or, preferably, in a 25 °C water bath) for 10 to 15 min.
6. With a fluorometer, read the fluorescence of the contents of the tubes, in the order in which copper reagent was added, at 365 nm primary wavelength (Corning filter no. 7–60 or 7–51, or Corning glass filter no. 3384) and 515 nm secondary wavelength (Corning no. 4303 plus no. 3384 or Wratten no. 8 plus Wratten no. 65A). Set the reagent blank to zero and adjust the instrument so that the highest standard reads more than 50% full-scale.

Analytical Procedure for Blood Soaked on Filter Paper

Analyze protein-free eluates of standards, blanks, controls, and unknown specimens in duplicate.

1. Add 0.2 mL of working reagent to each cuvet used for elution and shake.

Note: A 1-mL measuring pipette may be used because even a 10% error of volume measurement here will result in only 1% error in phenylalanine estimation.

2. Heat the cuvets at 60 °C in a water bath for 2 h.
3. Cool in an ice bath and add 1.0 mL of copper reagent.
4. Shake the cuvets well and allow to stand for 15 min before taking fluorometric readings.
5. Read the fluorescence of the cuvet contents with a fluorometer, in the order of addition of copper reagent, at 365 nm primary wavelength (Corning filter no. 7–60 or 7–51, or Corning glass filter no. 3384) and 515 nm secondary wavelength (Corning no. 4303 plus no. 3384 or Wratten no. 8 plus Wratten no. 65A).

Calculations

Plot the fluorescence values of the standards and blank against the equivalent serum concentrations, i.e., twice the real concentration, on graph paper. Draw the line of best fit and read the concentrations of the samples from this standard curve, or calculate the slope of the line and determine the concentration of each sample by multiplying the sample readings by the mean slope of the standards.

Slope of each standard
= fluorescence of standard/concn of phenylalanine in standard

Average the slopes of all the standards. Then:

Concn of unknown = average standard slope
× fluorescence of unknown specimen

Results

Data for standard curves for the liquid specimens and filter paper specimens are shown in Table 2. The reference intervals for phenylalanine at various ages are shown in Table 3.

Note: Evaluators E. W. N. and A. P. O. report a correlation coefficient of 0.98 with the Guthrie method.

Precision and recovery studies are shown in Tables 4 and 5.

Table 2. Data for Constructing Standard Curve[a]

Phenylalanine concn, mg/dL	Instrument (dial) readings Serum[b] (duplicates)	Filter paper samples[c] (duplicates)
0	0, 0	0, 0
1	3, 3	2, 2
2	4, 6	5, 3
3	9, 9	7, 9
5	14, 15	13, 11
10	29, 31	36, 38
15	43, 41	36, 38
20	59, 57	48, 46
30	86, 86	70, 68

[a] Submitter used a Turner Model II fluorometer with a Corning no. 7-60 (350 nm) filter in the primary and a Wratten no. 8 and 65A in the secondary (510 nm).
[b] Slit width, 1 ×; Wratten neutral density filters at 0.2 (= 63%) and 0.3 (= 50%).
[c] Slit width, 3 ×; Wratten neutral density filters at 0.6 (= 25%).

Table 3. Phenylalanine Concentrations in Blood of Normal and Phenylketonuric Infants (23)

	Phenylalanine range	
	μmol/L	mg/L
Premature infants	156–377	25–62
Infants, 0 to 7 days	104–127	17–21
Infants, 8 to 31 days	136–138	22–23
Children, 6 mo to 12 yr	40–210	7–35
PKU infants	1370–2380	223–388

Table 4. Within-Day Precision

Evaluator	Phenylalanine standard, mg/dL	N	Phenylalanine measured, mg/dL	SD, mg/dL	CV, %
E. W. N. and A. P. O.	2.0[a]	16	1.8	0.09	5.3
	6.0	16	5.22	0.22	4.0
M. H. C. and R. G. C.	2.30[b]	20	2.26	0.11	4.9
	11.18	20	11.00	0.45	4.1
Submitter	2.00[c]	20	2.65	0.15	5.7
	8.00	20	8.38	0.21	2.5

[a] Blood spots prepared by Robert Guthrie.
[b] Results from 20 replicate analyses with an AutoAnalyzer (Technicon).
[c] Results of duplicate analyses of phenylalanine added to whole blood, determined with a Beckman Amino Acid Analyzer.

Table 5. Analytical Recovery Studies

	Phenylalanine, mg/dL	Mean recovery (and range), %
Evaluators E. W. N. and A. P. O.	2–12	85 (79–97)
Evaluators M. H. C. and R. G. C.	3.7–7.1	97 (93–105)
Submitter	2.0–30	95 (94–105)

Discussion

The relative fluorescence produced by the phenylalanine/ninhydrin product in the presence of L-glycyl-L-leucine is nearly the same as that produced by L-leucyl-L-alanine (3) but the cost is one-third to one-fourth as much. Other usable dipeptides (and their relative fluorescences) are L-leucyl-L-alanine (100%), L-glycyl-L-alanine (95%); glycyl-D,L-phenylalanine (75%); and L-alanyl-L-alanine (70%) (23). The fluorescence-enhancing effects of the peptides are nonspecific, but no other constituent of normal plasma acts in the same manner. In the absence of a peptide, less than 0.5% of the expected fluorescence is developed.

Sensitivity

This method yields fluorescence proportional to phenylalanine concentrations from 0.1 to 3 μmol/L. Concentrations of ninhydrin between 6 and 8 mmol/L and glycyl-leucine between 0.6 and 1.1 mmol/L are optimum. Temperatures between 60 and 80 °C are acceptable, but outside this range little fluorescence will be produced. Shorter heating at 60 °C may be used, e.g., 60–90 min, but there will be some loss of sensitivity.

The final pH should be between 7 and 9. Lower pH values yield little fluorescence, and higher pH values produce very unstable fluorescence. The fluorescence is stable for at least 4 h in the procedure described.

Source of Errors

As noted earlier, the pH 5.8 ± 0.1 of succinate buffer is critical, the fluorescence decreasing markedly as the pH decreases, such that other amino acids can contribute to the fluorescence.

The major source of error is improper specimen collection or storage. Hemolyzed blood greatly increases the amount of phenylalanine measured. If not frozen, plasma or serum will yield higher values for phenylalanine within several hours.

For blood soaked on filter paper, errors arise from improper spotting of the specimen—i.e., the blood is not soaked all the way through the paper, or the spots are too small from improper storage (room temperature for a week or more)—or from contamination with fingers. Errors may also arise if the discs disintegrate in the tubes because of shaking too vigorously. Paper fibers interfere with reading the fluorescence.

Specificity

At pH 5.8 only leucine and arginine yield any appreciable fluorescence, about 5% of that of an equimolar amount of phenylalanine (24). Other amino acids (tyrosine and 5-hydroxytryptophan) yield less than 0.5% the fluorescence of an equimolar amount of phenylalanine. The maximum reaction of phenylalanine with ninhydrin occurs at pH 6.8, and is almost twice that at pH 5.8; however, other amino acids give 2 to 20% as much fluorescence as the same amount of phenylalanine.

Reference Intervals

The reference intervals for phenylalanine in blood vary with age and time of day. Values are greater two to six days after birth than before or after (25–27), and are significantly greater at 1900 hours than at 0700 hours for healthy infants (but the opposite occurs for babies with PKU) (28). These variations are relatively minor considering the concentrations expected in infants with PKU: between 1.00 and 2.50 mmol/L (160–400 mg/L). The present consensus is that values between 240 and 375 µmol/L (40–60 mg/L) should be considered normal during the first week of life. Values exceeding this should be investigated further (see Table 3).

Clinical Application

At present, the primary application of phenylalanine measurement in blood is in the diagnosis and dietary management of PKU. Detection of heterozygotes for this disorder is also possible (when combined with determinations of blood tyrosine), with or without the administration of phenylalanine load (29–32).

PKU is an autosomal recessive disorder that causes those affected to be mentally retarded. Most institutionalized patients have an IQ of 20, although about 2% of persons with true PKU have an IQ in the normal range.

The disorder is associated with the absence of activity of phenylalanine hydroxylase (EC 1.14.16.1), an enzyme present only in human liver. Phenylalanine hydroxylase functions to convert phenylalanine to tyrosine by *para*-hydroxylation. The lack of enzyme or enzyme activity results in greatly increased concentrations of phenylalanine in the blood, decreased tyrosine concentrations in the blood, and increased quantities of o-hydroxyphenylacetic acid, phenylpyruvic acid, and phenylacetic acid in the urine. Tyrosine becomes an essential amino acid when phenylalanine is deficient, resulting in decreased body concentrations of this amino acid and its metabolites, e.g., melanin and epinephrine. No relationship has been established between the values for any or all of the metabolites and the level of intelligence. The pathogenesis of mental deficiency in PKU is unknown.

Several clinical conditions are associated with high serum concentrations of phenylalanine during the newborn period, including disorders of tyrosine metabolism (33–36). When the diagnosis is in doubt, more detailed studies must be undertaken with the hospitalized infant so that nutritional and other factors can be adequately controlled. This is important because the low-phenylalanine diet used to control PKU can be damaging to an infant who does not have the disease, and several problems, including death, have been reported (37–40). Three causes of hyperphenylalaninemia severe enough to require treatment are listed in Table 6.

Table 6. Disorders Involving High Concentrations of Phenylalanine in Blood

Condition	Cause	Clinical aspects	Treatment	Ref.
Classical phenylketonuria I	Phenylalanine hydroxylase deficiency	Normal-appearing at birth	Dietary restriction	41
Phenylketonuria II	Dihydropteridine reductase (EC 1.6.99.7) deficiency	Mental retardation, progressive neurological degeneration	Dietary restriction (?), tetrahydrobiopterin administration	42
Phenylketonuria III	Dihydrobiopterin synthetase deficiency	Hypotonia		43

The caveat regarding PKU testing is that, although dietary treatment can be efficacious, it must be carried out judiciously. Frequent monitoring of blood for phenylalanine and careful assessment of growth and development are necessary in the management of patients with high concentrations of phenylalanine in the blood. There are many centers across the country that can help establish the diagnosis in a case of suspected PKU *(44)*.

The effort of Ada Falcon is gratefully acknowledged. This presentation was supported in part by Bureau of Community Health Services, MCH Training Grant, Project No. 288, and by DHEW Multidisciplinary Training Grant No. 903.

References

1. McCaman, M. W., and Robins, E., Fluorometric method for the determination of phenylalanine in serum. *J. Lab. Clin. Med.* **59**, 885–890 (1962).
2. Wong, P. W. K., O'Flynn, M. E., and Inouye, T., Micromethods for measuring phenylalanine and tyrosine in serum. *Clin. Chem.* **10**, 1098–1104 (1964).
3. Hill, J. B., Summer, G. K., and Hill, H. D., Modifications of the automated procedure for blood phenylalanine. *Clin. Chem.* **13**, 77–80 (1967).
4. Ambrose, J. A., Ross, C., and Whitfield, F., An ultramicro automated method (AutoAnalyzer) for the fluorometric determination of phenylalanine. In *Automation in Analytical Chemistry, Technicon Symposia (1967)*, Mediad, White Plains, NY, 1968, pp 13–19.
5. Ambrose, J. A., A shortened method for the fluorometric determination of phenylalanine. *Clin. Chem.* **15**, 15–23 (1969).
6. Terlingen, J. B. A., and Van Dreumel, H. J., Note on the fluorometric method for phenylalanine in serum. *Clin. Chim. Acta* **22**, 643–645 (1958).
7. Hill, J. B., Summer, G. K., Shavender, E. F., Scurletis, T. D., Robie, W. A., Maddry, L. G., Matheson, M. S., and Brooks, M. F., Early experience in the use of automation in screening for phenylketonuria in a newborn population. In *Automation in Analytical Chemistry, Technicon Symposia (1965)*, Mediad, New York, NY, 1966, pp 404–405.
8. Ambrose, J. A., Ultramicro automated dialyzer method (AutoAnalyzer) for the fluorometric determination of phenylalanine. In *Advances in Automated Analysis, Technicon International Congress (1970)*, Mediad, White Plains, NY, 1971, pp 25–32.
9. Hsia, D. Y., Berman, J. L., and Slatis, H. M., Screening newborn infants for phenylketonuria. *J. Am. Med. Assoc.* **188**, 203–206 (1964).
10. Irwin, H. R., Notrica, S., and Fleming, W., Blood phenylalanine levels of newborn infants—a routine screening program for the hospital newborn nursery. *Calif. Med.* **101**, 331–333 (1964).
11. Guthrie, R., and Susi, A., A simple phenylalanine method for detecting phenylketonuria in large populations of newborn infants. *Pediatrics* **32**, 338–343 (1963).
12. Hill, J. B., Summer, G. K., Pender, M. W., and Roszel, N. O., An automated procedure for blood phenylalanine. *Clin. Chem.* **11**, 541–546 (1965).
13. Searle, B., Mijuskovic, M. B., Widelock, D., and Davidow, B., A manual fluorometric paper disc method for detecting phenylketonuria. *Clin. Chem.* **13**, 621–625 (1967).
14. Bourdillon, J., and Vanderlinde, R. E., An improved screening procedure for blood phenylalanine. *Public Health Rep.* **81**, 991–995 (1966).
15. Faulkner, W. R., Phenylalanine. *Stand. Methods Clin. Chem.* **5**, 199–209 (1965).
16. Wu, J. T., Wu, L. H., Ziter, F. A., and Ash, K. O., Manual fluorometry of phenylalanine from blood specimens collected on filter paper. *Clin. Chem.* **25**, 470–472 (1979).
17. Andriaenssens, K., Vanheule, R., and Van Belle, M., A new simple screening method for detecting pathological aminoacidemias with collection of blood on paper. *Clin. Chim. Acta* **15**, 362–364 (1967).
18. Berry, H. K., Paper chromatographic method for estimation of phenylalanine. *Proc. Soc. Exp. Biol. Med.* **95**, 71–73 (1957).
19. Cawley, L. P., Goodwin, W. L., and Dibbern, P., One-dimensional paper chromatography of serum amino acids with particular reference to phenylalanine. *Am. J. Clin. Pathol.* **48**, 405–412 (1967).
20. Shen, R.-S., and Abell, C. W., Phenylketonuria: A new method for the simultaneous determination of plasma phenylalanine and tyrosine. *Science* **197**, 665–667 (1977).
21. La Du, B. N., and Michael, P. J., An enzymatic spectrophotometric method for the determination of phenylalanine in blood. *J. Lab. Clin. Med.* **55**, 491–496 (1960).
22. Levy, H. L., Baullinger, P. C., and Madigan, P. M., A rapid procedure for the determination of phenylalanine and tyrosine from blood filter paper specimens. *Clin. Chim. Acta* **31**, 447–452 (1971).
23. Robins, E., The measurement of phenylalanine and tyrosine in blood. *Methods Biochem. Anal.* **17**, 287–309 (1969).
24. Hsia, D. Y-Y., Litwack, M., O'Flynn, M., and Jakovcic, S., Serum phenylalanine and tyrosine levels in the newborn infant. *N. Engl. J. Med.* **267**, 1067–1070 (1962).
25. Holtzman, N. A., Mellits, D., and Kallman, C. H., Neonatal screening in phenylketonuria: II. Age dependence of initial phenylalanine in infants with PKU. *Pediatrics* **53**, 353–357 (1974).
26. Partington, M. W., and Lewis, E. J. M., Variations with age in plasma phenylalanine and tyrosine levels in phenylketonuria. *J. Pediatr.* **62**, 348–357 (1963).
27. Coburn, S. P., Seidenberg, M., and Fuller, R. W., Daily rhythm in plasma tyrosine and phenylalanine. *Proc. Soc. Exp. Biol. Med.* **129**, 338–343 (1968).
28. Guttler, F., Olesen, E. S., and Wamberg, E., Diurnal variations of serum phenylalanine in phenylketonuric children on low phenylalanine diet. *Am. J. Clin. Nutr.* **22**, 1568–1570 (1969).
29. Bessman, S. P., Williamson, M. L., and Koch, R., Diet, genetics and mental retardation interaction between phenylketonuric heterozygous mother and fetus to produce nonspecific diminution of IQ: Evidence in support of the justification hypothesis. *Proc. Natl. Acad. Sci. USA* **75**, 1562–1566 (1978).
30. Rampini, S., Anders, P. W., Curtius, H. D., and Marthaler, T., Detection of heterozygotes for phenylketonuria by column chromatography and discriminatory analysis. *Pediatr. Res.* **3**, 287–397 (1969).
31. Rosenblatt, D., and Scriver, C. R., Heterogeneity in genetic control of phenylalanine metabolism in man. *Nature* **218**, 677–678. (1968).
32. Griffin, R. F., Humienny, M. E., Hall, E. C., and Elsas, L. J., Classic phenylketonuria: Heterozygote detection during pregnancy. *Am. J. Hum. Genet.* **25**, 646–654 (1973).
33. Hsia, D. Y-Y., O'Flynn, M. E., and Berman, J. L., Atypi-

cal phenylketonuria with borderline or normal intelligence. *Am. J. Dis. Child.* **116,** 143–157 (1968).
34. Carpenter, G. G., Auerbach, V. H., and DiGeorge, A. M., Phenylalaninemia. *Pediatr. Clin. North Am.* **15,** 313–323 (1968).
35. Allen, R. J., Heffelfinger, J. C., Masotti, R. E., and Tsau, M. U., Phenylalanine hydroxylase activity in newborn infants. *Pediatrics* **33,** 512–525 (1964).
36. Jervis, G. A., Phenylketonuria. In *Amino Acid Metabolism and Genetic Variation,* McGraw-Hill, New York, NY, 1967, pp 5–9.
37. Birch, H. G., and Tigard, J., The dietary treatment of phenylketonuria not proven? *Dev. Med. Child. Neurol.* **9,** 9–12 (1967).
38. Knox, W. E., An evaluation of the treatment of phenylketonuria with diets low in phenylalanine. *Pediatrics* **26,** 1–11 (1960).
39. Anderson, J. A., Fisch, R., and Miller, E., Atypical phenylketonuric heterozygote. *J. Pediatr.* **68,** 351–360 (1966).
40. Castells, S., and Brandt, I. K., Phenylketonuria: Evaluation of therapy and verification of diagnosis. *J. Pediatr.* **72,** 34–40 (1968).
41. Jervis, G. A., Studies of phenylpyruvic oligophrenia: The position of the metabolic error. *J. Biol. Chem.* **169,** 651–656 (1947).
42. Kaufman, S., Holtzman, N. A., Milstien, S., Butler, I. J., and Krumholz, A., Phenylketonuria due to a deficiency of dihydropteridine reductase. *N. Engl. J. Med.* **293,** 785–790 (1975).
43. Kaufman, S., Berlow, S., Summer, G. K., Milstien, S., Schulman, J. D., Orloff, S., Spielberg, S., and Pueschel, S., Hyperphenylalanemia due to a deficiency of biopterin: A variant form of phenylketonuria. *N. Engl. J. Med.* **299,** 673–679 (1978).
44. Lynch, H. T., Guirgis, H., and Bergsma, D., *Birth Defects; International Directory of Genetic Services,* 5th ed., H. T. Lynch, H. Guirgis, and D. Bergsma, Eds., The National Foundation—March of Dimes, White Plains, NY, 1977, pp 1–58.

Phosphate, Inorganic

Submitters: Eugene S. Baginski and Slawa S. Marie, *Chemistry Laboratory, Doctors Hospital, Detroit, MI 48207*
Bennie Zak, *Department of Pathology, Wayne State University School of Medicine, Detroit Receiving Hospital and University Health Center, Detroit, MI 48201*

Evaluators: Elizabeth A. McLister and Philip G. Mullarky, *Doctors Hospital, Detroit, MI 48207*
Leonard Sideman and Irene Daza, *Division of Clinical Chemistry and Hematology, Bureau of Laboratories, Pennsylvania Department of Health, Lionville, PA 19353*

Reviewer: Basil T. Doumas, *Department of Pathology, The Medical College of Wisconsin, Milwaukee, WI 53226*

Introduction

The metabolism of phosphate is closely associated with that of calcium. The serum concentration of both elements is regulated by the parathyroid hormone (parathyrin), and a disturbance in the metabolism of one often affects the other. A determination of both elements in blood serum is useful to distinguish between a primary or a secondary cause of disturbance.

Most methods available for inorganic phosphate determination in clinical laboratories are based on the reaction between phosphate and molybdate in a reducing system. The blue complex formed under these conditions is then measured (1).

Principle

The reaction can be represented simply as follows:

$$12MoO_4^{2-} + H_3PO_4 + 24H^+ \longrightarrow H_3PMo_{12}O_{40} + 12H_2O$$

In an acid medium, molybdate is transformed into a heteropolyacid, which subsequently reacts with phosphate to produce a complex. It has also been postulated that in a strongly acid solution molybdate exists as a dimer, which polymerizes and reacts with phosphate to form duodecamolybdophosphoric acid (12 MPA) according to the following scheme:

$$6 \text{ Mo (VI)} \xrightarrow{H^+} \text{Mo (VI)} \xrightarrow{PO_4^{2-}} 12 \text{ MPA}$$
$$\text{(dimer)} \qquad \text{(polymer)}$$

Upon reduction the acid forms a blue product, referred to as "heteropoly blue." Mild reducing agents affect only some molybdate atoms, leaving others in a higher oxidation state. The complex exhibits two absorbance maxima, at 700 nm and 840 nm, either of which may be used to measure the blue color spectrophotometrically.

Ascorbic acid is used as a reducing agent. Sodium dodecyl sulfate keeps the protein in a dispersed form. The presence of arsenite in the system increases the sensitivity of the method. Citrate is used to complex the excess molybdate and to prevent the reduction of molybdate to insoluble oxides. Dimethylsulfoxide has a clearing effect in moderately lipemic specimens (2–4). However, in the case of a grossly lipemic specimen, a simple correction should be applied, as described later.

The presence of hemoglobin or bilirubin does not interfere because the absorbance peaks of those compounds are distant from that of the heteropoly blue complex. However, hemolyzed specimens are unsuitable because organophosphate compounds released from erythrocytes hydrolyze easily in the acid medium, thereby increasing the apparent inorganic phosphate concentration.

Materials and Methods

Specimen

Serum, plasma, cerebrospinal fluid, or acidified urine may be analyzed. Urine is acidified as follows: Add 2.0 mL of 2 mol/L HCl into the bottle containing the entire 24-h urine collection; mix well and check the pH with pH paper. The pH should be 3 or below. Add more acid if necessary. Because phosphate concentration in urine varies widely, use undiluted specimen first. If the color absorbance is too high, either make an appropriate dilution of the urinary aliquot with distilled water or use a smaller aliquot of the specimen.

Reagents

1. *Stock phosphate standard (1 g/L).* Dissolve

439.4 mg of potassium dihydrogen phosphate in phosphate-free water and dilute it to 100 mL. Stable.

2. *Phosphate working standard (50 mg/L).* Dilute 5.0 mL of the phosphate stock standard to 100 mL with phosphate-free water. Stable.

3. *Sulfuric acid, dilute.* Carefully add 16.0 mL of concd. sulfuric acid into a 1-L volumetric flask containing approximately 500 mL of water. Cool to room temperature and dilute to volume with water. Stable.

4. *Ascorbic acid–sodium dodecyl sulfate (A-SDS).* Dissolve 4.0 g of crystalline ascorbic acid in 100 mL of water. Then add 0.5 g of sodium metabisulfite and 4.0 g of sodium dodecyl sulfate (sodium lauryl sulfate); mix until all dissolves and transfer into an amber bottle. Stable at room temperature for at least two months. Do not refrigerate (see *Discussion*).

5. *Ammonium molybdate.* Dissolve 5.0 g of ammonium molybdate tetrahydrate in 1 L of water. Stable.

6. *Arsenite–citrate.* Dissolve 20.0 g of anhydrous sodium arsenite and 20.0 g of sodium citrate dihydrate in about 500 mL of water. Add 20.0 mL of glacial acetic acid followed by 400 mL of dimethyl sulfoxide. Cool to room temperature and dilute to 1 L with water. Stable.

Procedure

1. Pipet 50 µL of the specimen and the 50 mg/L working phosphate standard into separate 12 × 75 mm test tubes. Add 0.5 mL of the dilute sulfuric acid to each tube and to a third one to be used as the reagent blank. Mix.

2. Pipet 0.5 mL of A-SDS to all tubes and mix thoroughly. The presence of protein may cause turbidity at this stage; however, the mixture will clear upon agitation.

3. Add 0.5 mL of molybdate and mix. Allow 1 to 2 min to elapse before adding the next reagent.

4. Add 1.0 mL of arsenite–citrate and mix thoroughly. After 10 min, determine the absorbances of the sample and standard against the reagent blank with a spectrophotometer set at 700 nm.

Note: Absorbances may also be read at 840 nm.

Calculations

Calculate concentration from absorbance *(A)* as follows:

$$(A_s/A_{std}) \times 5 = \text{inorganic phosphate, mg/dL}$$

where: A_s = absorbance of sample
A_{std} = absorbance of standard

To convert mg/dL to mmol/L, multiply by 0.323.

Reference Intervals

Serum: 2.5–4.5 mg/dL

Cerebrospinal fluid: 0.9–2.0 mg/dL (0.29–0.65 mmol/L)

Urine (varies with dietary intake): 0.4–1.3 g/24 h (12.9–42.0 mmol/24 h)

The serum values for infants are as follows *(5)*:

Premature	7.6–8.1 mg/dL (2.46–2.61 mmol/L)
Newborn (full-term)	5.8–6.4 mg/dL (1.86–2.07 mmol/L)
1–10 years	4.4–4.8 mg/dL (1.43–1.54 mmol/L)

Discussion

Adhere *strictly* to the sequence of addition of reagents as described. The addition of each reagent *must* be followed by thorough mixing. When A-SDS reagent is added to the tube containing serum and sulfuric acid, turbidity develops but disappears immediately upon mixing. Mix the contents of the tube to dissolve the precipitate before the next reagent is added.

The addition of molybdate should be followed by thorough mixing to allow the reaction between phosphate and molybdate to go to completion before adding arsenite–citrate. If this is not done, values for phosphate may be lower than expected because molybdate binds citrate preferentially over phosphate.

Once molybdate is complexed with citrate, it is no longer available for the reaction with phosphate. The preferential binding of citrate by molybdate becomes particularly advantageous when the inorganic phosphate assay is performed in biological fluids containing organophosphate compounds such as ATP and creatine phosphate. These compounds slowly hydrolyze in an acid medium, releasing inorganic phosphate. However, when citrate is present it binds molybdate, preventing it from reacting with the inorganic phosphate released by hydrolysis. Indeed, if phosphate were to be added after citrate, it could not react with molybdate, owing to the complexation of molybdate by citrate.

This principle is applied to correct for opalescence in case a chylous specimen is encountered. In this circumstance, set up the test and the reagent blank in duplicate. In one set, follow the procedure exactly as described. In the other set, follow the procedure up to the step where molybdate is to be added; then add arsenite–citrate first and then add molybdate. The reverse sequence of addition of molybdate and arsenite–citrate causes citrate to tie up molybdate and prevents it from reacting with phosphate. The difference in absorbances between the two sets represents the corrected absorbance. This correction step for the background absorbance is valid because of the large sample dilution.

The reaction between phosphate and molybdate is apparently not instantaneous; approximately a 2-min delay is needed for its completion before arsenite–citrate can be added. The delay does not pose a problem when several specimens are determined at the same time because there is always a time interval between the addition of reagents. However, when only a single determination is performed, remember to allow a 2-min interval between the addition of molybdate and the arsenite–citrate.

All reagents are stable indefinitely except the A-SDS reagent, which is stable for about two months at room temperature. The reagent should not be stored under refrigeration because lauryl sulfate tends to precipitate at lower temperatures. Ascorbic acid should be of high purity. It is available in crystalline form from Fisher Scientific Co., Pittsburgh, PA 15219, or E. Merck Co., Rahway, NJ 07065. Sodium dodecyl sulfate (SDS), also known as sodium lauryl sulfate, is a common ionic detergent available from various sources, including Sigma Chemical Co., St. Louis, MO 63178, and Aldrich Chemical Co., Milwaukee, WI 53233. Because some lots of SDS have been found to be contaminated with phosphate, its purity should be checked as follows: set up the reagent blank as described under *Procedure* and determine its absorbance at 700 nm against water. The absorbance should not exceed 0.1. As the A-SDS reagent ages, the absorbance of the reagent blank may gradually increase; however, the results for inorganic phosphate are not affected as long as the phosphate standard is run simultaneously and its absorbance against the reagent blank remains constant. If the absorbance for the reagent blank exceeds 0.15, the A-SDS reagent should be discarded and a fresh one prepared.

Statistical Evaluations

The method described here was evaluated by two other laboratories. For correlation studies, Evaluators L. S. and I. D. used the method of Goldenberg and Fernandez (5), and Evaluators E. A. M. and P. G. M. used the Hycel Seventeen analyzer (Hycel, Inc., Houston, TX 77063).

Precision studies

Within-run	L.S. and I.D.	E.A.M. and P.G.M.
Mean, mg/dL	3.81[a]	3.88[b]
Range, mg/dL	3.7–3.9	
SD, mg/dL	0.56	0.078
CV, %	1.5	2.0
N	40	40

Day-to-day	No. 1[c]	No. 2[c]
Mean, mg/dL	3.8	8.34
Range, mg/dL	3.6–3.9	8.1–8.6
SD, mg/dL	0.90	1.24
CV, %	2.4	1.5
N	20	20

[a] Reference serum with manufacturer's value at 3.9 ± 0.3 mg/dL.
[b] Reference serum value not given.
[c] No. 1 = normal control, 3.9 ± 0.3 mg/dL; No. 2 = abnormal control, 8.5 ± 0.6 mg/dL.

Analytical recovery (Evaluators L. S. and I. D.)

Inorganic phosphate was added to a reference serum to increase the concentration by 30% and 70%. The sera were assayed in duplicate over a five-day period. The average recovery for the 30% addition was 97.9%, and at 70% was 95.4%. Recovery ranges were not given.

Correlation studies

Evaluators E. A. M. and P. G. M. found that with the Hycel method (y), the Submitters' method gave a correlation coefficient of 0.993, and the linear regression equation of $y = 1.03x - 0.045$, with 164 specimens. The t-value was 6.611, and the calculated F-value was 3.45 against a table value of 3.18. Although the t-value indicates statistical bias, the Evaluators do not consider this clinically significant. Evaluators L. S. and I. D. made 40 determinations on three control sera with both the Submitters' method and that of Goldenberg and Fernandez (5). They found t-values of 4.86, 10.5, and 2.39 and F-values of 1.11, 1.60, and 2.23.

Physiological and Clinical Considerations

Phosphate is found primarily in intracellular fluid, stored mostly in combination with calcium in the mineral matrix of bone; the rest is combined with organic compounds to form such compounds as adenosine triphosphate, creatine phosphate, and nucleic acids. Only a relatively small portion of total body phosphate exists in an inorganic form, and most of it is in blood plasma.

The metabolism of both phosphate and calcium is under the control of the parathyroid hormone, which lowers the concentration of serum phosphate by decreasing its reabsorption in renal tubules.

In a true, uncomplicated hyperparathyroidism, the serum phosphate concentration is decreased; it is increased in hypoparathyroidism. Vitamin D apparently stimulates renal reabsorption of phosphate, and it is for this reason that hypophosphatemia often develops in familial vitamin D-resistant rickets or in vitamin D deficiency (6). Renal excretion of phosphate also plays a role in the regulation of acid–base balance.

The blood seems to serve as a source of inorganic phosphate for bone, tissues, and plasma cells, including erythrocytes, where phosphate participates in the formation of glycolytic intermediates and other organophosphates, such as adenosine triphosphate and 2,3-diphosphoglycerate, both of which play a key role in the regulation of oxyhemoglobin function. It is not surprising, therefore, to observe a significant decrease in the synthesis of erythrocyte ATP and 2,3-diphosphoglycerate in severe hypophosphatemia, which is often precipitated by parenteral hyperalimentation (7–10).

Phosphate plays an important role as a cofactor in many enzymic reactions, and the fact that it is a part of the nucleic acid structure and provides energy through the synthesis of ATP and creatine phosphate means that a constant supply of inorganic phosphate is essential to maintain life. Under normal conditions diet provides sufficient sources of phosphate, and hypophosphatemia is a rather rare occurrence, unless it is caused by poor absorption, owing to ingestion of phosphate-binding antacids, or vitamin-D deficiency. A mild hypophosphatemia usually is not accompanied by clinical symptoms. However, a severe decrease in body phosphate is life-threatening. A moderate decrease in serum phosphate is often seen after insulin administration, in alcoholism, hyperthyroidism, diuretic therapy, renal tubular defect, etc. The decrease in serum inorganic phosphate observed after ingestion of carbohydrates or lipids is undoubtedly due to an increase in insulin release, which shifts the anion from plasma to tissues. Hypophosphatemia seems surprisingly common, even among an otherwise apparently normal hospital population (11).

Hyperphosphatemia in adult life is not common but may be noted in chronic renal failure, kidney shutdown due to hemolysis, or other causes. Growth is associated with a benign hyperphosphatemia (4.0–7.1 mg/dL or 1.3–3.4 mmol/L). Acromegalics may also have above-normal serum phosphate because of an increased concentration of growth hormone, which seems to increase the renal tubular reabsorption of phosphate.

The main value of serum phosphate determination seems to be in conjunction with determination of serum calcium, to assess the parathyroid status. The concentration of the two elements is usually in an inverse relationship. On the other hand, in carcinomas with metastases to the bone, for example, an increase of serum calcium may not be accompanied by a decrease in serum phosphate, a finding that is useful in ruling out the parathyroid involvement.

References

1. Baginski, E. S., Epstein, E., and Zak, B., Review of phosphate methodologies. *Ann. Clin. Lab. Sci.* 5, 399–416 (1975).
2. Baginski, E. S., and Zak, B., Microdetermination of serum phosphate and phospholipids. *Clin. Chim. Acta* 5, 834–838 (1960).
3. Baginski, E. S., Foa, P. P., and Zak, B., Microdetermination of inorganic phosphate, phospholipids, and total phosphate in biological materials. *Clin. Chem.* 13, 326–332 (1967).
4. Baginski, E. S., Marie, S. S., and Zak, B., Direct serum inorganic phosphate determination. *Microchem. J.* 19, 285–294 (1974).
5. Goldenberg, H., and Fernandez, A., Simplified method for the estimation of inorganic phosphorus in body fluids. *Clin. Chem.* 12, 871–882 (1966).
6. Harrison, H. E., Disturbances of calcium and phosphate metabolism. In *Pediatrics*, Appleton-Century-Crofts, 1968, p 371.
7. Kock, J. C., Williams, H. E., and Mentzer, W. C., Hemolytic anemia and somatic cell dysfunction in severe hypophosphatemia. *Arch. Intern. Med.* 134, 360–364 (1974).
8. Lichtman, M. A., Miller, D. R., Cohen, J., and Waterhouse, C., Reduced red cell glycolysis, 2,3-diphosphoglycerate and adenosine triphosphate concentration, and increased hemoglobin–oxygen affinity caused by hypophosphatemia. *Ann. Intern. Med.* 74, 562–568 (1971).
9. Craddock, P. R., Yawata, Y., VanSanten, L., Gilberstadt, S., Silvis, S., and Jacob, H. S., A complication of hypophosphatemia of parenteral hyperalimentation. *N. Engl. J. Med.* 290, 1403–1407 (1974).
10. Travis, S. F., Sugerman, H. J., Ruberg, R. L., Dudrick, S. J., Papadopoulos, M. D., Miller, L. D., and Oski, F. A., Alterations of red-cell glycolytic intermediates and oxygen transport as a consequence of hypophosphatemia in patients receiving intravenous hyperalimentation. *N. Engl. J. Med.* 285, 763–768 (1971)
11. Berto, M. G., and Pain, R. W., Hypophosphatemia and hyperphosphatemia in a hospital population, *Br. Med. J.* i, 273–276 (1972).

Protein (Total Protein) in Serum, Urine, and Cerebrospinal Fluid; Albumin in Serum

Submitters: Theodore Peters, Jr., and Gerald T. Biamonte,[1] *The Mary Imogene Bassett Hospital, Cooperstown, NY 13326*

Basil T. Doumas, *Department of Pathology, The Medical College of Wisconsin, Milwaukee, WI 53226*

Evaluators: Sandra M. Durnan and Richard M. Corcoran, *Emerson Hospital Concord, MA 01742*

Lindsay F. Hofman, *The Mason Clinic, Seattle, WA 98101*

Man M. Kochhar, *School of Pharmacy, Auburn University, Auburn, AL, 36380*

George T. Poole and Kap-Yong Hong, *Hospital of St. Raphael, New Haven, CT 06511*

Stephen E. Ritzmann, *Baylor University Medical Center, Dallas, TX 75246*

Reviewer: Richard J. Carter, *Technical Services, E. I. duPont de Nemours and Co., Wilmington, DE 19898*

Introduction

Proteins, the major dissolved solids of blood plasma or serum, provide for immune defenses, transport metabolites and hormones, and maintain the colloid osmotic pressure. "Total serum protein" consists of at least several hundred individual proteins, some occurring in trace amounts and others, such as albumin and immunoglobulin G, in concentrations of 10 to 50 g/L. The total protein concentration is useful as a rough indicator of nutrition and gastrointestinal function, as an indicator of plasma viscosity, and as a basis for quantitating fractions obtained by electrophoresis (see chapter on electrophoresis in this text). Of the individual proteins, albumin is the most widely measured, as a guide to nutritional status and to the function of the liver, kidney, and gastrointestinal tract. About 12 other individual proteins are frequently measured by immunochemical techniques, including the immunoglobulins, α_1-antitrypsin, transferrin, haptoglobin, and fibrinogen, but these procedures are not considered here. This chapter presents methods for total protein in serum, urine, and cerebrospinal fluid, and for albumin in serum.

TOTAL PROTEIN IN SERUM

Fasting or other preparation is not required, but may be desirable to decrease lipemia. Lipemic effects will be discussed later. Although plasma may be used for the determination of total protein, serum is the specimen of choice. Basically, the only difference in protein concentration between serum and plasma is that plasma contains the protein fibrinogen (normally only 2 to 4 g/L, but higher in many disease states). There is no change in the total protein or albumin of clear serum samples held for at least a week at −20 °C, 2–4 °C, or even at room temperature if kept in tightly stoppered containers (1). Thaw frozen samples and mix completely before sampling.

Note: Evaluators S. M. D. and R. M. C. confirmed that clear serum samples, tightly stoppered, are stable for a week or longer at room temperature, for a month at 2–4 °C, and up to two months at −20 °C.

Total Serum Protein by Refractometry

Principle

Dissolved solids increase the refraction of light by a solution, and upon reading the refractive index—a simple operation—one can obtain a reliable measure of total solids. (For reviews of refractometry for total serum proteins, see references 2–4.) Because serum normally contains about 16 g of nonprotein solids, mainly salts, per liter, a correction must be made for their contribution to the refractive index. In practice this is done by calibrating the refractometer with a serum of known protein content. The refractive index varies with temperature, but within the range of usual laboratory temperatures this effect has been overcome by proper refractometer design. High concentrations of nonprotein solids, such as urea in uremia or glucose in diabetes, are significant sources of error in disease states. An increase of 6000 mg of glucose per liter would cause an increase of

---

[1] Present address (G.T.B.): E. I. duPont de Nemours and Co., Wilmington, DE 19898.

6 g/L in the apparent protein (total solids) concentration; urea nitrogen of 2800 mg/L, corresponding to 6000 mg of urea per liter, would cause a similar 6 g/L increase.

Apparatus

The refractometer most widely used is the total solids meter (Model 10401; American Optical Instrument Co., Buffalo, NY 14215). It is used most conveniently when mounted on a companion stand with a built-in light source. The instrument is temperature-compensated and is precalibrated for serum protein as well as for specific gravity of urine. The zero reading can be checked with distilled water, but consult the manufacturer's instructions before attempting to reset the instrument.

Procedure

1. Allow a drop of serum to run between the plastic cover and the glass plate, or place a drop on the glass plate and then lower the plastic cover.

2. Press the cover gently and move the refractometer in relation to the light so that a dividing line between bright and dark fields is observed. Read the protein concentration from the scale at this line.

3. Clean the surfaces with water and wipe with lens tissue.

4. The scale is calibrated in g/dL; multiply by 10 to obtain g/L.

If the protein concentration is less than 35 g/L, the result is likely to be inaccurate and the biuret procedure (see below) should be used. If the concentration exceeds 110 g/L, dilute the specimen with an equal volume of 0.15 mol/L NaCl and repeat the determination.

Notes: Evaluator L. F. H. questioned the applicability of this procedure because of the effects of lipemia, icterus, and altered albumin/globulin ratio. These effects, however, have been shown not to cause errors greater than about 5% *(2)*. Evaluators S. M. D. and R. M. C. prefer this procedure over the biuret method because of its simplicity. They recommend verifying the calibration of the refractometer, recording the reading with distilled water daily, and taking particular care to keep the glass plate and plastic cover free of dust and debris. They also "prime" the device by placing a drop or two of control serum on the reading area then wiping it off before making a reading.

Evaluators S. M. D. and R. M. C. found within-run precision (CV) of 0.6–1.2%; G. T. P. and K. Y. H. reported 0.9% CV; and S. E. R. found 1.5%. Day-to-day CV was reported as 1.6–1.8% (S. M. D. and R. M. C.) and 2.0% (G. T. P. and K. Y. H.). Tests against other methods or laboratories showed a mean (g/L) of 69.9, compared with 68.7 by a reference laboratory (S. M. D. and R. M. C.); 72.6, compared with 68.0 by the American Monitor K. D. A. (G. T. P. and K. Y. H.); 72.7, compared with 71.5 by DuPont *aca* and 68.8 by Abbott ABA-100; and 119.2, compared with 144.3 by *aca* and 108.7 by ABA-100 (S. E. R.). Concentrated specimens such as the last one mentioned should be diluted before assay.

Total Serum Protein by Biuret Procedure

Principle

The biuret reaction is one of the oldest chemical methods for measuring protein, having been applied to serum in 1914 *(5)*; it remains one of the simplest and most accurate. This method depends on the formation of colored (purple) complexes between peptide chains and Cu(II) ions in strongly alkaline solution. The name comes from the reaction between Cu(II) and the compound biuret, $NH_2CONHCONH_2$, the condensation product formed by two molecules of urea at 180 °C. The reaction occurs with nitrogen atoms of peptide chains, the essential feature of proteins, and is not significantly influenced by differences in amino acid composition. Albumin, the most abundant protein of plasma, contains only peptide material and so is a desirable standard for this procedure.

There have been many formulations of the biuret reagent. The reagent presented by Reinhold in the first volume of this series *(6)* was one of the first to add tartrate to prevent precipitation of cupric hydroxide. Kingsley *(7)* offered a reagent containing sodium phosphate for this purpose. The reagent presented here is that of Doumas *(8)*, which yields a more extended range of linearity than the widely used reagent of Gornall et al. *(9)*. The latter is satisfactory, however, as long as the user demonstrates that results fall within the linear portion of the standard curve. Tartrate is present to complex Cu(II), and iodide is included as an antioxidant. The net absorbance at 540 nm of pure bovine serum albumin, 1 g/L, in the reagent of Doumas *(8)* is 0.297.

Reagents

Note: Satisfactory reagents may be purchased from several commercial sources. Biuret reagent prepared according to Gornall et al. *(9)* is available from Sigma Chemical Co., St. Louis, MO 63178, as no. 540-2, and from Fisher Scientific Co., Pittsburgh, PA 15219, as no. So-B-51; biuret blank may be obtained as Sigma no. 540-3.

1. *NaOH, 6.0 mol/L.* Dissolve 240 g of NaOH in freshly distilled water (to minimize carbon dioxide) with cooling, and dilute to 1 L. Use an unopened (or new) bottle of solid NaOH to assure that it is dry and free of carbonate. Store in a tightly closed polyethylene bottle at room temperature.

2. *Biuret reagent* ($CuSO_4$, 12 mmol/L; potassium sodium tartrate, 32 mmol/L; KI, 30 mmol/L; NaOH, 0.60 mol/L). Dissolve 3.00 g of copper sulfate ($CuSO_4 \cdot 5H_2O$) in about 500 mL of freshly distilled water. Add 9.0 g of potassium sodium tartrate [KOOC-$(CHOH)_2$COONa $\cdot$ 4H_2O] and 5.0 g of potassium iodide. After the solids have dissolved, add 100 mL of 6.0 mol/L NaOH and dilute to 1 L with distilled water. Store in a tightly closed polyethylene bottle

at room temperature. The reagent is stable for approximately six months.

3. *Biuret blank reagent.* Prepare exactly as the biuret reagent except omit the copper sulfate.

4. *Protein standard (albumin, 60–70 g/L).* Use a solution of pure bovine albumin that has been assayed by other means. Store the standard in a tightly stoppered bottle at 2–4 °C. There are several sources: *(a)* Standardized Protein Solution (Armour Pharmaceutical Co., Chicago, IL 60690) contains pure bovine albumin at about 10 g of nitrogen per liter (multiply the nitrogen value by 6.25 to obtain grams of albumin per liter). *(b)* Pentex® Albumin Monomer Solution (Miles Biochemicals, Elkhart, IN 46515) contains crystallized bovine albumin at about 70 g/L. *(c)* CAP Total Protein Standard (College of American Pathologists, Skokie, IL 60076) contains crystallized bovine albumin at about 62 g/L, standardized by reference laboratories. *(d)* Standard Reference Material 927 (National Bureau of Standards, Washington, DC 20760) contains pure bovine albumin, 70.45 g/L.

A standard may be prepared by adding 12 mL of sodium azide solution, 1.5 mmol/L (0.1 g/L), to 1.0 g of crystalline bovine serum albumin powder (from Armour Pharmaceutical Co. or Miles Biochemicals) in a small beaker and stirring gently to dissolve. If a properly calibrated ultraviolet spectrophotometer with bandwidth of 2 nm or less is available, add 100 µL to 10.0 mL of water and read absorbance at 280 nm in a 1-cm cuvet. Divide absorbance by 0.661 and multiply by 101 (dilution factor) to obtain the concentration of bovine albumin (g/L). If it is not possible to measure the absorbance at 280 nm as described, use the established absorptivity at 540 nm for the Doumas reagent (0.297 L · g^{-1} · cm^{-1} for bovine albumin) to calculate the albumin concentration. Determine the concentration of the bovine albumin standard by performing careful biuret assays in duplicate at several dilutions, using cuvets of accurately known pathlength, and calculating for each dilution:

absorbance (1-cm pathlength) × 51 / 0.297 = concn of that dilution of the standard (g/L).

(The factor of 51 corrects for the 5.1/0.1 dilution in the procedure.)

> *Note:* Evaluators G. T. P. and K. Y. H. claimed that the DADE (Div. American Hospital Supply, Miami, FL 33152) Human Protein Standard described later under the serum albumin method may also be used as a standard for the biuret procedure. Human serum albumin alone, however, is not an acceptable standard; its absorbance with this biuret reagent is about 3–4% higher than that of bovine serum albumin, according to Submitters T. P. and B. T. D.

Procedure

1. Pipet 5.00 mL of biuret reagent into a "test" series of tubes and 5.00 mL of biuret blank into a "blank" series.

2. Pipet 100 µL of of specimen (protein standard, serum, or control) into one tube of each series. Mix thoroughly by vortex-type action, or by repeated inversion after covering tubes with paraffin film (Parafilm®; American Can Co., Greenwich, CT 06830). Also prepare a reagent blank by adding 100 µL of water to 5.00 mL of biuret reagent.

3. Incubate tubes at room temperature for 30 min or at 37 °C for 10 min.

4. Set zero absorbance at 540 nm with biuret blank reagent. Read the absorbances of the blank series, then read the absorbances of the reagent blank and the test series.

5. Calculate as follows:

(a) Subtract both reagent blanks and sample blanks from the corresponding absorbances *(A)* obtained for the test series:

$$A_{net} = A_{test} - A_{reagent\ blank} - A_{sample\ blank}$$

The reagent blank should read 0.095–0.105 in a 1-cm optical path for the reagent described above. [The reagent of Gornall et al. *(9)* has somewhat less absorbance.] The blanks for specimens usually read less than $A = 0.010$ if the sera are clear.

(b) Compute:

$(A_{net\ unknown}/A_{net\ standard})$ × concn of std. (in g/L)
= concn of unknown (in g/L)

If a result exceeds 120 g/L, repeat the assay by diluting the specimen with an equal volume of 0.15 mol/L NaCl; multiply the result by 2.

6. Prepare a standard curve. The reaction has been demonstrated to be linear from 10 to 120 g/L with the reagent described above. To verify this, prepare a standard curve, using 100 µL of appropriate dilutions (up to sixfold) of the 60–70 g/L standard with 0.15 mol/L NaCl, to produce values from about 10 to 60 g/L, and increasing amounts of undiluted standard (150 and 200 µL) for values from 60 to 120 g/L. These increased amounts will increase the total volume slightly (from 5.1 to 5.15 and 5.2 mL, respectively); correct the observed net absorbances by multiplying them × (total volume/5.1) before plotting.

> *Notes:* For lipemic sera, the following modification of the procedure of Chrómy and Fischer *(10)* is recommended: In a test tube mix 100 µL of serum plus 0.5 mL of water with 10 mL of acetone. Perform the test in duplicate. Mix by inversion and centrifuge. Pour off the supernate, invert the test tube, touch the lip to an absorbent paper, and promptly pipet 5.00 mL of biuret reagent into one tube and 5.00 mL of biuret blank into the other. Vortex-mix; the sediment should dissolve promptly. Proceed with the assay as above (step 3 of *Procedure*).
>
> Evaluator L. F. H. obtained 100% analytical recovery with this treatment of lipemic specimens. Evaluators S. M. D. and R. M. C. recommend assaying only speci-

mens from fasting subjects to avoid the need for such treatment, although occasionally even these specimens may be extremely lipemic.

Evaluators S. M. D. and R. M. C. preferred the use of 10 min at 37 °C instead of 30 min at room temperature. They found the standard curve to be linear to 140 g/L under these conditions. They also found that, with careful pipetting and calibrated pipettes, the volumes of reagent and specimen could both be decreased by one-half.

Evaluator L. F. H. found the absorbance of most serum blanks to be 0.010 or less, with few as high as 0.030. Evaluators G. T. P. and K. Y. H. found the serum blank to be necessary only if the specimen is visibly colored (icteric or hemolyzed) or cloudy.

Discussion

Total protein is easy to measure by either refractometry or the biuret method. The accuracy of the refractometer is dependent solely on the calibration of the scale. Precision of reading is to 1 g/L. Surveys with relatively normal serum specimens show interlaboratory agreement of ± 3% for both biuret and turbidometric procedures (11). Results with the biuret reaction are highly dependent on the accuracy of the standard used; hence, the biuret standard has been specified in detail. A standard containing 51 g/L should give a net biuret absorbance near 0.300 in a 1-cm cuvet. Precision within a run is about ± 2% (CV) for the biuret procedure.

Notes: Within-run precision (CV) observed by Evaluators was as follows: 2.3% (S. M. D. and R. M. C.), less than 1% (L. F. H.), 2.6–4.3% (M. M. K.), and 0.3% (G. T. P. and K. Y. H.). Day-to-day precision was 3.0% and 3.7% (S. M. D. and R. M. C.), 3% (L. F. H.), 1.3–2.1% (M. M. K.), and 1.3% (G. T. P. and K. Y. H.). Evaluator S. E. R. found an average CV of 4.1% for replicates of 20 different specimens. Results (g/L) in comparison with other methods were good: 66.9, compared with 67.4 for a refractometric method (S. M. D. and R. M. C.); 71.9, compared with 73.0 with another biuret method (L. F. H.); and 64.6, compared with 64.4 by American Monitor K. D. A. (G. T. P. and K. Y. H.).

The chief source of error in the refractometric method is the presence of nonprotein solids such as urea or glucose in excess, as discussed in *Procedure*. Also, the refractometer should not be used to determine protein in fluids for which it has not been calibrated; drainage fluids, saliva, gastrointestinal fluids, etc. are not likely to contain the same amount of nonprotein solids as serum or plasma, and the calibration would not apply for them.

The chief source of error in the biuret procedure is variability in the blanks for lipemic specimens. The blank reagent containing no copper provides satisfactory compensation for the color of bilirubin and hemoglobin in sera, but not for excessive turbidity introduced by lipids. For lipemic sera an alternative blanking procedure is recommended. Also, in the unlikely event that dextrans (used rarely in the United States as a volume expander) are present in the blood specimen, precipitation may occur with the biuret reagent; if so, centrifuge the reaction mixture and read the clear supernate.

The reference mean for total serum protein is 72 g/L (range 62–82 g/L) for both men and women (12). Other sources report 66–82 g/L (3) and 60–80 g/L (4). Results are about 1 g/L higher in men than in women. The concentration decreases slightly with age as follows (13): mean protein concn (g/L) = $70.9 - 0.07y$ (males) or $68.6 - 0.05y$ (females), where y = ages in years. In pregnancy the value decreases from 69 to 61 g/L by term (14). In the newborn the protein concentration is 57 (SD 7) g/L, increasing to 60 (SD 4) g/L by six months (15), and to adult values by about three years. Premature infants show still lower values, 36–60 g/L (3). Diurnal variation is less than 5 g/L, and no differences have been noted after meals. Ambulatory individuals have higher total protein concentrations (about 4–8 g/L) than recumbent ones, which is a corollary of the postural shift of fluid from the vascular bed into the interstitial spaces. Vigorous exercise causes a similar increase for a few hours. Decreased concentrations of total protein are found in severe protein depletion (whether from dietary lack or intrinsic difficulties in digestion or absorption), in chronic wasting disease, or in prolonged protein loss such as in renal disease. Edema often occurs with serum protein concentrations less then 40 g/L. The protein concentration, however, is not a sensitive indicator of protein depletion, because in many diseases there is an accompanying increase in some of the serum globulins (immunoglobulins and "acute phase" proteins) that compensates for the usual decrease in albumin. This shift is detectable as a decreased albumin/globulin ratio (normally 1.5 to 2). Increased values of total protein occur with dehydration and in hyperimmune conditions or paraproteinemia, in which antibodies or monoclonal immunoglobulins occur in high concentrations, 100 g/L or more, often leading to a marked increase in plasma viscosity with impaired circulation to the extremities (12).

TOTAL PROTEIN IN URINE

Quantitative tests for urine protein are generally carried out on 24-h specimens, although 12-h and 18-h specimens are occasionally used. No preservative is required during collection, but the specimen, or a 100-mL aliquot, should thereafter be stored at 2–4 °C or frozen until assay.

Determination by Turbidometry

Principle

Proteins form fine precipitates when mixed with reagents such as trichloroacetic acid and sulfosalicylic acid. The precipitates can be quantitated by measuring their absorbance or their scattering of light (nephelometry). The reaction is a general one for proteins, and is apparently the result of salt formation between a complex anion such as trichloroacetate and the protein, which acquires a positive charge in acid medium. There is some difference in the behavior of different proteins, so the method must be regarded as an empiric one. The method given is similar to the one presented by Henry et al. *(3)*. A somewhat more complex procedure involving dextran gel and copper analysis has been published previously in this series *(16)*.

Reagents

1. *Trichloroacetic acid (TCA),*[2] *0.76 mol/L.* Dissolve 125 g of TCA in approximately 500 mL of water and dilute to 1 L in a volumetric flask. This solution is stable for several months at room temperature and indefinitely at 2–4 °C. Discard if any color or sediment is detectable.

2. *NaCl–azide reagent* (NaCl 150 mmol/L, NaN_3 7.5 mmol/L). Dissolve 8.5 g of NaCl and 0.5 g of NaN_3 in 1 L of water. Stability and storage are the same as the TCA solution.

3. *Stock serum pool.* Obtain a pool of composite normal serum specimens that are clear, nonicteric, and free of hemolysis. Add 0.50 g of sodium azide per liter (7.5 mmol/L). Determine the protein content of this pool by one of the methods for total protein in serum. Store tightly stoppered at 2–4 °C, and reanalyze about every two months. The pool is stable for about six months, or until sediment or mold appears.

Note: Evaluators G. T. P. and K. Y. H. prefer to use a fresh serum specimen as standard, the protein content of which is determined as above. They note that it should have normal albumin/globulin ratio.

4. *Working standards.* Dilute 3.00 mL of the stock serum pool to 200 mL with the NaCl–azide reagent in a volumetric flask, being careful to minimize foaming. Calculate the concentration of protein in this standard as follows: 0.015 × protein of serum pool (g/L) × 1000 = mg/L

For example, 70 g/L in serum pool = 1050 mg/L in standard. This solution should be stored at 2–4 °C and prepared freshly every month. For each assay dilute 1.0 and 5.0 mL of standard in NaCl–azide to make 10 mL of solution (do not use a volumetric flask because of foaming); for the example given, these standards would contain 105 and 525 mg of protein per liter.

Procedure

1. Allow refrigerated specimens to reach room temperature before analysis. Mix the urine thoroughly and record its total volume. Take a 100-mL aliquot for the analysis and filter through Whatman No. 1 filter paper (Whatman Inc., Clifton, NJ 07014).

2. Estimate the amount of protein in the urine by using any one of a number of commercially available reagent "dip sticks." If protein exceeds 1000 mg/L, dilute the urine with NaCl, 0.15 mol/L, to a concentration in the range of 500 to 1000 mg/L.

3. Set up two test tubes for each sample. Label one "test" and the other "blank." Include the two working standards in the range of 100 and 500 mg/L with each assay.

4. Into both the "test" tube and the "blank" tube pipet 4.00 mL of specimen (urine, diluted urine, standard, or control).

5. To each blank add 1.00 mL of NaCl–azide reagent to correct for pigments in the urine.

6. To each "test" tube, at timed intervals, add 1.00 mL of 0.76 mol/L TCA. Mix well by inversion.

7. Allow each tube to stand at room temperature for 10 min.

Note: Evaluators S. M. D. and R. M. C. recommend strongly that the specimens and reagents be allowed to come to room temperature before mixing; otherwise, errors as great as 35% were introduced. Evaluators L. F. H., G. T. P., K. Y. H., and S. E. R. found the use of timed intervals to be inconvenient, particularly if many specimens are assayed. Evaluator L. F. H. recommends handling this by adding reagents within a certain time interval (e.g., 2 min) and reading absorbance within 9 to 11 min later.

8. Gently mix the contents of each tube again and read absorbance at 620 nm. Set zero absorbance with the corresponding blank tube each time.

9. Calculate by using the standard having an absorbance near that of the specimen:

$(A_{unknown}/A_{standard}) \times$ concn of std. (in mg/L)
$\qquad =$ concn of unknown (in mg/L)

Multiply concentration of unknown by the total volume to calculate protein excreted (mg/d, for 24-h urine collections).

10. Prepare a standard curve to check the linearity of the assay every few months. Use working standards containing about 100, 200, 400, 600, 800, and 1000 mg of protein per liter. Check that the absorbance increases linearly through the range of concentration.

Notes: Within-run precision (CV) was 2.3–3.8% (Evaluators S. M. D. and R. M. C.) and 5.8% (Evaluators G. T. P. and K. Y. H.). Day-to-day precision was 5.4% at 35 mg/L and 27.7% at 19 mg/L (Evaluators S. M. D. and R. M. C.) and 9.0% at 41 mg/L (Evaluators G. T. P. and K. Y. H.). Note that these high CVs were obtained at low concentrations.

Mean values (mg/L) in a series of specimens were 437 compared with 406 by another turbidometric method (Evaluators S. M. D. and R. M. C.), 121 com-

---
[2] Nonstandard abbreviations: TCA, trichloroacetic acid; CSF, cerebrospinal fluid; BCG, bromcresol green.

pared with 126 by TCA–biuret procedure (Evaluator L. F. H.), and 177 compared with 92 by a sulfosalicylic acid procedure (Evaluators G. T. P. and K. Y. H., who felt that "both methods have problems"). Evaluators R. M. C. and S. M. D. found the standard curve to be linear with concentration to 1100 mg/L.

Discussion

Accuracy of recovery of pure protein, such as albumin, is probably about 10%, but the diverse nature of proteins in urines in different pathological conditions causes greater disparities. Results of interlaboratory surveys have been discouraging. For a urine specimen with added albumin in the range of 1200 mg/L, the interlaboratory CV was 59% in the initial survey of the College of American Pathologists *(17)*. Hence great care and attention to detail in the performance of this test is recommended, as is constant use of control urine preparations (available commercially from a number of suppliers). Interference may be caused by drugs that precipitate under acid conditions, including cephaloridine, cephalothin, chlorpromazine, promazine, penicillins in high dosage, sulfamethoxazole, tolbutamide, and many radiographic dyes *(18)*.

Notes: Evaluator L. F. H. wondered why we included this procedure when the discussion of it is "negative" and dip-stick tests are available. The answer is that dip sticks involve an indicator (bromphenol blue) that measures chiefly the binding of serum albumin, and can miss proteinurias of monoclonal immunoglobulins, of light chains (Bence Jones proteins), or of urinary tract proteins in lower tract disease. It is essential that a screening procedure detect proteins from any source.

Evaluator L. F. H. questioned the use of TCA for urinary protein and sulfosalicylic acid/sodium sulfate for cerebrospinal fluid (CSF) protein (next section). The latter reagent has been reported to give equal precipitation with albumin and globulins, which is particularly important for CSF, whereas we chose TCA for urinary protein because of the considerable experience with its use *(3)*. Evaluators S. M. D. and R. M. C., however, found they could use the sulfosalicylic acid/sodium sulfate reagent for urine as well as for CSF protein.

The reference interval for protein excretion given by Tietz is 25–75 mg/d *(4)*, but 100 mg/d is also widely used as the upper limit. The normal glomerulus for the most part bars the passage of albumin and high-M_r (relative molecular mass) proteins from plasma into the glomerular filtrate. Commonly, an increase in protein in urine to more than 150 mg/d is the result of damage to the glomerulus. In most kidney diseases involving proteinuria, albumin is the predominant protein. In a few instances the globulins predominate, notably in multiple myeloma and in macroglobulinemia *(19)*.

Note: Evaluators S. M. D. and R. M. C. find the suggested normal range of 25–75 mg/d too low, and prefer 100 mg/d (also preferred by Evaluators G. T. P. and L. F. H.), extending to 150 mg/d in pregnancy. As they noted, this is an area in need of research.

Proteinuria is often a manifestation of primary renal disease, although transient proteinuria may occur in the absence of renal disease with fevers, thyroid disorders, and heart disease. In 3 to 5% of young adults a small degree of proteinuria occurs during the day, but not at night when the subjects are in bed. This is termed postural or orthostatic proteinuria and is usually a benign condition.

TOTAL PROTEIN IN CEREBROSPINAL FLUID

A normal CSF sample is clear, colorless, and free of turbidity. Because of the difficulties in obtaining a specimen, the CSF should be handled with great care. Store any leftover portions at 2–4 °C (stable for several days) for possible repeat or additional determinations. Centrifuge CSF specimens before analysis, to remove erythrocytes or leukocytes. If any blood is present in the specimen because of a traumatic puncture, the results for CSF protein may be invalid. As little as 0.2 mL of blood can quadruple the apparent amount of protein present in 5.0 mL of CSF.

Determination by Turbidometry

Principle

The most common procedures used to estimate CSF proteins, like those for urine, measure turbidity when proteins are precipitated with acidic reagents. These procedures are simple and do not suffer from interference from amino acids, peptides, or other nonprotein substances but do require about 1 mL of CSF.

The main difficulty with the turbidometric procedures is the selection of the ideal precipitating reagent. Such a reagent should react similarly with both albumin and globulins. The precipitant least affected by changes in proportion of albumin and globulin present appears to be a combination of sulfosalicylic acid and sodium sulfate. The procedure given is that of Pennock et al. *(20)*, as modified by Tietz *(4)*. Two methods, one colorimetric *(21)* and one turbidometric with TCA *(22)*, have been published earlier in this series.

Reagents

Precipitating reagent. Sulfosalicylic acid, 0.12 mol/L, and sodium sulfate, 0.49 mol/L. Dissolve 30.0 g of sulfosalicylic acid ($C_7H_6O_6S \cdot 2H_2O$) and 70.0 g of sodium sulfate (Na_2SO_4) in water and dilute to 1 L in a volumetric flask. Mix and filter until clear. Store in a brown bottle. The reagent is stable; discard if any discoloration or turbidity is noted.

Stock serum pool, working standards, and NaCl–

azide reagent. Refer to the section on urine protein assay for preparation.

Procedure

1. Centrifuge the CSF to sediment any cells or particulate matter.

2. Check the clarity of the precipitating reagent by diluting 2.0 mL of reagent with 0.5 mL of NaCl–azide. Measure absorbance against NaCl–azide at 620 nm. If absorbance exceeds 0.010, prepare new reagent.

3. Set up two test tubes for each specimen (sample, standard, or control). Label one series as "test" and one as "blank."

4. To both the test and blank tubes add 0.50 mL of specimen.

5. Add 2.00 mL of NaCl–azide to all blank series tubes; mix.

6. At timed intervals add 2.00 mL of precipitating reagent to the test series; mix. Let stand at room temperature for 10 min.

7. At the end of 10 min, mix each "test" tube again by gentle inversion and read the absorbance at 620 nm. Set the absorbance to zero with the corresponding blank tube each time. Remember to observe the time interval used in step 6.

8. Calculate protein concentration with the standard having absorbance nearer to that of the specimen:

$$(A_{unknown}/A_{standard}) \times \text{concn of std. (in mg/L)} = \text{concn of unknown (in mg/L)}$$

Notes: Evaluators S. M. D. and R. M. C. have used this method for more than five years, and found it very reliable. They noted the importance of checking the reagent for color or turbidity, as described above. They did not find timing of intervals to be necessary, but emphasized the need to remix tube contents just before reading the turbidity.

Evaluator L. F. H. faults this method for the large volume of CSF required, a justifiable criticism. He recommends instead a highly sensitive procedure involving Coomassie Blue *(23).* However, we did not wish to recommend a procedure that has not had at least several years of trial.

Within-run precision (CV) was 1.3–3.8%, and between-day precision was 4.7–5.3% (Evaluators S. M. D. and R. M. C.). Evaluation with other procedures showed 665 (mg/L) by this method compared with 681 by a reference laboratory (S. M. D. and R. M. C.), and 640 compared with 640 by the Coomassie Blue method *(23)* (L. F. H.). Linearity extended to 600 (L. F. H.) or 1400 mg/L (S. M. D. and R. M. C.).

Discussion

Results with turbidometric analysis of spinal fluid protein are generally more satisfactory than those of urine protein, probably because of the absence in CSF of many interfering factors in urine, and because of the fewer protein species found in CSF. Precision is about 5% (CV) but deteriorates to about 10% at values below 150 mg/L. Agreement with other procedures is ± 7–10%, but, again, the proportional error is greater at lower values. Interference has been encountered from penicillins in large doses, sulfamethoxazole, tolbutamide, radiographic dyes, and after myelography *(18).*

The concentration of protein in CSF of adults is 150–450 mg/L. In the newborn (0 to 1 month of age) the concentration is higher, 600–900 *(22)* or 800–1200 *(21)* mg/L. In the elderly the upper limit of normal increases to 600 mg/L *(3).* Increases often signify injury to the brain or to the blood–brain barrier. Electrophoretic fractionation of the CSF proteins may be helpful in detecting certain conditions, particularly multiple sclerosis.

Note: Evaluator L. F. H. prefers a somewhat higher reference range for adults, 180–580 mg/L.

ALBUMIN IN SERUM

Specifications for specimen collection and handling are the same as for specimens for total protein.

Determination by Dye-Binding with Bromcresol Green

Principle

Albumin binds many negatively charged organic substances such as fatty acids, bilirubin, hormones, drugs, and dyes. Other proteins may bind some of these compounds but usually do so much less strongly and with lower capacity. The spectrophotometric method considered the most specific for albumin involves bromcresol green (BCG) at pH 4, a procedure introduced by Bartholomew and Delaney in 1966 *(24)* and modified by Doumas et al. in 1971 *(25).* A version of this procedure was published earlier in this series *(26).* Lately it has been recognized that interference by globulins is significant and is greater with longer reaction times *(27);* consequently, we now recommend an interval of only 30 s between mixing and reading absorbance *(28, 29).* The reagent of Doumas et al. *(25)* is used.

Reagents

1. *BCG reagent* (BCG, 150 µmol/L; sodium succinate buffer, 75 mmol/L; sodium azide, 1.5 mmol/L; Brij-35, 1.2 g/L; pH 4.20). Dissolve 105.0 mg of bromcresol green (3,3′,5,5′-tetrabromo-*m*-cresolsulfonphthalein) or 108.0 mg of its sodium salt, 8.85 g of succinic acid, 100 mg of sodium azide, and 4 mL of Brij-35 solution, 300 g/L (Technicon Instrument Corp., Tarrytown, NY 10591), in about 950 mL of water. Adjust to pH 4.15–4.25 with NaOH, 6 mol/L, then dilute to 1 L with water. (See section on biuret method reagents, above, for preparation of

NaOH solution.) The reagent may be stored six months at room temperature in a tightly closed polyethylene bottle.

Note: A similar BCG reagent may be purchased as no. 630-2 from Sigma, HyCel No. 149 from Fisher, no. 41292 from Pierce Chemical Co., Rockford, IL 61105, or "Albustrate" from General Diagnostics, Div. of Warner-Lambert Co., Morris Plains, NJ 07950. The Pierce reagent has citrate and the General Diagnostics reagent has lactate instead of succinate as buffer. Check extent of linearity before using any of these reagents.

2. *Human albumin standard* (about 60 g of albumin and 7.5 mmol of sodium azide per liter). Dissolve 6 g of human serum albumin (Cohn Fraction V) and 50 mg of sodium azide in 100 mL of water in a small beaker with gentle stirring. Measure the total protein concentration by the biuret procedure for total serum protein. This solution may be stored in a tightly closed bottle at 2–4 °C for six months. Human Protein Standard, 80 g/L (no. B5158; DADE), is an acceptable commerical substitute when properly diluted.

3. *BCG blank* (succinate buffer, 75 mmol/L; sodium azide, 1.5 mmol/L), needed for highly turbid specimens only. Dissolve 8.85 g of succinic acid and 100 mg of sodium azide in about 950 mL of water; add 4.0 mL of Brij-35 solution, adjust to pH 4.1–4.3, and make to 1 L with water. Store as for BCG reagent.

Procedure

1. Adjust absorbance at 628 nm to zero with water. Read the BCG reagent (usual absorbance is about 0.150) and reset the zero absorbance with this reagent.

2. Pipet 10.0 mL of BCG reagent into a series of tubes. *One at a time*, add 40.0 µL of human albumin standard (60 g/L) or of unknown serum to a tube, mix without delay and read absorbance at 628 nm at 30 s (± 3 s) after the addition.

Notes: Evaluators S. M. D. and R. M. C. emphasize the need for calibrated pipettes and careful pipetting technique. With care they are able to use half-volumes of reagent and specimen. They note that the absorbance may be read equally well at 625 nm, which may be more convenient for certain types of spectrophotometers than 628 nm.

Evaluators S. M. D. and R. M. C. stressed the importance of the time of reading. They used a stopwatch and read at 10 s after mixing; if the user can accomplish this, results will probably be improved. Evaluator L. F. H., on the other hand, found no difference between 30-s and 5-min readings for a series of samples. His results may be related to the choice of specimens; only specimens with very low albumin and high globulin concentrations show significant effects of time of reading.

3. Calculate protein as follows:

$(A_{unknown}/A_{net\ standard}) \times$ concn of std. (in g/L)
$=$ concn of unknown (in g/L)

4. Blanks are not required for hyperbilirubinemic or hemolyzed specimens; for highly lipemic (turbid) specimens, add 40 µL of sample to 10 mL of BCG blank; read the absorbance with BCG blank as a reference, and subtract this reading from the absorbance of the sample plus BCG reagent.

5. The reaction is linear from 10 to 60 g of albumin per liter. Verify this by diluting the human albumin standard to obtain concentrations in this range.

Notes: Evaluators S. M. D. and R. M. C. used this procedure for two years with excellent results. Evaluator M. M. K. found the color with the reagent described to be somewhat unstable near 30 s of reaction; Evaluators M. M. K. and L. F. H. recommend use of the Pierce reagent. Evaluators G. T. P. and K. Y. H. also recommend use of a commercial reagent, especially in small laboratories that do not have a pH meter and analytical balance available.

Precision (CV) found by the Evaluators was: within-run, 1.5% and 2.3% (S. M. D. and R. M. C.), 3.0–6.7% (M. M. K.), and 1.8% (G. T. P.); between-day, 2.7% and 3.1% (S. M. D. and R. M. C.), 3.5–8% (M. M. K.), and 3.4% (G. T. P. and K. Y. H.). Analytical recovery of added albumin was 98–101% (L. F. H.); linearity extended from 20 to 80 g/L (S. M. D. and R. M. C.). Comparison of results (g/L) with other procedures showed 37.1 vs 36.3 by another BCG method (S. M. D. and R. M. C.) and 34.3 vs 35.0 by American Monitor K. D. A. (G. T. P. and K. Y. H.).

Evaluator L. F. H. found some values to be lower than those obtained by cellogel electrophoresis.

Discussion

Correlations between the BCG-binding procedure and immunochemical or electrophoretic determination show that the BCG procedure gives satisfactory results with normal and hypoalbuminemic sera, provided that short reaction times are observed. Results of interlaboratory surveys with normal sera showed a CV of about 6% *(30);* day-to-day CVs within a laboratory ranged from 2.5 to 5.9%, depending on operator training *(25)*. The method is suitably free from interference by bilirubin and hemoglobin, and moderate lipemia may be compensated for by the serum blank. The absorptivity at 630 nm is about 3.1 L · g^{-1} · cm^{-1}, about 10 times as sensitive as the biuret reaction.

An alternative to the BCG procedure, should the workload in the laboratory permit, is to determine albumin by quantitation of protein electrophoresis patterns (described elsewhere in this volume). The assay of albumin by electrophoresis is free from the potential interference by globulins that bind slowly to BCG, but still suffers from nonlinearity and from albumin/globulin differences in dye-binding. Immunochemical assay of albumin is a promising approach but is little used at present.

Reference Intervals

The usual albumin concentration in an adult population is 42 g/L (range 35–50 g/L) *(12)*. Concentrations in women tend to be about 2 g/L less than those in men. This difference disappears at about 70 years of age, owing to a gradual decline in the concentration in men: g/L = 41.6 − 0.08y (men) and 39.1 − 0.04y (women), where y = age in years *(13)*. In late pregnancy the concentration is decreased by about 10 g/L *(14);* the newborn often has more albumin than the mother, although still less than the normal adult range. The mean for the newborn is 38 (SD 4) g/L, increasing to 47 (SD 4) g/L by six months *(15)*. Adult values are attained at about three months of age. The effects of posture, exercise, meals, and dehydration are similar to those observed with total protein.

Albumin concentrations in serum decrease in a large spectrum of diseases because of loss to the extravascular fluid through increased capillary permeability and impaired venous return, as well as decreased production of albumin. Because albumin synthesis is very sensitive to the supply of amino acids in the portal blood, albumin is a useful index of protein nutrition. The lowest albumin values are seen in protein-losing conditions such as nephrotic syndrome and protein-losing enteropathy, where concentrations of 10 to 20 g/L are common. Values of < 20 g/L are frequently observed in edema.

References

1. Wilson, S. S., Guillan, R. A., and Hocker, E. V., Studies of the stability of 18 chemical constituents of human serum. *Clin. Chem.* 18, 1498–1503 (1972).
2. Naumann, H. N., Determination of total serum proteins by refractometry. In *Serum Proteins and the Dysproteinemias,* F. W. Sunderman and F. W. Sunderman, Jr., Eds., Lippincott, Philadelphia, PA, 1964, pp 86–101.
3. Henry, R. J., Cannon, D. C., and Winkelman, J. W., *Clinical Chemistry, Principles and Technics,* 2nd ed., Harper and Row, Hagerstown, MD, 1974.
4. Tietz, N. W., *Fundamentals of Clinical Chemistry,* 2nd ed., W. B. Saunders Co., Philadelphia, PA, 1976.
5. Riegler, E., A colorimetric method for the determination of albumin. *Z. Anal. Chem.* 53, 242–245 (1914).
6. Reinhold, J. G., Total protein, albumin, and globulin. *Stand. Methods Clin. Chem.* 1, 88–97 (1953).
7. Kingsley, G. R., Procedure for serum protein determinations with a triphosphate biuret reagent. *Stand. Methods Clin. Chem.* 7, 199–207 (1972).
8. Doumas, B. T., Standards for total serum protein assays—a collaborative study. *Clin. Chem.* 21, 1159–1166 (1975).
9. Gornall, A. G., Bardawill, C. J., and David, M. M., Determination of serum proteins by means of the biuret reaction. *J. Biol. Chem.* 177, 751–766 (1949).
10. Chrómy, V., and Fischer, J., Photometric determination of total protein in lipemic sera. *Clin. Chem.* 23, 754–756 (1977).
11. Ross, J. W., and Fraser, M. D., Analytical clinical chemistry precision. State of the art for fourteen analytes. *Am. J. Clin. Pathol.* 68, 130–141 (1977).
12. Peters, T., Jr., Proteins. In *Chemical Diagnosis of Disease.* S. S. Brown, F. L. Mitchell, and D. S. Young, Eds., Elsevier, Amsterdam, 1979, pp 312–362.
13. Keating, F. R., Jr., Jones, J. D., Elveback, L. R., and Randall, R. V., The relation of age and sex to distribution of values in healthy adults of serum calcium, inorganic phosphorus, magnesium, alkaline phosphatase, total proteins, albumin, and blood urea. *J. Lab. Clin. Med.* 73, 825–834 (1969).
14. Elliott, J. R., and O'Kell, R. T., Normal clinical chemical values for pregnant women at term. *Clin. Chem.* 17, 156–157 (1971).
15. Trevorrow, V., Kaser, M., Patterson, J. P., and Hill, R. M., Plasma albumin, globulin and fibrinogen in healthy individuals from birth to adulthood. *J. Lab. Clin. Med.* 27, 471–486 (1941).
16. Doetsch, K., and Gadsden, R. H., Determination of urinary total protein by use of gel filtration and a modified biuret method. *Sel. Methods Clin. Chem.* 8, 179–182 (1977).
17. Glenn, G. C., and Hathaway, T. K., Urinary chemistry. A new CAP program. *Am. J. Clin. Pathol.* 68, 153–158 (1977).
18. Young, D. S., Thomas, D. W., Friedman, R. B., and Pestaner, L. C., Effects of drugs on clinical laboratory tests. *Clin. Chem.* 18, 1041–1303 (1972). Special issue.
19. Davidsohn, I., and Henry, J. B., Eds., *Todd–Sanford Clinical Diagnosis by Laboratory Methods,* 15th ed., W. B. Saunders Co., Philadelphia, PA, 1974.
20. Pennock, C. A., Passant, L. P., and Bolton, F. G., Estimation of cerebrospinal fluid protein. *J. Clin. Pathol.* 21, 518–520 (1968).
21. Friedman, H. S., Total proteins in cerebrospinal fluid (colorimetric). *Stand. Methods Clin. Chem.* 5, 223–230 (1965).
22. Rice, E. W., Total protein in cerebrospinal fluid (turbidometric). *Stand. Methods Clin. Chem.* 5, 231–236 (1965).
23. Johnson, J. A., and Lott, J. A., Standardization of the Coomassie Blue method for cerebrospinal fluid proteins. *Clin. Chem.* 24, 1931–1933 (1978).
24. Bartholomew, R. J., and Delaney, A. M., Sulphonphthaleins as specific reagents for albumin: Determination of albumin in serum. *Proc. Aust. Assoc. Clin. Biochem.* 1, 214–218 (1966).
25. Doumas, B. T., Watson, W. A., and Biggs, H. G., Albumin standards and the measurement of serum albumin with bromcresol green. *Clin. Chim. Acta* 31, 87–96 (1971).
26. Doumas, B. T., and Biggs, H. G., Determination of serum albumin. *Stand. Methods Clin. Chem.* 7, 175–188 (1972).
27. Webster, D., A study of the interaction of bromocresol green with isolated serum globulin fractions. *Clin. Chim. Acta* 53, 109–115 (1974).
28. Webster, D., The immediate reaction between bromocresol green and serum as a measure of albumin content. *Clin. Chem.* 23, 663–665 (1977).
29. Gustafsson, J. E. C., Automated serum albumin determination by use of the immediate reaction with bromcresol green reagent. *Clin. Chem.* 24, 369–373 (1977).
30. Batsakis, J. G., Aronsohn, R. S., Walker, W. A., and Barnes, B., Serum albumin, a CAP survey. *Am. J. Clin. Pathol.* 66, 238–243 (1976).

Prothrombin Time and Activated Partial Thromboplastin Time

Submitter: Leonard K. Dunikoski, Jr., *Department of Laboratories, Perth Amboy General Hospital, Perth Amboy, NJ 08861*

Evaluators: Walter E. Hordynsky, *Department of Biochemistry, St. Mary's Hospital, Orange, NJ 07050*

Paul F. A. Richter, *Presbyterian–University of Pennsylvania Medical Center, Philadelphia, PA 19104*

Henry G. Schriever, *Department of Laboratories, John F. Kennedy Medical Center, Edison, NJ 08817*

Catherine Sherry, *Department of Laboratories, St. Vincent's Hospital and Medical Center of New York, New York, NY 10011*

Leonard Sideman and Irene Daza, *Division of Clinical Chemistry and Hematology, Pennsylvania Department of Health, Lionville, PA 19353*

Reviewer: William D. Bostick, *Oak Ridge National Laboratory, Oak Ridge, TN 37830*

Introduction

In 1905, Paul Morawitz published what is now considered the "classic theory of coagulation" *(1)*. Unifying the work of several earlier investigators, he postulated that in the presence of calcium a "thromboplastin" factor could convert an inactive precursor (prothrombin) into thrombin, with subsequent conversion of fibrinogen into fibrin:

$$\text{prothrombin} \xrightarrow[\text{calcium}]{\text{thromboplastin}} \text{thrombin}$$

$$\text{fibrinogen} \xrightarrow{\text{thrombin}} \text{fibrin}$$

Later it was demonstrated that two distinct clotting systems exist in humans: an extrinsic system, activated by tissue enzymes, and an intrinsic system, activated by platelet thromboplastins. The gradual recognition that specific plasma proteins or factors are required to produce a thromboplastin capable of converting prothrombin into thrombin is reviewed in several texts *(2–4)* and makes interesting reading. In 1964, two similar theories were published postulating that most clotting factors exist in the circulation in an inactive precursor form. Once activated, each factor catalyzes the conversion of another inactive precursor into its active form, yielding a sequence of reactions similar to a "waterfall" *(5)* or "cascade" *(6)*. With some modifications, the "waterfall" or "cascade" hypothesis is still the most widely used concept of coagulation (Figure 1).

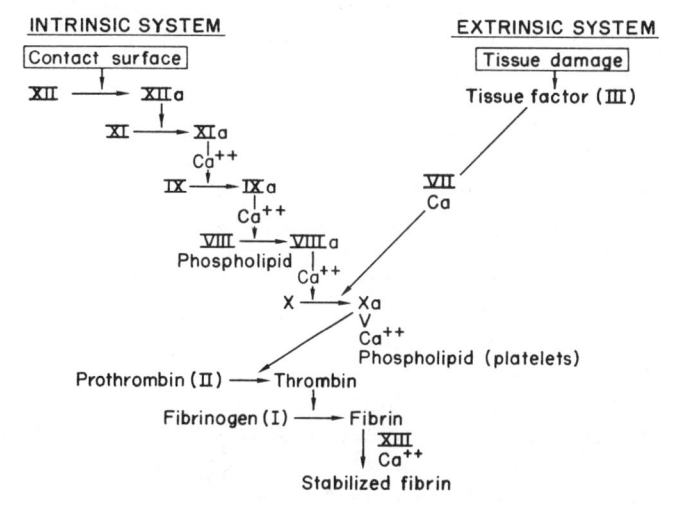

Fig. 1. A simplified version of the "waterfall" or "cascade" concept of coagulation
The APTT test standardizes contact activation, serves as a substitute for phospholipid (platelet factor 3), and is used as a monitor of the intrinsic system. The PT test provides an excess of tissue thromboplastin and is used as a monitor of the extrinsic system

The prothrombin time (PT) test, first described by Quick in 1935, measures the extrinsic pathway. It indicates factors that are decreased in some liver diseases, in vitamin K deficiency, and after administration of coumarin anticoagulants. Although the PT depends on the activity of several factors, a significant deficiency of a single factor (other than XII, XI, IX, or VIII) will prolong the prothrombin time.

The activated partial thromboplastin time (APTT) test measures the intrinsic pathway. As such, it depends on Factors XII, XI, IX, and VIII, as well as the factors common to the extrinsic pathway. The APTT is used in detecting hemophilias and in monitoring replacement therapy for hemophilia, as well as assessing anticoagulant therapy with heparin. The combination of PT and APTT tests will detect all clinically significant coagulation-factor deficiencies.

Two methods for both tests are presented: a manual "tilt-tube" technique and a semi-automated technique (Fibrometer® Precision Coagulation Timer; BBL, Div. of Becton, Dickinson and Co., Cockeysville, MD 21030). Although automated coagulation instruments are used in larger laboratories, recent surveys by the College of American Pathologists (7) show that a large majority of laboratories in the United States use the Fibrometer for routine coagulation testing procedures. The methods for the PT test conform to initial recommendations of the International Committee for Standardization in Hematology for a reference method for the one-stage PT test (8).

Principle

The PT is the time required for plasma to clot after a "complete" thromboplastin and an optimum amount of calcium have been added. The "complete" thromboplastin is a tissue extract of brain or lung origin, which reacts with Factors V, VII, and X to catalyze the conversion of prothrombin into thrombin.

The endpoint (or time) consists of the visual, optical, or mechanical detection of the clot formed by the conversion of fibrinogen to fibrin. An abnormal PT can result from deficiencies of fibrinogen, prothrombin, or Factors V, VII, or X, or from the presence of a coagulation inhibitor.

The APTT is the time required for a platelet-poor plasma to clot after the partial thromboplastin, an activator, and an optimum amount of calcium have been added. The activator maximizes contact activation and standardizes the activation of all plasmas. The partial thromboplastin serves as a substitute for platelet factor 3, and therefore the APTT is a measure of the intrinsic system.

In the tilt-tube techniques for PT and APTT, the fibrin clot is detected by eye, and a stopwatch is used to determine time of clot formation. With the Fibrometer, clot detection is electro-mechanical. The probe or clot-detection unit of the Fibrometer consists of two stainless-steel electrodes: one straight, stationary electrode and one moving electrode with a small hook at its tip. When a test is initiated, a timing device is activated, the stationary electrode drops into the reaction mixture, and the moving electrode sweeps through the reaction mixture, being lifted 1.27 mm out of the mixture every 0.5 s (9). As a fibrin strand forms, it is picked up on the small hook of the moving electrode, and an electrical detection circuit is completed between the stationary electrode, the liquid, the fibrin strand, and the moving electrode. Completion of this detection circuit stops the timing device.

Materials and Methods

Reagents

1. *Water.* Use reagent-grade distilled or de-ionized water for all reagents and controls. Water should be ammonia-free to prevent deterioration of factor V, and should be free from metal ion contamination.

2. *Sodium citrate, 109 mmol/L.* Place 32.0 g of trisodium citrate (dihydrate) in a 1-L volumetric flask and fill to volume with distilled water.

Note: This solution is not required if commercial evacuated collection tubes containing 32 g of trisodium citrate anticoagulant per liter are used for specimen collection. Although 38 g/L solution is used in many laboratories, the 32 g/L solution is the recommendation of the Expert Panel on Oral Anticoagulant Control (10). More concentrated solutions are hypertonic and reduce the hematocrit of blood anticoagulated with citrate.

3. *Calcium chloride, 25 or 20 mmol/L.* The molarity of calcium chloride depends on the commercial APTT reagent used. Evaluators P. F. A. R. and C. S. use DADE reagent (DADE, Div. American Hospital Supply Corp., Miami, FL 33152), which specifies 20 mmol/L calcium chloride. The Submitter and the other evaluators use the 25 mmol/L concentration, which is prepared by placing 3.68 g of $CaCl_2 \cdot 2H_2O$ in a 1-L volumetric flask, and filling to volume with de-ionized water. Alternatively, appropriate calcium chloride solutions may be purchased from the manufacturer of the APTT reagent used.

Note: This solution is not required for the PT test.

4. *Commercial thromboplastin/calcium reagent (for PT).* Prepare, store, and use this reagent according to the manufacturer's directions. In general, store the reconstituted reagent at 4–6 °C, mix it immediately before use, and do not keep at 37.5 °C longer than the manufacturer recommends. Although methods exist for preparing thromboplastin from autopsy tissues, the wide availability, controlled manufacture, and relative low cost of commercial products make them attractive. Because significant variability in PT test results is related to differences among thromboplastins (11–13), use a single type and lot number of the same manufacturer within each laboratory. The Submitter and Evaluators L. S. and H. G. S. use Simplastin reagent (General Diagnostics, Morris Plains, NJ 07950), while Evaluators C. S. and P. F. A. R. use Thromboplastin C (DADE), and Evaluator W. E. H. uses Ortho Brain Thromboplastin (Ortho Diagnostics, Raritan, NJ 08869). Stability for these reconstituted products averages five to seven days when refrigerated.

5. *Commercial APTT reagent.* Prepare, store, and use this reagent according to the manufacturer's instructions. The comments concerning in-laboratory preparation of PT reagent also apply to APTT re-

agent. Several types of activators may be used in these reagents, among them micronized silica, kaolin, Celite, or ellagic acid. Considerable differences among commercial APTT reagents have been reported *(14, 15)*, so a single lot number and type from the same manufacturer should be used within each laboratory. The Submitter and Evaluators L. S. and H. G. S. use Platelin Plus Activator or Automated APTT (General Diagnostics). Evaluators C. S. and P. F. A. R. use DADE Actin-Activated Cephaloplastin Reagent, and Evaluator W. E. H. uses Ortho Activated Thrombofax Reagent. Stability of these reconstituted reagents averages three to six days when refrigerated.

6. *Normal control material.* Commercial normal control products are available. Store, reconstitute, and use them according to the manufacturer's directions. Alternatively, prepare frozen plasma controls by pooling fresh citrated plasma from at least 10 individuals, both male and female. Do not use plasma from women using oral contraceptives. Freeze aliquots of the pooled plasma at −70 °C (if possible) and use them within four months. A −20 °C freezer may be used if it is not a self-defrosting type. Establish the mean and standard deviation with at least 20 aliquots assayed in duplicate.

7. *Abnormal control material.* Commercial abnormal control products are available. Use them in addition to a normal control if possible. Store, reconstitute, and use them according to the manufacturer's instructions. In general, these products are stable for 6–8 h after reconstitution when refrigerated.

Apparatus

1. *Water bath,* circulating, 37.5 ± 1 °C.

Note: Not required if a Fibrometer is used.

2. *Stopwatch with foot-pedal control,* such as Pedichron (Clay-Adams, Div. of Becton, Dickinson & Co., Parsippany, NJ 07054).

Note: Not required if a Fibrometer is used. Evaluator P. F. A. R. uses a stopwatch without a foot-pedal control and finds no difficulty.

3. *Culture tubes, 13 × 100 mm.* Clean, unscratched tubes should be used for coagulation testing. Disposable glass or plastic tubes are commercially available. When glass tubes must be reused, acid cleaning and thorough rinsing are essential.

Note: Not required if a Fibrometer is used. Evaluator L. S. uses 12 × 75 mm tubes instead, and finds them satisfactory.

4. *Fibrometer Precision Coagulation Timer.* Multiple timers and (or) companion heating blocks may be used to increase throughput.

5. *Sample cups.* Fibro tube® (BBL) disposable reaction cups or equivalent are used with the Fibrometer.

6. *Automatic pipette, 200 µL/100 µL, with disposable plastic tip.* The handheld pipetting device made especially for use with the Fibrometer may be used, or two separate automatic pipettes may be used (disposable tips are required). The Submitter uses MLA pipettes (Medical Laboratory Automation, Inc., Mt. Vernon, NY 10550); Evaluator P. F. A. R. uses Oxford pipettes (Oxford Labs., Foster City, CA 94404).

Collection and Handling of Specimens

Collection and handling techniques for PT and APTT specimens share some similarities but have several differences. For both procedures, citrate is used as an anticoagulant. Factors V and VIII are more stable in citrate than in oxalate *(16)*, and the APTT has been shown to be more sensitive in monitoring heparin therapy when citrated plasma is used instead of oxalated plasma *(17)*. Use a siliconized plastic or glass syringe, or a siliconized evacuated blood-collection tube, preferably with an 18- or 19-guage needle. Venipuncture must be clean, to avoid contamination with tissue fluid or hemolysis. There is still considerable disagreement about the necessity of using the so-called "two-syringe technique," where the first 1 or 2 mL of blood drawn is discarded and the contents of a second syringe or collection tube are used for coagulation studies. When drawing multiple samples (for serum as well as citrated plasma), however, it is just as easy not to use the initial blood for coagulation tests.

Add nine volumes of blood rapidly to one volume of sodium citrate (109 mmol/L) and mix by gentle inversion. Avoid frothing, foaming, excessive shaking, or turbulence. The ratio of blood to sodium citrate is critical, and the volume of blood added to the citrate is critical. The volume of blood added to the citrate should be within 10% of the 9/1 ratio; if evacuated collection tubes are used, the technologist should carry a reference tube containing a minimum acceptable total volume. In the Submitter's institution, such "reference tubes" became especially important when the manufacturer of evacuated tubes provided 4.5 mL-draw tubes that physically could accommodate up to 7 mL. The 9/1 ratio of blood to citrate is based on a normal hematocrit (0.45) and must be altered for grossly abnormal hematocrits so that citrate concentration in the final mixture is constant from sample to sample *(18)*. The amount of citrate required to achieve a constant ratio of citrate to plasma may be calculated from the following equation *(10)*:

$$V = (100 - H)/(595 - H)$$

where V = volume of citrate used for 1.0 mL of anticoagulated blood and H = hematocrit fraction × 100.

Examples:

1. To prepare 5 mL of citrated blood from a normal individual (hematocrit 0.45):
$V = (100 - 45)/(595 - 45) = 55/550 = 0.1$ mL of citrate per milliliter of citrated blood
Thus 5×0.1 mL = 0.5 mL of citrate would be added to 4.5 mL of whole blood to give 5.0 mL of citrated blood.

2. To prepare 7 mL of citrated blood from a patient with a hematocrit of 0.68:
$V = (100 - 68)/(595 - 68) = 32/527 = 0.06$ mL of citrate per milliliter of citrated blood
Thus 7×0.06 mL = 0.42 mL of citrate would be added to 6.58 mL of whole blood to give 7.0 mL of citrated blood.

For the PT test, keep the blood at room temperature to prevent cold-activation of kallikrein, which in some plasmas causes an increase in Factor VII *(19)*. Centrifuge the blood in a stoppered tube as soon as it is received, and keep it stoppered at room temperature. Grossly hemolyzed samples should not be analyzed, and lipemic samples may be difficult to assay by the tilt-tube procedure, because visibility of the clot will be impaired. Plasma need not be removed from the erythrocytes if the tube remains stoppered and if the test is performed within 4 h; if this cannot be done, separate and quick-freeze the plasma in alcohol/solid CO_2 and store at -70 °C if possible. A -20 °C freezer is satisfactory if it is not self-defrosting.

For the APTT test, centrifuge the citrated whole blood at 4–6 °C in a refrigerated centrifuge (if possible) and store it stoppered on melting ice or in ice water until the test is performed. If the assay cannot be completed within 4 h, quick-freeze an aliquot of the plasma in alcohol/solid CO_2 and store it at -70 °C. Again, a -20 °C freezer may be used, if it is not self-defrosting.

Prothrombin Time Test Procedure

Tilt-tube technique

1. Warm about 0.5 mL of citrated plasma at 37.5 °C for 5 min in a 13 × 100 mm culture tube in the water bath; do not allow it to stand for more than 10 min.

2. Likewise, warm about 0.5 mL of normal control at 37.5 °C for 5 min but not longer than 10 min.

3. Warm about 1 mL of commercial PT reagent in a 13 × 100 mm culture tube in the water bath for 5 min at 37.5 °C; do not exceed manufacturer's instructions for 37.5 °C storage.

4. Swirl the prewarmed PT reagent gently, then pipet 0.2 mL into a clean 13 × 100 mm culture tube in the 37.5 °C water bath.

5. Quickly but forcibly pipet 0.1 mL of prewarmed normal control plasma directly into the PT reagent, avoiding splattering onto the walls of the tube. Simultaneously, start a stopwatch, using the foot-pedal control.

6. After 6 s or half the expected clotting time (whichever is longer), remove the tube from the water bath, wipe it dry, and tilt the tube back and forth gently about once a second, permitting the mixture to flow up and down the wall of the tube and observing the contents for the first appearance of clot formation.

Note: A light source makes clot detection much easier. Evaluator P. F. A. R. uses an old Rh view box for this purpose. Removal of the tube from the water bath to a light source results in a rapid decrease in temperature and should be performed in a standard manner to attain maximum reproducibility. The Reviewer emphasizes that a decrease of 1 °C can prolong the clotting time by 1 s *(16)*. If the tube can be tilted and the clot formation monitored while keeping the liquid-containing portion of the tube in the water bath, a significant improvement in precision will result. Some laboratories use a glass-sided water bath or aquarium for this type of method.

7. At the first appearance of the fibrin clot, stop the stopwatch and record the time to the nearest 0.1 s.

8. Repeat steps 4–7 with the same normal control sample; results should agree within 1 s and should be within 2 SD of the laboratory mean for that control.

9. In a similar manner, test the patient's prewarmed plasma in duplicate. Record each value and report the average result. Miale and Kent *(19)* recommend that the difference between duplicates be no greater than 4.0% of the lower figure; otherwise, the test should be repeated in duplicate.

10. Periodically assay the abnormal control in duplicate in a similar manner, and record both values. Results should be within 2 SD of the laboratory mean for that control.

Fibrometer technique

The Fibrometer operates at a preset temperature of 37.2 ± 0.5 °C. The temperature can not be adjusted by the operator. Six prewarming wells and a reaction well are built into the heating block portion of the instrument. The prewarming wells will heat room-temperature liquids to 37.2 ± 0.5 °C within 3 min and refrigerated liquids within 5 min. Turn the instrument on for at least 10 min before beginning step 1, to allow the heating block to reach proper operating temperature. A small indicator light, located to the left of the digital readout, will glow as long as the instrument is at the correct operating temperature.

Note: Verify and record the instrument temperature at least daily with a sensitive thermocouple or thermometer.

1. Raise the probe arm to the up (rest) position above the center (reaction) well in the Fibrometer.
2. Place six sample cups into six of the warming wells. Using a felt tip marker, label one cup "patient," one cup "control," and four cups "reagent."
3. Pipet about 0.3 mL of citrated plasma into the "patient" cup and incubate for 3 min but not more than 10 min.
4. Pipet about 0.3 mL of normal control plasma into the "control" cup, and incubate 3 min if at room temperature or 5 min if refrigerated, but not more than 10 min in either case.
5. Swirl the PT reagent gently, and then pipet exactly 0.2 mL of it into each of the "reagent" cups. Incubate for 5 min (refrigerated reagent); do not exceed manufacturer's instructions for 37.5 °C storage.
6. Move one of the prewarmed "reagent" cups to the center reaction well of the Fibrometer.
7. Carefully pipet with a clean disposable plastic tip exactly 0.1 mL of prewarmed control plasma into the reagent cup; avoid splashing liquid onto sides of cup. Simultaneously press the timer bar.

Note: If you use the specially designed Fibrometer Automatic Pipette, ejection of the sample automatically triggers the instrument, and the timer bar need not be depressed.

8. At this point, the probe drops and the moving electrode begins to move through the reaction mixture. It is normal to hear a clicking sound.
9. When the instrument detects a fibrin strand, the electrode motion and the digital readout will stop. Record the time required for clot formation to the nearest 0.1 s.
10. Lift the probe arm to the up (rest) position. Wipe the electrode dry with a lint-free absorbent tissue such as a Kimwipes (Kimberly-Clark). Wash or clean the electrodes with distilled water from a wash bottle, then dry them with a lint-free tissue. We find the best precision is obtained by using a 1% phosphoric acid (10 g/L) wash followed by a distilled water rinse. Although other acids reportedly may pit the electrodes, Evaluator P. F. A. R. successfully uses a 0.1 mol/L HCl wash after each sample. In any case, clean the electrode after each sample, and be sure it is free of lint and all contamination, for proper fibrin strand detection.
11. Set the timer to zero by firmly pressing the red reset button for at least 1 s.
12. Repeat steps 6 through 9 for the prewarmed normal control plasma with another reagent cup and record both values. Results should agree within 4% of the lower figure *(19)* and should be within 2 SD of the laboratory mean for that control.
13. In a similar manner, test the patient's prewarmed sample in duplicate, following steps 6 through 11. Record both values and report the average result. Duplicate results should agree within 4% of the lower value *(19)*.
14. Periodically assay the abnormal control in duplicate in a similar manner and record both results. Results should be within 2 SD of the laboratory mean for that control.

Activated Partial Thromboplastin Time Test Procedure

Tilt-tube technique

1. Keep all plasma samples stoppered on melting ice until they are ready to test.
2. Equilibrate about 1 mL of 20 or 25 mmol/L (see above) calcium chloride solution in a 13 × 100 mm culture tube in the 37.5 °C water bath for at least 5 min, but not more than 60 min.
3. Equilibrate about 0.5 mL of APTT reagent for 5 min in a 13 × 100 mm culture tube in the 37.5 °C water bath; do not exceed the manufacturer's instructions for 37.5 °C storage.
4. Mix the APTT reagent gently, then pipet 0.1 mL into a clean 13 × 100 mm culture tube in the 37.5 °C water bath. Pipet 0.1 mL of normal control into the same test tube, swirl gently to mix, and incubate the mixture to activate for *exactly* 5 min at 37.5 °C.

Note: Evaluators P. F. A. R. and C. S. activate for exactly 3 min, as recommended by the DADE package insert.

5. After exactly 5 min, pipet 0.1 mL of prewarmed $CaCl_2$ solution into the mixture, simultaneously mix the reagents by tapping the tube gently, and start the stopwatch with the foot-pedal control.
6. After 30 s, remove the tube, wipe it dry, and tilt the tube back and forth gently about once a second, as in the PT test procedure (see also the *Note* accompanying step 6 of the PT test tilt-tube procedure).
7. At the first appearance of the fibrin clot, stop the stopwatch and record the time to the nearest 0.1 s.
8. Repeat steps 4–7 for the normal control; results should agree within 2 s and should be within 2 SD of the laboratory mean for that control.
9. In a similar manner, test the patient's plasma in duplicate, and record both times to the nearest 0.1 s. Results should agree within 3 s. Report the average of the two values.
10. Periodically assay the abnormal control in duplicate and record both results. Results should be within 2 SD of the laboratory mean for that control.

Fibrometer technique

The Fibrometer operates at a preset temperature of 37.2 ± 0.5 °C. The temperature can not be adjusted by the operator. Six prewarming wells and a reaction well are built into the heating block portion of the instrument. The prewarming wells will heat room temperature liquids to 37.2 ± 0.5 °C within 3 min, and refrigerated liquids within 5 min. Turn the instrument on for at least 10 min before beginning step 1, to allow the heating block to reach proper operating temperature. A small indicator light, located to the left of the digital readout, will glow as long as the instrument is at the correct operating temperature.

Note: Verify and record the instrument temperature at least daily with a sensitive thermocouple or thermometer.

1. Raise the probe arm to the up (rest) position above the center (reaction) well in the Fibrometer.
2. Place two sample cups into two of the warming wells. Using a felt tip marker, label one cup "CaCl$_2$," and one cup "reagent."
3. Keep all the plasma samples in an ice-water bath until ready to test.
4. Store the control plasma according to the manufacturer's instructions.
5. Pipet about 0.5 mL of APTT reagent into the "reagent" cup, and about 0.5 mL of calcium chloride into the "CaCl$_2$" cup. Incubate refrigerated liquid for 5 min; do not exceed manufacturer's instructions for 37.5 °C storage and do not heat the CaCl$_2$ longer than 60 min.
6. Place a new sample cup into the center "reaction" well of the Fibrometer, and carefully pipet exactly 0.1 mL of prewarmed APTT reagent into the cup. Avoid splashing reagent onto the sides of the cup.
7. Carefully pipet exactly 0.1 mL of control plasma into the reagent cup, swirl gently to mix, and allow the mixture to activate for *exactly* 5 min.

Note: Evaluators C. S. and P. F. A. R., using DADE reagent, incubate for exactly 3 min, according to the manufacturer's package insert.

8. After precisely 5 min, carefully pipet with a clean disposable plastic tip exactly 0.1 mL of prewarmed calcium chloride solution into the cup; simultaneously press the timer bar.

Note: If you use the specially designed Fibrometer Automatic Pipette, ejection of the sample automatically triggers the instrument, and the timer bar need not be depressed.

9. The probe will drop and the moving electrode will begin to move through the reaction mixture. It is normal to hear a clicking sound.
10. When the instrument detects a fibrin strand, the electrode motion and the digital readout will stop. Record the time required for clot formation to the nearest 0.1 s.
11. Lift the probe arm to the up (rest) position.

The electrode must be cleaned after each sample, and must be free of lint and all contamination for proper fibrin strand detection. The comments about electrode cleaning for PT specimens (see above) also apply here.

12. Set the timer to zero by firmly pressing the red reset button for at least 1 s.
13. Repeat steps 6 through 12 for the control plasma in duplicate. Results should agree within 5%, and should be within 2 SD of the laboratory mean for that control.
14. Similarly, test the patient's sample in duplicate, following steps 6 through 12, and record both results. Duplicate values should agree within 5%. Report the average of the two results.
15. Periodically assay the abnormal control in duplicate in a similar manner. Results should be within 2 SD of the laboratory mean for that control.

Results and Discussion

Precision Studies

Clotting assays, whether manual or automated, are extremely technique-dependent. Minor changes in specimen collection and handling, reagents, incubation time, temperature, or technique can yield large variations in the clotting times observed. Standardization of all variables leads to consistent results. This is especially true for the tilt-tube technique, where observer bias and temperature changes can be the greatest. Expected day-to-day and within-day precision data for PT results are shown in Table 1, and for APTT results in Table 2. To attain consistent day-

Table 1. Prothrombin Time Precision[a]

	Fibrometer CV,%		Tilt-tube CV,%	
Evaluator	Normal	Above-normal	Normal	Above-normal
		Within-run		
W. E. H	1.1	2.7		
P. F. A. R.	3.4	2.1	4.2	2.1
H. G. S.	3.6			
C. S.	1.0		3.6	
L. S.	3.2		5.0	
		Between-day		
W. E. H.	2.1	3.9		
P. F. A. R.	2.5	3.7	2.7	2.8
H. G. S.	5.4			
C. S.	4.1		6.2	
L. S., controls	3.5	4.6, 5.4	7.7	7.0, 5.7
patients[b]	4.3, 3.5		9.1, 8.6	

[a] Coefficient of variation (CV) listed for commercial coagulation controls unless otherwise noted. Where two values are listed, two different abnormal controls were used.

[b] Physiological specimens were included by Evaluator L. S. to evaluate collection technique as well as assay precision. Blood was drawn daily from one man and one woman volunteer for 20 working days of the study, with the two-syringe technic and Vacutainer Tubes (Becton-Dickinson) containing sodium citrate. Paired values listed are M, then F.

Table 2. Activated Partial Thromboplastin Time Precision[a]

	Fibrometer CV,%		Tilt-tube CV,%	
Evaluator	Normal	Above-normal	Normal	Above-normal
	Within-run			
W. E. H.	3.7	4.6		
P. F. A. R.	3.6	2.6	1.7	2.0
H. G. S.	5.9			
C. S.	4.2		2.1	
L. S.	6.0		5.8	
	Between-day			
W. E. H.	3.7	4.9		
P. F. A. R.	4.8	2.4	4.2	3.1
H. G. S.	5.9			
C. S.	3.8		6.8	
L. S., controls	8.1	15.5	3.6	11.9
patients[b]	7.1, 7.6		3.7, 5.1	

See Table 1 footnotes.

to-day precision, check all pipettes periodically for proper volume delivery, and have all technologists use the same set of pipettes. Change pipette tips after the addition of each reagent or test sample. Check and record the temperature of the Fibrometer or heating bath at least daily; deviations of test temperature to either side of 37.5 °C will prolong the coagulation times (19). Check the preincubation time needed to bring the individual reagents to 37.5 °C, and ensure that this temperature is maintained during the test. Verify the precision and accuracy of the stopwatch or timer periodically. Assay duplicate controls or patients' samples as random-blind samples to check operator bias. Maintain records for all routine controls, both normal and abnormal; record lot numbers for all reagents and blood-collection tubes as they are placed in use. Where possible, enroll in an external quality-control or proficiency-testing program to check your performance independently.

Owing to the great interlaboratory variations in PT and APTT results, many have proposed the standardization of test times. In the United States, reference plasmas have been advocated for this purpose (7, 20); in Europe, reference thromboplastins have been used (21–23). One or both of these approaches eventually may be successful in producing interlaboratory standardization to attain the best monitoring of anticoagulant therapy. Until such interlaboratory standardization is achieved, intralaboratory standardization can be addressed. Within each laboratory or institution, follow a single detailed procedure for each test on all days and on all work shifts. Use semiautomated (24) or automated coagulation instruments (25) instead of manual procedures whenever possible. However, if several automated instruments are available, do not interchange results from mechanical and photo-optical instruments.

Recent evidence suggests that precision for APTT testing may be related more to instrumentation than to the source of the partial thromboplastin (15). Also, different contact activators (Celite, kaolin, ellagic acid, etc.) produce significant variability in APTT results determined on the same plasma (14, 26, 27).

Interpretation of Results

Report PT and APTT results by giving the patient's "time" and the control time in seconds (to the nearest 0.1 s). Do not use dilution curves, percentile figures, or ratio of patient's result to control result (20), even though clinicians may use ratios in monitoring anticoagulant therapy. Establish a reference range for your laboratory for normal patients (males and females) as well as control ranges for normal and abnormal controls. If you use a commercial normal control, it is likely that the control range might be narrower than the reference range for normal patients. This resulting increase in precision reflects improved commercial standardization of control plasmas. Most clinical laboratories might expect to see reference ranges of 10 to 14 s for the PT and 30 to 45 s for the APTT in patients' samples.

When the PT and APTT are used as screening tests for bleeding disorders, both will be abnormally long for deficiencies of Factor X, V, prothrombin, or fibrinogen. An abnormal APTT with a normal PT indicates an intrinsic system deficiency (Factor XII, XI, IX, or VIII). An abnormal PT with a normal APTT indicates a Factor VII deficiency. In most laboratories, screening tests for suspected or actual bleeding problems will not be limited to coagulation tests, but will also include tests of vascular fragility and platelet function.

When the PT is used to monitor anticoagulant therapy with coumarin-type drugs (Coumadin, Panwarfin), the results are still reported in seconds, but the clinician uses a result 1.5- to 2.5-fold the patient's baseline PT (or 1.5- to 2.5-fold the normal control) as a commonly accepted therapeutic range. However, recent studies show that the institution of such a therapeutic range, while resulting in an excellent antithrombotic effect, also yields a high incidence of bleeding (28). Thus a lower therapeutic range may be indicated (1.5- to 2.0-fold the normal control). Differences in reagent composition also markedly affect the therapeutic range target value (20). Because of these differences and the substantial differences in technologists in the manual determination of PT, the Reviewer suggests that reporting data as a ratio of patient's result to control result should be given stronger emphasis. If this is done, the Submitter suggests the ratio should be given in addition to (and not instead of) the raw data in seconds. In any case, when monitoring oral anticoagulant therapy, note that coumarin-type drugs inhibit the utilization of vitamin K by the liver and therefore decrease the synthesis of vitamin-K-dependent factors (X, IX, VIII, and prothrombin). The peak effect on PT results occurs in 36–72 h, with full therapeutic effect requiring five to seven days of coumarin therapy *(29)*. After discontinuation of coumarin therapy, duration of effect lasts four to five days until vitamin-K-dependent factors are synthesized again *(29)*.

When the APTT is used to monitor heparin anticoagulation therapy, a result 1.5- to 2.5-fold the patient's baseline APTT (or 1.5- to 2.5-fold the normal control) is used as a commonly accepted therapeutic range. However, considerable differences from this guideline may exist in individual laboratories, because of differences in reagents used for APTT tests *(14)* and variable responses of the same APTT reagent to different types of heparin salts. A recent report confirmed that APTT results on a plasma specimen fortified with the sodium salt of heparin are consistently greater than those obtained on the specimen containing the calcium salt of heparin, even when both are present in the same concentrations (0.2 USP units/mL) *(30)*. In addition, a prolongation of the APTT depends on not only an adequate dose of heparin, but also an adequate functional plasma antithrombin III concentration. The only anticoagulant role of heparin appears to be the catalysis of the inactivation of thrombin by antithrombin III. In the absence of heparin, antithrombin III inactivates thrombin and Factors XI, IX, XIII, and X by a very slow process; in the presence of heparin, this reaction is markedly accelerated. If antithrombin III is decreased, a greater amount of heparin will be required to achieve the same antithrombotic effect that is achieved with less heparin in the presence of normal antithrombin III concentrations *(31)*.

The pharmacological half-life of heparin averages 90 min, but can range from 30 to 360 min and appears to increase with increasing dose *(29)*. Heparin preparations currently available are very heterogeneous, with only 25 to 35% of the bulk mucopolysaccharide being capable of catalyzing antithrombin III activity *(32)*. About 80% of heparin is metabolized in the liver and 20% excreted by the kidney.

For the APTT, a commonly accepted time for optimum monitoring of heparin is 1 h before the next heparin dose, when patients are receiving intermittent intravenous heparin bolus therapy. This "1 h before dose" regimen was instituted when heparin was given primarily by intravenous injection every 4 to 6 h. Many changes in sampling times now are occurring, and physicians often use constant infusion of heparin rather than the intermittent bolus. For prevention of postoperative venous thrombosis and pulmonary embolism, subcutaneous or "mini-dose" heparin therapy also may be used in certain patients. Clinical laboratories soon may perform direct plasma heparin and antithrombin III analyses to monitor these patients, because the APTT is insensitive to this low concentration of plasma heparin.

Some special precautions must be followed when an APTT is used for heparin monitoring. Once the specimen is drawn, it should be centrifuged at a high enough speed and long enough to render the plasma as platelet-free as possible *(33)*. Failure to use platelet-poor plasma can result in release of platelet factor 4 (a heparin-neutralizing factor), which can markedly alter APTT results for heparinized patients. Huseby and Shafer *(33)* suggest examination of plasmas for actual platelet count to determine optimum centrifugation conditions in each laboratory. Unusual APTT results in response to heparin therapy may be due to incorrect dose or concentration of administered heparin, in vivo platelet factor 4 release, decreased plasma antithrombin III values, or laboratory error *(34)*.

References

1. Morawitz, P., Die Chemie der Blutgerinnung. *Ergeb. Physiol.* **4**, 307–422 (1905).
2. Owen, C. A., Jr., Bowie, E. J. W., and Thompson, J. H., Jr., *The Diagnosis of Bleeding Disorders*, 2nd ed., Little, Brown and Co., Boston, MA, 1975.
3. Tullis, J. L., *Clot*, Charles C Thomas, Springfield, IL, 1976.
4. Macfarlane, R. G., The theory of blood coagulation. In *Human Blood Coagulation Haemostasis and Thrombosis*, R. Biggs, Ed., Blackwell Scientific Publ., Oxford, England, 1976, pp 1–31.
5. Davie, E. W., and Ratnoff, O. D., Waterfall sequence for intrinsic blood clotting. *Science* **145**, 1310–1313 (1964).
6. Macfarlane, R. G., An enzyme cascade in blood clotting mechanism and its function as a biochemical amplifier. *Nature (London)* **202**, 498–499 (1964).
7. Koepke, J. A., Gilmer, P. R., Triplett, D. A., and O'Sullivan, M. B., The prediction of prothrombin time system performance using secondary standards. *Am. J. Clin. Pathol.* **68**, Suppl. 1, 191–194 (1977).
8. Ingram, G. I. C., and Hills, M., Reference methods for the one-stage prothrombin time test on human blood; International Committee for Standardization in Hematology. *Thromb. Haemostas.* **36**, 237–238 (1976).
9. *The Fibro System® Manual*, 2nd ed., BBL, Division of Becton, Dickinson and Co., Cockeysville, MD, 1977.
10. Ingram, G. I. C., and Hills, M., The prothrombin time test; effect of varying citrate concentration. *Thromb. Haemostas.* **36**, 230–236 (1976).
11. Zucker, S., Brosious, E., and Cooper, G. R., One-stage prothrombin time survey. *Am. J. Clin. Pathol.* **53**, 340–347 (1970).
12. Bowyer, F. P., Stratos, M., Goldenfarb, P., and Hall, E. C., Reproducibility in the hematology laboratory: One-stage prothrombin times. *Am. J. Clin. Pathol.* **57**, 482–486 (1972).
13. Singer, J. W., and Sibley, C. A., Sensitivity of commercial thromboplastin to factor VII. *Am. J. Clin. Pathol.* **59**, 755–759 (1973).
14. Shapiro, G. A., Huntzinger, S. W., and Wilson, J. E., Variation among commercial activated partial thromboplastin time reagents in response to heparin. *Am. J. Clin. Pathol.* **67**, 477–480 (1977).

15. Triplett, D. A., Harms, C. S., and Koepke, J. A., The effect of heparin on the activated partial thromboplastin time. *Am. J. Clin. Pathol.* **70,** 556–559 (1978).
16. Barrington, J. D., and Peterson, E. W., The laboratory control of anticoagulation therapy: The one-stage prothrombin time-quality control in coagulation procedures. *Am. J. Med. Technol.* **33,** 296–305 (1967).
17. Lenahan, J. G., Frye, S., Jr., and Phillips, G. E., Use of the activated partial thromboplastin time in the control of heparin administration. *Clin. Chem.* **12,** 263–273 (1966).
18. Gjonnaess, H., Cold promoted activation of factor VII. I. Evidence for the existence of an activator. *Thromb. Diathes. Haemorrh.* **28,** 155–168 (1972).
19. Miale, J. B., and Kent, J. W., Standardization of the technique for the prothrombin time test. *Lab. Med.* **10,** 612–615 (1979).
20. Miale, J. B., and Kent, J. W., Standardization of the therapeutic range for oral anticoagulants based on standard reference plasmas. *Am. J. Clin. Pathol.* **57,** 80–88 (1972).
21. Leck, I., Thomson, J. M., and Poller, L., Quality control trials in the national reference thromboplastin scheme. *Br. J. Haematol.* **25,** 453–460 (1973).
22. Loelinger, E. A., van Dijk-wierda, C. A., and Roos, J., Quality control and standardization of the prothrombin time. Second International Symposium on Quality Control, Tokyo, Japan, June 1976.
23. The International Committee on Thrombosis and Haemostasis/The International Committee for Standardization in Hematology, Prothrombin time standardization: Report of the expert panel on oral anticoagulant control. *Thromb. Haemostas.* **42,** 1073–1131 (1979).
24. Beckala, H. R., Leavelle, D. E., and Didisheim, P., A comparison of five manually operated coagulation instruments. *Am. J. Clin. Pathol.* **70,** 71–75 (1978).
25. Klee, G. G., Didisheim, P., Johnson, R. J., and Giere, G. M., An evaluation of four automated coagulation instruments. *Am. J. Clin. Pathol.* **70,** 646–654 (1978).
26. Joist, J. H., Cowan, J. F., and Khan, M., Rapid loss of factor XII and XI activity in ellagic acid-activated normal plasma: Role of plasma inhibitors and implications for automated partial thromboplastin time recording. *J. Lab. Clin. Med.* **90,** 1054–1065 (1977).
27. Hathaway, W. E., Assmus, S. L., Montgomery, R. R., and Dubansky, A. S., Activated partial thromboplastin time and minor coagulopathies. *Am. J. Clin. Pathol.* **71,** 22–25 (1979).
28. Loelinger, E. A., The optimal therapeutic range in oral anticoagulation; history and proposal. *Thromb. Haemostas.* **42,** 1141–1152 (1979).
29. Watanabe, A. S., Anticoagulants. In *Handbook of Clinical Drug Data*, 4th ed., J. E. Knoben, P. O. Anderson, and A. S. Watanabe, Eds., Drug Intelligence Publ., Inc., Hamilton, IL, 1978, pp 291–294.
30. Banez, E. I., Triplett, D. A., and Koepke, J., Laboratory monitoring of heparin therapy—the effect of different salts of heparin on the activated partial thromboplastin time. *Am. J. Clin. Pathol.* **74** (Suppl.), 569–574 (1980).
31. Wessler, S., and Gitel, S. N., Heparin: New concepts relevant to clinical use. *Blood* **53,** 525–544 (1979).
32. Rosenberg, R. D., Heparin, antithrombin and abnormal clotting. *Ann. Rev. Med.* **29,** 367–378 (1978).
33. Huseby, R. M., and Shafer, D., Variables and questions related to a recent comparative heparin response evaluation. *Am. J. Clin. Pathol.* **69,** 99–100 (1978). Letter.
34. Soloway, H. B., Inappropriate response to heparin. *Diag. Med.* **2** (Sept/Oct), 31–33 (1979).

Salicylate

Submitter: Patricia E. Garrett, *Harvard Medical School, Boston, MA 02115*

Evaluators: John H. Kennedy, *Department of Clinical Chemistry, Queen Elizabeth Medical Centre, Birmingham B15 2TH, U.K.*

Marion Littleman, *Clinical Laboratory, G. N. Wilcox Memorial Hospital, Lihue, Kauai, HI 96766*

John Southgate and Charles A. Pennock, *Pediatric Chemical Pathology, Bristol Maternity Hospital, University of Bristol, Bristol BS2 8EG, U.K.*

Reviewer: Henry C. Nipper, *Laboratory Service, Veterans Administration Hospital, Baltimore, MD 21218*

Introduction

Worldwide use and abuse of salicylate as acetylsalicylate (aspirin) or of its derivatives probably exceed that of any other analgesic. Measurement of salicylate concentration in plasma and serum is useful in establishing and monitoring therapeutic dosage and in detecting and managing overdose or poisoning.

Several sophisticated methods are available for quantitating salicylate in body fluids, but the most commonly used is the simplest: the colorimetric Trinder method (1). Earlier workers, notably Brodie et al. (2) and Keller (3), measured the colored complex formed when salicylic acid is reacted with ferric ion in acid medium. Trinder used the same reaction but eliminated the need for extraction and a serum blank; protein precipitation and color formation were combined in a single step that minimized problems with nonspecific color formation. Micro and ultramicro modifications of the Trinder method have been devised by Caraway (4) and MacDonald et al. (5) for use with small volumes of sample, particularly for pediatric purposes, and were presented in volume 5 of this series (6).

Principle

In the procedures cited above, the principal reaction is the complexation of salicylic acid with ferric ion in acid medium to give a purple chromophore of undetermined structure. The reaction is not entirely specific for salicylic acid, but usual serum components do not interfere appreciably; the necessity for a serum blank is eliminated by incorporating mercuric salts in the reagent for the precipitation of protein. The color intensity is measured spectrophotometrically at 540 nm.

Materials and Methods

Reagents

1. *Trinder reagent:* 0.08 mol/L $HgCl_2$ and 0.10 mol/L $Fe(NO_3)_3$ in 0.12 mol/L HCl. Dissolve with stirring 20 g of mercuric chloride ($HgCl_2$; 0.04 mol) in 200 mL of hot distilled water, allow the solution to cool, and add 20 g of ferric nitrate [$Fe(NO_3)_3 \cdot 9H_2O$; 0.05 mol] and 60 mL of 1 mol/L hydrochloric acid (HCl; 0.06 mol) with stirring. Transfer the solution quantitatively to a 500-mL volumetric flask and dilute to the mark with distilled water. This solution is light yellow, and stable indefinitely.

Note: Evaluator M. L. recommends wearing a mask while weighing the mercuric chloride because of its extreme toxicity; the final reagent should be labeled: **Poison—do not pipet.**

2. *Stock salicylate standard.* Dissolve 116 mg of sodium salicylate (Aldrich Chemical Co., Inc., Milwaukee, WI 53233; 0.725 mmol) in distilled water (50 mL) in a 100-mL volumetric flask, add a drop of chloroform, and dilute to the mark with distilled water. The resulting solution (1 g of salicylic acid per liter) is stable for about six months when stored at 4–10 °C.

3. *Working salicylate standards.* Prepare working standards ranging in concentration from 1 to 50 mg/dL according to the following table by diluting with distilled water the indicated amount of stock standard to 50 mL in a volumetric flask.

Concn of final working standard, mg/dL	Vol of stock standard required, mL
1.0	0.5
5.0	2.5
10.0	5.0
20.0	10.0
30.0	15.0
50.0	25.0

Keep working and stock standard solutions in screw-cap bottles and refrigerate them except during use, to minimize evaporation and bacterial growth. Discard them if mold appears.

4. *Control.* Make a serum control by pooling specimens with high concentrations or by adding sodium salicylate to pooled serum; divide into aliquots and freeze. Do not use salicylic acid because it is insoluble. Use the conversion factor, 0.86,[1] to calculate the concentration of salicylic acid present after adding the sodium salicylate to serum. For example, suppose an addition of 100 mg of salicylic acid is desired: $0.86x = 100$; $x = 116$. Add 116 mg of sodium salicylate to the serum.

Collection and Handling of Specimens

The test may also be performed on spinal fluid, urine, or other fluids, but interpret results on these specimens, particularly urine, with caution because the presence of major salicylate metabolites such as the glucuronides will be undetected.

Plasma or serum may be used, although Evaluators J. S. and C. A. P. report a higher blank for plasma than for serum. Sodium ethylenediaminetetraacetate or heparin should be used as anticoagulant; oxalate interferes with the color formation. Specimens not analyzed immediately should be stored frozen. Do not add azide ion as preservative to either samples or standards; it forms a chromophore with the reagent that absorbs appreciably at 540 nm *(7)*.

Procedure

1. Label an appropriate number of tubes (to be centrifuged): reagent blank, Std. 1, Std. 5, Std. 10, Std. 20, Std. 30, Std. 50, Control, patient 1, patient 2, etc.
2. Place in each tube 2.5 mL of Trinder reagent.
3. To the reagent blank add 0.50 mL of distilled water. To each of the other tubes add 0.50 mL of the appropriate fluid. Shake or vortex-mix each tube without delay after addition of the sample.
4. Allow mixtures to stand at room temperature for approximately 10 min.
5. Centrifuge for 5–10 min at whatever speed is necessary to ensure good packing of the protein precipitate.
6. Carefully transfer the supernate to a cuvet with a 1-cm pathlength, and read the absorbance at 540 nm, setting the initial absorbance of the reagent blank at 0.
7. Plot absorbance *(y)* vs concentration *(x)* of the standards on regular graph paper. The result should be a straight line that passes through zero.
8. Read control and patients' sample values from the standard curve. Any result exceeding 30 mg/dL is considered to be in the toxic range and should be reported without delay.

---

[1] Relative molecular mass (M_r) of salicylic acid/M_r of sodium salicylate $= 138/160 = 0.86$.

Quality Control and Standardization

If the standards and samples are read under similar conditions with the same spectrophotometer for each run, it may not be necessary to rerun all standards for each set of patients' samples. One standard and one control specimen, the absorbances of which fall within specified limits (usually 2 SD about the mean for 20 determinations), will suffice to assure accuracy and precision for routine work. However, the entire group of standards should be rerun at predetermined regular intervals and whenever adjustments are made to the spectrophotometer that could affect its linearity. The data generated when the new curve is prepared should be recorded for ready reference and documentation.

Discussion

Acetylsalicylic acid (aspirin) is the salicylate derivative most commonly ingested, but salicylamide and methyl salicylate also have analgesic and antipyretic properties, and are often used. Esters of salicylic acid such as acetylsalicylic acid and methyl salicylate are rapidly hydrolyzed to the free acid after absorption through the gastrointestinal lumen *(8)*, so that the species being measured in serum or plasma by the Trinder method is salicylic acid or salicylate ion. Acetylsalicylic acid, in fact, does not react to give a purple color with acidified ferric ion, but because the half-life of this species is only approximately 15 min in the circulation *(9)*, its measurement is usually less important than that of salicylate ion, the principal active species. Salicylamide is not hydrolyzed in vivo, but is rapidly removed from the circulation both by absorption into tissue and by conversion to the sulfate or glucuronide derivatives, which are then excreted in urine along with some unchanged salicylamide and small amounts of other derivatives *(10)*. Thus, although salicylamide does complex with ferric ion under the reaction conditions to give the same molar absorptivity as sodium salicylate (P. E. Garrett, unpublished data), its measurement is of limited clinical usefulness.

The major metabolites of salicylate in humans are

acetylsalicylic acid   salicylic acid   salicylamide

salicyluric acid   acyl glucuronide   phenolic glucuronide salicyluric acid, formed by conjugation with glycine, and acyl and phenolic glucuronides. Of these, only salicyluric acid reacts with ferric ion to give the characteristic purple color, but because of rapid excretion, its concentration in serum or plasma is normally very low *(9)*.

The attractiveness of the Trinder procedure lies in its simplicity and economy. In addition, its accuracy has been established by comparison with extraction *(2)* and chromatographic *(11)* procedures. Precision is only minimally dependent on technique (mainly pipetting technique and the separation of the supernatant fluid from the precipitate). In the Submitter's hands, the coefficient of variation (CV) for frozen aliquots from a salicylate-supplemented serum pool, determined 20 times over as many days, was 7% at 20 mg/dL and 8% at 10 mg/dL; the CV was 17% for unsupplemented (blank) serum, with values ranging from 1.3 to 2.8 mg/dL (this serum was obtained from an outdated unit of donor blood, and thus was considered to be salicylate-free). CVs for plasma were slightly higher in these experiments. Recovery is nearly 100% with both serum and plasma. Reproducibility and recovery in the Evaluators' laboratories are given in Table 1. The reproducibility in the Evaluators' laboratory is apparently superior to that of the Submitter's, probably because of pipetting technique (according to the Submitter).

Table 1. Reproducibility and Analytical Recovery Studies

Precision studies		Recovery studies	
Mean (and SD), mg/dL	CV,%	Calculated concn, mg/dL	% recovery
Within-batch		5.9	93.4
8.28 (0.12)	1.4	11.6	94.3
31.95 (0.24)	0.8	17.0	99.1
		22.2	103.1
		34.0	107.0
Day-to-day		20	100
8.28 (0.15)	1.8	40	103
31.92 (0.73)	2.2		
28.2 (1.51)	5.4		
20	4		
40	8		

Interfering substances include acetoacetate, *p*-aminobenzoic acid, furosemide, and oxalate. Acetoacetate and oxalate complex with ferric ion under the reaction conditions to give a chromophore similar to salicylate, thus leading to falsely increased values. *p*-Aminobenzoic acid and furosemide, by acting in vivo to retard salicylate clearance, cause higher concentrations of salicylate in serum. Corticosteroids decrease serum salicylate values through an in vivo effect. Bilirubin, glucose, heparin, phenols, and urea have no measurable effect *(12)*.

The usual therapeutic dosage is 600 mg (10 grains), causing a maximum serum concentration of about 5 mg/dL. Concentrations as great as 20 mg/dL in serum are sometimes sought to treat refractory pain and inflammation. This can require dosages of 100 mg/kg body weight, and can cause tinnitus (ringing in the ears), usually the first clinical sign of toxicity. Clear toxicity occurs at serum salicylate concentrations of 30 mg/dL or more *(13)*.

A nomogram has been published for estimating the peak salicylate value when both the hours since ingestion and the current salicylate concentration are known *(14)*.

Caraway's micro procedure *(4)* involves a 1 to 10 (10-fold) dilution of the serum (0.10 mL) with water, and addition of diluted sample (1.0 mL) to an equal volume of Trinder reagent. MacDonald's ultramicro procedure *(5)* requires 25 μL of serum and 500 μL of Trinder reagent diluted with water (1/1 by vol). Each of these requires a cuvet suitable for the volume of supernatant fluid after centrifugation, and each presumably sacrifices some sensitivity to allow measurements on smaller samples. The procedure described here, incorporating the appropriate dilutions, may be used for these modifications.

Mercuric salts are toxic, and the reagent and wastes generated in these determinations should be treated with precautions appropriate to the handling of a nonvolatile toxic substance. Liquid waste should be washed down the drain with an excess of water, and solids should be buried in a place designated for the disposal of toxic wastes.

Note: Evaluators J. S. and C. A. P. state that, in their experience, the time since ingestion of salicylate is frequently unknown in patients who present with toxic symptoms; therefore, the nomogram described by Done *(14)* for estimation of the theoretical peak salicylate concentration is not particularly useful. It is much more important to recognize that a value for blood salicylate within the apparent therapeutic range in a patient who has taken an overdose may be misleading; in their laboratory they usually recommend that another blood sample be taken about 1–2 h after the initial test to determine whether the plasma salicylate concentration is increasing at a significant rate. This occurs particularly in patients who have received alkaline gastric lavage, which promotes rapid absorption of the salicylate bolus in the stomach; some patients have had salicylate concentrations of 20 mg/dL increase to between 80 and 120 mg/dL within 1 to 2 h.

References

1. Trinder, P., Rapid determination of salicylate in biological materials. *Biochem. J.* **57**, 301–303 (1954).
2. Brodie, B. B., Udenfriend, S., and Coburn, A. F., The determination of salicylic acid in plasma. *J. Pharmacol. Exp. Ther.* **80**, 114–117 (1944).
3. Keller, W. J., Jr., A rapid method for the determination of salicylates in serum or plasma. *Am. J. Clin. Pathol.* **17**, 415–418 (1947).
4. Caraway, W. T., *Microchemical Methods for Blood*

Analysis. Charles C Thomas, Springfield, IL, 1960, pp 105–106.
5. MacDonald, R. P., Ploompuu, J., and Knights, E. M., Jr., Ultramicro determination of salicylates in blood serum. *Pediatrics* **20**, 515–516 (1957).
6. MacDonald, R. P., Salicylate. *Stand. Methods Clin. Chem.* **5**, 237–243 (1965).
7. Frings, C. S., Effect of sodium azide on Trinder's method for the determination of salicylate. *Clin. Chim. Acta* **45**, 307–308 (1973).
8. Smith, J. J., and Smith, P. K., Eds., *The Salicylates*. Interscience, New York, NY, 1966, Chapters 1 and 2.
9. Davison, C., Salicylate metabolism in man. *Ann. N. Y. Acad. Sci.* **179**, 249–258 (1971).
10. Levy, G., and Matsuzawa, T., Pharmacokinetics of salicylamide elimination in man. *J. Pharmacol. Exp. Ther.* **156**, 285–293 (1966).
11. Rance, M. J., Jordan, B. J., and Nichols, J. D., A simultaneous determination of acetylsalicylic acid, salicylic acid and salicylamide in plasma by gas liquid chromatography. *Pharm. Pharmacol.* **27**, 425–429 (1975).
12. Young, D. S., Pestaner, L. C., and Gibberman, V., Effects of drugs on clinical laboratory tests. *Clin. Chem.* **21**, 360D (1975), and references contained therein.
13. Sunshine, I., *CRC Manual of Analytical Toxicology*. Chemical Rubber Co., Cleveland, OH, 1965, p 308.
14. Done, A. K., Salicylate intoxication: Significance of measurements of salicylate in blood in cases of acute ingestion. *Pediatrics* **26**, 800–807 (1960).

Sodium and Potassium

Submitter: Willie L. Ruff, *Clinical Laboratories, Howard University Hospital, Washington, DC 20060*

Evaluators: Ronald J. Byrnes, Aaron Sarafinas, Wayne Lambert, and Richard Brown, *Instrumentation Laboratory, Inc., Lexington, MA 02173*
Jack H. Edwards, *South Georgia Medical Center Laboratory, Valdosta, GA 31601*
Suzanne M. Garszczynski, *The Palmerton Hospital, Palmerton, PA 18071*
George Nix, *Wishard Memorial Hospital, Indianapolis, IN 46217*

Reviewer: Billy W. Perry, *Department of Pathology, The Medical College of Wisconsin, Milwaukee, WI 53226*

Introduction

Scientists have long known that when metals are heated to sufficiently high temperatures, they emit light of characteristic wavelengths. During the heating process, electronic transitions occur from relatively low, stable energy levels to higher energy levels. The transition of the electrons back to the low-energy states results in photoemissions that are characteristic for each element. By spectral isolation, e.g., with use of an interference filter, the various metals can be identified. With the advent of the atomizer for introducing solutions of metal into a flame, the technique of emission spectroscopy could be applied to the clinical laboratory via flame photometry. Richterich has traced the evolution of flame photometry from its inception to the present (1).

Sodium and potassium are determined in serum or plasma and urine in clinical laboratories on a routine or "stat" basis. Generally, though not always, sodium and potassium are ordered as part of an electrolyte panel (Na, K, Cl, CO_2). Values for the panel help in evaluating electrolyte imbalance and acid–base status. A primary use for sodium and potassium determinations is monitoring intravenous therapy, especially in children. Because of the role of sodium ion in osmotic pressure regulation, sodium balance is coupled to water balance (2); hypernatremia usually indicates water depletion, and hyponatremia refers more to water excess than to sodium depletion. Determining urinary sodium concentrations adds very little to the clinical picture or to the plasma findings. However, correlating urinary sodium with either aldosterone secretion or plasma renin values may be quite helpful.

Hyperkalemia, if severe, places a patient at risk for cardiac arrest. Hypokalemia may cause problems with neuromuscular transmission such as muscular weakness, hypotonia, and cardiac arrhythmias. Urinary potassium determinations are sometimes helpful in clinical diagnosis, e.g., in Conn's disease (3).

Ion-selective electrodes (4), atomic absorption spectrometry (4), and flame photometry (4, 5) are the most commonly used methods for the quantitative determination of sodium and potassium. Flame photometry is the most widely used of these. The instruments used for the data described in this chapter constitute "state-of-the-art" technology.

Principle

A serum or urine sample is usually diluted 50-, 100-, or 200-fold with a 15 mmol/L solution of LiCl. The diluted sample is nebulized into a spray chamber where it is mixed with air and a fuel, e.g., propane. Heavy droplets fall to the bottom of the chamber, but the sample aerosol is carried by the propane to a burner, where the mixture is ignited. The heat of the flame promotes electron excitation in some atoms of sodium, potassium, and lithium to a high-energy state that almost instantaneously reverts to a lower energy state. As the electrons return to the lower energy state, light is emitted at a frequency characteristic for the element. Sodium, potassium, and lithium atoms, when excited, characteristically emit light at the respective wavelengths of 591 nm (yellow flame), 768 nm (violet flame), and 671 nm (red flame). The emitted light passes through separate optical filters that permit passage of a narrow range of wavelengths. The filtered light is focused on a detector, the signal from the detector amplified, and the result displayed electronically. The ratio of the sodium or potassium signal to the lithium signal (serving as an internal standard) is used to calculate concentration.

Note: The IL 643 Flame Photometer (Instrumentation Laboratory, Lexington, MA 02173) involves use of a cesium internal standard, which not only works as well as lithium but also eliminates the switch-over to potassium as internal standard during lithium analysis.

For analysis of sodium and potassium by flame, dilute a serum or urine sample manually or with an automatic diluter. If manually, add a 50-μL sample to 5 mL of Li$^+$ diluent (15 mmol/L) and mix carefully. With the automatic diluter, aspirate a 50-μL sample from a sample cup and dispense into a mixing device with an appropriate volume of the Li$^+$ diluent. After the dilution and mixing, the sample is transported to a spray chamber, where it is nebulized and carried by air and fuel into the flame.

For standardization, place the instrument in the standardize mode. Dilute a 50-μL standard solution of 0 mmol/L Na$^+$ and 0 mmol/L K$^+$ with the Li$^+$ diluent either manually or with a diluter. Aspirate and burn the diluted sample in the flame. When the readout device stabilizes, adjust to zero with the zero knob. Then dilute 50 μL of 140 mmol/L Na$^+$ and 5 mmol/L K$^+$ standard with Li$^+$ diluent and aspirate into the flame. After stabilization, adjust the readout device to read 140/5 with the standard/calibrate knob. Dilute a serum or urine sample as described above and switch the flame photometer to an operate mode. Read the results from the display after the instrument has stabilized. Restandardize the instrument every 15 to 20 samples, or after it has been idle for 10 min or more.

Materials and Methods

A good flame photometer with or without a diluter is required.[1] The only solutions required are standards, controls, and lithium concentrate for dilutions.

Reagents

Lithium solutions: To prepare the stock solution, 150 mmol/L Li$_2$CO$_3$ (lithium carbonate), dissolve 11.8 g of Li$_2$CO$_3$ (previously dried at 110 °C and stored in a dessicator) in about 50 mL of water in a 1-L volumetric flask. Slowly add 100 mL of 3.5 mol/L HCl, avoiding foaming. Swirl gently until dissolved. Cool to room temperature and dilute the solution to 1 L with de-ionized water.

Prepare the working solution by diluting 100 mL of lithium stock solution to 1 L with de-ionized water; mix well. Store the working lithium solution (15 mmol/L) in a polyethylene or glass container of appropriate size (100–150 mL).

Standard solutions: Prepare sodium and potassium stock standards from National Bureau of Standards NaCl [Standard Reference Material (SRM) No. 919] and KCl (SRM No. 918). For a 1000 mmol/L NaCl solution, dissolve exactly 58.44 g of NaCl in de-ionized water and dilute to 1 L. Prepare a 100 mmol/L KCl solution by dissolving 7.46 g of KCl in de-ionized water; dilute to 1 L. Place 15 mL of the NaCl solution and 5 mL of the KCl solution into a 100-mL volumetric flask and dilute to 100 mL with de-ionized water. The final working solution contains 140 mmol of Na$^+$ and 5 mmol of K$^+$ per liter.

Alternatively, the working solution (140 mmol of Na$^+$ and 5 mmol of K$^+$ per liter) can be purchased directly from several companies, e.g., Instrumentation Laboratory, Inc.; Beckman Instruments, Inc., Fullerton, CA 92634; Technicon Instruments Corp., Tarrytown, NY 10591.

Controls: At least two serum controls (high and low ranges) are needed for each run of samples. Suggested ranges (mmol/L) for the controls are as follows: low control Na$^+$ 119–123, K$^+$ 2.8–3.3; high control Na$^+$ 149–159, K$^+$ 6.9–7.3. Use either in-house or commercially available controls. Reconstitute commercial controls strictly according to the manufacturer's instructions.

Apparatus

Use a good flame photometer. Several brands are commercially available. An automatic diluter is desirable, but manual dilutions can be used. You will also need a 50-μL micropipette, calibrated to ± 0.5% accuracy, and a 5-mL calibrated pipette for dispensing aliquots of lithium diluent.

Specimen Collection and Handling

Blood samples from adults, usually from the antecubital vein, are collected into suitable collection tubes. Using tubes containing an inert gel permits more thorough separation of cells and supernate after centrifugation, thus minimizing the risk of plasma/serum contamination with erythrocytes. Alternatively, serum separators such as Sur-Sep (General Diagnostics, Morris Plains, NJ 07950) may be added after clot formation but before centrifugation. Samples are allowed to stand for 15 to 30 min and then centrifuged for 15 min. For infants blood samples are collected by skin puncture of the heel or finger into microhematocrit tubes (1 to 1.2 mm i.d., 75 mm long) containing ammonium heparin or into heparinized polyethylene microcentrifuge tubes with caps *(6)*.

Serum must be separated from the cells promptly after centrifugation. Failure to do so may result in the diffusion of potassium from cells into the serum. Turbid samples, by virtue of their viscosity, may clog the flame photometer anywhere from the sample probe to the nebulizer.[2]

---

[1] The flame photometers used by the Evaluators are as follows: J. H. E., Beckman KLiNA Flame with DADE diluter; S. M. G., IL 343 with diluter; R. J. B., IL 643 with IL 144 diluter; G. N., IL 143 and Photovolt PVA-4.

[2] KLiNa Flame Technical Manual, Beckman Instruments, Inc., Fullerton, CA, 1972.

Note: Turbidity can be removed by high-speed centrifugation ($1.3 \times 10^5 \times g$).

Samples may be assayed promptly, or covered and refrigerated, or stored frozen for two weeks. These precautions reduce the risk of falsely high results related to sample evaporation.

Procedure

Sodium and potassium assay with an automatic diluter

Before starting a sample run, check the following:
1. Lithium diluent: is there sufficient volume?
2. Transfer tube (connects diluter and diluent): is it fully immersed in diluent?
3. Fuel (propane or methane): is there enough gas for the run?

Note: Natural gas from an in-house source may be used, if it is filtered to remove water condensate and particulate matter. Either glass wool or cotton is a satisfactory filter for this purpose. The gas pressure must be kept constant, to avoid erratic results.

After completing the pre-run checks, follow the operating procedure:
1. Turn on the air compressor and then the fuel. Ignite the air/fuel mixture. Ignition is often indicated by illumination of a panel light and a soft, characteristic sound of the burner. For greatest stability of operation before running samples, aspirate about 30 mL of lithium diluent directly into the spray chamber.
2. Select the correct potassium concentration readout mode, e.g., 0–20 (serum) or 0–200 (urine).
3. Aspirate de-ionized water or lithium diluent into the instrument. Permit three or four complete cycles (aspiration through readout) to occur. Then, use the appropriate knobs to adjust the sodium and potassium concentrations, respectively, to 000 and either 00.0 (serum) or 000 (urine).

Assay with a sample changer

Normally a diluter is used with the sample changer. If so, the sample probe must be adjusted to ensure that the tip enters the sample to a depth of approximately 6 mm (¼ in.) It is best to make the adjustment with de-ionized water in a 2.0-mL sample cup.

Load the sample tray as follows: standard, low control, high control, and 10 to 15 samples; repeat the sequence according to sample requirements.

With standard in the sample pick-up position, aspirate the standard and adjust the meter readings to 140 mmol of Na$^+$ and 5.0 mmol of K$^+$ per liter with the appropriate calibration control knobs. Verify the values for the 0 mmol/L Na$^+$ and 0 mmol/L K$^+$ control and the 140 mmol/L Na$^+$ and 5.0 mmol/L K$^+$ standard by repeating these procedures. Lock in the standard values by changing from calibrate mode to operate mode. Turn the sample changer on and begin the sampling process. Additional samples are automatically aspirated, mixed with diluent, and ignited in the burner; results are displayed by the readout device.

To end a run, switch the sample changer to the off position. If the instrument is to be used soon after a run, reload the sample changer with standard, low and high controls, and patients' samples, and begin the analysis process again as previously described. If no samples are to be run, place the instrument in standby mode.

Assays without a sample changer

If a sample changer is not used, the sample cups must be held in position manually to permit aspiration. Further, the sample cup must be held in position until the sample probe is completely withdrawn from it. Lowering the sample cup from the probe too quickly will cause erratic results.

Sodium and potassium assay without a diluter

For sodium and potassium without a diluter, manual dilutions must be made. Add a 25-µL sample to 2.5 mL of the Li$^+$ diluent (15 mmol/L) or add 50 µL of sample to 5 mL of diluent and mix carefully.

Note: If a diluter is an integral part of the flame photometer, it must be disconnected.

Zero adjustments, standardization, and sample determinations previously described are made by direct aspiration from hand-held cups into the spray chamber.

Quality Control

At least two controls should be run with each batch of samples. If only two controls are used, they should be in the low and high ranges for sodium and potassium. If three controls are used, they should be in the low, middle, and high ranges. Control values must be within a predetermined range, usually ± 2 SD of the mean. If the control is out of range, find the cause and correct it before accepting the result of any sample. Some possible causes include improper standardization, instrument malfunction, improper sample size, and improper reconstitution and handling of the control(s).

Calculations

No manual calculations are required. The instrument automatically calculates the values for the samples and controls.

Discussion

Studies were conducted independently in the laboratories of the Evaluators. Each laboratory analyzed

Table 1. Precision of the Method

Evaluator	n	Na Mean (and SD), mmol/L	CV, %	n	K Mean (and SD), mmol/L	CV, %
Within-day						
R. J. B., A. S., W. L., R. B.	20	143.57 (1.22)	0.85	20	4.30 (0.045)	1.04
J. H. E.	20	142.9 (0.7)	0.5	20	3.9 (0.1)	1.5
S. M. G.	20	140 (1.3)	0.9	20	4.7 (0.1)	2.1
G. N.	22	143.09 (1.15)	0.80	22	3.46 (0.16)	4.69
Day-to-day						
R. J. B., A. S., W. L., R. B.	40	129.9 (1.06)	0.81	40	4.56 (0.04)	0.81
S. M. G.	20	140 (1.0)	0.7	20	3.7 (0.1)	2.7

20 or more different patients' samples. The results were subjected to a basic statistical treatment to calculate mean, standard deviations, and coefficients of variation. Table 1 summarizes the precision data from the Evaluators.

Evaluators R. J. B., A. S., W. L., and R. B. reported data for accuracy and analytical recovery with the IL 643 flame photometer and the II 143 flame photometer with a 144 diluter. In the study, 157 paired samples were analyzed for sodium and potassium. The data from the study are given below:

	Na		K	
	IL 643	IL 143	IL 643	IL 143
Mean, mmol/L	138.5	139.2	4.10	4.09
Slope		1.00		1.01
y-intercept		−0.73		−0.01
r		0.971		0.991
Patients' values	121–152 mmol/L		3.0–5.8 mmol/L	

Without having a stringency requirement imposed, the results were reliable, accurate, and often precise. CVs of 0.5% or less for sodium and potassium doubtless are obtained only under the most stringent conditions (5). Proficiency testing samples from the Centers for Disease Control had CVs of 2.8% for sodium and 4.7% for potassium (5). The smaller CVs reported here may be attributed to improved flame photometers and proper sample handling. Probably the quality of today's instruments also is much greater than that of earlier models.

Hemolyzed and lipemic samples are not recommended for analysis. Hemolysis causes increased serum potassium readings because of leakage from erythrocytes, in which potassium concentration is about 20-fold that in plasma. Lipemic samples may obstruct the sample line or nebulizer, leading to complete line/nebulizer blockage or short sampling, resulting in falsely low values. In this study there was no chemical interference by lipemia in the results; however, it was occasionally necessary to clean the nebulizer.

Preventive Maintenance

A flame photometer performs best when it is properly maintained by a scheduled maintenance program. To ensure the best performance, follow the manufacturer's recommendations. Scheduled maintenance is preferred and recommended.

At the end of each day, or after running the equivalent of one day's operation, clean the sample lines by aspirating 30 to 50 mL of distilled water through the instrument. This facilitates the removal of loosely lodged particles and minimizes protein buildup.

Each week, remove the chimney, clean it with distilled water, and allow it to dry. Do not leave fingerprints on the chimney's optical surface!

Each month, clean the burner head and ignition block (if accessible and unless forbidden by the manufacturer) with a stiff brush; also remove and clean the air-compressor filter with soap and water.

Troubleshooting

If the instrument is not on after pressing the power switch, check to see that the power cord is connected and the fuse is not blown. If the instrument is on but ignition does not occur, check to see that:
1. The glass chimney is in place.
2. The chimney cover is dry and properly seated.
3. The spark electrode is not broken.
4. The fuel cylinder is not empty and the valve is open.
5. The compressor is running.

When the readout meters do not give values for sodium and potassium even though ignition has occurred:
1. Check the aspirator tube and clean with a stylet or stiff wire if necessary.
2. Check the lithium diluent and remake it as required.

3. Check the chimney alignment and change if needed.

If incorrect values are given for sodium and potassium:

1. Re-standardize the instrument with fresh standard(s).
2. Ensure that the chimney is cleaned, properly aligned, and positioned.
3. Check the aspiration rate and fuel/air ratio to ensure compliance with manufacturer's specifications.
4. Check the sample probe to verify its correct position in the sample cup.

Always check the operator's manual for detailed maintenance procedures and schedules.

Reference Intervals

The normal range for serum sodium in adults is 135 to 148 mmol/L *(4)*. Urinary sodium values are influenced markedly by diet and hence show wide variation. On an average diet, urinary sodium values range from 40–90 mmol/24 h to 43–217 mmol/24 h *(4)*. Potassium values range from 3.5 to 5.3 mmol/L in serum and 25 to 120 mmol/24 h in urine.

Evaluators R. J. B., A. S., W. L., and R. B. found the following serum values for a group of 50 adults (15 women and 35 men), ages 19 to 65: Na, 138.3–147.5 mmol/L; K, 3.71–4.95 mmol/L.

Evaluator S. M. G. derived the following ranges for serum values for a group of 20 apparently normal adults: Na, 134–145 mmol/L; K, 3.5–5.0 mmol/L.

Note: Evaluator S. M. G. uses the following "panic" values: Na, 128 and 155 mmol/L; K, 2.5 and 6.0 mmol/L.

References

1. Richterich, R., *Clinical Chemistry: Theory and Practice,* 2nd ed., translated by S. Raymond and J. H. Wilkinson, Academic Press, New York, NY, 1969, pp 121–124.
2. Zilva, J. F., and Pannall, P. R., *Clinical Chemistry in Diagnosis and Treatment,* Year Book Medical Publishers, Chicago, IL, 1975, pp 30–75.
3. Henry, J. B., and Krieg, A. F., Endocrine measurements. In *Clinical Diagnosis by Laboratory Methods,* 15th ed., I. Davidsohn and J. B. Henry, Eds., W. B. Saunders, Philadelphia, PA, 1974, pp 697–771.
4. Tietz, N. W., *Fundamentals of Clinical Chemistry,* 2nd ed., W. B. Saunders, Philadelphia, PA, 1976, pp 849–944.
5. Mason, W. B., Flame photometry. In *Clinical Chemistry: Principles and Technics,* 2nd ed., R. J. Henry et al., Eds., Harper & Row, Hagerstown, MD, 1974, pp 49–63.
6. Meites, S., and Levitt, M. J., Skin-puncture and blood-collecting techniques for infants. *Clin. Chem.* 25, 183–189 (1979).

Sweat Test for Cystic Fibrosis

Submitters: Keith B. Hammond and Brian J. Johnston, *Pediatric Microchemistry Laboratory, University of Colorado Health Sciences Center, Denver, CO 80262*

Evaluators: Melvin R. Glick, *Department of Clinical Pathology, University Hospital N-440, Indianapolis, IN 46223*

Gustavo Reynoso and William D. Slaunwhite, *Department of Pathology, Wilson Memorial Hospital, Johnson City, NY 13790*

Reviewer: Horace F. Martin, *Department of Pathology, Rhode Island Hospital, Providence, RI 02902*

Introduction

A positive sweat test, performed with pilocarpine iontophoresis, is necessary for the definitive diagnosis of cystic fibrosis (1). It is well documented that the sweat test is frequently poorly performed, leading to both false-positive and false-negative results (2, 3). Although it can be argued that the sweat test should be performed only in Cystic Fibrosis Centers, the cost of bringing patients from rural areas to these centers makes it desirable to have a reliable sweat test available in the smaller clinical laboratory.

A variety of techniques for testing sweat are described in the literature, one of which appeared in an earlier volume of this series (4). The sweat induction described here is by the classical Gibson–Cooke method (5), slightly modified in our laboratory. Either the sodium or the chloride content, or both, can be measured in collected sweat, although increases of these electrolytes in sweat appear to be due to a defect in *cation* transport (6). This is reinforced by the finding that certain anions can replace chloride in the sweat of patients with cystic fibrosis, which could lead to false-negative results (7). Because most laboratories rely on the measurement of sweat chloride, methods for both sodium and chloride will be described here.

Principle

Sweating is induced by forcing pilocarpine (a cholinomimetic alkaloid) into the sweat glands by iontophoresis. In this process a positive electrode is placed over a pad soaked with pilocarpine and a negative electrode is placed over a pad soaked with an electrolyte solution. A current is applied between the two electrodes for a fixed time. After the removal of the electrodes from the patient's skin, the sweat is collected onto a sodium- and chloride-free preweighed filter paper. The amount of sweat is calculated by reweighing the filter paper, and the sodium and chloride are leached out and quantified by standard analytical techniques.

Materials and Methods

Reagents

1. *Distilled or de-ionized water (Reagent water, Type I).*

2. *Pilocarpine nitrate, 5 g/L.* Dissolve 0.5 g of pilocarpine nitrate in water and dilute to 100 mL. Stored at room temperature, this solution is stable for at least one month.

Note: Evaluator M. R. G. finds the pilocarpine nitrate solution stable for at least one year when stored at 4 °C in a brown bottle.

3. *Potassium sulfate, 10 g/L.* Dissolve 1 g of K_2SO_4 in water and dilute to 100 mL. This solution is stable at room temperature indefinitely.

Note: Evaluator M. R. G. uses a sodium bicarbonate ($NaHCO_3$) solution. He comments that the choice of electrolyte should take into account the need to avoid the ion to be measured in the sweat, to preclude contamination from this source. If sodium is to be measured, then potassium sulfate is preferred to sodium bicarbonate.

4. *Lithium sulfate diluent, 15 mmol/L.* Dissolve 1.648 g of Li_2SO_4 in water and dilute to 1 L.

5. *Sodium chloride standard solution, 100 mmol/L.* Dissolve 5.85 g of NaCl in water and dilute to 1 L.

Apparatus

1. *Battery-operated power supply* (Model IPS-6 Gibson–Cooke Sweat Test Apparatus; Farrall Instruments, Grand Island, NE 68801).

Note: Evaluator M. R. G. comments that a suitable battery-operated power supply can be cheaply constructed from locally obtained components, given the information in the literature *(5, 8)*.

2. *Pliable copper electrodes*, 2.5 × 2.5 cm.

Note: Evaluator M. R. G. uses larger copper electrodes (approximately 4.0 × 4.0 cm). He comments that the use of chrome-plated or other coated copper should be avoided, as these substances may be driven into the skin by the current.

3. *Rubber electrocardiograph straps*, used to attach electrodes to the arm.

4. *Whatman No. 42 filter-paper circles*, 4.25-cm diameter (Whatman, Inc., Clifton, NJ 07014).

Note: Each new box of filter papers used must be checked to ensure that the papers are sodium- and chloride-free. This can be done by adding 4 mL of the lithium sulfate solution to a plastic vial containing a filter-paper circle, allowing to elute for 30 min, and assaying for sodium and (or) chloride as described below. A value of no more than 5 mmol/L should be obtained for either analyte.

5. *Plastic vials*, 16 mL, with caps (cat. no. 55–5; Thornton Plastics, Salt Lake City, UT 84104).

6. *Parafilm*, cut into 6.5 × 6.5 cm squares (American Can Corp., Neenah, WI 54956).

7. *Micropore surgical tape*.

8. *Flame photometer and (or) chloride-titrating apparatus*.

Note: The Submitters have used the KLiNa Flame Photometer (Beckman Instruments, Inc., Fullerton, CA 92634) for sodium determinations. Evaluators W. D. S. and G. R. have used the IL 143 and 343 flame photometers (Instrumentation Laboratory, Inc., Lexington, MA 02173), which consistently demonstrated linearity throughout the 0–200 meq/L range. They recommend the preparation of 50, 100, and 200 mmol/L sodium standards to verify the analytically acceptable range. For chloride determination Evaluator M. R. G. uses a coulometric–amperometric method as described by Cotlove (see also the chloride chapter by Dietz and Bond in this text), although a titrimetric method is equally satisfactory *(9)*.

Procedure

Iontophoresis

1. Using clean forceps, place a filter-paper circle in a capped, 16-mL plastic vial. Weigh on an analytical balance to within 0.1 mg and record the weight.

2. Position the copper electrodes on a rubber electrocardiograph strap so that the positive electrode can be placed on the volar surface of the forearm and the negative electrode on the posterior surface.

3. Wash the forearm with water and dry thoroughly with a soft paper tissue.

Note: Evaluator M. R. G. prefers to use Curity gauze squares (10 × 10 cm) rather than paper tissues. He finds that Curity gauze (distributed by Colgate-Palmolive Co., New York, NY 10022) is consistently free of sodium and chloride but recommends checking each freshly opened package.

4. Thoroughly moisten a filter-paper circle with the pilocarpine nitrate solution and position the paper on the volar surface of the forearm with the positive electrode over it.

5. Similarly, thoroughly moisten a filter-paper circle with the potassium sulfate solution and place on the posterior surface of the forearm with the negative electrode over it. Secure both electrodes with the rubber strap. Check that no part of the electrodes makes direct contact with the skin and that the strap is tight.

Note: Evaluators W. D. S. and G. R. feel that the use of 5.0 × 5.0 cm gauze pads instead of filter papers is technically superior because the gauze maintains increased surface contact with the infant's arm. Furthermore, the pads alleviate the discomfort and burns occasionally observed when filter papers are used.

6. Connect the electrodes to the power supply, being careful to connect the positive lead to the positive (pilocarpine) electrode and the negative lead to the negative (potassium sulfate) electrode.

7. Turn the switch on the power supply to the ON position.

8. Begin the iontophoresis by slowly adjusting the current to 2 mA.

Note: Some children may experience discomfort, which can usually be alleviated by adding a little of the pilocarpine nitrate and potassium sulfate solutions to the respective electrodes. Evaluator M. R. G. comments that on rare occasions an individual may show symptoms of hypersensitivity to pilocarpine. Should a patient complain of itching, dry throat, scratchy or watery eyes, or other allergic symptoms, the iontophoresis should be discontinued immediately.

9. At the end of 5 min turn the current off. Disconnect the leads and remove the strap and electrodes.

Collection of sweat

1. Using a soft paper tissue soaked with distilled or de-ionized water, thoroughly clean the pilocarpine-treated area on the forearm. Dry the area well with a dry tissue and proceed immediately to the next step.

Note: Evaluator M. R. G. recommends the use of Curity gauze squares (10 × 10 cm) in preference to tissues for this purpose.

2. Using forceps, remove the filter-paper circle from the preweighed vial and place over the exact area iontophoresed. Place the Parafilm square over the filter paper and secure it by wrapping surgical tape several times around the arm.

Note: Care must be taken to make an airtight seal between the skin and the Parafilm square. Evaluator M. R. G. uses waterproof plastic tape to seal all four edges to the skin.

3. Allow sweat to collect for 30 min. At the end of this time, press the Parafilm square down against the filter paper to include any water of condensation.

4. Carefully remove the tape and quickly remove the filter paper with clean forceps. Transfer without delay to the preweighed vial. Cap the vial immediately and re-weigh it to obtain the amount of sweat collected. Record this value.

Analysis of sweat sodium

1. Add 4 mL of the lithium diluent to the vial containing the filter paper and elute for 30–60 min.

2. Using clean forceps, remove the filter paper from the vial, squeezing any excess liquid from the paper into the vial.

3. Dilute 1 mL of the eluate with an additional 1 mL of lithium diluent and measure the sodium content of this sample by aspirating without further dilution into the flame photometer.

4. The remaining 3 mL of solution can be used for chloride determination as described elsewhere *(9)* (see also the alternative procedure for sweat chloride described below).

Standardization and Quality Control

1. Standardize the flame photometer, using the 100 mmol/L standard solution diluted 201-fold with lithium solution (20 µL of standard in 4 mL of diluent).

2. Perform appropriate quality-control procedures such as diluting control urine samples of known sodium content and assaying simultaneously with the standard and unknown.

Calculation

Calculate results (mmol/L) as follows:

$$\text{sodium concn} = R_{Na} \times 2 \times 20/W$$

where: R_{Na} = reading for sodium from flame photometer
2 = factor to compensate for dilution of eluate vs that of standard
20 = volume of standard solution used (µL)
W = weight of sweat collected (mg)

Note: For the Beckman KLiNa flame photometer operated in the urine mode, a sweat sample of 200 mg and a sodium concentration >40 mmol/L will exceed the operating range of the instrument. If the instrument indicates "out-of-range," then the 1 mL of the eluate should be diluted with a further 5 mL of the lithium diluent and re-assayed. The reading obtained from the flame photometer should then be multiplied by 3 before performing the calculation described above.

Alternative Procedure for Sweat Chloride

Analysis with Cotlove and Similar Chloride-Titrating Apparatus[1]

1. Add 5 mL of polyvinyl alcohol–acid reagent to the vial containing the filter paper and allow to elute for 30 min. Mix occasionally to facilitate extraction of the sweat electrolytes from the paper.

2. After elution is complete, pipet 4 mL of the polyvinyl alcohol–acid solution into a chloridometer cup and titrate chloride in the usual manner.

Standardization and quality control

1. Standardize the chloridometer, using the 100 mmol/L standard solution and a 100-µL pipette.

2. As a check of all the steps included in the sweat chloride analysis, pipet 100 µL of the 100 mmol/L chloride standard solution onto a filter-paper circle (using a preweighed filter paper and vial). Elute and titrate as for a patient's specimen. The weight should be 100 ± 0.5 mg and the calculated chloride concentration 100 ± 1 mmol/L.

Calculation

Calculate results (mmol/L) as follows:

$$\text{chloride concn} = R_{Cl} \times 1.25 \times 100/W$$

where: R_{Cl} = readout for chloride from chloride titrating apparatus (in mmol/L)
1.25 = factor to compensate for 5 mL of diluent used for unknown vs 4 mL actually analyzed
100 = volume (or weight) of standard solution used (µL or mg)
W = weight of sweat collected (mg)

Procedure with Corning and Similar Chloride-Titrating Apparatus

1. Add 15 mL of preconditioned "acid reagent plus chloride" to the vial and allow to elute for 30 min. Mix occasionally to facilitate extraction of the sweat electrolytes from the paper.

2. Pour all of the reagent (15 mL) back into the titration beaker.

3. Determine the chloride concentration in mmol/L.

Discussion

Despite commercially marketed systems for the *direct* measurement of sweat chloride with use of chloride electrodes or conductivity measurements, the Cystic Fibrosis Foundation considers the Gibson–Cooke method to be the method of choice *(1)*. In addition, many users of direct methods also use the

---

[1] Contributed by Evaluator M. R. G. Refer to the chapter on electrometric titration of chloride by Dietz and Bond for instructions on reagent preparation.

Gibson–Cooke method for confirmation or for validating "borderline" data.

Performing the sweat test is not particularly difficult, but requires careful attention to detail (2). The major problems associated with the method are: (a) cleaning and drying the skin after iontophoresis, (b) collecting an adequate sample, and (c) preventing evaporation of the sample after collection and before weighing. A paper tissue (or gauze) soaked with distilled water is used to cleanse the skin. Great care should be taken to ensure that the cleansed area is completely dry; residual water is absorbed by the filter paper, increasing the apparent volume of sweat.

Generally, there is little difficulty in collecting 100 to 200 mg of sweat, even in infants. Occasionally, however, smaller amounts are obtained. The concentration of sodium in sweat varies with the rate of sweat secretion (10). Because of this, it is conceivable that erroneous values are obtained if the sweating rate (and thereby the amount) is low. Although one cannot state categorically the minimum amount of sweat that will ensure an accurate result, it is generally accepted that a minimum of 50 mg should be collected. If less than 50 mg is collected, the sample should be discarded and the test repeated.

> *Note:* Evaluator M. R. G. comments that, in his experience, most of the patients who produced less than 50 mg of sweat per 30 min of collection time were less than two weeks old. He also confirms the observation of Hardy et al. (*Arch. Dis. Child.* **48**: 316, 1973) that very young children tend to have erroneously high sweat electrolyte concentrations; therefore, he tries to delay the initial sweat test until the child is at least two or three weeks old.

Evaporation of the sample occurs rapidly, especially in dry climates, leading to falsely high results. Care must be taken that the Parafilm used to cover the stimulated area of the arm makes an airtight seal over the filter-paper circle. This is best achieved by wrapping the surgical tape around the arm so as to completely cover the Parafilm, and is preferable to just taping down the edges. After the 30-min collecting period is complete, unwind the tape while holding the Parafilm in place over the filter paper.

Other precautions can be taken to reduce the possibility of false-positive and false-negative results:
1. The test should be performed by only one of the technologists on the laboratory staff.
2. The technologist should perform the entire test, from sweat stimulation and collection to determination of sodium and chloride concentration, and calculation.
3. All positive and borderline values should be repeated.

A modification of the chloride electrode method compares very favorably with the Gibson–Cooke method (11). The chloride electrode method offers certain advantages, notably, speed and convenience. However, the disadvantages of this method cannot be overlooked, especially when the test is performed in a smaller laboratory that performs the test only occasionally. The most likely source of error is in interpreting whether or not sweat has accumulated in the stimulated area. In the Gibson–Cooke method the exact amount of sweat generated is known; if it is insufficient, the test can be repeated.

Reference Intervals

Dr. H. Shwachman reports the following electrolyte ranges (mean ± 2 SD):

	No. of cases	mmol/L Na	Cl
Patients with cystic fibrosis	40	71–137	78–145
Controls:			
5 wk–11 mo	42	5–24	3–22
1– 9 yr	107	3–36	2–32
10–16 yr	17	6–52	2–38
17–60 yr	63	4–90	2–65

The concentration of sweat sodium and chloride tends to increase in normal individuals past adolescence, making it somewhat more difficult to interpret the test in adult patients. Increase in sweat sodium and chloride in diseases other than cystic fibrosis are rare but have been reported in malnutrition, hyperhydrotic ectodermal dysplasia, nephrogenic diabetes insipidus, and untreated adrenal insufficiency (8). False-positive results have also been reported in patients with hypothyroidism, glucose-6-phosphatase deficiency, mucopolysaccharidosis, and fucosidosis. Falsely low sweat sodium and chloride concentrations may be seen after administration of mineralocorticoids (aldosterone) and in edematous patients (12). Because cystic fibrosis may be concomitant with edema, the hypoalbuminemia should be corrected before performing the sweat test.

Ethnic variations have been reported in the response of the sweat glands to pilocarpine stimulation. Bantu children have been shown to have significantly lower sweating rates and sweat sodium concentrations than Caucasians and children of either Indian or Sudanese origin (13). These differences must be considered when the sweat test is applied to different ethnic groups. However, the incidence of cystic fibrosis among Asians and native African blacks is extremely rare.

References

1. Gibson, L. E., Position paper on sweat testing. In *GAP Conference Reports—Problems in Sweat Testing*, Cystic Fibrosis Foundation, Atlanta, GA 30326, 1975, pp 1–17.

2. Shwachman, H., and Mahmoodian, A., Quality of sweat test performance in the diagnosis of cystic fibrosis. *Clin. Chem.* **25,** 158–161 (1979).
3. Gibson, L. E., The decline of the sweat test. *Clin. Pediatr.* **12,** 450–453 (1973).
4. Ibbott, F. A., Chloride in sweat. *Stand. Methods Clin. Chem.* **5,** 101–111 (1965).
5. Gibson, L. E., and Cooke, R. E., A test for concentration of electrolytes in sweat in cystic fibrosis of the pancreas utilizing pilocarpine by iontophoresis. *Pediatrics* **23,** 545–549 (1959).
6. Kaiser, D., Excretion of protons, bicarbonate and cations by the single sweat gland of patients with cystic fibrosis. In *Fundamental Problems of Cystic Fibrosis and Related Diseases,* J. A. Mangos and R. C. Talamo, Eds., Intercontinental Medical Book Corp., New York, NY, 1973, pp 247–256.
7. Griffiths, A. D., and Bull, R. E., Anomalous sweat chloride levels in cystic fibrosis during antibiotic therapy. *Arch. Dis. Child.* **47,** 132–134 (1972).
8. Report of the committee for a study for evaluation of testing for cystic fibrosis. *J. Pediatr.* **88,** 711–750 (1976).
9. Gibson, L. E., diSant'Agnese, P. A., and Shwachman, H., Procedure for the quantitative iontophoretic sweat test for cystic fibrosis. Cystic Fibrosis Foundation, Atlanta, GA 30326, 1975.
10. Schwartz, I. L., and Thaysen, J. H., Excretion of sodium and potassium in human sweat. *J. Clin. Invest.* **35,** 114–120 (1956).
11. Finley, P. R., Dye, J. A., Lichti, D. A., Byers, J. M., and Williams, R. J., A modified ion-selective electrode method for measurement of sweat chloride. *Am. J. Clin. Pathol.* **69,** 615–618 (1978).
12. MacLean, W. C., Jr., and Tripp, R. W., Cystic fibrosis with edema and falsely negative sweat test. *J. Pediatr.* **83,** 86–88 (1973).
13. McCance, R. A., and Purohit, G., Ethnic differences in the response of the sweat glands to pilocarpine. *Nature (London)* **221,** 378–379 (1969).

Triglycerides in Serum, Colorimetric Method

Submitter: Christopher S. Frings, *Cunningham Pathology Associates and Medical Laboratory Associates, Birmingham, AL 35256*

Evaluators: Stephen H. Grossman and B. Patterson, *Department of Chemistry, University of Southwestern Louisiana, Lafayette, LA 70504*
Janet Y. Nutter and Robert F. Labbé, *Department of Laboratory Medicine, Harborview Medical Center, Seattle, WA 98104*
Shekhar Munavalli and J. Ben Flora, *Livingstone College, Salisbury, NC, and Biomedical Reference Laboratory, Burlington, NC 28114*
F. Goodland and C. P. Price, *Department of Chemical Pathology, General Hospital, Southampton S09 4XY, U.K.*

Reviewer: Gerald R. Cooper, *Metabolic Biochemistry Branch, Center for Environmental Health, Centers for Disease Control, Atlanta, GA 30333*

Introduction

Measurement of serum triglycerides is useful for diagnosis of hyperlipidemia and hyperlipoproteinemia. The correct diagnosis and treatment of hyperlipoproteinemia depends on accurate and precise methods for triglyceride determination.

Presented here is a manual colorimetric method described by Neri and Frings (1). This method is relatively simple and requires only equipment that is usually available even in small clinical chemistry laboratories. It is ideally suited for laboratories that do not have a large number of requests for triglyceride assays and for those wanting to make their own reagents.

Principle

Isopropanol extracts of serum are prepared and then treated with alumina to remove glycerol, phospholipids, and glucose. Triglycerides are then saponified to yield glycerol, which is oxidized by sodium metaperiodate to formaldehyde. Formaldehyde reacts with acetylacetone to form a yellow dihydrolutidine derivative. The absorbance of this product at 405 nm is proportional to the serum concentration of triglycerides.

Materials and Methods

Reagents

1. *Isopropanol.* "Aldehyde-free," analytical grade reagent.

2. *Alumina (Al_2O_3), washed.* Wash alumina (Woelm, neutral, activity grade I for chromatography; Waters Associates, Inc., Framingham, MA 01701) until fines are removed. This usually requires eight to 10 washings, with approximately four bed-volumes of de-ionized water for each wash. Dry the alumina in an oven for 15 to 18 h at 100–110 °C. *Do not keep the alumina in the oven longer than 18 h.* It can be used satisfactorily for at least two months when stored in a tightly stoppered container at room temperature.

3. *Saponification reagent.* Dissolve 10.0 g of KOH in 75 mL of water and 25 mL of isopropanol. This reagent is stable for at least two months when stored at room temperature in a brown glass bottle.

4. *Sodium metaperiodate reagent.* Dissolve 77 g of anhydrous ammonium acetate in 700 mL of de-ionized water; add 50 mL of glacial acetic acid and 650 mg of sodium metaperiodate ($NaIO_4$). Dissolve the sodium metaperiodate in this solution and dilute to 1 L with de-ionized water. *Do not change the order of addition of chemicals.* This reagent is stable for at least two months when stored at room temperature in a brown glass bottle.

5. *Acetylacetone reagent.* Add 0.40 mL of 2,4-pentanedione to 100 mL of isopropanol. This reagent is stable for at least two months when stored at room temperature in a brown glass bottle.

Note: Reviewer G. R. C. notes that discolored yellow reagent should be discarded.

6. *Standard.* The stock standard contains 1.000 g of triolein (A grade; Calbiochem, San Diego, CA 92112) per 100 mL of isopropanol. Prepare a working standard of 200 mg/dL by diluting 2.0 mL of the stock standard to volume with isopropanol in a 10-mL volumetric flask. The stock standard is stable for at least two months at 4–7 °C in a tightly sealed

container. Prepare working standard freshly each week and store in a brown glass bottle at 4–7 °C.

Note: Evaluators S. H. G. and B. P. found that weekly preparation of the working standard was inadequate and recommended preparing fresh working standard every other day.

Reviewer G. R. C. states that the accuracy of the triolein standards depends on the purity of the triolein and isopropanol solvent. The triolein should be 99% chromatographically pure. The isopropanol should be peroxide free by KI test. Each concentration of triolein should be weighed to four significant places and volumes measured at one temperature. Solutions of the standards are dispensed into amber bottles and sealed with tape soaked with melted paraffin. Solutions are stable for at least six months at 4 °C.

Apparatus

The procedure requires 16 × 125 mm screw-capped culture tubes, a mechanical rotator, a vortex-type mixer, a water bath at 65–70 °C, and a spectrometer or spectrophotometer to measure absorbance at 405 nm.

Specimen Collection

Because serum triglyceride concentrations are increased after an intake of most foods, blood samples must be taken after a 10–12 h fast. This method requires 0.2 mL of serum, but plasma from blood collected with ethylenediaminetetraacetate (EDTA) as an anticoagulant is also satisfactory. Triglycerides in serum are stable at least three days when stored at room temperature (24 °C) or refrigerated at 4–7 °C *(2)*.

Procedure

1. To a 16 × 125 mm screw-capped culture tube (blank) containing 0.7 (±0.1) g of washed alumina, add 5.0 mL of isopropanol and 0.20 mL of water.

Note: A "calibrated scoop" may be used to add the washed alumina.

2. To another tube (standard) containing 0.7 (±0.1) g of washed alumina, add 5.0 mL of isopropanol and 0.20 mL of working standard.

3. To another tube (unknown) containing 0.7 (±0.1) g of washed alumina, add 5.0 mL of isopropanol and 0.20 mL of serum.

4. Place all tubes on a mechanical rotator for 15 min.

Note: Evaluators J. Y. N. and R. F. L. had problems getting a tight seal with used screw-top extraction tubes. To get a better extraction, they tried mixing the tubes and contents with a vortex-type mixer three or four times for 10 s each during a 15-min period. The results correlated very well with 15 min of rotation.

5. Centrifuge for approximately 10 min at high speed to pack the alumina.

6. Transfer 2.0 mL of the clear supernatant fluid to appropriately labeled test tubes.

7. Add 0.60 mL of saponification reagent to all tubes and mix with a vortex-type mixer.

8. Let all tubes stand for at least 5 min, but no longer than 15 min at room temperature.

9. Add 1.5 mL of periodate reagent to all tubes and mix with a vortex-type mixer.

10. Add 1.5 mL of acetylacetone reagent to all tubes and mix with a vortex-type mixer. Tightly cover each tube with Parafilm.

11. Place all tubes in a 65–70 °C water bath for approximately 15 min.

12. Remove all tubes from the water bath, allow to cool at room temperature for 10–15 min, and vortex-mix for 5 s.

13. Within 1 h measure the absorbance of standard and unknowns against the blank at 405 nm in 10- or 12-mm cuvets.

14. Calculate values for unknowns by comparing the absorbances *(A)* of the unknown and standard as follows:

$$(A \text{ unknown}/A \text{ standard}) \times 200 = \text{triglycerides, mg/dL}$$

Clinical Interpretation

Normal concentrations of serum triglycerides for fasting (10–12 h) adults are approximately 30–150 mg/dL.

Discussion

The absorbance at 405 nm of the final product is linearly related to triglyceride concentrations up to 500 mg/dL. The absorbance of the chromogen at 405 nm does not change measurably for at least 1 h. If the absorbance of the final solution is greater than 0.8, the sample should be diluted with water and reassayed.

Glycerol does not interfere, having been removed with alumina in the extraction step *(3)*.

Analytical recovery of triglycerides is essentially 100% (Table 1). The within-run and day-to-day precision is given in Table 2.

Table 1. Added Triglycerides Analytically Recovered in Serum

Evaluators	mg/dL Expected	mg/dL Observed	Recovery, %
F. G. and C. P. P.	294	295	100
	451	478	106
	243	239	98
J. Y. N. and R. F. L.	50	49	97
	100	101	101
	200	198	99

Table 2. Precision Studies

Evaluators	Within-run N	Mean, mg/dL	CV, %	Day-to-day N	Mean, mg/dL	CV, %
S. H. G. and B. P.	20	17.6	3.4	20	168	3.4
J. Y. N. and R. F. L.		89.8	9.6		97.2	12.7
F. G. and C. P. P.	20	126	2.9	15	130	5.6
	20	312	2.5	15	263	3.8

References

1. Neri, B. P., and Frings, C. S., Improved method for determination of triglycerides in serum. *Clin. Chem.* **19**, 1201–1202 (1973).
2. Frings, C. S., Neri, B. P., Freeman, K., and Fendley, T. W., Stability of triglycerides in serum. *Clin. Chem.* **20**, 87–88 (1974). Letter.
3. Frings, C. S., and Queen, C. A., Use of alumina for removing phospholipids, glycerol and glucose from isopropanol extracts prior to assaying triglycerides: Comparison with zeolite mixtures. *Clin. Chem.* **18**, 709 (1972). Abstract.

Urea in Serum, Urease–Berthelot Method

Submitters: Alex Kaplan and Lin-Nar L. Teng, *Departments of Laboratory Medicine and Pharmaceutical Sciences, University of Washington, Seattle, WA 98195*

Evaluators: Jack H. Edwards, Jr., *South Georgia Medical Center Laboratory, Valdosta, GA 31601*
Betty J. Lash, *Animal Reference Laboratory, Houston, TX 77086*

Reviewer: David L. Witte, *Department of Pathology, University of Iowa, Iowa City, IA 52242*

Introduction

The measurement of the serum concentration of urea has been used for decades as one of several valuable tests for the assessment of renal function. Serum urea (or urea-N) may be determined quantitatively (a) directly, by reacting diacetyl monoxime with urea to form a colored complex, or (b) indirectly, by splitting urea with the specific enzyme urease (EC 3.5.1.5) and measuring the hydrolysis product, ammonia. The method described here deals with the latter approach.

In 1914, Van Slyke and Cullen (1) increased the specificity of blood urea analysis by utilizing urease to convert urea into ammonia and CO_2, transferring the ammonia to an acid trap by aeration, and then measuring the trapped ammonia titrimetrically. Over the years, several different chemical methods (gasometric, titrimetric, and colorimetric) have been used to measure the ammonia evolved, in attempts to improve the sensitivity, precision, and simplicity of the blood urea determination. The large number of modifications of these approaches attests to the failure to reach a consensus concerning a satisfactory method.

The Berthelot reaction (2), in which ammonia reacts with hypochlorite, phenol, a catalyst, and alkali to produce an intensely blue stable dye (indophenol), seemed to be an excellent reaction for determining the amount of ammonia liberated from urea by urease. More than 100 years passed, however, before the reaction was incorporated into a satisfactory, stable method for serum urea. The many variables of the Berthelot reaction included: the order of addition of reagents, the particular combination of multiple reagents in the various solutions, the optimum concentration of reagents, the choice of catalyst, the pH of the reactions, temperature of color development, and others.

In 1962, Chaney and Marbach (3) modified the Berthelot reaction of Fawcett and Scott (4) by combining the latter's catalyst (disodium nitrosylpentacyanoferrate III, commonly called sodium nitroprusside) with phenol into a single reagent that was added before the alkaline sodium hypochlorite. When combined with the prior hydrolysis of urea by urease, the Chaney–Marbach method for the determination of serum urea was rapid, highly sensitive, and precise; their results were quickly verified and confirmed by Searcy et al. (5). The method requires only 10 µL of serum and can readily be scaled down to 1 or 2 µL, if necessary, because the final volume of solution, 10 mL, produces more color than required by most spectrophotometers.

Note: The catalyst added to the reaction mixture is nitrosylpentacyanoferrate III but it may be converted by the alkali to a ferrous nitritopentacyano complex. See discussion of the reaction mechanism.

The Chaney–Marbach method, with a minor modification, has been used manually in the Submitters' laboratory for several years and was described in an earlier volume of this series (6). The method described below is essentially the same as in the previous protocol (6), but the concentrations of several reagents have been changed to optimize conditions, and more recent views concerning the reaction mechanism are presented.

Principle

Serum urea is determined in a two-step procedure:
1. Urease hydrolyzes urea specifically to form ammonia and carbamic acid; the carbamic acid spontaneously decomposes into ammonia and carbon dioxide, so that the net result is the formation of 2 mol of NH_3 and 1 mol of CO_2 from 1 mol of urea, as follows:

$$H_2N{-}C({=}O){-}NH_2 + HOH \xrightarrow{urease} NH_3 + HO{-}C({=}O){-}NH_2 \rightarrow 2\,NH_3 + CO_2$$

2. The ammonia is reacted with phenol, a catalyst, and an alkaline solution of sodium hypochlorite to form an intensely blue dye, indophenol. There are many different ways of carrying out this reaction, the Berthelot reaction. The mechanism is complex and not completely understood, but may be as follows:

$$NH_4^+ + OH^- \rightarrow NH_3 + H_2O$$

$$NH_3 + OCl^- \rightarrow \underset{\text{chloramine}}{NH_2Cl} + OH^-$$
(hypochlorite)

$$NH_2Cl + \underset{\text{phenol}}{\text{C}_6H_4\text{-OH}} + OH^- \xrightarrow{\text{catalyst}} Cl^- + H_2O + \underset{p\text{-aminophenol}}{HO\text{-C}_6H_4\text{-}NH_2}$$

$$HO\text{-C}_6H_4\text{-}NH_2 + \text{C}_6H_5\text{-OH} + OH^- + O_2 \rightarrow \underset{\text{indophenol}}{O=\text{C}_6H_4=N\text{-C}_6H_4\text{-}O^-} + 3H_2O$$

The final product of the reaction, indophenol, is beyond dispute, although several different pathways have been proposed (7–9) for the conversion of p-aminophenol to indophenol (see Discussion).

Materials and Methods

Reagents

For all reagents, use only distilled water that has been passed through a cation-exchange resin[1] to remove any possible traces of ammonia. This will help keep absorbance of the reagent blanks low (less than 0.040).

Note: The best system for producing water for general laboratory use involves a mixed-bed resin for removing cations and anions, followed by a bed of activated charcoal to remove organic contaminants. The further addition of a 0.2-µm pore-size membrane will remove any bacteria that might be in the purification system.

1. *Stock phenol–catalyst.* To approximately 800 mL of de-ionized water in a 1-L volumetric flask, add 50 g of reagent-grade phenol and 0.25 g of sodium nitrosylpentacyanoferrate (III),[2] $Na_2Fe(CN)_5NO \cdot 2H_2O$. The solution contains 0.53 mol of phenol and 0.8 mmol of the catalyst per liter. The reagent is stable for at least two months at 4–8 °C.

2. *Stock alkaline hypochlorite solution.* In approximately 600 mL of de-ionized water in a 1-L volumetric flask, place 20 g of NaOH pellets. When cool, add 43 mL of hypochlorite solution containing approximately 50 g of NaOCl per liter. Dilute to volume with water. The solution contains 0.50 mol of NaOH and 30 mmol of NaOCl per liter. The reagent is stable for at least three months, when protected from light and stored at 4–8 °C.

Note: Four brands of commercial bleach solutions, Clorox, F/S, White, and West Best, were tested and found to be satisfactory hypochlorite solutions; other brands probably can be used.

Note: This solution is less alkaline than the Chaney–Marbach reagent. Lorentz (7) has shown that maximum color develops in the indophenol reaction between pH 10.4 and 11.2. The optimum pH is achieved in the described method when the color-producing reagents are added to the urease solution containing ethylenediaminetetraacetate (EDTA) and phosphate buffer, pH 6.9. The concentration of NaOCl is not critical. The same results were obtained for solutions of urea or ammonium sulfate even though the concentration of NaOCl was increased or decreased by 25%.

3. *Phosphate buffer–EDTA solution* (per liter, 20 mmol of phosphate and 10 mmol of EDTA), pH 6.9 at 25 °C. To approximately 900 mL of water add 3.36 g of EDTA disodium salt and 3.48 g of K_2HPO_4. Stir until dissolved and then adjust the pH to 6.90–6.95 with HCl. Make up to 1000 mL volume with water and store at 4–8 °C.

Note: Urease is readily inactivated by traces of heavy metals. EDTA forms complexes with ions of heavy metals and protects against this type of inactivation. When adjusted to pH 6.9, the EDTA and phosphate make up a buffer near the optimum pH for urease. The incubation medium should be slightly acid, to trap volatile NH_3 as an ammonium salt. Watson (10) demonstrated that urease was more active in a potassium phosphate buffer than in either sodium phosphate or EDTA. One milliliter of 20 mmol/L phosphate buffer is equivalent to the 0.2 mL of 0.1 mol/L phosphate used by Watson.

4. *Stock urease,* approximately 40 Sigma units/mL (456 U/mL). Dissolve 2000 Sigma units of enzyme powder (approximately 500 mg of Type III urease; Sigma Chemical Co., St. Louis, MO 63178) in 25 mL of de-ionized water and add 25 mL of glycerol. The enzyme solution is stable for at least four months at 4–8 °C.

Note: Sigma defines its unit as that amount of urease liberating 1 mg of ammonia nitrogen from urea in 5 min at 30 °C under their assay conditions. The IUB unit (U) is the activity of urease liberating 1 µmol of NH_3 from urea per minute at 25 °C. According to the manufacturer's literature, one Sigma unit at 30 °C is equivalent to 11.4 U at 25 °C.

5. *Working urease in phosphate buffer–EDTA solution.* Transfer 1 mL of stock urease to a 100-mL volumetric flask and make up to volume with phosphate buffer–EDTA solution. The solution is at pH 6.9 and contains 0.4 Sigma units (4.6 U) per milliliter. The solution is stable for at least three weeks at 4–8 °C.

---

[1] Dowex 50 × 8 (Dow Chemical Co., Midland, MI 48640) or Amberlite IR 120 or 122 (Rohm and Haas, Philadelphia, PA 19105).

[2] This is the correct name for the chemical frequently listed by chemical suppliers as sodium nitroferricyanide or in the literature as sodium nitroprusside; it is a nitroso- and not a nitro-compound.

Note: Pour into a small flask the amount of enzyme solution sufficient for the day's work. Discard the excess at the end of the day.

6. *Stock urea standard, 100 mmol/L.* Dry a quantity of National Bureau of Standards urea (or reagent grade urea if the other is not obtainable) in a desiccator at reduced pressure. Transfer 601 mg to a 100-mL volumetric flask, add 50 mg of sodium azide as a preservative, and make up to volume with de-ionized water. The solution contains 100 mmol/L urea (6.01 g/L); in terms of urea-N, the corresponding concentration is 2.80 g/L (280 mg/dL).

7. *Working urea standards:*

   a. 5 mmol/L (per liter, 300 mg of urea, or 140 mg of urea-N). Dilute 5 mL of stock urea solution to 100 mL with 0.5 g/L sodium azide.

   b. 20 mmol/L (per liter, 1.20 g of urea or 560 mg of urea-N). Dilute 10 mL of stock urea solution to 50 mL with 0.5 g/L sodium azide.

8. *Stock standard ammonium sulfate, 100 mmol/L.* Transfer 1.321 g of desiccated $(NH_4)_2SO_4$ to a 100-mL volumetric flask, dissolve in 0.5 g/L sodium azide, and make up to volume. The concentration is 100 mmol/L, equivalent to 2.80 g/L of nitrogen, the same as the stock urea solution. The solution is stable at 4–8 °C.

9. *Working ammonium sulfate standard, 20 mmol/L.* Dilute 10 mL of stock ammonium sulfate standard to 50 mL with 0.5 g/L sodium azide solution. Store at 4–8 °C.

Procedure

Note: Serum is usually the fluid of choice for measuring urea but plasma may be substituted, provided the anticoagulant used contains no ammonium salts. Sodium, potassium, or lithium salts of heparin, EDTA, or oxalate are satisfactory anticoagulants. It is not necessary to precipitate proteins. However, whole blood cannot be used without prior precipitation of proteins.

The procedure should be carried out in a room free of traces of ammonia vapors. It is inadvisable to store urine samples in the same room. All operations involving the use of NH_4OH or its generation by the addition of alkali to ammonium salts (e.g., in the sodium nitroprusside test for ketones) should be conducted in a fume hood or in another room. It is imperative that no ammonia-containing solutions be used for washing the floor or windows.

1. Pipet into a series of 16 × 100 or 16 × 150 mm test tubes 1.0 mL of working urease solution (reagent no. 5). Add to the appropriate tubes 10 μL of either serum, control serum, standard, or water, according to the protocol in Table 1.

Note: The greatest source of error in the procedure is imprecision in pipetting the 10-μL sample of serum. We have obtained the greatest manual precision by using an SMI pipet with glass tip (Micro/Pettor®; Scientific Manufacturing Industries, Emeryville; CA 94608). The careful use of a pipetter-dilutor would also improve precision.

Table 1. Protocol for Serum Urea Determination

Reagent added	Sera Unknown	Control	Urea standard (no. 7)	Reagent blank
	mL			
Urease (no. 5)[a]	1.0	1.0	1.0	1.0
Serum	0.010	0.010		
Working standard[b]			0.010	
Water				0.010
Incubate 5 min at 55 °C				
Phenol–catalyst (no. 1)	1.0	1.0	1.0	1.0
Alkaline hypochlorite (no. 2)	1.0	1.0	1.0	1.0
Incubate 5 min at 55 °C				
Water	7.0	7.0	7.0	7.0

Read absorbance at 630 nm

[a] Reagent no. refers to reagents listed in text. The directions given here are for the separate pipetting of serum and urease solution. If a good pipetter-dilutor is available, it is more convenient to transfer 10 μL of serum plus 0.79 mL of water washout to a test tube and then add by a dispensing pipette 0.2 mL of a working urease solution that is fivefold as concentrated as reagent no. 5. The working urease solution is kept in a refrigerator in its dispensing bottle when not in use.

[b] Include both urea standards, nos. 7a and 7b, on each run.

Note: The 1-mL and 7-mL volumes of reagents and water, respectively, are most conveniently measured and transferred by means of a dispensing pipette. Many companies make dispensing pipettes: Brinkmann Instruments, Westbury, NY 11590; Labindustries, Berkeley, CA 94710; Oxford Labs, Foster City, CA 94404; and others.

Note: For rigid control, include an ammonium sulfate standard, 20 mmol/L, as well as the urea standard in each run, to simplify troubleshooting if something goes wrong. If the ammonium sulfate standard gives full color but the urea standard is low, the urease solution has deteriorated. A decreased color in all tubes, including both types of standards, indicates that either or both of the color reagents (reagents no. 1 and no. 2, above) have deteriorated; this happens rarely, but the hypochlorite solution is the first one to suspect.

2. Incubate the tubes for 5 min at 55 °C after addition of the urease; 15 min at 37 °C will suffice if a 55 °C water bath is not available.

3. Add 1.0 mL of stock phenol–catalyst solution (reagent no. 1) to each by means of a dispensing pipette and mix.

4. Immediately add 1.0 mL of alkaline hypochlorite solution (reagent no. 2) to each tube and mix.

Note: With the Chaney–Marbach procedure, full color development will take place even after a 30-min delay between the addition of the phenol–catalyst and alkaline hypochlorite reagents (11), but this is not true for those procedures in which the hypochlorite reagent is added first. The timing is so critical when the reagents are added in reverse order that a delay of several minutes in the addition of phenol–catalyst to the alkaline hypochlorite results in a measurable decrease in color.

5. Develop color by placing the tubes in a 55 °C water bath for 5 min. The same may be accomplished in 15 min at 37 °C.

6. To each tube, add 7.0 mL of de-ionized water from a dispensing pipette and mix.

Note: It is possible to avoid this step by performing the Berthelot reaction with 4 mL of phenol–catalyst solution that has been diluted 1+3 (fourfold) with water plus 5 mL of an alkaline hypochlorite solution that has been diluted 1+4 (fivefold). The color develops more quickly, however, with the stock solutions, which is an advantage in an emergency laboratory. The color is stable for hours.

7. Read the absorbance of each tube against a reagent blank at 630 nm.

Note: Always check the absorbance of the blank vs water to see whether any of the reagents have absorbed NH_3. The absorbance of the blank vs water should not exceed 0.040 if special purity urease was used, or 0.100 if a less potent preparation was used.

The absorbance follows Beer's Law up to 75 mmol of urea (2.1 g of urea-N) per liter, but the color is too intense to read at this high concentration in cuvets with a 10-mm light path. High concentrations of serum urea (up to 75 mmol/L) may be measured, however, by appropriate dilution of the developed color in both sample and blank. Dilute each 1+4 (fivefold) with de-ionized water, read at 630 nm, and multiply the results by 5. If a dispensing pipette is set to deliver 7.0 mL of water, one to six extra dispensings of 7-mL portions of water can be dispensed into both the sample and blank before absorbance is read. For example, if five extra portions of 7-mL each are taken for diluting the sample and blank, the sample volume is 10 mL + 35 mL = 45 mL; the dilution factor is 45/10 = 4.5.

Calculation

Serum urea concentration (mmol/L) = $(A_u/A_s) \times C$, where A_u and A_s are the respective absorbances of unknown and standard, and C is the concentration of urea (mmol/L). To convert to *urea*, mg/L, multiply by 60. To convert to *urea-N*, mg/L, multiply urea mmol/L by 28.

Note: The weight of 1 mmol of urea is 60 mg. Each millimole of urea contains 2 mmol, or 28 mg, of nitrogen.

Urine Urea and Preformed Ammonia

Urine urea (or urea-N) may be determined in a manner similar to that described for serum urea but with some modifications made necessary by the high concentration of urea in urine and by the presence of ammonium salts *(6)*. Measure the concentration of urea + ammonia by diluting the urine 1+9 (10-fold) with de-ionized water and treating the mixture the same as a serum sample for the determination of urea. In addition, to measure the preformed ammonia, run a urine blank determination in which water is substituted for the urease solution. Subtract the value for preformed ammonia from that for urea + ammonia and multiply by 10, to obtain the urine urea concentration.

Discussion

Reaction mechanism. It is generally agreed that in the Berthelot reaction, ammonia is converted by hypochlorite to chloramine, which then reacts with alkaline phenol to form *p*-aminophenol. The catalyst must be involved in this latter reaction because *p*-aminophenol is rapidly converted into indophenol, in the absence of any catalyst, when treated with phenol and alkali. The disagreement centers upon the pathway taken by *p*-aminophenol to form indophenol. Bolleter et al. *(8)* and Fenton *(9)* believed that *p*-aminophenol is oxidized by hypochlorite to quinonechloroimine, which then condenses with alkaline phenol to yield the indophenol ion. It is true that *p*-aminophenol is oxidized by hypochlorite *in acid solution* to quinonechloroimine *(13)*, but the Berthelot reaction takes place in alkaline solution. Furthermore, *p*-aminophenol is converted by alkaline phenol alone (without hypochlorite or catalyst) to indophenol *(6–8)*, proving that quinonechloroimine is not an obligatory intermediate. If any quinonechloroimine had formed in the early reaction, however, it would condense with phenol in alkaline solution to form indophenol.

Lorentz *(7)*, in his thorough study of the reaction, thought that the major portion of the *p*-aminophenol oxidatively condensed with phenol to yield the leuco form of indophenol (reaction 1), which is then oxidized by oxygen in the solution to indophenol (reaction 2). To us, it seems more likely that *p*-aminophenol is oxidized to quinonemonoimine, a highly reactive compound *(13)*, which is then oxidatively condensed with phenol to form indophenol (reactions 3 and 4, respectively). The alternative pathways appear below.

$$HO-\langle\rangle-NH_2 + \langle\rangle-OH + 1/2\ O_2 \rightarrow HO-\langle\rangle-\overset{H}{N}-\langle\rangle-OH + H_2O \quad (1)$$
leuco form (reduced) of indophenol

$$HO-\langle\rangle-\overset{H}{N}-\langle\rangle-OH + 1/2\ O_2 + OH^- \rightarrow O=\langle\rangle=N-\langle\rangle-O^- + 2H_2O \quad (2)$$
indophenol (blue, dissociated form)

$$HO-\langle\rangle-NH_2 + 1/2\ O_2 \rightarrow O=\langle\rangle=NH + H_2O \quad (3)$$
quinonemonoimine (*p*-benzoquinonemonoimine)

$$O=\langle\rangle=NH + \langle\rangle-OH + 1/2\ O_2 + OH^- \rightarrow O=\langle\rangle=N-\langle\rangle-O^- + 2H_2O \quad (4)$$
indophenol (blue, dissociated form)

We agree with Lorentz *(7)* that in the Berthelot reaction as carried out in alkaline solution, quinonechloroimine is not the active intermediate. There is little evidence to support his contention, however, that *p*-aminophenol oxidatively condenses with phenol to yield the reduced (leuco) form of indophenol, which is then oxidized to phenol. It is far more likely that *p*-aminophenol is oxidized to quinonemonoimine (reaction 3), a compound that has been crystallized and identified in such a reaction *(14)*. The next step (reaction 4) is the alkaline oxidative condensation of quinonemonoimine with phenol to produce the blue indophenol dye, another reaction known to occur *(13)*.

The nature of the catalyst, sodium nitrosylpentacyanoferrate III, $Na_2Fe(CN)_5NO$ (listed in chemical catalogs as sodium nitroferricyanide and formerly known as sodium nitroprusside), is not known with certainty. Although Fernelius *(15)* states that sodium nitroprusside is a nitrosylo complex of Fe III, there is evidence that it is converted by alkali to a ferrous nitritopentacyano complex, $Na_4[Fe(CN)_5ONO]$ *(16–18)*.

So even though the Berthelot reaction is not completely elucidated, it nevertheless is a sensitive, specific, and precise method for the determination of urea-N in body fluids. The blue-colored reaction product (indophenol) is stable for hours, and few physiologic substances or drugs interfere with the test.

Precision

Within-run: When a control serum containing a urea-N concentration of 380 mg/L was used, one individual in our laboratory obtained coefficients of variation (CVs) of 2.12 and 2.29%, respectively, on two runs of 20 samples each; two others obtained CVs of 2.60 and 3.90%, respectively, on 20 sample runs of the same control serum. With a serum urea-N concentration of 160 mg/L, the CVs were 1.65, 2.65, and 2.72%, respectively, by the same three individuals on runs containing 14 to 20 samples each.

Day-to-day: The CV obtained by one person analyzing the 160 mg/L serum upon 20 different occasions was 2.65%.

The most critical factor affecting the precision of the method is the pipetting of 10 µL of serum.

Note: Evaluator B. J. L. analyzed a control serum for urea for 20 days by the method described and by the Gilford system (Gilford Instrument Labs., Inc., Oberlin, OH 44074). At a urea concentration of 170 mg/L (2.8 mmol/L), the day-to-day CV was 5.4% by the evaluated method and 3.7% by the semiautomated method on the Gilford. This demonstrates the critical nature of the sample pipetting because the serum was pipetted manually (Helena pipettes; Helena Labs., Beaumont, TX 77704) in the evaluated method, whereas the Gilford has a built-in pipette that is operated semiautomatically.

Evaluator J. H. E. found that the method described for serum urea-N gave very reproducible results in a patient population that varied from normal patients to those with sufficiently severe renal disease to require periodic dialysis. He compared the serum urea-N for 18 patients determined by the present method (*y*) with that of the Beckman BUN Analyzer (Beckman Instruments, Inc., Fullerton, CA 92634) and obtained a correlation coefficient of 0.998, a slope of 0.994, an intercept of 1.4, and a standard error of estimate of 1.36. The urea-N values ranged from 20 to 710 mg/L. In a comparison of 27 patients' values by the present method with those obtained with the American Monitor KDA (American Monitor Corp., Indianapolis, IN 46268), J. H. E. obtained a correlation coefficient of 0.998, a slope of 0.944, an intercept of 0.45, and a standard error of estimate of 1.11. The urea-N values varied from 40 to 800 mg/L. J. H. E. also found that the most critical step in the method was the precise pipetting of the 10-µL serum sample. His laboratory encountered no pipetting problems with either the Oxford or Beckman (manufactured for Beckman by Sherwood Medical Industries, St. Louis, MO 63103) pipetters.

Numerous studies have shown that the determination of serum urea or urea-N by the Berthelot reaction agrees closely with those obtained by other urease methods, or by the diacetyl monoxime methods *(19–21)*, in which urea reacts directly with a coupling reagent.

Interfering Substances

Richterich and Kuffer *(12)* tested a number of preservatives that might inhibit the activity of urease, and found that formaldehyde at 20 g/L or greater was the only inhibitor. Aldosan, ammonium fluoride, and thymol–isopropanol had no influence upon the enzymic activity. Heavy metals such as lead and mercury strongly inhibit urease activity, but the presence of EDTA in the urease solution should offer protection against accidental contamination with these ions.

Some authors who have used variants of the Berthelot reaction (adding hypochlorite in acid solution or in alkaline solution before the phenol) claimed that certain amino acids and other nitrogenous constituents also produced a blue color in the reaction. Kaplan *(6)* reviewed this work previously and showed that there was negligible interference from amino acids, glucosamine, citrulline, guanidine, bilirubin, glutamine, histidine, uracil, uric acid, creatinine, salicylic acid, and a number of salts. Richterich and Kuffer *(12)* tested and found no interference by *N*-acetyl-*p*-aminophenol, creatine, glutamine, sulfathiazole, chloramphenicol, uric acid, phosphate, salicylate, or streptomycin. Although there are reports in the literature that these substances interfere with the Berthelot reaction, they are always tested by some *variant of the reaction* in which the hypochlorite, with or without alkali, is added first; the substances do not interfere when the phenol–catalyst is added first, followed by the alkaline hypochlorite (the Chaney–Marbach procedure). Richterich and Kuffer also showed that hyperlipidemia required a serum blank correction and that extensive hemolysis inhibited the Berthelot reaction; hemoglobin at 117 and 230 mmol/L (188 and 373 mg/dL) inhibited the development of the indophenol blue color by 4.5 and 13.8%, respectively.

> *Note:* Evaluator J. H. E. reports that concentrations of NaF as great as 25 mg/mL in the blood-collection tube do not inhibit the action of urease in this method. He further found that heparin salts used as an anticoagulant do not interfere.

As discussed above, *p*-aminophenol, if present in serum, would produce the same blue color as an equimolar amount of ammonia. Fortunately, there are no medications with *p*-aminophenol so this is not a laboratory problem yet. Acetaminophen (4-hydroxyacetanilide; Tylenol®), a common drug widely used as an alternative to aspirin, does not interfere with the Berthelot reaction because it is conjugated in the liver with glucuronate or sulfate *(22)*; *p*-aminophenol is not one of its catabolic products. Richterich and Kuffer *(12)* tested acetaminophen at a concentration of 15 mg/dL and found that it gave no color in the Berthelot reaction.

Normal and Pathologic Values for Serum Urea-N

The serum urea-N concentration of an individual is so dependent upon the amount of protein consumed daily and upon physiologic factors affecting protein catabolism that the distribution curve of the urea concentration of apparently healthy individuals is skewed toward the high side. There is a fairly large overlapping area that includes "high normals" and mild renal disease. Reed et al. *(23)* studied 1419 clinically normal adults and reported a normal range (2.5 to 97.5 percentiles) for urea of 100–250 mg/L in men (1.7–4.2 mmol/L) and 70–210 mg/L in women (1.2–3.5 mmol/L). Cole *(24)*, in his retrospective study of hospital patients, demonstrated that a large overlap exists in the area of 180–270 mg/L (3.0–4.5 mmol/L) between patients with renal or prerenal problems and those without.

> *Note:* Evaluator B. J. L. reports that the normal serum urea concentration in several animal species are close to that in humans. In her laboratory, the normal value for dogs is 60–280 mg/L (1.0–4.7 mmol/L), for cats 120–320 mg/L (2.0–5.3 mmol/L) and for horses 100–200 mg/L (1.7–3.3 mmol/L). As in humans, the serum urea concentration in animals varies with the protein content of the diet.

Above-normal serum concentrations of urea or urea-N may be caused by impaired excretion, by an accelerated production of urea as a consequence of increased protein catabolism, or by a combination of both factors. Processes that decrease urea excretion are: renal damage that decreases the glomerular filtration rate, obstruction to the flow of urine (calculi, carcinomas obstructing the ureters or urethra, prostatic hypertrophy), or an insufficient blood pressure or renal blood flow to form urine at a normal rate. Increased protein catabolism may occur as a result of high protein intake, prolonged fever, increased secretion of adrenal glucocorticoids, therapeutic administration of cortisol-like steroids, or bleeding peptic or duodenal ulcers. Thus, an increased serum urea-N concentration may arise from prerenal, renal, or postrenal causes.

A decreased concentration of serum urea-N usually occurs in starvation, protein deprivation, late pregnancy, and late in hepatic failure, when there is malfunction of the urea cycle.

References

1. Van Slyke, D. D., and Cullen, G. E., A permanent preparation of urease and its use in the determination of urea. *J. Biol. Chem.* **19**, 211–228 (1914).
2. Berthelot, M. P. E., *Repert. Chim. Appl.* **1**, 284 (1859).
3. Chaney, A. L., and Marbach, E. P., Modified reagents for determination of urea and ammonia. *Clin. Chem.* **8**, 130–132 (1962). Presented at Research Symposium, Southern California Section of American Association for Clinical Chemists, Los Angeles, CA, May 1961.
4. Fawcett, J. K., and Scott, J. E., A rapid and precise method for the determination of urea. *J. Clin. Pathol.* **13**, 156–159 (1960).
5. Searcy, R. L., Gough, G. S., Korotzer, J. L., and Bergquist, L. M., Evaluation of a new technique for estimation of urea nitrogen in serum. *Am. J. Med. Technol.* **27**, 255–262 (1961).
6. Kaplan, A., Urea nitrogen and urinary ammonia. *Stand. Methods Clin. Chem.* **5**, 245–256 (1965).
7. Lorentz, K., Mechanismus und Specifität der Indophenolreaktion zur Ammoniakbestimmung. II. NH$_2$-Radikale, *p*-Aminophenol, Reaktion von Aminen und aromatischen Hydrazinen. *Z. Klin. Chem. Klin. Biochem.* **6**, 293–295 (1967).
8. Bolleter, W. T., Bushman, C. J., and Tidwell, P. T., Spectrophotometric determination of ammonia as indophenol. *Anal. Chem.* **33**, 592–594 (1961).
9. Fenton, J. C. B., The estimation of plasma ammonia by ion exchange. *Clin. Chim. Acta* **7**, 163–175 (1962).
10. Watson, D., A note on urease-catalyzed hydrolysis of urea. *Clin. Chim. Acta* **14**, 571–572 (1966).
11. Weatherburn, M. W., Phenol–hypochlorite reaction for determination of ammonia. *Anal. Chem.* **39**, 971–974 (1967).
12. Richterich, R., and Kuffer, H., Die Bestimmung des Harnstoffs in Plasma und Serum (Urease/Berthelot-Methode) mit dem Greiner Electronic Selective Analyzer GSA II. *Z. Klin. Chem. Klin. Biochem.* **11**, 553–564 (1973).
13. Rodd, E. H., *Chemistry of Carbon Compounds*, 3B, Elsevier, Amsterdam, 1956, pp 715–722.
14. Smith, P. A. S., *The Chemistry of the Open-Chain Nitrogen Compounds*, 1, Benjamin, Inc., New York, NY, 1965, p 125.
15. Fernelius, W. C., Nomenclature of coordination compounds and its relation to general inorganic nomenclature. *Adv. Chem. Ser.* **8**, 9–37 (1953).
16. Sidgwick, N. V., and Bailey, R. W., Structures of metallic carbonyl and nitrosyl compounds. *Proc. R. Soc. London* **144**, 521–537 (1934).
17. Horn, D. B., and Squire, C. R., An improved method

for the estimation of ammonia in blood. *Clin. Chim. Acta* 17, 99–105 (1967).
18. Weichselbaum, T. E., Hagerty, J. C., and Mark, H. B., Jr., A reaction rate method for ammonia and blood urea nitrogen utilizing a pentacyanonitrosyloferrate catalyzed Berthelot reaction. *Anal. Chem.* 41, 848–850 (1969).
19. Searcy, R. L., and Cox, F. M., A modified technique for ultramicro estimation of urea nitrogen. *Clin. Chim. Acta* 8, 810–812 (1963).
20. Wilcox, A. A., Carroll, W. E., Sterling, R. E., Davis, H. A., and Ware, A. G., Use of the Berthelot reaction in the automated analysis of serum urea nitrogen. *Clin. Chem.* 12, 151–157 (1966).
21. Wybenga, D. R., Di Giorgio, J., and Pileggi, V. J., Manual and automated methods for urea nitrogen measurement in serum. *Clin. Chem.* 18, 891–895 (1971).
22. Goodman, L. S., and Gilman, A., *The Pharmacological Basis of Therapeutics*, 5th ed., MacMillan, New York, NY, 1975, pp 343–347.
23. Reed, A. H., Cannon, D. C., Winkelman, J. W., Bhasin, Y. P., Henry, R. J., and Pileggi, V. J., Estimation of normal ranges from a controlled sample survey. 1. Sex- and age-related influence on the SMA 12/60 screening group of tests. *Clin. Chem.* 18, 57–66 (1972).
24. Cole, G. W. The clinical usefulness of multiple result ranges for a laboratory test: The serum urea nitrogen. *Pathologist* 32, 101–104 (1978).

Urea in Serum, Direct Diacetyl Monoxime Method

Submitter: Frank W. Fales, *Department of Biochemistry, Emory University School of Medicine, Atlanta, GA 30322*

Evaluators: June D. Boyett, *St. Vincent Hospital and Medical Center, Portland, OR 97225*
Epperson E. Bond, Ralston W. Reid, and Albert A. Dietz, *Veterans Administration (Edward Hines, Jr.) Hospital, Hines, IL 60142*
Robert C. Elser, *York Hospital Department of Pathology, York, PA 17405*
Michael McAneny, *Chemistry Laboratory, Kadlec Hospital, Richland, WA 99352*

Reviewer: David L. Witte, *Department of Pathology, The University of Iowa, Iowa City, IA 52242*

Introduction

This direct diacetyl monoxime method for determining serum urea nitrogen has the advantages of being simple, sensitive, and accurate. It does not require the preparation of protein-free filtrate, and the entire procedure can be carried out in colorimeter tubes. This manual method has somewhat less sensitivity than the urease–Berthelot reaction method presented in the previous chapter, but it has the advantage of not being affected by ammonia fumes from laboratory reagents, floor and window cleaners, or cigarette smoke. Its disadvantages, however, include the use of viscous, orthophosphoric acid reagent, which is difficult to pipet, and the release of unpleasant-smelling fumes during the heating of the reaction mixture in a boiling water bath. These disadvantages are minimized by using a dispenser to deliver the phosphoric acid reagent and by carrying out the heating in a fume hood.

Principle

The reaction of diacetyl monoxime with urea was first described by Fearon (1). The sample was heated with diacetyl monoxime in a highly acidic solution, and then an oxidizing reagent was added. A yellow product formed with compounds having the structure $R_1NH-CO-NHR_2$, where R_1 was H or a single aliphatic group, and R_2 was not an acyl group. Although urea was the only compound at an appreciable concentration that reacted in protein-free filtrates from blood, it was difficult to adapt the reaction to blood because the colored product was unstable. The color faded rapidly, especially when exposed to light (2), and the time required to reach the maximum absorbance increased with increasing urea concentration (3). These difficulties were overcome as described below.

A plausible reaction postulated by Dickenman et al. (4) is as follows (see also *Addendum* to this chapter): 1 mol of urea reacts with diacetyl monoxime (2-nitroso-3-butanone; also called 2,3-butanedione monoxime) to form the yellow product 4,5-dimethyl-2-imidazolone, with the release of 1 mol of water and 1 mol of hydroxylamine.

$$\begin{array}{c} CH_3 \\ | \\ C=O \\ | \\ C=NOH \\ | \\ CH_3 \end{array} + \begin{array}{c} H_2N \\ C=O \\ H_2N \end{array} \longrightarrow \begin{array}{c} CH_3 \\ | \\ C=N \\ C=O + H_2O + H_2NOH \\ C=N \\ | \\ CH_3 \end{array}$$

Diacetyl Monoxime Urea 4,5-Dimethyl-2-Imidazolone

Natelson et al. (5) postulated that diacetyl monoxime first hydrolyzed to diacetyl, which reacted with urea. However, this was disputed by Beale and Croft (6). Regardless of which is the reactant, the reaction with diacetyl monoxime is smoother than with diacetyl (6–8). In the Technicon AutoAnalyzer method (9, 10) some problems were eliminated by adding ferric salt and thiosemicarbazide. The former removed the hydroxylamine by oxidation and the latter presumably formed a Schiff base with the imidazolone. This both stabilized the structure and produced a red product of much higher absorbance than the original yellow compound. This procedure was not readily adaptable to the manual method, however, because there was no linear relationship between the absorbance and the urea concentration.

The reagent proposed in the modification of the Coulombe and Favreau (11) appeared to overcome this obstacle. They reported a linear relationship, which was confirmed by Moore and Sax (12). However, the reagent had to be used at greater dilution for the product to obey Beer's Law over an extended range of concentration. Moreover, light catalyzes the formation of the colored product, so direct rays from lights must be avoided during heating. The distinctive features were the use of phosphoric acid as the sole acid, and omission of the oxidizing agent.

Wybenga et al. *(13)* proposed a diacetyl monoxime method that could be used directly on serum without removal of protein. This direct use of serum combined with the modified procedure of Coulombe and Favreau provides the simple, accurate, manual method for the determination of serum urea described here. The photometric determination of the serum urea concentration is accomplished by measuring absorbance at 550 nm, a wavelength of high sensitivity, free from protein interference.

Materials and Methods

Reagents

Diacetyl monoxime/thiosemicarbazide reagent. Dissolve 3.00 g of diacetyl monoxime, m.p. 76–77 °C, and 0.15 g of thiosemicarbazide, m.p. 180–182 °C, in distilled water and dilute to 500 mL; heat to dissolve the thiosemicarbazide. Store in an amber bottle at room temperature. Do not store in a refrigerator. In a cool room a small amount of the thiosemicarbazide may crystallize and settle out of the solution without apparent harm to the reagent. The reagent loses little sensitivity in one month.

Phosphoric acid reagent. Mix 1200 mL of reagent grade H_3PO_4 (85–87%; 850–870 g/L) with distilled water and dilute to 2000 mL. The reagent is stable.

Urea stock standard, 600 mg/dL (100 mmol/L) urea. Dissolve 600 mg of desiccated urea (ACS grade) in distilled water saturated with chloroform and dilute to volume in a 100-mL volumetric flask.

Urea working standards. (a) Dilute 2.5 mL and 5 mL of stock urea standard to 100 mL in volumetric flasks with water saturated with chloroform: final concentration, 15 and 30 mg/dL (2.5 and 5 mmol/L). (b) Dilute 10 mL of stock urea standard to 50 mL with water saturated with chloroform: final concentration, 120 mg/dL (20 mmol/L). The chloroform acts as preservative. Store the standards in a refrigerator in tightly stoppered vessels. If care is taken to avoid contamination, the standards will keep for several months.

Additional reagents for the protein-free filtrate method:

Sodium tungstate, 100 g/L. Dissolve 100 g of sodium tungstate ($Na_2WO_4 \cdot 2H_2O$) in distilled water and dilute to 1 L. Stable.

Sulfuric acid, 333 mmol/L. Weigh 35 g of sulfuric acid (reagent grade, 95–98% purity, sp. grav. 1.84) in a small tared beaker. Pour the acid into a 1-L volumetric flask about half filled with distilled water and rinse the residual acid into the flask with several portions of water. Dilute to volume and mix thoroughly. Make final volume adjustment when solution has cooled to room temperature.

Urea standards, 1.5, 3, and 12 mg/dL. Dilute 5 mL of the 15, 30, and 120 mg/dL standards to 50 mL each.

Apparatus

Dispenser. Either the Dispensette® (Brinkmann Instruments, Westbury, NY 11590) or the Repipet® (Labindustries, Berkeley, CA 94710) is satisfactory.

Collection and Handling of Specimens

Collect the blood without stasis into an evacuated collection tube not containing anticoagulant and allow it to clot. Separate the clot from the tube with a stirring rod and then remove the serum after centrifugation.

Note: After taking samples for the analysis, the remainder may be stored in a tightly stoppered tube in a refrigerator for one week so that the determination can be repeated, should questions arise concerning the analysis.

Contrary to an earlier report *(14)*, the urea is equally distributed throughout blood fluid *(15)*, and there is no change in extracellular urea concentration during clotting. Therefore, plasma treated with oxalate, EDTA, or heparin can be substituted for serum in the determination of urea.

Procedure[1]

1. To labeled, calibrated colorimeter tubes—one tube for the blank; duplicate tubes for 15, 30, and 120 mg/dL urea standards; and separate tubes for each unknown—add 1.0 mL of diacetyl monoxime/thiosemicarbazide reagent.

2. With TC pipettes or with an accurate dispensing device, add 20 µL of standard or serum to the labeled tubes.

3. To each tube, including the blank, add with the dispenser 8.0 mL of the phosphoric acid reagent directly into the solution. This aids in mixing the viscous, phosphoric acid with the diacetyl monoxime reagent. Mix thoroughly.

4. Place the tubes in an appropriate test-tube holder such as a wire basket sectioned at two levels with perpendicular strands of rubber-covered wire, and place the basket of tubes in a boiling water bath, positioned to avoid direct rays from lights.

5. After 20 min, transfer the basket of tubes to a tap-water bath to cool to room temperature. Transfer the tubes from the basket without agitation. There is no dilution error at this stage because the few drops of low-density condensate remain at the surface of the high-density solution. Partial mixing

---

[1] See footnote to Table 1 for method to use with protein-free filtrate.

Table 1. Protocol for Serum Urea Determination[a]

		Standards (in duplicate)			
Sequence	Blank	15 mg/dL	30 mg/dL	120 mg/dL	Serum unknowns
Diacetyl monoxime/ thiosemicarbazide	1.0	1.0	1.0	1.0	1.0
Serum					0.02
15.0 mg/dL standard		0.02			
30.0 mg/dL standard			0.02		
120 mg/dL standard				0.02	
Water	0.02				
Phosphoric acid	8.0	8.0	8.0	8.0	8.0

Mix thoroughly
Heat in a boiling water bath for 20 min
Cool in a tap water bath for about 8 min
Read the absorbance at 550 nm

[a] In the protocol for determination of serum urea by the tungstic acid/protein-free filtrate method use filtrate instead of serum; use 1.5, 3, and 12 mg/dL standards instead of 15, 30, and 120 mg/dL standards; and use 0.20 mL each of filtrate standards and water, instead of the 0.02 mL indicated above.

of the low-density condensate with the high-density solution may cause Schlieren lines to extend deep enough from the surface to interfere with the absorbance measurements, but this is much less of a problem with the modified than with the original procedure because of the greater depth of the solution.

Note: With the unmodified procedure, Evaluator J. D. B. performed the above steps in borosilicate tubes, 25 × 150 mm, covered with glass marbles to act as condenser. This avoided condensate on the sides of the tubes after boiling. After boiling and cooling, the solutions were vortex-mixed before decanting into colorimeter tubes. Precision was improved.

6. Read the absorbance of the standards and unknowns at 550 nm after setting the instrument to zero with the blank. See outline in Table 1.

Sera from highly lipemic and badly hemolyzed blood are not suitable for urea determination by the direct method. With these samples, prepare a protein-free filtrate as follows: To 1.40 mL of water add 0.20 mL of serum and 0.20 mL of sodium tungstate reagent. Mix and then add 0.20 mL of 333 mmol/L sulfuric acid reagent. Mix thoroughly and filter on 5.5-cm diameter Whatman No. 1 filter paper (Whatman, Inc., Clifton, NJ 07014). The modified protocol is outlined in the footnote to Table 1.

7. Calculate the concentration of urea in serum using the average factor, F, of the three standards as follows:

$$F = \frac{1}{3}\left(\frac{15}{A_{15}} + \frac{30}{A_{30}} + \frac{120}{A_{120}}\right)$$

$$C_{unk} = F \times A_{unk}$$

where A_{unk} is the absorbance of the unknown and A_{15}, A_{30}, and A_{120} are the absorbances of the 15, 30, and 120 mg/dL standards, respectively. C_{unk} is the urea concentration of the unknown in mg/dL; to convert to mmol/L, divide by 6.

Discussion

Three features of the diacetyl monoxime method were examined critically: the stability of the colored product, agreement with Beer's Law, and suitability of the direct use of whole serum, there being reports of difficulties in these areas. Wybenga et al. (13) reported that their diacetyl monoxime/thiosemicarbazide reagent faded as much as 7% in 30 min when oxidizing agent was not added. However, Coulombe and Favreau (11) observed a maximum color development that was stable for several hours with their reagent, which did not contain oxidizing agent. The Submitter, using the reagent of Coulombe and Favreau, observed a gradual increase in absorbance. Initially the absorbance increased at a rate of about 1 to 1.5% in 15 min; it continued to increase for many hours but at a decreasing rate. Readings may increase by 15 to 20% if left standing overnight. The ratios remain constant for several hours; therefore, no error results if absorbance of the standards is read not too long after the reading of the unknown.

It was necessary to dilute the reagent of Coulombe and Favreau because the range over which absorbance was linearly related to concentration was insufficient. The range of linearity increased as the sensitivity of the reagent decreased with age. The Submitter found that the calibration was nearly linear up to 150 mg/dL (25 mmol/L).

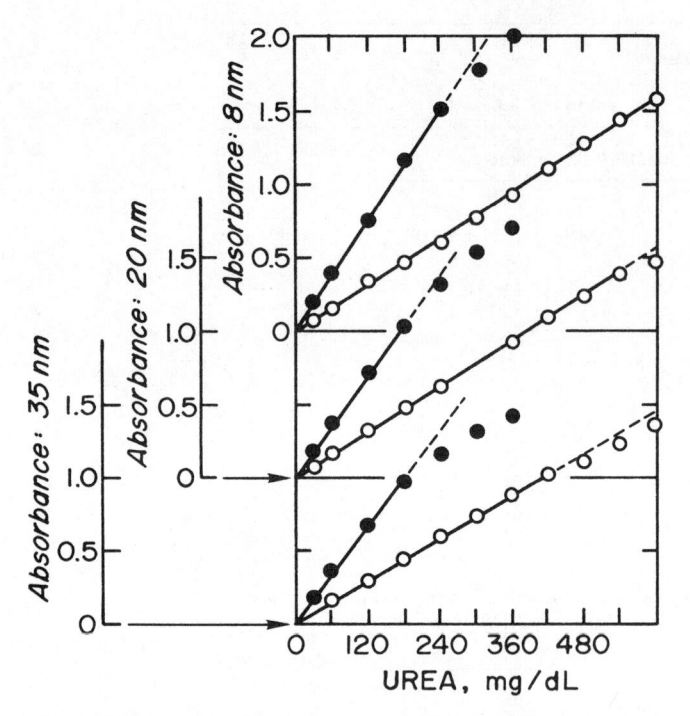

Fig. 1. The increase in absorbance with increasing urea concentration measured with spectrophotometers having absorbance bandwidths (spectral bandpass) of 35 nm (Coleman Junior, Model 6D), 20 nm (Junior II, Model 6/20), and 8 nm (Junior III, Model 6/8)

●, original, unmodified procedure, with 5 mL of the phosphate reagent and absorbance read at 530 nm; ○, the modified procedure, with 8 mL of phosphate reagent and absorbance read at 550 nm

Other evaluations of the unmodified procedure are given in Tables 2 and 3. The precision studies were acceptable and the recoveries were satisfactory at the concentrations studied. However, Evaluators E. E. B., R. W. R., A. A. D., and J. D. B. found a limited concentration range of linearity, and noted that their automated comparison method gave significantly higher values for a series of samples than did the unmodified, manual procedure. Evaluator R. C. E. observed a much broader range of linearity, and he and Evaluator M. M. found satisfactory agreement between the comparison and manual procedures. These findings suggested that the unsatisfactory agreement with comparison method and the narrow range of linearity might be related. A study of the effect of spectral bandwidth on linearity, suggested by Reviewer D. L. W., is shown in Figure 1. With bandwidths of 35, 20, and 8 nm, the absorbance was linear up to 150, 180, and 240 mg/dL, respectively, as determined with the unmodified procedure. It is probable that the linearity would extend to higher concentrations with an expensive spectrophotometer, especially with solutions of high absorbance. However, methods for the small clinical laboratory should be designed to give valid results with an inexpensive spectrophotometer. With the modified reagent, the linearity extended to a urea concentration of 420 mg/dL, even with a spectrophotometer having a 35-nm spectral bandwidth.

Nonspecific absorbance became a problem with the diluted reagent. In Table 4 the apparent urea concentration of a serum with a mean urea concentration of 29 (SD 1.2) mg/dL was determined by measurements of the absorbance of the sample and a 60 mg/dL standard at each wavelength. With the original reagent (1 mL of diacetyl monoxime/thiosemicarbazide and 5 mL of phosphoric acid solution) the nonspecific absorbance was evident at wavelengths shorter than 530 nm, the wavelength of maximum absorbance. However, dilution of the reagent by adding 10 mL rather than 5 mL of phos-

Table 2. Precision and Recovery with the Unmodified Procedure[a]

		Within-run			Between-run			Urea, mg/dL		
Evaluator	n	Mean (SD), mg/dL	CV, %	n	Mean (SD), mg/dL	CV, %	Evaluator	Expected	Measured	Recovery, %
J. D. B.		32.6 (1.1)	3.4		32.4 (1.3)	4.0	J. D. B.	102.2	103.1	100.9
		122.6 (4.3)	3.5		124.1 (5.4)	4.4		86.1	84.2	97.8
M. M.	20	21.8	4.3	20	22.5	8.6		72.2	73.3	101.5
	20	49.5	3.5	20	48.4	5.3		70.7	71.1	100.6
	20	95.4	3.2	20	95.4	3.9		63.6	58.9	92.6
R. C. E.	20	59.8	0.5	20	40.7	1.0		47.6	44.8	94.1
				21	96.6	1.9		34.1	34.3	100.6
								33.0	31.5	95.5
							M. M.			92–104
							R. C. E.	58.3	55.7	95.6
								73.1	72.9	99.7
								100.7	100.7	100.0
								128.6	132.9	103.3

[a] Using 1 mL of diacetyl monoxime/thiosemicarbazide and 5 mL of phosphoric acid reagent and reading absorbance at 530 nm.

Table 3. Comparison with Other Methods

Evaluator	Comparison method (n = 40)		Unmodified[a] mg/dL	System I mg/dL	Upper limit of linearity, unmodified method, mg/dL
R. C. E.	Gilford System I	Mean	28.6	28.7, F-test = 1.04 (n.s.)	265
		SD	26.8	26.2, t-test = 0.017 (n.s.)	
		(SD)²	718.4	684.4	
		Regression formula: $y = 1.0073\,x - 0.135$ ($r = 0.987$)			
J. D. B.	SMA 6/60	Regression formula: $y = 0.88\,x - 0.8$ ($r = 0.896$) $S_b = 0.35$, $S_a = 0.8$. Corrected values:[b] $b = 0.9012$, $a = -0.78$. Significant proportional error was observed, the SMA 6/60 giving higher values.			180
M. M.	SMA 12/60[c]	Regression formula: $y = 0.972\,x - 0.32$ ($r = 0.985$)			
E. E. B., R. W. R., and A. A. D.	SMA 6	Higher values with the comparison method, with average differences of 3.8[d] and 3.5[e] mg of urea nitrogen per deciliter.			65[d] 130[e]

n.s., not significant.
[a] 1 mL of diacetyl monoxime/thiosemicarbazide, 5 mL of phosphoric acid reagent, absorbance read at 530 nm.
[b] By the Deming regression model described by Cornbleet and Gochman (*Clin. Chem.* **25**: 432–438, 1979).
[c] Spectrophotometer used: Gilford Stasar II, [d] Coleman Jr. Model 6A, [e] Gilford Model 340.

phoric acid solution brought about a more pronounced nonspecific absorbance, extending to 540 or 550 nm when measured with a spectrophotometer with a 20-nm bandpass. The nonspecific absorbance was eliminated when a tungstic acid protein-free filtrate was used. Moreover, when the urea was removed by action of urease, the nonspecific absorbance remained. Thus protein appeared to be responsible for the nonspecific absorbance.

The use of 1 mL of diacetyl monoxime/thiosemicarbazide plus 10 mL rather than 5 mL of the

Table 4. Apparent Urea Concentration Determined at Different Wavelengths

	Urea concn, mg/dL		
Wavelength, nm	Whole serum plus 5 mL of phosphate reagent	Whole serum plus 10 mL of phosphate reagent	Protein-free filtrate plus 10 mL of phosphate reagent
430		228	
440	66	110	
450	49	73	16
460		55	28
470	37	47	26
480		42	28
490	33	38	28
500	32	36	26
510	31	36	27
520	30	33	27
530	28	31	26
540	28	29	26
550	28	28	27
560	27	28	28
570	28	29	28

Table 5. Concentration and Recovery of Urea in Serum as Determined at 530 and 550 nm

Sample no.[a]	No. of replicates	Urea, mg/dL Measured	Added	Recovery, %	CV, %
		530 nm			
1	18	32	0	—	5.0
2	14	77	0	—	3.6
3	16	123	0	—	3.3
4	16	132	0	—	3.6
4	14	246	112	101.8	3.5
4	14	365	224	104.0	3.1
4	13	471	337	100.6	3.7
		550 nm			
1	16	29	0	—	4.7
2	12	74	0	—	3.5
3	16	118	0	—	3.4
4	16	129	0	—	3.7
4	14	237	112	96.4	2.9
4	14	354	224	100.4	3.3
4	13	465	337	99.7	3.8

[a] Sample 4 was a lyophilized serum control reconstituted with 10 mL of water or with 10 mL of 120, 240, and 360 mg/dL urea solutions. The final volume of the control samples was 10.7 mL.

phosphoric acid solution yielded a straight-line relationship between absorbance and concentration of standards up to at least 480 mg/dL (80 mmol/L); with 8 mL of phosphoric acid solution, the calibration line was linear up to 360 mg/dL (60 mmol/L). The precision and recovery of urea determined with the 10-mL phosphoric acid modification are shown in Table 5. This study indicated that urea at high concentration was accurately determined with use of the diluted reagent. Consistently lower values were obtained when measurements were at 550 rather than 530 nm, but the average difference was only 6.0 mg/dL (1.0 mmol/L). The chief drawback of this modified procedure is its relatively low sensitivity in the diagnostically important region of 16–27 mg of urea nitrogen per 100 mL. The modification in which 8 mL of phosphoric acid solution was used gives valid results for a great majority of samples with high urea concentrations and has the advantage of increased sensitivity, as shown in the following regression formulas (x = urea concentration):

$$y = 0.00255x - 0.001 \quad (530 \text{ nm}; 10 \text{ mL } H_3PO_4)$$
$$y = 0.00207x + 0.001 \quad (550 \text{ nm}; 10 \text{ mL } H_3PO_4)$$
$$y = 0.00283x + 0.01 \quad (530 \text{ nm}; 8 \text{ mL } H_3PO_4)$$
$$y = 0.00262x + 0.01 \quad (550 \text{ nm}; 8 \text{ mL } H_3PO_4)$$

In all four cases, the correlation coefficient exceeded 0.9995. The values obtained with the 8- and 10-mL phosphoric acid modification were in excellent agreement with urea results for serum samples (Table 5) ranging from 30 to 360 mg/dL.

The inclusion of three standards serves two useful functions. Calculations based on the average factor reduce the day-to-day variation, and recording the absorbance and ratio of absorbance of the three standards is useful in quality control. Large decreases in absorbances suggest deterioration of the reagent, whereas a drastic change in the average ratio suggests deterioration of a standard.

The within-run precision is generally better than the between-run precision. In the former, the readings of the standards establish a calibration constant and only deviations in test solutions are included in the estimate of precision, but deviations in both tests and standards are included in the estimate of the between-run precision. This factor of between-run variability can be minimized by use of standards having absorbances between 0.3 and 0.9, the range of least instrument error, as shown in Table 6. However, changes in the reagents, standards, and the environment may also contribute to the day-to-day variations. Changes in atmospheric pressure, room temperature, and room illumination may be particularly bothersome in the present procedure because of the development of the colored product in a boiling water bath, the change of absorbance at room temperature, and the sensitivity of the product to light. The effect of some of these environmental factors may not be independent of urea concentration, so one standard is chosen within the normal range (where results for most samples will occur) and a second standard is within the range of minimum instrument error.

To test the reproducibility of the method in the critical region bridging normal and above-normal concentrations, seven serum samples with concentrations of 15–66 mg/dL (2.5–11 mmol/L), mean 46.2 mg/dL (7.7 mmol/L), were analyzed in duplicate on three consecutive days. The within-day CV was 3.06% and the day-to-day CV was 3.16% for the individual samples. The day-to-day variation given here may not be comparable to the day-to-day variations shown in Table 2 because in that case each analysis was carried out on a separate day.

Groups of serum samples previously analyzed with the Technicon AutoAnalyzer were selected to yield a variety of urea concentrations for comparison with the manual method. Urea was determined both directly in serum and in tungstic acid protein-free filtrates. The results of the study are shown in Table 7. Statistical comparison of the values obtained by the three methods by paired differences revealed no bias. The straight-line regression formula for automated (x) vs direct method and for filtrate (z) vs direct method were respectively:

$$y = 0.98x - 1.5, \; r^2 = 0.997$$
$$y = 1.01z - 1.3, \; r^2 = 0.996$$

The following statistical values were obtained in the comparison of the direct, manual method (y) with the automated procedure: slopes $b_{y \cdot x} = 0.9776$ and $1/b_{x \cdot y} = 0.9808$; standard deviation of the regression $S_{y \cdot x} = 2.98$; standard deviation of slope $S_b = 0.012$ with fiducial limits of slope at 95% confidence, 0.9776 ± 0.0248; and standard deviation of the intercept $S_a = 2.20$. The excellent agreement supported the validity of the direct manual method.

The only foreseen difficulty was interference during analysis of serum from icteric, lipemic, and hemolyzed bloods. No interference was observed with increase in icteric index. However, hemolysis sufficient to color the serum a deep red caused a positive error with the direct method, giving an apparent urea concentration 11 mg/dL higher than with the protein-free filtrate. Surprisingly, a significant negative error was observed with highly lipemic samples. A lipid control serum with high concentrations of triglycerides and cholesterol gave an apparent urea concentration 9 mg/dL lower with the direct

Table 6. Variation of Instrument Error in Absorbance of an Inexpensive Spectrophotometer[a]

Absorbance	Instrument error, %	Absorbance	Instrument error, %
0.05	9.22	0.90	2.16
0.10	4.90	1.00	2.39
0.20	2.81	1.10	2.69
0.30	2.16	1.20	3.05
0.40	1.91	1.30	3.51
0.50	1.81	1.40	4.07
0.60	1.80	1.50	4.76
0.70	1.86	1.60	5.61
0.80	1.98	1.70	6.67

[a] A reading error of 0.5% transmission was assumed for both blank and test in calculation of the instrument error.

Table 7. Serum Urea in 25 Samples, as Determined by Manual and Automated Methods

Urea concn, mg/dL

Manual		Automated
Direct	Filtrate[a]	(AutoAnalyzer)
15	16	15
20	20	17
22	21	21
27	26	28
35	36	34
44	47	43
57	58	56
77	82	81
89	94	86
151	158	150
13	11	11
20	19	15
22	22	21
25	26	24
36	36	32
39	40	39
49	50	51
55	56	62
173	167	180
205	193	204
16	16	15
28	26	26
34	34	34
42	44	41
50	51	49
53.76[b]	53.96[b]	53.40[b]

[a] Tungstic acid protein-free filtrate.
[b] Mean.

method than with the filtrate. With an authentic lipemic serum with milk-like appearance, the difference between the two types of specimens was 15 mg/dL.

Reference Intervals

The "normal" range of urea in blood is widened by variation in protein and water intake, even after an overnight fast (16). Healthy individuals with high protein intake and low water consumption will have serum urea concentrations much higher than similar individuals on a low-protein diet and high water intake. In a study of 278 normal subjects, MacKay and MacKay (16) observed a total blood urea nitrogen range of 5 to 23 mg/dL with a "great majority" between 8 and 18 mg/dL. Schales (17) in the first volume of this series reported a reference interval of 10–18 mg/dL in conjunction with his urease–Nessler reagent method, and in a later volume Kaplan (18) reported 7–18 mg/dL with the urease–Berthelot reaction method.

Note: Evaluator R. C. E. states that, in his experience, the range of values in the literature is too narrow. In his laboratory a study on 204 ambulatory individuals admitted for elective surgical procedures showed a nongaussian distribution skewed towards higher values. The 95% central range extended from 4 to 27 mg/dL.

Evaluator J. D. B. observed in 51 adults a central 95% percentile range of 7 to 27 mg/dL with a median of 14, and a mean of 14.9 mg/dL. She used the original reagent of 1 mL of the diacetyl monoxime/thiosemicarbazide plus 5 mL of the phosphoric acid solution.

Evaluator M. M. reports 8–25 mg/dL as the normal range determined with a commercial version of the diacetyl monoxime method.

However, Hoffman (19) pointed out that serum urea nitrogen concentrations greater than 16 mg/dL may be suspect in patients after several days on the usual hospital regimen high in fluids and low in proteins. Therefore, use of 8–18 mg/dL as the "normal" range for serum urea nitrogen is advisable. In terms of urea concentration, the range is 17–39 mg/dL (2.8–6.7 mmol/L).

References

1. Fearon, W. R., The carbamido diacetyl reaction: A test for citrulline. *Biochem. J.* **33**, 902–907 (1939).
2. Andersen, C. J., and Strange, B., Colorimetric determination of urea. *Scand. J. Clin. Lab. Invest.* **11**, 122–127 (1959).
3. Ormsby, A. A., A direct colorimetric method for the determination of urea in blood and urine. *J. Biol. Chem.* **146**, 595–604 (1942).
4. Dickenman, R. C., Crafts, S., and Zak, B., Use of alpha diketones for analysis of urea. *Am. J. Clin. Pathol.* **24**, 981–984 (1954).
5. Natelson, S., Scott, M. L., and Beffa, C., A rapid method for the estimation of urea in biologic fluids by means of the reaction between diacetyl and urea. *Am. J. Clin. Pathol.* **21**, 275–281 (1951).
6. Beale, R. N., and Croft, D., A sensitive method for the colorimetric determination of urea. *J. Clin. Pathol.* **14**, 418–424 (1961).
7. Friedman, H. S., Modification of the determination of urea by the diacetyl monoxime method. *Anal. Chem.* **25**, 662–664 (1953).
8. Kitamura, M., and Iuchi, I., An improved diacetyl monoxime method for the determination of urea in blood and urine. *Clin. Chim. Acta* **4**, 701–706 (1953).
9. Marsh, W. H., Fingerhut, B., and Miller, H., Automated and manual direct methods for the determination of blood urea. *Clin. Chem.* **11**, 624–627 (1965).
10. Evans, R. T., Manual and automated methods for measuring urea based on a modification of its reaction with diacetyl monoxime and thiosemicarbazide. *J. Clin. Pathol.* **21**, 527–531 (1968).
11. Coulombe, J. J., and Favreau, L., A new simple semimicro method for colorimetric determination of urea. *Clin. Chem.* **9**, 102–108 (1963).
12. Moore, J. J., and Sax, S. M., A revised automated procedure for urea nitrogen. *Clin. Chim. Acta* **11**, 475–477 (1965).
13. Wybenga, D. R., Di Giorgio, J., and Pileggi, V. J., Manual and automated methods for urea nitrogen measurements in whole serum. *Clin. Chem.* **17**, 891–895 (1971).
14. Ralls, J. O., Urea is not equally distributed between the water of the blood cells and that of the plasma. *J. Biol. Chem.* **151**, 529–541 (1943).

15. Fales, F. W., Water distribution in blood during sickling of erythrocytes. *Blood* **51**, 703–709 (1978).
16. MacKay, E. M., and MacKay, L. L., The concentration of urea in the blood of normal individuals. *J. Clin. Invest.* **4**, 295–306 (1927).
17. Schales, O., Urea nitrogen. *Stand. Methods Clin. Chem.* **1**, 118–122 (1953).
18. Kaplan, A., Urea nitrogen and urinary ammonia. *Stand. Methods Clin. Chem.* **5**, 245–256 (1965).
19. Hoffman, W. S., *The Biochemistry of Clinical Medicine*, 4th ed., Year Book Medical Publisher, Chicago, IL, 1970, p 353.

Addendum: Speculations Concerning the Mechanisms of the Fearon Reaction

The proposal by Dickenman et al. *(1)* that 4,5-dimethyl-2-imidazolone is the colored product of the Fearon reaction was based on the analogous reaction between *o*-diphenylene diamine with glyoxal and α-diketones. The former yields quinoxaline, and the latter, substituted quinoxalines *(2)*. Furthermore, urea reacts with 1-phenyl-1,2-propanedione *(3)* or with α-isonitrosopropiophenone *(4)* to form a colored product with greater maximum absorbance and longer wavelength of maximum absorbance than with diacetyl monoxime. The postulated product would be 4-methyl-5-phenyl-2-imidazolone and the observed changes in absorbance would be expected from the increased extent of the conjugated system.

Beale and Croft *(5)* proposed that the six-membered ring structure, 3-hydroxy-5,6-dimethyl-1,2,4-triazine, was the colored product from reaction between urea and diacetyl monoxime. However, this structure does not seem possible because Lugosi et al. *(6)* have conclusively shown that the identical colored product was obtained when urea reacted either with diacetyl or with diacetyl monoxime. Interestingly, this same dimethyl triazine structure was proposed for the variant purple product obtained by reaction of urea with the dimethylglyoxime/thiosemicarbazide reagent *(7, 8)*.

Others *(3, 6)* have postulated that a relatively insoluble, crystalline material isolated from the acid-catalyzed reaction between urea and diacetyl was the immediate precursor of colored product of Fearon reaction. Schiff *(9)* isolated a white, crystalline product from reacting urea, glyoxal, and hydrochloric acid at 100 °C, with elemental analysis corresponding to the following structures:

Schiff considered that there was little probability that two different urea molecules would react simultaneously with glyoxal to form structure **1**. Furthermore, the compound was hardly affected by boiling with acetic anhydride, a stability not expected for a diamide. Therefore, structure **2**, the glycoluril, was chosen. Franchimont and Klobbie *(10)* were the first to isolate the similar product from reaction between diacetyl and urea. Its elemental analysis agreed with structure **4** below. The initial step of the synthesis might be similar to the reaction of benzamidine [C_6H_5—C(=NH_2)—NH_2] with diacetyl *(11)* with a tautomeric form of urea being the reactant as shown below:

Compound **3** may then add a second urea to form compound **4**, dimethyl glycoluril. Compound **3** (4,5-dihydroxy-4,5-dimethyl-2-imidazolidone) is produced in the base-catalyzed reaction between urea and diacetyl, and its ring structure has been proved by isolation of the expected *cis* and *trans* dihydroxy isomers *(12)*. Also, considerable data support the given structure of compound **4**. Rheineck *(13)* found that allantoin was converted to glycoluril (compound **2**) by treatment with sodium amalgam, a reaction involving a two-electron reduction and ring closure with release of a water molecule. Furthermore, two isomers resulted from the reaction of methyl urea with glyoxal *(14–16)*.

Nematollahi and Ketcham *(16)* found that the NMR spectrum of the major, more symmetrical, product showed two sharp peaks with a ratio of 1:3, consistent with six identical methyl hydrogens and two identical tertiary hydrogens; the spectrum of the minor product showed one peak of methyl hydrogens and a four-line resonance pattern ascribed to the spin-spin splitting of the two adjacent, non-identical, tertiary hydrogens.

Vaniamin and Vakirtzi-Lemonias *(3)* proposed that compound **3** was an intermediate and compound **4** was the immediate precursor of the yellow pigment of Fearon reaction. When compound **4** was dissolved in hot acid reagent, they found that during the disappearance of the dimethyl glycoluril, as determined by absorbance measurements at 256 nm, there was a concomitant increase in yellow pigment measured at 478 nm.

Despite the strong evidence in support of the dimethyl glycoluril structure, Lugosi et al. *(6)* proposed that the compound was actually diacetyl diureide, the dimethyl derivative of compound 1, and proposed it as the precursor of the Fearon reaction colored product. In addition to the cited chemical evidence against this, their diamide structure does not seem to fit the physical characteristic of very low solubility in water. They based their supposition chiefly upon release of a colorless, odorless, noncombustible gas with nitrous acid. However, the sample had been dissolved in 3 mol/L HCl and heated in a boiling water bath for 10 min before the nitrous acid test. When diacetyl reacts with urea in hot acid solution, a number of products and intermediates may form, and the eventual separation from solution of the dimethyl glycoluril crystals by no means indicates that it was the chief product, but rather that it was the least soluble product. Thus when the glycoluril was heated with the 3 mol/L HCl, it is likely that the equilibria between the reactants and products were reestablished. The observed gas production may have been CO_2 and N_2, derived from the action of nitrous acid on urea, or N_2 derived from some amide intermediate. By similar logic, the formation of the yellow pigment from diacetyl glycoluril *(3)* does not prove that the glycoluril was a precursor of the colored product under the conditions of the Fearon reaction. When urea and diacetyl monoxime are heated with added HCl at concentrations recommended for dimethyl glycoluril synthesis *(6)*, first the soluble, yellow pigment appears and increases in concentration with time; much later, the insoluble diacetyl glycoluril gradually separates out from the solution, while the intensity of the yellow Fearon-reaction pigment continues to increase. Thus the production of the yellow pigment and the glycoluril seem to be independent of each other. Although there is no direct evidence that the yellow pigment has the 4,5-dimethyl-2-imidazolone structure proposed by Dickenman et al. *(1)*, its conjugated ring structure make it a plausible colored product.

References

1. Dickenman, R. C., Crafts, S. B., and Zak, B., Use of alpha diketones for analysis of urea. *Am. J. Clin. Pathol.* **24**, 981–986 (1954).
2. Heterocyclic compounds. In *Chemistry of Carbon Compounds*, **4B**, E. A. Rodd, Ed., Elsevier Press, Amsterdam, Holland, 1959, p 1346.
3. Vaniaman, M. P., and Vakirtzi-Lemonias, C., Chemical basis of the carbamido-diacetyl micromethod for estimation of urea, citrulline, and carbamyl derivatives. *Clin. Chem.* **16**, 3–6 (1970).
4. Archibald, R. M., Colorimetric determination of urea. *J. Biol. Chem.* **157**, 507–518 (1945).
5. Beale, R. N., and Croft, D., A sensitive method for colorimetric determination of urea. *J. Clin. Pathol.* **14**, 418–424 (1961).
6. Lugosi, R., Theibert, W. J., Holland, W. J., and Lam, L. K., A study of the reaction of urea with diacetyl monoxime and diacetyl. *Clin. Biochem.* **5**, 171–181 (1972).
7. Khramov, V. A., and Galaev, Y. V., Thiosemicarbazide modification of the Fearon reaction and its use for quantitative determination of urea. *Vopr. Med. Khim.* **15**, 435–439 (1969) [cited in *Chem. Abs.* **71**, 98880a (1969)].
8. Khramov, V. A., and Narbutovich, N. I., Determination of urea in blood by color reaction with dimethyl glyoxime. *Lab. Delo* **7**, 439–440 (1969) [cited in *Chem. Abs.* **71**, 1805b (1969)].
9. Schiff, H., Ueber Acetyleneharnstoff. *Ann.* **189**, 157–162 (1878).
10. Franchimont, A. P. N., and Klobbie, E., Sur quelques uréides et leurs derives nitres. *Recl. Trav. Chim.* **7**, 236–257 (1888).
11. Heterocyclic compounds. In *Chemistry of Carbon Compounds* (see ref. *2*), **4A**, 1957, p 287.
12. Vail, S. L., Barker, R. H., and Mennitt, P. G., Formation and identification of *cis* and *trans* dihydroxyimidinones from urea and glyoxal. *J. Org. Chem.* **30**, 2179–2182 (1962).
13. Rheineck, H., Ueber das Verhalten des Allantoins zu Natrum. *Ann. Chem.* **134**, 219–228 (1865).
14. Franchimont, A. P. N., and Klobbie, E., Quelques nouveaux derives de l'ureae. *Recl. Trav. Chim.* **7**, 12–24 (1888).
15. Weitzner, E., IV. Ueber Dimethylglykolurile und β-Methylhydrantoin. *Ann. Chem.* **362**, 125–131 (1908).
16. Nematollahi, J., and Ketcham, R., Imidazoimidazole. I. The reaction of ureas with glyoxal. Tetrahydroimidazo-(4,5-d)-imidazole-2,5-diones. *J. Org. Chem.* **28**, 2378–2380 (1963).

Uric Acid

Submitter: Wendell T. Caraway, *Laboratories of McLaren General Hospital and St. Joseph Hospital, Flint, MI 48502*

Evaluators: Lester I. Burke, *Toledo Hospital Laboratory, Toledo, OH 43623*
Samuel Y. Chu, *Ottawa General Hospital, Ottawa, Ontario K1N 5C8, Canada*
George M. Maruyama, *Medical Associates, Dubuque, IA 52001*

Reviewer: Nathan Gochman, *Clinical Chemistry Section, Laboratory Service, Veterans Administration Medical Center, San Diego, CA 92161*

Introduction

Numerous methods have been described for the determination of uric acid in biological fluids. Earlier techniques were applied to whole blood and required the isolation of uric acid as its silver salt. For colorimetric estimation the silver urate was dissolved in cyanide solution and measured by its reducing action on phosphotungstate in mildly alkaline solution. Because cyanide also increased the intensity of the color produced, it was retained in some methods after the silver precipitation technique had been abandoned. The introduction of the use of uricase permitted the development of highly specific methods for measurement of uric acid and provided criteria for evaluating the older colorimetric procedures.

Colorimetric methods for the most part are based on the reduction of phosphotungstate at pH 9–10 to produce a blue color. Uricase/colorimetric methods depend on the colorimetric measurement of chromogens before and after treatment of the sample with uricase; the difference is assumed to represent "true" uric acid.

Uricase/spectrophotometric methods are based on direct measurement of the decrease in absorbance at 292 nm after treatment with uricase. Uric acid may also be separated by ion-exchange chromatography and its absorbance measured directly at 292 nm.

Various reagents have been used to adjust the alkalinity or to intensify the color of the final reaction mixture. Buffers include carbonate, phosphate, tungstate, silicate, and borax. Cyanide, hydrazine, and hydroxylamine improve sensitivity and specificity of the reaction, inasmuch as the color produced by uric acid is increased as much as 10-fold in the presence of these reagents.

Principle

The method presented is based on the reduction of phosphotungstate by urate in carbonate buffer at pH 10.2. The blue-colored reduced oxides of phosphotungstate are readily measured with a spectrophotometer in the red portion of the spectrum (1).

Materials and Methods

Reagents

1. *Stock phosphotungstic acid reagent.* Transfer 50 g of molybdate-free sodium tungstate, $Na_2WO_4 \cdot 2H_2O$, to a 500-mL Erlenmeyer flask and dissolve in 75 mL of water. Add 40 mL of 85% (850 g/kg) o-phosphoric acid and warm the solution to dissolve any precipitate. Cover with an inverted beaker and place in an autoclave at 120–125 °C for 2 h. Cool to room temperature, transfer with rinsing to a 500-mL volumetric flask, and dilute to the mark with water. The solution has a slight green color. Store in a brown bottle at room temperature. The solution is stable indefinitely.

Because the acidity of the reagent is important in the reaction, other phosphotungstic acid reagents may not be substituted indiscriminantly. An equivalent reagent is available commercially (no. 3033, Uric Acid Reagent; Harleco, Div. of EM Industries, Inc., Gibbstown, NJ 08027).

Note: Evaluator L. I. B. found that the reagent for the Technicon SMA 12/60 (CS 280–1; Fisher Scientific Co., Pittsburgh, PA 15219) also gave equivalent results.

2. *Dilute phosphotungstic acid.* Dilute 10.0 mL of stock reagent to 100 mL with water. This is stored in a brown glass bottle and is stable for at least one year.

3. *Sodium carbonate solution, 100 g/L.* Dissolve 100 g of anhydrous Na_2CO_3 in water and dilute to 1000 mL. Filter the solution if it is not as clear as water. It is stable when stored in a tightly stoppered alkali-resistant bottle or polyethylene container.

4. *Sodium tungstate, 100 g/L.* Dissolve 100 g of molybdate-free $Na_2WO_4 \cdot 2H_2O$ in water and dilute to 1000 mL.

5. *Sulfuric acid, 0.33 mol/L*. Dilute 18.5 mL of concd. sulfuric acid (sp. grav. 1.84) to 1000 mL with water.

6. *Tungstic acid solution.* To 800 mL of water add, with mixing, 50 mL of 100 g/L sodium tungstate, 0.05 mL of 850 g/kg *o*-phosphoric acid, and 50 mL of 0.33 mol/L sulfuric acid. Stored in a borosilicate or polyethylene bottle, this solution will usually be stable for months; discard if it becomes cloudy. A fine yellow sediment may form occasionally on glass surfaces, but this does not interfere with the use of the reagent.

7. *Uric acid stock standard, 100 mg/dL*. Transfer 1.000 g of uric acid to a 1000-mL volumetric flask. Dissolve separately 600 mg of lithium carbonate, Li_2CO_3, in 150 mL of warm water, filter, and heat the filtrate to 60 °C. Ignore a slight turbidity at this point. Add the warm lithium carbonate solution to the flask and mix until the uric acid is completely dissolved (about 5 min); the flask may be warmed under running hot water if necessary, then held under running cold water to cool. The solution may remain slightly turbid. Add 20 mL of 400 g/L formaldehyde (formalin) solution. Dilute to about 500 mL with water, then add slowly and with mixing 25 mL of 0.5 mol/L sulfuric acid (prepared by diluting concd. acid 36-fold (1 + 35). Dilute this solution to 1 L with water, mix well, and store in a brown bottle in the cold. This standard is stable for five years or longer at 2–10 °C.

8. *Dilute uric acid working standard, 0.5 mg/dL*. Transfer 0.50 mL of uric acid stock standard to a 100-mL volumetric flask and dilute to the mark with benzoic acid solution (2 g/L). This working standard is stable for at least six months at room temperature.

Collection and Handling of Specimens

Serum is preferred. Plasma is also satisfactory if heparin or ethylenediaminetetraacetate (EDTA) (sodium or lithium salt) is used as an anticoagulant. Potassium salts should not be used because they will produce turbidity with phosphotungstate under the reaction conditions. Slight hemolysis does not interfere significantly, but grossly hemolyzed specimens should not be used. Erythrocytes contain glutathione and ergothioneine, which act as reducing substances and produce falsely higher results. Uric acid in serum is stable for at least seven days if the serum is removed from the clot and stored at 4 °C.

Procedure for Determination in Serum

1. Add 9.0 mL of tungstic acid solution to 1.0 mL of serum. Mix thoroughly and centrifuge or filter.
2. Transfer 5.0 mL of clear supernate or filtrate to a test tube or cuvet. Transfer 5.0 mL of water to a second tube for a blank and 5.0 mL of dilute standard (0.5 mg/dL) to a third tube.
3. To each tube add 1.0 mL of sodium carbonate solution (100 g/L) and mix; let stand 10 min.
4. Add 1.0 mL of dilute phosphotungstic acid, *mix immediately*, and let stand 30 min.
5. Measure the absorbances of the standard and unknown within the next 20 min vs the blank with a spectrophotometer set at 700 nm. Any photoelectric colorimeter equipped with a red filter may be used (nominal wavelength 640–720 nm).

Note: Evaluator G. M. M. performs analyses in duplicate by using 2 mL of supernate, water, or dilute standard, followed by 0.4 mL each of sodium carbonate and phosphotungstic acid solutions.

Procedure for Determination in Urine

If the urine is cloudy, warm a portion of the well-mixed specimen at 60 °C for 10 min to dissolve any precipitated urates. Dilute 1.0 mL of urine to 100 mL with water. Use 5.0 mL of the diluted urine and proceed as described for the protein-free filtrate of serum.

Calculations

Prepare a standard curve to determine the extent of agreement with Beer's Law for the particular spectrophotometer or colorimeter used. When observed absorbances *(A)* fall on the linear portion of the curve, the concentration of uric acid in the unknown may be calculated from the equation:

$$(A_{unknown}/A_{standard}) \times 5 = \text{mg of uric acid per 100 mL of serum}$$

For urines initially diluted 100-fold (1 + 99),

$$(A_{unknown}/A_{standard}) \times 50 = \text{mg of uric acid per 100 mL of urine}$$

The standard calibration curve usually follows Beer's Law to a concentration equilavent to 10 mg of uric acid per 100 mL of serum. Higher values may be read from the curve, or the analysis may be repeated by starting with 2.0 mL of filtrate and 3.0 mL of water. With the latter procedure, final results are multiplied by 5/2.

Calibration

Prepare a standard calibration curve as follows: Transfer 3.00 mL of uric acid stock standard (100 mg/dL) to a 200-mL volumetric flask and dilute to the mark with water to provide a dilute standard containing 1.5 mg/dL. Prepare working standards by diluting 1.0, 2.0, 3.0, 4.0, and 5.0 mL of the dilute standard to 5.0 mL with water. Proceed with color development of the standards and a blank, as described under *Procedure*. These standards correspond to 3, 6, 9, 12, and 15 mg of uric acid per 100 mL of serum. On linear graph paper, plot the absorbance for each standard vs the equivalent concentration of uric acid, expressed as milligrams per 100 mL of serum. If desired, all readings may be taken in percent transmittance; in that case, set the blank equal to 100% transmittance and plot the transmittance readings for the standards on semilog paper.

Discussion

Analytical recovery of uric acid added to serum before precipitation of protein ranged from 96 to 102%. For these studies a stock standard was prepared in lithium carbonate solution only and was used the same day. Recoveries will be much lower if the standard containing formaldehyde is used because some uric acid binds to protein in the presence of formaldehyde.

> *Note:* Evaluators found mean recoveries ranging from 94 to 98%.

Precision studies conducted on a single pool of serum for 20 days showed a mean value of 6.1 (SD 0.21) mg/dL and a coefficient of variation (CV) of 3.4%.

> *Note:* Evaluators found day-to-day CVs ranging from 2.9 to 4.4% and within-day CVs ranging from 1.2 to 3.5%.

Accuracy was evaluated on 53 serum samples analyzed before and after incubation with uricase. Values for total chromogens ranged from 2.4 to 12.6 mg/dL. Values for non-urate chromogens (after treatment with uricase to destroy uric acid) ranged from 0 to 0.4 mg/dL, with a mean value of 0.10 mg/dL (2). The major non-urate chromogen in fresh serum appears to be ascorbic acid (3), which is effectively destroyed during the 10-min incubation with sodium carbonate. Other substances that may produce color in the reaction are methyldopa, levodopa, and metabolites of caffeine and theophylline. These have a negligible effect at concentrations normally found in serum, but may cause falsely high values for uric acid in urine (4).

> *Note:* Evaluators L. I. B. and S. Y. C. compared the above method with the SMA 12/60 saline method for serum and found no significant differences ($p > 0.05$). L. I. B. also found no significant difference between this method and a uricase procedure (no. 292; Sigma Chemical Co., St. Louis, MO 63178). Evaluator G. M. M., however, found that the colorimetric method gave results about 0.6 mg/dL higher than another commercial uricase procedure (American Monitor Corp., Indianapolis, IN 46268). Reviewer N. G. noted that patients on chronic hemodialysis may exhibit spuriously high values for uric acid, because of the accumulation of non-urate chromogens.

Full color development in the procedure described depends on the presence of sufficient carbonate to bring the pH of the final reaction mixture to 10.0–10.4. This requires a balance between the sodium carbonate and the acidity contributed by the protein-free filtrate and the phosphotungstic acid.

An increase in the concentration of sodium carbonate results in a more rapid color development but decreases the period of constant absorbance. Increasing the concentrations of phosphotungstic acid and carbonate will result in better agreement with Beer's Law at higher concentrations of urate, but turbidities tend to develop.

Temperature control is not critical. Final reaction mixtures were incubated for 30 min at 20, 25, or 30 °C, with equal results. The final absorbance obtained was also the same after 30 min for mixtures incubated in the dark or exposed to sunlight.

Reference Intervals

Normal values obtained for serum uric acid by various investigators are reasonably consistent (1). Typical ranges are: men, 3–7 mg/dL; women, 2–6 mg/dL. Children have values in the same range as women. Values in adults show a strong positive correlation with body weight (6). The daily variation in the serum uric acid of an individual is approximately 0.4–0.5 mg/dL. Normal concentrations in serum are slightly lower in subjects receiving a low-purine diet.

Urinary excretion on an average diet ranges from 250 to 750 mg/24 h but varies with the purine content of the diet (low-purine diet, less than 450 mg/24 h; high-purine diet, as much as 1200 mg/24 h).

Increased values for serum uric acid (hyperuricemia) are observed in all forms of nephritis with nitrogen retention. Values as high as 10 mg/dL are frequently observed; values exceeding 15 mg/dL are unusual.

Hyperuricemia occurs commonly in chronic leukemia, particularly myeloid, and also occasionally in polycythemia and multiple myeloma. This is probably a reflection of increased production of uric acid associated with an increase in the number of short-lived cells and the resulting increase in nucleic acid turnover and metabolism.

The hyperuricemia of primary gout is caused by excessive production of purines and by renal retention of uric acid. Increase in serum uric acid, often as great as 10–15 mg/dL, occurs almost always in acute attacks of gout.

The Lesch–Nyhan syndrome presents in childhood and is associated with an almost complete deficiency of the enzyme hypoxanthine phosphoribosyl transferase (EC 2.4.2.8). This leads to increased uric acid production and hyperuricemia. The amount of uric acid in the urine is five- to sixfold normal.

Other conditions in which hyperuricemia may occur are lead poisoning, alcoholism, Down's syndrome, cancer chemotherapy, postirradiation, and toxemia of pregnancy.

Low values for serum uric acid are less commonly observed but may occur after administration of probenecid, cortisone, allopurinol, coumarins, and high doses of salicylate. Hereditary xanthinuria is a relatively benign and rare disorder in which there is a deficiency of xanthine oxidase (EC 1.2.3.2). This results in low uric acid production and excretion of the precursor, xanthine. Serum uric acid is typically 1 mg/dL or less, and urinary excretion is less than 50 mg/24 h.

Alternative Micromethod for Serum

The carbonate method described above does not produce sufficient color to permit its use in a simple, scaled-down micromodification. The following method is based on that of Brown (5) and depends on the reduction of phosphotungstic acid by uric acid in the presence of cyanide and urea.

Reagents

1. *Sodium cyanide, 120 g/L* (**POISON**). Dissolve 12 g of sodium cyanide in water and dilute to 100 mL. This is stable for at least one month when stored in a dark bottle at 2–10 °C. Dispense with an automatic dispenser or from a pipette equipped with a safety pipetting device. Use analytical reagent grade NaCN to lessen the color of the reagent blank. **CAUTION:** Reagents or solutions containing cyanide should be disposed of in an acid-free drain, in a well-ventilated hood, and flushed with a large amount of water.
2. *Urea, 500 g/L*. Dissolve 50 g of urea in water and dilute to 100 mL. This solution is stable at room temperature.
3. *Stock phosphotungstic acid reagent*. Same as in the macro procedure.
4. *Tungstic acid solution*. Same as in the macro procedure.
5. *Uric acid stock standard, 100 mg/dL*. Same as in the macro procedure.
6. *Uric acid working standard, 5 mg/dL*. Transfer 5.0 mL of uric acid stock standard to a 100-mL volumetric flask and dilute to the mark with benzoic acid solution (2 g/L). This solution is stable at least six months at room temperature.

Procedure

1. Pipet 50 µL of serum or working standard to small test tubes, add 2.0 mL of tungstic acid solution, mix thoroughly, and centrifuge.
2. Transfer 1.0 mL of clear supernatant liquid from each test tube to clean test tubes or cuvets.
3. Pipet 1.0 mL of tungstic acid solution into another test tube for a reagent blank.
4. To each tube add 1.0 mL of NaCN solution (**poison**) and 1.0 mL of urea solution; mix. Add 0.5 mL of stock phosphotungstic acid reagent; mix.
5. Let stand 45 min at room temperature.
6. Measure the absorbance of the standard and unknown against the blank in a spectrophotometer set at 600 nm.

Note: Evaluator L. I. B. found that the color was stable for at least 50 min after the 45-min incubation period.

Calculations

Calculate the concentration of uric acid in the unknown the same as with the macro method. The standard curve is usually linear to 15.0 mg/dL. For higher concentrations, repeat the analysis by starting with 0.5 mL of protein-free supernate and 0.5 mL of tungstic acid solution; multiply the final results by 2.

Calibration

Dilute exactly 0.25, 0.50, 0.75, 1.00, 1.25, and 1.50 mL of uric acid stock standard (100 mg/dL) to 10.0 mL with water and mix well. These working standards contain, respectively, 2.5, 5.0, 7.5, 10.0, 12.5, and 15.0 mg of uric acid per 100 mL. Pipet 50 µL of each working standard to test tubes and proceed exactly as described for the microprocedure for serum, including the blank. On linear graph paper plot the absorbance for each standard against the corresponding concentration.

Discussion

With this micromethod, Evaluators S. Y. C. and L. I. B. found average recoveries of added uric acid to be 96 and 97%. In precision studies, CVs averaged about 5% for within-day and day-to-day studies. Because maximum absorbance for this reaction occurs near 700 nm, this wavelength may be used to improve precision. The wavelength of 600 nm is suggested for obtaining a working range up to 15 mg/dL.

Evaluator S. Y. C. found no significant difference between the micromethod and the SMA 12/60 procedure ($p > 0.05$). Evaluator L. I. B. found no significant difference between the micromethod and a commercial uricase procedure (no. 292; Sigma Chemical Co.).

References

1. Caraway, W. T., Uric acid. *Stand. Methods Clin. Chem.* **4**, 239–247 (1963).
2. Caraway, W. T., and Marable, H., Comparison of carbonate and uricase–carbonate methods for the determination of uric acid in serum. *Clin. Chem.* **12**, 18–24 (1966).
3. Caraway, W. T., Non-urate chromogens in body fluids. *Clin. Chem.* **15**, 720–726 (1969).
4. Caraway, W. T., and Kammeyer, C. W., Chemical interference by drugs and other substances with clinical laboratory test procedures. *Clin. Chim. Acta* **41**, 395–434 (1972).
5. Brown, H., The determination of uric acid in human blood. *J. Biol. Chem.* **158**, 601–608 (1945).
6. Goldberg, D. M., Handyside, A. J., and Winfield, D. A., Influence of demographic factors on serum concentrations of seven chemical constituents in healthy human subjects. *Clin. Chem.* **19**, 395–402 (1973).

Urinalysis

Submitters: K. Owen Ash, *Department of Pathology, College of Medicine, University of Utah, Salt Lake City, UT 84132*

Gordon P. James, *Department of Pathology, University of Texas Medical Branch, Galveston, TX 77550*

Evaluators: William L. Gyure, *Department of Pathology, Jersey City Medical Center, Jersey City, NJ 07304*

Michael Guarnieri, *Automated Analytical Laboratory Services, Medical Products Division, 3M Company, Baltimore, MD 21215*

Charles A. Pennock, *Paediatric Chemical Pathology, Bristol Maternity Hospital, University of Bristol, BSZ 8EG, U.K.*

Reviewer: Rosanne M. Savol, *Scientific and Regulatory Affairs, Professional Products Group, Miles Laboratories, Inc., Elkhart, IN 46515*

Introduction

More tests are performed on urine than on any other body fluid; urine testing is performed not only in the clinical laboratory but also on the wards, in the clinics, in the physician's office *(1, 2)*, and even by patients in their homes. Almost two decades have passed since an informative, well-written text on urinalysis published the statement, "Modern tablet, tape, and dipstick tests of urinalysis are useful because they require no laboratory and they can be run and read correctly by intelligent patients as well as by the physician at the bedside" *(3)*. This attitude that dipstick urinalysis is the near ultimate in analytical simplicity still persists today, but the data do not justify such a casual approach *(4–10)*. When dipstick results obtained on hospital wards are compared with laboratory results on the same sample by identical tests, the false-positive and false-negative rates are alarmingly high; these rates were 15 and 17%, respectively, in a study by Simpson and Thompson *(11)*. Even though dipsticks, tablets, and tapes may involve "state of the art" chemistry, those analytical tools must be used by trained personnel who understand the limitations of the tests, or the results can be misleading. The use of any test without thought as to interpretation of the results is a useless exercise. A negative test result (i.e., the absence of an abnormal metabolite) does not necessarily exclude the disorder. For example, infants not receiving milk cannot excrete galactose even if they have galactosemia, and infants on low-protein intake will not excrete excess phenylketones even if they have phenylketonuria. Urine screening tests alone are inadequate for detection of these two important, treatable, inherited disorders. It may be important to question why a negative test is obtained when a positive result is expected.

This chapter includes tests having wide application in small clinical laboratories. Heavy emphasis is placed on routine urinalysis with dipsticks, and on the backup tests necessary to confirm significant results. Selected reference works provide details for additional laboratory tests, not covered in this chapter, that one might need to consider *(12–15)*. Unless circumstances require otherwise, tests infrequently ordered should be referred to a reputable reference laboratory. The difficulty of maintaining fresh reagents and technical competence usually dictates that infrequent tests and tests requiring special technical expertise can be done more accurately and economically in a large laboratory setting.

Quality Control

No test should be performed in the clinical laboratory until an adequate control program has been established. Too frequently, quality-control programs for urinalysis have been weak or even absent. Yet, quality control in urinalysis is as urgent as in any other area of the laboratory. Quality-control programs for the small clinical laboratory need not be complex, but they must be clearly defined, written down, and followed carefully to assure high-quality results. Commercial control products that might be used include QC-U®, TEK-CHEK®, and URINTROL® (Products of General Diagnostics, Div. of Warner-Lambert Co., Morris Plains, NJ 07950; Ames Co., Div. of Miles Laboratories, Inc., Elkhart, IN 46515; and DADE Div., American Hospital Supply Corp., Miami, FL 33152, respectively). Without quality control, the analyst has no basis for confidence in test results. The essential features of a quality assurance program for small clinical laboratories are discussed in other chapters in this book.

Specimen Collection

The clinical value of urinalysis is very much dependent on the quality of the specimen. The results may be useless or even misleading unless proper attention is given to collection of the specimen. The second morning specimen, which provides freshly formed urine, is generally considered to be most suitable for a random specimen sample.

For routine urinalysis, preservatives are generally not required. The specimen should be properly collected and analyzed within 60 min *(14)*. If longer delays are expected, the specimen can be refrigerated for up to 12 h. Some changes are to be expected when specimens are refrigerated. When a urine sample is to be sent to a reference laboratory for testing, it is important to consult with the reference laboratory concerning sample preservation. The more common urine preservatives include boric, glacial acetic, and hydrochloric acids *(13)*.

When urine specimens are intended for microbiological workup, "clean-catch" specimen collection is very important. Children and infants will require the assistance of trained personnel to obtain proper urine specimens. Adult patients can be given simple instructions for "clean-catch" urine specimen collection (see Table 1) *(16)*.

Table 1. Instructions for Obtaining Clean-Catch Urine *(16)*

Male patients
1. Set the collecting material on back of the toilet, and remove the cover from the sterile, disposable specimen cup. DO NOT touch the inside of cup.
2. Wash hands.
3. Expose penis.
4. Wash penis with soaped cotton balls. Retract the foreskin if not circumcised.
5. Rinse with sponge. Put the used sponge in a waste basket, NOT in the toilet.
6. Void a small amount of urine into the toilet to cleanse the urethral canal.
7. Hold the cup by its outside and pass urine into it.
8. Replace the cover on the cup and return cup to the laboratory.

Female patients
1. Place the collecting tray of materials on the wash basin.
2. Remove underpants completely.
3. Wash hands carefully.
4. Sit facing the back of the toilet; legs will spread appropriately in this position.
5. Take one wet cotton ball from the bowl of cotton balls soaking in green soap.
6. Spread self with the other hand and wash with the wet cotton ball. Do this well, from high up front toward the back. When finished, drop the used cotton ball into the waste container.
7. Continue to keep spread and wash the same way with a second wet cotton ball, then with a third, fourth, and a fifth, and discard each cotton ball as you finish with it. DO NOT use any cotton ball more than once.
8. Remain spread and start urinating. Once started, catch in midstream some urine in a sterile, disposable, clear plastic cup, held so that it does not touch skin or clothing. DO NOT TOUCH the inside of the cup.
9. Replace the lid and return the cup to laboratory.

Table 2. Routine Urinalysis

General properties and appearance
Clarity
Color
Specific gravity
pH

Chemical determinations
Protein
Glucose and (or) other reducing substances
Ketones
Bilirubin
Blood
Urobilinogen

Microscopic examination of urine sediment
Casts
Cells
Crystals
Miscellaneous

Routine Analysis Procedure

Routine urinalysis generally includes the tests listed in Table 2. A suggested procedure for completing the routine urinalysis follows. This procedure requires 7 to 10 min for complete analysis of one sample, 15 min for a batch of three samples, and about 30 min for 12 samples.

1. Mix the urine well; note color and appearance.
2. Using combination reagent strips, perform assays precisely as instructed by the manufacturer (be certain the reagent strips have not expired).
3. When possible, perform confirmation tests on positive results.

Note: Dipsticks may sometimes register false-negative results; vitamin C may interfere with urine glucose and occult blood tests, but usually is not present in urine at concentrations that cause interference. Because protein tests are sensitive only to albumin, proteins such as Bence Jones protein and gamma-globulin will not be detected. Negative as well as positive results should be carefully scrutinized.

4. Pour 10 to 12 mL into a 15-mL conical centrifuge tube. Centrifuge for 5 min at 2000 rpm (300 × g).
5. Determine the specific gravity of the supernate.
6. Pour off the supernate and resuspend the sediment in the residual fluid, approximately 1 mL.
7. Place a drop of the sediment suspension on a microscope slide and cover with a coverslip. Using reduced light, examine under low power (100×) for

epithelial cells, mucus, and casts. In the examination of the urine deposit, it may be helpful to use a nonspecific background stain (e.g., iodine) against which to view cells and casts, but this is usually not necessary if one sets the microscope up correctly and reduces the light sufficiently to identify cells and casts against a darkened background.

Casts often migrate to the edges of the coverslip. Turn the magnification to high-dry (450×), increase the light, and examine for leukocytes, erythrocytes, yeast, bacteria, crystals, and *Trichomonas vaginalis*. A meticulous search of almost any sediment suspension under the microscope will exhibit some unusual or abnormal structures. Such occasional abnormal findings are sometimes insignificant, but the analyst is often not prepared to determine their significance. Therefore, the laboratory should adopt a simple scheme, such as given in Table 3, which permits the analyst to report all observations and yet does not overemphasize nonrepresentative aberrant structures *(16)*.

General Properties and Appearance of Urine

Clarity

Urine is normally clear, but in certain conditions freshly voided urine may appear hazy, cloudy, or even turbid. A cloudy urine may reflect increased numbers of bacteria and leukocytes, and thus indicate urinary tract infection. The cloudy urine might also be related to the precipitation of amorphous or stellar phosphate crystals, urates, or oxalate crystals. Urine containing gross amounts of uric acid crystals can be clarified by warming the sample under tap water until the precipitate dissolves. Other possible causes for cloudy or turbid urine include carbonate and nitrate crystals, erythrocytes, spermatozoa, prostatic fluid, mucin, calculi, fecal contamination, roentgenogram (x-ray) media, and fat *(13)*. Deposition of these substances is not itself significant, but may justify estimation of true 24-h excretion of the substances in certain patients.

Color

Normal urine color ranges from straw-yellow to amber, depending upon the urine composition and concentration. Within this normal range, the color of a specific specimen is not generally informative. However, abnormal variations in urine color may provide valuable clues to the patient's condition. Various medications and other causes for abnormal urine coloration have been reported *(13)*; such colorations may be very alarming to a patient who has not been informed of these phenomena. Red urine may be due to hemoglobin, myoglobin, porphyrin, phenindione, aminopyrine, fuchsin (aniline dye), beets, menstrual contamination, or erythrocytes. Urine may appear quite bloody and yet contain very little blood. Methylene blue, present in various medications, may produce a blue-green urine. Blue-green colorations have also been reported with indigo-carmine, indican, and *Pseudomonas* infections. Brown urine may result from phenacetin, phenols, and biological products such as hemoglobin, myoglobin, bilirubin, homogentisic acid, or melanin. Dark-yellow to brown urine can be caused by conjugated bilirubin, biliverdin, acriflavine, mepacrine, nitrofurantoin, riboflavin, phenazopyridine, urobilin, senna, rhubarb, or cascara *(13)*.

Specific Gravity

Specific gravity is the relationship of the weight of a substance (urine) to the weight of an equal volume of water at a specified temperature, and is related to the number of dissolved particles in the substance *(14)*. Unlike osmolarity, which is a colligative measurement, each substance in the urine contributes somewhat differently to the specific gravity. A variety of substances, including roentgenogram contrast media, intravenous fluids, high concentrations of glucose, and, to some extent, abnormal

Table 3. Reporting Microscopic Observations of Urine Sediment *(16)*

Low-power field (100×)	High-power field (450×)
Squamous	*Erythrocytes*
rare = 1 with 4–5 fields empty	rare = 1 with 4–5 fields empty
occ = 1 with 2–3 fields empty	occ = 1 with 2–3 fields empty
few = less than 25% of field	few = 1–5
1+ = 25% of field	1+ = 6–20
2+ = 50% of field	2+ = 21–50
3+ = 75% of field	3+ = 51–100
4+ = 100% of field	4+ = more than 100
pkd = packed	
Round (transitional)	*Leukocytes*
rare = 1 with 4–5 fields empty	rare = 1 with 4–5 fields empty
occ = 1 with 2–3 fields empty	occ = 1 with 2–3 fields empty
few = less than 25% of field	few = 1–5
1+ = 25% of field	1+ = 6–20
2+ = 50% of field	2+ = 21–50
3+ = 75% of field	3+ = 51–100
4+ = 100% of field	4+ = more than 100
Casts	*Bacteria*
rare = 1 with 4–5 fields empty	1+ = 25% of field
occ = 1 with 2–3 fields empty	2+ = 50% of field
1+ = 1–5	3+ = 75% of field
2+ = 6–10	4+ = 100% of field
3+ = 11–30	
4+ = more than 30	
Crystals	*Renal tubular cells*
rare = 1 with 4–5 fields empty	rare = 1 with 4–5 fields empty
occ = 1 with 2–3 fields empty	occ = 1 with 2–3 fields empty
few = less than 25% of field	1+ = 1–5
1+ = 25% of field	2+ = 6–10
2+ = 50% of field	3+ = 11–30
3+ = 75% of field	4+ = more than 30
4+ = 100% of field	

protein concentrations, influence specific gravity.

Urinary specific gravity is generally determined either directly, with a hydrometer (urinometer), or indirectly, with a Total Solids Meter (American Optical Corp., Buffalo, NY 14215), which measures the refractive index of urine and is calibrated to read specific gravity (17, 18). Refractive index measurements generally correlate sufficiently with the direct methods and require less specimen. Before using a hydrometer, calibrate it with distilled water at a specific temperature (e.g., sp. grav. 0.998 at 20 °C) and then routinely check it with a known solution, possibly saturated ammonium oxalate. Temperature is important. To correct for temperature, add (or subtract) 0.001 for every 3 °C above (or below) the calibration temperature. Make readings at the bottom level of the meniscus; the hydrometer should not touch the side of the container.

The physiological range for specific gravity is between 1.002 and 1.035 (12). Normal values for a 24-h specimen range between 1.015 and 1.025. The specific gravity fluctuates considerably from day to day, but even so, this measurement has clinical value in assessing the concentrating ability of the renal tubules. If antidiuretic hormone (ADH) activity is absent, the specific gravity of excreted urine will be near 1.002; with maximum ADH activity, its specific gravity will approach 1.035. In renal tubular damage, the concentrating function is frequently the first to be lost, resulting in a low specific gravity. Urine specific gravity may increase in dehydration or uncontrolled diabetes.

pH

Acid–base balance is regulated primarily by the kidneys and lungs. Nonvolatile organic acids of metabolic origin are excreted by the kidneys. These acids are excreted as salts, with the kidneys selectively regulating cations, i.e., potassium, sodium, ammonium, or calcium. The kidneys maintain acid–base balance by reabsorbing variable amounts of sodium in exchange for hydrogen and ammonium ions.

A healthy person on an ordinary diet will excrete urine with a pH of approximately 6.0. However, normal urine pH can range from 4.6 to 8, depending on metabolic and dietary considerations (12). Urine produced during the night is usually more acidic than daytime urine. Metabolism of fats generates more acid residues than metabolism of carbohydrates. Normal postprandial urine is more alkaline, presumably as a result of increased acid secretion into the stomach; this has been termed the "alkaline tide."

Urine pH values greater than the normal range may suggest certain pathologic conditions. In renal tubular acidosis, the impaired exchange of hydrogen ions for cations can cause a relatively alkaline urine; the urine pH cannot be lowered below 6 to 6.5 by administration of acid-loading substances (12). Therefore, a urine pH of 5.0 or lower would eliminate renal tubular acidosis as a possible diagnosis. Alkaline urine is also produced in metabolic or respiratory alkalosis. Other reported causes of alkaline urine include ingestion of certain fruits (especially citrus fruits) and vegetables, sodium bicarbonate, potassium citrate, or acetazolamide. Urinary tract infection with urea-splitting bacteria also may result in alkaline urine.

Acid urines may result from metabolic or respiratory acidosis. A paradoxical aciduria may result from prolonged vomiting or in hypercorticism, in which the urine may be somewhat more acidic than normal even in the presence of metabolic alkalosis. Diets high in protein produce acid urine. Ingested ammonium chloride, methionine, and methenamine mandelate have been used therapeutically to acidify urine; pH is a convenient and important monitor of this therapy.

An estimate of urine pH is usually sufficient. Adequate pH measurements can be made in the laboratory by using simple disposable reagent strips that incorporate pH indicators, i.e., methyl red, phenolphthalein, and bromthymol blue. These formulations allow the skilled, trained technologist to estimate urine pH confidently to within 1.0 pH unit (7). If greater accuracy is required, one must use a glass pH electrode. pH determinations should always be made on fresh urine at room temperature. After standing at room temperature, urine pH may drift higher or lower because of bacterial metabolism or loss of CO_2. Urine with a high pH will give a false-positive protein result.

Odor

Characteristic urine odors are produced as a result of certain abnormal conditions. Urine from infants with phenylketonuria may have a distinctly "mousey" smell, and the odor in maple syrup urine disease is unmistakable. Ketones are often detectable, and δ-ketobutyric acid has a distinct "beery" smell. Cystine and homocysteine in high concentration give a sulfurous smell, and the "sweaty feet" odor in the rare patient with isovaleric acidemia is pungently obvious. Certain foods, i.e., asparagus and garlic, impart their own characteristic odor. The smell of ammonia suggests that gross infection is present or that the urine sample is not fresh. Always report any unusual odor, because it could provide a clue to an abnormal condition. Record such observations in the log book.

Chemical Determinations by Dipstick Analysis

Multi-reagent test strips or "dipsticks" are commonly used in performing routine urinalysis because of their simplicity and convenience but, where possible, the laboratory should have "backup" procedures to confirm findings.

In dipstick urinalysis procedures, results depend on the visual acuity as well as the concentration and attitude of the analyst. The ability to discern small changes in color shade can easily be compromised by the anxiety of the patient or inexperience of the staff. Therefore, untrained staff or patients should not perform these tests. Caution and skill must be applied to estimate analyte concentrations with dipstick procedures (4–11). Only widely differing concentrations can be distinguished with confidence, with the possible exception of determining pH (5). Some inconsistencies can be decreased by reading the reagent strips spectrophotometrically. Two such instruments have been introduced: the CliniTek® Reflectance Photometer (Ames Co.) and the Urotron® System (Bio-Dynamics/bmc, Indianapolis, IN 46350). Both instruments are designed to assay and record data from only those reagent strips manufactured by their respective companies. These instruments may in time be commonly used in the larger clinical laboratories, but in most smaller laboratories visual interpretation by comparison with a color chart is likely to remain.

This chapter provides the information necessary for the intelligent use of urinalysis dipsticks, test tapes, tablets, and other supporting urinalysis procedures. Suitable commercial controls, used regularly, can be useful. However, the tests must be performed precisely as specified by the manufacturers. Timing is especially important in the dipstick methods. Package insert instructions are supplied with each product. The procedures are simple and can usually be completed in 2 min or less.

Materials

Various dipstick products for performing urinalysis are available from commercial sources. The products from Ames Co. and Bio-Dynamics/bmc have been extensively studied and compared. Appropriately used, these products are suitable (19). Test sticks are available for single tests or several different combinations of multiple tests. Expiration dates and storage conditions are important.

Procedure

1. Briefly, but completely, immerse the dipstick test areas in fresh, well-mixed, uncentrifuged urine.

   Note: Some urines may contain particulate matter that can interfere with dipstick readings. In this case, centrifuge first and then perform the dipstick tests, but remember that occult blood readings may be altered because whole erythrocytes will be removed by centrifugation.

2. Remove the dipstick from the urine immediately and remove the excess urine from the strip by drawing the edge of the test strip along the urine container rim.

3. The test strip should be kept in a horizontal position to avoid mixing the chemicals from adjacent test areas.

4. Allow the reaction to proceed for the specified time (see respective package insert for details for each test). Timing can be critical.

5. After the specified reaction time, compare the test-strip color with the corresponding color charts on the test-strip container. Estimate the nearest color and record the associated results; borderline results should be read in a consistent manner.

6. Check quality control for each dipstick test. Use at least two quality-control urines, one negative for all analytes with neutral pH, and one positive for all analytes with abnormal pH. Record all quality-control results. Quality-control materials can be made in the laboratory or purchased from laboratory supply companies, e.g., General Diagnostics, Ames, DADE.

Specific Analytes

Glucose and other reducing substances

The detection and subsequent monitoring of diabetes mellitus are the primary objectives of urinary glucose determinations. Even though a number of different sugars have been detected in urine, only glucose has pathological significance in adults (14). Normal adults excrete an average of about 130 mg of urine glucose per 24 h (20). The reabsorptive capacity of the tubules for glucose is approximately 160 mg/dL. Therefore, when blood glucose concentrations exceed this capacity, glucose appears in the urine. Glucose in the urine in the presence of normal blood glucose concentrations is called renal glucosuria, a condition in which the renal threshold for glucose is abnormally low.

Urine glucose is generally determined either by an enzyme assay specific for glucose, or by a nonspecific assay that determines total reducing substances with cupric ion in an alkaline solution. Both types of assay are relatively simple and should be available in the urinalysis laboratory.

Dipstick assays are based on glucose oxidase (EC 1.1.3.4) and peroxidase (EC 1.11.1.7) reactions, as follows:

$$\text{Glucose} + O_2 \xrightarrow{\text{glucose oxidase}} \text{gluconic acid} + H_2O_2$$

$$H_2O_2 + \text{reduced indicator} \xrightarrow{\text{peroxidase}} \text{oxidized indicator} + H_2O$$

o-Tolidine and potassium iodide have been used as indicators for the peroxidase reaction. In their oxidized forms they are green and brown, respectively. The glucose oxidase reaction is specific for glucose: there are no known urine constituents that will react with glucose oxidase to give a false-positive reaction. However, false-positive results can occur if urine is collected in a container with trace amounts of peroxides. False-negative results can occur if large amounts of ascorbic acid, as might result from therapeutic doses of vitamin C, are present in the urine.

For proper interpretation, the sensitivity of dipstick glucose assays must be appreciated. It is doubtful that slightly increased urine glucose concentrations (20–60 mg/dL) can be detected with confidence by current glucose dipstick assays *(4)*. The ability to distinguish between different glucose concentrations with dipsticks is also likely to be severely limited *(4)*. Therefore, very little confidence should be placed in the difference between trace, 1+, 2+, etc. In one study *(4)*, technologists assigned the values of negative, trace, 1+, and 2+ (with N-Multistix®) and 1+, 2+, and 3+ (with ChemStrip®-8) to a urine with a glucose concentration of 200 mg/dL. If the dipstick glucose assay is performed correctly, it has qualitative value, but nothing more *(4, 8–10, 19)*.

Galactose appears in the urine of infants with the rare disease galactosemia, and will be detected as only a reducing substance. All patients under 10 years old should be tested by both the glucose-specific determination and the assay for total reducing substances. A negative glucose test accompanied by a positive test for reducing substances will alert the physician to the need for possible follow-up to determine the nature of the reducing substances. In addition to galactose, many other reducing substances might be present in urine, including ascorbic acid, uric acid, creatinine, cysteine, lactose, fructose, ribose, and sulfanilamide *(14)*.

Total reducing substances can be estimated with the Clinitest® tablets (Ames Co.). These tablets contain anhydrous cupric sulfate, sodium hydroxide, citric acid, and sodium bicarbonate and are used as follows:

1. Mix five drops of urine with 10 drops of water in a test tube.
2. Add one Clinitest tablet and allow to stand undisturbed for 15 s; observe the color reaction.
3. After mixing, compare with color chart provided by the manufacturer.

Careful observation after addition of the tablet is necessary because sugar concentrations greater than 2 g/dL will cause color changes that rapidly pass through the usual endpoint colors (i.e., green, tan, and orange) to a dark greenish-brown. If the pass-through phenomenon is observed with the five-drop Clinitest procedure, the analyst should perform a two-drop test and compare the color of the reaction with the two-drop color chart provided by Ames. These tablets absorb moisture and darken. When discoloration is evident or when the quality-control results are not within limits, the tablets should not be used.

Proteins

Normal urine contains 20–200 mg of protein per liter *(12–15)* with a mean of about 150 mg of protein excreted per 24 h *(21)*. The small amount of protein in normal urine includes globulins, albumins, and mucoproteins. Proteinuria, which is not always clinically significant, exists when normal urine concentrations of protein are exceeded *(22)*. Four factors may cause proteinuria *(23)*: glomerular leakage of plasma protein (pathologically, the most important cause), overflow of filtered proteins, failure of tubular reabsorption of filtered proteins, or renal proteins derived from kidney tissue. Proteinuria may be observed in a wide variety of abnormalities, including primary renal disease, fevers, thyroid disorders, heart disease, pyelonephritis, acute glomerulonephritis, nephrotic syndrome, and hypertensive vascular disease.

The primary application of the dipstick urine protein test is in screening for early renal disease. Dipstick methods used for screening purposes are insensitive to protein concentrations usually found in normal urine; sensitivity may not be great enough to pick up low levels of proteinuria.

Because of the grave significance of a positive urine-protein test, all positive dipstick protein results should first be confirmed by a backup qualitative test such as the sulfosalicylic acid method *(23, 24)*. If abnormal proteinuria is still indicated, the result should be followed up with a quantitative protein assay on a 24-h urine.

Dipstick protein assays are based on the so-called "protein error of indicators" principle *(25)*, wherein protein displaces hydrogen ion and thereby changes the color of a pH indicator. For example, with tetrabromphenol blue as the indicator, colors range from green-yellow for negative proteinuria through green-blue for positive reactions. If the dipstick is left in the urine sample too long, the buffer may be washed out and the over-exposed test strip will appear blue, irrespective of any protein that may or may not be present.

The dipstick tests for proteins should be used and interpreted with caution. The ability to distinguish between normal urine and mild proteinuria is severely compromised in current dipstick protein assays. Several problems combine to produce this lack of discernment. Dipstick assays are sensitive to albumin but not globulins or Bence Jones proteins *(24)*, and variations in urine salt concentration affect the dipstick protein readings *(24)*. False-positive results may be obtained with alkaline, highly buffered urine, or with contamination by quaternary ammonium compounds such as Zephiran used in the collection process for clean-catch urines. However, the most important problem with the dipstick protein assay is poor precision attributable to variation of the technologists *(6)*. Urine protein concentrations less than

20 mg/L can be distinguished from concentrations greater than 300 mg/L with confidence with dipstick procedures. Attempts to distinguish protein concentrations closer than this are progressively less reliable *(6)*.

Ketones

The expression "ketone bodies" refers collectively to acetone, acetoacetic acid, and β-hydroxybutyric acid. Increased ketone concentrations can develop in either blood (ketonemia) or urine (ketonuria) when carbohydrate metabolism is limited. An estimate of the degree of ketonuria is useful in diagnosing and monitoring diabetic ketoacidosis and in the differential diagnosis of coma.

Acetone and acetoacetic acid react with nitroprusside in an alkaline medium to form a purple chromogen. Sensitivities of the dipstick assays have been set so that normal concentrations of ketone bodies in urine will not be detected. Some technologists can see a color change in the dipstick with as little as 2 mg of acetoacetate per 100 mL, but all technologists should see a color change with 10 mg/dL. The dipstick chromogen is not stable; therefore, package-insert instructions regarding the reading time after wetting the strip must be respected. This assay is much more sensitive to acetoacetic acid than acetone, and it is insensitive to β-hydroxybutyric acid. A false-positive ketone reaction can be obtained if large concentrations of L-dopa or its metabolites are present.

The nitroprusside tests for ketones have generally replaced the ferric chloride (Gerhardt's test) method because of the many substances that interfere with the latter. However, for other reasons the ferric chloride test is included in this chapter.

Ferric chloride test

In the past, the ferric chloride test has been used as a screening test for phenylketonuria, but many constituents of urine give colored products that confuse the test results *(13)*. More specific and sensitive tests are now available for phenylketonuria *(26)*. However, a wide variety of urine constituents with possible clinical significance give colored products with a ferric chloride test. Therefore, this simple test can help to identify those patients who may need follow-up studies to identify the specific urine constituent causing the color change in the ferric chloride test; details are given in Table 4.

Bilirubin

Bilirubin is a degradation product of hemoglobin metabolism in the reticuloendothelial system. Normal urine does not contain detectable concentrations of bilirubin, but conjugated bilirubin, the only form found in urine, may be increased to detectable concentrations in cases of heptocellular disease or biliary obstruction. When liver function is normal, hemolytic jaundice generally does not increase urine bilirubin; nor does liver damage increase urine bilirubin unless some obstruction is involved *(12)*.

Table 4. Ferric Chloride Test *(13)*

Reagents

FeCl₃ solution.[a] Dissolve 10 g of FeCl₃·6H₂O in 100 mL of distilled or de-ionized water.

Phosphate-precipitating agent.[b] Mix 2.3 g of MgCl₂·6H₂O, 1.4 g of NH₄Cl, and 2.0 mL of concd. NH₄OH in 100-mL container and dilute to volume.

Procedure

1. Add, with mixing, 1 mL of phosphate precipitant to 4.0 mL of fresh urine.
2. Filter, then acidify the filtrate with two or three drops of concd. HCl.
3. Slowly add two or three drops of FeCl₃ solution and note any color formation.

Results

Urine constituent[c]	Observation
Phenylpyruvic acid, bilirubin, or imidazole pyruvic acid	Green to blue-green, persists 2 to 4 min
Homogentisic acid	Green-blue, fades rapidly
p-Hydroxyphenylpyruvic acid	Green, fades rapidly
Lactic acid	Gray
Maple syrup urine disease (valine, leucine, isoleucine)	Blue
δ-Ketobutyric acid	Purple, fades to red-brown
Salicylates	Stable purple
Phenothiazines	Pink-purple
Cyanates	Red

[a] When stored in a clear bottle at room temperature, the FeCl₃ solution develops a precipitate within about one to two days. Even though the supernatant solution continues to yield the described colors with compounds listed, the solution should be prepared freshly each day.

[b] This solution is stable for at least four months at room temperature.

[c] This list represents only part of the possible constituents that react in the ferric chloride test.

Several qualitative tests have been developed for detecting bilirubin in urine *(12, 14)*. The diazotization methods are most common, but a simple "shake test" provides a crude way to detect bilirubin; 3–4 mL of urine is placed in a 10-mL test tube and shaken vigorously. If the foam is distinctly yellow, the urine likely contains conjugated bilirubin. If the foam is white, high concentrations of bilirubin are not present. The shake test must always be followed up with a more definitive method for urine bilirubin.

In dipstick assays bilirubin reacts with diazonium salts in an acid environment. The intensity of the resulting color is roughly proportional to the bilirubin concentration. The Ames method involves diazotized dichloroaniline and the Bio-Dynamics method involves 2,6-dichlorobenzene–diazonium–tetrafluoroborate. Both manufacturers provide detailed directions and color comparison charts to aid in evaluating results. Even trace amounts of bilirubin detected by the dipstick methods require further investigation. Most technologists using dipstick assays can detect as little as 3 mg of bilirubin per liter.

The Ictotest® reagent tablets (Ames Co.) are more sensitive, detecting as little as 1 mg/L, and can be used to follow-up questionable findings from the dipstick methods. In addition to the reagent tablets and absorbent mats provided by the manufacturer, the only additional equipment required is a dropper. The reaction takes 30 s and the test can be completed in about 60 s. A blue or purple color indicates a positive reaction, and the intensity of the color approximates the concentration of bilirubin present in the urine.

The analyst should keep in mind that bilirubin is labile in urine; assays must be performed on fresh urine samples protected from light.

Urobilinogen

After bilirubin enters the intestine in the bile, the intestinal flora reduce the conjugated bilirubin to colorless urobilinogen (which includes mesobilirubinogen, stercobilinogen, and urobilinogen) (13). Most urobilinogen is eliminated with the stool, but a small amount is absorbed from the intestine into the blood. This absorbed urobilinogen is re-excreted, in part, by the liver into the bile, and the remainder is excreted in the urine. Conditions leading to increased intestinal bilirubin will generally increase urine urobilinogen, but certain liver disorders may also substantially increase urine urobilinogen. With biliary obstruction, urine concentrations of urobilinogen are decreased because the precursor, bilirubin, is being at least partly blocked from entering the intestinal tract (27).

Dipstick urobilinogen assays are based on the p-dimethylaminobenzaldehyde reaction, or alternatively, with 4-methoxybenzene–diazonium–tetrafluoroborate. The sensitivities of both assay reactions (approximately 5 mg/L) are such that normal urine urobilinogen concentrations are detectable; 1 Ehrlich unit equals 1 mg of urobilinogen (12). Tests for urobilinogen must be carried out on fresh urine: urobilinogen is rapidly converted to urobilin, which does not react with p-dimethylaminobenzaldehyde but imparts an orange color to the urine. Numerous interfering substances in urine have been reported (28). Increased or decreased concentrations may have clinical significance.

Nitrite

An association between urine nitrite and urinary tract infection was first established in 1914 (29). Subsequently, many investigators studied urine nitrite as an indicator of urinary tract infection. Nitrite appears in the urine as a result of nitrate-reducing organisms in the urinary tract or bladder. False-positive results are rare and any detectable amount of nitrite in urine is presumed to indicate urinary tract infection (30).

In dipstick procedures, aromatic amines react with nitrite to produce a diazonium salt, which in turn is coupled to N-(1-naphthyl)ethylenediamine or 3-hydroxyl-1,2,3,4-tetrahydro-7,8-benzoquinoline to produce a pink chromogen. These reactions are very rapid and the dipstick can be read within 30 s after wetting. These dipstick assays are very specific for nitrite, with no known urine constituent interfering to produce a false-positive result. However, false-negative results may be obtained if urine contains ascorbic acid, abnormal amounts of urobilinogen, or its pH is below 6. Abnormal amounts of bilirubin, ketones, hemoglobin, protein, glucose, creatinine, urea, uric acid, ferric ion, sulfate, or phosphate do not interfere with the assay. These dipstick assays are sensitive and dependable indicators of nitrite. However, they are not dependable indicators of urinary tract infections.

Note: Because many false-negative results are observed (30), a negative dipstick nitrite result should never be considered evidence for the absence of urinary tract infection.

Instructions that accompany dipstick nitrate assays may suggest that the preferred sample for this specific test is the first morning voiding. This allows at least a 4-h urine retention in the bladder in contact with the nitrate-reducing organisms, which provides sufficient time for nitrate reduction to nitrite.

Note: Some manufacturers of dipsticks suggest that inadequate dietary nitrate may be another cause of false-negative results. The Submitters doubt that either bladder-retention time or diet are serious considerations. The primary problem with the assay is the instability of nitrite in some urines (30). Urine samples collected by the patient at home and brought to the physician's office are not suitable for nitrite analysis because false-positive results (due to incubation with contaminating bacteria) and false-negative results (due to in vitro oxidation of nitrite) could be obtained from such samples (30).

Blood

Normal urine contains minute quantities of blood, an average of about 1000 erythrocytes per milliliter (31). During menses special care must be taken to avoid contaminating urine specimens with blood. When free hemoglobin or erythrocytes are present in urine in abnormal concentrations, they are referred to as hemoglobinuria or hematuria, respectively. The detection of hemoglobinuria and (or) hematuria is an important diagnostic clue that requires further investigation. Hemoglobinuria suggests intravascular hemolysis, such as a transfusion reaction. A finding of hematuria indicates possible bleeding from a lesion in the urinary tract.

Blood, when present in the urine, is in the form of intact erythrocytes and (or) hemoglobin (i.e., lysed cells). Because erythrocytes may lyse before urinalysis, the dipstick assay alone cannot distinguish between hematuria, hemoglobinuria, and myoglobinuria (the latter condition resulting from muscle trauma or disease). The appropriate test for hematuria is microscopic examination. The presence of erythrocytes in urine is detected by microscopic examination when they exceed about 60/μL of urine (32).

The laboratory tests generally used to screen urine for occult blood are based on the pseudoperoxidase activity of hemoglobin, which catalyzes a reaction between peroxide and a chromogen such as o-tolidine; myoglobin also catalyzes this reaction. When myoglobinuria is suspected, it may be important to determine whether a positive reaction for occult blood is due to hemoglobin or to myoglobin. Careful observation of the patient's plasma may help to make this distinction. Intravascular hemoglobin binds to haptoglobin, forming a complex too large to be filtered by the glomerulus. Even before the haptoglobin-binding sites become saturated with hemoglobin, the plasma has a distinct pink coloration. On the other hand, myoglobin, a much smaller molecule, is rapidly cleared by the kidneys and the plasma is not discolored. Electrophoresis (14) is suitable for distinguishing between myoglobin and hemoglobin in urine. The ammonium sulfate precipitation method (13) has not been reliable in our experience.

Another source of error in urine assays for blood is the peroxidases occasionally found in urinary-tract-infecting organisms. These can cause a false-positive blood assay.

Most technologists can detect hemoglobin from 10 to 20 lysed erythrocytes per microliter. The best sensitivity for intact erythrocytes varies from five to 50 erythrocytes per microliter, depending on the dipstick product used (7). Do not attempt to estimate the quantity of blood present on the basis of a dipstick assay because there is a generally poor association between the dipstick procedures and the hemocytometer or sedimentation count. The quantity of blood in urine is not diagnostic of a particular disease process (33); however, even a trace amount of blood appearing repeatedly in urine is abnormal and should lead to urologic examination. The dipsticks are generally satisfactory for screening purposes. However, these methods are based on free hemoglobin and are usually less sensitive to intact erythrocytes; each routine urinalysis should include a microscopic examination of the urine sediment.

Microscopic Examination of Urine

A comprehensive treatment of the microscopic examination of urine is beyond the scope of this chapter but a summary review will serve to emphasize this important aspect of urinalysis. Each laboratory must obtain suitable reference materials (13, 34) containing pictures to help identify the formed elements that may be observed in the urine sediment. Microscopic examination of the urine can be an invaluable aid to detection and evaluation of urinary and kidney disorders. Microscopic observations should be made on fresh urine samples. Those structural elements most often identified during the microscopic examination of urine have been grouped as casts, cells, crystals, and a miscellaneous category (16). A brief summary follows.

Casts

Casts are cylindric proteinaceous structures that can be most easily observed under the microscope with subdued illumination. They are formed in the distal convoluted tubules. Identification of casts can be important because anything contained within the cast has come from the renal parenchyma. For example, the presence of erythrocytes in the urine sediment does not determine the origin of bleeding, but erythrocytes found within a cast are from the kidney. Casts have been classified by several different schemata. Some of the more common designations are as follows:

Hyaline casts are colorless, homogeneous, transparent, and contain very few formed elements. They may be seen in urine of patients with or without renal disease. The shape of the casts is determined by the site of their formation.

Erythrocyte casts appear as solid cylinders of tightly packed erythrocytes. These casts appear only in active glomerulonephritis. The erythrocytes within these casts often do not retain their cellular integrity but the pinkish color of hemoglobin persists. They are then designated as blood or hemoglobin casts, and have the same significance as erythrocyte casts.

Leukocyte casts contain packed leukocytes; these are usually seen in acute pyelonephritis but may also occur with active renal inflammation. Identification of these casts in urine sediment should always be followed by bacteriological investigation.

Tubular epithelial casts are seen with acute tubular necrosis as well as other chronic or acute renal parenchymal diseases. In unstained sediment, they may be difficult to distinguish from leukocyte casts.

Granular casts are thought by some investigators (35) to be composed of degenerated epithelial cells, leukocytes, or erythrocytes, along with albumin and fat. As the cells within the cast degenerate, the internal structure first becomes coarsely granular, with cellular and granular elements often observed in the same cast. As deterioration continues, the texture of the cast changes to fine granular and then to waxy; the structure within the cast may be entirely lost. The presence of granular casts has the same pathological significance as the renal epithelial casts.

Casts are formed in the renal tubule and therefore are usually 25 to 30 μm in diameter. However, with renal disease, abnormal tubules may have greater diameters and thus produce *broad casts;* this designation should be used when casts are significantly wider than four erythrocyte diameters. Any of the above casts might appear as broad casts, e.g., a broad erythrocyte cast. Broad casts should be specifically noted.

Cells

Erythrocytes. The quantity of erythrocytes in urine may be so great that the urine resembles blood. Small quantities are apparent only when the concentrated sediment is examined under the microscope; from one to five erythrocytes per high-power (450×) field should not be considered abnormal. Hematuria should be followed up by a more complete urologic investigation, except when it is due to contamination of urine by menstrual discharge.

The erythrocytes in urine may vary widely in size and configuration because of their exposure to nonisotonic urine. At times it may be difficult to distinguish them from yeast or fatty droplets; however, mixing a drop of acetic acid solution with a drop of sediment suspension will lyse the erythrocytes but will generally not change the other, confusing structures. The acetic acid solution (3 mL of glacial acetic acid in 97 mL of water) is stable for 12 months at room temperature.

Leukocytes. Large numbers of leukocytes in urine generally indicate a suppurative process in the kidney, bladder, or urethra *(36)*. More than five cells per high-power field constitutes pyuria. Pus cells are somewhat larger than erythrocytes but generally smaller than most epithelial cells. The leukocytes are recognized by their characteristic nuclei and granular cytoplasm. The cytoplasmic granulates of these cells in urine may show brownian movement and be referred to as "glitter cells."

Renal epithelial cells. Unstained renal tubular epithelial cells are about the size of leukocytes or slightly larger, but have a large, round nucleus that occupies 50–80% of the cell. They may look cuboidal or columnar.

Squamous epithelial cells. These cells, which originate principally in the urethra and vagina, are large relative to other cellular components and are therefore quite easy to identify. Except when present in abundance, they have limited diagnostic significance, usually indicating a contaminated urine specimen.

Oval fat bodies. At low magnification (100×), these bodies may appear as black specks on the slide. Under higher magnification (450×), they can be seen to be epithelial cells laden with refractile fat droplets. The larger fatty globules are anisotropic to polarized light.

Other cells. Histiocytes, epithelial cell inclusion bodies, malignant cells, yeast cells, and spermatozoa are among the other cellular structures that might occasionally be detected in urine sediment.

Crystals

Crystalluria may be entirely asymptomatic or may be associated with the formation of calculi in the urinary tract. Generally, crystals have limited diagnostic significance but may provide clues to certain pathologic conditions. For example, hexagonal cystine crystals in urine are rarely observed but are very significant because they indicate cystinuria. Urate crystals can indicate gout, but are not conclusive. Identification of the crystalline materials in urine may be quite simple, but in some cases the identification can require analytic capabilities beyond those available in many laboratories. If pathologic conditions are suggested, the crystal identity should be confirmed, where possible, by specific chemical tests. The more common crystals observed in urine are listed in Table 5 *(16);* the presence of a good illustrated reference book *(13, 34)* greatly facilitates their identification. Because a variety of metabolic products and drugs may crystallize in urine when present in increased concentrations, careful review of a patient's dietary and medication habits may aid in determining the nature of urine crystals.

Table 5. Common Urinary Crystals

Composition	Crystal forms
Calcium oxalate	Envelopes or octahedrals, birefringent hourglass
Calcium phosphate (dissolves in acetic acid)	Amorphous, granular, wedge-shaped, rosettes
Triple (ammonium magnesium phosphate)	Prisms, feathery crystals
Calcium sulfate	Thin needles, prisms
Calcium carbonate (dissolves in acetic acid with CO_2 evolution)	Anhedral, dumbbell, rhombohedral, needles
Uric acid and urates (sparingly soluble in acetic acid)	Rhombic, sheaves, prismatic rosettes, rectangular and hexagonal plates, amorphous

Miscellaneous

Many kinds of foreign materials are occasionally observed in the urine sediment. Pollen particles, fecal contaminants, dandruff, hair, food particles, fabric fibers, starch particles, glass chips, cosmetic powders, and ointments are only a few of the artifacts that have been seen in urine specimens. Even to the experienced observer these foreign materials may, under the microscope, be mistaken for pathologic crystals or formed elements. These artifacts, which obscure proper examination of the sediment, generally arise from unclean receptacles and (or) improper collection procedures. The presence of large numbers of bacteria in the urine may signify infection, but can also indicate that a specimen was stored improperly. Nonbacterial organisms such as trichomonads may also be found in urine of infected individuals. Whenever the microscopic examination of the urine reveals a significant amount of unidentified material, further investigation is recommended.

References

1. Cowen, J. B., Results of a survey of physicians' offices in northwest Illinois. *Il. Med. J.* 149, 377–380 (1976).
2. National Survey of Physicians' Office Laboratories. HSM 110-72-341, Bureau of Quality Assurance. United States Public Health Service, Rockville, MD, 1975.
3. Kark, R. M., Lawrence, J. R., Pollak, V. E., Pirani, C. L., Muehrcke, R. C., and Silva, H., *A Primer of Urinalysis*, 2nd ed., Harper and Row, New York, NY, 1963, p 3.
4. James, G. P., and Bee, D. E., Glucosuria: Accuracy and precision of laboratory diagnosis by dipstick analysis. *Clin. Chem.* 25, 996–1001 (1979).
5. James, G. P., Bee, D. E., and Fuller, J. B., Accuracy and precision of urine pH determinations using two commercially available dipsticks. *Am. J. Clin. Pathol.* 70, 368–374 (1978).
6. James, G. P., Bee, D. E., and Fuller, J. B., Proteinuria: Accuracy and precision of laboratory diagnosis by dipstick analysis. *Clin. Chem.* 24, 1934–1939 (1978).
7. Bee, D. E., James, G. P., and Paul, K. L., Hemoglobinuria and hematuria: Accuracy and precision of laboratory diagnosis. *Clin. Chem.* 25, 1696–1699 (1979).
8. Simpson, E., and Thompson, D., An assessment of hospital routine urinalysis. *Clin. Chem.* 24, 389–390 (1978).
9. Assa, S., Evaluation of urinalysis methods used in 35 Israeli laboratories. *Clin. Chem.* 23, 126–128 (1977).
10. Brereton, D. M., Sontrop, M. E., and Fraser, C. G., Timing of urinalysis reactions when reagent strips are used. *Clin. Chem.* 24, 1420–1421 (1978).
11. Simpson, E., and Thompson, D., An assessment of hospital routine urinalysis. *Ann. Clin. Biochem.* 15, 241–242 (1978).
12. Henry, J. B., Ed., *Clinical Diagnosis and Management by Laboratory Methods*, 16th ed., W. B. Saunders Co., Philadelphia, PA, 1979.
13. Davidsohn, I., and Henry, J. B., *Clinical Diagnosis by Laboratory Methods*, 15th ed., W. B. Saunders Co., Philadelphia, PA, 1974.
14. Tietz, N., *Fundamentals of Clinical Chemistry*, W. B. Saunders Co., Philadelphia, PA, 1976.
15. Henry, R. J., Cannon, D. C., and Winkelman, J. W., Eds., *Clinical Chemistry: Principles and Technics*, Harper and Row, New York, NY, 1974.
16. Ash, K. O., Matsen, J. M., and Rothstein, G., *Family Medicine Principles and Practice*, R. B. Taylor, Ed., Springer-Verlag, New York, NY, 1978, p 411.
17. Rubini, H. E., and Wolfe, A. V., Refractionmetric determinations of total solids and water of serum and urine. *J. Biol. Chem.* 225, 869–876 (1957).
18. Wolfe, A. V., Urinary concentrative properties. *Am. J. Med.* 32, 329–332 (1962).
19. Smith, B. C., Michael, J. P., and Fraser, C. G., Urinalysis by use of multi-test reagent strips: Two dipsticks compared. *Clin. Chem.* 23, 2337–2340 (1977).
20. Date, J. W., Quantitative determination of some carbohydrates in normal urine. *Scand. J. Clin. Lab. Invest.* 10, 155–162 (1958).
21. Relman, A. S., and Levinsky, N. G., Clinical examination of renal function. In *Diseases of the Kidney*, M. C. Strauss and L. G. Welt, Eds., Little-Brown, Boston, MA, 1971.
22. Manuel, Y., Revillard, J. P., and Betnul, H., Eds., *Proteins in Normal and Pathological Urine*, Karger, Basel, 1970.
23. Grant, G. H., and Kachmar, J. F., The proteins of body fluids. In *Fundamentals of Clinical Chemistry* (see ref. *14*), pp 298–376.
24. Gyure, W. L., Comparison of several methods for semiquantitative determination of urinary protein. *Clin. Chem.* 23, 876–879 (1977).
25. Sorenson, S., The measurement of the hydrogen ion concentration and its importance for enzymatic processes. *Biochem. Z.* 21, 131 (1909).
26. Wu, J. T., Wu, L. H., Ziter, F. A., and Ash, K. O., Manual fluorometry of phenylalanine from blood specimens collected on filter paper: A modified procedure. *Clin. Chem.* 25, 470–472 (1979).
27. Free, A. H., and Free, H. M., *Urinalysis in Clinical Laboratory Practice*, CRC Press, Inc., Cleveland, OH, 1976.
28. Young, D. S., Pestaner, L. C., and Gibberman, V., Effects of drugs on clinical laboratory tests. *Clin. Chem.* 21, 385D–386D (1975).
29. Cruickshank, J., and Moyes, J., The presence and significance of nitrite in urine. *Br. Med. J.* ii, 712 (1914).
30. James, G. P., Paul, K. L., and Fuller, J. B., Urinary nitrite and urinary-tract infection. *Am. J. Clin. Pathol.* 70, 671–678 (1978).
31. Wright, W. T., Cell counts in urine. *Arch. Intern. Med.* 103, 76–78 (1959).
32. Cook, M. H., Free, H. M., and Free, A. H., The detection of blood in urine. *Am. J. Med. Technol.* 22, 218–231 (1956).
33. Higgins, C. C., The clinical significance of hematuria. *J. Am. Med. Assoc.* 166, 203–206 1958.
34. *Urine under the Microscope*, ROCOM Press, Div. of Hoffmann-LaRoche, 1973.
35. Lippman, R. W., *Urine and the Urinary Sediment*, 2nd ed., Charles C Thomas, Springfield, IL, 1957.
36. Miller, S. E., Examination of urine. In *A Textbook of Clinical Pathology*, 7th ed., S. E. Miller, Ed., Williams and Wilkins, Baltimore, MD, 1966.

Appendix I: Conversion Factors, Units for Analytes in This Book

Submitter: H. Peter Lehmann, *Department of Pathology, Louisiana State University Medical Center, New Orleans, LA 70112*

Constituent	Common unit	Factor	Recommended unit[a]
Acid phosphatase	Bessey–Lowry Brock[b] unit/L	16.67	U/L
Albumin	g/dL	10	g/L
Alanine aminotransferase	U/L	1	U/L
Alcohol (ethanol)	mg/dL	0.217	mmol/L
Alkaline phosphatase	Bessey–Lowry Brock[b] unit/L	16.67	U/L
Ammonia	µg/dL	0.587	µmol/L
Amylase	Somogyi unit/dL	1.85	U/L
Aspartate aminotransferase	U/L	1	U/L
Barbiturates	mg/dL	10	mg/L
Bilirubin	mg/dL	17.1	µmol/L
Calcium	mg/dL	0.25	mmol/L
	meq/L	0.5	mmol/L
	mg/24 h	0.025	mmol/24 h
Carbon dioxide (bicarbonate)	meq/L	1	mmol/L
Carbon monoxide	% saturation	0.01	1 (saturated fraction)
Chloride	mg/dL	0.282	mmol/L
	meq/L	1	mmol/L
Cholesterol	mg/dL	0.0259	mmol/L
Creatine kinase	U/L	1	U/L
Creatinine	mg/dL	88.4	µmol/L
	mg/24 h	8.84	µmol/24 h
Estriol	µg/24 h	3.47	nmol/24 h
Glucose	mg/dL	0.0555	mmol/L
	mg/24 h	0.00555	mmol/24 h
Hemoglobin	g/dL	10	g/L
Hematocrit	%	0.01	1 (volume fraction)
Iron and iron-binding capacity	µg/dL	0.179	µmol/L
Lactate dehydrogenase	U/L	1	U/L
Magnesium	mg/dL	0.411	mmol/L
	meq/L	0.5	mmol/L
	mg/24 h	0.0411	mmol/24 h
Osmolality	milli-osmol/kg	1.0	mmol/kg
Phenylalanine	mg/dL	0.061	mmol/L
p_{CO_2}	mmHg	0.133	kPa
p_{O_2}	mmHg	0.133	kPa
Phosphate	mg/dL	0.323	mmol/L
	g/24 h	32.3	mmol/24 h
Protein, total	g/dL	10	g/L
Prothrombin time	s	1	s
Salicylate	mg/dL	72.4	µmol/L
Triglyceride	mg/dL	0.0113[c]	mmol/L
Urea	mg/dL	0.167	mmol/L
	g/24 h	16.7	mmol/24 h
Urea nitrogen	mg/dL	0.714	mmol/L
	g/24 h	71.4	mmol/L

[a] Recommended by: International Union of Pure and Applied Chemistry and International Federation of Clinical Chemistry. Approved Recommendation 1978. Quantities and Units in Clinical Chemistry, 1979. *Clin. Chim. Acta* **96**: 157F–183F, 1979. List of Quantities in Clinical Chemistry. *Clin. Chim. Acta* **96**: 185F–204F, 1979. Commonly used unit × factor = recommended unit.

[b] *p*-Nitrophenyl phosphate substrate.

[c] Factor based on relative molecular mass of triolein = 885.

Appendix II: Analyte Reference Intervals ("Normal" Ranges) Reported in This Book

	Interval	Page no.
Acid phosphatase		
Thymolphthalein monophosphate substrate	0.5–1.9 U/L	62
p-Nitrophenyl phosphate substrate		67
Total	2–11 U/L	
Prostatic (tartrate-sensitive)	2–3 U/L	
Albumin, serum	35–50 g/L	324–325
Alanine aminotransferase (ALT)	5–19 U/L	72
Alkaline phosphatase		82
Newborns, 1 wk	up to 250 U/L	
Girls		
1–12 yr	up to 350 U/L	
>15 yr	25–100 U/L	
Boys		
1–12 yr	up to 350 U/L	
12–15 yr	up to 500 U/L	
>20 yr	25–100 U/L	
Adults	25–100 U/L	
Ammonia		89
Newborns	up to 171 μmol/L (290 μg/L)	
"Postnatal"	up to 43 μmol/L (73 μg/L)	
Infants and children	up to 91 μmol/L (155 μg/L)	
Adults	12–47 μmol/L (20–80 μg/L) Also see page 89	
Amylase		
Amyloclastic		94
Serum	40–180 units/dL	
Urine	<300 units/h	
Starch chromogen		100
Serum	45–200 dye units/dL	
Urine	0.66–5.43 dye units/min	
Aspartate aminotransferase (AST)	10–29 U/L	106
Bilirubin		
Modified Jendrassik–Grof		118
Total	0.1–1.2 mg/dL	
Malloy–Evelyn, modified		122
Total	0.1–1.5 mg/dL	
Conjugated	0.0–0.3 mg/dL Also see page 122	
Calcium		126
Adults		
Serum	8.5–11.0 mg/dL (2.1–2.8 mmol/L)	
Cerebrospinal fluid	4.2–5.4 mg/dL (1.1–1.4 mmol/L)	
Urine	50–150 mg/24 h (12.5–37.5 mmol/24 h)	

	Interval	Page no.
Children, serum		
Premature, first wk	6.0–10.0 mg/dL (1.5–2.5 mmol/L)	
Full-term, first wk	7.0–12.0 mg/dL (1.8–3.0 mmol/L)	
Child	8.0–11.0 mg/dL (2.0–2.8 mmol/L)	
Carbon dioxide		135
Adults		
Venous plasma	25.1–28.3 mmol/L	
Arterial plasma	22.5–26.7 mmol/L	
Fingertip plasma	24.3–27.4 mmol/L	
Newborns		
Venous plasma	24.3–26.9 mmol/L	
Arterial plasma	21.9–25.5 mmol/L	
Fingertip plasma	23.1–26.2 mmol/L	
Adults and children, total CO_2		302
Arterial	20–24 mmol/L	
Venous	21–28 mmol/L	
Carbon monoxide		141
Adults		
Non-smoking	0–2%	
Smoking	5–14%	
Chloride		
Colorimetric	95–106 mmol/L	146
Coulometric-amperometric		
Serum	101–109 mmol/L	151
Cerebrospinal fluid	122–128 mmol/L	
Cholesterol	See pages 158–163	
Creatine kinase		189
Men	63–460 U/L	
Women	16–206 U/L	
Creatine kinase isoenzymes	See page 196	
Creatinine		
Fuller's earth method	See page 204	
Column method	5.0–13.0 mg/L	210
Estrogens, total pregnancy (TPE), placental	See pages 239–240	
Glucose		
Hexokinase	70–105 mg/dL (3.9–5.8 mmol/L)	248
o-Toluidine	70–122 mg/dL (3.9–6.8 mmol/L)	253
Glucose-6-phosphate dehydrogenase	6.54 ± 0.79 U/g of hemoglobin	257
Hematocrit		261
Men	0.47 ± 0.07	
Women	0.42 ± 0.05 Also see page 261	
Hemoglobin, blood		266
Children	12–14 g/dL	
Men	14.5–17 g/dL	
Women	13–15 g/dL Also see page 266	

(continued on next page)

	Interval	Page no.
Iron, serum		270
Men	70–168 µg/dL	
Women	38–153 µg/dL	
Iron-binding capacity, total		270
Men	264–394 µg/dL	
Women	293–478 µg/dL	
Lactate dehydrogenase	211–518 U/L	275
Magnesium		278
Magon dye		
Newborn	0.75–1.15 mmol/L	
Children	0.70–0.95 mmol/L	
Adults	0.65–1.25 mmol/L	
Titan Yellow	0.7–2.1 mmol/L	285
Osmolality		290
Serum	280–290 mmol/kg of water	
Urine	300–1400 mmol/kg of water	
Free water clearance	−100 to −25 mL/h	
pH		302
Newborns	7.33–7.49[a]	
2 mo–2 yr	7.34–7.46[a]	
Children and adults		
Arterial	7.35–7.45	
Venous	7.32–7.42	
p_{CO_2}		302
Newborns	27–40 mmHg[a]	
2 mo–2 yr	26–41 mmHg[a]	
Children and adults		
Arterial	33–46 mmHg	
Venous	40–50 mmHg	
p_{O_2}		302
Newborns	60–76 mmHg[a]	
Children and adults		
Arterial	80–105 mmHg	
Venous	25–47 mmHg	
Phenylalanine		308
Premature infants	156–377 µmol/L (25–62 mg/L)	
Infants, 0–7 days	104–127 µmol/L (17–21 mg/L)	
Infants, 8–31 days	136–138 µmol/L (22–23 mg/L)	
Children, 6 mo–12 yr	40–210 µmol/L (7–35 mg/L)	
Phenylketonuric infants	1370–2380 µmol/L (223–388 mg/L) Also see page 309	
Phosphate (inorganic)		314
Serum		
Premature infants	7.6–8.1 mg/dL (2.46–2.61 mmol/L)	
Newborns (full-term)	5.8–6.4 mg/dL (1.83–2.07 mmol/L)	
Children (1–10 yr)	4.4–4.8 mg/dL (1.43–1.54 mmol/L)	

	Interval	Page no.
Phosphate, serum (cont.)		
Adults	2.5–4.5 mg/dL (0.8–1.45 mmol/L)	
Cerebrospinal fluid	0.9–2.0 mg/dL (0.29–0.65 mmol/L)	
Urine	0.4–1.3 g/24 h (12.9–42.0 mmol/24 h)	
Potassium		345
Serum	4.5–5.3 mmol/L	
Urine	25–120 mmol/24 h	
Proteins, serum, by electrophoresis		227
Total protein	6.6–8.2 g/dL	
Albumin	3.5–4.8 g/dL	
Alpha$_1$-globulin	0.2–0.4 g/dL	
Alpha$_2$-globulin	0.5–0.9 g/dL	
Beta-globulin	0.6–1.2 g/dL	
Gamma-globulin	0.7–2.0 g/dL	
Protein, total		
Serum	62–82 g/L	320
Cerebrospinal fluid		323
Newborns	600–1200 mg/L	
Adults	150–450 mg/L	
Urine	25–100 mg/24 h	322
Prothrombin time	10–14 s Also see pages 333–334	333
Salicylate	See page 339	
Sodium		345
Serum	135–148 mmol/L	
Urine (average diet)	40–90 to 43–217 mmol/24 h	
Sweat test results		350
	Na, mmol/L Cl, mmol/L	
5 wk–11 mo	5–24 3–22	
1 yr– 9 yr	3–36 2–32	
10 yr–16 yr	6–52 2–38	
17 yr–60 yr	4–90 2–65	
Cystic fibrosis patients	71–137 78–145	
Thromboplastin time, activated partial	30–45 s Also see pages 333–334	333
Triglycerides	30–150 mg/dL Also see pages 159, 161–162	354
Urea, serum		
Berthelot reaction		362
Men	10–25 mg/dL (1.7–4.2 mmol/L)	
Women	7–21 mg/dL (1.2–3.5 mmol/L)	
Diacetyl monoxime	4–27 mg/dL (0.7–4.6 mmol/L)	371
Uric acid		377
Men	3–7 mg/dL	
Women	2–6 mg/dL Also see pages 377–378	

[a] Arterial or arterialized.

Index

Absorbance calibration, 37–38, 72, 105, 272
"Absurd" values, 34
Accuracy in quality assurance, 21, 29
Acetylsalicylic acid (aspirin), 337–339
Acid–base status, 293, 294, 299, 301–303, 315, 382
Acidemia (acidosis), 302–303, 382
Acid phosphatase, 13, 59–63, 65–67
    p-Nitrophenyl phosphate substrate, 65–67
    Prostatic, 59, 62, 65–67
    Tartrate inhibition of, 59, 62, 65–67
    Thymolphthalein monophosphate substrate, 59–63
Activated partial thromboplastin time (APTT), 327–334
Alanine aminotransferase (ALT), 11, 69–72
Albumin, 11, 127, 194, 223, 227–231, 317–318, 320, 322–325, 380, 384, 387
Albumin/globulin ratio, 318, 320
Alcohol abuse, 185, 191, 280, 283, 289, 291, 316, 377
Alcohol dehydrogenase, 75–78
Alcohol, serum, 13, 75–78, 213
    See also Ethanol
Alkalemia (alkalosis), 302–303, 382
Alkaline phosphatase, 11, 79–83, 189
Aminotransferase, see Alanine aminotransferase, Aspartate aminotransferase
Ammonia, plasma, 8, 85–90
Amphetamines, see Drug screening
Amylochrome, 91
Amylase, 11, 91–100
    Amyloclastic (starch) procedure, 91–94
    Saccharogenic (reducing sugars) procedure, 92, 94
    Starch chromogen, 95–100
    Units, 93, 98
Analytical balance, maintenance, 39–40
Anemia, 255, 258, 261–262, 266, 271
Anticoagulants and preservatives, 5–7, 18, 380
    See also Citrate, EDTA, Fluoride, Heparin, Oxalate, Sodium azide
Antidiuretic hormone (ADH), 289, 382
α_1-Antitrypsin, 223
Arterial blood collection, 4, 7, 302

Arterialization, 6, 303
Aspartate aminotransferase (AST), 11, 69, 71, 90, 101–107
Aspirin (acetylsalicylate), 337–339
Azotemia, 85

Barbiturates, 13, 88, 109–110, 214, 216–218, 220
Beer's law stated, 37
Bence Jones protein, 230, 322, 380, 384
Berthelot reaction for urea, 357, 360–362
Bias, systematic error, 20–21
Bicarbonate, 294, 301–303
Bilirubin, 11, 14, 81, 90, 113–124, 144, 147, 171–173, 175, 177, 182, 192, 204, 209–210, 253, 279, 313, 323, 381, 385–386
    Direct-reacting (conjugated), 113–115, 119–120, 122–124, 381, 385–386
    Indirect-reacting (unconjugated), 122
    Jendrassik–Grof procedure, 113–118, 121–123
    Meites–Hogg modification, 113, 116–117, 119–124
    Modified Malloy–Evelyn procedure, 119–124
    Total, 113–124
    Urobilinogen, 386
Biohazards, 50–51, 53–54
Bisalbuminemia, 231
Biuret determination of total serum protein, 318–320
Blood
    Ammonia, 8, 85–90
    Collection, 3–9, 302–303
    Gases, 4, 293–303
    Glucose, 241–253
    Hemoglobin (BHb), 261–266
    pH, 4–8, 280, 293–303
Bromide, 143, 147, 149

Caffeine
    Effects on tests, 3, 377
    Reagent, 113–114, 117
Calcium, 6, 7, 11, 14, 92, 99, 125–130, 277, 280, 283, 313, 315–316, 327
    Metabolism and clinical aspects, 127–128

Calibration, 18, 37–40, 72, 93, 98–99, 105, 115–116, 158, 168–169, 264, 272–273, 296–297
Calibrators and standards, 18, 34, 38, 51, 296
Cancer, 128, 271, 362
    Prostatic, 59, 62, 67, 191, 195
Capillary blood collection, 3, 167, 259, 264, 295
Carbon dioxide, 11, 131–136, 146, 260, 300, 303
    See also Blood gases
Carbon monoxide, 139–142, 191, 195, 265
Carboxyhemoglobin (HbCO), 4, 139, 141–142, 263, 265
Cardiovascular disease, 162–163, 170
    See also Heart damage
Centrifugal force, 8, 260
Centrifuge, 8, 33, 40, 47, 260–261
Cell indices (Wintrobe's) and hemoglobin content, 262
Cerebrospinal fluid, analysis with, 89, 125–126, 134, 145, 146, 278, 313–314, 322–323, 338
Chloride, 11, 143–155
    Colorimetric determination, 143–148
    Coulometric-amperometric (electrometric) determination, 143, 149–152, 348, 349
    Interferences with, 143, 147, 149
    Mercurimetric determination, 143, 153–155
    Sweat, 347–350
    Urine, 147
Cholesterol, 157–183
    Clinical aspects, 162–163, 170
    Enzymic determination, 157–158, 165–174
    Esters, 165, 175
    Free, see Cholesterol, unesterified
    High-density lipoprotein (HDL), 13, 157–158, 160, 162, 170
    Iron–uranyl acetate determination, 157, 165, 179–182
    Liebermann–Burchard determination, 157–158, 165, 175–178
    Lipoproteins, 13, 157–158, 160, 162, 165, 170
    Low-density lipoprotein (LDL), 157–158, 160, 170

Cholesterol *(Continued)*
  Parekh–Jung determination, 179–182
  Reference values, 158–163, 170
  Specimen collection, 3, 157
  Tissue, 181
  Total, 157–159, 161–163, 165, 170–171, 179
  Unesterified, 165, 171
Citrate anticoagulant, 7, 13, 62, 66, 92, 99, 187, 243, 256, 268, 284, 288, 295, 328–330
Clearance
  Creatinine, 100, 201, 204, 209
  Free water, 289–290
  Osmolal, 290
Coagulation testing, 327–334
Collection of specimens, 3–9, 294–295, 329–330, 380
Colligative properties, 287, 381
Conjugated bilirubin, *see* Bilirubin
Control charts, 17, 21–31, 188, 274
Control limits, 22–30
Conversion factors, 391
Cortisol, 3, 7, 14
Coumarin anticoagulants, 327, 333–334, 377
Creatine kinase (CK), 4, 7, 11, 14, 185–196
  Evaluation of methods for, 195–196
  Interferences with assay for, 187, 189, 194
  Isoenzymes, 187, 191–196
Creatinine, 11, 14, 34, 201–204, 207–211, 235–239, 253, 384
  Clearance, 100, 201, 204, 209
  Column-chromatographic determination, 207–211
  Fuller's earth (Lloyd's reagent) determination, 201–204, 207
Cyanmethemiglobin, 263
Cystic fibrosis, 347–350

Delta check, 34
Design, laboratory, 18, 54–55
Diabetes, 235–236, 246–247, 277, 280, 283, 289, 303, 317, 350, 382, 383
Dipstick analysis, *see* Urinalysis
Direct-reacting bilirubin, *see* Bilirubin
Drains and waste disposal, 53
Drug screening, 213–221

EDTA (ethylenediaminetetraacetate) anticoagulant, 6, 7, 13–14, 70, 85, 91, 99, 103, 114, 145, 187–189, 243, 247, 256, 259–260, 264, 268, 273, 284, 288, 295, 338, 354, 359, 366
Electrical equipment and safety, 52–53
Electrodes, 301–302, 304–309, 347–348
  Clot-detection, 328
  Ion-selective, 8, 143, 144, 341
  Maintenance of, 150–151, 297–298

Electrolytes, 11, 34, 280, 341, 347, 349–350
Electrophoresis
  Creatine kinase isoenzymes, 191–196
  Proteins, 223–233, 387
Emphysema, 85
Enzyme immunoassay (EIA), 213–214, 218–220
Enzyme-multiplied immunoassay 213, 219
Epilepsy, 110
Error, sources of, 19–21, 33, 35
Estriol, glucuronates, sulfates, 14, 235
Estrogen/creatinine (E/C) ratio, 235–236, 238–240
Estrogen-receptor protein, 12, 13
Estrogens
  Serum cholesterol and, 161–162
  Total pregnancy (TPE), placental, 235–240
Estrone, estradiol-17β, 235
Ethanol (ethyl alcohol), 3–4, 88, 157
  See also Alcohol, serum
Ethylenediaminetetraacetate, *see* EDTA
Evacuated tubes, blood collection, 3, 5–6, 85, 135, 157

Fat, urinary, 381, 387–388
Fatty acids, 3, 13, 165, 328
Fibrinogen, 13, 224, 231, 317, 327–328, 333
Fire and explosion hazards, 50
First aid, 46
Flame photometry, 286, 341–345
  Maintenance of instrumentation, 40, 344–345
Fluoride, 14, 147, 189
  Anticoagulant/preservative, 7, 13–14, 66, 70, 76, 103, 145, 243–244, 247, 250, 361–362
Free-water clearance, 289–290
Freezer maintenance, 39
Freezing point osmometry, 288–290
Fructose interference with glucose determination, 246, 253, 384
Fuller's earth (Lloyd's reagent), 201–204
Fume hoods, 50, 53–54
Furniture, laboratory, 54

Galactosemia, 379, 384
Gases, compressed, cryogenic, 51–52
Gaussian distribution, 20, 309
Genetic abnormalities, 89–90, 246, 255, 257–258, 262, 305, 308–310, 315, 377–379
Globulin
  Alpha$_1$-, 223, 227–231
  Alpha$_2$-, 223, 227–231
  Beta-, 223–224, 227–231
  Gamma-, 223–224, 227–231, 380
Glomerular filtration, 201, 362
Glucose, 5, 8, 11, 88, 135, 239, 241–253, 289, 317, 320, 380, 381, 383–384

Glucose *(Continued)*
  Hexokinase method, 241–248
  Tolerance test, 3, 7
  o-Toluidine method, 249–253
Glucose-6-phosphate dehydrogenase (G-6-PD) activity, 13, 241–246, 255–258
Glutamine in spinal fluid, 89
γ-Glutamyltransferase, 4, 11
Gout, 377, 388
Guthrie bacterial inhibition assay, 305

Halide, *see* Chloride
Halogen, *see* Chloride
Hazards, laboratory, 44–55
HbCO, *see* Carboxyhemoglobin
Heart damage, 85, 101, 162, 185, 191, 194–196, 271, 280, 341, 384
  See also Cardiovascular, Myocardial
Heavy metals, 14, 35, 358, 361
  Wastes, 49
Hematocrit (HCT), packed cell volume (PCV), 34, 259–262, 328
Hemodialysis, 377
Hemoglobin, 13, 34, 81, 139–142, 257, 261–266, 381, 385–387
Hemoglobinometry, 263–266
Hemolysis, 5, 7, 9, 12, 19, 66, 85, 97, 103, 113, 114, 117, 120, 123, 126, 168, 172–173, 178, 182, 187, 192, 204, 209–210, 231, 243, 247, 255, 258, 264, 267, 273, 278, 279, 284, 295, 309, 313, 320, 322, 329–330, 344, 362, 367, 371, 376, 386
Hemophilia, 327
Henderson–Hasselbalch equation, 294, 301–302
Heparin anticoagulant, 6–7, 13–14, 61, 62, 70, 81, 92, 99, 103, 114, 126, 133, 135, 141, 145, 150, 187, 243, 250, 256, 260–261, 264, 268, 273, 278, 284, 288, 294–295, 299, 328, 329, 334, 338, 342, 359, 362, 366, 376
Hepatic disease, 71, 79, 85, 89, 101, 113, 223, 231, 271, 280, 362, 385
Hydrogen ion concentration, *see* pH
Hyperammonemia, 85
Hyper- and hypocalcemia, 125, 127–128, 280
Hypergammaglobulinemia, 264
Hyperlipidemia, 353, 361
Hypertension, 235, 384
Hyperthyroidism, 128, 280, 316
Hypothyroidism, 185, 195, 350

Immunoglobulins (IgG, IgM, IgA, IgD), 223, 228–231, 317, 320, 322
Indirect-reacting (unconjugated), bilirubin, *see* Bilirubin
Infectious agents, handling of, 7, 50–51
Inorganic phosphate, *see* Phosphate
Iodide (in chloride determination), 147, 149
Iodine (in amylase determination), 91–93

Iodoacetate, 7
Iontophoresis, 347–350
Iron, 7, 263, 267–270
Iron-binding capacity, total (TIBC), 267–270

Jaffé reaction, 201, 239
Jaundice, 113, 122, 126, 271, 385
Jendrassik–Grof bilirubin method, 113–118, 121–123

Ketoacidosis, 201, 303, 385
Kernicterus, 113
Kidney disorders, see Renal
Kits, method selection and evaluation, 18, 30

Laboratory design and safety, 43–55
Lactate dehydrogenase (LD) activity, 11, 69, 102, 271–275
  Isoenzymes, 271, 273
Lactose tolerance test, 3
Lactic acid, 4, 13
Lesch–Nyhan syndrome, 377
Leukemia, 264, 271, 377
Lipase, 11
Lipemia, 3, 19, 123, 125–127, 172–173, 187, 192, 204, 273, 279, 318–320, 324, 330, 344, 367, 371
Lipids, 34, 144, 157–163, 170, 187, 253, 279
Lipoproteins, see Cholesterol
Lipoproteinemia, 165
Liver disease, see Hepatic
Lloyd's reagent (Fuller's earth), 201–204

Macroamylasemia, 94
Macroglobulinemia, 230, 322
Magnesium, 277–280, 283–285
  Magon dye method, 277–280
  Titan Yellow method, 283–285
  Physiological and clinical aspects, 279–280, 283
Maintenance of equipment, 17, 33, 37–42, 54, 298–299
Malloy–Evelyn bilirubin determination, modified, 119–124
Meites–Hogg modification, bilirubin determination, 113, 116–117, 119–124
Mecurimetric titration of chlorides, 143, 153–155
Mercury, 14
  Carbon dioxide determination with, 131–136
  Chloride determination with, 143–148, 153–155
  Salicylate determination with, 337
  Toxic hazard, 49; see also Heavy metals
Methemoglobin, 141
Microcapillary centrifuge, 133, 261
Microgasometer, 131–136
Microhematocrit, 260–261
Multiple myeloma, 128, 230, 264, 322, 377

Multiple sclerosis, 323
Muscular dystrophy, 185, 191, 194–195
Myocardial infarction, 185, 191, 194–196, 271, 280
Myoglobulinuria, 195, 381, 387

Natelson microgasometer, 131–136
Nephrosis, 223, 231, 325, 384
Neoplastic disease and hypercalcemia, 127
Nicotine, see Smoking
Normal values, see Reference values
Nuclear Regulatory Commission, 44, 51, 52

Occupational Safety and Health Act (OSHA), 43–44
Osmolality, 14, 287–292, 381
  Clinical significance, 289
Osmometry, 288
  Freezing point, 288–290
  Vapor pressure, 290–292
  Volatile substances, interference by, 291
Osmotic pressure, 287–288, 341
Osteomalacia and hypocalcemia, 127–128
Oxalate anticoagulant, 7, 66, 70, 76, 92, 99, 103, 150, 187, 243, 250, 260, 264, 268, 273, 284, 295, 329, 338–339, 359, 366
Oxyhemoglobin (HbO_2), 139, 141, 263–265, 315

Pancreatitis, 91, 94, 95, 99
Parathyroid hormone (parathyrin, parathormone), 14, 127, 280, 313, 315–316
Parotitis, 95
Partial thromboplastin time, see Thromboplastin
p_{CO_2}, 293–303
Pediatrics, 5, 8, 79–80, 85, 88, 90, 113, 120, 125, 127, 129, 182, 208, 257, 266, 267, 277, 292, 294–295, 308–310, 314, 325, 342, 347–350
pH, 4–8, 35, 293–303, 382
Phenol hazard, 49
Phlebotomy, see Blood collection
Phenylalanine, 305–310
Phenylketonuria (PKU), 305, 309–310, 379, 382, 385
Phosphatase, acid, see Acid phosphatase
Phosphatase, alkaline, see Alkaline phosphatase
Phosphate, inorganic, 5, 127, 313–316
Pilocarpine, 347–350
Pipetting hazards, 45–46
"Plasma-crit," 259
Polycythemia, 260–261, 266, 377
p_{O_2}, 293–303
Potassium, 4, 5, 7, 8, 11, 40, 88, 145, 280, 341–345, 376
Pregnancy, 235–240, 262, 266, 280, 320, 325, 362, 377

Preventive maintenance, 37–42, 54
Prostatic carcinoma, 59, 62, 67, 191, 195
Protein
  Cerebrospinal fluid, 322–323
  Electrophoresis, 223–233
  Serum, 223–233, 317–320
  Total, 11, 317
  Urine, 230–234, 239, 320–322, 380, 381, 384–385
"Protein effect," 125
Proteinuria, 322, 384
Prothrombin time (PT), 6, 327–334
  Interpretation of results, 333–334

Quality assurance, quality control, 17–31, 33–35, 379

Radioactive materials, use and disposal of, 51
Range, use in quality assurance, 20
Reference values, 62, 67, 72, 82, 88–89, 94, 100, 106, 118, 122, 126, 135, 141, 146, 151, 158–163, 170, 189, 196, 204, 209–210, 227, 239, 248, 253, 257, 261, 266, 270, 275, 278, 285, 290, 302, 308–309, 314, 317, 320, 322–325, 333, 345, 350, 354, 362, 371, 377–378, 393–394
Refractometry and total serum proteins, 317–318, 320
Refrigerator maintenance, 39
Renal damage, 195, 210, 236, 271, 283, 289, 293, 302–303, 316, 322, 362, 384, 387–388
Reye's syndrome, 85, 90, 185, 191

Saccharogenic amylase determination, 92, 94
Safety, laboratory, 43–55
Salicylate (acetylsalicylate, aspirin), 213, 253, 303, 337–339, 378, 385
Sarcoidosis, 128
Serum
  Albumin, 127, 194, 223, 227–231, 317–318, 320, 322–325
  Alcohol (ethanol), 13, 75–78, 213
  Iron, 7, 263, 267–270
  Iron-binding capacity, total, 267–270
  Protein, 223–233, 317–320
  Urea, 34, 317–318, 357–362, 365–373
  Uric acid, 5, 34, 375–378
Shipment of specimens, 11–14, 18
Skin puncture, 3–7, 259, 264
Smoking, 3–4, 46, 50, 87, 141, 365
Sodium, 8, 11, 40, 88, 145, 146, 265, 341–345, 347–350
Sodium azide, 48, 50, 117, 338
Solvents, storage and disposal of, 49–50
Somogyi unit, amylase, 98
Specimen
  Collection, 3–9, 18, 302–303, 329–330, 380
  Identification, 7, 18–19

397

Specimen (Continued)
　Source (arterial, skin-puncture, venous), 3–4, 302
　Transport, 8, 11–14, 18
Specific gravity, 287, 381–382
Spectrophotometer maintenance, 37–38
Splenomegaly, 262
Starch chromogen procedure for amylase, 95–100
Starch–iodine procedure for amylase, 91–94
Sterols, 171
Storage, laboratory, 48–51, 54–55
Sugar, see Glucose
Sulfhydryl ions, 149–150, 187–188, 192
Sulfides, 149
Sweat test, 347–350

Thermometer maintenance, 38–39
Thin-layer chromatography (TLC), 213–218, 220
Thiocyanate, 143–149
Thrombin, 6, 327, 334
Thromboplastin time, activated partial (APTT), 327–334
　Interpretation of results, 333–334
Tobacco, see Smoking
Tonometry, 299–300
Tourniquet use in blood collection, 4, 81, 167, 176, 295
Toxemia, 235, 377
Toxic and hazardous substances, 48–49
Toxicology, 6, 213
Toxic Substances Control Act (TOSCA), 43
Trace elements, 5–6, 14

Transaminases, see Alanine aminotransferase, Aspartate aminotransferase
Transportation of specimens, 8, 11–14, 18
Triglycerides, 3, 147, 157–162, 170, 353–355
Trinder method for salicylate, 337–339
Turbidimetry, 320–323

Unconjugated bilirubin, see Bilirubin
Units, conversation table, 391
Urea-nitrogen, 11, 34, 201, 317–318, 320, 357, 360–362, 371
Urea, serum, 34, 317–318, 320, 357–373
　Berthelot reaction method, 357–362, 365, 371
　Dimethyl monoxime method, 357, 365–373
Urea, urinary, 360
Uremia, 127, 317
Uric acid, 5, 34, 189, 253, 375–378, 381, 384, 388
Urinalysis, 379–389
　Bilirubin, 381, 385–386
　Blood, 380, 386–387
　Casts, cells, crystals, 380–381, 387–389
　"Clean-catch" specimen collection, 380
　Dipstick analysis, 322, 379–387
　Drugs, 213–221
　Glucose and other reducing substances, 383–384
　Ketones, 382, 385
　Microscopic examination, 387–389

Urinalysis (Continued)
　Nitrite, 386
　pH, 382, 383
　Pregnancy estrogens, 235–240
　Proteins, 230–234, 239, 320–322, 380, 381, 384–385
　Specific gravity, 381–382
　Timed specimens, 87, 92, 94, 96, 126, 203, 235–240, 278, 290, 313, 320–322, 381, 384, 386
　Urobilinogen, 386

Vacuum systems and laboratory design, 53
Vapor pressure osmometry, 290–292
Variability in measurement, 19–21
Venipuncture, 3–5, 7, 75, 76, 81, 85, 167, 260–261
Ventilation and laboratory design, 53–54
Vitamin C, 380, 384
Vitamin D, 127–128, 280, 283, 315–316
Vitamin K, 327, 333–334
Volatile substances in osmometry, 291

Waldenstrom's macroglobulinemia, 230
Wastes, storage and disposal, 47–49, 51
Water-bath maintenance, 33, 40
Wavelength calibration, 18, 33, 37–38, 72, 105, 272
Wintrobe hematocrit tube, 260

Xanthinuria, 378

Zone electrophoresis, 223